[barcode label] ...CO-ANR-293

Quantity	Symbol	Value
Hubble constant†	H_0	$100\ h\ km\ s^{-1}\ Mpc^{-1}$
Normalized Hubble constant†	$h = H_0/100$	$0.50 < h < 0.85$
Hubble time†	$1/H_0$	$9.778\ 13 \times 10^9\ h^{-1}$ years
Age of Universe†	t_0	$15(5) \times 10^9$ years
Critical density of Universe†	$\rho_c = 3H_0^2/8\pi G$	$1.878\ 82(24) \times 10^{-29}\ h^2\ g\ cm^{-3}$
		$2.775\ 366\ 27 \times 10^{11}\ h^2\ M_\odot\ Mpc^{-3}$
		$1.053\ 94(13) \times 10^{-5}\ h^2\ GeV\ cm^{-3}$
Density parameter of Universe†	$\Omega_0 = \rho_0/\rho_c$	$0.1 < \Omega_0 < 2$
(ρ_0 = present mass density of Universe)		
Local disk density	ρ disk	$3 - 12 \times 10^{-24}\ g\ cm^{-3}$
Local halo density	ρ halo	$3 - 7 \times 10^{-25}\ g\ cm^{-3}$
Cosmic background radiation (CBR) temperature†	T_0	2.726 ± 0.005 K
Solar velocity with respect to CBR		$369.5 \pm 3.0\ km\ s^{-1}$
Energy density of CBR	ρ_γ	$4.647\ 7 \times 10^{-34}\ (T/2.726)^4\ g\ cm^{-3}$
		$0.260\ 71\ (T/2.726)^4\ eV\ cm^{-3}$
Number density of CBR photons	n_γ	$410.89\ (T/2.726)^3\ cm^{-3}$
Conversion Factors		
Parsec	pc = 1A/1 arc second	$3.085\ 677\ 582 \times 10^{18}$ cm = 3.2616 ly
Megaparsec	Mpc = 10^6 pc	$3.085\ 677\ 582 \times 10^{24}$ cm
Light year	ly	$9.460\ 730\ 47 \times 10^{17}$ cm = 0.306 601 pc
Day	d	86 400 seconds
Julian century	JC	36 525 days
Radian	rad	$2.062\ 648\ 062\ 5 \times 10^5$ seconds of arc
Steradian*	str	$3283\ (degree)^2 = 4.25 \times 10^{10}\ (arcsec)^2$
Jansky	Jy = 1 flux unit	$10^{-23}\ erg\ cm^{-2}\ s^{-1}\ Hz^{-1}$
Solar flux unit	sfu = 10^4 flux units	$10^{-19}\ erg\ cm^{-2}\ s^{-1}\ Hz^{-1}$

† Subscript 0 denotes present-day values

* A sphere subtends 4π steradians = 41253 square degrees

[1] Adapted from LANG, K.R.: *Astrophysical Data: Planets and Stars* (New York: Springer-Verlag, 1992), Siedelmann, K.P.: *Explanatory Supplement to the Astronomical Almanac* (Mill Valley, CA: University Science Books, 1992), and Physical Review **D50**, 1234 (1994). The figures in parenthesis after the values give the one-standard-deviation uncertainties in the last digits.

ASTRONOMY AND ASTROPHYSICS LIBRARY

Series Editors:
I. Appenzeller, Heidelberg, Germany
G. Börner, Garching, Germany
M. Harwit, Washington, DC, USA
R. Kippenhahn, Göttingen, Germany
J. Lequeux, Paris, France
P. A. Strittmatter, Tucson, AZ, USA
V. Trimble, College Park, MD, and Irvine, CA, USA

Springer
Berlin
Heidelberg
New York
Barcelona
Hong Kong
London
Milan
Paris
Singapore
Tokyo

ASTRONOMY AND ASTROPHYSICS LIBRARY

A&A LIBRARY

Series Editors: I. Appenzeller · G. Börner · M. Harwit · R. Kippenhahn
J. Lequeux · P. A. Strittmatter · V. Trimble

Theory of Orbits (2 volumes)
Volume 1: Integrable Systems and Non-perturbative Methods
Volume 2: Perturbative and Geometrical Methods
By D. Boccaletti and G. Pucacco

Galaxies and Cosmology
By F. Combes, P. Boissé, A. Mazure and A. Blanchard

The Solar System 2nd Edition By T. Encrenaz and J.-P. Bibring

Nuclear and Particle Physics of Compact Stars
By N. K. Glendenning

The Physics and Dynamics of Planetary Nebulae By G. A. Gurzadyan

Astrophysical Concepts 2nd Edition By M. Harwit

Stellar Structure and Evolution By R. Kippenhahn and A. Weigert

Modern Astrometry By J. Kovalevsky

Observational Astrophysics 2nd Edition
By P. Léna, F. Lebrun and F. Mignard

Supernovae Editor: A. Petschek

General Relativity, Astrophysics, and Cosmology
By A. K. Raychaudhuri, S. Banerji and A. Banerjee

Tools of Radio Astronomy 2nd Edition
By K. Rohlfs and T. L. Wilson

Atoms in Strong Magnetic Fields
Quantum Mechanical Treatment and Applications
in Astrophysics and Quantum Chaos
By H. Ruder, G. Wunner, H. Herold and F. Geyer

The Stars By E. L. Schatzman and F. Praderie

Gravitational Lenses By P. Schneider, J. Ehlers and E. E. Falco

Relativity in Astrometry, Celestial Mechanics and Geodesy By M. H. Soffel

The Sun An Introduction By M. Stix

Galactic and Extragalactic Radio Astronomy 2nd Edition
Editors: G. L. Verschuur and K. I. Kellermann

Reflecting Telescope Optics (2 volumes)
Volume I: Basic Design Theory and its Historical Development
Volume II: Manufacture, Testing, Alignment, Modern Techniques
By R. N. Wilson

Galaxy Formation By M. S. Longair

K. R. Lang

Astrophysical Formulae

Volume I: Radiation, Gas Processes and High Energy Astrophysics

Third Enlarged and Revised Edition

With 36 Figures and 39 Tables

RECEIVED
AUG 25 1999
MSU - LIBRARY

Springer

QB
461
.L36
1999
v.1

Kenneth R. Lang
Tufts University
Department of Physics and Astronomy
Robinson Hall
Medford, MA 02155, USA

Series Editors

Immo Appenzeller
Landessternwarte, Königstuhl
D-69117 Heidelberg, Germany

Gerhard Börner
MPI für Physik und Astrophysik
Institut für Astrophysik
Karl-Schwarzschild-Str. 1
D-85748 Garching, Germany

Martin Harwit
511 H Street SW
Washington, DC 20024, USA

Rudolf Kippenhahn
Rautenbreite 2
D-37077 Göttingen, Germany

James Lequeux
Observatoire de Paris
61, avenue de l'Observatoire
F-75014 Paris, France

Peter A. Strittmatter
Steward Observatory
The University of Arizona
Tucson, AZ 85721, USA

Virginia Trimble
Astronomy Program
University of Maryland
College Park, MD 20742, USA
and
Department of Physics
University of California
Irvine, CA 92717, USA

Cover picture: Composite image of the Sun taken by two instruments aboard the Solar and Heliospheric Observatory (SOHO) and joined at the black circle. It reveals the Sun's outer atmosphere from the base of the corona to millions of kilometers above the solar surface. Raylike structures appear in the ultraviolet light emitted by oxygen ions flowing away from the Sun to form the solar wind (*outside the black circle*). The solar wind with the highest speed originates in coronal holes, which appear as dark regions at the north pole (*top*) and across the solar disk (*inside the black circle*). Courtesy of the SOHO EIT and SOHO UVCS consortium.

Library of Congress Cataloging-in-Publication Data

Lang, Kenneth R. Astrophysical formulae: a compendium for the astronomer, astrophysicist and physicist / Kenneth R. Lang. – 3rd enl. and rev. ed. p. cm. – (Astronomy and astrophysics library, ISSN 0941-7834) Includes bibliographical references and index. Contents: v. 1. Radiation, gas processes, and high energy astrophysics – v. 2. Space, time, matter and cosmology. ISBN 3-540-61267-X (v. 1: alk. paper). – ISBN 3-540-64664-7 (v. 2: alk. paper)
1. Astrophysics–Formulae. I. Title. II. Series. QB461.L36 1998 523.01'02'12–dc21 89-27747 CIP

The 2nd Edition was published as a single volume. This 3rd Edition has 2 volumes.

ISSN 0941-7834
ISBN 3-540-61267-X 3rd Edition (Vol. I) Springer-Verlag Berlin Heidelberg New York
ISBN 3-540-55040-2 2nd Edition Springer-Verlag Berlin Heidelberg New York

This work is subject to copyright. All rights are reserved, whether the whole or part of the material is concerned, specifically the rights of translation, reprinting, reuse of illustrations, recitation, broadcasting, reproduction on microfilm or in any other way, and storage in data banks. Duplication of this publication or parts thereof is permitted only under the provisions of the German Copyright Law of September 9, 1965, in its current version, and permission for use must always be obtained from Springer-Verlag. Violations are liable for prosecution under the German Copyright Law.

© Springer-Verlag Berlin Heidelberg 1978, 1992, 1999
Printed in Germany

The use of general descriptive names, registered names, trademarks, etc. in this publication does not imply, even in the absence of a specific statement, that such names are exempt from the relevant protective laws and regulations and therefore free for general use.

Typesetting: SPS Madras, India
Cover design: *design & production* GmbH, Heidelberg
10039996 55/3120-5 4 3 2 1 0 - Printed on acid-free paper

3980529
v1
9.28.99

To Marcella

Ebbene, forse voi credete
che l'arco senza fondo della volta spaziale
sia un vuoto vertiginoso di silenzi.
Vi posso dire allora che verso
questa terra sospettabile appena
l'universo giá dilaga di pensieri.

(Mario Socrate, *Favole paraboliche*)

Well, maybe you think
That the endless arch of the space vault
Is a giddy, silent hollowness.
But I can tell you that,
Overflowing with thought,
the universe is approaching
this hardly guessable earth.

(Mario Socrate, *Parabolic Fables*)

Preface to the Third Edition

When the first edition of **Astrophysical Formulae** was published in 1974, it was designed as a fundamental reference for the student and researcher in the field of astrophysics. Here the reader could find that long-forgotten formula or reference to the more comprehensive article. Anyone interested in learning about a new area of astrophysics could first turn to **Astrophysical Formulae** to obtain an introductory background and guidance to the relevant literature. In all these ways, the book has succeeded far beyond my initial expectations, becoming a widely–used, standard reference.

I am grateful to the many astronomers and astrophysicists who have told me how useful this book has been to them. Such comments have helped sustain my writing and led to this revised, updated version. It could not have been written without the caring, steadfast support of my wife, Marcella. This third edition remains dedicated to her; we courted and were married during the writing and publication of the first edition.

It is more disappointing than astonishing that today's astronomers and astrophysicists so quickly forget the exciting moments of yesterday's science, and often inadequately acknowledge the important work of their predecessors. This can be partly attributed to the accelerated pace of our celestial science; it is difficult to stay abreast of rapid developments in one's own field, let alone others. By providing a comprehensive reference for fundamental formulae and original articles, **Astrophysical Formulae** provides a foundation for overcoming these difficulties. It can also help researchers in all areas to enjoy the development and exciting new discoveries of astronomy and astrophysics.

This completely revised, third edition of **Astrophysical Formulae** more than doubles the number of formulae and references found in the previous versions. There are now more than 4 000 formulae. The references are now gathered together alphabetically for the entire book, rather than by chapter as in previous versions. The authors, titles and journal information have now been provided for more than 5 000 journal articles!

Astrophysical Formulae has now been divided into two volumes. The first **Volume I: Radiation, Gas Processes and High Energy Astrophysics,** contains the classical physics that underlies astrophysics, updated with recent work and references in their applications to the cosmos. **Volume II: Space, Time, Matter and Cosmology** includes many of the topics that are under active scrutiny by contemporary astronomers and astrophysicists.

The first **Volume I: Continuum Radiation, Monochromatic (Line) Radiation, Gas Processes, and High Energy Astrophysics** retains the classical material of previous versions, including the suggestions of the Russian translators of the first edition. Some new sections have now been provided, such as those on helioseismology, solar neutrinos, neutrino oscillations, neutrino emission from stellar collapse and supernovae, and energetic particles and radiation from solar flares. These chapters have additionally been updated with references to important review articles and books. Many of the tables have been removed from these chapters, since more comprehensive tabular information can now be found in Kenneth R. Lang's **Astrophysical Data: Planets and Stars** (New York: Springer-Verlag 1992).

New formulae and supporting data in **Volume I** include, in order of presentation, those dealing with: synchrotron radiation, gyrosynchrotron radiation, synchrotron self-absorption, bremsstrahlung, dispersion measures and time delays of pulsar emission, longitudinal waves, Čerenkov radiation, inverse Compton radiation, interstellar dust, interstellar extinction, interstellar polarization, interstellar scintillation, transition probabilities, emission lines from planetary nebulae and H II regions, recombination lines, LS coupling, 21-cm line of interstellar hydrogen, Fraunhofer lines, solar emission lines, Zeeman effect and cosmic magnetic fields, gravitational redshifts, impact broadening, polytropes, Strömgren radius and its expansion, white dwarf stars and neutron stars, equations of state for degenerate gases, limiting masses for compact objects, rotation of fluid masses, Roche limit, Roche lobes, solar and stellar coronas, solar and stellar winds, heating of the Sun's chromosphere and corona, sound waves and solar oscillations, depth of the solar convection zone, rotation and gas currents inside the Sun, shock waves, gravitational collapse in the presence of a magnetic field, star formation in molecular clouds, magnetohydrodynamic waves, accretion, variable stars, nonradial stellar oscillations, fluid instabilities, pinch instability, magnetic reconnection, fundamental particles, nucleosynthesis, atomic mass excesses, thermonuclear reaction rates, electron screening, neutrinos, neutron star cooling, origin and abundance of the elements, spallation reactions of cosmic ray particles with interstellar matter, formation of lithium, beryllium, and boron, supernova explosions, supernova remnants, supernova SN 1987A, the solar neutrino problem, neutrino astrophysics, gamma-ray lines from solar flares, nonthermal hard X-ray bremsstrahlung of solar flares, particle acceleration in solar flares, observed cosmic rays, ultra-high-energy cosmic rays, shock wave acceleration of cosmic rays, and gamma ray bursts.

New advances, both observational and theoretical, have come at an ever increasing rate, including the discovery of the cosmic microwave background radiation with its anisotropy; improved tests of the special and general theories of relativity; new measurements of the extragalactic distance scale and age of the Universe; the discovery of, and search for, invisible dark matter; the discovery of the binary pulsar and its implications for gravitational radiation; the mature development of radio and X-ray astronomy; the discovery of gravitational lenses; the development of theories for accretion onto compact objects; a new knowledge of white dwarfs, neutron stars and pulsars; the discovery of candidate black holes at the centers of galaxies and the establishment of a comprehensive theory for black holes; a new inflationary model for the earliest stages of the Universe; the discovery of superluminal motions in quasars and nearby micro-quasars; the development of theories for active galactic nuclei; improved computations of big bang nucleosynthesis; new observations of deuterium and helium throughout the cosmos; and dramatic improvements in our knowledge of the shape, structure, content and formation of the Universe. All of these comparatively recent developments are included in a completely new **Volume II: Space, Time, Matter and Cosmology**. I am indebted to my colleagues for the opportunity to learn about so many captivating new ideas and discoveries.

Over one hundred of the seminal contributions to twentieth century astronomy and astrophysics, through to the year 1975, are now reproduced in, **A Source Book in Astronomy and Astrophysics 1900–1975** by Kenneth R. Lang and Owen Gingerich (Cambridge, Mass.: Harvard University Press 1979). It contains a dozen new English translations of articles originally in German, French or Dutch, and includes a historical introduction to each. It is heartening to realize that every article in the **Source Book** was previously referenced in the first edition of **Astrophysical Formulae**, and that several of these articles include the work of subsequent winners of the Nobel Prize in Physics – including Subramanyan Chandrasekar, William A. Fowler, Antony Hewish, Arno A. Penzias, Sir Martin Ryle, and Robert F. Wilson. When the **Source Book** is updated to the turn of the millennium in the year 2000, it will surely consist of articles referenced in this third edition of **Astrophysical Formulae**.

The library at the Harvard-Smithsonian Center for Astrophysics has been an invaluable resource, and special thanks go to Seth Redfield for helping to locate material there.

I have profited from the advice of many experts who have read individual sections of this version. Those who have read portions of the new material and have supplied critical comment include: Neil Ashby, John N. Bahcall, Roger D. Blandford, Margaret Geller, Jack Harvey, Mark Haugan, Bruce Partridge, Martin Rees, David Schramm, and Kenneth Seidelmann. Persons who have provided suggestions for new formulae and references include: E. H. Avrett, K. Borkowski, R. Catchpole, K. P. Dere, T. Forbes, J. Franco, V. G. Gurzadyan, H. J. Haubold, M. Hernanz, R. W. John, K. I. Kellermann, K. Krisciunas, W. Kundt, M. A. Lee, J. Linsley, G. P. Malik, W. C. Martin, J.-M. Perrin, V. A. Razin, D. O. Richstone, N. N. Shefov, B. V. Somov, J. Terrell, R. Woo, S. Woosley, and J. Yang.

Medford
October, 1998

Kenneth R. Lang

Preface

This book is meant to be a reference source for the fundamental formulae of astrophysics. Wherever possible, the original source of the material being presented is referenced, together with references to more recent modifications and applications. More accessible reprints and translations of the early papers are also referenced. In this way the reader is provided with the often ignored historical context together with an orientation to the more recent literature. Any omission of a reference is, of course, not meant to reflect on the quality of its contents. In order to present a wide variety of concepts in one volume, a concise style is used and derivations are presented for only the simpler formulae. Extensive derivations and explanatory comments may be found in the original references or in the books listed in the selected bibliography which follows. Following the convention in astrophysics, the c.g.s (centimeter-gram-second) system of units is used unless otherwise noted. To conserve space, the fundamental constants are not always defined, and unless otherwise noted they have the meaning and value given in the tables of physical constants and astrophysical constants provided at the beginning and end of this book.

A substantial fraction of this book was completed during two summers as a visiting fellow at the Institute of Theoretical Astronomy, Cambridge, and I am especially grateful for the hospitality and courtesy which the members of the Institute have shown me. I am also indebted to the California Institute of Technology for the freedom to complete this book. The staff of the scientific periodicals library of the Cambridge Philosophical Society and the library of the Hale observatories are especially thanked for their aid in supplying and checking references. Those who have read portions of this book and have supplied critical comment and advice include Drs. L. H. Aller, H. Arp, R. Blandford, R. N. Bracewell, A. G. W. Cameron, E. Churchwell, D. Clayton, W. A. Fowler, J. Greenstein, H. Griem, J. R. Jokipii, B. Kuchowicz, M. G. Lang, A. G. Michalitsanos, R. L. Moore, E. N. Parker, W. H. Press, M. J. Rees, J. A. Roberts, J. R. Roy, W. L. W. Sargent, W. C. Saslaw, M. Schmidt, I. Shapiro, P. Solomon, E. Spiegel, and S. E. Woosley.

Medford
October, 1974

Kenneth R. Lang

Contents

Contents of Volume II

1. Continuum Radiation

"What is light? Since the time of Young and Fresnel we know that it is wave motion. We know the velocity of the waves, we know their lengths, and we know that they are transverse; in short, our knowledge of the geometrical conditions of the motion is complete. A doubt about these things is no longer possible; a refutation of these views is inconceivable to the physicist. The wave theory of light is, from the point of view of human beings, certainty."

Hertz, 1889

"It is now, I believe, generally admitted that the light which we receive from the clear sky is due in one way or another to small suspended particles which divert the light from its regular course... There seems to be no reason why the color of the compound light thus scattered should not agree with that of the sky... Suppose for distinctness of statement, that the primary ray is vertical, and that the plane of vibration is that of the meridian. The intensity of the light scattered by a small particle is constant, and a maximum for rays which lie in the vertical plane running east and west, while there is no scattered ray along the north and south line. If the primary ray is unpolarized, the light scattered north and south is entirely due to that component which vibrates east and west, and is therefore perfectly polarized."

Lord Rayleigh, 1871

1.1 Static Electric Fields

The experimentally determined Coulomb force, $\boldsymbol{F}$, between two static point charges, q_1 and q_2 is (Coulomb, 1785)

$$\boldsymbol{F} = \frac{q_1 q_2}{R^2} \boldsymbol{n}_{\mathrm{r}} \ , \tag{1.1}$$

where R is the distance between the charges, and $\boldsymbol{n}_{\mathrm{r}}$ is a unit vector directed from one charge to the other. The static electric field, $\boldsymbol{E}$, of a point charge q_1 is defined so that

$$\boldsymbol{F} = q_2 \boldsymbol{E}, \quad \text{where} \quad \boldsymbol{E} = \frac{q_1}{R^2} \boldsymbol{n}_{\mathrm{r}} \tag{1.2}$$

and R is the distance from q_1. Integrating equation (1.2) over a closed spherical surface we obtain Gauss's law

$$\oint_{\mathrm{S}} \boldsymbol{E} \cdot \boldsymbol{n} \, ds = 4\pi q = 4\pi \int_{\mathrm{V}} \rho \, dv \ , \tag{1.3}$$

where ρ is the charge density, $\oint_{\mathrm{S}}$ denotes the closed surface integral, $\boldsymbol{E} \cdot \boldsymbol{n}$ is the component of $\boldsymbol{E}$ which is normal to the surface element ds, and $\int_{\mathrm{V}} \rho \, dv$ is the amount of charge within the closed surface. Using the equations of vector analysis, Eq. (1.3) may be expressed as Poisson's equation (Poisson, 1813)

$$\nabla \cdot \boldsymbol{E} = -\nabla^2 \varphi = 4\pi\rho \ , \tag{1.4}$$

where

$$\boldsymbol{E} = -\nabla\varphi \ , \tag{1.5}$$

and φ is called the scalar electric potential. For one static point charge, q, we have

$$\varphi = q/R \ , \tag{1.6}$$

where R is the radial distance from the charge. For an electrostatic dipole consisting of two charges, $+q$ and $-q$, separated by a distance a along the z axis

$$\varphi = aq \cos\theta / R^2 \tag{1.7}$$

so that

$$\boldsymbol{E}_{\mathrm{r}} = \frac{2d}{R^3} \cos\theta \boldsymbol{n}_{\mathrm{r}} \tag{1.8}$$

$$\boldsymbol{E}_{\theta} = \frac{d \sin\theta}{R^3} \boldsymbol{n}_{\theta}$$

and

$$\boldsymbol{E}_{\varphi} = 0 \ ,$$

where the dipole moment $d = aq$, the angle θ is the angle between the z axis and the radial direction, R is the radial distance from the dipole, and $\boldsymbol{n}_{\theta}$ and $\boldsymbol{n}_{\mathrm{r}}$ are unit vectors in the θ and the radial direction.

1.2 Static Magnetic Fields

The experimentally determined magnetic force, $\boldsymbol{F}$, between two current elements $I_1 \boldsymbol{dl}_1/c$ and $I_2 \boldsymbol{dl}_2/c$ is (Ampére, 1827)

$$\boldsymbol{F} = \frac{I_1 I_2}{c^2 R^2} \boldsymbol{dl}_2 \times (\boldsymbol{dl}_1 \times \boldsymbol{n}_{\mathrm{r}}) \ , \tag{1.9}$$

where I_1 and I_2 are the two currents, $\boldsymbol{dl}_1$ and $\boldsymbol{dl}_2$ are unit vector elements of length whose directions are the same as that of the current flow, R is the distance between the two elements, and $\boldsymbol{n}_{\mathrm{r}}$ is a unit vector directed from length element $\boldsymbol{dl}_1$ towards element $\boldsymbol{dl}_2$. The attractive force per unit length between two long straight parallel wires with currents flowing in the same direction is, for example,

$$F = \frac{2 I_1 I_2}{c^2 D} \ , \tag{1.10}$$

where D is the perpendicular distance between the wires. The static magnetic field, $\boldsymbol{B}$, of current element $I_1 \boldsymbol{dl}_1/c$ may be defined so that (Biot and Savart, 1820)

$$\boldsymbol{F} = \frac{I_2 \boldsymbol{dl}_2}{c} \times \boldsymbol{B} \quad \text{where} \quad \boldsymbol{B} = \frac{I_1}{c} \frac{\boldsymbol{dl}_1 \times \boldsymbol{n}_{\mathrm{r}}}{R^2} \ . \tag{1.11}$$

Integrating equation (1.11) over a closed contour we obtain Ampére's law

$$\oint_{c} \boldsymbol{B}\cdot d\boldsymbol{l} = \frac{4\pi}{c} I = \frac{4\pi}{c}\int_{s} \boldsymbol{J}\cdot\boldsymbol{n}\, ds \ , \tag{1.12}$$

where $\oint_c$ denotes a closed contour integral, I is the current passing through the closed contour, $\int_s$ is an open surface integral over any surface bounded by the closed curve, and $\boldsymbol{J}\cdot\boldsymbol{n}$ is the component of the current density, $\boldsymbol{J}$, which is normal to the surface element ds. Using the equations of vector analysis, Eq. (1.12) may be expressed in the differential form

$$\nabla\times\boldsymbol{B} = \frac{4\pi}{c}\boldsymbol{J} \ . \tag{1.13}$$

As an example, it follows directly from Eq. (1.12) that the magnetic field of a long wire of current, I, extending along the z axis is

$$\boldsymbol{B}_{\varphi} = \frac{2I}{cR}\boldsymbol{n}_{\varphi} \ , \tag{1.14}$$

where R is the perpendicular distance from a point on the z axis and $\boldsymbol{n}_{\varphi}$ is unit vector in the φ direction of a spherical coordinate system whose positive z axis is in the direction of current flow. It follows directly from Eq. (1.11) that we may express the static magnetic field $\boldsymbol{B}$ in terms of a magnetic vector potential $\boldsymbol{A}$ such that

$$\boldsymbol{B} = \nabla\times\boldsymbol{A} \quad \text{where} \quad \boldsymbol{A} = \frac{I d\boldsymbol{l}}{cR} \ , \tag{1.15}$$

and therefore

$$\nabla\cdot\boldsymbol{B} = \nabla\cdot\nabla\times\boldsymbol{A} = 0 \ . \tag{1.16}$$

For each Cartesian coordinate of the vector $\boldsymbol{A}$, we have

$$\nabla^2\boldsymbol{A} = \frac{-4\pi}{c}\boldsymbol{J} \ , \tag{1.17}$$

and

$$\boldsymbol{A} = \frac{1}{c}\int_{v}\frac{\boldsymbol{J}}{R}dv \ ,$$

where R is the distance to the volume element dv. As an example, consider a closed loop of current, I, and radius, a, placed in the xy plane and centered at the z axis. Using Eq. (1.15) we obtain (Jackson, 1962)

$$\boldsymbol{A} = \frac{\pi a^2 I}{cR^2}\sin\theta\,\boldsymbol{n}_{\varphi} \ , \tag{1.18}$$

so that

$$\boldsymbol{B}_{\mathrm{r}} = \frac{2m\cos\theta}{R^3}\boldsymbol{n}_{\mathrm{r}} \quad \text{for} \quad R \gg a \ , \tag{1.19}$$

$$\boldsymbol{B}_{\theta} = \frac{m\sin\theta}{R^3}\boldsymbol{n}_{\theta} \quad \text{for} \quad R \gg a$$

and

$$\boldsymbol{B}_\varphi = 0 \ ,$$

where the magnetic dipole moment $m = \pi a^2 I/c$, the angle θ is the angle between the z axis and the radial direction, R is the radial distance from the center of the loop, and $\boldsymbol{n}_\varphi, \boldsymbol{n}_\theta$, and $\boldsymbol{n}_\mathrm{r}$ are unit vectors in the directions of the spherical coordinate system centered in the current loop and with the z axis perpendicular to the plane of the loop. For a spherical mass of radius, r, the latitude $\lambda = \pi/2 - \theta$ if often used, and the dipole moment of the static magnetic field is given by

$$m = r^3 B_\mathrm{p}/2 \ , \tag{1.20}$$

where B_p is the polar magnetic field strength at the surface of the sphere. The magnetic moments and radii of the planets are given by Lang (1992). The dynamo theory for the Earth's varying magnetic field is reviewed by Inglis (1981), and the status of the dynamo theory for cosmic magnetism is discussed by Cowling (1981).

1.3 Electromagnetic Fields in Matter – Constitutive Relations

Experiments on the flow of current in metals (Davy, 1821) which are subjected to an applied electric potential were explained by Ohm (1826). Ohm's law states that the current density, $\boldsymbol{J}$, is given by

$$\boldsymbol{J} = \sigma \boldsymbol{E} \ , \tag{1.21}$$

where the constant σ is called the conductivity and $\boldsymbol{E}$ is the electric field strength. Kelvin (1850) suggested that the magnetic induction $\boldsymbol{B}$ induced in matter by a magnetizing force, $\boldsymbol{H}$, is given by

$$\boldsymbol{B} = \mu \boldsymbol{H} \ , \tag{1.22}$$

where μ is called the magnetic permeability. Similarly, Maxwell (1861) suggested that the electric displacement, $\boldsymbol{D}$, caused by the application of an electromotive force to matter is given by

$$\boldsymbol{D} = \varepsilon \boldsymbol{E} \ , \tag{1.23}$$

where ε is called the dielectric constant or permittivity. In most astrophysical situations $\mu = 1$, its free space value. In free space, $\sigma = 0$ and $\varepsilon = 1$. The constants ε and μ define the index of refraction, n, of a medium with zero conductivity and no magnetic field

$$n = \sqrt{\varepsilon\mu} \ . \tag{1.24}$$

The indices of refraction for a conductive medium and for a medium with a magnetic field are discussed in Sects. 1.12 and 1.33, respectively.

The macroscopic constant, ε, was related to the atomic properties of matter by Lorentz (1880) and Lorenz (1881). They argued that the mean or observed field, $\boldsymbol{E}$, coming from a region containing a great number of atoms or molecules will be different from the effective field, $\boldsymbol{E}'$, acting on each individual molecule. It was assumed that under the influence of $\boldsymbol{E}'$ a molecule will polarize itself forming an electric dipole moment

$$\boldsymbol{d} = \alpha \boldsymbol{E}' \ , \tag{1.25}$$

where the constant, α, is called the polarizability. From Eq. (1.8) the resulting dipole field at the surface of a spherical molecule will be

$$\boldsymbol{E}_\mathrm{s} = \frac{\boldsymbol{d}}{a^3} = \frac{\alpha \boldsymbol{E}'}{a^3} \ , \tag{1.26}$$

where a is the molecular radius. Thus the observed field, $\boldsymbol{E}$, from an isotropic medium with a number density of N molecules will be

$$\boldsymbol{E} = \boldsymbol{E}' - N\left[\frac{4}{3}\pi a^3 \boldsymbol{E}_\mathrm{s}\right] = \left[1 - \frac{4\pi}{3} N\alpha\right] \boldsymbol{E}' \ . \tag{1.27}$$

The total electric polarization of the medium, $\boldsymbol{P}$, is given by

$$\boldsymbol{P} = N\alpha \boldsymbol{E}' = \chi_\mathrm{e} \boldsymbol{E} = \left[\frac{N\alpha}{1 - \frac{4}{3}\pi N\alpha}\right] \boldsymbol{E} \ , \tag{1.28}$$

where $\boldsymbol{P}$ is the volume density of the electric dipole moment, and the constant χ_e is called the dielectric susceptibility. The dielectric constant, ε, is introduced by noting that

$$\boldsymbol{D} = \varepsilon \boldsymbol{E} = \boldsymbol{E} + 4\pi \boldsymbol{P} = (1 + 4\pi\chi_\mathrm{e}) \boldsymbol{E} \ , \tag{1.29}$$

and the polarizability, α, is given by the Lorentz–Lorenz formula

$$\alpha = \frac{3}{4\pi N} \frac{\varepsilon - 1}{\varepsilon + 2} = \frac{3}{4\pi N} \frac{n^2 - 1}{n^2 + 2} \ , \tag{1.30}$$

where N is the number density of molecules, and n is the index of refraction.

The macroscopic version of Eq. (1.4) which excludes the contribution of electric dipole charges is Coulomb's law

$$\nabla \cdot \boldsymbol{D} = 4\pi\rho \ . \tag{1.31}$$

In a similar way, a magnetic susceptibility, χ_m, may be defined so that the magnetic field $\boldsymbol{H}$ is given by

$$\boldsymbol{M} = \chi_\mathrm{m} \boldsymbol{H} \ , \tag{1.32}$$

where the magnetization, $\boldsymbol{M}$, is the volume density of the dipole magnetic moment. In this case we have

$$\boldsymbol{B} = \mu \boldsymbol{H} = \boldsymbol{H} + 4\pi \boldsymbol{M} \ , \tag{1.33}$$

and the macroscopic version of Eq. (1.13) which excludes the current density of the magnetic dipoles is Ampére's law

$$\nabla \times \boldsymbol{H} = \frac{4\pi \boldsymbol{J}}{c} \ . \tag{1.34}$$

1.4 Induced Electromagnetic Fields

Faraday (1843) first showed that an electric potential was induced in a circuit moving through the lines of force of a magnet, and his law for the induced electric potential is Faraday's law

$$\oint_{c} \boldsymbol{E} \cdot d\boldsymbol{l} = -\frac{1}{c}\frac{\partial}{\partial t}\int_{s} \boldsymbol{B} \cdot \boldsymbol{n}\, ds \ , \tag{1.35}$$

where $\oint_c$ denotes the closed contour integral, $\boldsymbol{B}$ is called the magnetic induction, $\int_s$ denotes an open surface integral over any surface bounded by the closed contour, and $\boldsymbol{B} \cdot \boldsymbol{n}$ denotes the component of $\boldsymbol{B}$ which is normal to the surface element ds. The line integral in Eq. (1.35) is called the induced electromotive force, and the magnetic flux (or number of lines of force) is defined as

$$\Phi = \int_{s} \boldsymbol{B} \cdot \boldsymbol{n}\, ds \ .$$

Using the equations of vector analysis, the differential form of Eq. (1.35) is

$$\nabla \times \boldsymbol{E} + \frac{1}{c}\frac{\partial \boldsymbol{B}}{\partial t} = 0 \ . \tag{1.36}$$

Here the electric field vector, $\boldsymbol{E}$, and the magnetic induction, $\boldsymbol{B}$, are defined in the same reference frame.

1.5 Continuity Equation for the Conservation of Charge

The conservation of electric charge is expressed by the continuity equation

$$\frac{d}{dt}\int_{v} \rho\, dv + \oint_{s} \boldsymbol{J} \cdot \boldsymbol{n}\, ds = 0 \ , \tag{1.37}$$

where ρ is the charge density, $\boldsymbol{J} \cdot \boldsymbol{n}$ is the component of current density normal to the surface element ds, $\oint_s$ denotes a closed surface integral, and $\int_v \rho\, dv$ is a volume integral within the closed surface. This equation states that the current flowing into a closed surface area is equal to the time rate of change of charge within the area, and was first confirmed experimentally by Faraday (1843). Using the equations of vector analysis, the differential form of Eq. (1.37) is

$$\frac{\partial \rho}{\partial t} + \nabla \cdot \boldsymbol{J} = 0 \ . \tag{1.38}$$

1.6 Maxwell's Equations

Maxwell (1861, 1873) first noted that the electrostatic equations (1.31) and (1.34) were inconsistent with the continuity equation (1.38) for time varying charges and current. He suggested that a displacement current $(1/4\pi)\partial D/\partial t$ be added so that Eqs. (1.16), (1.31), and (1.34) for the static fields together with Eq. (1.35) for time varying fields define the electromagnetic field equations:

$$\nabla \times \boldsymbol{H} - \frac{1}{c}\frac{\partial \boldsymbol{D}}{\partial t} = \frac{4\pi}{c}\boldsymbol{J} \quad \text{or} \quad \oint_{\mathrm{C}} \boldsymbol{H} \cdot d\boldsymbol{l} - \frac{1}{c}\frac{\partial}{\partial t}\int_{\mathrm{S}} \boldsymbol{D} \cdot \boldsymbol{n}\, ds = \frac{4\pi}{c}\int_{\mathrm{S}} \boldsymbol{J} \cdot \boldsymbol{n}\, ds \ , \tag{1.39}$$

$$\nabla \times \boldsymbol{E} + \frac{1}{c}\frac{\partial \boldsymbol{B}}{\partial t} = 0 \quad \text{or} \quad \oint_{\mathrm{C}} \boldsymbol{E} \cdot d\boldsymbol{l} + \frac{1}{c}\frac{\partial}{\partial t}\int_{\mathrm{S}} \boldsymbol{B} \cdot \boldsymbol{n}\, ds = 0 \ , \tag{1.40}$$

$$\nabla \cdot \boldsymbol{D} = 4\pi\rho \quad \text{or} \quad \oint_{\mathrm{S}} \boldsymbol{D} \cdot \boldsymbol{n}\, ds = 4\pi \int_{\mathrm{V}} \rho\, dv \ , \tag{1.41}$$

and

$$\nabla \cdot \boldsymbol{B} = 0 \quad \text{or} \quad \oint_{\mathrm{S}} \boldsymbol{B} \cdot \boldsymbol{n}\, ds = 0 \ . \tag{1.42}$$

Thus five vectors describe the electromagnetic field and its effect on material objects. They are the electric vector, $\boldsymbol{E}$, the magnetic vector, $\boldsymbol{H}$, the electric displacement, $\boldsymbol{D}$, the magnetic induction, $\boldsymbol{B}$, and the electric current density, $\boldsymbol{J}$. The continuity equation (1.38) for the conservation of electric charge is assumed to hold, and the current density $\boldsymbol{J}$ is given by

$$\boldsymbol{J} = \sigma \boldsymbol{E} + \rho \boldsymbol{v} = \boldsymbol{J}_\sigma + \boldsymbol{J}_{\mathrm{c}} \ , \tag{1.43}$$

where the conduction current $\boldsymbol{J}_\sigma = \sigma \boldsymbol{E}$ for material with conductivity, σ, and the convection current $\boldsymbol{J}_{\mathrm{c}} = \rho \boldsymbol{v}$ for charges of density ρ moving with velocities $\boldsymbol{v}$.

1.7 Boundary Conditions

Maxwell's equations (1.39)–(1.42) apply for material in which the dielectric constant, ε, and the magnetic permeability, μ, are continuous. At surfaces of discontinuity in ε and/or μ, the electromagnetic fields at each side of the discontinuity are related by the following equations which may be obtained directly from the integral form of Maxwell's equations

$$\boldsymbol{n}_{12} \cdot (\boldsymbol{B}_2 - \boldsymbol{B}_1) = 0 \ , \tag{1.44}$$

$$\boldsymbol{n}_{12} \cdot (\boldsymbol{D}_2 - \boldsymbol{D}_1) = 4\pi\rho_{\mathrm{s}} \ , \tag{1.45}$$

$$\boldsymbol{n}_{12} \times (\boldsymbol{E}_2 - \boldsymbol{E}_1) = 0 \ , \tag{1.46}$$

$$\boldsymbol{n}_{12} \times (\boldsymbol{H}_2 - \boldsymbol{H}_1) = 4\pi \boldsymbol{J}_{\mathrm{s}}/c \ , \tag{1.47}$$

where $\boldsymbol{n}_{12}$ is a unit vector normal to the surface of discontinuity and pointing from medium 1 into medium 2. Eqs. (1.44) and (1.45) apply to components of $\boldsymbol{B}$ and $\boldsymbol{D}$ which are normal to the surface, whereas Eqs. (1.46) and (1.47) apply to components of $\boldsymbol{E}$ and $\boldsymbol{H}$ which are tangential to the surface. The ρ_{s} and $\boldsymbol{J}_{\mathrm{s}}$ denote, respectively, the charge density and current density at the boundary surface.

1.8 Energy Density of the Electromagnetic Field

The total energy density, U, of the electromagnetic field is (Kelvin, 1850; Maxwell, 1861, 1873)

$$U = U_E + U_H = \frac{\varepsilon}{8\pi}E^2 + \frac{\mu}{8\pi}H^2 = \frac{1}{8\pi}[\boldsymbol{E}\cdot\boldsymbol{D} + \boldsymbol{B}\cdot\boldsymbol{H}] \ , \tag{1.48}$$

where U_E and U_H are, respectively, the electric and magnetic energy densities, ε and μ are, respectively, the dielectric constant and magnetic permeability of the medium, and E and H are, respectively, the magnitudes of the electric and magnetic vectors, $\boldsymbol{E}$ and $\boldsymbol{H}$.

1.9 Poynting Energy Flux and Poynting's Theorem

The amount of energy per unit time which crosses a unit area normal to the direction of the electric and magnetic vectors, $\boldsymbol{E}$ and $\boldsymbol{H}$, is given by the magnitude of the Poynting vector (Poynting, 1884)

$$\boldsymbol{S} = \frac{c}{4\pi}(\boldsymbol{E}\times\boldsymbol{H}) \ , \tag{1.49}$$

for time varying electromagnetic fields. The conservation of energy is expressed by Poynting's theorem:

$$\partial U/\partial t + \nabla\cdot\boldsymbol{S} = -\boldsymbol{J}\cdot\boldsymbol{E} \ , \tag{1.50}$$

for a total energy density, U, and current density $\boldsymbol{J}$.

1.10 Electromagnetic Momentum and Radiation Pressure

The momentum density of the electromagnetic field is (Maxwell, 1861)

$$g = \frac{\boldsymbol{S}}{c^2} = \frac{(\boldsymbol{E}\times\boldsymbol{H})}{4\pi c} \ .$$

The radiation pressure, $\boldsymbol{P}$, is given by the net rate of momentum transfer per unit area.

$$\boldsymbol{P} = \frac{\boldsymbol{S}}{c} = \frac{\boldsymbol{E}\times\boldsymbol{H}}{4\pi} \ . \tag{1.51}$$

1.11 Lorentz Force Law

The vector force, $\boldsymbol{F}$, exerted by an electromagnetic field on a charge, q, moving at the velocity, v, is (Heaviside, 1889; Lorentz, 1892)

$$\boldsymbol{F} = q\left(\boldsymbol{E} + \frac{\mu}{c}\boldsymbol{v}\times\boldsymbol{H}\right) \ , \tag{1.52}$$

where the magnitude of $\boldsymbol{v} \times \boldsymbol{H}$ is the component of velocity in the direction perpendicular to the magnetic field when μ is unity and $\boldsymbol{B} = \mu \boldsymbol{H}$.

Let us suppose that a charged particle of charge, q, has initial velocity components $v_\perp$ and $v_\parallel$ perpendicular and parallel to the direction of a uniform constant magnetic field, $\boldsymbol{H}$. It follows from Eq. (1.52) that the particle motion in the presence of a $\boldsymbol{H}$ will be the superposition of a constant velocity, $v_\parallel$, in the direction of $\boldsymbol{H}$ together with a circular motion in the plane perpendicular to $\boldsymbol{H}$. As first pointed out by Heaviside (1904) the circular radian frequency (the gyrofrequency) is given by

$$\omega_{\mathrm{H}} = \frac{q\mu H}{\gamma mc} = \frac{qB}{\gamma mc} , \tag{1.53}$$

and the gyroradius (the Larmor radius) is

$$r_{\mathrm{H}} = \frac{v_\perp}{\omega_{\mathrm{H}}} = \frac{mc\gamma v_\perp}{qB} , \tag{1.54}$$

where m is the mass of the particle, and the particle momentum is $\boldsymbol{p} = \gamma m \boldsymbol{v}$ where $\gamma = [1 - (v^2/c^2)]^{-1/2}$. For low velocities, $v \ll c$, we have $\gamma \approx 1$. In general, the path of the charged particle is a helix of radius, r_H, and pitch angle $\psi = \tan^{-1}(v/\omega_{\mathrm{H}} r_{\mathrm{H}})$.

In the presence of a uniform constant electric field, $\boldsymbol{E}$, directed along the x axis, a charged particle of charge q, momentum γmv, total energy γmc^2, and kinetic energy $(\gamma - 1)mc^2$ will follow a trajectory specified by Eq. (1.52).

$$\begin{aligned} x &= \frac{\gamma mc^2}{qE} \cosh\left(\frac{eEy}{\gamma mvc}\right) \\ &\approx \frac{qE}{2mv^2} y^2 \quad \text{for } v \ll c , \end{aligned} \tag{1.55}$$

where the initial velocity, v, is in the xy plane.

The average motion of a charged particle in the presence of uniform constant electric and magnetic fields, $\boldsymbol{E}$ and $\boldsymbol{H}$, will be in a direction perpendicular to both fields. The average drift velocity is

$$\boldsymbol{v} = \frac{c\boldsymbol{E} \times \boldsymbol{H}}{H^2} \tag{1.56}$$

where it has been assumed that $v \ll c$ and $\boldsymbol{E} \ll \boldsymbol{H}$.

1.12 Electromagnetic Plane Waves

Maxwell's equations (1.39)–(1.42) in a charge and current free region (zero conductivity and no magnetic field) assume the form

$$\nabla^2 E - \frac{\varepsilon\mu}{c^2}\frac{\partial^2 E}{\partial t^2} = 0 , \tag{1.57}$$

and

$$\nabla^2 H - \frac{\varepsilon\mu}{c^2}\frac{\partial^2 H}{\partial t^2} = 0 ,$$

where ε and μ are, respectively, the dielectric constant and the magnetic permeability. Eqs. (1.57) are wave equations for electromagnetic waves which propagate with the velocity

$$v = c/\sqrt{\varepsilon\mu} \ . \tag{1.58}$$

Of special interest is the harmonic solution to Eq. (1.57) for which

$$\boldsymbol{E} = \boldsymbol{E}_0 \exp[i(\boldsymbol{k} \cdot \boldsymbol{r} - \omega t)] \ , \tag{1.59}$$

where t is the time variable and the harmonic frequency, ω, is related to the propagation or wave vector, $\boldsymbol{k}$, by

$$\boldsymbol{k} = \frac{2\pi}{\lambda}\boldsymbol{s} = \frac{\omega}{c} n\boldsymbol{s} \ , \tag{1.60}$$

where λ is the wavelength, $\boldsymbol{s}$ is the direction of wave propagation which is perpendicular to $\boldsymbol{E}$, and n is the index of refraction. The phase velocity of the wave is

$$v_p = \frac{\omega}{k} = \frac{c}{n} = \frac{c}{\sqrt{\varepsilon\mu}} \ , \tag{1.61}$$

whereas the group velocity is given by the differential

$$v_g = \frac{d\omega}{dk} \ . \tag{1.62}$$

It follows directly from Maxwell's equations that when the electric field vector $\boldsymbol{E}$ is given by Eq. (1.59), the magnetic field vector $\boldsymbol{H}$ is given by

$$\boldsymbol{H} = \left(\frac{\varepsilon}{\mu}\right)^{1/2} \boldsymbol{s} \times \boldsymbol{E} \ ,$$

where $\boldsymbol{s}$ is a unit vector denoting the direction of propagation of the wave. It follows that the field vectors of a plane wave are perpendicular to each other and perpendicular to the direction of propagation, and hence they are called transverse plane waves.

From Eq. (1.49), the energy flux in the plane wave is

$$S = \frac{c}{4\pi}\sqrt{\frac{\varepsilon}{\mu}}E^2 = \frac{c}{4\pi}\sqrt{\frac{\mu}{\varepsilon}}\boldsymbol{H}^2 = \frac{c}{8\pi}\sqrt{\frac{\varepsilon}{\mu}}E_0^2 \tag{1.63}$$

which is directed along the direction of wave propagation. The energy density, U, in the plane wave is, from Eq. (1.48),

$$U = \frac{\varepsilon E^2}{4\pi} = \frac{\mu H^2}{4\pi} \ . \tag{1.64}$$

When the medium has a non-zero conductivity, σ, the propagation vector, k, is complex and is given by (Jackson, 1962)

$$\begin{aligned} k &= \left[\varepsilon\mu\frac{\omega^2}{c^2}\left(1 + i\frac{4\pi\sigma}{\omega\varepsilon}\right)\right]^{1/2} \\ &\approx \sqrt{\mu\varepsilon}\frac{\omega}{c} + i\frac{2\pi}{c}\sqrt{\frac{\mu}{\varepsilon}}\sigma \quad \text{for } \sigma \ll \omega\varepsilon \end{aligned}$$

$$\approx (1+i)\frac{\sqrt{2\pi\omega\mu\sigma}}{c} \quad \text{for } \sigma \gg \omega\varepsilon \ , \tag{1.65}$$

and the waves are attenuated exponentially with distance. The phase velocity of the wave is $v_p = \omega/\alpha = c/n$ where α is the real part of k and n is the index of refraction. When a magnetic field is present there are additional real and imaginary terms for k, and these terms are given in Sect. 1.33.

1.13 Polarization of Plane Waves – The Stokes Parameters

If the propagation direction of a harmonic plane wave is the z axis of a Cartesian coordinate system, the components of the electric field vector are

$$E_x = a_1 \cos(\omega t - \boldsymbol{k} \cdot \boldsymbol{r} + \delta_1)$$

$$E_y = a_2 \cos(\omega t - \boldsymbol{k} \cdot \boldsymbol{r} + \delta_2) \tag{1.66}$$

and $E_z = 0$,

where a_1 and a_2 are scalar amplitudes, ω is the harmonic frequency, t is the time variable, $\boldsymbol{k}$ is the wave vector, $\boldsymbol{r}$ is the vector radius, and δ_1 and δ_2 are arbitrary phases. It follows directly from Eqs. (1.66) that E_x and E_y are related by

$$\left(\frac{E_x}{a_1}\right)^2 + \left(\frac{E_y}{a_2}\right)^2 - \frac{2E_xE_y}{a_1a_2}\cos\delta = \sin^2\delta \ , \tag{1.67}$$

where $\delta = \delta_2 - \delta_1$. Eq. (1.67) describes an ellipse whose tilt angle, ψ, with respect to the x axis, and the length of the major and minor axis, $2a$ and $2b$, are related by the equations (Born and Wolf, 1964)

$$\tan 2\psi = (\tan 2\alpha)\cos\delta \ ,$$

$$\tan \chi = \pm b/a \ ,$$

$$a^2 + b^2 = a_1^2 + a_2^2 \ ,$$

and

$$\sin 2\chi = (\sin 2\alpha)\sin\delta \ , \tag{1.68}$$

where

$$\tan\alpha = a_2/a_1 \ .$$

In general, the wave is said to be elliptically polarized. When $\delta = \delta_2 - \delta_1 = m\pi$ where m is an integer, the ellipse reduces to a straight line and the wave is said to be linearly polarized. When $a_1 = a_2 = a$ and $\delta = \pm\pi/2 + 2m\pi$, where m is an integer, the wave is said to be circularly polarized. By definition the wave is right hand or left hand circularly polarized according as the $\pm$ sign is plus or minus.

The Stokes parameters for the wave are (Stokes, 1852)

$$I = a_1^2 + a_2^2 \ ,$$

$$Q = a_1^2 - a_2^2 = I\cos 2\chi \cos 2\psi \ ,$$

$$U = 2a_1a_2\cos\delta = I\cos 2\chi \sin 2\psi \ ,$$

and

$$V = 2a_1 a_2 \sin\delta = I \sin 2\chi \ , \tag{1.69}$$

where

$$I^2 = Q^2 + U^2 + V^2 \ ,$$

and I represents the intensity of the wave.

1.14 Reflection and Refraction of Plane Waves

By considering the boundary conditions for the electromagnetic fields (Eqs. (1.44)–(1.47)) the laws of reflection and refraction of the harmonic plane wave may be obtained. If a plane wave falls on a boundary between two homogeneous media, the angle θ_i, between the direction of propagation of the incident wave and the normal to the surface is related to the angle θ_t, between the direction of propagation of the transmitted wave and the surface normal by Snell's law of refraction (Snell, 1621)

$$\frac{\sin\theta_i}{\sin\theta_t} = \left(\frac{\varepsilon_2\mu_2}{\varepsilon_1\mu_1}\right)^{1/2} = \frac{n_2}{n_1} = \frac{v_1}{v_2} \ , \tag{1.70}$$

where ε is the dielectric constant, μ is the magnetic permeability, $n = \sqrt{\varepsilon\mu}$ is called the index of refraction, v is the phase velocity, and subscripts 1 and 2 denote, respectively, the incident medium and the medium of the transmitted wave. The similar relation for the angles θ_i and θ_r which the incident and reflected waves make with the surface normal is

$$\sin\theta_i = \sin\theta_r \ , \tag{1.71}$$

which follows from Fermat's (1627) principle that light follows that path which brings it to its destination in the shortest possible time.

The ratio of the amount of energy in the reflected wave to that in the incident wave is given by (Fresnel, 1822)

$$R_{\parallel} = \frac{\tan^2(\theta_i - \theta_t)}{\tan^2(\theta_i + \theta_t)} \ ,$$
$$R_{\perp} = \frac{\sin^2(\theta_i - \theta_t)}{\sin^2(\theta_i + \theta_t)} \ , \tag{1.72}$$

where the reflection coefficients $R_{\parallel}$ and $R_{\perp}$ are, respectively, for waves which are linearly polarized in the plane of incidence and normal to the plane of incidence. Using Snell's law, Eq. (1.70), and assuming $\mu = 1$, these equations become

$$R_{\perp} = \left[\frac{\cos\theta_i - (\varepsilon - \sin^2\theta_i)^{1/2}}{\cos\theta_i + (\varepsilon - \sin^2\theta_i)^{1/2}}\right]^2 \ ,$$

and (1.73)

$$R_{\parallel} = \left[\frac{\varepsilon\cos\theta_i - (\varepsilon - \sin^2\theta_i)^{1/2}}{\varepsilon\cos\theta_i + (\varepsilon - \sin^2\theta_i)^{1/2}}\right]^2 \ ,$$

where ε is the dielectric constant of the medium. Use of Eq. (1.73) in determining the dielectric constant of the Moon and Venus is discussed by Heiles and Drake (1963). Clark and Kuzmin (1965), Berge and Greisen (1969), and Muhleman (1969). The degree of polarization, Π, of the reflected light is given by:

$$\Pi = \left|(R_{\|} - R_{\perp})/(R_{\|} + R_{\perp})\right| \ . \tag{1.74}$$

The ratio of the energy in the transmitted wave to that in the incident wave is given by the transmission coefficients

$$T_{\|} = 1 - R_{\|} \text{ and } T_{\perp} = 1 - R_{\perp} \ . \tag{1.75}$$

For normal incidence, Eqs. (1.73) and (1.75) reduce to

$$R = \left(\frac{n-1}{n+1}\right)^2 \text{ and } T = \frac{4n}{(n+1)^2} \ , \tag{1.76}$$

where

$$n = \frac{n_2}{n_1} = \left(\frac{\varepsilon_2 \mu_2}{\varepsilon_1 \mu_1}\right)^{1/2} \ .$$

When a wave from free space is normally incident upon a medium with $\mu = 1$, the reflection coefficient becomes

$$R_0 = \left[\frac{1 - \sqrt{\varepsilon}}{1 + \sqrt{\varepsilon}}\right]^2 \ . \tag{1.77}$$

If the medium is partially conducting with conductivity, σ, then the term ε appearing in Eq. (1.77) must be replaced by

$$\frac{1}{\mu}\left[\varepsilon + \frac{i\sigma}{2\pi\nu}\right] \ , \tag{1.78}$$

where ν is the wave frequency. In this case the amplitude of the transmitted wave will fall to $1/e$ of its initial value in the skin depth distance

$$\delta = c/\sqrt{4\pi^2\mu\nu\sigma} = c/\sqrt{2\pi\mu\omega\sigma} \ . \tag{1.79}$$

Radar astronomers often measure the radar cross section given as

$$\sigma_r = gR_0\pi r^2 \ , \tag{1.80}$$

where r is the radius of the planet, and the coefficient, g, accounts for surface roughness. Early radar cross section measurements of Venus, Mars and the Moon are reviewed by Evans (1969); detailed radar mapping of the surface of Venus from the orbiting Magellan spacecraft can be found in the series of papers introduced by Saunders et al. (1992).

Of special interest is the Brewster angle of incidence for which the electric vector of the reflected radiation is linearly polarized in a plane normal to the plane of incidence. This angle is given by (Brewster, 1815)

$$\tan\theta_i = n = \frac{n_2}{n_1} = \left(\frac{\varepsilon_2 \mu_2}{\varepsilon_1 \mu_1}\right)^{1/2} \ , \tag{1.81}$$

which is $\sqrt{\varepsilon}$ for incidence from free space upon a medium with $\mu = 1$. The Brewster angle θ_i is also called the polarizing angle. At this angle, the reflected and transmitted rays are perpendicular to each other with $\theta_i + \theta_t = \pi/2$, and the electric vector of the reflected radiation has no component in the plane of incidence.

Eqs. (1.51), (1.64), (1.71), and (1.72) can be combined to give the net radiation pressure at the surface of reflection.

$$P_{\mathrm{N}} = \frac{E^2}{4\pi}(1+R)\cos^2\theta_i \ ,$$

and

$$P_{\mathrm{T}} = \frac{E^2}{4\pi}(1-R)\sin\theta_i\cos\theta_i \ , \tag{1.82}$$

where $E^2/4\pi$ is the energy density of the wave, P_{N} and P_{T} denote the normal and tangential components of pressure, R is the reflection coefficient, and θ_i is the angle of incidence.

1.15 Dispersion Relations

An electron of charge e and mass m which is part of an atomic system acts like a harmonic oscillator whose motion in the x direction is given by

$$x(t) = x_0\cos(\omega_0 t) \ ,$$

where t is the time variable and ω_0 is the radian frequency of the harmonic oscillation. When this oscillator is subjected to a harmonic plane wave of radian frequency, ω, its equation of motion as specified by the Lorentz force law (Eq. (1.52)) is

$$m\ddot{x} + m\omega_0^2 x = eE(t) = eE_0\exp[i\omega t] \ , \tag{1.83}$$

where E_0 is the amplitude of the electric field vector $\boldsymbol{E}$ which is assumed to have the x direction, and $\ddot{}$ denotes the second derivative with respect to time. Eq. (1.83) has the solution

$$x(t) = \frac{eE(t)}{m(\omega_0^2 - \omega^2)} \ ,$$

so that the dipole moment is

$$d(t) = ex(t) = \frac{e^2E(t)}{m(\omega_0^2 - \omega^2)} \ . \tag{1.84}$$

Using Eqs. (1.25), (1.30), and (1.84), the index of refraction, n, of a density of N such bound electrons is given by the dispersion relation

$$\frac{n^2-1}{n^2+2} = \frac{4\pi Ne^2}{3m(\omega_0^2 - \omega^2)} \ . \tag{1.85}$$

For a gas, $n \approx 1$ and we have

$$n \approx \left[1 - \frac{4\pi Ne^2}{m(\omega_0^2 - \omega^2)}\right]^{1/2} \ , \tag{1.86}$$

a relation first derived by Maxwell (1899). For a system of N atoms, the quantum mechanical form of Eq. (1.86) is (Ladenburg, 1921; Kramers, 1924)

$$n \approx \left[1 - \frac{Ne^2}{\pi m} \sum_k \frac{f_{1k}}{\nu_{1k}^2 - \nu^2}\right]^{1/2} , \tag{1.87}$$

where f_{1k} and ν_{1k} are, respectively, the oscillator strength and the frequency of the transition from the ground state to the k level. The oscillator strength is defined in terms of the Einstein transition probabilities in Sect. 2.4.

For a plasma with free electron density, N_e, Eq. (1.86) becomes

$$n \approx \left[1 - \left(\frac{\omega_p}{\omega}\right)^2\right]^{1/2} , \tag{1.88}$$

where the plasma frequency, ω_p, is given by

$$\omega_p = \left(\frac{4\pi N_e e^2}{m}\right)^{1/2} ,$$

and the depth of penetration, δ_p, of the plasma is

$$\delta_p = c/(\omega_p^2 - \omega^2)^{1/2} \approx c/\omega_p .$$

When absorption must be taken into account, the expression for the index of refraction becomes more complex as shown in Sect. 1.33. There is a general relation between refraction and absorption, however, which was first derived by Kronig and Kramers (1928). If the electric susceptibility $\chi = \chi_1 + i\chi_2$, where χ_1 and χ_2 are real, then the Kramers–Kronig relations are

$$\chi_1(\omega_0) = \frac{2}{\pi} \int_0^\infty \frac{\omega \chi_2(\omega) d\omega}{\omega^2 - \omega_0^2} \tag{1.89}$$

and

$$\chi_2(\omega_0) = -\frac{2\omega_0}{\pi} \int_0^\infty \frac{\chi_1(\omega) d\omega}{\omega^2 - \omega_0^2} ,$$

where ω_0 is the frequency at which χ_1 or χ_2 are being evaluated. When a medium has a total extinction cross section, σ_e, then Eq. (1.89) becomes

$$\chi_1(\omega_0) = \frac{2c}{\pi V} \int_0^\infty \frac{\sigma_e(\omega) d\omega}{\omega^2 - \omega_0^2} , \tag{1.90}$$

where V is the volume of the medium. As first derived by Feenberg (1932), the extinction cross section for matter with an absorption cross section, σ_a, and scattering cross section, σ_s, is given by $\sigma_e = \sigma_a + \sigma_s$. The Kramers–Kronig relations for interstellar polarization are given by Martin (1975).

1.16 Lorentz Coordinate Transformation

The Lorentz transformation is a means by which the coordinates x, y, z, t of an event in one inertial system (the K system) can be transferred into the coordinates x', y', z', t' of the same event in another inertial system (the K' system). An inertial system is defined as one in which a freely moving body proceeds with constant velocity, and according to the first postulate of special relativity the laws of nature are invariant with respect to the Lorentz transformation of the coordinates from one inertial system to another. If the K' system is moving at a uniform velocity, v, along the x axis and away from the K system, then the Lorentz transformation (Lorentz, 1904) is

$$x' = \gamma[x - vt], \quad y' = y, \quad z' = z, \quad \text{and } t' = \gamma[t - (xv)/c^2] \ , \tag{1.91}$$

where

$$\gamma = \left[1 - \frac{v^2}{c^2}\right]^{-1/2} .$$

In his 1904 paper Lorentz showed that Maxwell's equations are invariant when subjected to his coordinate transformation.

1.17 Lorentz Transformation of the Electromagnetic Field

If the K system has electromagnetic fields, $\boldsymbol{E}$ and $\boldsymbol{H}$, then the fields $\boldsymbol{E}'$ and $\boldsymbol{H}'$ in the K' system moving at the velocity v with respect to the K system may be calculated by using the Lorentz coordinate transformations. The components of E and H parallel to the direction of motion remain unchanged, whereas the components perpendicular to $\boldsymbol{v}$ are given by

$$\boldsymbol{E}' = \gamma[\boldsymbol{E} + (\boldsymbol{v}/c) \times \boldsymbol{H}] \tag{1.92}$$

and

$$\boldsymbol{H}' = \gamma[\boldsymbol{H} - (\boldsymbol{v}/c) \times \boldsymbol{E}] \ ,$$

where $\gamma = [1 - (v^2/c^2)]^{-1/2}$, and we assume that $\mu = 1$ as it is for free space. In many astrophysical situations, $\boldsymbol{E} \ll \boldsymbol{H}$ and $v \ll c$ so that Eqs. (1.92) become

$$\boldsymbol{E}' \approx \boldsymbol{E} + (\boldsymbol{v}/c) \times \boldsymbol{H} \tag{1.93}$$

and

$$\boldsymbol{H}' \approx \boldsymbol{H} \ .$$

1.18 Induced Electric Fields in Moving or Rotating Matter (Unipolar [Homopolar] Induction)

From Eq. (1.93), and the boundary conditions given by Eqs. (1.45) and (1.47), a uniform slab of perfectly conducting matter moving at velocity $\boldsymbol{v}$ through a magnetic field $\boldsymbol{B}$ will have an induced electric field given by

$$\boldsymbol{E} = -\frac{\boldsymbol{v}}{c} \times \boldsymbol{B} \ , \tag{1.94}$$

whose lines of force terminate in the surface charge density

$$\rho_s = E/4\pi \ . \tag{1.95}$$

A surface current density

$$J_s = cB/4\pi \tag{1.96}$$

must also be present.

Similarly, the electric field, $\boldsymbol{E}$, induced on the surface of a sphere of radius, $\boldsymbol{R}$, rotating at an angular velocity, $\boldsymbol{\Omega}$, through a magnetic field $\boldsymbol{B}$ is

$$\boldsymbol{E} = -\frac{\boldsymbol{\Omega} \times \boldsymbol{R}}{c} \times \boldsymbol{B} \ , \tag{1.97}$$

which also follows from the Lorentz force law (Eq. (1.52)). When the magnetic field is a magnetic dipole (Eqs. (1.19)), it follows from Eq. (1.97) that the potential difference induced between latitude λ and the equator is

$$V = \frac{\Omega B_p R^2}{2} (\cos^2 \lambda - 1) \ , \tag{1.98}$$

where R is the sphere's radius, and B_p is the polar magnetic field strength.

1.19 Electromagnetic Field of a Point Charge Moving with a Uniform Velocity

When a point charge, q, is at rest it has no magnetic field and an electric field, E, given by Eq. (1.2)

$$\boldsymbol{E} = \frac{q}{R^2} \boldsymbol{n}_r \ ,$$

where R is the radial distance from the charge and $\boldsymbol{n}_r$ is a unit vector in the radial direction. When the charge is moving at the uniform velocity, $\boldsymbol{v}$, the Lorentz transformations give (Poincaré, 1905)

$$\boldsymbol{E} = \frac{q}{R^2} \frac{1 - \frac{v^2}{c^2}}{\left(1 - \left(\frac{v^2}{c^2}\right) \sin^2 \theta\right)^{3/2}} \boldsymbol{n}_r \ , \tag{1.99}$$

and

$$\boldsymbol{H} = -(\boldsymbol{v} \times \boldsymbol{E})/c \ ,$$

where θ is the angle between the direction of motion and the unit radial vector, $\boldsymbol{n}_r$.

1.20 Vector and Scalar Potentials (The Retarded and Liénard–Wiechert Potentials)

Because the divergence of the curl of any vector is zero, it follows from Eq. (1.42) that the magnetic induction, $\boldsymbol{B}$, may be expressed as

$$\boldsymbol{B} = \nabla \times \boldsymbol{A} \ , \tag{1.100}$$

where $\boldsymbol{A}$ is called the magnetic vector potential. It then follows from Eq. (1.40) that

$$\boldsymbol{E} = -\frac{1}{c}\frac{\partial \boldsymbol{A}}{\partial t} - \nabla\varphi \ , \tag{1.101}$$

where φ is called the scalar electric potential. The $\boldsymbol{A}$ and φ may be related by the Lorentz condition (Lorentz, 1892)

$$\nabla \cdot \boldsymbol{A} + \frac{1}{c}\frac{\partial \varphi}{\partial t} = 0 \ . \tag{1.102}$$

When the Lorentz condition is used, it follows from Maxwell's Eqs. (1.39)–(1.42) that in a vacuum ($\varepsilon = \mu = 1$) the potentials satisfy the wave equations

$$\nabla^2 \boldsymbol{A} - \frac{1}{c^2}\frac{\partial^2 \boldsymbol{A}}{\partial t^2} = -\frac{4\pi}{c}\boldsymbol{J} \tag{1.103}$$

and

$$\nabla^2 \varphi - \frac{1}{c^2}\frac{\partial^2 \varphi}{\partial t^2} = -4\pi\rho \ ,$$

where $\boldsymbol{J}$ is the current density and ρ is the charge density. These equations have the retarded solutions (Lorentz, 1892)

$$\boldsymbol{A}(R, t) = \frac{1}{c}\int \frac{\boldsymbol{J}\left(t - \frac{R}{c}\right)}{R} dv$$

and

$$\varphi(R, t) = \int \frac{\rho\left(t - \frac{R}{c}\right)}{R} dv \ , \tag{1.104}$$

where the integrals are volume integrals of charge and current density at the retarded time $t' = t - R/c$, and R is the radial distance.

If each volume element of charge is moving at the velocity, v, the volume integrals in Eqs. (1.104) must be corrected for the changing charge density. Consider a spherical shell of thickness dr at the distance R from the origin of the coordinate system used in the integration. During the time $dt = dr/c$, the amount of charge flowing through the inner surface of the shell is

$$\rho \boldsymbol{n}_{\mathrm{r}} \cdot \boldsymbol{v}\, dt = \rho \frac{\boldsymbol{n}_{\mathrm{r}} \cdot \boldsymbol{v}}{c} dr \ , \tag{1.105}$$

where $\boldsymbol{n}_{\mathrm{r}}$ is a unit vector in the radial direction and $\boldsymbol{v}$ is the velocity vector. The sum of the charge elements, dq, to be accounted for in the volume element, dv, is

$$dq = \rho\left[1 - \frac{\boldsymbol{n}_{\mathrm{r}} \cdot \boldsymbol{v}}{c}\right] dv \ . \tag{1.106}$$

For a point charge, q, moving at the velocity, v, Eqs. (1.104) and (1.106) give the Liénard–Wiechert potentials (Liénard, 1898; Wiechert, 1901)

$$\varphi(R,t) = \left.\frac{q}{R\left[1 - \left(\boldsymbol{n}_{\mathrm{r}} \cdot \frac{\boldsymbol{v}}{c}\right)\right]}\right|_{t'=t-R/c} \tag{1.107}$$

$$\boldsymbol{A}(R,t) = \left.\frac{q\boldsymbol{v}}{cR\left[1 - \left(\boldsymbol{n}_{\mathrm{r}} \cdot \frac{\boldsymbol{v}}{c}\right)\right]}\right|_{t'=t-R/c},$$

where R is the distance from the charge.

1.21 Electromagnetic Radiation from an Accelerated Point Charge

It follows directly from Eqs. (1.101) and (1.107) that the electric field vector $\boldsymbol{E}$ and magnetic induction $\boldsymbol{B}$ of a point charge q moving along the z axis at a velocity $v \ll c$ are given by

$$\boldsymbol{E} = \left.\frac{q\dot{v}}{c^2 R} \sin\theta \boldsymbol{n}_\theta\right|_{t'=t-R/c} \tag{1.108}$$

$$\boldsymbol{B} = \left.\frac{q\dot{v}}{c^2 R} \sin\theta \boldsymbol{n}_\varphi\right|_{t'=t-R/c},$$

where $\dot{}$ denotes the first derivative with respect to time, $\dot{v}$ is the acceleration, θ is the angle between the z axis and the radial vector to the point of observation, $\boldsymbol{n}_\theta$ and $\boldsymbol{n}_\varphi$ are unit vectors in the θ and φ directions defined by the spherical coordinate system, and it is assumed that the radial distance, R, is large. It then follows from Eqs. (1.63) and (1.64) that the total energy radiated per unit time per unit solid angle, $d\Omega$, is

$$\frac{dP}{d\Omega} = \left.\frac{q^2 \dot{v}^2}{4\pi c^3} \sin^2\theta\right|_{t'=t-R/c}, \tag{1.109}$$

and the total energy radiated per unit time in all directions is

$$P = \left.\frac{2}{3}\frac{q^2}{c^3}\dot{v}^2\right|_{t'=t-R/c}, \tag{1.110}$$

a result first obtained by Larmor (1897).

When the velocities are large, the correct results are (Liénard, 1898; Jackson, 1962)

$$\boldsymbol{E} = \left[q\frac{(\boldsymbol{n}-\boldsymbol{\beta})(1-\beta^2)}{(1-\boldsymbol{n}\cdot\boldsymbol{\beta})^3 R^2}\right]\Bigg|_{t'=t-R/c} + \frac{q}{c}\left[\frac{\boldsymbol{n}\times\{(\boldsymbol{n}-\boldsymbol{\beta})\times\dot{\boldsymbol{\beta}}\}}{(1-\boldsymbol{n}\cdot\boldsymbol{\beta})^3 R}\right]\Bigg|_{t'=t-R/c} \tag{1.111}$$

$$\boldsymbol{H} = \boldsymbol{n} \times \boldsymbol{E}$$

$$\frac{dP}{d\Omega} = \left.\frac{q^2}{4\pi c}\frac{\left|\boldsymbol{n}\times\{(\boldsymbol{n}-\boldsymbol{\beta})\times\dot{\boldsymbol{\beta}}\}\right|^2}{(1-\boldsymbol{n}\cdot\boldsymbol{\beta})^5}\right|_{t'=t-R/c}$$

and

$$P = \frac{2}{3}\frac{q^2}{c}\gamma^6[(\dot{\boldsymbol{\beta}})^2 - (\boldsymbol{\beta} \times \dot{\boldsymbol{\beta}})^2]\Big|_{t'=t-R/c},$$

where $\boldsymbol{\beta} = \boldsymbol{v}/c$, the unit vector in the direction of observation is $\boldsymbol{n}$, and $\gamma = [1 - (v/c)^2]^{-1/2}$.

1.22 Electromagnetic Radiation from Electric and Magnetic Dipoles

The electric dipole moment $d(t)$ of a charge q is related to its acceleration $\dot{v}(t)$ by $\dot{v}(t) = \ddot{d}(t)/q$. It therefore follows from Eqs. (1.108) that for an electric dipole oriented along the z axis, the electric and magnetic field vectors are

$$\boldsymbol{E} = \frac{\ddot{d}(t)}{c^2 R}\sin\theta \boldsymbol{n}_\theta\Big|_{t'=t-R/c} \tag{1.112}$$

$$\boldsymbol{H} = \frac{\ddot{d}(t)}{c^2 R}\sin\theta \boldsymbol{n}_\varphi\Big|_{t'=t-R/c},$$

where the distance R is large, θ is the angle between the z axis and the radial vector to the point of observation, and $\boldsymbol{n}_\theta$ and $\boldsymbol{n}_\varphi$ are unit vectors in the θ and φ direction. It also follows from Eqs. (1.63) and (1.64) that the total energy radiated per unit time per unit solid angle is

$$\frac{dP}{d\Omega} = \frac{[\ddot{d}(t)]^2}{4\pi c^3}\sin^2\theta\Big|_{t'=t-R/c} \tag{1.113}$$

and the total energy radiated per unit time in all directions is

$$P = \frac{2}{3}\frac{[\ddot{d}(t)]^2}{c^3}\Big|_{t'=t-R/c}. \tag{1.114}$$

As an example, for an electric dipole rotating in the x–y plane with angular velocity, $\boldsymbol{\Omega}$, the total energy radiated per unit time per unit solid angle is (Landau and Lifshitz, 1962)

$$\frac{dP}{d\boldsymbol{\Omega}} = \frac{d_0^2\Omega^4}{8\pi c^3}(1 + \cos^2\theta) \tag{1.115}$$

and the total energy radiated per unit time in all directions is

$$P = \frac{2d_0^2\boldsymbol{\Omega}^4}{3c^3}, \tag{1.116}$$

where d_0 is the maximum value of the dipole moment $d(t)$ and quantities have been averaged over the period of the rotation. As another example, for harmonic oscillation of a charge, q, at frequency, ν_0, the total energy radiated per unit time is

$$P = \frac{64\pi^4\nu_0^4}{3c^3}\left|\frac{qx_0}{2}\right|^2, \tag{1.117}$$

where $|qx_0/2|$ is the rms value of the dipole moment, and x_0 is the peak value of the harmonic displacement. The corresponding quantities for a magnetic dipole with magnetic dipole moment $m(t)$ are given by Eqs. (1.112)–(1.117) with $d(t)$ replaced with $m(t)$, with the exception that the roles of $\boldsymbol{E}$ and $\boldsymbol{H}$ are reversed. That is, the magnetic dipole formula for $\boldsymbol{H}$ is given by the electric dipole formula for $\boldsymbol{E}$ with $d(t)$ replaced by $m(t)$. The equations for dipoles moving at speed v comparable to c were first obtained by Heaviside (1902).

Eqs. (1.112)–(1.117) are solutions for the far field ($R \gg c\dot{d}/\ddot{d}$ and $R > cd/\dot{d}$). The complete solution for the electromagnetic fields of a dipole was obtained by Hertz (1889). The components of the fields in the spherical coordinate directions are

$$E_{\mathrm{r}} = 2\left(\frac{[d]}{R^3} + \frac{[\dot{d}]}{cR^2}\right)\cos\theta \tag{1.118}$$

$$E_\theta = \left(\frac{[d]}{R^3} + \frac{[\dot{d}]}{cR^2} + \frac{[\ddot{d}]}{c^2R}\right)\sin\theta$$

$$H_\varphi = \left(\frac{[\dot{d}]}{cR^2} + \frac{[\ddot{d}]}{c^2R}\right)\sin\theta\ ,$$

where [ ] denotes an evaluation at the retarded time $t' = t - (R/c)$, the dipole direction is along the z axis, θ is the angle from the z axis to the point of observation, and R is the distance to the point of observation. The fields for a magnetic dipole of magnetic dipole moment, $m(t)$, are given by Eqs. (1.118) with $d(t)$ replaced by $m(t)$, E replaced by H and H replaced by E.

1.23 Thermal Emission from a Black Body

The brightness, $B_\nu(T)$, of the radiation from a black body, a perfect absorber, in thermodynamic equilibrium at the temperature T is (Planck, 1901, 1913; Wien, 1893, 1894; Rayleigh, 1900, 1905; Jeans, 1905, 1909; Milne, 1930)

$$B_\nu(T) = \frac{2h\nu^3}{c^2}\frac{n_\nu^2}{[\exp(h\nu/kT) - 1]} \qquad \text{Planck's law} \tag{1.119}$$

$$= \frac{2h\nu^3 n_\nu^2}{c^2}\exp(-h\nu/kT) \qquad \text{if } h\nu \gg kT \qquad \text{Wien's law} \tag{1.120}$$

$$= \frac{2n_\nu^2 \nu^2 kT}{c^2} \quad \text{if } h\nu \ll kT \qquad \text{Rayleigh–Jeans law} \tag{1.121}$$

where $n_\nu \approx 1$ is the index of refraction of the medium at frequency ν, c is the velocity of light, and h and k are, respectively, Planck's and Boltzmann's constants. The Rayleigh–Jeans approximation is used at radio frequencies, $\nu \approx 10^9$ Hz, for the criterion $h\nu \ll kT$, or $\nu < 2 \times 10^{10}\ T$, is valid. Wien's approximation is sometimes used at optical frequencies, where $\nu \approx 10^{15}$ Hz. The temperature, T, is called the equivalent brightness temperature and the units of brightness are erg sec^{-1} cm^{-2} Hz^{-1} rad^{-2}. Plots of $B_\nu(T)$ for various T are shown in Fig. 1.1. In wavelength units, the brightness, $B_\lambda(T)$, is

$$B_\lambda(T) = \frac{2hc^5}{\lambda^5} \frac{1}{[\exp(-hc/k\lambda T) - 1]} \tag{1.122}$$

where the constant appearing in the exponential function is called the second radiation constant, given by Lang (1992)

$$c_2 = hc/k = 1.438769(12) \text{cm} \ ^\circ\text{K} \ .$$

The first radiation constant is $c_1 = 2\pi hc^2 = 3.7417749(22) \times 10^{-5}$ erg s^{-1} cm^2.

The wavelength, $\lambda_{\max}$, for which the brightness is a maximum depends on the temperature, T, according to (Wien, 1894)

$$\lambda_{\max} = 0.2897756(24) T^{-1} \text{ cm} \qquad \text{Wien displacement law} \ , \tag{1.123}$$

where the temperature is in degrees Kelvin, or °K, and the Wien displacement law constant $b = \lambda_{\max}\mathrm{T} = c_2/4.96511423 = 0.2897756(24)$ cm °K.

When $B_\nu(T)$ is integrated over all frequencies, the total emittance, πB, of the black body is found to be (Stefan, 1879; Boltzmann, 1884; Milne, 1930)

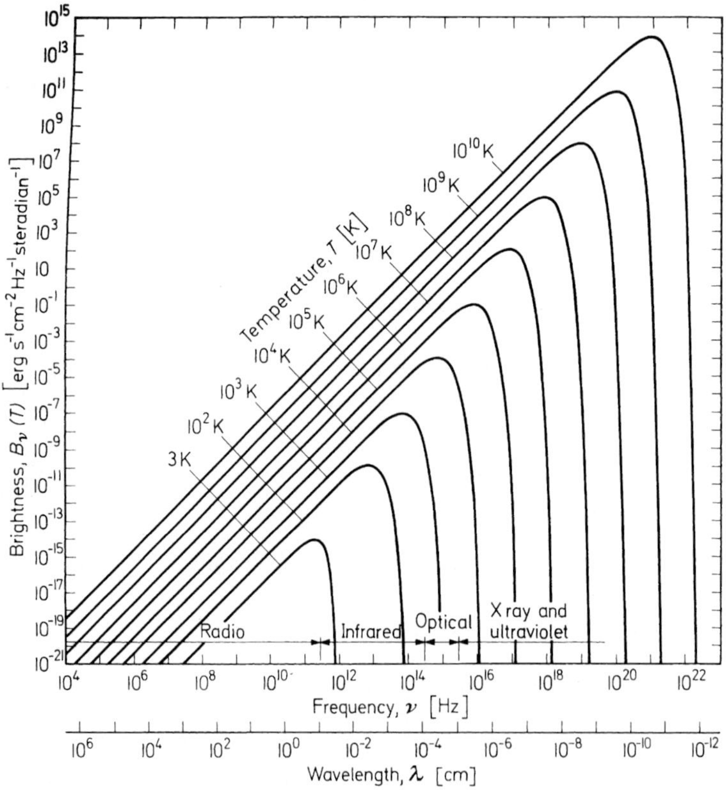

Fig. 1.1. The brightness, $B_\nu(T)$, of a black-body radiator at frequency, ν, and temperature, T. The Planck function $B_\nu(T)$, is given by Eq. (1.119)

$$\pi B = \sigma T^4 n^2 \quad \text{Stefan–Boltzmann law} , \tag{1.124}$$

where $n \approx 1$ is the index of refraction of the medium and the Stefan–Boltzmann constant $\sigma = 2\pi^5 k^4/(15c^2h^3) = 5.670 \times 10^{-5}$ erg cm^{-2} sec^{-1} °K^{-4}. The total black body intensity, I, is therefore given by

$$I = \frac{\sigma n^2}{\pi} T^4 = 1.8046 \times 10^{-5}\, T^4 \text{ erg cm}^{-2} \text{ sec}^{-1} \text{ steradian}^{-1} , \tag{1.125}$$

and the radiation energy density, U, is

$$U = aT^4 n^3 , \tag{1.126}$$

where the radiation density constant, $a = 4\sigma/c = 7.564 \times 10^{-15}$ erg cm^{-3} °K^{-4}.

The absolute luminosity, L, of a thermal radiator, or black body, of radius, R, and effective temperature, T_{eff}, is:

$$L = 4\pi\sigma R^2 T_{\text{eff}}^4 \text{ erg s}^{-1} \qquad \text{Stefan–Boltzmann law.} \tag{1.127}$$

The effective temperature is the surface temperature that the object would have if it were a perfect black body radiating at absolute luminosity L. For the Sun, we have $L_{\odot} = 3.85 \times 10^{33}$ erg s^{-1}, $R_{\odot} = 6.96 \times 10^{10}$ cm, and $T_{\text{eff}} = 5780$ °K.

The radiant flux, S, of a black body observed at a distance, D, is:

$$S = L/(4\pi D^2) = \sigma R^2 T^4/D^2 \text{ erg s}^{-1} \text{ cm}^{-2} . \tag{1.128}$$

For the Sun, observed at the Earth's mean distance from the Sun of one astronomical unit $= A =$ a.u. $= 1.49598 \times 10^{13}$ cm, the measured $S_{\odot} = 1.368(4) \times 10^6$ erg s^{-1} cm^{-2}.

Optical astronomers measure flux in terms of apparent magnitude, m. For two stars whose apparent magnitudes are m_1 and m_2, the ratio of the measured total flux from the two stars, s_1/s_2, is given by the relation

$$m_1 - m_2 = -2.5 \log(s_1/s_2) . \tag{1.129}$$

One magnitude is equivalent to minus four decibels, and a calibration visual magnitude for the Sun is $m_{\odot} = -26.73 \pm 0.03$ and $s_{\odot} = 1.368 \times 10^6$ erg sec^{-1} cm^{-2}.

For a gray body in the region of frequencies where the Rayleigh–Jeans law is valid, the effective temperature, T_{eff}, is given by

$$T_{\text{eff}} = [1 - R_0(\lambda)]T , \tag{1.130}$$

where $R_0(\lambda)$ is the reflection coefficient at the wavelength, λ, of observation given by Eqs. (1.77) or (1.78), $[1 - R_0(\lambda)]$ is called the emissivity, and T is the black body temperature.

For a planet which has no heat conductivity and no atmosphere the effective temperature is

$$T_{\text{eff}} = T_{\odot} \left(\frac{R_{\odot}}{a}\right)^{1/2} (1 - A)^{1/4} \approx 392(1 - A)^{1/4} a^{-1/2} , \tag{1.131}$$

where the temperature of the Sun, $T_{\odot} \approx 5780$ °K, the solar radius $R_{\odot} \approx 6.96 \times 10^{10}$ cm, the semi-major axis of the planetary orbit is a and is given in a.u. in the numerical approximation, and the Bond albedo (Bond,

1863), A, of a planet is defined (Russell, 1916) as the ratio of the total amount of radiant energy reflected by the planet in all directions to the amount it receives from the Sun. If the planet rotates slowly and radiates predominantly from its sunlit side, the effective temperature is:

$$T_{\text{eff}} = \frac{T_\odot}{(2)^{1/4}}\left(\frac{R_\odot}{a}\right)^{1/2}(1-A)^{1/4} \approx 330(1-A)^{1/4}a^{-1/2} ,$$

and for the rapidly rotating major planets,

$$T_{\text{eff}} = \frac{T_\odot}{\sqrt{2}}\left(\frac{R_\odot}{a}\right)^{1/2}(1-A)^{1/4} \approx 277(1-A)^{1/4}a^{-1/2} .$$

The effective and observed temperatures of the giant planets are given by Lang (1992). The measured temperatures of Jupiter, Saturn and Neptune are about twice the effective temperatures because these giant planets generate their own internal heat. The observed surface temperatures of the terrestrial planets, also given by Lang (1992), can differ substantially from the effective temperature because of the greenhouse warming effect of their atmospheres. For example, the "natural" greenhouse effect of the terrestrial atmosphere raises the Earth's surface temperature to about 288 °K or about 33 °K above its effective temperature of 255 °K, which is below the freezing temperature of water, and the surface temperature of Venus is 730 ± 5 °K, or hundreds of degrees above the temperature due to sunlight if there was no atmosphere on Venus.

For a black body whose radius is r and whose distance is D, the flux density $S_\nu(T)$ incident at the Earth is:

$$S_\nu(T) = \boldsymbol{\Omega}_{\text{s}} B_\nu(T) \approx \frac{\pi r^2}{D^2} B_\nu(T) , \tag{1.132}$$

where $\boldsymbol{\Omega}_{\text{s}}$ is the solid angle subtended by the source and $B_\nu(T)$ is the brightness of the source at frequency ν and temperature T. The flux observed, $S_{0\nu}(T)$, with an antenna of efficiency, η_A, and beam area, Ω_A, is given by

$$S_{0\nu}(T) = \eta_A \Omega_{\text{s}} B_\nu(T) \quad \text{if } \Omega_{\text{s}} < \Omega_A \tag{1.133}$$
$$\text{or } \eta_A \Omega_A B_\nu(T) \quad \text{if } \Omega_{\text{s}} \geq \Omega_A .$$

At radio frequencies, the Rayleigh–Jeans approximation is used with Eq. (1.133) to obtain

$$S_{0\nu}(T) = 3.07 \times 10^{-37}\, \eta_A \Omega_{\text{s}} \nu^2 T \quad \text{erg sec}^{-1}\ \text{cm}^{-2}\ \text{Hz}^{-1} \quad \text{if } \Omega_{\text{s}} < \Omega_A . \tag{1.134}$$

Radio astronomers measure flux density in flux units, where one flux unit = 10^{-23} erg sec^{-1} cm^{-2} Hz^{-1} = 10^{-26} watt m^{-2} Hz^{-1}.

At optical frequencies, the Wien approximation may be used with Eq. (1.133) to obtain

$$S_{0\nu}(T) = 1.47 \times 10^{-47} \eta_A \Omega_{\text{s}} \nu^3 \exp\left(-4.8 \times 10^{-11} \frac{\nu}{T}\right) \quad \text{erg sec}^{-1}\ \text{cm}^{-2}\ \text{Hz}^{-1} . \tag{1.135}$$

at the Earth. Further formulae relating flux densities and magnitudes to luminosity, and formulae for magnitude corrections, are given in Volume II.

The Lorentz transformations given in Eqs. (1.91) were first used by Mosengeil (1907) to calculate the radiation from a black body moving at a velocity, v, away from an observer. At the radiation source, the total flux, s, of the radiation in the frequency range $d\nu$ and solid angle, $d\Omega$, is

$$s = B_\nu(T) d\nu \, d\Omega \ , \tag{1.136}$$

where $B_\nu(T)$ is the brightness of the black body at temperature T and frequency, ν. The observed frequency, ν_{obs}, is given by (Lorentz, 1904)

$$\nu_{\text{obs}} = \frac{\nu}{(1-\beta^2)^{1/2}}(1 - \beta\cos\theta) \ , \tag{1.137}$$

which accounts for the frequency shift of a moving source which was first observed by Doppler (1842). Here $\beta = v/c$ and θ denotes the angle between the velocity vector and the wave vector of the radiating source. The observed solid angle, $d\Omega_{\text{obs}}$, is given by

$$d\Omega_{\text{obs}} = \frac{1-\beta^2}{(1-\beta\cos\theta)^2} d\Omega \ . \tag{1.138}$$

It follows from Eqs. (1.92) and (1.119) that the total flux density seen by the observer, S_{obs}, and the observed brightness, $B_{\nu\text{obs}}(T_{\text{obs}})$, are given by

$$S_{\text{obs}} = B_\nu(T)\frac{(1-\beta\cos\theta)^2}{(1-\beta^2)} d\nu d\Omega = B_{\nu\text{obs}}(T_{\text{obs}}) d\nu_{\text{obs}} d\Omega_{\text{obs}} \tag{1.139}$$

and

$$B_{\nu\text{obs}}(T_{\text{obs}}) = \frac{B_\nu(T)(1-\beta\cos\theta)^3}{(1-\beta^2)^{3/2}} = \frac{2h\nu_{\text{obs}}^3}{c^2 \exp\{[h\nu_{\text{obs}}/(kT)][(1-\beta^2)^{1/2}/(1-\beta\cos\theta)]\} - 1} \ . \tag{1.140}$$

The angular distribution of temperature, given by Peebles and Wilkinson (1988), is $T(\theta) = (1-\beta^2)^{1/2}/(1-\beta\cos\theta)$; such an anisotropy has been used to infer our velocity with respect to the cosmic microwave background radiation (see Volume II).

1.24 Radiation Transfer and Observed Brightness

If a beam of radiation of intensity, I_0, passes through an absorbing cloud of thickness, L, the intensity of the radiation when leaving the cloud is given by

$$I_{0\nu} = I_0 \exp(-\tau_\nu) \ , \tag{1.141}$$

where the optical depth $\tau_\nu = \int_0^L \alpha_\nu dx$, the absorption coefficient per unit length in the cloud is α_ν, and the subscript ν denotes the frequency dependence of the variables.

An absorber also emits radiation, and the emission coefficient, ε_ν, is defined as the amount of energy a unit volume of material emits per second per unit solid angle in the frequency range ν to $\nu + d\nu$. The total intensity I_ν emitted by a column of gas of unit cross sectional area and length, L, is therefore

$$I_\nu = \int_0^L \varepsilon_\nu \exp[-\alpha_\nu x] dx \ . \tag{1.142}$$

For matter in thermodynamic equilibrium at temperature, T, we have (Kirchhoff, 1860; Milne, 1930)

$$\varepsilon_\nu = n_\nu^2 \alpha_\nu B_\nu(T) \qquad \text{(Kirchhoff's law)} \ , \tag{1.143}$$

where $B_\nu(T)$ is the vacuum brightness of a black body at temperature, T, and n_ν is the index of refraction of the medium.

It follows from Eqs. (1.142) and (1.143) that

$$\begin{aligned} B_{C\nu}(T) &= B_\nu(T)[1 - \exp(-\tau_\nu)] \\ &= B_\nu(T) \quad \text{if } \tau_\nu \gg 1 \text{ (optically thick)} \\ &= \tau_\nu B_\nu(T) \quad \text{if } \tau_\nu \ll 1 \text{ (optically thin)} \ , \end{aligned} \tag{1.144}$$

where $B_{C\nu}(T)$ is called the brightness of the cloud at frequency, ν, and the index of refraction is assumed to be unity.

If a source of brightness, $B_{S\nu}(T_S)$, at frequency, ν, and temperature, T_S, is irradiating a cloud of temperature, T_C, the total brightness, $B_{O\nu}(T_{TOT})$ of the cloud will be

$$\begin{aligned} B_{O\nu}(T_{TOT}) &= B_{S\nu}(T_S)\exp(-\tau_\nu) + B_{C\nu}(T_C) \ , \\ \text{or } \ T_{TOT} &= T_S \exp(-\tau_\upsilon) + T_C[1 - \exp(-\tau_\upsilon)] \ , \end{aligned} \tag{1.145}$$

where $B_{C\nu}(T_C)$ is given by Eq. (1.144). For the optically thin case $\tau \ll 1$, the total temperature, T_{TOT}, will be:

$$T_{TOT} = T_S(1 - \tau) + \tau T_C \ .$$

It follows from Eqs. (1.121) and (1.133) that the radio flux density, S, of the source plus cloud will be:

$$S = 2kT_S\lambda^{-2}\exp(-\tau)\Omega_S + 2kT_C\lambda^{-2}[1 - \exp(-\tau)]\Omega_C \ . \tag{1.146}$$

1.25 Magnetobremsstrahlung or Gyroradiation (Gyromagnetic and Synchrotron Radiation) of a Single Electron

Theoretical studies by Schott (1912) indicated that a charged particle moving in the presence of a magnetic field will emit radiation. Electrons accelerated to nearly the velocity of light in the magnetic fields of the General Electric Company's synchrotron particle accelerator were subsequently found to emit linearly polarized radiation (Elder, Langmuir and Pollock, 1948), stimulating further theoretical treatments of this type of nonthermal emission (Schwinger, 1949). Radiation from electrons moving at nearly the velocity of light, c, or at relativistic speeds, in a magnetic field has subsequently become known as synchrotron radiation. When the electrons are moving at mildly relativistic velocities $v \approx 0.1c$, the emission is called gyrosynchrotron radiation, while gyromagnetic radiation is emitted by slow electrons of $v \ll c$ spiraling about magnetic fields.

In the meantime, radio emission from the Milky Way had been detected by Jansky (1935) and confirmed by Reber (1944). The high radio brightness temperature was attributed to the nonthermal synchrotron radiation of high-speed, cosmic ray electrons spiraling about either stellar magnetic fields (Alfvén and Herlofson, 1950) or interstellar magnetic fields (Kiepenheuer, 1950; Ginzburg, 1956). Both the optical and radio emission of the Crab Nebula supernova remnant were attributed to synchrotron emission (Shklovskii, 1953), providing an explanation for its nonthermal spectrum (Minkowski, 1942). Following Shklovskii's prediction, polarized radiation was detected from the Crab Nebula at both optical wavelengths (Dombrovskii, 1954) and radio wavelengths (Oort and Walraven, 1956), confirming its synchrotron origin. Strong radio emission from the planet Jupiter was also explained by synchrotron or gyromagnetic emission of electrons trapped in its magnetic field (Burke and Franklin, 1955).

The long-wavelength radio emission of our Galaxy is now attributed to relativistic electrons, with velocities v approaching the velocity of light c, moving in interstellar magnetic fields with a magnetic field strength $H \approx 10^{-6}$ Gauss; this radiation has been reviewed by Salter and Brown (1988). Powerful, discrete extragalactic radio sources, named radio galaxies and quasars, also emit radio radiation by the synchrotron process [see Kellermann and Owen, 1988 for a review]; these objects are discussed in greater detail in Volume II. The synchrotron radiation of compact and extended extragalactic radio sources are respectively reviewed by Kellermann and Pauliny-Toth (1981) and Miley (1980). Saikia and Salter (1988) discuss the polarization properties of extragalactic radio sources. Good reviews of various synchrotron radiation formulae have been provided by Pacholczyk (1970), Moffet (1975), Jackson (1975), Rybicki and Lightman (1979), Salter and Brown (1988), Kellermann and Owen (1988), Shu (1991) and Longair (1994).

When an electron moves linearly at the velocity, v, and with the acceleration, $\dot{v}$, the total power radiated per unit solid angle is (Liénard, 1898)

$$\frac{dP}{d\Omega} = \frac{e^2\dot{v}^2}{4\pi c^3}\frac{\sin^2\theta}{(1-\beta\cos\theta)^5} \; , \tag{1.147}$$

where $\beta = v/c$ and θ is the angle between the line of sight and the common direction of v and $\dot{v}$ (here we assume the acceleration vector is parallel to the velocity vector). At low velocities this distribution becomes the Larmor distribution (Eq. (1.109)). At high velocities the distribution becomes a narrow cone of half-angle, θ, given by

$$\theta = \gamma^{-1} = [1-\beta^2]^{1/2} = mc^2E^{-1} = 8.2 \times 10^{-7}\, E^{-1} \text{ radians} \; , \tag{1.148}$$

where $\beta = v/c$ and E is the total energy of the electron. This directed beam of radiation is typical regardless of the relation between the acceleration and velocity vectors.

When the electron moves about a circular orbit of radius, ρ, the observer sees a pulse of radiation of approximate duration $\rho/(\beta c\gamma)$ in the electron frame, and $\rho/(\beta c\gamma^3)$ in his own frame. Thus each pulse of radiation contains frequencies up to the critical frequency, ν_c, given by

$$2\pi\nu_c = \left(\frac{c\beta}{\rho}\right)\gamma^3 \approx 6 \times 10^{28}\frac{E^3}{\rho}\,\mathrm{Hz} \ . \tag{1.149}$$

When the electron motion is assumed to be periodically circular, the observed radiation will consist of harmonics of the rotation frequency $\beta c/\rho$, up to the critical harmonic, γ^3.

The total power radiated, P_r, is given by (Jackson, 1962)

$$P_{r_{\|}} = \frac{2}{3}\frac{e^2}{c^3}\dot{v}^2\gamma^6 \ , \tag{1.150}$$

where $\dot{v}$ denotes the first derivative of the velocity, v, with respect to time. The subscript $_{\|}$ in Eq. (1.150) denotes the case where the acceleration and velocity vectors are parallel. When they are perpendicular, the total power radiated is

$$P_{r_{\perp}} = \frac{2}{3}\frac{e^2}{c^3}\dot{v}^2\gamma^4 \tag{1.151}$$

a result first obtained by Liénard (1898).

For a charge in a uniform magnetic field, the entire (nonradiative) acceleration occurs perpendicular to the velocity vector, and we have

$$\partial v/\partial t = \omega_{\mathrm{H}} v \sin\psi \tag{1.152}$$

where the frequency of gyration, ω_{H}, is given by Eq. (1.53); for an electron of energy E, velocity v, and Lorentz factor $\gamma = [1-(v/c)^2]^{-1/2}$,

$$\omega_{\mathrm{H}} = eH/(\gamma mc) = 1.76 \times 10^7\, H/\gamma\ \mathrm{Hz} = 14.4\, H/E\ \mathrm{Hz} \ ,$$

the pitch angle ψ is the angle between the magnetic field vector and the velocity vector, and the radius of gyration $r_{\mathrm{H}} = \rho$, given by Eq. (1.54) for an electron, is:

$$\rho = \frac{v}{\omega_{\mathrm{H}}} \approx 2 \times 10^9\, EH^{-1}\ \mathrm{cm\ for}\ v \approx c \ . \tag{1.153}$$

The resulting gyroradiation is called gyromagnetic when $v \ll c$. When the electron velocity is relativistic, the radiation is called synchrotron radiation. In this case, the radiation is primarily directed in the direction transverse to the magnetic field, and in a narrow beam of width γ^{-1}. The idea that cosmic radio sources might be radiating synchrotron radiation was first suggested by Alfvén and Herlofson (1950). Subsequently, Ginzburg (1953) and Shklovskii (1953) suggested that the optical radiation of the Crab Nebula was synchrotron radiation and would be found to be polarized–as it was at optical wavelengths by Dombrovskii (1954) and in the radio wavelength region by Oort and Walraven (1956).

As pointed out by Epstein and Feldman (1967) the conventional formulae for the synchrotron radiation of a single electron must be corrected for the Doppler shifted gyrofrequency, ω_{DH}, given by $\omega_{\mathrm{DH}} = \omega_{\mathrm{H}}/(1-\beta\cos^2\psi)$ $\approx \omega_{\mathrm{H}}/\sin^2\psi$.

As these effects cancel out when considering the radiation of an ensemble of electrons (Scheuer, 1968), the Doppler shift effect will be ignored in what follows.

The frequency near which the synchrotron emission is a maximum is called the critical frequency and is defined as

$$\nu_c = \frac{3}{4\pi}\frac{eH}{mc}\gamma^2 \sin\psi \approx 6.266 \times 10^{18}\ HE^2 \sin\psi \quad \text{Hz} \ . \tag{1.154}$$

The constant in Eq. (1.154) is 16.1 when H is given in microgauss, E is in GeV and ν is in MHz. For radio emission in typical astronomical magnetic fields of $H = 10^{-5}$ to 10^{-6} Gauss, the electrons must have Lorentz factors of $\gamma = 10^3$ to 10^5. From Eqs. (1.150) and (1.151) the total radiated power, P_r, is obtained

$$\begin{aligned} P_r &= \frac{2}{3}\frac{e^4}{m^2c^3}\beta^2\gamma^2H^2\sin^2\psi \approx 2 \times 10^{-3}\ H^2E^2\sin^2\psi \text{ erg sec}^{-1} \\ &\approx 1.6 \times 10^{-15}\ H^2\gamma^2\sin^2\psi \text{ erg sec}^{-1} \ . \end{aligned} \tag{1.155}$$

or

$$P_r = 2\sigma_T c U_H \beta^2\gamma^2 \sin^2\psi \ ,$$

where $\beta = v/c$ the ratio of the velocity, v, to the speed of light, c, the Thomson cross section $\sigma_T = 8\pi r_e^2/3 = 6.65 \times 10^{-25}$ cm^2, the classical electron radius $r_e = e^2/(mc^2) = 2.82 \times 10^{-13}$ cm, and the magnetic energy density $U_H = H^2/(8\pi)$. For an isotropic distribution of electrons of energy $E = \gamma mc^2$, the angle averaged value of $\langle \sin^2\psi \rangle = 2/3$ can be used. For nonrelativistic gyroradiation, with $v \ll c$ and $\gamma \approx 1$, the expression for the loss of energy by the electron is given by Eq. (1.155) with $\gamma = 1$, and the radiation is emitted at the gyrofrequency ω_H of the electron, given in Eq. (1.152) with $\gamma = 1$.

O'Dell and Sartori (1970) note that Eq. (1.155) and similar equations to follow are only valid if $\gamma \sin\psi \gg 1$ or if the radian frequency $\omega \gg 10^7\ H/\sin\psi$. From Eq. (1.155) the total power radiated per unit frequency interval ν to $\nu + d\nu$ centered at the maximum frequency is

$$P(\nu_m)d\nu \approx \frac{P_r}{\nu_c} \approx 3 \times 10^{-22}\ H \text{ erg sec}^{-1}\text{ Hz}^{-1} \ . \tag{1.156}$$

The lifetime, τ_r, of the electron due to radiation damping is

$$\tau_r \approx \frac{E}{P_r} \approx 500\ E^{-1}\ H^{-2} \text{ sec} \ . \tag{1.157}$$

Detailed calculations for the radiation from a relativistic particle in circular motion were first obtained by Schott (1912). When these equations are applied to the ultrarelativistic motion ($E \gg mc^2$) of an electron in a magnetic field (Vladimirskii, 1948; Schwinger, 1949; Westfold, 1959) the following formulae are obtained.

The average power radiated per unit frequency interval in all directions may be divided into two components $P_1(\nu)$ and $P_2(\nu)$ according as the component is parallel or perpendicular to the projection of the H field line in the plane normal to the direction of observation.

$$P_1(\nu)d\nu = \frac{\sqrt{3}e^3H}{2\pi mc^2}\frac{\nu}{2\nu_c}\left[\int_{\nu/\nu_c}^{\infty} K_{5/3}(\eta)d\eta - K_{2/3}\left(\frac{\nu}{\nu_c}\right)\right]d\nu \ , \tag{1.158}$$

$$P_2(\nu)d\nu = \frac{\sqrt{3}e^3H}{2\pi mc^2}\frac{\nu}{2\nu_c}\left[\int_{\nu/\nu_c}^{\infty} K_{5/3}(\eta)d\eta + K_{2/3}\left(\frac{\nu}{\nu_c}\right)\right]d\nu \ , \tag{1.159}$$

where it is assumed that the electron has the appropriate pitch angle to radiate towards the observer, and K is the modified Bessel function (a Bessel function of the second kind with imaginary argument). The relevant integrals have been evaluated by Westfold (1959) and Le Roux (1961); more recent treatments are provided by Rybicki and Lightman (1979) and Longair (1994).

The total power radiated in frequency interval between ν and $\nu + d\nu$ is therefore given by

$$\begin{aligned} P(\nu)d\nu &= \frac{\sqrt{3}e^3}{mc^2}H_\perp\frac{\nu}{\nu_c}\int_{\nu/\nu_c}^{\infty} K_{5/3}(\eta)d\eta \, d\nu \\ &\approx 5.04\times 10^{-22}\, H_\perp\left(\frac{\nu}{\nu_c}\right)^{1/3} d\nu \text{ erg sec}^{-1}\text{ Hz}^{-1} \quad \text{for } \nu \ll \nu_c \\ &\approx 2.94\times 10^{-22}\, H_\perp\left(\frac{\nu}{\nu_c}\right)^{1/2} \exp\left(\frac{-\nu}{\nu_c}\right) d\nu \text{ erg sec}^{-1}\text{ Hz}^{-1} \quad \text{for } \nu \gg \nu_c \ , \end{aligned} \tag{1.160}$$

where $H_\perp$ is the component of H which is perpendicular to the velocity vector. The function $P(\nu)d\nu$ has its maximum at $\nu = 0.3\,\nu_c$, and describes the spectral distribution of power shown in Fig. 1.2. These asymptotic expressions are derived by Scheuer (1968), Rybicki and Lightman (1979) and Longair (1994); they indicate that there is very little synchrotron radiation power at frequencies $\nu \gg \nu_c$ and that the radiation spectrum varies as $\nu^{1/3}$ for $\nu \ll \nu_c$.

For ultrarelativistic motion ($E \gg mc^2$), the total angular spectrum can also be divided into two components $P_1(\psi)$ and $P_2(\psi)$ according as the direction of polarization is parallel or perpendicular to the projection of the magnetic field in the plane normal to the direction of observation (Westfold, 1959; Ginzburg and Syrovatskii, 1965)

$$P_2(\psi) = \frac{3}{4\pi^2 r^2}\frac{e^3H}{mc^2\zeta}\left(\frac{\nu}{\nu_c}\right)^2\left(1+\frac{\psi^2}{\zeta^2}\right)^2[K_{2/3}(g_\nu)]^2$$

and

$$P_1(\psi) = \frac{3}{4\pi^2 r^2}\frac{e^3H}{mc^2\zeta}\left(\frac{\nu}{\nu_c}\right)^2\frac{\psi^2}{\zeta^2}\left(1+\frac{\psi^2}{\zeta^2}\right)[K_{1/3}(g_\nu)]^2 \ , \tag{1.161}$$

where

$$g_\nu = \frac{\nu}{2\nu_c}\left(1+\frac{\psi^2}{\zeta^2}\right)^{3/2} \ .$$

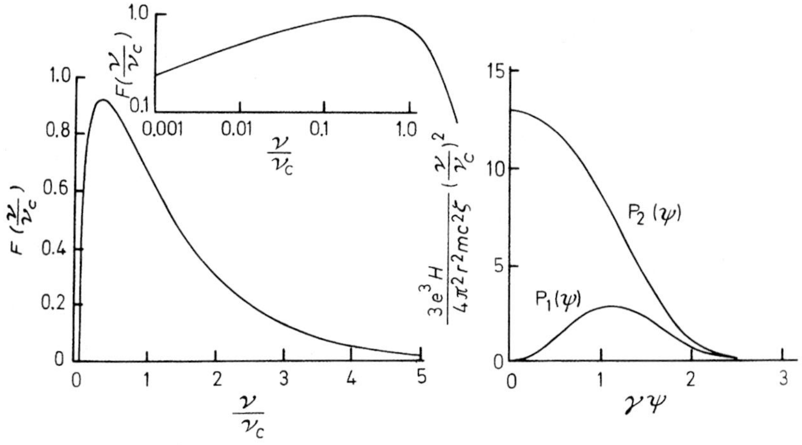

Fig. 1.2. The function $F(\nu/\nu_c) = (\nu/\nu_c) \int_{\nu/\nu_c}^{\infty} K_{5/3}(\eta)d\eta$, which characterizes the spectral distribution of synchrotron radiation from a single electron, is shown in both linear and logarithmic plots [cf. Vladimirskii, 1948; Schwinger, 1949]. The total synchrotron power radiated per unit frequency interval is related to $F(\nu/\nu_c)$ by Eq. (1.160), and critical frequency, ν_c, is given by Eq. (1.154). Also shown is the angular spectrum for the synchrotron radiation of a single electron in directions parallel, $P_1(\psi)$, and perpendicular, $P_2(\psi)$, to the projection of the magnetic field on the plane of the figure [after Ginzburg and Syrovatskii, 1965, by permission of Annual Reviews, Inc.]. The angle, ψ, is the angle between the direction of observation and the nearest velocity vector of the radiation cone, H is the magnetic field intensity, r is the distance from the radiating electron, $\gamma = \zeta^{-1} = \left[1 - (v/c)^2\right]^{-1/2}$ where v is the velocity of the electron, and $P_1(\psi)$, and $P_2(\psi)$ are given by Eqs. (1.161). The angular spectrum plots are for $\nu/\nu_c = 0.29$

The r^2 term has been added to denote the dependence of the radiated power on the distance, r, from the source, $\zeta = \gamma^{-1} = mc^2/E$, and here ψ is not the pitch angle but the angle between the direction of observation and the nearest velocity vector of the radiation cone. The $P_1(\psi)$ and $P_2(\psi)$ are illustrated in Fig. 1.2. In general, then, the polarization is elliptical with the axes parallel and perpendicular to the projection of the magnetic field on the plane transverse to the direction of observation. The direction of the ellipse is right or left hand according as $\psi >$ or < 0. The polarization is seen to be linear only when $\psi = 0$. The degree of polarization of the total power per unit frequency interval at a given frequency ν, is (Westfold, 1959)

$$\begin{aligned} \Pi &= \frac{K_{2/3}(\nu/\nu_c)}{\int_{\nu/\nu_c}^{\infty} K_{5/3}(\eta)d\eta} \\ &\approx \frac{1}{2} \quad \text{for } \nu \ll \nu_c \\ &\approx 1 - \frac{2\nu_c}{3\nu} \quad \text{for } \nu \gg \nu_c \ . \end{aligned} \tag{1.162}$$

Here the degree of polarization, Π, is related to the maximum and minimum observable values of intensity, P_{max} and P_{min}, by $\Pi = (P_{max} - (P_{min}/(P_{max} + P_{min})$. Detailed formulae for the Stokes parameters of the synchrotron radiation from an ultrarelativistic electron are given by Legg and Westfold (1968) and Longair (1994).

The rate at which an electron loses its energy, E, by synchrotron radiation is given by

$$\frac{dE}{dt} = -\int_0^\infty P(\nu)d\nu = -2.368 \times 10^{-3} H_\perp^2 E^2 \text{erg sec}^{-1} \ , \tag{1.163}$$

where $P(\nu)d\nu$ is given by Eq. (1.160). The electron energy, $E(t)$, as a function of time t, is given by

$$E(t) = \frac{E_0}{(1 + t/t_{1/2})} \ , \tag{1.164}$$

where the time required for the electron to lose half its initial energy, E_0, is

$$t_{1/2} = \frac{4.223 \times 10^2}{H_\perp^2 E_0} \text{sec} \ . \tag{1.165}$$

The constant 4.223×10^2 has the more practical units of 8.352×10^9 years $(\mu \text{ Gauss})^2$ GeV. As pointed out by Takakura (1960), the radiation lifetime may not be so small as the collision lifetime, t_c, when thermal electrons are present. For example, the basic relaxation time, τ_r, for an electron of energy, E, moving through singly charged ions of density, N_i is given by (Trubnikov, 1965)

$$\begin{aligned} \tau_r &\approx \frac{\sqrt{m}}{15\pi\sqrt{2}e^4} \frac{E^{3/2}}{N_i} \\ &\approx 10^8 E^{3/2} N_i^{-1} \text{sec} \ , \end{aligned}$$

where the numerical approximation is for E in keV and N_i in cm^{-3}.

Detailed calculations of the angular spectrum and the frequency spectrum of the synchrotron radiation for moderate electron energies, $E \approx mc^2$, include motions parallel to the H field (Trubnikov, 1958; Takakura, 1960; Bekefi, 1966). The total power radiated per unit solid angle in the n th harmonic at the angle θ to the H field line is

$$\frac{dP_n}{d\Omega} = \frac{e^4 H^2 (1-\beta)^2 \gamma^2}{2\pi m^2 c^3 (1 - \beta_{||} \cos\theta)^2} \left\{ \left(\frac{\cos\theta - \beta_{||}}{\beta_\perp \sin\theta} \right)^2 J_n^2(\gamma \sin\theta) + J_n'^2(\gamma \sin\theta) \right\} \ , \tag{1.166}$$

where $\beta^2 = \beta_{||}^2 + \beta_\perp^2 = (v_{||}/c)^2 + (v_\perp/c)^2 = (v/c)^2$, the $v_{||}$ and $v_\perp$ are the instantaneous particle velocities along and perpendicular to the magnetic field,

$$\gamma = \frac{n\beta_\perp}{1 - \beta_{||}\cos\theta} \ ,$$

J_n is a Bessel function of order n, and $'$ denotes the first derivative with respect to the argument. This radiation is at the critical radian frequency

$$\omega = \frac{neH(1-\beta^2)^{1/2}}{mc(1 - \beta_{||}\cos\theta)} \ .$$

The angular distribution of the radiation from a mildly relativistic electron is shown in Fig.1.3.

The total power radiated in the n th harmonic is

$$P_n = \frac{2e^4H^2}{m^2c^3}\frac{1-\beta_0^2}{\beta_0}\left[n\beta_0^2 J'_{2n}(2n\beta_0) - n^2(1-\beta_0)^2 \int_0^{\beta_0} J_{2n}(2nt)dt\right] ,$$

where $\beta_0 = \beta_\perp/(1-\beta_{||}^2)^{1/2}$. When the electron velocity $v \ll c$, the nonrelativistic case, there is little synchrotron power radiated at high harmonics, but when $v \geq 0.1\ c$ the emission at higher harmonics becomes important.

The total power radiated over all harmonics is

$$P_r = \frac{2e^4H^2}{3m^2c^3}\left[(1-\beta_{||}^2)(1-\beta^2)^{-1} - 1\right] = 1.59 \times 10^{-15}\frac{\beta_\perp^2 H^2}{1-\beta^2}\text{erg sec}^{-1} . \quad (1.167)$$

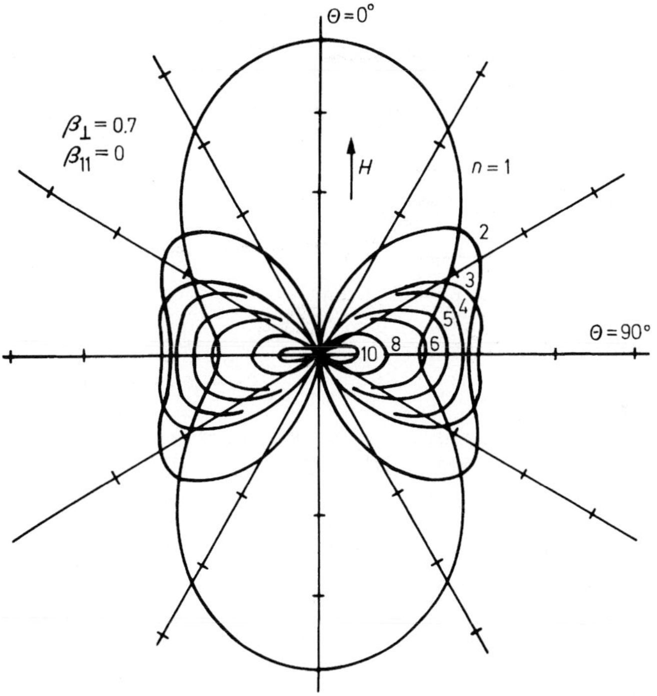

Fig. 1.3. The angular distribution of the radiation intensity from a mildly relativistic electron [after Oster, 1960]. The intensity distribution is shown for various harmonics, n, as a function of the angle, θ, between the observer and the direction of the magnetic field, H. The figure is for a fixed particle energy corresponding to $\beta_\perp = V_\perp/c = 0.7$ where the electron velocity is V and $\perp$ denotes the component perpendicular to the direction of the magnetic field. The short bars on the radial lines denote equal increments of intensity; and the complete polar surface is obtained by rotating the figure about the vertical axis. The figure illustrates the fact that higher frequencies are emitted in a narrow angular range about the electron orbital plane defined by $\theta = 90°$

The first and second terms in the brackets { } of Eq.(1.166) respectively represent polarized components of emission with the electric vector parallel, $\|$, and perpendicular, $\perp$, to the magnetic field. Analytic approximations for the Bessel function terms of Eq.(1.166) are given by Wild and Hill (1971), and they obtain

$$P_{\|} = Az^2\left(\frac{1.5}{n_c} + \frac{0.5033}{n}\right)^{-1/3} n \exp\left(\frac{-n}{n_0}\right) \tag{1.168}$$

$$P_{\perp} = A\left(\frac{1.5}{n_c} + \frac{1.193}{n}\right)^{1/3}\left(1 - \frac{1}{5n^{2/3}}\right)^2 n \exp\left(\frac{-n}{n_0}\right) ,$$

where P is the power emitted by a single relativistic or sub-relativistic electron per unit solid angle per unit frequency interval in the n th harmonic,

$$z = \frac{(\cos\theta - \beta\cos\varphi)}{(1 - \beta\cos\varphi\cos\theta)} ,$$

$$A = \frac{e^3 H}{2\pi mc^2 \sin^2\theta} \frac{(1-\beta^2)^{1/2}}{(1 - \beta\cos\varphi\cos\theta)} ,$$

φ is the angle between the electron velocity vector and the magnetic field direction, θ is the angle between the magnetic field direction and the observer's line of sight, and

$$n_c = \frac{3}{2}(1 - X^2)^{-3/2}$$

and

$$(2n_0)^{-1} = \ln\left[1 + (1 - X^2)^{1/2}\right] - \ln X - (1 - X^2)^{1/2} ,$$

where

$$X = \frac{\beta\sin\varphi\sin\theta}{(1 - \beta\cos\varphi\cos\theta)} .$$

The polarization of the radiation from mildly relativistic electrons will be elliptical, with the ratio of the axes of the ellipse given by

$$R_n = -\frac{(\cos\theta - \beta_{\|})J_n(\gamma\sin\theta)}{\beta_{\perp}\sin\theta J_n'(\gamma\sin\theta)} . \tag{1.169}$$

In the case of gyroradiation where the electron kinetic energy $E \ll mc^2$, Eq.(1.166) for the total power radiated per unit solid angle in the n th harmonic can be simplified to the expression

$$\frac{dP_n}{d\Omega} = \frac{e^4 H^2}{2\pi m^2 c^3} \frac{(n\beta_{\perp}/2)^{2n}}{[(n-1)!]^2} (\sin\theta)^{2n-2}(1 + \cos^2\theta) , \tag{1.170}$$

where θ is the angle between the direction of wave propagation and the magnetic field, and $\beta_{\perp} = (v_{\perp}/c)$ corresponds to the component of velocity, $v_{\perp}$, perpendicular to the magnetic field. For the case of a Maxwellian distribution of electron kinetic energies, the absorption coefficient per unit length, α_ν, due to gyroradiation is given by (Sitenko and Stepanov, 1957; Wild, Smerd, and Weiss, 1963)

$$\alpha_\nu = \frac{4\pi^{3/2}}{c}\frac{\nu_P^2}{\nu}\frac{(n/2)^{2n}}{n!}\frac{(\sin\theta)^{2n-2}(1+\cos^2\theta)}{\cos\theta}\beta_0^{2n-3}\exp\left[-\frac{(1-n\nu_{\boldsymbol{H}}/\nu)^2}{\beta_0^2\cos^2\theta}\right] , \tag{1.171}$$

where the plasma frequency $\nu_P = \left[e^2N_e/(\pi m)\right]^{1/2}$, the frequency of the radiation is ν, the $\beta_0^2 = 2kT/(mc)^2$, and the gyrofrequency $\nu_H = e\boldsymbol{H}/(2\pi mc)$. For optical depth calculations, the effective thickness, L, of a layer of thermal electrons is given by $L = 2L_H\beta_0\cos\theta$ where the scale of the magnetic field $L_H = H/\nabla H$, is the characteristic distance over which a change in ν_H is significant.

When an electron loses a nonnegligible amount of kinetic energy in its orbit in a magnetic field, quantum effects must be taken into account. This condition is fulfilled when (Canuto and Chiu, 1971)

$$\left(\frac{e^2}{\hbar c}\right)\gamma^2\left(\frac{H}{H_q}\right) \gg 1$$

or

$$\gamma^2 H \gg 6\times 10^{15}\ \text{Gauss} .$$

Here the total energy of the electron is $E = \gamma mc^2$ and $H_q = m^2c^3/(e\hbar) \approx 4.414\times 10^{13}$ Gauss. Quantum effects also become important for $E - mc^2 \approx H/H_q$ or for $H/H_q \gg 1$ (Erber, 1966). When the electron orbits become quantized, the energy of the electron in a magnetic field is given by (Rabi, 1928)

$$E = mc^2\left[1 + \left(\frac{p_z}{mc}\right)^2 + \frac{2nH}{H_q}\right]^{1/2} , \tag{1.172}$$

where p_z is the component of electron momentum in the direction of the magnetic field, H is the magnetic field intensity, and the principle quantum number $n = 0, 1, 2, \ldots$. The transition between two quantum states n and n' gives rise to a photon of energy $h\nu$ given by (Canuto and Chiu, 1971)

$$h\nu = \frac{E - p_z c\cos\theta}{\sin^2\theta}\left\{1 - \left[1 - \frac{m^2c^4\sin^2\theta(n-n')}{(E - p_z c\cos\theta)^2}\right]^{1/2}\right\}$$
$$\approx mc^2\left\{(n-n')\frac{H}{H_q} + \theta\left[\left(\frac{h\nu}{mc^2}\right)^2\cos^2\theta\right]\right\} \text{ for } H \ll H_q ,$$

where θ is the angle of the emitted photon with respect to the axis of the magnetic field. For the $n = 1$ to $n' = 0$ transition, all of the radiation will have an energy of $1.16\times 10^{-8}H$ eV and the total power radiated per unit volume, P, is given by (Chiu and Fassio-Canuto, 1969)

$$P = \frac{1}{2}\frac{e^2mc}{\hbar^2}mc^2N_e\left(\frac{h\nu}{mc^2}\right)^2\left(\frac{H}{H_q}\right)^3(1+\cos^2\theta)$$

$$\approx 2.85 \times 10^{-5} N_e H_8^3 (1 + \cos^2 \theta) \quad \text{erg sec}^{-1} \text{ cm}^{-3} \ ,$$

where N_e is the number density of electrons in quantum state $n = 1$, and $H_8 = H/10^8$. The lifetime of the electron in this quantum regime is given by

$$\tau = \left[\frac{2}{3}\frac{e^2}{\hbar c}\frac{mc^2}{\hbar}\left(\frac{H}{H_q}\right)^2 \gamma\right]^{-1} \approx 2.6 \times 10^{-19} \left(\frac{H_q}{H}\right)^{-2} \gamma^{-1} \text{ sec} \ .$$

1.26 Synchrotron Radiation from an Ensemble of Particles

The volume emissivity (power per unit frequency interval per unit volume per unit solid angle) of the ultrarelativistic radiation of a group of electrons is

$$\varepsilon(\nu) = \int P(\nu) N(E) dE \ , \tag{1.173}$$

where $P(\nu)$ is the total power radiated per unit frequency interval by one electron (Eq. (1.160)), and $N(E)dE$ is the number of electrons per unit volume per unit solid angle along the line of sight which are moving in the direction of the observer and whose energies lie in the range E to $E + dE$.

If all the ultrarelativistic electrons possess the same energy, E, then Eqs. (1.160) and (1.173) give the total intensity

$$I(\nu) \approx 5.04 \times 10^{-22} H_\perp \left(\frac{\nu}{\nu_c}\right)^{1/3} \int N(E) dl \text{ erg sec}^{-1} \text{cm}^{-2} \text{Hz}^{-1} \text{rad}^{-2} \text{ for } \nu \ll \nu_c$$

$$\approx 2.94 \times 10^{-22} H_\perp \left(\frac{\nu}{\nu_c}\right)^{1/2} \exp\left(\frac{-\nu}{\nu_c}\right) \int N(E) dl \text{ erg sec}^{-1} \text{cm}^{-2} \text{Hz}^{-1} \text{rad}^{-2} \text{ for } \nu \gg \nu_c \ ,$$

where ν is the frequency, $\nu_c \approx 6 \times 10^{18} H_\perp E^2$Hz, the component of H along the line of sight is $H_\perp$, and $\int N(E)dl$ is the number of electrons per unit solid angle per square centimeter along the line of sight whose velocities are directed towards the observer. The degree of polarization in this case is given by Eq. (1.162)

$$\Pi = \frac{1}{2}\left\{1 + \frac{\Gamma(1/3)}{2}\left(\frac{\nu}{2\nu_c}\right)^{2/3}\right\} \text{ for } \nu \ll \nu_c$$

$$= 1 - \frac{2}{3}\frac{\nu_c}{\nu} \qquad \text{for } \nu \gg \nu_c \ .$$

Because the observed cosmic rays have a power law spectrum, and because the spectra of many observed radio sources are power law, it is often assumed that

$$N(E)dE = KE^{-\gamma} dE \ , \tag{1.174}$$

where K is a constant. The power law index γ is *not* the Lorentz factor; it is also specified by the symbol γ. To avoid confusion the symbols p or δ have been used by some authors to describe the power law index. The intensity of the ultrarelativistic synchrotron radiation of a homogeneous and isotropic distribution of electrons whose $N(E)$ are given by Eq. (1.174), and which are imbedded in a homogeneous magnetic field of strength H, may be obtained from Eqs. (1.160) and (1.173)

$$\begin{aligned} I(\nu) &= l\varepsilon(\nu) = Kl\alpha(\gamma)\frac{\sqrt{3}}{8\pi}\frac{e^3}{mc^2}\left[\frac{3e}{4\pi m^3c^5}\right]^{(\gamma-1)/2} H_\perp^{(\gamma+1)/2}\nu^{-(\gamma-1)/2} \\ &\approx 0.933\times 10^{-23}\alpha(\gamma)KlH_\perp^{(\gamma+1)/2}\left(\frac{6.26\times 10^{18}}{\nu}\right)^{(\gamma-1)/2} \\ &\quad \mathrm{erg\,sec^{-1}cm^{-2}Hz^{-1}rad^{-2}} , \end{aligned} \tag{1.175}$$

where ν is the frequency, l is the dimension of the radiating region along the line of sight, and $\alpha(\gamma)$ is a slowly varying function of γ which is of order unity and is given by

$$\alpha(\gamma) = 2^{(\gamma-3)/2}\frac{\gamma+7/3}{\gamma+1}\Gamma\left(\frac{3\gamma-1}{12}\right)\Gamma\left(\frac{3\gamma+7}{12}\right)$$

for $\gamma > \frac{1}{3}$.

The degree of linear polarization, Π, of the synchrotron radiation from ultrarelativistic electrons with velocities $v \approx c$ and a power law spectrum of index γ in a uniform magnetic field is:

$$\prod = (\gamma+1)/(\gamma+7/3) \ . \tag{1.176}$$

If the magnetic field is randomly distributed and $N(E)$ is given by Eq. (1.174) then I(ν) is given by Eq. (1.175) with a new constant of order unity substituted for $\alpha(\gamma)$. In this case, however, the degree of polarization is zero.

Measurements of the degree of circular polarization, $\prod_0$, of the synchrotron radiation from an ensemble of ultrarelativisitic electrons ($E \gg mc^2$) may be used to determine the magnetic field intensity, H, of the radiating source (Legg and Westfold, 1968; Melrose, 1971).

For an ensemble of electrons with a power law energy distribution of index, γ, as given by Eq. (1.174), the degree of circular polarization is given by

$$\begin{aligned} \prod_0 &= \frac{4}{\sqrt{3}}\frac{\gamma+1}{\gamma(\gamma+\frac{7}{3})}\cot\theta\left(\frac{\nu_H\sin\theta}{\nu}\right)^{1/2}\frac{\Gamma\left(\frac{3\gamma+4}{12}\right)\Gamma\left(\frac{3\gamma+8}{12}\right)}{\Gamma\left(\frac{3\gamma-1}{12}\right)\Gamma\left(\frac{3\gamma+7}{12}\right)} \\ &\times\left[\gamma+2+\tan\theta\frac{\varphi'(\theta)}{\varphi(\theta)}\right] \end{aligned} \tag{1.177}$$

for $\gamma > \frac{1}{3}$. Values of the gamma functions, Γ, are tabulated by Legg and Westfold (1968) for values of γ in the range 0.4–9.0. The degree of circular polarization given by Eq. (1.177) is the ratio of the fourth Stokes parameter, V, to the first Stokes parameter, I. Melrose (1971) has extended these formulae to include the effects of reabsorption and differential Faraday rotation on the degree of circular polarization. Both effects can cause the circular polarization to become slightly smaller and to reverse in sign. The polarization of gyrosynchrotron radiation in a magnetoactive plasma is given by Ramaty (1969).

From Eq. (1.175) we see that the synchrotron radiation from electrons with a power law distribution of index $-\gamma$ has a power law frequency spectrum proportional to ν^{α}, where the spectral index, α, is given by

$$\alpha = -\frac{(\gamma - 1)}{2} \ . \tag{1.178}$$

Most radio sources are observed to have a power law frequency spectrum, and some examples are shown in Fig. 1.4.

It follows from Eqs.(1.164) and (1.174) that the total electron energy, U_{e}, is given by (Burbidge, 1959; Kellermann and Owen, 1988)

$$U_{\mathrm{e}} = 422 \frac{L}{H^2} \frac{3-\gamma}{2-\gamma} \left(\frac{E_2^{2-\gamma} - E_1^{2-\gamma}}{E_2^{3-\gamma} - E_1^{3-\gamma}} \right) \mathrm{erg} \ , \tag{1.179}$$

where E_1 and E_2 denote, respectively, the lowest and highest electron energies, it is assumed that the electron energy distribution index $\gamma \neq 2$ or 3, and the total luminosity, L, of the source is given by

$$L = 4\pi r^2 \left(1 + \frac{z}{2}\right)^2 \frac{S_{\nu_{\mathrm{r}}}}{\nu_{\mathrm{r}}^{\alpha}} \int_{\nu_1}^{\nu_2} \nu^{\alpha} d\nu \ , \tag{1.180}$$

where $r = cz/H_0$ is the distance to the source, H_0 is the Hubble constant, z is the redshift, $S_{\nu_{\mathrm{r}}}$ is the source flux density measured at some reference frequency, ν_{r}, and ν_1 and ν_2 are the lower and upper cutoff frequencies beyond which the source radiation is negligible, Assuming that each electron radiates only at its critical frequency given by Eq.(1.154), Eq.(1.179) may be written as

$$U_{\mathrm{e}} = 10^{12} \frac{L}{H^{3/2}} \frac{\nu_2^{1/2+\alpha} - \nu_1^{1/2+\alpha}}{\nu_2^{1+\alpha} - \nu_1^{1+\alpha}} \frac{(2\alpha + 2)}{(2\alpha + 1)} \mathrm{erg} \ , \tag{1.181}$$

provided that $\alpha \neq -\frac{1}{2}$ or -1. As discussed in Sect. 1.38, the luminosity of synchrotron radiation is limited by the inverse Compton effect.

The total energy in the source, U_{T}, is given by (Burbidge, 1959, Burbidge, Jones and O'Dell (1974))

$$U_{\mathrm{T}} = U_{\mathrm{p}} + U_{\mathrm{m}} = aU_{\mathrm{e}} + \frac{V\,\mathrm{H}^2}{8\pi} \ , \tag{1.182}$$

where the total particle energy is U_{p}, the constant $a \approx 100$ denotes the increase in energy of the more energetic baryons above that of the relativistic electrons, the energy stored in the magnetic field is U_{m}, and V is the volume of the source. The value of $a = 100$ is typical for cosmic rays. Luminosities of radio galaxies and quasars range from 10^{40} to 10^{44}erg sec^{-1}, whereas those for normal galaxies are between 10^{34} and 10^{38}erg sec^{-1}. A typical field of $H \approx 10^{-4}$ gauss gives $U_{\mathrm{m}} \approx U_{\mathrm{p}} \approx 10^{60}$ erg if $a = 100$.

The total energy, $U_{\mathrm{e}} + U_{\mathrm{m}}$, in high-speed electrons and magnetic fields is minimized when dE/dH = 0, and the minimum magnetic field strength is given by its "equipartition" value (Burbidge, 1959; Moffet, 1975; Kellermann and Owen, 1988):

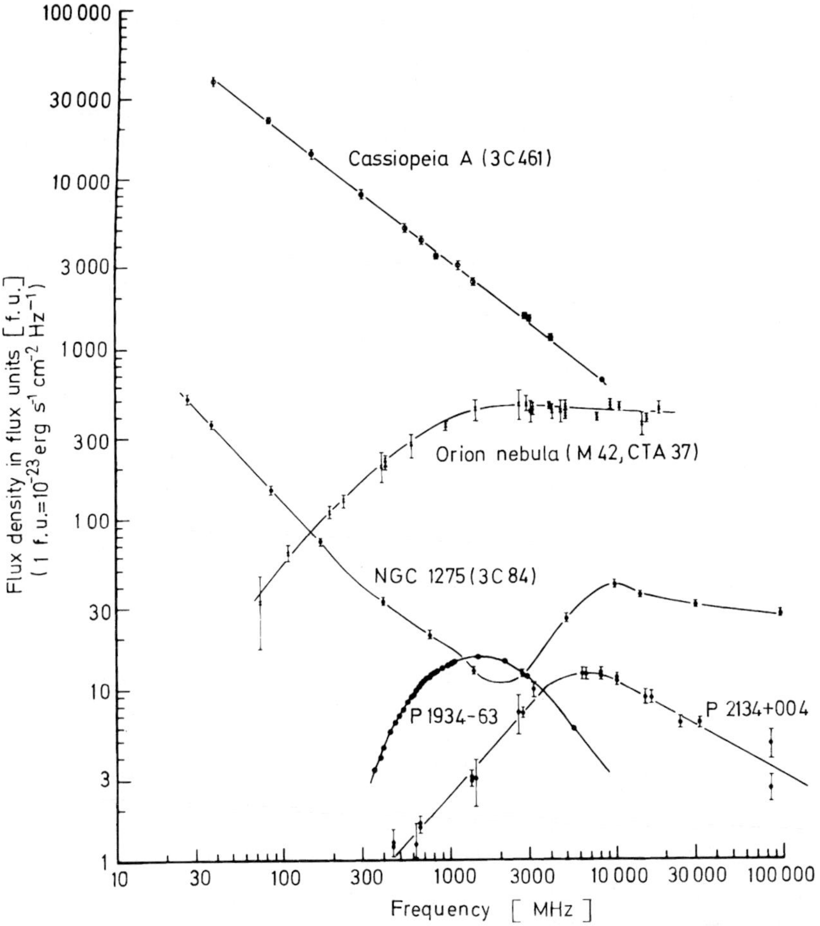

Fig. 1.4. Radiofrequency spectra of sources exhibiting the power law spectrum of synchrotron radiation (Casseopeia A), the flat spectrum of thermal bremsstrahlung radiation with low frequency self absorption (Orion Nebula), unusual high frequency radiation (NGC 1275), and low frequency absorbtion processes (P1934 − 63 and P2134 + 004). The data for P2134 + 004 are from E. K. Conklin, and the other data are from Kellermann [1966], Hjellming, and Churchwell [1969], Terzian and Parrish [1970], and Kellermann, Pauliny-Toth, and Williams [1969].

$$H_{min} \approx (10^{13} L/V)^{2/7} \approx 1.5 \times 10^{-4} \theta^{9/7} z^{-2/7} S^{2/7} \text{ Gauss} , \tag{1.183}$$

for a source of angular size θ in units of arcseconds, redshift z, and flux density S in Janskys or flux units. The minimum electron energy, U_{emin}, is given when the energy is nearly equally distributed between the relativistic electrons and the magnetic fields, so that

$$\begin{aligned} U_{emin} \approx U_{Hmin} &\approx VH^2_{min}/(8\pi) \approx (10^{12} L)^{4/7} V^{3/7} \\ &\approx 10^{59} \theta^{9/7} z^{17/7} S^{2/7} \text{ ergs} . \end{aligned} \tag{1.184}$$

Miley (1980) provides formulae for the minimum energy conditions in extended radio sources that emit synchrotron radiation.

Gyrosynchrotron radiation from mildly relativistic electrons, with velocities $v \approx 0.1c$, is discussed by Dulk and Marsh (1982) who give simplified expressions for the emissivity, ε_ν, in units of erg cm^{-3} s^{-1} Hz^{-1} rad^{-2}, absorptivity, α_ν, in units of cm^{-1}, the effective temperature, T_{eff}, the degree of circular polarization, $\prod_c$, and the frequency of peak brightness, ν_{peak}. They consider an isotropic pitch angle distribution with both nonthermal (power-law) and thermal (Maxwellian) distributions of electron energy. These formulae are reproduced by Dulk (1985) in his review of radio emission from the Sun and stars, with a correction for the numerical expression for the degree of circular polarization. For an electron distribution that is power law in energy, E, with a number density $n(E) = KE^{-\delta}$ above a cutoff $E_O = 10\,\mathrm{keV} = 1.6 \times 10^{-8}$erg, Dulk (1985) gives the following approximation formulae for the gyrosynchrotron radiation from mildly relativistic electrons:

$$\begin{aligned}
\varepsilon_\nu/(NH) &\approx 3.3 \times 10^{-24} 10^{-0.52\delta} (\sin\theta)^{-0.43+0.65\delta} (\nu/\nu_H)^{1.22-0.90\delta} ,\\
\alpha_\nu H/N &\approx 1.4 \times 10^{-9} 10^{-0.22\delta} (\sin\theta)^{-0.09+0.72\delta} (\nu/\nu_H)^{-1.30-0.98\delta} ,\\
T_{eff} &\approx 2.2 \times 10^9 10^{-0.31\delta} (\sin\theta)^{-0.36-0.06\delta} (\nu/\nu_H)^{0.50+0.085\delta} ,\\
\prod\nolimits_c &\approx 1.26\ 10^{0.035\delta}\ 10^{-0.071\cos\theta} (\nu/\nu_H)^{-0.782+0.545\cos\theta} (\tau_\nu \ll 1) ,\\
\nu_{peak} &\approx 2.72 \times 10^3 10^{0.27\delta} (\sin\theta)^{0.41+0.03\delta} (NL)^{0.32-0.03\delta} H^{0.68+0.03\delta} ,
\end{aligned} \tag{1.185}$$

where H is the magnetic field strength, N is the total number of electrons per cm^3 above the energy E_O, the constant $K = (\delta - 1)E_0^{\delta-1}N$. The viewing angle is θ, and the gyrofrequency $\nu_H = 2.8 \times 10^6 H$.

Dulk (1985) also gives formulae for the same variables in the case of synchrotron radiation of ultrarelativistic electrons with velocities $v \approx c$ and a power-law energy distribution.

$$\begin{aligned}
\frac{\varepsilon_\nu}{HN} &= \frac{1}{2}(\delta - 1)E_0^{\delta-1} g(\delta) \frac{\sqrt{3}e^3}{8\pi mc^2} \sin\theta \left[\frac{2m^2c^4}{3\sin\theta}\frac{v}{v_H}\right]^{-(\delta-1)/2}\\
&\approx 8.6 \times 10^{-24}(\delta - 1)\sin\theta \left[\frac{0.175}{\sin\theta}\left(\frac{E_O}{1\mathrm{MeV}}\right)^{-2}\frac{v}{v_H}\right]^{-(\delta-1)/2}
\end{aligned} \tag{1.186}$$

where $g(\delta)$ involves a product of gamma functions and is within 15% of 1.85 over the range $2 \le \delta \le 5$;

$$\begin{aligned}
\frac{\alpha_v H}{N} &= (\delta - 1)E_0^{\delta-1} h(\delta) \frac{2\pi e m^5 c^{10}}{9} \frac{1}{\sin\theta} \left[\frac{m^2c^4}{3\sin\theta}\frac{\nu}{\nu_H}\right]^{-(\delta+4)/2}\\
&\approx 8.7 \times 10^{-12} \frac{\delta - 1}{\sin\theta} \left(\frac{E_O}{1\mathrm{MeV}}\right)^{\delta-1} \left[\frac{8.7 \times 10^{-2}}{\sin\theta}\frac{\nu}{\nu_H}\right]^{-(\delta+4)/2} ,
\end{aligned}$$

where $h(\delta)$ is within 15% of 0.74 over the range $2 \le \delta \le 5$;

$$T_{\rm eff} \approx 2.6 \times 10^9 2^{-\delta/2} \left[\frac{\nu}{\nu_{\rm H} \sin\theta}\right]^{1/2} ;$$

$$\nu_{\rm peak} \approx 3.2 \times 10^7 \sin\theta \left(\frac{\rm E_O}{\rm 1MeV}\right)^{(2\delta-2)/(\delta+4)} \times \left[8.7 \times 10^{-12} \frac{\delta - 1}{\sin\theta} {\rm NL}\right]^{2/(\delta+4)} {\rm H}^{(\delta+2)/(\delta+4)} . \qquad (1.187)$$

The degree of linear polarization, Π_l, for a uniform, homogeneous, optically thin source is $\Pi_l = (\delta + 1)/(\delta + 7/3)$, circular polarization of synchrotron radiation is small, and both the linear and circular polarization approach zero if the magnetic field of the source is randomly distributed.

The principal characteristics of the radio synchrotron radiation and bremsstrahlung of thermal electrons in solar active regions are given by Zlotnik (1968) and Zheleznyakov (1970) as a function of the known electron concentration, kinetic temperature and magnetic field strength, explaining the slowly varying component of solar radio emission.

1.27 Synchrotron Radiation in a Plasma

The total power radiated per unit frequency interval by a single relativistic electron in a plasma is given by (Ginzburg and Syrovatskii, 1965)

$$P(\nu) = \sqrt{3}\frac{e^3 H \sin\psi}{mc^2}\left[1 + \left(\frac{\nu_p E}{\nu mc^2}\right)^2\right]^{-1/2} \frac{\nu}{\nu'_{\rm c}} \int_{\nu/\nu'_{\rm c}}^{\infty} K_{5/3}(\eta) d\eta , \qquad (1.188)$$

where the magnetic field intensity is H, the pitch angle is ψ, the total energy of the electron is E, the frequency of the synchrotron radiation is ν, the modified Bessel function is K, the critical frequency, $\nu'_{\rm c}$, is given by

$$\nu'_{\rm c} = \frac{3eH \sin\psi}{4\pi mc}\frac{mc^2}{E}\left[1 - \left(\frac{n_\nu v}{c}\right)^2\right]^{-3/2}$$

or

$$\nu'_{\rm c} \approx \nu_{\rm c}\left[1 + \left(\frac{\nu_{\rm p} E}{\nu mc^2}\right)^2\right]^{-3/2} = \nu_{\rm c}\left[1 + \left(1 - n_\nu^2\right)\left(\frac{E}{mc^2}\right)^2\right]^{-3/2} ,$$

where the critical frequency, $\nu_{\rm c}$, in a vacuum is given in Eq. (1.154), the electron velocity is v, the index of refraction, n_ν, is assumed to be less than unity, and the plasma frequency, $\nu_{\rm p}$, is given by

$$\nu_{\rm p} = \nu(1 - n_\nu^2)^{1/2} = \left(\frac{e^2 N_{\rm e}}{\pi m}\right)^{1/2} \approx 9 \times 10^3 N_{\rm e}^{1/2}\ {\rm Hz} ,$$

where $N_{\rm e}$ is the number density of thermal electrons. The general formulae for the synchrotron radiation emitted in the nth harmonic by an electron in an isotropic plasma were first derived by Eidman (1958), slightly modified by Liemohn (1965), and numerically computed by Ramaty (1969). Wild and Hill (1971) give these formulae together with an approximate expression for the power radiated in the parallel component by a relativistic or subrelativistic electron.

As first pointed out by Tsytovich (1951) and rigorously derived by Eidman (1958) and Razin (1960), the synchrotron radiation in a plasma will be appreciably suppressed at frequencies, ν, below the critical frequency, ν_R, given by

$$\nu_R = \frac{4ecN_e}{3H\sin\psi} \approx 20\frac{N_e}{H\sin\psi} \approx \frac{2\nu_p^2}{3\nu_H} . \quad (1.189)$$

This Tsytovich-Razin effect suppresses synchrotron radiation at frequencies where the index of refraction in the plasma becomes significantly less than unity. The total power radiated, P, by a single relativistic electron in a plasma is obtained by integrating Eq. (1.188) over all frequencies and is given by (Simon, 1969)

$$P = \pi\frac{3^{3/4}}{\sqrt{2}}\frac{e^3H\sin\psi}{mc^2}\frac{E}{mc^2}\nu_p \exp\left[\frac{-2\sqrt{3}\pi mc\nu_p}{\gamma eH\sin\psi}\right] , \quad (1.190)$$

where $\gamma = E/(mc^2)$. The asymptotic form for the volume emissivity, $\varepsilon(\nu)$, of the synchrotron radiation from a group of electrons with a power law energy distribution is given by (Simon, 1969)

$$\varepsilon(\nu) = 6.9N_0\left(\frac{\nu}{\nu_R}\right)^{3/2}\nu_p\left(\frac{\nu}{\sqrt{2}\nu_p}\right)^{-\gamma}\exp\left(-3.7\frac{\nu_R}{\nu}\right) \quad \text{erg sec}^{-1}\ \text{Hz}^{-1}\text{cm}^{-3} \text{ for } \nu < \nu_R , \quad (1.191)$$

where $\varepsilon(\nu)$ is the power per unit frequency per unit volume, and it is assumed that the number density of electrons per unit solid angle along the line of sight, which are moving in the direction of the observer and whose energies lie in the range E to $E + dE$, is given by

$$N(E)dE = N_0\left(\frac{E}{mc^2}\right)^{-\gamma} dE . \quad (1.192)$$

It follows from Eqs. (1.175) and (1.191) that the spectral index $\alpha = -(\gamma - 1)/2$ for frequencies $\nu > \nu_R$ and is roughly $\alpha = \frac{3}{2} - \gamma$ for $\nu < \nu_R$.

Synchrotron radiation will also be suppressed by thermal absorption in a plasma at those frequencies for which the optical depth is greater than unity. The absorption coefficient and optical depth of a plasma are discussed in detail in Sect. 1.30. The formula for the absorption coefficient at radio frequencies $\nu > \nu_p$ leads to the conclusion that radiation will be thermally absorbed at frequencies below the critical frequency, ν_T, given by

$$\nu_T = \left[lT^{-3/2}N_e^2\right]^{-2.1} , \quad (1.193)$$

where l is the extent of the plasma along the line of sight to the radiating source, and T and N_e are, respectively, the kinetic temperature and the number density of the thermal electrons. When the thermal electrons are mixed with the relativistic electrons the spectral index will be $\alpha_0 = -(\gamma - 1)/2$ for $\nu > \nu_T$ and $\alpha = 2.1 + \alpha_0$ for $\nu < \nu_T$. If the synchrotron radiation is created in a vacuum but subsequently passes through a plasma, the intensity of the radiation will fall off as $\exp[-(\nu/\nu_T)^{-2.1}]$ for $\nu < \nu_T$.

1.28 Additional Modifications of the Synchrotron Radiation Spectrum

As first suggested by Twiss (1954) and calculated by Le Roux (1961), McCray (1969), Wild, Smerd and Weiss (1963), Ginzberg and Syrovatskii (1965), Shu (1991), and Longair (1994) synchrotron radiation may become self absorbed with an absorption coefficient per unit length, α_ν, given by

$$\alpha_\nu = -\frac{c^2}{8\pi\nu^2}\int_0^\infty E^2 \frac{d}{dE}\left[\frac{N(E)}{E^2}\right] P(\nu)dE \ , \tag{1.194}$$

where ν is the frequency, $N(E)dE$ is the number density of electrons per unit solid angle along the line of sight which are moving in the direction of the observer and whose kinetic energies lie in the range E to $E + dE$, the total power radiated per unit frequency interval by a single relativistic electron is $P(\nu)$ and is given by Eq. (1.160), and it is assumed that the electrons are distributed isotropically. For electrons which have the power law energy distribution given by Eq. (1.174), Eq. (1.194) becomes (Le Roux, 1961)

$$\begin{aligned}\alpha_\nu &= g(\gamma)\frac{e^3}{2\pi m}\left(\frac{3e}{2\pi m^3 c^5}\right)^{\gamma/2} K(H\sin\psi)^{(\gamma+2)/2}\nu^{-(\gamma+4)/2} \\ &\approx 0.019(3.5\times 10^9)^\gamma K(H\sin\psi)^{(\gamma+2)/2}\nu^{-(\gamma+4)/2} \quad \mathrm{cm}^{-1} \ ,\end{aligned} \tag{1.195}$$

where $g(\gamma)$ is a constant of order unity given by

$$g(\gamma) = \frac{\sqrt{3}}{16}\left(\gamma + \frac{10}{3}\right)\Gamma\left(\frac{3\gamma+2}{12}\right)\Gamma\left(\frac{3\gamma+10}{12}\right) \ .$$

Here γ is the index of the power law energy distribution given in Eq. (1.174), H is the magnetic field intensity, ψ is the average pitch angle, and the constant K which also appears in Eq. (1.174) is a function of the intensity I_ν of the source and is given by

$$K \approx 7.4 \times 10^{21}\frac{I_\nu}{lH}\left(\frac{\nu}{6.26\times 10^{18}H}\right)^{(\gamma-1)/2} \ ,$$

where l is the extent of the radiating source along the line of sight.

For a very compact source, we have the possibility that the source will be optically thick due to synchrotron self absorption at low frequencies $\nu < \nu_s$, even if it is optically thin at high frequencies $\nu > \nu_s$. The physics of compact nonthermal sources is discussed by Jones, O'Dell and Stein (1974), and the transfer of polarized radiation in self-absorbed synchrotron radio sources is given by Jones and O'Dell (1977). The synchrotron radiation will become appreciably suppressed at frequencies, ν, below the critical frequency, ν_s, for which the optical depth $\tau_\nu = \alpha_\nu l$ is unity. For a source of maximum flux density S_{ν_s}, magnetic field intensity, H, angular size, θ, and redshift, z, the critical frequency, ν_s, is given by (Slish, 1963; Williams, 1963; Kellermann and Pauliny-Toth, 1981):

$$\nu_s \approx f(p) H^{1/5} S_{\nu s}^{2/5} \theta^{-4/5} (1+z)^{1/5} \delta^{-1/5} \quad \text{GHz} , \tag{1.196}$$

where the function f(p) depends only weakly on source geometry and the value of the electron energy power-law index is p, $f(p) \approx 8$ for $p = \gamma = 2.5$, H is the magnetic field strength in Gauss, θ is the angular size in milliarcseconds or 10^{-3} seconds of arc, the flux density, $S_{\nu s}$, is in units of Janskys or flux units, the quantity δ is a correction for the relativistic Doppler shift of the entire radiating source if it is moving with high velocity and $\delta = 1$ for source velocities less than the velocity of light, c (also see Volume II), and ν_s is in units of $\text{GHz} = 10^3$ MHz $= 10^9$Hz. The similar expression for ν_s given by Kellermann and Owen (1988) is in error because the flux density and angular size parmeters were inadvertently transposed. Synchrotron radiation losses in a compact radio source lead to a characteristic half-life, t, at frequency ν_m of (Kellermann and Pauliny-Toth, 1981; Kellermann and Owen, 1988):

$$t \approx H^{-3/2} \nu_m^{-1/2} \approx 10^7 S^3 \theta^6 \nu_s^{-8} (1+z)^{3/2} \delta^{3/2} \text{ years} . \tag{1.197}$$

The spectral index of the radiation intensity will be $\alpha = -(\gamma - 1)/2$ for $\nu > \nu_s$ and $\alpha = 2.5$ for $\nu < \nu_s$. Observations of the maximum flux density, critical frequency, and angular size of many of the compact radio sources (Kellermann and Pauliny-Toth, 1969; Clarke et al., 1969) indicate that $H = 10^{-4\pm1}$ Gauss if the observed low frequency cutoff in their spectra is due to synchrotron self-absorption.

It follows from Eq. (1.163) that high energy electrons lose their energy by synchrotron radiation faster than low energy electrons and therefore the spectrum of the radiation will become steeper with increasing time. When synchrotron radiation losses predominate, a source which is initially supplied with electrons with a power law energy distribution of index $-\gamma$ will develop a spectral index of $\alpha = -(2\gamma + 1)/3$ for frequencies higher than the critical frequency, ν_K, given by (Kardashev, 1962)

$$\nu_K = 3.4 \times 10^8 H^{-3} t^{-2} \quad \text{Hz} \tag{1.198}$$

where H is the magnetic field intensity in Gauss, and t is the age of the radio source in years. Spectral changes are also induced by energy losses due to the expansion of the source, collisions of electrons and heavy particles, the inverse Compton effect, and ionization. The details of these changes are given by Kardashev (1962) and de la Beaujardiére (1966). The emission spectrum can also be altered by energy cutoffs in the electron distribution, but the steepest low energy cutoff that is possible has a spectral index of only one third. The evolution of expanding nonthermal sources has been discussed by Cocke (1974) and Vitello and Pacini (1977, 1978).

O'Dell and Sartori (1970) have pointed out that there is a natural low frequency cutoff in the spectrum of the synchrotron radiation of a single electron (cf. Fig. 1.2) which falls off as the one-third power of the frequency for frequencies, ν, less than the critical frequency

$$\nu_c = \frac{eH}{2\pi mc\psi} \approx 3 \times 10^6 \frac{H}{\psi} \quad \text{Hz} \tag{1.199}$$

where ψ is the average pitch angle. This cyclotron turnover induces a low frequency spectral index of $\alpha = \gamma$ for a group of electrons with a power law energy distribution of index $-\gamma$, whereas the high frequency spectral index is $\alpha = -(\gamma - 1)/2$.

1.29 Bremsstrahlung (Free–Free Radiation) of a Single Electron

Electromagnetic waves are radiated when an electron is accelerated in the electrostatic fields of the ions or the nuclei of atoms in the substance through which it passes. This radiation is known as braking radiation, or in German bremsstrahlung. It is also called free–free emission since the radiation corresponds to transitions between unbound states in the field of a nucleus. A relatively recent discussion of nonrelativistic, thermal bremsstrahlung and relativistic bremsstrahlung is given by Longair (1992).

When a fast charged particle encounters an atom, molecule, or ion it is acccelerated and emits radiation called bremsstrahlung (braking radiation). Consider a nonrelativistic electron of mass, m, charge, e, and velocity, v. When the electron enters the Coulomb force field of a charge Ze, the angular deflection θ, of the particle is given by Rutherford (1911).

$$\tan\frac{\theta}{2} = \frac{Ze^2}{mv^2 b} \ , \tag{1.200}$$

where the impact parameter, b, designates the perpendicular distance from the Ze charge to the original path of the electron.

It follows from the Larmor (1897) formula for the total power radiated by an accelerated electron (Eq. (1.110)), and from Eq. (1.200) that the total bremsstrahlung energy radiated in frequency interval between ν and $\nu + d\nu$ by a non-relativistic electron during a collision with a charge Ze is given by

$$\begin{aligned} W(\nu)d\nu \approx \frac{2e^2}{3\pi c^3}|\Delta v|^2\, d\nu \approx \frac{2e^2 v^2}{3\pi c^3}(1-\cos\theta)d\nu \approx \frac{8}{3\pi}\left(\frac{e^2}{mc^2}\right)^2 \frac{Z^2e^2}{c}\left(\frac{c}{vb}\right)^2 d\nu \\ \text{for } \nu < \frac{v}{b} \\ \approx 0 \quad \text{for } \nu > \frac{v}{b} \ , \end{aligned} \tag{1.201}$$

where Δv is the change in electron velocity caused by the collision. The frequency spectrum is flat out to the critical frequency $\nu_c \approx v/b$, and the total energy radiated during one collision is $[vW(\nu)]/b$.

The total number of encounters per unit time between an electron and a volume density of N_i ions in the parameter range b to $b + db$ is

$$N_i v 2\pi b db = N_i v 2\pi \frac{d\sigma_s}{d\Omega} \sin\theta d\theta \ , \tag{1.202}$$

where the differential scattering cross section, $d\sigma_s/d\Omega$, is given by

$$\frac{d\sigma_s}{d\Omega} = \frac{b}{\sin\theta}\left|\frac{db}{d\theta}\right| = \frac{Z^2}{4}\left(\frac{e^2}{mv^2}\right)^2 \frac{1}{\sin^4\left(\frac{\theta}{2}\right)} \ . \tag{1.203}$$

Using Eqs. (1.201) and (1.202), the total bremsstrahlung power, $P_i(v, \nu)d\nu$, radiated in frequency interval between ν and $\nu + d\nu$ in collisions with N_i ions is given by

$$P_i(v, \nu)d\nu = N_i v Q_r(v, \nu)d\nu \ , \tag{1.204}$$

where the radiation cross section, $Q_r(v, \gamma)$ is given by

$$Q_r(v, \nu) = \frac{16}{3}\frac{Z^2e^6}{m^2c^3v^2}\int_{b_{min}}^{b_{max}}\frac{db}{b} = \frac{16}{3}\frac{Z^2e^6}{m^2c^3v^2}\ln\left(\frac{b_{max}}{b_{min}}\right) \ , \tag{1.205}$$

which has the dimensions of area energy/frequency. The photon cross section $\sigma_r(v, \hbar\omega)$, is also often used, and is defined by

$$\sigma_r(v, \hbar\omega) = \frac{Q_r(v, \nu)}{\hbar^2\omega} = \frac{16}{3}Z^2\frac{e^2}{\hbar c}\left(\frac{e^2}{mc^2}\right)^2\frac{c^2}{\hbar\omega v^2}\ln\left(\frac{b_{max}}{b_{min}}\right) \ , \tag{1.206}$$

which has the dimensions of area per unit photon energy. The fine structure constant $\alpha = e^2/(\hbar c) = 1/137$, and the classical electron radius $r_e = e^2/(mc^2) = 2.82 \times 10^{-13}$ cm. The classical limits to the impact parameter are (Jackson, 1962, 1975; Longair, 1992):

$$b_{max} = v/\omega \tag{1.207}$$

$$b_{min} = Ze^2/(mv^2) \text{ for low velocity}$$

and

$$b_{min} = \hbar/(mv) \text{ for high velocity} \tag{1.208}$$

where low velocities are $v \ll \alpha Zc$ and at high velocities $v \gg \alpha Zc$ a quantum mechanical approach is required.

The logarithmic Gaunt factor $\ln(b_{max}/b_{min})$ depends on the radian frequency, ω, of the bremsstrahlung radiation and the velocity, v, of the electron according to the formulae (Bohr, 1915; Bethe, 1930)

$$\begin{aligned}
\ln\left(\frac{b_{max}}{b_{min}}\right) &= \ln\left(\frac{\gamma^2 mv^3}{Ze^2\omega}\right) && \text{for } v < \frac{Ze^2}{\hbar} \text{ and } \omega < \frac{mv^3}{Ze^2} \\
&= 0 && \text{for } v < \frac{Ze^2}{\hbar} \text{ and } \omega > \frac{mv^3}{Ze^2} \ , \\
\ln\left(\frac{b_{max}}{b_{min}}\right) &= \left[\ln\left(\frac{2\gamma^2 mv^2}{\hbar\omega}\right) - \frac{v^2}{c^2}\right] && \text{for } v > \frac{Ze^2}{\hbar} \text{ and } \omega < \frac{mv^2}{\hbar} \\
&= 0 && \text{for } v > \frac{Ze^2}{\hbar} \text{ and } \omega > \frac{mv^2}{\hbar} \ ,
\end{aligned} \tag{1.209}$$

where $\gamma = E/(mc^2)$ is not $\gg 1$, the kinetic energy of the electron is E, and it is assumed that ω is larger than the plasma frequency $\omega_p = (4\pi N_e e^2/m)^{1/2} \approx 5.64 \times 10^4 N_e^{1/2}$, where N_e is the electron density. When $\omega < \omega_p$, Eqs. (1.209) are appropriate with ω_p substituted for ω. As suggested by Bethe and Heitler (1934), Eqs. (1.209) are best used with the average velocity $v = [E^{1/2} + (E - \hbar\omega)^{1/2}]/\sqrt{2m}$, where E is the initial energy of the electron and $\hbar\omega$ is the photon energy. For the special case of electron-electron scattering at high velocities but in the nonrelativistic case, we have (Bethe, 1930)

$$\ln\left(\frac{b_{max}}{b_{min}}\right) = \ln\left[\frac{(\gamma-1)(\gamma+1)^{1/2}}{\sqrt{2}}\frac{mc^2}{\hbar\omega}\right] \quad \text{for } \omega < \frac{mv^2}{\hbar}$$
$$\approx 0 \quad \text{for } \omega > \frac{mv^2}{\hbar} \ . \qquad (1.210)$$

When the electron is relativistic, $\gamma \gg 1$, the differential scattering cross section, $d\sigma_s/d\Omega$, is given by the Mott formula (Mott and Massey, 1965)

$$\frac{d\sigma_s}{d\Omega} = \frac{Z^2e^4}{4m^2v^4}\frac{1}{\sin^4\left(\frac{\theta}{2}\right)}\frac{1-\beta^2\sin^2\left(\frac{\theta}{2}\right)}{\gamma^2} \ , \qquad (1.211)$$

where $\beta = v/c$, the electron velocity is v, and θ is the angular deflection of the electron. Approximate relativistic formulae for the total bremsstrahlung power radiated per unit frequency interval, $P_i(v, \nu)$ are given by Eqs. (1.204) and (1.205) when $b_{max} = \gamma^2 v/\omega$ and

$$\ln\left(\frac{b_{max}}{b_{min}}\right) \approx \ln\left(\frac{\gamma^2 mc^2}{\hbar\omega}\right) \quad \text{for } \gamma \gg 1 \quad \text{and } \omega < \frac{(\gamma-1)mc^2}{\hbar}$$
$$\approx 0 \quad \text{for } \gamma \gg 1 \quad \text{and } \omega > \frac{(\gamma-1)mc^2}{\hbar} \ . \qquad (1.212)$$

The exact formula for the photon cross section, σ_r, of a high energy electron incident on an unshielded static charge Ze is given by (cf. Bethe, 1930; Bethe and Heitler, 1934; Wheeler and Lamb, 1939; Heitler, 1954; Jauch and Rohrlich, 1955)

$$\sigma_r = 4Z^2\frac{e^2}{\hbar c}\left(\frac{e^2}{mc^2}\right)^2\frac{1}{\hbar\omega}\frac{1}{E_i^2}\left(E_i^2 + E_f^2 - \frac{2}{3}E_iE_f\right)\left[\ln\left(\frac{2E_iE_f}{mc^2\hbar\omega}\right) - \frac{1}{2}\right] \ , \qquad (1.213)$$

where the initial and final electron energy are, respectively, E_i and E_f, and $E_i - E_f = \hbar\omega$, the photon energy. Eqs. (1.212) and (1.213) are equally valid for the collisions of a relativistic electron with free electrons, free protons, or with hydrogen atoms (cf. Blumenthal and Gould, 1970). As first pointed out by Fermi (1940), however, the screening effects of atoms limits the maximum impact parameter to the atomic radius, a. For the collision of relativistic electrons with atoms we have the additional condition that

$$\ln\left(\frac{b_{max}}{b_{min}}\right) = \ln\left(\frac{amv}{\hbar}\right) = \ln\left(\frac{1.4}{Z^{1/3}}\frac{\hbar c}{e^2}\frac{v}{c}\right) \quad \text{for } \omega < \omega_s \ , \qquad (1.214)$$

$$\text{where } \omega_s = \frac{Z^{1/3}}{1.4}\frac{e^2}{\hbar c}\frac{\gamma^2 mc^2}{\hbar} ,$$

b_{max} is the atomic radius $a = 1.4\hbar^2/(Z^{1/3}me^2)$, and the fine structure constant $\alpha = e^2/\hbar c = 1/137$.

For relativistic electrons ($\gamma \gg 1$), energy losses by radiation predominate over collisional energy losses. It follows from Eqs. (1.204), (1.205), (1.212), and (1.214) that the radiative energy loss, dE_r/dx, of a relativistic electron when traversing a unit distance of matter is given by

$$\begin{aligned}\frac{dE_r}{dx} &\approx \frac{16}{3}N_iZ^2\left(\frac{e^2}{\hbar c}\right)\left(\frac{e^2}{mc^2}\right)^2 E \ln \gamma \qquad \text{for } \omega > \omega_s \\ &\approx \frac{16}{3}N_iZ^2\left(\frac{e^2}{\hbar c}\right)\left(\frac{e^2}{mc^2}\right)^2 E \ln\left(\frac{1.4\,\hbar c}{Z^{1/3}\,e^2}\right) \qquad \text{for } \omega < \omega_s ,\end{aligned} \tag{1.215}$$

where the electron energy $E = \gamma mc^2$, the fine structure constant $\alpha = e^2/\hbar c \approx 1/137$, and the classical electron radius $e^2/mc^2 \approx 2.8 \times 10^{-13}$ cm.

For nonrelativistic electrons, collisional energy loss predominates over radiative energy loss. The total energy lost per unit length, dE_c/dx, in collisions with N atoms with Z electrons is (Bohr, 1915; Bethe, 1930)

$$\frac{dE_c}{dx} = 4\pi NZ\frac{e^4}{mv^2}\ln\left(\frac{b_{max}}{b_{min}}\right) , \tag{1.216}$$

where $\ln(b_{max}/b_{min})$ is given by Eqs. (1.209).

1.30 Bremsstrahlung (Free–Free Radiation) from a Plasma

The total bremsstrahlung power radiated from a plasma per unit volume per unit solid angle per unit frequency interval ν to $\nu + d\nu$ is called the volume emissivity, ε_ν, and is given by

$$\varepsilon_\nu = n_\nu \frac{N_e}{4\pi}\int P_r(v,\nu)f(v)dv , \tag{1.217}$$

where n_ν is the index of refraction, N_e is the electron density, $f(v)$ is the electron velocity distribution, and $P_r(v,\nu)$ is the total power radiated per unit frequency interval in the collision of an electron of velocity, v, with N_i ions and is given by Eqs. (1.204) to (1.214). The velocity criterion $v \lessgtr Ze^2/\hbar$ may be converted to the temperature criterion $T \lessgtr 3.16 \times 10^5 Z^2$ through the relation

$$v \approx \left(\frac{kT}{m}\right)^{1/2} \approx 3.89 \times 10^5 T^{1/2}\ \text{cm sec}^{-1} . \tag{1.218}$$

When $f(v)$ is Maxwellian (Eq. (3.114)), Eq. (1.217) becomes

$$\begin{aligned}\varepsilon_\nu d\nu &= \frac{8}{3}\left(\frac{2\pi}{3}\right)^{1/2}\frac{n_\nu Z^2 e^6}{m^2c^3}\left(\frac{m}{kT}\right)^{1/2} N_iN_e g(\nu,T)\exp(-h\nu/kT)d\nu \\ &\approx 5.4 \times 10^{-39} n_\nu Z^2 \frac{N_iN_e}{T^{1/2}} g(\nu,T)\exp(-h\nu/kT)d\nu \ \text{erg sec}^{-1}\ \text{cm}^{-3}\ \text{Hz}^{-1}\ \text{rad}^{-2} ,\end{aligned} \tag{1.219}$$

and the free-free Gaunt factor, $g(v, T)$ is given by

$$g(v, T) = \frac{\sqrt{3}}{\pi} \ln \Lambda \approx 0.54 \ln \Lambda \ , \tag{1.220}$$

where $\ln \Lambda$ is given in Table 1.1. Discussions of the appropriate Gaunt factor in various domains are given by Gaunt (1930), Sauter (1933), Sommerfeld (1931), Elwert (1954), Scheuer (1960), Oster (1961), Brussard and van de Hulst (1962), Mercier (1964), and Oster (1970). Values have been tabulated by Karzas and Latter (1961).

For photon energies $hv \ll kT$, the logarithmic term is $\ln \Lambda \approx \ln(4.7 \times 10^{10} T/v)$ at X-ray wavelengths and $\ln \Lambda \approx \ln[4.9 \times 10^{10} T^{3/2}/(Zv)]$ at radio frequencies (Longair, 1992). The free-free Gaunt factor at radio frequencies is $g(v, T) \approx 11.962\, T_e^{0.15}\, v^{-0.1}$ (Brown, 1987). For $hv \gg kT$, the Gaunt factor $g(v, T) \approx [hv/(kT)]^{1/2}$. Thermally averaged, nonrelativistic free–free Gaunt factors have been computed by Hummer (1988) over a wide range of temperatures and frequencies using the exact expressions given by Karzas and Latter (1961). The relativistic free–free Gaunt factor for a dense, high-temperature plasma was calculated by Nakagawa, Kohyama and Itoh (1987).

The exact value for the bremsstrahlung cross section is given by Bethe and Heitler (1934) see Eq. (1.213) and Eq. (4.388). Bremsstrahlung cross-section formulas and related data are reviewed by Koch and Motz (1959); at high values of the ratio of photon energy to electron energy, the cross section is highly anisotropic (Gluckstern and Hull, 1953, Tseng and Pratt, 1973).

The free–free absorption coefficient per unit length for electron-ion bremsstrahlung is given by (Kirchhoff, 1860; Margenau, 1946; Smerd and Westfold, 1949)

$$\alpha_v = \frac{\varepsilon_v}{n_v^2} \frac{c^2}{2hv^3} \left[\exp\left(\frac{hv}{kT}\right) - 1\right] = \frac{\varepsilon_v}{n_v^2 B_v(T)} \ , \tag{1.221}$$

Table 1.1 The logarithm factor, Λ, where the Gaunt factor $g(v, T) = (\sqrt{3}/\pi) \ln \Lambda$. The frequency is v, the constants $\gamma = 1.781$ and $e' = 2.718$, the plasma frequency is $\omega_p \approx 5.64 \times 10^4 N_e^{1/2}$ Hz, the electron density is N_e, the electron temperature is T, and the ion charge is Ze

	$T < 3.6 \times 10^5\, Z^2$ °K	$T > 3.6 \times 10^5\, Z^2$ °K
$\omega \gg \omega_p$	$\Lambda = \left(\frac{2}{\gamma}\right)^{5/2} \left(\frac{kT}{m}\right)^{1/2} \left(\frac{kT}{2\pi Z e^2 v}\right)$	$\Lambda = \frac{4kT}{\gamma hv}$
	$\approx 5.0 \times 10^7 (T^{3/2}/Zv)$	$\approx 4.7 \times 10^{10} (T/v)$
$\omega \le \omega_p$	$\Lambda = \left(\frac{2}{\gamma}\right)^2 \frac{1}{e'^{1/2}} \left(\frac{kT}{m}\right)^{1/2} \frac{kT}{Ze^2 \omega_p}$	$\Lambda = \left(\frac{8}{e'\gamma}\right)^{1/2} \left(\frac{kT}{m}\right)^{1/2} \frac{(mkT)^{1/2}}{\hbar \omega_p}$
	$\approx 3.1 \times 10^3 (T^{3/2}/Z N_e^{1/2})$	$\approx 3.0 \times 10^6 (T/N_e^{1/2})$

where $B_\nu(T)$ is the vacuum brightness of a black-body radiator. For frequencies $\nu \ll 10^{10}T$, the Rayleigh–Jeans approximation for $B_\nu(T)$ can be used with Eqs. (1.219) and (1.221) to obtain

$$\begin{aligned}\alpha_\nu &= \frac{N_e N_i}{n_\nu(2\pi\nu)^2}\left[\frac{32\pi^2 Z^2 e^6}{3(2\pi)^{1/2} m^3 c}\right]\left(\frac{m}{kT}\right)^{3/2} \ln \Lambda \\ &\approx \frac{9.786 \times 10^{-3}}{n_\nu \nu^2}\frac{N_e N_i}{T^{3/2}} \ln[4.954 \times 10^7 (T^{3/2}/Z\nu)] \quad \mathrm{cm}^{-1} \\ &\quad \text{for } T < 3.16 \times 10^5 \ {}^\circ\mathrm{K} \\ &\approx \frac{9.786 \times 10^{-3}}{n_\nu \nu^2}\frac{N_e N_i}{T^{3/2}} \ln[4.7 \times 10^{10} (T/\nu)] \quad \mathrm{cm}^{-1} \\ &\quad \text{for } T > 3.16 \times 10^5 \ {}^\circ\mathrm{K}\ .\end{aligned} \tag{1.222}$$

The optical depth, $\tau_\nu = \int \alpha_\nu dl$, is often expressed as the approximation (Altenhoff, Mezger, Strassl, Wendker, and Westerhout, 1960)

$$\tau_\nu \approx 8.235 \times 10^{-2} T_e^{-1.35} \nu^{-2.1} \int N_e^2 dl\ , \quad \text{for } \nu \ll 10^{10}T, \text{and } T < 9 \times 10^5 \ {}^\circ\mathrm{K}\ , \tag{1.223}$$

where T_e is the electron temperature in °K, the frequency ν is in GHz or 10^9 Hz, and the emission measure $\int N_e^2 dl$ is in parsec cm^{-6}, where one parsec $= 3 \times 10^{18}$ cm. Eq. (1.223) is the same as the more exact result found by using Eq. (1.222) if the former is multiplied by a factor $a(\nu, T)$ which is very nearly unity, and has been tabulated by Mezger and Henderson (1967).

It follows from Eq. (1.223) that the optical depth, τ_ν, becomes unity at the critical frequency, ν_T, given by

$$\nu_T \approx 0.3(T_e^{-1.35} N_e^2 l)^{1/2}\ , \tag{1.224}$$

where ν_T is in GHz and the extent, l, of the plasma along the line of sight is in parsecs. At radio frequencies where $h\nu \ll kT$, or $\nu \ll 10^{10}T$, the exponential factor in Eq. (1.219) becomes unity and the radiation spectrum is nearly flat for frequencies $\nu > \nu_T$. For frequencies $\nu < \nu_T$, the radiation is self-absorbed and the radiation spectrum is that of a black–body falling off as ν^2 at lower frequencies. Thermal bremsstrahlung from regions of ionized hydrogen, HII regions, exhibit such a spectrum, and an example is shown in Fig. 1.4 for the Orion nebula. The energy density, $U(\nu)d\nu$, of the radiation of a thermal plasma in the frequency range ν to $\nu + d\nu$ is given by (Smerd and Westfold, 1949)

$$U(\nu)d\nu = \frac{4\pi}{c} n_\nu B_\nu(T) d\nu\ , \tag{1.225}$$

where the brightness of a black–body radiator, $B_\nu(T)$ is given by Eq. (1.119) and the index of refraction, n_ν, is related to the plasma frequency and the electron density just below Eq. (1.188).

The total bremsstrahlung luminosity, L, of a thermal plasma is given by integrating Eq. (1.219) over all frequencies to obtain

$$L = 4\pi V \int \varepsilon_\nu d\nu = \frac{32\pi}{3}\left(\frac{2\pi}{3}\right)^{1/2} \frac{e^6}{mc^3 h}\left(\frac{kT}{m}\right)^{1/2} N_e N_i V Z^2 g$$
$$\approx 1.43 \times 10^{-27}\, N_e N_i T^{1/2} V Z^2 g \ \text{erg sec}^{-1} \ , \tag{1.226}$$

where the volume of the source is V, and it has been assumed that the index of refraction is unity.

The radio flux density, S_ν, arising from the bremsstrahlung of an ionized, expanding stellar atmosphere is derived by Panagia and Felli (1975). Since the relevant formula involves the mass loss rate of the star, it has been given in Volume II.

In the presence of an intense magnetic field ($H > 10^{13}$ Gauss) the electron bremsstrahlung differs from the ordinary bremsstrahlung process in that in a magnetic field the electrons are one dimensional particles free to move in the direction of the field, but bound in the plane perpendicular to the magnetic field in quantized circular orbits with energy (in the nonrelativistic limit) in multiples of $1.19 \times 10^{-8} H$ eV, where H is the magnetic field intensity in Gauss. Formulae for the emission rate and absorption coefficients of electron bremsstrahlung in intense magnetic fields are given by Canuto, Chiu, and Fassio-Canuto (1969) for a vacuum, and by Canuto and Chiu (1970) for a plasma. The emissivity for the zero level quantum state is given by (Chiu, Canuto, and Fassio-Canuto, 1969)

$$\varepsilon = 3.01 \times 10^{-29} N_i N_e H_8 Z^2 E_5^{-5/2} \nu_8^{-2} \ \text{erg sec}^{-1}\ \text{cm}^{-3} (10^8\ \text{Hz})^{-1} \ , \tag{1.227}$$

where N_e and N_i are the electron and ion number densities, respectively, Z is the charge of the nucleus, E_5 is the electron energy measured in $k \cdot 10^5\ ^\circ\text{K} =$ 8.63 eV, and ν_8 is the frequency of the radiation in 10^8 Hz.

1.31 Photoionization and Recombination (Free–Bound) Radiation

In the recombination of an electron of mass, m, charge, e, and velocity, v, with an ion of charge eZ to form a bound atom in a state with principal quantum number, n, or, for the converse process of photoionization from the level, n, the frequency, ν, of emission or absorption is given by (Einstein, 1905)

$$h\nu = \frac{1}{2} m v^2 + E_i - E_n \ , \tag{1.228}$$

where E_i is the ionization energy, and E_n is the energy of the n th state. For hydrogen-like atoms, Eq. (1.228) becomes

$$h\nu = \frac{1}{2} m v^2 + Z^2 E_H / n^2 \ , \tag{1.229}$$

where $E_H = 2\pi^2 e^4 m / h^2 = 13.6\ \text{eV} \approx 2.2 \times 10^{-11}$ erg is the ionization energy of hydrogen, and $Z^2 E_H$ is the ionization energy of a hydrogenic ion of charge $e(Z-1)$.

For the photoionization of a hydrogen-like system consisting of one electron in the field of charge, eZ, of one ion, the total bound–free absorption cross section, $\sigma_a(\nu, n)$ for light of frequency, ν, is (Kramers, 1923)

$$\sigma_{\rm a}(\nu, n) = \frac{32\pi^2 e^6 R_\infty Z^4}{3^{3/2} h^3 \nu^3 n^5} \approx 2.8 \times 10^{29} \frac{Z^4}{\nu^3 n^5} \text{ cm}^2 \ , \tag{1.230}$$

where $R_\infty = 2\pi^2 e^4 m/(h^3 c)$ is the Rydberg constant, and n is the principle quantum number of the initial state. The more exact expression for $\sigma_{\rm a}(\nu, n)$ is given by (Gaunt, 1930; Menzel and Pekeris, 1935; Burgess, 1958)

$$\sigma_{\rm a}(\nu, n) g_{\rm fb}(T, \nu) \ ,$$

where the bound-free Gaunt factor, $g_{\rm fb}(T, \nu)$, is nearly unity at optical frequencies. Gaunt factors have been tabulated by Menzel and Pekeris (1935) and Karzas and Latter (1961).

A contribution to bound–free absorption will occur whenever there are enough atoms in the nth excited states. Denoting their number density by N_n, the total opacity for bound free transitions is:

$$\kappa_{\rm bf} = \sum_n N_n \sigma_{\rm a} g_{\rm bf}/\rho \ .$$

For stellar atmospheres the Rosseland mean opacity for bound–free absorption is (Bowers and Deeming, 1984):

$$\kappa_{\rm bf} = \kappa_0 \rho T^{-3.5} \ , \tag{1.231}$$

where

$$\kappa_0 = 4.32 \times 10^{25} Z(1 + X)\bar{g}_{\rm bf}/t \ ,$$

where Z and X respectively denote the mass fraction of heavy metals and hydrogen, $\bar{g}_{\rm bf}$ is the frequency averaged Gaunt factor, and the quantity t is the guillotine factor.

The recombination cross section, $\sigma_{\rm r}(\nu, n)$, for the recombination radiation of a single electron–ion encounter may be obtained by using the Milne (1921) relation

$$\frac{\sigma_{\rm a}}{\sigma_{\rm r}} = \frac{m^2 c^2 v^2 g_Z(1)}{\nu^2 h^2 g_{Z-1}(n)} \ , \tag{1.232}$$

where $g_Z(1)$ is the statistical weight of the ion in the ground state and $g_{Z-1}(n)$ is the statistical weight of the nth level of the atom. For hydrogen-like systems $g_Z(1) = 1$ and $g_{Z-1}(n) = 2n^2$. Thus Eqs. (1.230) and (1.232) give

$$\sigma_{\rm r}(\nu, n) = \frac{128\pi^4 e^{10} Z^4 g_{\rm fb}}{3\sqrt{3} m c^3 h^4 \nu n^3 v^2} \approx 3 \times 10^{10} \frac{Z^4 g_{\rm fb}}{\nu n^3 v^2} \text{ cm}^2 \ . \tag{1.233}$$

For a plasma with electron density, $N_{\rm e}$, and ion denisty, $N_{\rm i}$, the total number of recombinations between ground-state ions and electrons per unit volume from the velocity range v to $v + dv$ is

$$N_{\rm i} N_{\rm e} f(v) \sigma_{\rm r}(\nu, n) v dv \ , \tag{1.234}$$

where $f(v)$ is the electron velocity distribution and $\sigma_{\mathrm{r}}(v, n)$ is the single electron–ion cross section for recombination to level n (given by Eq. (1.233) for hydrogen-like atoms). When the free electrons are in local thermodynamic equilibrium at temperature, T, the velocity distribution is Maxwellian. In this case, the recombination power, $P_{\mathrm{r}}(n, \nu)d\nu$, radiated per unit volume in the frequency interval ν to $\nu + d\nu$ when recombining to level n is, for hydrogen like atoms, given by:

$$P_{\mathrm{r}}(n, \nu)d\nu = N_{\mathrm{e}}N_{\mathrm{i}}\frac{Z^4K}{(kT)^{3/2}}\frac{g_{\mathrm{fb}}}{n^3}\exp\left[-\frac{h\nu}{kT}+\frac{Z^2E_{\mathrm{H}}}{n^2kT}\right]d\nu \ , \qquad (1.235)$$

where

$$K = \frac{64\pi^{1/2}e^4h}{3^{3/2}m^2c^3}(E_{\mathrm{H}})^{3/2} \approx 3.4 \times 10^{-40}(E_{\mathrm{H}})^{3/2}\ \mathrm{erg}^{5/2}\ \mathrm{cm}^3$$

and $E_{\mathrm{H}} = 2\pi^2e^4m/h^2 \approx 2.2 \times 10^{-11}$ erg is the ionization energy of hydrogen. The total recombination power $P_{\mathrm{r}}(\nu)d\nu$, radiated per unit volume in the frequency interval ν to $\nu + d\nu$ is obtained by summing $P_{\mathrm{r}}(n, \nu)$ over all permitted quantum levels.

$$P_{\mathrm{r}}(\nu)d\nu = N_{\mathrm{e}}N_{\mathrm{i}}\frac{Z^4K}{(kT)^{3/2}}\exp\left(\frac{-h\nu}{kT}\right)\sum_{n_0}^{\infty}n^{-3}\exp\left(\frac{Z^2E_{\mathrm{H}}}{n^2kT}\right)g_{\mathrm{fb}}d\nu \ , \qquad (1.236)$$

where $n_0 \geq (Z^2E_{\mathrm{H}}/h\nu)^{1/2}$.

When detailed values of $P_{\mathrm{r}}(\nu)d\nu$ are required, the approximation formulae of Brussard and van de Hulst (1962) may be used to evaluate the summation in Eq. (1.236). The total volume emissivity, ε_ν, of both thermal bremsstrahlung and recombination radiation is given by Brussard and van de Hulst (1962).

$$\begin{aligned}\varepsilon_\nu d\nu &= \frac{8}{3}\left(\frac{2\pi}{3}\right)^{1/2}\frac{Z^2e^6}{m^2c^3}\left(\frac{m}{kT}\right)^{1/2}N_{\mathrm{i}}N_{\mathrm{e}}bd\nu \\ &\approx 5.4 \times 10^{-39}Z^2\frac{N_{\mathrm{i}}N_{\mathrm{e}}}{T^{1/2}}bd\nu\ \mathrm{erg\ sec^{-1}cm^{-3}\ Hz^{-1}\ rad^{-2}}\end{aligned} \qquad (1.237)$$

where

$$b = [g(\nu, T) + f(\nu, T)]\exp\left(\frac{-h\nu}{kT}\right) \ , \qquad (1.238)$$

the free-free Gaunt factor $g(\nu, T)$ is given by Eqs. (1.220) and Table 1.1, and the free-bound Gaunt factor, $f(\nu, T)$, is given by

$$f(\nu, T) = 2\theta\sum_{n=m}^{5}g_n\frac{e^{\theta/n^2}}{n^3} + 2\theta g_\infty\sum_{n=6}^{\infty}\frac{e^{\theta/n^2}}{n^3} + 50\theta(g_5 - g_\infty)\sum_{n=6}^{\infty}\frac{e^{\theta/n^2}}{n^5} \ , \qquad (1.239)$$

where m is the principal quantum number of the lowest level to which emission at the frequency, ν, can occur, $g_n(\nu)$ is the Gaunt factor for the transition from the free to bound state, $\theta = 2\pi^2me^4Z^2/(h^2kT)$. A plot of the spectral function b

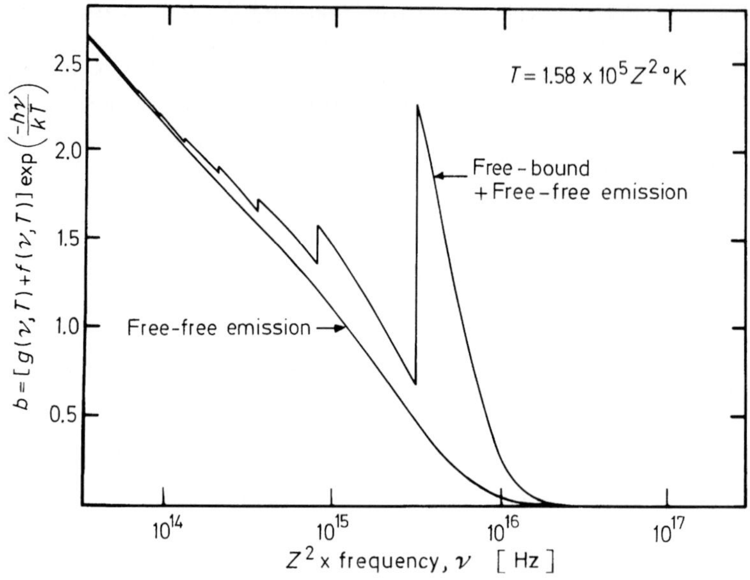

Fig. 1.5. The emission spectrum of a Maxwellian distribution of electrons at a temperature of $T = 2\pi^2 me^4 Z^2/(h^2k) = 1.58 \times 10^5 Z^2\,°K$ with exact Gaunt factors taken into account [after Brussard and van de Hulst, 1962]. The lower curve illustrates the spectrum of thermal bremsstrahlung emission, whereas the upper curve illustrates the combined spectrum of thermal bremsstrahlung and recombination radiation. The volume emissivity, ε_ν, is related to b by Eq. (1.237)

for $T = 1.58 \times 10^5\,°K$ is shown in Fig. 1.5 together with the spectrum of bremsstrahlung radiation alone. These plots indicate that recombination radiation is not important when compared with bremsstrahlung for $h\nu \ll kT$ or $\nu \ll 10^{10} T$. At optical frequencies recombination radiation of neutral and ionized hydrogen and helium become important, as does two photon emission of hydrogen. These effects are discussed in detail in Sect. 2.11, and by Chandrasekhar and Breen (1946) and Brown and Mathews (1970). The continuous spectra arising from free–free and free–bound transitions at X-ray wavelengths (1 Å to 30 Å) are calculated by Culhane (1969) for temperatures between 0.8×10^6 and $10^8\,°K$.

A lower limit to the total luminosity, L, of the recombination radiation may be obtained by integrating Eq. (1.236) over all frequencies resulting from capture into the ground state only ($n = 1, \nu \geqslant Z^2 E_H/h$). This term has the value (Cooper, 1966)

$$L_{min} \approx 10^{-21} N_e N_i T^{-1/2} V Z^4 \text{ erg sec}^{-1} , \qquad (1.240)$$

where the volume of the source is V. A comparison of Eqs. (1. 226) and (1.240) indicates that the total luminosity due to recombination radiation may be greater than that due to thermal bremsstrahlung for temperatures less than $10^6\,°K$.

Table 1.2 Electron densities, N_e, molecular density, N_m, + atomic density, N_a, electron temperature, T_e, and magnetic field strength, H, for astrophysical plasmas.

Plasma	N_e	$N_m + N_a$	T_e	H
	(cm^{-3})	(cm^{-3})	(°K)	(Gauss)
Ionosphere:				
$D \sim 70$ km	10^3	2×10^{15}	2×10^2	$\sim 3 \times 10^{-1}$
$E \sim 100$ km	10^5 day, 10^3 night	6×10^{12}	$2\text{–}3 \times 10^2$	$\sim 3 \times 10^{-1}$
$F_1 \sim 200$ km	10^5 day, absent night	10^{10}	10^3	$\sim 3 \times 10^{-1}$
$F_2 \sim 300$ km	10^6 day, 10^5 night	10^9	$1\text{–}3 \times 10^3$	$\sim 3 \times 10^{-1}$
Interplanetary Space	$1\text{-}10^4$	≈ 0	$10^2\text{–}10^3$	$10^6\text{–}10^{-5}$
Solar Corona	$10^4\text{–}10^8$	≈ 0	$10^3\text{–}10^6$	$10^{-5}\text{–}100$
Stellar Chromosphere	10^{12}	≈ 0	10^4	10^3
Stellar Interiors	10^{27}	≈ 0	$10^{7.5}$	–
Planetary Nebulae	$10^3\text{–}10^5$	≈ 0	$10^3\text{–}10^4$	$10^{-4}\text{–}10^{-3}$
HII Regions	$10^2\text{–}10^3$	≈ 0	$10^3\text{–}10^4$	10^{-6}
White Dwarfs	10^{32}	≈ 0	10^7	10^6 (surface)
Pulsars	10^{42} center, 10^{12} surface	≈ 0		10^{12} (surface)
Interstellar Space	$10^{-3}\text{–}10$ (av ~ 0.03)	$10^{-2}\text{–}10^5$ (av ~ 1)	10^2	10^{-6}
Intergalactic Space	$< 10^{-5}$	≈ 0	$10^5\text{–}10^6$	$\lesssim 10^{-8}$

1.32 Astrophysical Plasmas

Almost all matter in space can be considered to be a plasma, an ionized gas consisting of electrons (e), ions (i), and neutral atoms (a) or molecules (m) with respective densities, N_e, N_i, N_a, N_m; respective masses, m, M_i, M_a, M_m; and respective charges, e, Ze, and zero for the neutral particles. The plasma is often taken to be in local thermodynamic equilibrium at the temperature, T; and often contains a magnetic field of strength, H. Typical values of N_e, $N_a + N_m$, T_e, and H are given in Table 1.2 for well known astrophysical plasmas.

For a plasma in thermodynamic equilibrium at temperature, T, the electron velocity distribution is Maxwellian (Maxwell, 1860) Eq. (3.114), and the rms velocity, v_{th}, and the average kinetic energy, K.E., are given by

$$v_{th} = (3kT/m)^{1/2} \approx 6.7 \times 10^5\, T^{1/2} \text{ cm sec}^{-1}$$
$$\text{K.E.} = 3kT/2 \approx 2.1 \times 10^{-16}\, T \text{ erg} . \qquad (1.241)$$

For an electron density, N_e, the average separation between the electrons, $\langle r \rangle$, and the average Coulomb energy, C.E., are given by

$$\langle r \rangle = (3N_e/4\pi)^{-1/3} \approx N_e^{-1/3} \text{ cm}$$
$$\text{C.E.} = e^2/\langle r \rangle \approx 2.3 \times 10^{-19}\, N_e^{1/3} \text{ erg} . \qquad (1.242)$$

It follows from Poisson's equation (Poisson, 1813, Eq. (1.4)) that the electric field intensity, E, due to a linear displacement, $x(t)$, of the electrons is given by

$$E = 4\pi e N_e x(t) . \qquad (1.243)$$

By using the Lorentz force equation (Lorentz, 1892) Eq. (1.52) with Eq. (1.243), the equation of motion for the displaced charge is obtained

$$m\ddot{x}(t) + 4\pi e^2 N_e x(t) = 0 \ , \tag{1.244}$$

where $\ddot{}$ denotes the second derivative with respect to time, t. Eq. (1.244) is the equation of motion of a harmonic oscillator whose oscillating frequency is the plasma frequency, ω_p, given by

$$\omega_p = \left(\frac{4\pi e^2 N_e}{m}\right)^{1/2} \approx 5.64 \times 10^4 N_e^{1/2} \quad \text{rad sec}^{-1}$$

or

$$\nu_p = \omega_p / 2\pi \approx 8.977 \times 10^3 N_e^{1/2} \quad \text{Hz} \ , \tag{1.245}$$

a result first obtained by Tonks and Langmuir (1929). It also follows directly from the Lorentz force equation (1.52) that a charged particle moving in a magnetic field of intensity, H, will gyrate about the field at the cyclotron frequency, ω_H, given by

$$\omega_H = \frac{ZeH}{Mc\gamma} \approx 1.76 \times 10^7 \frac{ZmH}{M\gamma} \quad \text{rad sec}^{-1}$$

or

$$\nu_H \approx 2.8 \times 10^6 H\gamma^{-1} \quad \text{Hz} \quad \text{for} \quad M = m \ , \tag{1.246}$$

a result first obtained by Heaviside (1904). Here the particle mass is M, its charge is Ze, the velocity is v, and the $\gamma = [1 - (v/c)^2]^{-1/2} = E/mc^2$ where E is the particle energy. The radius of gyration, r_H, is given by

$$r_H = \frac{v}{\omega_H} \approx \frac{v_{th}}{\omega_H} = \frac{c(3kT/m)^{1/2} M\gamma}{ZeHm} \approx 3.8 \times 10^{-2} \frac{T^{1/2} M\gamma}{ZHm} \ \text{cm} \ . \tag{1.247}$$

When a test charge, Ze, is placed in a plasma in thermodynamic equilibrium its electrostatic potential, $\varphi(r)$, will cause the electrons to move into a spatial density distribution $N(r)$ given by Boltzmann's equation (3.4) and Maxwell's distribution (Eq. 3.114)

$$\begin{aligned} N(r) &= N_e \exp\left[-\frac{e\varphi(r)}{kT}\right] \\ &\approx N_e \left[1 - \frac{e\varphi(r)}{kT}\right] \quad \text{for } e\varphi(r) \ll kT \ , \end{aligned} \tag{1.248}$$

where r is the distance from the test charge. Using the approximate form of Eq. (1.248) together with Poisson's equation (Poisson, 1813) Eq. (1.4) we obtain the equation

$$\varphi(r) = \frac{Ze}{r} \exp\left(-\frac{r}{r_D}\right) \ , \tag{1.249}$$

where the Debye radius, r_D, is given by

$$r_D = 3^{-1/2}\left(\frac{v_{th}}{\omega_p}\right) = \left(\frac{kT}{4\pi e^2 N_e}\right)^{1/2} \approx 6.69 \left(\frac{T}{N_e}\right)^{1/2} \ \text{cm} \ . \tag{1.250}$$

Eqs. (1.249) and (1.250) were first obtained by Debye and Hückel (1923) and they showed that the electrons of a plasma move in such a way as to screen out the Coulomb field of a test charge for distances $r > r_D$. Landau (1946) has shown that the wavelength of electron plasma oscillations must be greater than the Debye length in order for the waves to exist.

It follows from Maxwell's equations (Maxwell, 1860) (Eqs. (1.39)–(1.42)) that the magnetic field intensity, H, the electric field intensity, E, and the current density, J, in a vacuum are related by the equation

$$c\nabla \times H = 4\pi J + \frac{\partial E}{\partial t} \quad . \tag{1.251}$$

When electrons of density, N_e, are introduced into the vacuum, $J = N_e e v$, where the velocity, v, is given by $v = eE_0/(i\omega m)$, the latter equation following from the Lorentz force law for an electron subjected to a oscillatory electric field given by $E = E_0 \exp(i\omega t)$. Eq. (1.251) may then be written as

$$c\nabla \times H = i\omega \left[1 - \frac{4\pi e^2 N_e}{m\omega^2}\right] E_0 = i\omega \varepsilon_v E_0 \ , \tag{1.252}$$

where ε_v is the dielectric constant. It then follows that the index of refraction, $n_{v\perp}$, for transverse ($\perp$) electromagnetic waves of frequency, v, is given by

$$n_{v\perp} = \sqrt{\varepsilon_v} = \left[1 - \left(\frac{v_p}{v}\right)^2\right]^{1/2} \approx \left[1 - 8.06 \times 10^7 \frac{N_e}{v^2}\right]^{1/2} \ , \tag{1.253}$$

an equation first derived by Maxwell (1899).

Eq. (1.253) is valid if the thermal speed of the electrons is sufficiently low that the mean distance between most of the interacting particles does not change appreciably during the period of an oscillation. When the thermal motion of the electrons is taken into account it can be shown that longitudinal plasma oscillations are possible. If φ is the average potential due to the motion of plasma particles, Poisson's equation (1.4) may be written as

$$\begin{aligned} -\nabla^2 \varphi = 4\pi\rho &= 4\pi e N_e \left[1 - \int \frac{f(v)dv}{\left(\frac{1+2e\varphi}{mU^2}\right)^{1/2}}\right] \\ &\approx \frac{4\pi K^2 e^2 N_e \varphi}{m} \int \frac{f(v)dv}{(\omega - \boldsymbol{K}\cdot\mathbf{v})^2} \ , \end{aligned} \tag{1.254}$$

where U is the wave velocity in the wave coordinate system, v is the velocity in the laboratory system, $f(v)$ is the velocity distribution function, ω is the wave frequency, and the wave number $K = 2\pi/\lambda$, where λ is the wavelength. Eq. (1.254) was first obtained by Vlasov (1945). Assuming a Maxwellian velocity distribution for a plasma in thermodynamic equilibrium (Eq. (3.114)), expanding the integral, and solving for ω we obtain

$$\omega^2 \approx \frac{4\pi e^2 N_e}{m} + \left(\frac{3kT}{m}\right) K^2 = \omega_p^2 + v_{th}^2 K^2 \ , \tag{1.255}$$

to give an index of refraction $n_{v\|}$ for longitudinal waves of

$$n_{\nu\|} = \frac{cK}{\omega} = \left(\frac{c}{v_{\mathrm{th}}}\right)\left[1-\left(\frac{\nu_{\mathrm{p}}}{\nu}\right)^2\right]^{1/2} \approx 4.5\times10^4\ T^{-1/2}\left[1-8.06\times10^7\frac{N_{\mathrm{e}}}{\nu^2}\right]^{1/2} . \tag{1.256}$$

Eqs. (1.255) and (1.256) were first derived by Bohm and Gross (1949), and have been shown to be valid for longitudinal waves with frequencies $\nu \approx \nu_{\mathrm{p}}$.

Black body radiation intensity for a plasma (Planck, 1901; Milne, 1930)

$$\begin{aligned} B_{\nu\mathrm{p}}(T) &= n_\nu^2 \frac{2h\nu^3}{c^2}\left[\exp\left(\frac{h\nu}{kT}\right)-1\right]^{-1} \\ &\approx 2n_\nu^2\nu^2 kT/c^2 \text{ for } \nu < 10^{10}T . \end{aligned} \tag{1.257}$$

Radiation energy density in frequency interval between ν and $\nu + d\nu$ for a plasma (Smerd and Westfold, 1949)

$$\begin{aligned} U(\nu)d\nu &= \frac{4\pi}{c}B_{\nu\mathrm{p}}(T)\left|\frac{\partial\omega n_\nu}{\partial\omega}\right|d\nu \\ &\approx 8\pi\frac{\nu^2 kT}{c^3}\left[1-\left(\frac{\nu_{\mathrm{p}}}{\nu}\right)^2\right]^{1/2} d\nu \\ &\qquad \text{for } \nu < 10^{10}\ T,\ \text{transverse waves} \\ &\approx \frac{8\pi\nu^2 kT}{3\sqrt{3}c^3}\left(\frac{mc}{kT}\right)^3\left[1-\left(\frac{\nu_{\mathrm{p}}}{\nu}\right)^2\right]^{1/2} d\nu \\ &\qquad \text{for } \nu < 10^{10}\ T,\ \text{longitudinal waves } \nu \approx \nu_{\mathrm{p}} . \end{aligned} \tag{1.258}$$

Kirchhoff's law for a plasma (Kirchhoff, 1860; Milne, 1930)

$$\varepsilon_\nu = \alpha_\nu B_{\nu\mathrm{p}}(T) , \tag{1.259}$$

where ε_ν is the emission coefficient (power per unit volume per unit frequency interval per unit solid angle), α_ν is the absorption coefficient per unit length, and $B_{\nu\mathrm{p}}(T)$ is the black body radiation intensity for a plasma (n_ν^2 times that for free space).

Absorption coefficient per unit length (Kramers, 1923; Gaunt, 1930)

$$\alpha_\nu = \frac{1}{n_\nu}\left(\frac{\nu_{\mathrm{p}}}{\nu}\right)^2\frac{\nu_{\mathrm{eff}}}{c} \quad \text{for } \nu < 10^{10}\ T . \tag{1.260}$$

Effective collision frequency for electrons with ions, i, or molecules, m,

$$\begin{aligned} \nu_{\mathrm{eff}}(i) &= \frac{4}{\sqrt{2}}\left(\frac{\pi}{3}\right)^{3/2} Z^2\frac{e^4}{m^2}\left(\frac{m}{kT}\right)^{3/2} N_{\mathrm{i}}g(\nu,T) \approx 50\frac{N_{\mathrm{i}}}{T^{3/2}}\ \mathrm{Hz} \\ \nu_{\mathrm{eff}}(m) &= \frac{4\pi}{3}a^2 v_{\mathrm{th}}N_{\mathrm{m}} \approx 9\times10^5\pi a^2 T^{1/2}N_{\mathrm{m}}\ \mathrm{Hz} , \end{aligned} \tag{1.261}$$

where the Gaunt factor $g(\nu,T) = (\sqrt{3}/\pi)\ln\Lambda$. The logarithmic factors are given in Table 1.1. The effective collision cross section for molecules of radius a is πa^2 which has the value 4.4×10^{-16} cm^2 for air.

Mean free path:

$$r_{\mathrm{m}} = \frac{v_{\mathrm{th}}}{\nu_{\mathrm{eff}}} \approx 10^4(T^2/N_{\mathrm{i}})\ \mathrm{cm} . \tag{1.262}$$

1.33 Propagation of Electromagnetic (Transverse) Waves in a Plasma

Transverse electromagnetic waves in a cold plasma have been reviewed by Benz (1993) and Sturrock (1994), including the Faraday rotation and dispersion of radio waves that are used to infer the electron density and magnetic field strength of interstellar space. Both authors also discuss longitudinal waves in a warm plasma and Čerenkov radiation described in the next Sect. 1.34. Radiation from electron beams, including related observations of solar radio bursts, are discussed by Dulk (1985) and Benz (1993). Collective plasma radiation processes leading to coherent radio emission with brightness temperatures $T_B \gg 10^{10}$ K are reviewed by Melrose (1991), including the electron cyclotron maser emission mechanism.

Following Ratcliffe (1959) we define the dimensionless parameters

$$X = \left(\frac{\nu_p}{\nu}\right)^2, \quad Y = \frac{\nu_H}{\nu} \text{ and } Z = \frac{\nu_{eff}}{2\pi\nu} \ , \tag{1.263}$$

where ν is the frequency of the electromagnetic wave, and ν_p, ν_H, and ν_{eff} are defined in Sect. 1.32. The complex index of refraction, ε_ν, for any arbitrary direction of the magnetic field is then given by

$$\varepsilon_\nu^2 = [n_\nu - iq_\nu]^2 = 1 - \frac{X}{\frac{1 - iZ - \frac{1}{2}Y_T^2}{1 - X - iZ} \pm \left[\frac{\frac{1}{4}Y_T^4}{(1 - X - iZ)^2} + Y_L^2\right]^{1/2}} \ , \tag{1.264}$$

where the + and − signs of the ± denote, respectively, the ordinary and extraordinary waves, $Y_T = Y \sin\psi$ and $Y_L = Y \cos\psi$, where the angle ψ is the angle between the propagation of the electromagnetic wave and the magnetic field vector. The index of refraction is n_ν and the absorption coefficient, α_ν, is $4\pi q_\nu \nu/c$.

For the important special case of quasi-longitudinal propagation, where the wave nearly propagates along the magnetic field line, we have the index of refraction

$$n_\nu \approx \{1 - \nu_p^2[\nu(\nu \pm \nu_H)]^{-1}\}^{1/2} = \left[1 - \frac{X}{1 \pm Y}\right]^{1/2} \tag{1.265}$$

and the absorption coefficient

$$\alpha_\nu \approx \frac{\nu_{eff}}{n_\nu c}\left(\frac{\nu_p}{\nu}\right)^2\left[1 \pm \left(\frac{\nu_H}{\nu}\right)\right]^{-2} = \frac{\pi\nu}{n_\nu c}\frac{XZ}{(1 \pm Y)^2 + Z^2} \ . \tag{1.266}$$

The dispersion relation in this case is that of a "cold" plasma given by Appleton (1932) and Hartree (1931). The ordinary and extraordinary waves are both nearly completely circularly polarized, the sense of the ordinary wave being left hand. The phase difference, $\Delta\varphi$, between the extraordinary and ordinary waves after traversing the length dl of the plasma is (Rybicki and Lightman, 1979; Shu, 1991; Sturrock, 1994):

$$\Delta\varphi = \frac{\omega}{2c}(n_{\nu+} - n_{\nu-})dl \approx \frac{\pi\nu}{c}\left(\frac{\nu_p}{\nu}\right)^2\left(\frac{\nu_H}{\nu}\right)dl = \frac{\pi\nu}{c}X\,Y\,dl \ , \tag{1.267}$$

where $n_{\nu+}$ and $n_{\nu-}$ are given by Eq. (1.265) with the + and − signs. Thus a linearly polarized electromagnetic wave, which is incident on a plasma, will be rotated through Ω radians, where Ω is given by

$$\Omega = \int_0^L \Delta\varphi dl = \frac{e^3}{2\pi m^2 c^2 \nu^2} \int_0^L N_e H \cos\psi dl \approx \frac{2.36 \times 10^4}{\nu^2} \int_0^L N_e H \cos\psi dl \text{ radians} \tag{1.268}$$

after traversing a thickness, L, of the plasma. The angle ψ is the angle between the magnetic field and the line of sight. This effect is called Faraday rotation after its discovery by Faraday (1844). As long as $\nu \gg \nu_H$, the rotation due to components of H which are perpendicular to the line of sight is negligible compared with Ω. The line integral $\int_0^L N_e H \cos\psi dl$ is often measured by observing the periodic rise and fall of intensity at closely spaced frequencies received with a linearly polarized feed. The period, P, of this sinusoidal oscillation is given by

$$P = 1.33 \times 10^{-4} \nu^3 \left[\int_0^L N_e H \cos\psi dl \right]^{-1} \text{Hz} \approx 10^9 [\lambda^3 \text{R.M.}]^{-1} \text{ Hz} , \tag{1.269}$$

where the rotation measure, R.M., is given by

$$\text{R.M.} = 8.1 \times 10^5 \int N_e H \cos\psi dl \text{ rad m}^{-2} , \tag{1.270}$$

where the electron density, N_e, is in units of cm^{-3}, the magnetic field strength, H, is in Gauss and dl in parsecs. The rotation angle $\Omega = \lambda^2$ R.M. radians if λ is in meters. Rotation measures for pulsars are tabulated by Lang (1992) with values ranging between 10 and 1,000 rad m^{-2}.

For transverse propagation, the index of refraction is given by

$$n_{\nu 0} = [1 - (\nu_p/\nu)^2]^{1/2} = [1 - X]^{1/2} \tag{1.271}$$

whereas for quasi-transverse propagation

$$n_{\nu 0} = \{1 - \nu_p^2 [\nu^2 + (\nu^2 - \nu_p^2) \cot^2\theta]^{-1}\}^{1/2} ,$$

where θ represents the angle between the direction of propagation and the magnetic field. For both cases

$$n_{\nu e} = \left\{ 1 - \left(\frac{\nu_p}{\nu}\right)^2 \frac{[1 - (\nu_p/\nu)^2]}{1 - (\nu_p/\nu)^2 - (\nu_H/\nu)^2} \right\}^{1/2} = \left\{ 1 - \frac{X(1-X)}{1 - X - Y^2} \right\}^{1/2} , \tag{1.272}$$

and the absorption coefficient is given by

$$\alpha_{\nu 0} = \frac{\nu_{\text{eff}}}{n_\nu c} \left(\frac{\nu_p}{\nu}\right)^2 = \frac{\nu_{\text{eff}} X}{n_\nu c} , \tag{1.273}$$

$$\alpha_{\nu e} = \frac{\nu_{\text{eff}}}{n_\nu c} \left(\frac{\nu_p}{\nu}\right)^2 \frac{[1 + (\nu_H/\nu)^2]}{[1 - (\nu_H/\nu)^2]^2} = \frac{\nu_{\text{eff}} X}{n_\nu c} \frac{1 + Y^2}{(1 - Y^2)^2} , \tag{1.274}$$

where the subscripts 0 and e denote, respectively, the ordinary and extraordinary waves. The two waves in this case are nearly completely linearly polarized, the electric vector of the ordinary wave being parallel to the magnetic field.

In the absence of a magnetic field, the dielectric constant, ε_ν, and the conductivity, σ_ν, of a plasma are given by (Ginzburg, 1961)

$$\varepsilon_\nu = 1 - \frac{\nu_p^2}{\nu^2 + (\nu_{\mathrm{eff}}/2\pi)^2} \tag{1.275}$$

and

$$\sigma_\nu = \frac{\nu_p^2 \nu_{\mathrm{eff}}}{4\pi[\nu^2 + (\nu_{\mathrm{eff}}/2\pi)^2]} \approx \frac{\nu_{\mathrm{eff}}}{4\pi}\left(\frac{\nu_p}{\nu}\right)^2 . \tag{1.276}$$

The index of refraction, n_ν, and the absorption coefficient α_ν, are given by

$$\begin{aligned} n_\nu &= \left\{\frac{\varepsilon_\nu}{2} + \left[\left(\frac{\varepsilon_\nu}{2}\right)^2 + \left(\frac{\sigma_\nu}{\nu}\right)^2\right]^{1/2}\right\}^{1/2} \\ &\approx \left[1 - \frac{\nu_p^2}{\nu^2 + (\nu_{\mathrm{eff}}/2\pi)^2}\right]^{1/2} \quad \text{for } \varepsilon_\nu \gg \frac{\sigma_\nu}{\nu} , \end{aligned} \tag{1.277}$$

and

$$\begin{aligned} \alpha_\nu &= 4\pi\frac{\nu}{c}\left\{-\frac{\varepsilon_\nu}{2} + \left[\left(\frac{\varepsilon_\nu}{2}\right)^2 + \left(\frac{\sigma_\nu}{\nu}\right)^2\right]^{1/2}\right\}^{1/2} \\ &\approx \frac{1}{n_\nu}\left(\frac{\nu_p}{\nu}\right)^2 \frac{\nu_{\mathrm{eff}}}{c} \approx \frac{4\pi\sigma_\nu}{n_\nu c} \quad \text{for } \varepsilon_\nu \gg \frac{\sigma_\nu}{\nu} . \end{aligned} \tag{1.278}$$

When a wave of frequency, ν, propagates through a plasma with $\nu > \nu_H$. its group velocity is

$$u = c n_\nu \approx c\left[1 - \left(\frac{\nu_p}{\nu}\right)^2\right]^{1/2} . \tag{1.279}$$

For a plasma with index of refraction $n \approx [1 - (\nu_p/\nu)^2]^{1/2}$ for frequency ν with plasma frequency ν_p, the phase velocity $v_{ph} = 2\pi\nu/k = \omega/k = c/n$, and the group velocity $u = v_{gr} = \partial\omega/\partial k = cn$. For radio observations of the interstellar medium the observed frequency $\nu \gg \nu_p$, and we can expand Eq. (1.279) by the binomial theorem, showing that a pulse of radiation will be delayed by the time, τ_D, given by (Rybicki and Lightman, 1979; Shu, 1991; Sturrock, 1994):

$$\begin{aligned} \tau_D &\approx \frac{L}{c} + \frac{e^2}{m\nu^2}\int_0^L N_e dl \\ &\approx 0.3 \times 10^{-10}\, L + \frac{1.35 \times 10^{-3}}{\nu^2}\int_0^L N_e dl \text{ seconds} , \end{aligned} \tag{1.280}$$

when propagating through a plasma of thickness, L. The differential dispersion will be

$$\Delta\tau_D = \frac{e^2}{2\pi mc}\left[\frac{1}{\nu_1^2} - \frac{1}{\nu_2^2}\right]\int_0^L N_e dl \approx 1.35 \times 10^{-3}\left[\frac{1}{\nu_1^2} - \frac{1}{\nu_2^2}\right]\int_0^L N_e dl \text{ sec} \ , \tag{1.281}$$

which is the delay between a pulse received at a high frequency, ν_1, and that received at a low frequency, ν_2, after propagating through a plasma of thickness, L. The parameter $\int_0^L N_e dl$ is called the dispersion measure, denoted DM, and has c.g.s. units of cm^{-2}. The units of parsec cm^{-3} are also often used for dispersion measure (one parsec = 3.0857 $\times 10^{18}$ cm). The dispersion measures, DM, for pulsars are given by Lang (1992) in units of pc cm^{-3}, with values ranging between 10 and 1,000 pc cm^{-3}. In these units, the delay $\Delta\tau_D = 4.148 \times 10^{15}$ DM ν^{-2} seconds for a frequency ν in Hz. For DM = 30 pc cm^{-3}, corresponding to a column of 10^{20} electrons, the delay is 12 seconds for a signal at 100 MHz relative to infinite frequency.

The galactic distribution of free electrons has been modeled using the dispersion, distance and scattering measurements of pulsars and other radio sources (Cordes, Weisberg, Frail, Spangler and Ryan, 1991). The rotation measures of pulsars can be combined with their dispersion measures to obtain the weighted estimate $\langle H_\parallel \rangle$ of the interstellar magnetic field strength parallel to the signal path using Eqs. (1.270) and (1.280) to obtain $\langle H_\parallel \rangle = 1.23 \times 10^{-6}$ RM/DM Gauss, where RM is in units of rad m^{-2} and DM is in units of pc cm^{-3}.

1.34 Propagation of Longitudinal (P mode) Waves in a Plasma: Plasma Line Radiation and Čerenkov Radiation

Because a plasma is macroscopically neutral, local perturbations in the electron or ion density set up electron or ion oscillations near the plasma frequency, ω_p. As Tonks and Langmuir (1929) pointed out, these oscillations will be propagated as longitudinal plasma waves of frequency, ω_p. The dispersion relation for a longitudinal wave of wave number $k = 2\pi/\lambda$ at wavelength λ is given by Eq. (1.255), which was derived by Bohm and Gross (1949), and has been described by Jackson (1962, 1975), Shu (1992) and Benz (1993). For radian frequency $\omega = 2\pi\nu$ and plasma frequency $\omega_p = 2\pi\nu_p$, the approximate dispersion relation valid at small k and long wavelengths is:

$$\omega^2 = \omega_p^2 + 3\langle v^2 \rangle k^2 = \omega_p^2 + 3\, k_B T\, k^2/m \ , \tag{1.282}$$

where the root mean square thermal velocity in one direction (parallel to the electric field) is $\langle v^2 \rangle^{1/2} = (k_B T/m)^{1/2}$ for Boltzmann's constant k_B and electron mass m. For a Maxwellian distribution, the root mean square distribution, in three dimensions, is $\langle v^2 \rangle^{1/2} = (3k_B T/m)^{1/2}$.

The index of refraction for a longitudinal wave in a plasma is given by Eq. (1.256), which indicates that longitudinal waves with frequencies, ν, larger than the plasma frequency, ν_p, can propagate, as they will for transverse electromagnetic waves – see Eq. (1.253). Longitudinal waves will be seriously damped for wave numbers $k \gg k_D$, where the Debye wave number, k_D, is given by (Debye and Hückel, 1923; Landau, 1946)

$$k_D^2 = \omega_p^2 / \langle v^2 \rangle \ . \tag{1.283}$$

The mechanism of such Landau damping has been described by Jackson (1962, 1975) and Shu (1992). Longitudinal waves are damped for wavelengths, λ, smaller than λ_D, or for

$$\lambda \ll \lambda_D = 2\pi / k_D = \langle v^2 \rangle^{1/2} / \omega_p \approx 10 \ (T/N_e)^{1/2} \ \text{cm} \ , \tag{(1.284)}$$

for a temperature T in °K and electron density N_e in cm^{-3}. For small $k \ll k_D$ or for long wavelengths $\lambda \gg \lambda_D$ the longitudinal plasma oscillations are virtually undamped. Their phase velocity, v_{ph}, and group velocity, v_{gr}, are given by:

$$u = v_{ph} \approx \omega_p / k \gg \ \langle v^2 \rangle^{1/2}$$

and

$$u = v_{gr} \ = 3 \ \langle v^2 \rangle / v_{ph} \ll \ \langle v^2 \rangle^{1/2} \ . \tag{1.285}$$

The longitudinal plasma waves may be converted into electromagnetic waves when the plasma is nonuniform (cf. Zheleznyakov, 1970, for details of several such processes). Of special interest is the electromagnetic radiation produced by the scattering of longitudinal plasma waves on the small-scale thermal fluctuations in plasma density, N (Ginzburg and Zheleznyakov, 1958, 1959; Salpeter, 1960; Smerd, Wild, and Sheridan, 1962). Rayleigh scattering by the quasi-neutral random fluctuations due to the thermal motion of the ions produces electromagnetic radiation at the ion frequency $\nu_i = (e^2 N_i / \pi m_i)^{1/2}$ and at the plasma frequency $\nu_p = (e^2 N_e / \pi m_e)^{1/2}$. The total flux of energy, $W(\nu_p)$, scattered at the plasma frequency in a volume, V, is given by (Ginzburg and Zheleznyakov, 1959)

$$W(\nu_p) \approx n_\nu \frac{e^4 N_e V}{6 m^2 c^3} \mathrm{E}_0^2 \ , \tag{1.286}$$

where the index of refraction $n_\nu = \sqrt{3} v_{th} / v_\varphi$, the phase velocity of the plasma wave is v_φ, and E_0 is the electric field amplitude of the plasma wave. The efficiency parameter, Q, for the change of plasma wave energy into the energy of electromagnetic waves at the plasma frequency is given by

$$Q = \frac{W(\nu_p)}{L^2 S_p} \approx \frac{4\pi e^4 N_e L}{3 m^2 c^3 v_{th}} \ ,$$

where L is the linear extent of the scattering region, and S_p is the plasma wave energy flux density. If the emission takes place over a frequency range $\Delta\nu$, across an area A, and into a solid angle $\Delta\Omega$, then an effective temperature, T_b, defined by using the Rayleigh–Jeans law for the brightness of a black body, is given by

$$T_b = \frac{c^2 W(\nu_p) e^{-\tau}}{2 \nu_p^2 k \Delta\nu \Delta\Omega A} \ , \tag{1.287}$$

where τ denotes the optical depth between the observer and the source.

Combination scattering by the space-charge fluctuations due to the thermal motion of the electrons produces electromagnetic radiation at the frequency $2\nu_p$ whose total flux of energy, $W(2\nu_p)$, is given by (Zheleznyakov, 1970)

$$W(2\nu_p) \approx \frac{4\sqrt{3}}{5}\frac{e^4 N_e V}{m^2 c^3}\frac{kT}{mc^2}E_0^2 , \tag{1.288}$$

provided that $v_\varphi \ll c/\sqrt{3}$. This emission is beamed strongly in the direction normal to that of the plasma wave, and has an angular spectrum of intensity, $I(\theta)$, given by (Smerd, Wild, and Sheridan, 1962)

$$I(\theta) \propto \sin^3\theta[1 + 3(v_0/c)^2 - 2\sqrt{3}(v_0/c)\cos\theta] , \tag{1.289}$$

where θ is the angle from the direction of the plasma wave, and v_0 is the mean velocity of the stream of particles exciting the plasma wave.

For a stream of particles of density $N_s \ll N_e$ and velocity $u \gg v_\varphi$ where v_φ is the phase velocity of the plasma, the amplitude of the steady-state plasma wave is given by (Bohm and Gross, 1949)

$$E_0 = \frac{m}{2eK^3}\left\{\frac{16}{3}v_\varphi(2\pi\nu_{ps})^2\frac{\nu_s}{\nu}\left(\frac{df(u)}{du}\right)_{u=v_\varphi}\right\}^2$$
$$\approx \frac{\pi m \nu_p}{e v_0}\left\{\frac{16}{3}\frac{v_0^3 m}{kT_s}\left(\frac{N_s}{N_e}\right)^2\right\}^2 , \tag{1.290}$$

where $K = 2\pi\nu_p/v_\varphi$, the plasma frequency $\nu_p = [e^2N_e/\pi m]^{1/2}$, the plasma frequency of the stream $\nu_{ps} = [e^2N_s/\pi m]^{1/2}$, the collision frequency of the stream, ν_s, and the plasma, ν, are related by $\nu_s/\nu \approx N_s/N_e$, and the velocity distribution of the stream, $f(u)$, is assumed to be Maxwellian with a mean value of $v_0 \approx v_\varphi$ and $(df(u)/du)_{u=v_\varphi} \approx v_{th}^{-2}$.

Longitudinal waves may also be generated as a consequence of the accelerated motions of electrons scattered by ions in analogy with thermal bremsstrahlung (cf. Bekefi, 1966). In this case the frequency of the longitudinal waves are again concentrated near ν_p, but in this frequency range the volume emissivity is greater than that of the thermal bremsstrahlung by the factor (c/v_{th}^2).

If a charged particle of velocity, u, is injected into a plasma for which the phase velocity, v_φ, is less than u, the particle will excite longitudinal waves in a manner which is analogous to the optical Čerenkov effect (Čerenkov, 1937). Optical Čerenkov radiation is concentrated into a cone centered on the trajectory of the charged particle and given by Jackson (1962)

$$\cos\psi = \left[\frac{c}{un_\nu}\right] , \tag{1.291}$$

where n_ν is the refractive index of the medium. The total energy radiated per unit frequency interval per unit length of the medium is, in the optical case, given by

$$I_\nu = \frac{2\pi e^2\nu}{c^2}\left[1 - \frac{c^2}{n_\nu^2 u^2}\right] , \tag{1.292}$$

where it is understood that $u > c/n_\nu$ and the index of refraction $n_\nu = (\varepsilon_\nu)^{1/2}$ for a dielectric constant ε_ν and a permeability $\mu = 1$. Čerenkov radiation of a particle moving in longitudinal electrostatic waves is discussed by Benz (1993).

For a highly relativistic particle of velocity $v \approx c$, entering the Earth's upper atmosphere, the Čerenkov radiation, often written Cherenkov, is emitted in a small forward angle, $\Delta\theta$, given by (Harwit, 1988)

$$\Delta\theta \approx 2[1 - (v/c)^2]^{1/2} \ . \tag{1.293}$$

This radiation can be used to identify the existence and direction of arrival of cosmic ray particles and neutrinos.

The frequency and angular spectrum for the Čerenkov radiation of an electron in a plasma have been calculated by Cohen (1961). The radiation is symmetric about the trajectory of the charged particle. If θ denotes the angle from this trajectory, the angular spectrum of the radiated energy is

$$I_\theta = 6.3 \times 10^5 \frac{e^2 \nu_p^2}{u^2} \frac{\cos\theta}{\cos^2\theta - (v_{th}/u)^2} \quad \text{erg rad}^{-2}\,\text{cm}^{-1} \ . \tag{1.294}$$

Here u is the velocity of the incident electron and $v_{\text{th}} = (3kT_e/m)^{1/2}$ where T_e is the electron temperature. The total energy radiated per unit frequency interval per unit path of the electron is

$$\begin{aligned} I_\nu &= 2\pi \frac{e^2 \nu_p}{u^2} \frac{(\nu/\nu_p)}{(\nu/\nu_p)^2 - 1} \quad \text{for } \nu > \nu_c \\ &= 0 \text{ for } \nu < \nu_c \ , \end{aligned} \tag{1.295}$$

where the low frequency cutoff is

$$\nu_c = \nu[1 - (v_{\text{th}}/u)^2]^{-1/2} \ .$$

Plots of the spectral form of I_ν and the angular form of I_θ are shown in Fig. 1.6. The high frequency cutoff is assumed to be limited to $\sqrt{2}\omega_p$ by Landau damping ($v_\varphi \geq \sqrt{2}v_{\text{th}}$) and the total radiated energy per unit length of path is

$$\begin{aligned} W &= 2\pi^2 \frac{e^2 \nu_p^2}{u^2} \ln\left[\left(\frac{u}{v_{\text{th}}}\right)^2 - 1\right] \\ &\approx 5.1 \times 10^{-39} \nu_p^2 \left(\frac{c}{u}\right)^2 \ln\left[\left(\frac{u}{v_{\text{th}}}\right)^2 - 1\right] \quad \text{erg cm}^{-1} \ . \end{aligned} \tag{1.296}$$

1.35 Scattering from a Harmonic Oscillator

The equation of motion of a harmonic oscillator which is driven by a harmonic plane wave is

$$m\ddot{x} + m(2\pi\nu_0)^2 x = eE_0 \exp(i2\pi\nu t) - m\gamma_{\text{cl}}\dot{x} \ , \tag{1.297}$$

where $\dot{}$ and $\ddot{}$ denote, respectively, the first and second derivatives with respect to time, x is the direction of the electric field vector, E_0 and ν are, respectively, the strength and frequency of the electric field, m and e are, respectively, the

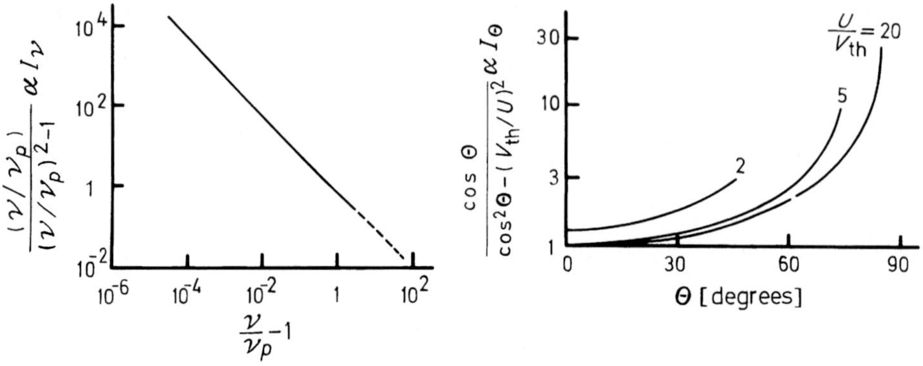

Fig. 1.6. The angular, I_θ, and frequency, I_ν, power spectrum of the Čerenkov radiation of a single charged particle in a plasma (after Cohen, 1961). The electron velocity is U, the thermal velocity is $V_{th} = (3kT/m)^{1/2}$, and the plasma frequency is $\nu_p = (e^2N_e/\pi m)^{1/2}$, where the temperature is T and the electron density is N_e. The radiation is symmetrical about the trajectory of the electron, and θ is the angle from this trajectory. The I_θ, and I_ν, are given, respectively by Eqs. (1.294) and (1.295). The frequency spectrum has a low frequency cutoff at $\nu_c = \nu_p[1 - V_{th}/U)^2]^{-1/2}$, and in the dashed region Landau damping inhibits wave propagation. The Čerenkov angular spectrum also exhibits a high angle cutoff which results from Landau damping

mass and charge of the oscillator, ν_0 is the resonance frequency of the oscillator, and the classical damping constant is

$$\gamma_{cl} = \frac{8\pi^2 e^2 \nu_0^2}{3mc^3} \approx 2.5 \times 10^{-22}\nu_0^2 \quad \sec^{-1} . \tag{1.298}$$

This equation has the steady state solution

$$x(t) = \frac{e}{4\pi^2 m}\mathcal{R}\left[\frac{E_0 \exp[i2\pi\nu t]}{(\nu^2 - \nu_0^2) + i(\gamma_{cl}\nu/2\pi)}\right] , \tag{1.299}$$

where $\mathcal{R}$ denotes the real part of the term in parenthesis.

From Eq.(1.109) the total energy radiated per unit time per unit solid angle, $d\Omega$, is

$$\frac{dP}{d\Omega} = \frac{3}{8\pi}\sigma_s c\, U \sin^2\theta , \tag{1.300}$$

where $U = E_0^2/4\pi$ is the energy density of the incident field, θ is the angle between the x axis and the direction of observation, and the scattering cross section, σ_s, is given by

$$\begin{aligned}\sigma_s &= \frac{8\pi}{3}\left(\frac{e^2}{mc^2}\right)^2 \frac{\nu^4}{(\nu^2 - \nu_0^2)^2 + (\gamma_{cl}^2\nu^2/4\pi^2)} \\ &\approx 6.65 \times 10^{-25}\frac{\nu^4}{(\nu^2 - \nu_0^2)^2 + (\gamma_{cl}^2\nu^2/4\pi^2)} \quad \text{cm}^2 .\end{aligned} \tag{1.301}$$

Similarly, the total energy radiated per unit time in all directions, P, is obtained from Eq. (1.110)

$$P = \sigma_s c U \ . \tag{1.302}$$

The total scattering cross section obtained by integrating σ_s over all frequencies is

$$\sigma_{tot} = \frac{\pi e^2}{mc} \approx 0.0265 \text{ cm}^2 \ . \tag{1.303}$$

The corresponding result for a quantum mechanical oscillator is (Ladenburg, 1921; Kramers, 1924)

$$\sigma_{tot} = \frac{\pi e^2}{mc} f_{nm} \ , \tag{1.304}$$

where f_{nm} is the oscillator strength of the $m - n$ transition.

1.36 Rayleigh Scattering by Bound Electrons

When plane wave radiation of frequency, ν, is incident upon atoms or molecules whose characteristic transition frequencies, ν_0, are much larger than ν, the total energy radiated per unit time per unit solid angle, $dP/d\Omega$, and the total energy radiated per unit time in all directions, P, are given by Eqs. (1. 300) and (1.302) with

$$\begin{aligned} \sigma_s = \sigma_R &= \frac{8\pi}{3}\left(\frac{e^2}{mc^2}\right)^2 \frac{\nu^4}{(\nu^2 - \nu_0^2)^2} \quad \text{for } \nu < \nu_0 \\ &\approx 6.65 \times 10^{-25}\left(\frac{\nu}{\nu_0}\right)^4 \text{ cm}^2 \text{ for } \nu \ll \nu_0 \ . \end{aligned} \tag{1.305}$$

This result, which was first obtained by Rayleigh (1899), follows from Eq. (1.301) when we assume $\nu \gg \gamma_{cl}$ when $\nu \ll \nu_0$.

The corresponding quantum mechanical expression for the Rayleigh scattering cross section is (Ladenburg, 1921; Kramers, 1924)

$$\sigma_R \approx 6.65 \times 10^{-25}\left(\sum_k f_{1k}\left[\frac{\nu_{1k}^2}{\nu^2} - 1\right]^{-1}\right)^2 \text{ cm}^2 \ , \tag{1.306}$$

where f_{1k} is the oscillator strength of the $1 - k$ transition.

From Eq. (1.86), the index of refraction, n, of a gas of density N is

$$n = \left[1 - \frac{4\pi N e^2}{4\pi^2 m(\nu_0^2 - \nu^2)}\right]^{1/2} \ . \tag{1.307}$$

Substituting Eq. (1.307) into Eq. (1.305) we obtain

$$\sigma_R = \frac{8\pi^3}{3}\left(\frac{\nu}{c}\right)^4\left(\frac{n^2 - 1}{N}\right)^2 \approx \frac{128\pi^5 \alpha^2}{3\lambda^4} \ . \tag{1.308}$$

where λ is the wavelength of the incident radiation and α is the polarizability.

1.37 Thomson and Klein–Nishina Scattering by a Free Electron

The total energy radiated per unit time per unit solid angle by a free electron is (Larmor, 1897, Eq. (1.109))

$$\frac{dP}{d\Omega} = \frac{e^2}{4\pi c^3}\dot{v}^2 \sin^2\theta \ , \tag{1.309}$$

where $\dot{v}$ is the acceleration and θ is the angle between the direction of acceleration and the direction of observation. From the Lorentz force equation (1.52) an incident electric field of strength E_0 gives the electron an acceleration $\dot{v} = eE_0/m$. Substituting into Eq. (1.309) we obtain

$$\frac{dP}{d\Omega} = \frac{3}{8\pi}\sigma_{\mathrm{T}} c U \sin^2\theta \ , \tag{1.310}$$

where $U = E_0^2/4\pi$ is the energy density of the incident plane wave and the Thomson cross section (Thomson, 1903) is given by

$$\sigma_{\mathrm{T}} = \frac{8\pi}{3}\left(\frac{e^2}{mc^2}\right)^2 \approx 6.65 \times 10^{-25}\mathrm{cm}^2 \ . \tag{1.311}$$

This result could have been obtained from Eq. (1.301) with $\nu_0 = 0$. The total energy radiated per unit time in all directions, P, is obtained by integrating Eq. (1.310) over all angles

$$P = \sigma_{\mathrm{T}} c U \ . \tag{1.312}$$

When the photon energy $h\nu$ is greater than or near mc^2, or $\nu \gtrsim 10^{20}$ Hz, the correct scattering cross section for free electrons is (Klein and Nishina, 1929)

$$\sigma_{\mathrm{KN}} = \frac{3}{4}\sigma_{\mathrm{T}}\left\{\frac{1+\alpha}{\alpha^2}\left[\frac{2(1+\alpha)}{1+2\alpha} - \frac{1}{\alpha}\ln(1+2\alpha)\right] + \frac{1}{2\alpha}\ln(1+2\alpha) - \frac{1+3\alpha}{(1+2\alpha)^2}\right\} \ , \tag{1.313}$$

where $\alpha = h\nu/(mc^2)$. At low and high photon energies this equation becomes

$$\sigma_{\mathrm{KN}} = \sigma_{\mathrm{T}}\left\{1 - \frac{2h\nu}{mc^2} + \frac{26}{5}\left(\frac{h\nu}{mc^2}\right)^2 + \cdots\right\} \quad \text{for} \quad \nu \ll 10^{20}\ \mathrm{Hz} \tag{1.314}$$

$$= \frac{3}{8}\sigma_{\mathrm{T}}\left(\frac{mc^2}{h\nu}\right)\left[\ln\left(\frac{2h\nu}{mc^2}\right) + \frac{1}{2}\right] \quad \text{for} \quad \nu \gg 10^{20}\ \mathrm{Hz}$$

$$\approx \frac{3 \times 10^{-5}}{\nu}\left[\ln(1.6 \times 10^{-20}\nu) + \frac{1}{2}\right]\ \mathrm{cm}^2 \quad \text{for} \quad \nu \gg 10^{20}\ \mathrm{Hz} \ . \tag{1.315}$$

The total energy radiated per unit time is, in this case, given by

$$P = \sigma_{\mathrm{KN}} c U \ , \tag{1.316}$$

where U is the energy density of the incident wave. The general formula for the angular distribution of the scattered radiation is, however, given by

$$I = \left(\frac{r_0}{R}\right)^2 \frac{\sin^2\psi I_0}{[1+\alpha(1-\cos\theta)]^3}\left\{1 + \frac{\alpha^2(1-\cos\theta)^2}{2\sin^2\psi[1+\alpha(1-\cos\theta)]}\right\} \ , \tag{1.317}$$

where I and I_0 are, respectively, the intensities of the incident and scattered radiation, $r_0 = e^2/mc^2 \approx 2.8 \times 10^{-13}$ cm is the classical electron radius, R is the distance from the free electron to the observation point, and ψ and θ are, respectively, the angles that the incident electric vector and the direction of incident wave propagation make with the direction of observation.

In the presence of a magnetic field of intensity, H, the Thomson scattering cross section is reduced below the classical value given by Eq. (1.311) when the photon frequency, ν, becomes smaller than the gyrofrequency, $\nu_H \approx 2 \times 10^6 H$ Hz. The reduction factor is roughly $(\nu/\nu_H)^2$ for $\nu \ll \nu_H$ and is discussed in detail by Canuto, Lodenquai, and Ruderman (1971).

1.38 Compton Scattering by Free Electrons and Inverse Compton Radiation

In the case of Thomson scattering, there is no change in the frequency of the radiation. For Compton scattering, the incident radiation transfers energy to the electron, and the scattered photon has less energy and a longer wavelength than the incident one. Thomson scattering applies for photon energies much less than the rest mass energy of the electron, or for $h\nu \ll mc^2$, whereas the Compton effect becomes important for larger incident photon energies (Compton, 1923). In the inverse Compton effect, high energy electrons scatter low energy photons so that in the Compton interaction the photons now gain energy and the electrons lose energy. Relatively recent discussions of these effects are given by Rybicki and Lightman (1979) and Longair (1992). The inverse Compton energy loss of electrons that also emit synchrotron radiation is reviewed by Kellermann and Owen (1988). Calculations of the energy exchange between free electrons and electromagnetic radiation during Compton scattering are given by Barbosa (1982), who also demonstrates the relationship of the Klein–Nishina cross section to that of classical electromagnetic radiation theory. Lieu and Axford (1993) consider synchrotron radiation from the viewpoint of an inverse Compton effect.

Let a photon of energy, $h\nu_1$, collide with an electron which is at rest. If the incident photon energy is large enough, the frequency of the scattered photon, ν_2, will be smaller than ν_1 by an observable amount. The effect is named after Compton who first observed and explained the effect (Compton, 1923). If the photon is deflected by an angle, φ, and if the electron recoils with a velocity, v, at an angle, θ, with respect to the initial trajectory of the photon, the relativistic equations for conservation of energy and momentum are

$$h\nu_1 = h\nu_2 + mc^2\left\{\left(1 - \frac{v^2}{c^2}\right)^{-1/2} - 1\right\} ,$$

$$\frac{h\nu_1}{c} = \frac{h\nu_2}{c}\cos\varphi + \frac{mv\cos\theta}{\left[1 - \frac{v^2}{c^2}\right]^{1/2}} , \quad (1.318)$$

and

$$\frac{h\nu_2}{c}\sin\varphi = \frac{m\,v}{\left[1-\left(\frac{v^2}{c^2}\right)\right]^{1/2}}\sin\theta \ .$$

It follows from Eqs. (1.318) that

$$\nu_2 = \nu_1\left[1+\frac{h\nu_1}{m\,c^2}(1-\cos\varphi)\right]^{-1} \tag{1.319}$$

and that the increment in wavelength is

$$\Delta\lambda = \lambda_2 - \lambda_1 = h\,(1-\cos\varphi)/(mc) \ ,$$

which is termed the Compton shift; the ratio $\lambda_c = h/(mc) = 2.42631058$ (22) $\times 10^{-10}$ cm is called the electron Compton wavelength.

These formulae assume that the electron is at rest. If α is the angle which the incident photon makes with the trajectory of a moving electron whose total energy $E = \gamma mc^2$, the energy of a photon in the rest frame of the electron is

$$h\nu_1 = \gamma h\nu(1+\beta\cos\alpha) \ ,$$

where $h\nu$ is the energy of the photon in its own frame, and $\beta = v/c$ for an electron of velocity, v.

For the inverse Compton effect, when the velocity of the electron, v, and $\beta = v/c$, are both large, we have the scattered frequency:

$$\nu_2 \approx \gamma^2\nu \quad \text{for } \gamma h\nu \ll m\,c^2 \ . \tag{1.320}$$

The scattering cross section for this case is (Feenberg and Primakoff, 1948; Donahue, 1951).

$$\sigma_s \approx \gamma^2\sigma_T \quad \text{for } \gamma h\nu \ll m\,c^2 \ , \tag{1.321}$$

where $\sigma_T \approx 6.65 \times 10^{-25}$ cm^2, and the total energy radiated per unit time, P, by an electron passing through radiation of energy density, U, is given by

$$P = \gamma^2\sigma_T c\,U \quad \text{for } \gamma h\nu \ll m\,c^2 \ . \tag{1.322}$$

A more precise relation derived by Longair (1992) for $\gamma h\nu \ll mc^2$ is:

$$P = (4/3)\sigma_T\,c\,U(\gamma^2-1) = (4/3)\,\sigma_T\,c\,U(v^2/c^2)\gamma^2 \ .$$

When the electron velocity is high and $\gamma h\nu \gg m\,c^2$, all of the electron energy, $\gamma m\,c^2$, is transferred into scattered radiation regardless of the incident photon frequency. Hence,

$$\nu_2 \approx \gamma m\,c^2/h \quad \text{for } \gamma h\nu \gg m\,c^2 \ , \tag{1.323}$$

and the Klien-Nishina cross section is

$$\sigma_s \approx \frac{3}{8}\sigma_T\left(\frac{m\,c^2}{\gamma h\nu}\right)\left[\ln\left(\frac{2\gamma h\nu}{m\,c^2}\right)\right] \quad \text{for } \gamma h\nu \gg m\,c^2 \ , \tag{1.324}$$

so that the total energy radiated per unit time by one electron in passing through radiation of energy density, U, is

$$P = \frac{3}{8}\sigma_T c\,U\left[\frac{m\,c^2}{\gamma h\nu}\right]\ln\left(\frac{2\gamma h\nu}{m\,c^2}\right) \quad \text{for } \gamma h\nu \gg m\,c^2 \ . \tag{1.325}$$

The Compton scattering of relativistic particles is further discussed in Chap. 4, where it is noted that other processes such as pair production are also important. Applications of scattering formulae to astronomical objects are given by Felten and Morrison (1963, 1966), Goldreich and Morrison (1964), and Gould and Schréder (1966). Scattering of high energy particles has been reviewed by Blumenthal and Gould (1970).

In order to maintain synchrotron radiation against inverse Compton losses, the magnetic field energy density in the region where the electrons are radiating must dominate over the local radiation energy density. For a homogeneous, isotropic source of luminosity, L, and radius, R, this condition means that (Felten and Morrison, 1966; Hoyle, Burbidge, and Sargent, 1966)

$$H^2 > \frac{8L}{R^2 c} , \tag{1.326}$$

where H is the magnetic field intensity, and the luminosity of the source is given by $L \approx 4\pi R^2 \nu S/\theta^2$, where ν is the frequency, S is the flux density, and θ is the angular size of the source. This means that the Compton effect will set in for sources which would otherwise have a frequency, $\nu_{\rm max}$, of maximum flux density given by

$$\nu_{\rm max} \gtrsim 10^{13} H \text{ Hz} . \tag{1.327}$$

For radio sources with an upper cutoff frequency, ν_c, and maximum effective brightness temperature, $T_{\rm max}$, it follows from Eq. (1.121) that the brightness $B \approx 2\nu_c^2 kT/c^2$. Using Eq. (1.154) with $E \approx kT$, we obtain $H \approx 10^{-19}\nu_c(kT)^{-2}$ and substituting these values of B and H into Eq. (1.326) we obtain (Kellermann and Pauliny-Toth, 1969)

$$10^{-72}\, T_{\rm max}^5 \nu_c < 1 . \tag{1.328}$$

Probably $\nu_c \approx 10^{11}$ Hz so that $T_{\rm max} < 10^{12}\,^\circ$K for an incoherent synchrotron radiator.

It follows that the minimum observable angular size, $\theta_{\rm IC}$, for a compact synchrotron source undergoing quenching by inverse Compton radiation is given by Kellermann and Owen (1998)

$$\theta_{\rm IC} \approx 10^{-3} S_{\rm m}^{1/2} \nu_{\rm c}^{-1} \text{seconds of arc} , \tag{1.329}$$

where the frequency $\nu_{\rm c}$ is the self-absorption cutoff frequency in GHz $= 10^9$ Hz. The observed angular sizes of compact radio sources are generally in good agreement with those expected from this equation and the measured peak flux densities, $S_{\rm m}$ in Janskys, and upper cutoff frequency, $\nu_{\rm c}$ in GHz, and there is no evidence that the peak brightness temperature ever exceeds 10^{12} K. This is strong evidence that the compact radio sources indeed radiate by the synchrotron process and that the radio emission is limited by the inverse Compton effect.

These angular sizes may also be compared with that caused by interstellar scattering, $\theta_{\rm scat}$, which follows from Eq. (1.385) together with measurements of Lang (1971) and Harris, Zeissig, and Lovelace (1970).

$$\theta_{\rm scat} \approx 25 D^{1/2} \nu^{-2} \text{ seconds of arc} , \tag{1.330}$$

where D is the source distance in parsecs (one parsec $\approx 3 \times 10^{18}$ cm) and ν is the frequency in MHz.

The inverse Compton scattered flux density, S_C, at an electron energy, E, is given by Marscher (1983) and Kellermann and Owen (1988) as:

$$S_C \approx 10^{-6} \ln(\nu/\nu_m)\theta^{-2(p+2)}\nu_m^{-0.5(3p+7)}S_m^{(p+3)}E^{-0.5(p-1)}(1+z)^{(p+3)}\delta^{-(p+3)}\text{Jy} \ , \tag{1.331}$$

where ν_m is the upper cutoff frequency of the synchrotron radiation spectrum, the p is the power-law index of the electron energy distribution, z is the redshift, and the quantity δ is a correction for the relativistic Doppler shift if the source is moving with high velocity and $\delta \approx 1$ if $v \ll c$.

Near an energy $E = 1$ keV appropriate for soft X-ray observations:

$$S_C \approx 10^{-6} T_B^{p+2} S\nu^{(p+1)/2}(1+z)^{p+3}\delta^{-(p+3)}\text{Jy} \ , \tag{1.332}$$

where the effective brightness temperature, T_B, is approximately $\nu^2\theta^2 S/1.22$ when T_B is expressed in units of 10^{12} K. Observations at millimeter wavelengths, where the effect of self-absorption is small, do indeed show a correlation between measured radio and X-ray flux density in the sense expected if the X-ray emission is due to inverse Compton scattering from the radio photons (Owen et al., 1981).

1.39 Rayleigh Scattering from a Small Sphere

Let a harmonic plane wave which is linearly polarized in the x direction and propagating in the z direction be incident upon a dielectric sphere of radius, a. Small spheres for which $a < 0.05\lambda_0$, where λ_0 is the wavelength of the incident radiation, will oscillate along the x axis in synchronism with the electric field of the plane wave. The sphere will therefore radiate as an electric dipole whose dipole moment, $d(t)$, is given by Eqs. (1.25) and (1.30).

$$d(t) = a^3\left(\frac{n^2-1}{n^2+2}\right)E_0(t) \approx a^3\left[\frac{\varepsilon_2-\varepsilon_1}{\varepsilon_2+2\varepsilon_1}\right]E_0(t) \ , \tag{1.333}$$

where $E_0(t)$ is the strength of the electric field vector of the plane wave, ε_2 and ε_1 are, respectively, the dielectric constants of the sphere and the surrounding medium, and the relative index of refraction is $n = [\varepsilon_2\mu_2/(\varepsilon_1\mu_1)]^{1/2}$ where μ is the magnetic permeability. Using a spherical co-ordinate system centered at the center of the sphere, the total energy radiated per unit time per unit solid angle, $dP/d\Omega$, may be obtained from Eq. (1.113)

$$\frac{dP}{d\Omega} = \frac{4\pi^3ca^6}{\lambda_0^4}\left(\frac{n^2-1}{n^2+2}\right)^2 \sin^2\theta E_0^2(t) \ , \tag{1.334}$$

where θ is the angle between the x axis and the radial vector to the point of observation. From Eq. (1.114), the total energy radiated per unit time in all directions is

$$P = \sigma_s cU \ , \tag{1.335}$$

where $U = E_0^2(t)/4\pi$ is the energy density of the plane wave, and the scattering cross section, σ_s, is given by (Rayleigh, 1871)

$$\sigma_s = \frac{128\pi^5 a^6}{3\lambda_0^4}\left(\frac{n^2-1}{n^2+2}\right)^2 \quad \text{for } a < 0.05\lambda_0 \ . \tag{1.336}$$

The efficiency factor, Q_s, for scattering is defined as the ratio of σ_s to the geometric cross section

$$Q_s = \frac{\sigma_s}{\pi a^2} = \frac{128\pi^4 a^4}{3\lambda_0^4}\left(\frac{n^2-1}{n^2+2}\right)^2 \ . \tag{1.337}$$

From Eq. (1.113) the intensity of the scattered radiation, I, for an incident wave of unit intensity is

$$I = \frac{16\pi^4 a^6}{R^2\lambda_0^4}\left(\frac{n^2-1}{n^2+2}\right)^2 \sin^2\theta \ , \tag{1.338}$$

where R is the distance from the sphere to the observation point.

When the incident plane wave is linearly polarized, the scattered radiation is also linearly polarized. When the incident wave is unpolarized, both the incident and scattered waves may be regarded as a superposition of two linearly polarized waves. When the plane of observation is defined as the plane that contains both the direction of propagation of the incident wave and the direction of observation, the intensity of the scattered light may be resolved into components $I_\parallel$ and $I_\perp$ which are parallel and perpendicular to this plane. For unit intensity of unpolarized incident radiation, the scattered intensities are

$$I_\perp = \frac{16\pi^4 a^6}{R^2\lambda_0^4}\left(\frac{n^2-1}{n^2+2}\right)^2 \ , \tag{1.339}$$

and

$$I_\parallel = \frac{16\pi^4 a^6}{R^2\lambda_0^4}\left(\frac{n^2-1}{n^2+2}\right)^2 \cos^2\theta \ , \tag{1.340}$$

and the total scattered intensity is

$$I = \frac{I_\perp + I_\parallel}{2} = \frac{8\pi^4 a^6}{R^2\lambda_0^4}\left(\frac{n^2-1}{n^2+2}\right)^2 (1+\cos^2\theta) \ , \tag{1.341}$$

where θ is the angle between the direction of propagation of the incident wave and the direction of observation.

Even when the incident radiation is unpolarized, the scattered radiation is partly polarized. The degree of polarization, P, is defined as

$$P = \frac{I_\perp - I_\parallel}{I_\perp + I_\parallel} = \frac{\sin^2\theta}{1+\cos^2\theta} \ , \tag{1.342}$$

where here again θ is the angle between the directions of propagation and observation. What is actually observed at optical frequencies is a magnitude difference, Δm_p, given by

$$\Delta m_{\mathrm{p}} = 2.50 \log \frac{I_\perp}{I_\parallel} \approx 2.172P \quad \text{for } P \ll 1 \ , \tag{1.343}$$

where $I_\perp$ and $I_\parallel$ are, respectively, the maximum and minimum observed intensities.

When the scattering sphere is not a perfect dielectric, but is partially conducting, part of the incident radiation is absorbed as well as scattered. If σ is the conductivity of a conducting sphere in a dielectric medium, then the index of refraction of the sphere, m_2, is given by

$$m_2 = \sqrt{\varepsilon_2 \mu_2} + i \frac{2\pi\sigma}{\omega} \sqrt{\frac{\mu_2}{\varepsilon_2}} \quad \text{for } \sigma \ll \omega\varepsilon \tag{1.344}$$

and

$$m_2 = \left(\frac{2\pi\mu_2\sigma}{\omega} \right)^{1/2} (1+i) \quad \text{for } \sigma \gg \omega\varepsilon \ , \tag{1.345}$$

where ε_2 and μ_2 are, respectively, the dielectric constant and magnetic permeability of the sphere, and ω is the radian frequency of the harmonic plane wave. When the surrounding space is a perfect dielectric, its refractive index is $m_1 = \sqrt{\varepsilon_1 \mu_1}$ where ε_1 and μ_1 are, respectively, the dielectric constant and magnetic permeability of the medium. The relative dielectric constant for a badly conducting sphere is then given by

$$m = \frac{m_2}{m_1} = \left(\frac{\varepsilon_2}{\varepsilon_1} \right)^{1/2} + i \frac{2\pi\sigma}{\omega\sqrt{\varepsilon_1 \varepsilon_2}} \tag{1.346}$$

or

$$m^2 \approx \frac{\varepsilon_2}{\varepsilon_1} + i \frac{4\pi\sigma}{\omega\varepsilon_1} \ . \tag{1.347}$$

It follows directly from a limiting case of the Mie scattering theory discussed in Sect. 1.41 that the scattering cross section in this case is

$$\sigma_{\mathrm{s}} = \frac{128\pi^5 a^6}{3\lambda_0^4} \mathcal{R} \left(\frac{m^2 - 1}{m^2 + 2} \right) \quad \text{for } a < 0.05\lambda_0 \ , \tag{1.348}$$

where $\mathcal{R}$ denotes the real part of the term within parenthesis.

1.40 Interstellar Dust, Extinction and Reddening of Stars

The idea that interstellar matter might be composed of small solid particles, called dust, as well as gas atoms had its origin in Vesto Slipher's studies of reflection nebulae, Edward Barnard's photographs of dark markings in the sky, and Max Wolf's investigations of dark nebulae. Slipher showed that some diffuse "reflection" nebulae exhibit absorption lines similar to those of nearby stars, and he reasoned that these nebulae are composed of solid particles that reflect starlight (Slipher, 1916, 1918, 1919). Barnard used telescopes with a wide field of view to photograph dark regions (Barnard, 1919, 1927); the close resemblance in the form and size of the dark markings suggested to Barnard that they were nebulous. By counting stars in dark and nearby bright regions, Wolf (1923) was able to show that the dark areas are obscuring dust clouds that absorb the light of distant stars. Because Wolf could not detect any substantial difference in the colors of stars that lie outside and behind the dark nebulae, he concluded that the dark regions must be composed of solid dust particles, whose scattering properties depend weakly on wavelength, λ, rather than gas atoms, which scatter light much more effectively at shorter wavelengths, by Rayleigh scattering that varies as λ^{-4}.

By assuming that all open star clusters have the same linear diameter and that the magnitudes of the individual stars are overestimated because of interstellar absorption of starlight, Trumpler (1930) obtained an average photographic absorption of 0.67 magnitude per kiloparsec, where 1 kpc $= 3.2 \times 10^{21}$ cm. When Joy (1939) compared two estimates of the distances for Cepheid variable stars, inferred from their period-luminosity relation and differential galactic rotation, he found that the light from these stars is absorbed at the rate of 0.85 magnitude per kiloparsec of interstellar matter. These results meant that the distance to the galactic center, and the total extent of our Galaxy, had been underestimated by a factor of almost 2. An average extinction of ≈ 1 magnitude per kiloparsec at visible wavelengths will be produced by solid dust particles with a mass density of $\approx 10^{-26}$ g cm^{-3}, which is roughly 0.5 percent of the density of the interstellar medium (Aannestad and Purcell, 1973).

The combined effects of scattering and absorption of electromagnetic radiation by interstellar dust is called extinction. The observed intensity of light, I_λ, at wavelength, λ, is given by:

$$I_\lambda = I_{\lambda 0} \exp(-\tau_\lambda) \tag{1.349}$$

where $I_{\lambda 0}$ is the intensity that would be received at the Earth in the absence of interstellar extinction along the line of sight and τ_λ is the optical depth at the wavelength observed. Photoelectric measurements indicate that the approximate interstellar extinction varies inversely with the wavelength, or as λ^{-1}, in the visual wavelength region, and that this variation is approximately the same for all stars (Stebbins, Huffer and Whitford, 1940; Whitford, 1958; Savage and Mathis, 1979). In the optical region,

$$\tau_\lambda = Cf(\lambda) \approx C/\lambda \ ,$$

where the constant C depends on the star, but the function $f(\lambda) \approx \lambda^{-1}$ is the same for all stars. The extinction is therefore strongest toward the shorter, blue end of the visible spectrum, and a star's colour appears reddened when viewed through interstellar dust. Detailed examination of the extinction curve, plotted as a function of $1/\lambda$ from ultraviolet to infrared wavelengths, shows departures from the λ^{-1} rule (Schild, 1977; Savage and Mathis, 1979).

The observed extinction of light at optical wavelengths is explained by Mie scattering (see the next Sect. 1.41) of solid interstellar dust grains comparable in size to the wavelength of light or about 10^{-5} cm across (van de Hulst, 1949). By showing that the shape of the curve of starlight extinction as a function of wavelength gives information about the size of the interstellar dust grains, van de Hulst set the stage for decades of speculation on the exact nature of interstellar dust. For example, when observations were subsequently extended to the ultraviolet spectral region, the amount of interstellar extinction was found to continue to increase, suggesting that some dust grains also have sizes that are on the order of the shorter wavelengths. A bump in the extinction curve observed at wavelengths near 2,175 Å can be attributed to absorption by small graphite particles. Absorption features found in the infrared spectra of cool but luminous stars indicate the presence of solid silicate particles that were presumably formed out of gas that was originally free of solids. Interstellar dust grains may therefore also be composed of silicates. Ices, carbon, and silicates, or some combinations of these, are probably the principal ingredients of dust grains, whose sizes may range from 10^{-4} to 5×10^{-5} cm.

A complete study of the physics of dust particles in interstellar, circumstellar and extragalactic space is given by Evans (1993). Cosmic dust has been discussed by Martin (1978); successive reviews of interstellar dust grains were given by Greenberg (1963), Lynds and Wickramasinghe (1968), Aannestad and Purcell (1973), and Savage and Mathis (1979). Then current knowledge of interstellar dust, including its extinction, polarization, distribution and relation to interstellar molecules, was provided in the symposium edited by Greenberg and van de Hulst (1973). The formation of molecules on interstellar dust grains is discussed in Sect. 2.5.13. An atlas of dust and H II regions in galaxies is provided by Lynds (1974). Dust in T Tauri stars and in galaxies are respectively reviewed by Bertout (1989) and Stein and Soifer (1987). Dust-gas interactions and the infrared emission from hot astrophysical plasmas are reviewed by Dwek and Arendt (1992), while the IRAS infrared view of the cosmos is discussed by Beichman (1987). Formulae for the extinction and polarization of starlight due to interstellar dust are given by Bowers and Deeming (1984); Osterbrock (1989) describes dust in H II regions. Small grains and large aromatic molecules are reviewed by Puget and Leger (1989).

The total extinction cross section, σ_e, of a sphere of radius a is given by (Feenberg, 1932; van de Hulst, 1949, 1957)

$$\sigma_e = \sigma_s + \sigma_a = -\frac{8\pi^2 a^3}{\lambda_0} \mathscr{I}\left(\frac{m^2-1}{m^2+2}\right) \quad \text{for } a < 0.05\lambda_0 \ , \tag{1.350}$$

where the complex index of refraction is m, the σ_s is given by Eq. (1.348), the $\mathscr{I}$ denotes the imaginary part of the term within the parenthesis, and σ_a is the cross section for the absorption of radiation by the sphere. The extinction cross section is the ratio of the total radiated energy scattered and absorbed per unit time to the incident radiant energy per unit time per unit area. An extinction coefficient is defined as $Q_e = \sigma_e/(\pi a^2)$, and the albedo, A, is defined as $A = Q_s/Q_e$. For a totally reflecting sphere of high conductivity

$$Q_s \approx Q_e \approx \frac{10}{3}\left(\frac{2\pi a}{\lambda_0}\right)^4 \quad \text{for } a < 0.05\lambda_0 \ , \tag{1.351}$$

whereas for a nonconducting sphere $\sigma_e \approx \sigma_s$ where σ_s is given by Eq. (1.348).

Extinction at wavelength, λ, is usually expressed in magnitudes as the extinction $A_\lambda = \Delta m = m(\lambda) - m_0(\lambda)$ where m_0 is the apparent magnitude if no dust existed along the line of sight. For a star of absolute magnitude $M(\lambda)$ at wavelength λ and distance D in parsecs, the apparent magnitude, $m(\lambda)$, is given by:

$$m(\lambda) = M(\lambda) + 5\log D - 5 + A_\lambda \ , \tag{1.352}$$

so the extinction A_λ measures how much interstellar dust diminishes starlight; it increases the apparent magnitude by an amount $\Delta m = A_\lambda$. For spherical grains of radius, a, and column density, N_c, the optical depth, τ_λ, for extinction is

$$\tau_\lambda = \pi a^2 Q_e N_c \ , \tag{1.353}$$

so that

$$A_\lambda = \Delta m = -2.5\log I_\lambda/I_{\lambda 0} = 1.086\pi a^2 Q_e N_c = 1.086 N\sigma_e D \ ,$$

where the extinction cross section $\sigma_e = \pi a^2 Q_e$, the efficiency Q_e specifies the efficiency of starlight attenuation, N is the volume density of dust grains, $N = N_c/L$ for length L along the line of sight, and D is the distance in cm.

The color excess, $E(B-V)$, due to the difference in the interstellar extinction in the blue, B, and visual, V, passbands is related to the absorption, A_V, in the visual band by $A_V = R \times E(B-V)$. The reddening factor R, or ratio of total to selective absorption, has a value of $R \approx 3.1$ to 3.3 (Schalen, 1975; Sandage, 1975; Olson, 1975); an exact formula for R is given by Olson (1975) and reproduced in Volume II. Values of $A_\lambda/E(B-V)$ are tabulated by Savage and Mathis (1979) for other wavelengths, λ, and reproduced by Rowan-Robinson (1985). Formulae for the absorption A_V and A_B in terms of the interstellar hydrogen concentration are also given in Volume II.

Interstellar dust plays an important role in absorbing energetic ultraviolet light that would otherwise destroy interstellar molecules. As discussed in Sects. 2.15.1 and 2.15.3, interstellar molecules are found within dark dust clouds, and dust also acts as a catalyst in forming complex molecules. Considerable atoms and molecules are deposited on dust grain mantles in dark molecular clouds.

How and where do the solid interstellar dust particles form? Bertil Lindblad reasoned that particles of the appropriate size might grow by the sublimation of interstellar gas around particles in a process analogous to that in which nucleating particles are formed (Lindblad, 1935). Early publications about the origin of interstellar dust are those of Kramers and Ter Haar (1946) and Oort and van de Hulst (1946). The formation and destruction of dust grains were reviewed by Salpeter (1977), and discussed by Salpeter (1974) and Draine and Salpeter (1979). Dust grains are most likely to originate in dark interstellar clouds, as well as in the outflowing gas of cool stars, where the density is high enough for atoms to form molecules that nucleate into larger particles.

The existence of interstellar grains is also supported by observations of linearly polarized light from reddened stars (Hiltner, 1949; Hall, 1949). The linearly polarized starlight can be explained in terms of dust particles about 10^{-5} cm in size aligned by an interstellar magnetic field of about 10^{-6} Gauss in strength (Davis and Greenstein, 1951; Spitzer and Tukey, 1951). Polarization observations indicate that the interstellar magnetic fields are aligned along the galactic plane (Mathewson and Ford, 1970). The polarization, Π, is defined by

$$\prod = (I_1 - I_2)/(I_1 + I_2) \ , \tag{1.354}$$

where I_1 is the intensity in the plane of polarization, or the maximum observed through a polarizing filter, and I_2 is the intensity in the perpendicular plane, corresponding to the minimum intensity observed through a polarizing filter. Expressed as a magnitude difference, Δm_Π, the polarization is given by

$$\Delta m_\Pi = 2.50 \log(I_1/I_2) = 2.717 \ \Pi \ .$$

Observations provide an upper limit of $\Delta m_\Pi/A_V = \Delta m_\Pi/\Delta m \leq 0.065$. The Kramers–Kronig relations for interstellar polarization are given by Martin (1975).

1.41 Mie Scattering from a Homogeneous Sphere of Arbitrary Size

The formulae for scattering from a sphere of radius a, have been derived from Maxwell's equations by Mie (1908) and Debye (1909). When the plane of observation is defined as that plane that contains both the direction of propagation of the incident wave and the direction of observation, the intensity of the scattered light may be resolved into two components, $I_\parallel$ and $I_\perp$, which are parallel and perpendicular to this plane. For unit intensity of unpolarized incident radiation, Mie's results for the scattered intensities are (Born and Wolf, 1970).

$$I_\perp = \frac{\lambda^2}{4\pi^2 R^2} |S_1|^2 \sin^2 \varphi \ , \tag{1.355}$$

and

$$I_\parallel = \frac{\lambda^2}{4\pi^2 R^2} |S_2|^2 \cos^2 \varphi \ ,$$

where R is the distance from the sphere, and

$$S_1 = \sum_{n=1}^{\infty} \frac{2n+1}{n(n+1)} \{a_n \pi_n(\cos\theta) + b_n \tau_n(\cos\theta)\} \ ,$$

$$S_2 = \sum_{n=1}^{\infty} \frac{2n+1}{n(n+1)} \{a_n \tau_n(\cos\theta) + b_n \pi_n(\cos\theta)\} \ ,$$

$$a_n = \frac{\psi_n(\alpha)\psi'_n(\beta) - m\psi_n(\beta)\psi'_n(\alpha)}{\zeta_n(\alpha)\psi'_n(\beta) - m\psi_n(\beta)\zeta'_n(\alpha)} \ ,$$

$$b_n = \frac{m\psi_n(\alpha)\psi'_n(\beta) - \psi_n(\beta)\psi'_n(\alpha)}{m\zeta_n(\alpha)\psi'_n(\beta) - \psi_n(\beta)\zeta'_n(\alpha)} \ ,$$

$$\alpha = 2\pi m_2 a/\lambda_0 \ ,$$

$$\beta = 2\pi m_1 a/\lambda_0 \ ,$$

λ_0 is the wavelength in vacuum, m_2 and m_1 are, respectively, the complex indices of refraction in the sphere and the surrounding medium, $m = m_1/m_2$, and the Riccati-Bessel functions are related to the Bessel, J, Neumann, N, and Hankel, H, functions by

$$\psi_n(Z) = (\pi Z/2)^{1/2} J_{n+1/2}(Z)$$
$$\zeta_n(Z) = (\pi Z/2)^{1/2} H^{(1)}_{n+1/2}(Z) = \psi_n(Z) + i\chi_n(Z) \ ,$$
$$\chi_n(Z) = -(\pi Z/2)^{1/2} N_{n+1/2}(Z) \ ,$$

and

$$\pi_n(\cos\theta) = \frac{P_n^{(1)}(\cos\theta)}{\sin\theta} \ ,$$
$$\tau_n(\cos\theta) = \frac{d}{d\theta} P_n^{(1)}(\cos\theta) \ ,$$

where $P_n^{(1)}(\cos\theta)$ is the associated Legendre function of the first kind, θ is the angle between the direction of propagation of the incident wave and the direction of observation, and φ is the other angle of the spherical coordinate system.

The scattering cross section, σ_s, and the extinction cross section, σ_e, are given by

$$\sigma_s = \left(\frac{\lambda^2}{2\pi}\right) \sum_{n=1}^{\infty} (2n+1)\{|a_n|^2 + |b_n|^2\} \ , \tag{1.356}$$

and

$$\sigma_s = \left(\frac{\lambda^2}{2\pi}\right) \sum_{n=1}^{\infty} (2n+1)\{\mathscr{R}(a_n + b_n)\} \ ,$$

where λ is the wavelength in the medium outside the sphere, and $\mathscr{R}$ ( ) denotes the real part of the term in parenthesis. The absorption cross section, σ_a, is given by

$$\sigma_a = \sigma_e - \sigma_s \ . \tag{1.357}$$

By definition, the extinction cross section is related to the total energy absorbed and scattered per unit time, P, by

$$P = \sigma_e c U \ , \tag{1.358}$$

where U is the energy density of the incident radiation and cU is the incident energy per unit time per unit area. The associated extinction efficiency, Q_e, is given by $\sigma_e/(\pi a^2)$ where πa^2 is the geometrical cross section of the sphere.

When the radius a is less than $0.05\,\lambda_0$,

$$a_1 = -\frac{2}{3} i \left(\frac{m^2-1}{m^2+2}\right) \left(\frac{2\pi a}{\lambda_0}\right)^3 \ , \tag{1.359}$$

where the square of the relative index of refraction, m^2, is given by

$$m^2 = \left(\frac{m_2}{m_1}\right)^2 \approx \frac{\varepsilon_2}{\varepsilon_1} + i\left(\frac{4\pi\sigma}{\omega\varepsilon_1}\right)$$

for a badly conducting sphere of conductivity, σ, and dielectric constant ε_2. The medium surrounding the sphere has dielectric constant, ε_1, and the radian frequency of the radiation is ω. In this case, we obtain the formulae for Rayleigh scattering (Rayleigh, 1871) which are given in the previous section.

When the radius, a, is much larger than the wavelength, λ, the effective extinction cross section becomes

$$\sigma_e = 2\pi a^2 \ . \tag{1.360}$$

In this case, for small angles, θ, between the direction of propagation of the incident wave and the direction of observation, the observed intensity for unit incident intensity is given by

$$I \approx \frac{Q_e}{4}\left[\frac{J_1\left(\frac{2\pi a}{\lambda}\sin\theta\right)}{\sin\theta}\right]^2 ,$$

where J is the Bessel function, and the intensity pattern is the Fraunhofer diffraction pattern of a circular aperture, first derived by Airy (1835).

1.42 Radar Backscatter

The radar cross section, σ_R, is defined to be 4π times the ratio of the reflected power per unit solid angle in the direction of the source to the power per unit area in the incident wave. That is, the radar cross section of a target is the projected area of a perfectly conducting sphere which, if placed in the same position as the real target, would scatter the same amount of energy to the observer.

For a planet at a distance, R, with a radar cross section, σ_R, the echo power, P_R, observed with a radar transmitting a signal of power, P_T, is given by the radar equation

$$P_R = \frac{GA}{(4\pi R^2)^2}\sigma_R P_T \ , \tag{1.361}$$

where G is the gain of the transmitting antenna over an isotropic radiator in the direction of the planet, and A is the effective collecting area of the receiving antenna for signals arriving from the direction of the planet. The backscatter cross section, σ_b, is the radar cross section for the case when the observer and transmitter occupy the same point. It follows from the previous sections that the backscatter cross section of a free electron is given by

$$\sigma_b = \frac{3}{2}\sigma_T = 4\pi\left(\frac{e^2}{m c^2}\right)^2 , \tag{1.362}$$

where the Thomson (1903) scattering cross section $\sigma_T \approx 6.65 \times 10^{-25}$ cm^2. For Rayleigh scattering from a poorly conducting sphere of radius, a,

$$\sigma_b = 4\pi a^2\left(\frac{2\pi a}{\lambda_0}\right)^4\left(\frac{m^2-1}{m^2+2}\right)^2 \quad \text{for } a < 0.05\lambda_0 \ , \tag{1.363}$$

where m is the relative index of refraction and λ_0 is the wavelength. For a planet of radius, r_0, the cross section is

$$\sigma_b = g R_0 \pi r_0^2 \ , \tag{1.364}$$

where the reflection coefficient for normal incidence, R_0, is given by Eqs. (1.77) and (1.78), and the coefficient g is a measure of the surface roughness. For a smooth surface with a Gaussian distribution of surface slopes, for example, $g = 1 + s^2$ where s is the r.m.s. slope (Hagfors, 1964).

For spherical targets it is often of interest to measure the backscatter cross section $\sigma_b(\varphi)$ as a function of the angle φ between the direction of illumination and an arbitrary normal to the surface. When a pulse of radiation of duration, τ, is used, it illuminates a surface annulus of linear width $c\tau/(2\tan\varphi)$ at the angle

$$\varphi = \cos^{-1}\left(1 - \frac{ct}{2r_0}\right) \ , \tag{1.365}$$

where the range delay, t, is the time after which the pulse first strikes the surface, and r_0 is the radius of the sphere (cf. Fig. 1.7). Range delay mapping of planetary surfaces provides a measure of $\sigma_b(\varphi)$ which can then be compared with theory to determine the roughness of the surface. As first pointed out by Rayleigh (1879) the image of a scattered wavefront is not seriously affected unless the wavefront deformation is greater than $\lambda/8$, where λ is the wavelength. Radar signals will, therefore, detect surface features whose sizes are larger than $\lambda/8$. Hagfors (1961, 1964) has assumed that the probability that the height of the true surface will depart from the mean by an amount h is proportional to $\exp[-1/2(h/h_0)^2]$, where h_0 is the r.m.s. height variation. The horizontal structure of the surface is then specified by an autocorrelation function, $\rho(\Delta r)$, of the height, $h(r)$, as a function of the distance, r, from any arbitrary point on the surface. Here

$$\rho(\Delta r) = \langle h(r)h(r + \Delta r)\rangle / h_0^2 \ . \tag{1.366}$$

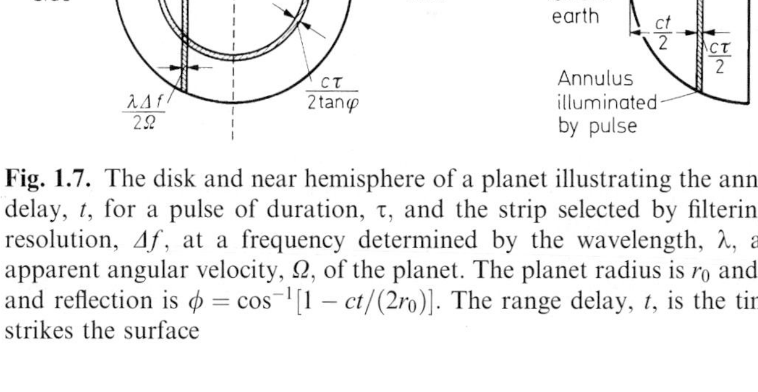

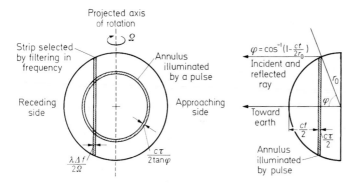

Fig. 1.7. The disk and near hemisphere of a planet illustrating the annulus of constant range delay, t, for a pulse of duration, τ, and the strip selected by filtering in frequency with a resolution, Δf, at a frequency determined by the wavelength, λ, and the instantaneous apparent angular velocity, Ω, of the planet. The planet radius is r_0 and the angle of incidence and reflection is $\phi = \cos^{-1}[1 - ct/(2r_0)]$. The range delay, t, is the time after the pulse first strikes the surface

Hagfors finds that the backscatter cross section normalized to a unit surface area is given by

$$\sigma_b(\varphi) \propto \frac{1}{\cos^4\varphi}\exp\left[\frac{-r_0^2\tan^2\varphi}{2h_0^2}\right] \quad \text{for } \rho(\Delta r) = \exp\left(\frac{-\Delta r^2}{2r_0^2}\right) \tag{1.367}$$

and

$$\sigma_b(\varphi) \propto \frac{1}{\left[\cos^4\varphi + \left(\frac{r_0\lambda}{4\pi h_0^2}\right)^2 \sin^2\varphi\right]^{3/2}} \quad \text{for } \rho(\Delta r) = \exp\left(\frac{-\Delta r}{r_0}\right) .$$

If the surface were white so that its brightness is the same in all directions, independently of the direction from which it is illuminated, then $\sigma_b(\varphi)$ is given by Lambert's law (Lambert, 1760)

$$\sigma_b(\varphi) \propto \cos^2(\varphi) . \tag{1.368}$$

The sidereal rotational velocity, Ω_s, of a planet can be determined by measuring its Doppler effect in radar echoes. For an annulus defined by the range delay, t, the total bandwidth, B, of the echo is given by Shapiro (1967)

$$B = \frac{2\nu_0}{c}[\Omega_s + \Omega_a][c\,t(4r_0 - c\,t)]^{1/2}[1 - (\boldsymbol{\Omega}\cdot\boldsymbol{r})^2]^{1/2} , \tag{1.369}$$

where ν_0 is the transmitted frequency, Ω_a is the apparent angular rotational velocity of the planet caused by the relative motion of the radar site and the planet, r_0 is the radius of the planet, $\boldsymbol{\Omega}$ is a unit vector in the direction of the total rotation, $\Omega_s + \Omega_a$, $\boldsymbol{r}$ is a unit vector in the direction from the Earth to the planet at the time of reflection, and Ω_s is the intrinsic rotational velocity of the planet.

When radar echoes are viewed in a narrow frequency interval, $\Delta\nu$, centered about a frequency, ν, different from the transmitted frequency, ν_0, the received radiation comes from a narrow strip of surface of linear width $c\Delta\nu/(2\Omega_s\nu)$ which is parallel to the axis of rotation. The combination of both delay and Doppler measurements then define two sections on the planet's surface which are equidistant from its equator, and which have the backscatter cross section $\sigma_b(t,\nu)$, given by Green (1968)

$$\sigma_b(t,\nu) = \sigma(t,\nu)\sec\varphi = \frac{c^2\sec\varphi\,\sigma(\varphi)}{2\nu_0\Omega_s\cos\alpha}\left[\frac{ct}{2r_0}\left(2 - \frac{ct}{2r_0}\right) - \left(\frac{\nu c}{2\nu_0 r_0\Omega_s\cos\alpha}\right)^2\right]^{-1/2} \tag{1.370}$$

provided that the planet is a uniform sphere of radius, r_0, and rotational velocity, Ω_s, with an angle α between its polar axis and the plane perpendicular to the line of sight. Here $\sigma(t,\nu)$ is the cross section per unit area of the surface, and $\sigma(\varphi)$ is the same cross section at the angle $\varphi = \cos^{-1}[1 - (ct/2r_0)]$. The delay – Doppler mapping technique is illustrated in Fig. 1.7.

For a plasma with plasma frequency $\nu_p = [e^2N_e/(\pi m)]^{1/2} \approx 8.97 \times 10^3 N_e^{1/2}$ Hz, the backscatter cross section $\sigma_b dv$ per unit volume, dv, is given by Booker and Gordon (1950) Booker (1956)

$$\sigma_b d\,v = \frac{4\pi^3}{c^4}\left|\frac{\Delta N_e}{N_e}\right|^2 \nu_p^4 P(2kl, 2km, 2kn) , \tag{1.371}$$

where N_e is the electron density, ΔN_e is the fluctuating component of electron density, the frequency of transmission, ν, is assumed to be $\gg \nu_p$, and the three-dimensional wave number spectrum of the density variations, $P(2kl, 2km, 2kn)$ is a function of the propagation constant $k = 2\pi\nu/c$, and l, m, n is a unit vector in the direction of observation. For spherically symmetric irregularities, $\Delta N_e/N_e$, of scale size, a,

$$P(2kl, 2km, 2kn) = (2\pi)^{3/2} a^3 \exp\left(-\frac{8\pi a^2}{\lambda^2}\right) , \tag{1.372}$$

where λ is the wavelength of observation. For a plasma which is in thermal equilibrium, $\sigma_b dv$ is given by Fejer (1960)

$$\sigma_b dv = 4\pi\left(\frac{e^2}{mc^2}\right)^2 N_e \approx \frac{3}{2}\sigma_T N_e \qquad \text{for } \lambda < r_D , \tag{1.373}$$

where the Thomson (1903) scattering cross section $\sigma_T \approx 6.65 \times 10^{-25}\,\text{cm}^2$, and the Debye radius $r_D = [kT_e/(4\pi e^2 N_e)]^{1/2} \approx 6.9(T_e/N_e)^{1/2}$ cm, where T_e is the electron temperature of the plasma. For wavelengths λ larger than r_D ions begin to contribute to the backscatter cross section and we have the relations (Buneman 1962; Farley, 1966)

$$\sigma_b dv = 4\pi\left(\frac{e^2}{mc^2}\right)^2 N_e\left[\frac{1}{(1+\alpha^2)(2+\alpha^2)}\right] \qquad \text{for } T_e = T_i , \tag{1.374}$$

and

$$\sigma_b dv \approx 4\pi\left(\frac{e^2}{mc^2}\right)^2 N_e\left[\frac{\alpha^2}{1+\alpha^2} + \frac{1}{(1+\alpha^2)(1+\alpha^2+T_e/T_i)}\right] \qquad \text{for } T_e \lesssim 3T_i ,$$

where the first and last terms in the brackets of the second expression denote, respectively, the electronic and ionic contributions, and $\alpha = 4\pi r_D/\lambda$. As shown by Dougherty and Farley (1960), Salpeter (1960), and Hagfors (1961), the ratio T_e/T_i may be determined by observing the frequency spectrum of the radar echo signal. The power spectrum $W(\omega)$ is given by

$$W(\omega) = \frac{\left|1+\left(\frac{\lambda}{4\pi}\right)^2 \sum_i \left(\frac{1}{r_{Di}}\right)^2 F_i(\omega)\right|^2 \overline{|N_e^0(\omega)|^2} + \left(\frac{\lambda}{4\pi r_{De}}\right)^4 |F_e(\omega)|^2 \sum_i \overline{|N_i^0(\omega)|^2}}{\left|1+\left(\frac{\lambda}{4\pi}\right)^2\left\{\left(\frac{1}{r_{De}}\right)^2 F_e(\omega) + \sum_i \left(\frac{1}{r_{Di}}\right)^2 F_i(\omega)\right\}\right|^2} , \tag{1.375}$$

where

ω = angular radio wave frequency displacement from the transmitted frequency
λ = radio wavelength
r_{De} = electron Debye length = $[kT_e/(4\pi e^2 N_e)]^{1/2}$
r_{Di} = ion Debye length = $[kT_i/(4\pi Z^2 e^2 N_i)]^{1/2}$

$$F_e(\omega) = 1 - \omega \int_0^\infty \exp\left(-\frac{16\pi^2 kT_e}{\lambda^2 m_e}\tau^2\right) \sin(\omega\tau) d\tau$$

$$- i\omega \int_0^\infty \exp\left(-\frac{16\pi^2 kT_e}{\lambda^2 m_e}\tau^2\right) \cos(\omega\tau) d\tau \ ,$$

$$F_i(\omega) = 1 - \omega \int_0^\infty \exp\left(-\frac{16\pi^2 kT_i}{\lambda^2 m_i}\tau^2\right) \sin(\omega\tau) d\tau$$

$$- i\omega \int_0^\infty \exp\left(-\frac{16\pi^2 kT_i}{\lambda^2 m_i}\tau^2\right) \cos(\omega\tau) d\tau \ .$$

The term $\overline{|N_e^0(\omega)|^2}$ is the fluctuation spectrum of the independent electrons. That is,

$$\overline{|N_e^0(\omega)|^2} = 2N \int_0^\infty \exp\left(-\frac{16\pi^2 kT_e}{\lambda^2 m_e}\tau^2\right) \cos(\omega\tau) d\tau \ , \tag{1.376}$$

and $\overline{|N_i^0(\omega)|^2}$ is the corresponding spectrum for the ions obtained by replacing N_e by N_i, T_e by T_i, and m_e by m_i. Here N_e is the number density of electrons, T_e is the electron temperature, m_e is the electron mass, and with the subscript i, they represent the corresponding quantities for ions.

1.43 Phase Change and Scattering Angle Due to Fluctuations in Electron Density

The electric field vector, $\boldsymbol{E}$, of an electromagnetic plane wave is given by

$$\boldsymbol{E} = E_0 \exp[i(\boldsymbol{k} \cdot \boldsymbol{r} - \omega t))]\boldsymbol{\eta}_\perp \ , \tag{1.377}$$

where E_0 is the amplitude of the wave, ω is its radian frequency, $\boldsymbol{\eta}_\perp$ is a unit vector which is perpendicular to the direction of propagation, and the wave vector $\boldsymbol{k}$ is given by

$$\boldsymbol{k} = \frac{\omega}{c} n\boldsymbol{s} \ , \tag{1.378}$$

where n is the index of refraction of the medium and $\boldsymbol{s}$ is a unit vector in the direction of propagation. When the plane wave is propagating in an ionized gas, the index of refraction is given by Eq. (1.253)

$$n = \left[1 - \frac{4\pi N_e e^2}{m\omega^2}\right]^{1/2} \ , \tag{1.379}$$

where N_e is the free electron density. A small change, Δn, in the index of refraction will give rise to a small phase change, $\Delta\phi$, of the electric field vector. From Eqs. (1.377) to (1.379), a change in electron density, ΔN_e, occurring in a turbule of size, a, will give rise to the phase change.

$$\Delta\varphi = \frac{\omega \Delta n a}{c} = \frac{e^2 a \Delta N_e}{m v c} = r_0 a \Delta N_e \lambda \ , \tag{1.380}$$

where $r_0 = e^2/(mc^2) \approx 2.8 \times 10^{-13}$ cm is the classical electron radius, and ν and λ are, respectively, the frequency and wavelength of the plane wave.

Let the spatial distribution of the gas turbules be Gaussian, and describe an isotropic Gaussian phase autocorrelation function, $\rho(r)$, given by

$$\rho(r) = \langle\varphi(x)\varphi(x+r)\rangle/\varphi_0^2 = \exp[-r^2/(2a^2)] \ , \tag{1.381}$$

where $\langle \quad \rangle$ denotes a spatial average, $\varphi(r)$ describes the phase distribution across the wave front, and the r.m.s. value of this distribution is $\varphi_0 = \langle[\varphi(x)]^2\rangle^{1/2}$. Then from Eq. (1.380), a thickness, L, of a gas with turbules distributed according to Eq. (1.381) will cause an r.m.s. phase shift, φ_0, given by

$$\begin{aligned}\varphi_0 &= (2\pi)^{1/4}\left(\frac{L}{a}\right)^{1/2}\langle\Delta\varphi^2\rangle^{1/2} = (2\pi)^{1/4}\frac{e^2}{mc\nu}(La)^{1/2}\langle\Delta N_e^2\rangle^{1/2} \\ &\approx 10^{-2}\frac{(La)^{1/2}}{\nu}\langle\Delta N_e^2\rangle^{1/2} \quad \text{radians} \ .\end{aligned} \tag{1.382}$$

When studying the ionosphere, φ_0 is often expressed in the alternate form

$$\varphi_0 \approx \pi(La)^{1/2}\nu_p^2\frac{\lambda}{c^2}\left\langle\left(\frac{\Delta N_e}{N_e}\right)^2\right\rangle^{1/2} \ , \tag{1.383}$$

where the plasma frequency, ν_p, is defined by the condition that the index of refraction given by Eq. (1.379) becomes zero

$$\nu_p = \left(\frac{e^2 N_e}{\pi m}\right)^{1/2} \approx 8.97 \times 10^3 N_e^{1/2}\,\text{Hz} \ , \tag{1.384}$$

When a ray of the plane wave radiation has traversed a thickness, L, of a gas, the r.m.s. value, θ_{scat}, of its angular deviation, θ, from a straight line path is (Chandrasekar, 1952; Fejer, 1953)

$$\begin{aligned}\theta_{scat} &\approx (2\pi)^{-3/4}\left(\frac{L}{a}\right)^{1/2}\langle\Delta n^2\rangle^{1/2} \approx (2\pi)^{-3/4}\left(\frac{L}{a}\right)^{1/2}\frac{c}{\omega a}\langle\Delta\varphi^2\rangle^{1/2} \\ &\approx 10^8\left(\frac{L}{a}\right)^{1/2}\frac{\langle\Delta N_e^2\rangle^{1/2}}{\nu^2} \quad \text{radians} \ ,\end{aligned} \tag{1.385}$$

where the index of refraction change, Δn, and the phase change, $\Delta\varphi$, are given by Eq. (1.380), and it has been assumed that the turbules are distributed according to Eq. (1.381). The parameter θ_{scat} determines the minimum detectable size of an astronomical source. That is, even if a source is smaller than θ_{scat}, it will appear to have the angular size θ_{scat}. Values of θ_{scat} are given in Table 1.3 together with other scintillation parameters for the atmosphere, the ionosphere, and the interplanetary and interstellar medium.

Table 1.3. Representative values of scintillation parameters for the atmosphere, the ionosphere, the interplanetary medium, and the interstellar medium[1]. The rms scattering angle is θ_{scat}, the scale size is a, the wind velocity is v, the decorrelation time scale of the intensity fluctuations is τ_v, the r.m.s. electron density is $\langle N_e^2 \rangle^{1/2}$, the r.m.s. fluctuation electron density is $\langle \Delta N_e^2 \rangle^{1/2}$, the thickness of the scintillating screen is L, and the effective screen distance is D

Medium	θ_{scat}, (seconds of arc)	a (cm)	v (km/sec)	τ_v (sec)	$\frac{\langle \Delta N_e^2 \rangle^{1/2}}{\langle N_e^2 \rangle^{1/2}}$	$\langle N_e^2 \rangle^{1/2}$ (cm^{-3})	L (cm)	D (cm)
Atmosphere (at 4000 Å)	0.75	10	10^{-4}	1	10^{-4}	10^{19}	10^4	10^5
Ionosphere (at 45 MHz)	5	10^4–10^5	0.1–0.3	30	10^{-3}	10^5	5×10^6	3×10^7
Interplanetary medium (at 200 MHz)	10^{-3}–10^4	10^7 *	250–700	0.2	10^{-2}	10–10^8	10^{13}	10^{13}
Interstellar medium (at 318 MHz)	10^{-3}	10^{11}	30–200	10^3	10^{-2}	0.03	10^{22}	10^{22}

[1] Values are for the frequencies given in parenthesis and the L and D which are tabulated. The references are Chandrasekhar [1952] for the atmosphere, Hewish [1951] and Aarons, Whitney, and Allen [1971] for the ionosphere, Dennison and Hewish [1967] and Cohen, Gundermann, Hardebeck, and Sharp [1967] for the interplanetary medium, and Lang and Rickett [1970], Rickett [1970] and Lang [1971] for the interstellar medium.
*Spacecraft observations indicate a characteristic scale or correlation length of 10^{11} cm for the turbulence of the interplanetary medium.

1.44 The Scintillation Pattern

As reviewed by Hewish (1975), early studies of radio galaxies led to the identification of ionospheric scintillation, and then to interplanetary scintillation that was subsequently used to probe the structure of radio sources and led to the serendipitous discovery of pulsars (Hewish et al., 1968). The radio scintillation effects of pulsar radiation were in turn used to probe the interstellar medium. Early papers on the interstellar scintillation of pulsar radiation include those of Rickett (1970), Lang (1971), Lang and Rickett (1970) and Rickett and Lang (1973); interstellar scintillation of extragalactic radio sources is discussed by Condon and Backer (1975). Ratcliffe (1956) gives a good review of the theory applicable to ionospheric scintillation at radio frequencies, while Tatarski (1961) pioneered the theory of optical scintillation in the Earth's atmosphere. Reviews of the interplanetary scintillation of radio waves are provided by Coles, Rickett and Rumsey (1974) and Jokipii (1973). The theory of strong scintillations in astrophysics is given by Lee and Jokipii (1975, 1976). Papers that describe spacecraft radio scintillation and propagation in the solar wind are given by Woo (1993, 1996). The observations and theory of interstellar scattering and scintillation were reviewed by Rickett (1977). Radio propagation through the turbulent interstellar plasma has been additionally reviewed by Rickett (1990).

Booker, Ratcliffe, and Shinn (1950) first showed that the autocorrelation function of a wave disturbance over any plane parallel to a thin phase changing screen is the same as that over the wave front just after it has emerged from the screen. The observable statistical properties of the variations in wave intensity across the Earth are therefore directly related to the spatial variations of the wave phase across the wave front at the screen. In particular, the spatial autocorrelation function, $\varphi_0^2\rho(r)$, of the phase, $\varphi(x)$, across the wave front may be directly related to the spatial autocorrelation function, $M(r)$, of the fluctuation $\Delta I(x) = I(x) - \langle I(x)\rangle$ of the wave intensity, $I(x)$, at the Earth. Here

$$M(r) = \langle \Delta I(x) \Delta I(x+r)\rangle \ , \tag{1.386}$$

where $\langle\ \rangle$ denotes a spatial average. The $M(r)$ is usually specified by computing its Fourier transform, $M(q)$, given by

$$M(q) = \int_{-\infty}^{+\infty} M(r) \exp(-irq) dr \ . \tag{1.387}$$

The $M(q)$ is called the power spectrum of the intensity fluctuations because it is equal to the square of the Fourier transform of $\Delta I(x)$. That is,

$$M(q) = \left| \int_{-\infty}^{+\infty} \Delta I(x) \exp(-ixq) dx \right|^2 \ . \tag{1.388}$$

When a source is not a point source, the intensity fluctuations due to different parts of the source will overlap and the net result will be to smear out the intensity fluctuations. That is, intensity fluctuations due to a point source will be deeper than those due to an extended source, and very extended sources will not be observed to scintillate. This condition is illustrated by the formula for the power spectrum, $M_s(q)$, of an extended source (Salpeter, 1967)

$$M_s(q) = M(q)|V(qD)|^2 \ , \tag{1.389}$$

where D is the distance to the scintillating screen, $M(q)$ is the power spectrum for a point source, and the visibility function, $V(qD)$, is the Fourier transform of the source brightness distribution.

Assuming that the intensity fluctuations in time, $I(t)$, observed by a fixed antenna are caused by the motion of a thin phase changing screen across the observer's line of sight, the autocorrelation function, $M(\tau)$, of the intensity fluctuations, $I(t)$, is given by

$$M(\tau) = M\left(\frac{r}{v}\right) = \frac{\langle [I(t) - \langle I(t)\rangle][I(t+\tau) - \langle I(t+\tau)\rangle]\rangle}{\sigma^2} \ , \tag{1.390}$$

where the screen moves with the velocity, v, perpendicular to the line of sight to the source, the angular brackets denote a time average, and the mean square deviation of $I(t)$ is σ^2.

If measurements of intensity fluctuations are made with an antenna pointing ON a source and then OFF of it, the depth of modulation of the intensity fluctuations may be measured. This modulation or scintillation index is given by Cohen et al. (1967)

$$m = \{[(\sigma_{\mathrm{ON}})^2 - (\sigma_{\mathrm{OFF}})^2]/[\langle I_{\mathrm{ON}}(t)\rangle - \langle I_{\mathrm{OFF}}(t)\rangle]^2\}^{1/2} \ , \qquad (1.391)$$

where $I\ (t)$ is the observed intensity variation in time, σ^2 denotes the mean square value of the intensity, and the angular brackets denote a time average. As mentioned before, an extended source decreases the modulation index below its value for a point source. When a source is broader than the critical angle, ψ_c, the modulation depth is considerably reduced (Salpeter, 1967)

$$\psi_{\mathrm{c}} \approx a/D \quad \text{for small } m \ ,$$
$$\psi_{\mathrm{c}} \approx a/(\varphi_0 D) \quad \text{for } m \approx 1 \ . \qquad (1.392)$$

Another statistical parameter which may be measured is the probability distribution, $P(I)$, of the intensity. If distributions are measured while pointing ON and OFF a source, they are related by the convolution equation

$$\mathrm{P}_{\mathrm{ON}}(x) = \int_{-\infty}^{+\infty} P(I) P_{\mathrm{OFF}}(x - I) dI \ , \qquad (1.393)$$

where $P(I)$ is the probability distribution due to the scintillating screen.

A decorrelation time, τ_v, may be defined as the equivalent width of the autocorrelation function, $M(t)$, given by Eq. (1.390)

$$\tau_v = \frac{\int_{-\infty}^{+\infty} M(t) dt}{M(t=0)} \approx \frac{a}{\varphi_0 v} \approx \frac{c}{\nu v \theta_{\mathrm{scat}}} \approx 10^3 \left(\frac{a}{L}\right)^{1/2} \frac{\nu}{v\langle \Delta N_{\mathrm{e}}^2\rangle^{1/2}} \ \sec \ , \qquad (1.394)$$

where v is the velocity at which the scattering turbules are moving transverse to the line-of-sight. The decorrelation time is roughly the quarter period of the intensity fluctuations, and representative values are given in Table 1.3. A decorrelation frequency, f_v, may be similarly defined as the equivalent width of a frequency correlation function. If two radiometers, 1 and 2, record intensities $I_1(t)$ and $I_2(t)$, respectively, at two different frequencies, then the cross-correlation function, $\Gamma_{12}(\tau)$, is given by

$$\Gamma_{12}(\tau) = \frac{1}{\sigma_1 \sigma_2} \langle [I_1(t) - \langle I_1(t)\rangle][I_2(t+\tau) - \langle I_2(t+\tau)\rangle]\rangle \ , \qquad (1.395)$$

where the angular brackets denote a time average and σ_1^2 and σ_2^2 are the mean square deviations of $I_1(t)$ and $I_2(t)$, respectively. The f_v may then be defined as the equivalent width of a plot of $\Gamma_{12}(0)$ versus frequency separation of the radiometers. For strong phase fluctuations where geometric optics applies, the path difference between a direct and scattered ray is $D\theta_{\mathrm{scat}}^2/4$ if the screen distance, D, is half the distance to the source. Consequently, the frequency difference, f_v, for which the two waves interfere is given by

$$f_v \approx \frac{4c}{D\theta_{\mathrm{scat}}^2} \approx \frac{10^{-5} a \nu^4}{\langle \Delta N_{\mathrm{e}}^2\rangle D^2} \ \mathrm{Hz} \quad \begin{array}{l}\text{for strong scintillations}\\ \text{in an extended medium.}\end{array} \qquad (1.396)$$

A pulse of radiation will be broadened in time by f_ν^{-1}. That is, the time profile of a scattered pulse will be the convolution of the emitted pulse with an exponential function whose $1/e$ decay time is f_ν^{-1}. Theoretical formulae for f_ν in different conditions of scattering have been derived together with formulae for $M(q)$, m, and $P(I)$ by Hewish (1951), Fejer (1953), Mercier (1962), Salpeter (1967), Jokipii and Hollweg (1970), and Lovelace et al. (1971). A few of these formulae are given in Table 1.4.

If a stable diffraction pattern is moving with a velocity, v, at an angle θ to the projected baseline, B, between two antennas, then an upper limit to the velocity is given by

$$v = B|\cos\theta|/T \leq B/T \ , \tag{1.397}$$

where T is the time displacement between the arrival of the pattern at the two antennas. If the intensity fluctuations seen at the two antennas are highly correlated, then we have the lower limit (Lang and Rickett, 1970)

$$v \geq B[T^2 + \tau_\nu^2]^{-1/2} \ , \tag{1.398}$$

where τ_ν is the decorrelation time. If the stable diffraction pattern is observed simultaneously with three antennas, then a unique velocity may be specified (cf. Dennison and Hewish, 1967). If, however, the scintillating structure rearranges itself as it moves, this apparent velocity overestimates the true drift velocity (cf. Briggs, Phillips, and Shinn, 1950). Finally, Lovelace et al. (1971) have shown that under certain circumstances the Bessel transform of the time autocorrelation function measured by a single antenna may give the drift velocity, v. Values of v for the atmosphere, ionosphere, and the interplanetary and interstellar medium were given in Table 1.3.

Rickett (1977) provides additional formulae for the scattering angle and the pulse broadening of pulsar radiation, as well as the functional dependence of interstellar intensity scintillation in time, space and frequency. Refractive interstellar scintillations of compact radio sources are reviewed by Rickett (1990), including its importance in the interpretation of variable radio sources and Very Long Baseline Interferometry (VLBI) images at low galactic latitudes and long wavelengths.

Table 1.4. Formulae for the power spectrum, $M(q)$, the modulation index, m, the probability distribution of intensity, $P(I)$, and the decorrelation frequency, f_v, for three regions of scintillation. Here a thin phase scattering screen is assumed to be located at a distance, D, to have turbules of scale size, a, and to cause an r.m.s. phase fluctuation, ϕ_0. For a Gaussian screen, $\phi^2(q) = (2\pi)^{-1/2} a\phi_0^2 \exp(-q^2a^2/2)$

Region	$M(q)$	m	$P(I)$	f_v
$\phi_0 \ll 1$ and $D \ll 2\pi a^2/\lambda$ or $\phi_0 > 1$ and $D \ll 2\pi a^2/(\lambda\phi_0)$	$4\phi^2(q)\sin^2\left(\frac{q^2D\lambda}{4\pi}\right) \approx \left(\frac{q^2D\lambda}{2\pi}\right)^2 \phi^2(q)$	$\frac{\phi_0 D\lambda}{2\pi a^2} \propto \lambda^2$	$\propto \exp\left[\frac{-I^2}{2m^2}\right]$	$\to \nu$
$\phi_0 \ll 1$ and $D \gg 2\pi a^2/\lambda$	$4\phi^2(q)\sin^2\left(\frac{q^2D\lambda}{4\pi}\right) \approx 2\phi^2(q)$	$\sqrt{2}\phi_0 \propto \lambda$	$\propto \exp\left[\frac{-I^2}{2m^2}\right]$	$\frac{2\pi a^2 c}{\lambda^2 D} \propto \lambda^{-2}$
$\phi_0 \gg 1$ and $D \gg 2\pi a^2/(\lambda\phi_0)$	$\exp\left[-\left(\frac{qa}{2\phi_0}\right)^2\right]$	1	$\propto \exp[-I]$	$\frac{2\pi a^2 c}{\phi_0^2\lambda^2 D} \propto \lambda^{-4}$

2. Monochromatic (Line) Radiation

"They (atoms) move in the void and catching each other up jostle together, and some recoil in any direction that may chance, and others become entangled with one another in various degrees according to the symmetry of their shapes and sizes and position and order, and they remain together and thus the coming into being of composite things is effected."

Simplicus (6th Century A. D.)

"I write about molecules with great diffidence, having not yet rid myself of the tradition that atoms are physics, but molecules are chemistry, but the new conclusion that hydrogen is abundant seems to make it likely that the above-mentioned elements H, O, and N will frequently form molecules."

Sir Arthur Eddington, 1937

"Modern improvements in optical methods lend additional interest to an examination of the causes which interfere with the absolute homogeneity of spectrum lines. So far as we know these may be considered under five heads, and it appears probable that the list is exhaustive."

(I) The translatory motion of the radiating particles in the line of sight, operating in accordance with Doppler's principle.
(II) A possible effect of the rotation of the particles.
(III) Disturbance depending on collisions with other particles either of the same or another kind.
(IV) Gradual dying down of the luminous vibrations as energy is radiated away.
(V) Complications arising from the multiplicity of sources in the line of sight. Thus if the light of a flame be observed through a similar one, the increase of illumination near the centre of the spectrum line is not so great as towards the edges and the line is effectively widened."

Lord Rayleigh, 1915

2.1 Parameters of the Atom

Classical electron radius. Consider the orbital motion of an electron of mass, m, and charge, e, about a nucleus of charge Ze. Equating the Coulomb force of attraction to the force from centripetal acceleration we obtain

$$\frac{Ze^2}{r^2} = \frac{mv^2}{r} , \tag{2.1}$$

where r is the radius of the orbit, and v is the velocity of the electron. Solving for r,

$$r = r_0 \left(\frac{c}{v} \right)^2 Z , \tag{2.2}$$

where the classical electron radius

$$r_0 = e^2/(mc^2) \approx 2.818 \times 10^{-13}\mathrm{cm} \ . \qquad (2.3)$$

Radius of the first Bohr orbit. Following Bohr (1913) we may assume that the angular momentum of the electron is quantized so that

$$mvr = \frac{hn}{2\pi} \ , \qquad (2.4)$$

where $h \approx 6.625 \times 10^{-27}$ erg sec is Planck's constant and n is an integer. Using Eqs. (2.1) and (2.4) we obtain the radius r_n of the nth Bohr orbit.

$$r_n = a_0 \frac{n^2}{Z} \ , \qquad (2.5)$$

where the radius of the first Bohr orbit of hydrogen ($Z = 1$) is

$$a_0 = \frac{h^2}{4\pi^2 me^2} \approx 0.529 \times 10^{-8}\mathrm{cm} \ . \qquad (2.6)$$

Line frequency. For atomic or molecular radiation resulting from the transition between two levels of energy, E_m and E_n, the frequency of radiation, ν_{mn}, is given by (Planck, 1910; Bohr, 1913)

$$\nu_{mn} = |E_m - E_n|/h \ , \qquad (2.7)$$

where $||$ denotes the absolute value, and h is Planck's constant.

Rydberg constant for infinite mass. The total energy, E_n, of an electron whose velocity is v and orbital radius is r, is given by

$$E_n = -\frac{Ze^2}{r} + \frac{1}{2}mv^2 = \frac{-2\pi^2 me^4 Z^2}{h^2 n^2} \ , \qquad (2.8)$$

where n is an integer and it has been assumed that the angular momentum is quantized according to Eq. (2.4). Using Eq. (2.8) in Eq. (2.7) the frequency corresponding to this energy is

$$v_{mn} = cR_\infty Z^2 \left| \frac{1}{n^2} - \frac{1}{m^2} \right| \ , \qquad (2.9)$$

where m and n are integers, the Rydberg constant for infinite mass is (Rydberg, 1890),

$$R_\infty = \frac{2\pi^2 m_e e^4}{ch^3} \approx 1.097 \times 10^5 \mathrm{cm}^{-1} \ , \qquad (2.10)$$

and m_e is the electron mass.

Compton frequency and wavelength. The photon energy, $h\nu$, becomes equal to the electron rest mass energy, mc^2, at the Compton frequency

$$\nu_0 = mc^2/h \approx 1.23 \times 10^{20} \ \mathrm{Hz} \ , \qquad (2.11)$$

which corresponds to the Compton wavelength (Compton, 1923)

$$\lambda_0 = \frac{h}{2\pi mc} \approx 3.862 \times 10^{-11}\mathrm{cm} \ . \qquad (2.12)$$

Zeeman displacement–Larmor frequency of precession. Assume that the moving electron is a harmonic oscillator which in the absence of electromagnetic fields exhibits rectilinear oscillation given by

$$\vec{r}(t) = \vec{r}_0 \cos(2\pi\nu_0 t) \quad , \tag{2.13}$$

where ν_0 is the frequency of oscillation, $\vec{r}_0$ is the peak linear displacement, and t is the time variable. The equation of motion of the forced vibration of the oscillator in a magnetic field of strength, $\vec{H}$, directed along the z axis is (Lorentz, 1897)

$$m\ddot{\vec{r}} + m(2\pi\nu_0)^2\vec{r} = \frac{e}{c}\dot{\vec{r}} \; x \; \vec{H} \quad , \tag{2.14}$$

where $\cdot$ and $\cdot\cdot$ denote, respectively, the first and second derivatives with respect to time, and the magnetic force term is the term on the right side of the equation. Solving Eq. (2.14) for the frequency of oscillation, ν, we obtain

$$\nu = \nu_0 + o \text{ for } o \ll \nu_0 \quad , \tag{2.15}$$

where the Zeeman displacement (Zeeman, 1896, 1897; Lorentz, 1897)

$$o = \frac{eH}{4\pi mc} \approx 1.4 \times 10^6 H \text{ Hz} \quad , \tag{2.16}$$

when the magnetic field intensity, H, is in Gauss. The latter constant is also called the Larmor frequency of precession because Larmor (1897) showed that a superimposed H field leaves the motion of an electron in its orbit alone except for a uniform precession of the orbit about the direction of the lines of force, the precessional velocity being $2\pi o$.

Bohr magneton and nuclear magneton. The magnetic dipole moment of an electric current is equal to the product of the current strength, the area enclosed by the circulating current, and c^{-1}. For an electron of charge, e, and angular velocity, ω, the current strength is $e\omega/2\pi$. Multiplying this strength by the area πr^2 where r is the radius of the orbit, and using Eq. (2.4) for the quantization of angular momentum, we obtain the magnetic moment

$$M = L\mu_{\text{B}} \quad , \tag{2.17}$$

where L is an integer, and the Bohr magneton, μ_{B}, is given by

$$\mu_{\text{B}} = \frac{eh}{4\pi mc} \approx 9.274\,054(31) \times 10^{-21} \text{ erg Gauss}^{-1} \quad . \tag{2.18}$$

The corresponding magneton for the hydrogen nucleus is the nuclear magneton

$$\mu_{\text{K}} = \mu_{\text{B}}\left(\frac{m}{m_{\text{p}}}\right) \approx 5.050\,7866(17) \times 10^{-24} \text{ erg Gauss}^{-1} \quad , \tag{2.19}$$

where m_{p} is the proton mass.

From Eqs. (2.7), (2.15), and (2.16) we also see that the energy associated with the magnetic field of strength H is

$$E = \mu_{\text{B}} H \quad . \tag{2.20}$$

Fine structure constant. From Eqs. (2.4) and (2.6) we obtain the ratio of the velocity of the first Bohr orbit to the velocity of light. This fine structure constant is

$$\alpha = \frac{v}{c} = \frac{2\pi e^2}{hc} = \frac{2\mu_{\text{B}}}{ea_0} = \frac{r_0}{\lambda_0} = \frac{\lambda_0}{a_0} \approx \frac{1}{137.036} \approx 7.3 \times 10^{-3} \quad , \tag{2.21}$$

where μ_B is the Bohr magnetron, a_0 is the radius of the first Bohr orbit, r_0 is the classical electron radius, and λ_0 is the Compton wavelength.

Sommerfeld (1916) showed that the total energy of an electron in an elliptical orbit is given by Eq. (2.8) unless relativity corrections are considered. In this case, the energy of level n is given by

$$E_n = \frac{-2\pi^2 me^4}{h^2}\frac{Z^2}{n^2}\left[1 + \frac{\alpha^2 Z^2}{n}\left(\frac{1}{k} - \frac{3}{4n}\right)\right] , \tag{2.22}$$

where α is the fine structure constant, the ratio of major to minor axes of the ellipse is n/k, and the integer k is called the azimuthal quantum number.

2.2 Einstein Probability Coefficients, Bound–Bound Photoprocesses

When considering spontaneous and radiation induced transitions between atomic energy levels, the probability per unit time, P_{mn}, that an atom will undergo a transition from a high state of energy, E_m, to a lower state of energy, E_n, is (Einstein, 1917)

$$P_{mn} = A_{mn} + B_{mn}U_\nu \quad , \tag{2.23}$$

where A_{mn} is the Einstein coefficient for spontaneous transition between the two states (spontaneous emission), and B_{mn} is the Einstein stimulated emission coefficient for a transition induced by radiation of energy density $U_\nu\, d\nu$ in the frequency range ν to $\nu + d\nu$. The probability per unit time for a radiation induced absorption is

$$P_{nm} = B_{nm}U_\nu \quad , \tag{2.24}$$

where B_{nm} is the Einstein coefficient for photoabsorption. From Eq. (1.64), $U_\nu = E_\nu^2/4\pi$ for plane wave radiation with electric field strength, E_ν, at frequency, ν. From Eq. (1.119), the energy density of black body radiation at the temperature, T, in the frequency range ν to $\nu + d\nu$ is (Planck, 1901)

$$U_\nu = \frac{8\pi h\nu^3}{c^3}\left[\exp\left(\frac{h\nu}{kT}\right) - 1\right]^{-1} = \frac{4\pi}{c}B_\nu(T) \quad , \tag{2.25}$$

where $B_\nu(T)$ is the brightness of the radiator at frequency, ν, and temperature, T. Because the number of downward transitions must equal the number of upperward transitions, and because in thermodynamic equilibrium each state has a population determined by the Boltzmann distribution, Eq. (3.126), it follows that

$$\frac{g_n}{g_m}\exp[-E_n/kT]B_{nm}U_\nu = \exp[-E_m/kT](A_{mn} + B_{mn}U_\nu) \quad . \tag{2.26}$$

Choosing $h\nu = E_m - E_n$ in Eq. (2.25), it follows from Eq. (2.26) that

$$A_{mn} = \frac{8\pi h\nu^3}{c^3}B_{mn} \quad , \tag{2.27}$$

and

$$g_m B_{mn} = g_n B_{nm} \quad , \tag{2.28}$$

where g_m is the statistical weight of the mth level. Some authors define P_{nm} as $B_{nm}B_\nu(T)$ where $B_\nu(T)$ is given by Eq. (2.25). In this case, Eq. (2.27) becomes $A_{mn} = 2h\nu^3 B_{mn}/c^2$. Although relations (2.27) and (2.28) were determined under the assumption of thermodynamic equilibrium, the Einstein coefficients must be properties of the atom only and therefore Eqs. (2.27) and (2.28) are true regardless of the nature of the radiation.

The A_{mn} is the probability per second that an atom with an electron in level m will spontaneously emit a photon of energy $h\nu_{mn} = E_m - E_n$, so the energy emitted per unit time is $h\nu_{mn}A_{mn}$ and the energy emitted per unit volume by the spontaneous bound-bound transition is:

$$dP/dv = N_m h\nu_{mn} A_{mn} \quad , \tag{2.29}$$

where N_m is the volume density of atoms in level m and $h\nu_{mn} = E_m - E_n$.

A bibliographic database on atomic transition probabilities, containing approximately 3,000 references, is maintained by J. R. Fuhr and H. R. Felrice of the National Institute of Standards and Technology (NIST) at its World Wide Web Site http://physics.nist.gov/under Physical Reference Data. Their Atomic Spectroscopic Database also includes atomic energy levels and wavelengths. The Chianti atomic database provides atomic data needed for calculating astrophysical emission line spectra at wavelengths greater than 50 Angstroms as a function of both density and temperature (Dere et al., 1997). It is available on the World Wide Web at http://wwwsolar.nrl.navy.mil/chianti.html and includes atomic energy levels, wavelengths, radiative data and electron excitation data for ions which are abundant in cosmic plasmas.

2.3 Einstein Probability Coefficient for Spontaneous Emission from an Electric Dipole

The electric dipole moment, $d(t)$, of an electron of charge, e, which is constrained to move harmonically in the x direction is

$$d(t) = ex(t) = ex_0 \cos(2\pi\nu_0 t) \quad , \tag{2.30}$$

where ν_0 is the frequency of the oscillation and x_0 is the peak displacement of the oscillation. Using the Larmor (1897) relation for the total energy radiated per unit time in all directions by a dipole (Eq. (1.114)), we obtain the average emitted power:

$$P = \frac{2}{3}\frac{\langle|\ddot{d}(t)|^2\rangle}{c^3} = \frac{64\pi^4\nu_0^4}{3c^3}\left(\frac{ex_0}{2}\right)^2 \quad , \tag{2.31}$$

where $\ddot{}$ denotes the second derivative with respect to time, $||$ denotes the absolute value, $\langle\rangle$ denotes a time average, and $(ex_0/2)$ is the mean dipole moment.

By comparing Eqs. (2.29) and (2.31), the spontaneous emission coefficient, A_{mn}, for electric dipole radiation is obtained.

$$A_{mn} = \frac{64\pi^4 \nu_{mn}^3}{3hc^3} |\mu_{mn}|^2 \approx 1.2 \times 10^{-2} \, \nu_{mn}^3 |\mu_{mn}|^2 \ \text{sec}^{-1} \ , \tag{2.32}$$

where the electric dipole matrix element, μ_{mn}, is often given by

$$\mu_{mn}^2 = S_{mn}/g_m \approx (e^2 x_0^2) \ , \tag{2.33}$$

where S_{mn} is called the strength of the electric dipole and g_m is the statistical weight of the mth level.

For an estimate of the value of the A_{mn} of an electric dipole transition, let

$$\nu_{mn} = cR_\infty \approx 3.3 \times 10^{15} \ \text{Hz} \tag{2.34}$$

be the radiation frequency of an electron in the first Bohr orbit, whose squared electric dipole matrix element is

$$|\mu_{mn}|^2 \approx e^2 a_0^2 \approx 6.46 \times 10^{-36} \ \text{cm}^2 \ \text{e.s.u.}^2 \ , \tag{2.35}$$

where a_0 is the radius of the first Bohr orbit. Substituting Eqs. (2.34) and (2.35) into Eq. (2.32) we obtain

$$A_{mn} \approx 10^9 \text{sec}^{-1} \ . \tag{2.36}$$

The dipole moments are sometimes given in Debye units, where one Debye $= 10^{-18}$ e.s.u. cm.

2.4 Relation of the Electric Dipole Emission Coefficient to the Classical Damping Constant and the Oscillator Strength

The kinetic energy, E, of a harmonic oscillator is given by

$$E = \frac{m[\dot{x}(t)]^2}{2} = 2\pi^2 m \nu_0^2 x_0^2 \cos^2(2\pi\nu_0 t) \ , \tag{2.37}$$

where $x(t)$ is given by Eq. (2.30). It follows from Eq. (2.31) for the time rate of change of energy, dE/dt, and from Eq. (2.37), that

$$\frac{dE}{dt} = -\gamma_{\text{cl}} E \ , \tag{2.38}$$

where the classical damping constant

$$\gamma_{\text{cl}} = \frac{8\pi^2 e^2 \nu_0^2}{3mc^3} \approx 2.47 \times 10^{-22} \nu_0^2 \ \text{sec}^{-1} \ .$$

Comparing Eq. (2.38) with Eq. (2.29), we obtain

$$A_{mn} = -3\gamma_{\text{cl}} f_{mn} \ , \tag{2.39}$$

where the oscillator strength, f_{mn}, is the effective number of electrons per atom.

The oscillator strength has the property

$$g_m f_{mn} = -g_n f_{nm} \quad , \tag{2.40}$$

where g_m is the statistical weight of the mth level. The Einstein probability coefficient A_{mn} for spontaneous bound–bound emission from level m to level n, with $n < m$, is:

$$A_{mn} = 8\pi^2 e^2 \nu_{mn}^2 (g_n/g_m) f_{nm}/mc^3 = 3\gamma_{cl}(g_n/g_m) f_{nm} \quad , \tag{2.41}$$

the Einstein probability coefficient for bound–bound photoabsorption from level n to level m is:

$$B_{nm} = \pi e^2 f_{nm}/(mh\nu_{nm}) = (g_m/g_n) B_{mn} \quad ,$$

where B_{mn} is the Einstein probability coefficient for stimulated bound–bound emission from level m to level n, and

$$g_n f_{nm} = [mc^3/(8\pi^2 e^2 \nu_{nm}^2)] g_n A_{mn} = 1.499 \lambda_{nm}^2 g_m A_{mn} \quad ,$$

for λ_{mn} in cm. For a hydrogen atom $g_n = 2n^2$, and the Kramer's formula for the oscillator strength, f_{nm}, is given by Eq. (2.118) in Section 2.12. An asymptotic expression for the dipole oscillator strengths of helium at large n is provided by Khandelwal, Khan and Wilson (1989), and oscillator strengths of ionized silicon (Si II) resonance lines are given by Luo, Pradhan and Shull (1988). A bibliographic database on atomic transition probabilities, containing approximately 3,000 references is maintained by J. R. Fuhr and H. R. Felrice of the National Institute of Standards and Technology at its World Wide Web site http://physics.nist.gov/under Physical Reference Data.

When m is larger than n, as we have assumed in the previous discussion, f_{mn} and f_{nm} are called, respectively, the emission and absorption oscillator strengths. The oscillator strength obeys the Reiche–Thomas–Kuhn sum rule (Thomas, 1925; Kuhn, 1925)

$$\sum_{n<m} f_{mn} + \sum_{n>m} f_{mn} = \sum_{n<m} f_{mn} - \sum_{m<n} \frac{g_n}{g_m} f_{nm} = Z \quad , \tag{2.42}$$

for an atom with Z optical electrons. Typical values of the oscillator strength at optical frequencies are on the order of unity. Shu (1991) provides a quantum mechanical proof that the sum of f-values for a single electron, with a given initial state, over all possible final states (both higher and lower) exactly equals 1.

2.5 Probability Coefficient for Spontaneous Emission from a Magnetic Dipole

Let a magnetic dipole have the oscillatory magnetic dipole moment, $m(t)$, given by

$$m(t) = m_0 \cos(2\pi\nu_0 t), \tag{2.43}$$

where ν_0 is the harmonic frequency. As the time averaged energy radiated per unit time is given by Eq. (2.31) with $d(t)$ replaced by $m(t)$, it follows from

Eq. (2.32) that the spontaneous emission coefficient, A_{mn}, for magnetic dipole radiation is

$$A_{mn} = \frac{64\pi^4 \nu_{mn}^3}{3hc^3} |\mu_{mn}|^2 \approx 1.2 \times 10^{-2} \nu_{mn}^3 |\mu_{mn}|^2 \ \mathrm{sec}^{-1} \ , \tag{2.44}$$

where the magnetic dipole matrix element, μ_{mn}, is given by

$$\mu_{mn}^2 = S_{mn}/g_m \ , \tag{2.45}$$

where S_{mn} is the strength of the magnetic dipole and g_m is the statistical weight of the level m.

For an estimate of the value of A_{mn} of a magnetic dipole transition, let

$$\nu_{mn} = cR_\infty \approx 3.3 \times 10^{15} \ \mathrm{Hz} \tag{2.46}$$

be the radiation frequency of an electron in the first Bohr orbit, whose magnetic moment is

$$|\mu_{mn}|^2 = \mu_B^2 = \left(\frac{eh}{4\pi mc}\right)^2 \approx 8.6 \times 10^{-41} \ \mathrm{erg}^2 \ \mathrm{Gauss}^{-2} \ , \tag{2.47}$$

where μ_B is the Bohr magneton. Substituting Eqs. (2.46) and (2.47) into Eq. (2.44), we obtain

$$A_{mn} \approx 10^4 \ \mathrm{sec}^{-1} \ . \tag{2.48}$$

2.6 Probability Coefficient for Spontaneous Emission from an Electric Quadrupole

The quadrupole moment of a set of charges of magnitude, e, is (Landau and Lifshitz, 1962)

$$D_{\alpha\beta} = \sum e(3x_\alpha x_\beta - \delta_{\alpha\beta} r^2) \ , \tag{2.49}$$

where x_α and x_β are the x coordinates of the radius vector, r, between two charges α and β, $\delta_{\alpha\beta}$ is the Kronecker delta, and the summation is performed over all the charges e. The electric, E, and magnetic, H, field strengths of the quadrupole radiation at a far distance, R, from the set of charges are given by

$$E = H = \frac{\dddot{\vec{D}}}{6c^3 R} \sin\theta \ , \tag{2.50}$$

where $\dddot{}$ denotes the third derivative with respect to time, and θ is the angle between the direction of observation and the vector $\vec{D}$ with the components $D_\alpha = D_{\alpha\beta} n_\beta$ where $\vec{n}$ is the unit vector, $\vec{n} = \vec{R}/R$. The total energy emitted per unit time in all directions, P, is obtained by integrating the Poynting flux $cE^2/4\pi$ over all angles and noting that $D_{\alpha\beta}$ is a symmetric tensor with five independent components.

$$P = \frac{\dddot{D}_{\alpha\beta}^2}{180c^5} \ . \tag{2.51}$$

If we assume that $D_{\alpha\beta}$ is oscillatory in time and is given by

$$D_{\alpha\beta} = 3ex_0^2\cos(2\pi\nu_0 t) \ , \tag{2.52}$$

where ν_0 is the oscillatory frequency, Eq. (2.51) gives a time averaged power

$$P = \frac{32\pi^6\nu_0^6}{5c^5}\left(\frac{ex_0^2}{2}\right)^2 \ , \tag{2.53}$$

where $ex_0^2/2$ is the average electric quadrupole moment.

By comparing Eq. (2.29) with Eq. (2.53), the spontaneous emission coefficient, A_{mn}, for electric quadrupole radiation is obtained

$$A_{mn} = \frac{32\pi^6\nu_{mn}^5}{5hc^5}\left(\frac{S_{mn}}{g_m}\right) \approx 3.8\times10^{-23}\nu_{mn}^5\left(\frac{S_{mn}}{g_m}\right)\ \mathrm{sec}^{-1} \ , \tag{2.54}$$

where S_{mn} is the strength of the electric quadrupole, and g_m is the statistical weight of the level m.

For an estimate of the value of A_{mn} of an electric quadrupole transition, let

$$\nu_{mn} = cR_\infty \approx 10^{15}\ \mathrm{Hz} \ , \tag{2.55}$$

be the radiation frequency of an electron in the first Bohr orbit whose electric quadrupole moment is

$$\frac{S_{mn}}{g_m} = a_0^4e^2 \approx 1.8\times10^{-52}\ \mathrm{cm}^4\ \mathrm{e.s.u.}^2 \ , \tag{2.56}$$

where a_0 is the radius of the first Bohr orbit. Substituting Eqs. (2.55) and (2.56) into Eq. (2.54), we obtain

$$A_{mn} \approx 10\ \mathrm{sec}^{-1} \ . \tag{2.57}$$

2.7 Radiation Transfer

The equation of radiation transfer for the intensity, $I(\nu,x)$, of radiation at the frequency, ν, and at the point, x, measured along the line of sight is (Kirchhoff, 1860; Woolley, 1947; Smerd and Westfold, 1949)

$$\frac{d}{d\tau_\nu}\left[\frac{I(\nu,x)}{n_\nu^2}\right] = S_\nu - \frac{I(\nu,x)}{n_\nu^2} \ , \tag{2.58}$$

where the source function, S_ν, is given by

$$S_\nu = \frac{\varepsilon_\nu}{\alpha_\nu n_\nu^2} \ , \tag{2.59}$$

and the optical depth, τ_ν, is defined by

$$d\tau_\nu = \alpha_\nu(x)dx \ . \tag{2.60}$$

The volume emissivity, ε_ν, is the power emitted per unit volume per unit frequency interval per unit solid angle, the absorption coefficient per unit length is α_ν, and n_ν is the index of refraction. For a plasma, n_ν will be unity as long as the frequency $\nu \gg \nu_\mathrm{p} = 8.9\times10^3 N_\mathrm{e}^{1/2}$ Hz, where ν_p is the plasma frequency and N_e is the free electron density.

For an isotropic and homogeneous plasma of thickness, L, the optical depth may be written

$$\tau_\nu(L) = \int_0^L \alpha_\nu(x)dx = \alpha_\nu L = N\sigma_\nu L \ , \tag{2.61}$$

where N is the volume density of absorbing atoms and σ_ν is called the absorption cross section.

Under conditions of local thermodynamic equilibrium (LTE) at temperature, T, the source function, S_ν, is given by (Planck, 1901; Milne, 1930)

$$\begin{aligned} S_\nu = \frac{\varepsilon_\nu}{\alpha_\nu n_\nu^2} = B_\nu(T) &= \frac{2h\nu^3}{c^2}\left[\exp\left(\frac{h\nu}{kT}\right) - 1\right]^{-1} \\ &\approx \frac{2h\nu^3}{c^2}\exp\left[-\left(\frac{h\nu}{kT}\right)\right] \quad \text{for } \nu \gg 10^{10}T \\ &\approx \frac{2\nu^2 kT}{c^2} \quad \text{for } \nu \ll 10^{10}T \ , \end{aligned} \tag{2.62}$$

which is the expression for the vacuum brightness, $B_\nu(T)$, of a black body radiator at temperature, T, and frequency, ν.

The general solution to Eq. (2.58) for the intensity at some point A on the line of sight in terms of some other point B on the line of sight is

$$\frac{I(\nu, A)}{n_\nu^2(A)} = \frac{I(\nu, B)}{n_\nu^2(B)}\exp[\tau_\nu(B) - \tau_\nu(A)] - \int_{\tau_\nu(A)}^{\tau_\nu(B)} S_\nu(t)\exp[t - \tau_\nu(A)]\, dt \ , \tag{2.63}$$

where $\tau_\nu(A)$ is the optical depth from $\tau = 0$ to the point A and similarly for $\tau_\nu(B)$. It follows from Eqs. (2.62) and (2.63) that the emergent intensity from an isothermal radiator of thickness L is

$$\begin{aligned} I(\nu, 0) &= n_\nu^2(0)B_\nu(T)\{1 - \exp[-\tau_\nu(L)]\} \\ &\approx n_\nu^2(0)B_\nu(T) \quad \text{for } \tau_\nu(L) \gtrsim 1 \text{ (optically thick)} \\ &\approx n_\nu^2(0)B_\nu(T)\tau_\nu(L) \quad \text{for } \tau_\nu(L) < 1 \text{ (optically thin)} \ . \end{aligned} \tag{2.64}$$

If there is a bright background source, of specific intensity $I(\nu, 0)$, in addition to an emitting and absorbing medium, or cloud, in local thermodynamic equilibrium at uniform temperature T_c, then the observed intensity $I(\nu)$ will be [see Eq. (1.145)]:

$$I(\nu) = I(\nu, 0)\exp[-\tau_\nu(L)] + B_\nu(T_c)\{1 - \exp[-\tau_\nu(L)]\} \ , \tag{2.65}$$

where $B_\nu(T_c) = 2kT_c\nu^2/c^2$ at radio frequencies and the Rayleigh-Jeans approximation can be used. The observed brightness and brightness temperatures are given by Eq. (1.145). Both equations assume no significant scattering; the formal solution of the equation of radiative transfer including scattering in stellar atmospheres is given by Mihalas (1970) and Shu (1991).

2.8 Resonance Absorption of Line Radiation

The quantum mechanical absorption coefficient per unit length, α_ν, for the self absorption of the line radiation resulting from a transition from a high energy level m to a lower level n is (Einstein, 1917)

$$\alpha_\nu = \frac{c^2 N_n g_m}{8\pi\nu^2 g_n n_\nu^2} A_{mn} \left[1 - \frac{g_n N_m}{g_m N_n}\right] \varphi_{mn}(\nu) \ , \tag{2.66}$$

where N_n is the population of level n, A_{mn} is the Einstein coefficient for spontaneous transition from level m to level n, the index of refraction of the medium, n_ν, usually does not differ appreciably from unity, the statistical weight $g_n = 2J + 1$ where J is the total angular quantum number of the level, and $\varphi_{mn}(\nu)$ is the spectral intensity distribution over the line. Detailed descriptions of the $\varphi_{mn}(\nu)$ for various processes are given in Sects. 2.18 to 2.22.

The absorption coefficient must be normalized to satisfy the criteria (Ladenburg, 1921; Kramers, 1924)

$$\int_0^{+\infty} \alpha_\nu d\nu = \frac{c^2 N_n g_m}{8\pi\nu^2 g_n n_\nu^2} A_{mn} \left[1 - \frac{g_n N_m}{g_m N_n}\right] \ . \tag{2.67}$$

It follows from Eqs. (2.40) and (2.41) that Eq. (2.66) may also be expressed in the form

$$\int_0^{\infty} \alpha_\nu d\nu = \frac{\pi e^2}{mc} f_{nm} N_n \left[1 - \frac{g_n N_m}{g_m N_n}\right] \ ,$$

where f_{nm} is the absorption oscillator strength for the $m - n$ transition, and the number of oscillators is $N_n f_{nm}$. These criteria are the same as the criterion that

$$\int_0^{\infty} \varphi_{mn}(\nu) d\nu = 1 \ . \tag{2.68}$$

For a Gaussian line profile of full width to half maximum, $\Delta\nu_{\mathrm{D}}$, this criterion means that $\varphi_{mn}(\nu_{mn}) = (\ln 2/\pi)^{1/2}(2/\Delta\nu_{\mathrm{D}})$. For a Lorentz line profile of full width to half maximum, $\Delta\nu_{\mathrm{L}}$, the $\varphi_{mn}(\nu_{mn}) = (2/\pi)\Delta\nu_{\mathrm{L}}^{-1}$. In the following we will pick $\varphi_{mn}(\nu_{mn}) = \Delta\nu_{\mathrm{L}}^{-1}$, noting there must be a small correction depending on the exact line profile.

Under conditions of local thermodynamic equilibrium (LTE) at temperature, T, the populations of levels m and n are given by Boltzmann's equation (3.126). This equation specifies that $\exp[-h\nu_{mn}/(kT)] = g_n N_m/(g_m N_n)$, and that $N_n = g_n N_{\mathrm{tot}} \exp[-\chi_n/(kT)]/U$, where N_{tot} is the total density of particles, U is the partition function (cf. Chap. 3), and χ_n is the excitation energy of level n from the ground state. It follows that at the line frequency, ν_{mn}, Eq. (2.66) becomes, in LTE conditions,

$$\alpha_{\nu_{mn}}(\mathrm{LTE}) = \frac{c^2}{8\pi\nu_{mn}^2} \frac{N_m}{\Delta\nu_{\mathrm{L}}} \left[\exp\left(\frac{h\nu_{mn}}{kT}\right) - 1\right] A_{mn} \ . \tag{2.69}$$

At radio frequencies the factor $[1 - \exp(-h\nu_{mn}/kT)]$ becomes $h\nu_{mn}/kT$, whereas it is approximately unity at optical frequencies. Thus, for example, Eq. (2.67) becomes

$$\int_0^\infty \alpha_\nu d\nu \approx \frac{\pi e^2}{mc} f_{nm} N_n \approx 2.15 \times 10^{-2} f_{nm} N_n \, \mathrm{cm^2\,s^{-1}} \quad \text{for } h\nu_{mn} \gg kT \ . \tag{2.70}$$

This result shows that the integrated absorption coefficient is a constant if N_n is constant, irrespective of the physical process responsible for the line broadening. Using Eqs. (2.32), (2.33), and (2.40), the self absorption coefficient, $\alpha_{\nu_{mn}}$(LTE), at the line frequency, ν_{mn}, under conditions of local thermodynamic equilibrium may also be written in the forms

$$\begin{aligned} \alpha_{\nu_{mn}}(\mathrm{LTE}) &= \frac{8\pi^3 \nu_{mn}}{3hc} |\mu_{mn}|^2 \frac{N_m}{\Delta\nu_\mathrm{L}} \left[\exp\left(\frac{h\nu_{mn}}{kT}\right) - 1\right] \\ &= \frac{8\pi^3 \nu_{mn}}{3hc} \left|\frac{S_{mn}}{g_m}\right| \frac{N_m}{\Delta\nu_\mathrm{L}} \left[\exp\left(\frac{h\nu_{mn}}{kT}\right) - 1\right] \\ &= \frac{\pi e^2}{mc} f_{nm} \frac{N_n}{\Delta\nu_\mathrm{L}} \left[1 - \exp\left(\frac{-h\nu_{mn}}{kT}\right)\right] \ , \end{aligned} \tag{2.71}$$

where μ_{mn} is the dipole matrix element, S_{mn} is the dipole strength, g_m is the statistical weight of the mth level, and we recall that the population density, N_m, of the mth level may also be given by $N_m = g_m N_\mathrm{tot} \exp[-\chi_m/(kT)]/U$, where N_tot is the total density of particles, χ_m is the excitation energy of the mth level above the ground state, and U is the partition function.

2.9 Line Intensities Under Conditions of Local Thermodynamic Equilibrium

Under conditions of local thermodynamic equilibrium, the optical depth at the center of a line coming from a homogeneous and isotropic plasma of thickness, L, is

$$\tau_{\nu_{mn}}(L) = L\alpha_{\nu_{mn}}(\mathrm{LTE}) \ , \tag{2.72}$$

where only self absorption of the line has been considered, and $\alpha_{\nu_{mn}}$(LTE) is given by Eqs. (2.69) or (2.71). Using the solution of the radiation transfer equation (2.64), together with Eqs. (2.71) and (2.72), the intensity $I(\nu_{mn}, 0)$ at the center of the line emerging from an optically thin plasma is obtained.

$$I(\nu_{mn}, 0) = B_{\nu_{mn}}(T) \frac{\pi e^2 f_{nm}}{mc\Delta\nu_\mathrm{L}} \left[1 - \exp\left(\frac{-h\nu_{mn}}{kT}\right)\right] L N_n \ , \tag{2.73}$$

where the brightness of a black body at temperature, T, is $B_{\nu_{mn}}(T)$ and is given by Eqs. (2.62). Once the oscillator strength, f_{nm}, is calculated, and the intensity, $I(\nu_{mn}, 0)$, and the line width, $\Delta\nu_\mathrm{L}$, are measured, the column density, LN_n, may be obtained using Eq. (2.73).

In actual practice, in addition to the line radiation, an antenna also sees continuum radiation from the cloud itself, c, from discrete sources, s, behind the cloud, and from the general background, B, of the Galaxy and the isotropic 3 °K radiation. If the respective brightness temperatures are T_c, T_s, and T_B, and the solid angles of the two discrete sources are Ω_s and Ω_c, an antenna with efficiency, η_A, and beam solid angle, Ω_A, will observe the antenna temperature

$$T_A = \eta_A T_{0c} + \eta_A T_{0e}[1 - \exp(-\tau_{\nu_{mn}})] + \eta_A[T_{0s} + T_B]\exp(-\tau_{\nu_{mn}}) \ , \tag{2.74}$$

where

$$T_{0e} = T_{exc} \quad \text{or} \quad T_{exc}\Omega_c/\Omega_A \quad \text{according as } \Omega_c > \text{ or } < \Omega_A$$
$$T_{0c} = T_c \quad \text{or} \quad T_c\Omega_c/\Omega_A \quad \text{according as } \Omega_c > \text{ or } < \Omega_A$$

and

$$T_{0s} = T_s \quad \text{or} \quad T_s\Omega_s/\Omega_A \quad \text{according as } \Omega_s > \text{ or } < \Omega_A \ ,$$

where T_{exc} is the excitation or spin temperature of the line radiation. When the receiver is frequency switched between ν_{mn} and another frequency different from ν_{mn}, the difference in the observed antenna temperature is

$$\begin{aligned} \Delta T_A &= \eta_A[T_{0e} - (T_{0s} + T_B)][1 - \exp(-\tau_{\nu_{mn}})] \\ &\approx \eta_A[T_{0e} - (T_{0s} + T_B)]\tau_{\nu_{mn}} \quad \text{for } \tau_{\nu_{mn}} \ll 1 \ . \end{aligned} \tag{2.75}$$

When $T_{0e} \gg T_{0s} + T_B$, we see the line in emission and when $T_{0e} \ll T_{0s} + T_B$ we see it in absorption.

For the special case of an optically thin gas ($\tau_{mn} \ll 1$) in local thermodynamic equilibrium, and seen in emission at radio frequencies ($h\nu_{mn} \ll kT$), Eqs. (2.71), (2.72), and (2.75) give

$$\Delta T_A \Delta\nu_L = \eta_A \frac{\pi e^2 h}{mc} \frac{\nu_{mn}}{k} \frac{LN_{tot}}{U} g_n f_{nm} \ , \tag{2.76}$$

where we assume $\Omega_c > \Omega_A$. In this case, the area of the line profile leads to a direct measurement of LN_{tot}, the number of atoms in a cylinder of unit cross sectional area and length L, which are radiating at the frequency ν_{mn}.

For the special case of an optically thin gas ($\tau_{mn} \ll 1$) in local thermodynamic equilibrium, and seen in absorption at optical frequencies ($h\nu_{mn} \gg kT$), Eqs. (2.71), (2.72), and (2.75) give

$$\frac{\Delta T_A \Delta\nu_L}{T_0} = -\eta_A \frac{\pi e^2}{mc} \frac{LN_{tot}}{U} g_n f_{nm} \exp\left(\frac{-\chi_m}{kT}\right) \ , \tag{2.77}$$

where T_0 is the temperature of the background continuum, and we assume $\Omega_s > \Omega_A$. In this case, the normalized area of the line profile (the equivalent width) leads to a measurement of LN_{tot}. Details depend on the exact form of the line broadening function. Nevertheless, comparisons of the equivalent widths of lines from the same object do lead to the relative densities of the elements in the object. Calculations of line intensities in stellar atmospheres must include the effects of both scattering and absorption, and these effects are discussed in detail in Sect. 2.16. Absorption of line and continuum radiation by photoionization, free-free transitions, and scattering is described by Eqs. (2.61) and (2.72) together with the absorption and scattering cross sections given in Sects. 1.30, 1.31, and 1.37.

2.10 Line Intensities When Local Thermodynamic Equilibrium Does Not Apply

Let b_n designate the ratio of the actual population of level n to the population in conditions of local thermodynamic equilibrium (obtainable from the Saha-Boltzmann equation (3.129)). The absorption coefficient at the frequency ν_{mn} of the $m-n$ transition is (Goldberg, 1966)

$$\alpha_{\nu_{mn}} = b_n \beta_{nm} \alpha_{\nu_{mn}}(\mathrm{LTE}), \tag{2.78}$$

where

$$\beta_{nm} = \frac{1-(b_m/b_n)\exp(-h\nu_{mn}/kT)}{1-\exp(-h\nu_{mn}/kT)} ,$$

and the absorption coefficient, $\alpha_{\nu_{mn}}(\mathrm{LTE})$, in conditions of local thermodynamic equilibrium is given by Eqs. (2.69) or (2.71). The emission coefficient is given by

$$\varepsilon_{\nu_{mn}} = b_m \alpha_{\nu_{mn}}(\mathrm{LTE}) B_{\nu_{mn}}(T) , \tag{2.79}$$

where the black body brightness, $B_{\nu_{mn}}(T)$, at temperature, T, is given by

$$B_{\nu_{mn}}(T) = \frac{2h\nu_{mn}^3}{c^2} \frac{1}{[\exp(h\nu_{mn}/kT)-1]} .$$

Tables of departure coefficients b_n for hydrogenic ions are given by Salem and Brocklehurst (1979).

Using Eq. (2.71) for $\alpha_{\nu_{mn}}(\mathrm{LTE})$ in Eq. (2.78), we obtain

$$\alpha_{\nu_{mn}} = \frac{\pi N_n e^2 f_{nm} b_n}{mc\Delta\nu_\mathrm{L}} \left[1-\left(\frac{b_m}{b_n}\right)\exp\left(\frac{-h\nu_{nm}}{kT}\right)\right] , \tag{2.80}$$

where f_{nm} is the oscillator strength of the $n-m$ transition, $\Delta\nu_\mathrm{L}$ is the width of the line, and N_n is the population of the nth level. This equation reduces to

$$\alpha_{\nu_{mn}} \approx \alpha_{\nu_{mn}}(\mathrm{LTE}) \quad \text{for } \nu \gg 10^{10} T$$

and

$$\alpha_{\nu_{mn}} \approx N_n \frac{\pi e^2 f_{nm}}{mc\Delta\nu_\mathrm{L}} \frac{h\nu_{mn}}{kT} b_m \left\{1-\left(\frac{kT}{h\nu_{mn}}\right)\left(\frac{b_m-b_n}{b_m}\right)\right\} \quad \text{for } \nu \ll 10^{10} T , \tag{2.81}$$

where

$$\frac{b_m-b_n}{b_n} \approx (m-n)\frac{d\ln b_n}{dn} .$$

The absorption coefficient is very sensitive to small changes in level populations at low frequencies. If $b_m - b_n$ exceeds $b_m h\nu_{mn}/(kT)$, the absorption coefficient becomes negative providing maser amplification of the line.

The population, N_n, of level n may be determined from the equation of statistical equilibrium

$$N_n \sum_{m=1}^{\infty} P_{nm} = N_c P_{cn} + \sum_{m=1}^{\infty} N_m P_{mn} \ , \tag{2.82}$$

where P_{mn} denotes the probability per unit time of a transition from level m to level n, and the subscript c denotes the continuum. A few of the possible processes which determine level populations are given below together with their probabilities.

a) *Spontaneous bound–bound transition between levels.* The probability per unit time of a spontaneous transition from a high level m to a lower level n is A_{mn}, the Einstein coefficient for spontaneous emission, which is given in Sect. 2.12 for hydrogen-like atoms.

b) *Radiation induced bound–bound transitions between levels.* The probability per unit time for a radiation induced transition between a high level m and a lower level n is

$$B_{mn} U_\nu \quad \text{or} \quad B_{nm} U_\nu \ , \tag{2.83}$$

where the B_{mn} and B_{nm} are the Einstein stimulated emission coefficients for a transition induced by radiation of energy density $U_\nu d\nu$ in the frequency range ν to $\nu + d\nu$. The interrelationships between A_{mn}, B_{mn}, and B_{nm} are given by Eqs. (2.27) and (2.28).

c) *Radiation induced ionizations.* It follows from Eq. (1.230) for the absorption cross section, σ_ν, for photoionization by radiation of frequency, ν, that the probability per unit time of a radiation induced ionization is

$$P_{nc} = \int_0^\infty \frac{c U_\nu \sigma_\nu d\nu}{h\nu} \approx 1.3 \times 10^{66} Z^4 \int_0^\infty \frac{U_\nu d\nu}{\nu^4 n^5} \ \sec^{-1} \tag{2.84}$$

for a hydrogen-like atom. Here the photon energy is $h\nu$, U_ν is the energy density of the radiation, and n is the energy level of the unionized atom.

An overview of the photoionization equilibrium and thermal equilibrium of gaseous nebulae is given by Osterbrock (1989). He applies the theory to planetary nebulae, active galactic nuclei, and H II regions. Davidson and Netzer (1979) also review photoionization equilibrium and apply it to the emission lines of quasars, while Osterbrock and Mathews (1986) review the emission-line regions of active galaxies and quasars. Photoabsorption cross sections for positive atomic ions with $Z \leq 30$ are presented by Reilman and Manson (1979) for the energy range 5 eV $\leq h\nu \leq$ 5 keV. Brown (1971) gives cross sections for the photoionization of helium in the soft X-ray region, and Henry (1970) provides photoionization cross-sections for atoms and ions of carbon, nitrogen, oxygen and neon. The ground states and ionization energies of the elements are provided in the Atomic Spectroscopic Database of the National Institute of Standards and Technology (NIST) at their World Wide Web site http://physics.nist.gov/under Physical Reference Data.

d) *Radiative recombination of ions with electrons*. The probability per unit time of a radiative recombination is

$$P_{cn} = N_{\mathrm{e}} \int_0^\infty \sigma_{\mathrm{r}}(v, n) \left[1 + \frac{c^3 U_\nu}{8\pi h \nu^3}\right] v f(v) dv \ , \tag{2.85}$$

where N_{e} is the free electron density, U_ν is the radiation energy density at frequency, ν, the distribution of velocity, v, is $f(v)$, and $\sigma_{\mathrm{r}}(v, n)$ is the cross section for the recombination of a free electron of velocity, v, with an ion to form an atom with level, n. Using the Milne (1921) relation,

$$\sigma_{\mathrm{r}}(v, n) = \frac{h^2 g(n) \nu^2}{m^2 c^2 g(1) v^2} \sigma_\nu \ , \tag{2.86}$$

where $g(n)$ is the statistical weight of level, n, the statistical weight of the ground state of the ion is $g(1)$, and σ_ν is the cross section for photoionization from level n and is given by Eq. (1.230) for hydrogen-like atoms, for which $g(n) = 2n^2$.

Photoionization equilibria that include radiative recombination have been reviewed by Osterbrock (1974, 1989); such equilibrium calculations are given by Landini and Monsignori Fossi (1972), Jacobs et al. (1977, 1979), and Shull and Van Steenberg (1982) for astrophysically abundant elements. Recombination spectra and line intensities of hydrogenic ions are calculated by Pengelly (1964) and Hummer and Storey (1987). Recombination coefficients for neutral helium atoms are provided by Brown and Mathews (1970). Radiative recombination of complex ions is discussed by Gould (1978).

e) *Collision induced transitions between levels*. The probability per unit time of a collision induced transition between levels m and n is

$$P_{mn} = N \gamma_{mn} = N \int_0^\infty \sigma_{mn}(v) v f(v) dv \ , \tag{2.87}$$

where N is the density of colliding particles, γ_{mn} is called the rate coefficient for collisional excitation, $\sigma_{mn}(v)$ is the collisional cross section for collisions with a particle of velocity, v, and $f(v)$ is the velocity distribution. For example, for a Maxwellian distribution at temperature, T, we have (Eq. (3.114))

$$f(v) dv = \left(\frac{2}{\pi}\right)^{1/2} v^2 \left(\frac{m_{\mathrm{r}}}{kT}\right)^{3/2} \exp\left[-\frac{m_{\mathrm{r}} v^2}{2kT}\right] dv \ , \tag{2.88}$$

where the reduced mass, $m_{\mathrm{r}} = m_{\mathrm{a}} m_{\mathrm{p}} / (m_{\mathrm{a}} + m_{\mathrm{p}})$ and m_{a} and m_{p} are, respectively, the masses of the atom and the colliding particle. The equation of detailed balancing for collisional excitation is

$$g_m \gamma_{mn} = g_n \gamma_{nm} \exp\left[\frac{-h\nu_{mn}}{kT}\right] \ , \tag{2.89}$$

where g_m is the statistical weight of the mth level, and γ_{mn} is the rate coefficient for collisional deexcitation.

The collision cross section, $\sigma_{mn}(v)$, is often expressed in terms of the collision strength, $\Omega(m,n)$, by the expression (Hebb and Menzel, 1940; Seaton, 1958; Osterbrock, 1974, 1979)

$$\sigma_{mn}(v) = \frac{\pi}{g_m}\left(\frac{h}{2\pi m_e v}\right)^2 \Omega(m,n) \ , \tag{2.90}$$

where $\Omega(m,n)$ is on the order of unity for most collisions, and v is the velocity of the particle before collision. Another estimate of the collision cross section is, of course, the geometric cross section, πa^2, where a is the atomic radius. For electric dipole moment transitions of oscillator strength f_{mn}, the collision strength $\Omega(m,n)$ is given by Seaton (1962)

$$\Omega(m,n) \approx \frac{1.6\pi}{3^{1/2}} \frac{g_n f_{nm} E_1}{h\nu_{mn}} \ , \tag{2.91}$$

where $h\nu_{mn}$ is the energy of the $m-n$ transition, and E_1 is the binding energy of the neutral hydrogen atom.

An approximate expression for P_{nm} of hydrogen-like atoms in the case of a Maxwellian velocity distribution of free electrons is therefore (Seaton, 1962)

$$P_{nm} \approx 4\pi a_0^2 \left(\frac{8kT}{m}\right)^{1/2} \left(\frac{E_1}{kT}\right)^2 \left[\frac{\exp(-\chi_{nm})}{\chi_{nm}} - 1\right] 3 f_{nm} N_e \ , \tag{2.92}$$

where $\pi a_0^2 = 8.797 \times 10^{-17}\mathrm{cm}^2$, $\chi_{nm} = h\nu_{mn}/(kT)$, and P_{nm} and P_{mn} are related by the equation

$$n^2 \exp(\chi_n) P_{nm} = m^2 \exp(\chi_m) P_{mn} \ . \tag{2.93}$$

Numerical values for the electron-hydrogen collisional cross sections were given by Callaway (1985) and Callaway, Unnikrishnan and Oza (1987).

When the transitions have no electric dipole moment, $\Omega(m,n)$ is independent of velocity, and Eqs. (2.87) and (2.90) lead to

$$P_{mn} \approx \frac{N_e h^2 \Omega(m,n)}{g_m (2\pi m_e)^{3/2} (kT)^{1/2}} \approx 8.63 \times 10^{-6} T^{-1/2} N_e \frac{\Omega(m,n)}{g_m} \mathrm{sec}^{-1} \tag{2.94}$$

for collisions with free electrons of density, N_e, whose velocity distribution is Maxwellian. This rate is appropriate for collisional deexcitation of positive ions, and the appropriate collisional excitation rate is given by Eqs. (2.93) and (2.94). Collision strengths for the forbidden transitions of cosmically abundant ions are tabulated by Mendoza (1983) and Osterbrock (1989).

f) *Collisional ionization by thermal electrons and the inverse process of three body recombination.*

The semi-empirical formula for the probability per unit time of a collisional induced ionization of a hydrogen-like atom is (Seaton, 1960; Jeffries, 1968)

$$P_{nc} \approx 7.8 \times 10^{-11} T_e^{1/2} n^3 \exp(-\chi_n) N_e \,\mathrm{sec}^{-1} \ , \tag{2.95}$$

where n is the principal quantum number of the energy level, χ_n is the ionization potential of the nth level in units of kT_e, and N_e is the free electron density. Here the principle of detailed balance gives

$$P_{cn} = \frac{N_e h^3}{(2\pi m k T_e)^{3/2}} n^2 \exp(\chi_n) P_{nc} \ , \tag{2.96}$$

where the statistical weight of the nth level is $g_n = 2n^2$.

Atomic collision processes in gaseous nebulae and the electron impact of positive ions have been respectively reviewed by Seaton (1968) and Seaton (1975). Electron-impact excitation cross sections for complex ions are given by Sampson (1974). Y.-K. Kim and M. E. Rudd provide a database of electron-impact ionization cross sections of atoms and molecules at the National Institute of Standards and Technology (NIST) World Wide Web site http://physics.nist.gov/under Physical Reference Data. Shull and Van Steenberg (1982) discuss the ionization equilibria of astrophysically abundant elements, including rates of collisional ionization, radiative recombination and dielectronic recombination; Landini and Monsignori Fossi (1972) include these processes as well as autoionization. The Chianti database includes atomic energy levels, wavelengths, radiative data and electron excitation data for ions which are abundant in cosmic plasmas (Dere et al., 1997); it is on the World Wide Web at http://wwwsolar.nrl.navy.mil/chianti.html.

g) *Dielectronic recombination of an ion with an electron to give a doubly excited atom followed by a radiative transition to a singly excited state.* A complex ion with one or more electrons may recombine with an electron to form an atom in level i, nl by way of the intermediate doubly excited level j, nl. In this case the term

$$\sum_j \sum_l N_{j,nl} b(j, nl) A_{ji} \tag{2.97}$$

must be added to the right-hand side of the equation of statistical equilibrium (Eq. (2.82)). Here the factor $b(j, nl)$ by which the population of j, nl differs from its value in thermodynamic equilibrium, $N_{j,nl}$, is given by (Bates and Dalgarno, 1962; Davies and Seaton, 1969)

$$b(j, nl) = \frac{A_a}{A_a + A_r} \ , \tag{2.98}$$

where A_a is the probability coefficient for autoionization from level j, nl and $A_r = A_{ji}$ is the radiative probability coefficient for the transition $j, nl \to i, nl$. The equilibrium population, $N_{j,nl}$, is given by the Saha–Boltzmann equation (Eq. (3.127))

$$\begin{aligned} N_{j,nl} &= N_i N_e \frac{h^3}{(2\pi m k T)^{3/2}} \frac{g_{j,nl}}{2 g_{i,nl}} \exp\left(\frac{-h\nu_{ij}}{kT}\right) \\ &\approx 2.1 \times 10^{-16} N_i N_e T^{-3/2} \frac{g_{j,nl}}{g_{i,nl}} \exp\left(\frac{-h\nu_{ij}}{kT}\right) \mathrm{cm}^{-3} \ , \end{aligned} \tag{2.99}$$

where N_i and N_e are, respectively, the ion and electron densities, $g_{i,nl}$ is the statistical weight of the level i, nl, and $h\nu_{ij}$ is the energy of the $i - j$ transition.

Dielectronic recombination coefficients have been calculated by Nussbaumer and Storey (1983) for ions of carbon, nitrogen and oxygen at temperatures of 10^3 to 6×10^{4}°K; this process was included in the ionization equilibrium calculations of astrophysically abundant elements by Jacobs *et al.* (1977, 1979) and Shull and Van Steenberg (1982). Dielectronic recombination and resonances in excitation cross sections are discussed by Raymond (1978).

Dielectronic recombination has been employed to explain the temperatures of the solar corona (Burgess, 1964, 1965; Burgess and Seaton, 1964), and to explain the anomalous intensity of the radio frequency carbon line (Palmer et al., 1967; Goldberg and Dupree, 1967; Dupree, 1969).

The level populations, N_n, or equivalently, b_n, for high temperature ($5000°\mathrm{K} < T < 10{,}000°\mathrm{K}$) conditions in ionized hydrogen regions have been calculated by Sejnowski and Hjellming (1969) and Brocklehurst (1970) for electron densities in the range $10\ \mathrm{cm}^{-3} < N_e < 10^4\ \mathrm{cm}^{-3}$ (also see Salem and Brocklehurst, 1979). These b_n values, which include effects from processes $(a - f)$, have been used to calculate the ratio of hydrogen line temperatures to continuum temperatures (Andrews and Hjellming, 1969; Hjellming, Andrews, and Sejnowski, 1969). In general, the ratio is enhanced beyond that of the LTE case for values of n larger than 60, and the amount of enhancement depends upon the electron density, the temperature, and the size of the region. The most detailed comparison of observed hydrogen recombination line intensities and non-LTE theory has been carried out for the Orion nebula (Hjellming and Gordon, 1971), and these observations are consistent with theoretical non-LTE results. Observationally, the ratio, ρ, of the central line intensities of the $n + 1 \rightarrow n$ transition and a nearby higher order $m + \Delta m \rightarrow m$ transition is compared with the LTE value

$$\rho_{\mathrm{LTE}} = \frac{n^2 f_{n,n+1}}{m^2 f_{m,m+\Delta m}} , \qquad (2.100)$$

where the oscillator strength $f_{m,m+\Delta m}$ is given by Menzel (1969) $f_{m,m+\Delta m}/m \approx 0.194$, 0.0271, 0.00841, 0.00365, and 0.00191 for $\Delta m = 1, 2, 3, 4$, and 5, respectively.

2.11 Planetary Nebulae, Forbidden Lines, Recombination Spectra and the Balmer Decrement

When William Herschel began to "sweep the heavens" with his telescopes in the 1780s, he found many nebulous patches of light. Some, because of their uniform disks and bluish color, resembling the distant planet Uranus, he dubbed planetary nebulae. Nearly a century later, William Huggins showed that the spectra of planetary nebulae exhibit the emission lines of hydrogen and an unidentified substance designated "nebulium". The central stars of these nebulae radiate an extraordinary amount of ultraviolet light (Wright, 1918), suggesting that this radiation provides the source of nebular luminosity (Russell, 1921; Hubble, 1922). The energetic photons of ultraviolet starlight will ionize the hydrogen atoms, whose free electrons and protons will subsequently recombine, providing the hydrogen emission spectra (Menzel, 1926; Zanstra, 1927).

The previously unidentified nebulium emission lines were explained as forbidden transitions of ionized oxygen and nitrogen excited by collisions with free-electrons before recombination (Bowen, 1928). The green nebular lines were, for example, identified with doubly ionized oxygen, O^{++}, or [O III] in modern notation, where the square brackets designate forbidden transition; Bowen also matched other previously unidentified nebular lines with singly ionized oxygen and nitrogen, or [O II] and [N II]. These solutions depended on the rarity of atomic collisions in the low density nebulae; permitting "forbidden" transitions that seldom take place in a higher-density laboratory situation, where an atom will almost always be jostled into a different state before the forbidden radiation can be emitted. (This set the stage for a similar explanation of the Sun's "coronium" emission lines in terms of forbidden transitions in a tenuous, unexpectedly hot (millions of degrees) solar corona (Grotrian, 1939; Edlén, 1941)). Bowen (1935) additionally described the unusual strength of the allowed transitions of O III and N III by a remarkable resonance-fluorescence pumping mechanism that depends on a series of chance coincidences in wavelengths between ionized helium and doubly ionized oxygen and nitrogen. Osterbrock (1989) discusses the Bowen resonance-fluorescence mechanism for O III from He II Lyman α photons. A simplified theory for this mechanism is given by Kallman and McCray (1980), who compare it to the observed intensities of O III and N III lines in the spectra of planetary nebulae.

Following Bowen's pioneering article, the theory of both recombination and forbidden line spectra was then developed, enabling astronomers to combine observations with theory to derive the temperatures and electron densities of the nebular gas, as well as the temperatures of their associated stars. Many of the pioneering papers dealing with physical processes in planetary nebulae are reproduced by Menzel (1962). A good general summary of theory and observation of gaseous nebulae, written in the context of planetary nebulae, is given by Seaton (1960); reviews of collision and electron impact processes are respectively given by Seaton (1968, 1975). The astrophysics of gaseous nebulae is reviewed in the books by Osterbrock (1974, 1989), who includes references to fundamental papers and comparisons with observations of planetary nebulae and ionized hydrogen (H II) regions (see Section 2.12), as well as active galactic nuclei (see Volume II). The chemical composition of planetary nebulae is discussed by Kaler (1970, 1978); planetary nebulae and their central stars have been reviewed by Kaler (1985).

The electron densities, electron temperatures, stellar temperatures, Hβ flux, radio flux density, radial velocities and expansion velocities of planetary nebulae are given by Lang (1992), together with their celestial positions, angular sizes, distances, radii and ages. These and additional data for planetary nebulae have also been compiled by Acker et al. (1990). Bright, named planetary nebulae are also compiled by Lang (1992).

Certain very hot stars radiate sufficient energy in the ultraviolet wavelength region to ionize the surrounding gas according to the cross section given in Eq. (1.230). After photoionization, the ejected electrons lose energy by inelastic collisions followed by radiation. As a result a Maxwellian distribution of electron velocities with electron temperature, T_e, is established (Böhm and Aller, 1947). As first suggested by Bowen (1928), the forbidden lines observed in the nebulae are excited by electron impact. It follows from Eq. (2.94) that for a Maxwellian distribution of electrons the probability per second, a_{mn}, for collisional deactivation from level m to level n is given by

$$a_{mn} = \frac{8.63 \times 10^{-6}}{g_m T_e^{1/2}} N_e \Omega(n,m) \sec^{-1} \quad \text{for } E_n < E_m \ , \tag{2.101}$$

where g_m is the statistical weight of level m, the electron density is N_e, and the collision strength $\Omega(n,m)$ has been assumed to be independent of energy, as it is for most positive ions. For collisional excitation, the probability per second, b_{mn}, is given by

$$b_{nm} = \frac{g_m}{g_n} a_{mn} \exp\left[-\frac{(E_m - E_n)}{kT_e}\right] \quad \text{for } E_n < E_m \ . \tag{2.102}$$

The intensity, I_{mn}, of the $m \rightarrow n$ transition is given by

$$I_{mn} = N_m A_{mn} h\nu_{mn} \ , \tag{2.103}$$

where N_m is the population of level m, the spontaneous transition probability is A_{mn}, and ν_{mn} is the frequency of the $m \rightarrow n$ transition. The population N_m is determined by the equation of statistical equilibrium (Eq. (2.82)) with both collision induced and spontaneous transitions being taken into account. The populations N_2 and N_3 are related, for example, by the equation

$$\frac{N_2}{N_3} = \frac{b_{12}(A_{31} + a_{31}) + (b_{12} + b_{13})(A_{32} + a_{32})}{b_{12}(b_{12} + b_{13}) + b_{13}(A_{21} + a_{21})} \ . \tag{2.104}$$

References for collisional excitation and de-excitation of ions and the methods to calculate the collision strengths are given by Seaton (1968) and Seaton (1975). Wavelengths of the forbidden transitions are given by Bowen (1960) and Osterbrock (1974, 1989). Transition probabilities of forbidden transitions of abundant ions were clarified by a fully relativistic quantum mechanical treatment (Zeippen, 1982; Mendoza and Zeippen, 1982; Zeippen, 1987). Numerical values of collision strengths and transition probabilities of cosmically abundant ions are given by Mendoza (1983) and Osterbrock (1989). Using these numerical values, Kaler (1986) and Osterbrock (1989) obtain the following relations for optical temperature determinations of planetary nebulae, as well as H II regions. Here R_O and R_N, for [O III] and [N II], respectively, are the ratios of the intensities of the nebular lines to that of the auroral line, corrected for interstellar extinction.

For [O III]

$$R_O = \frac{I(\lambda 4959) + I(\lambda 5007)}{I(\lambda 4363)} = \frac{7.73 \exp[(3.29 \times 10^4)/T]}{1 + 4.5 \times 10^{-4}(N_e/T^{1/2})} \ ,$$

and (2.105)

$$T_e[\text{O III}] = \frac{14320}{\log R_O - 0.890 + \log(1 + 0.046x)} \ .$$

For [N II]

$$R_N = \frac{I(\lambda 6548) + I(\lambda 6583)}{I(\lambda 5755)} = \frac{6.91 \exp[(2.50 \times 10^4)/T]}{1 + 2.5 \times 10^{-3}(N_e/T^{1/2})} ,$$
$$T_e[\text{N II}] = \frac{10860}{\log R_N - 0.841 + \log(1 + 0.251x)} , \quad (2.106)$$

where T is the temperature, N_e is the electron density, and $x = 10^{-2}N_e/\sqrt{T_e}$. Electron temperatures for 107 planetary nebulae calculated from the [O III] or [N II] line intensities or both are given by Kaler (1986). Measured optical values of temperature and electron density are given by Lang (1992), who also identifies bright planetary nebulae with well-known names.

The electron temperature, T_e, may be approximated by equating the thermal energy, kT_e, to the excitation energy of approximately 2 eV for the forbidden lines to obtain $T_e \approx 10^4$°K. A more exact method is to measure the relative intensities of the allowed transitions of the emission lines of hydrogen and helium. It follows from Eqs. (2.82), and (2.101) to (2.103) that the observed intensity, I_{mn}, of the $m \rightarrow n$ transition is given by

$$I_{mn} = \varepsilon h \nu_{mn} N_e N_+ \alpha_{mn} \frac{R}{3\theta^2} \text{erg cm}^{-2} \text{sec}^{-1} \text{rad}^{-2} , \quad (2.107)$$

where the filling factor, ε, denotes the fraction of the observed volume of the nebula which is occupied by line emitting atoms or ions, N_e is the electron density, N_+ is the atom or ion density, ν_{mn} is the frequency of the $m \rightarrow n$ transition, R and θ denote, respectively, the radius and angular extent of the nebula, and α_{mn} is the effective recombination coefficient.

The recombination coefficient may be calculated using the equation of statistical equilibrium, Eq. (2.82), together with the recombination cross section $\sigma_r(\nu, n)$ given by Milne (1921), Eq. (1.233), or the photoionization cross section, $\sigma_a(\nu, n)$ given by Kramers (1923), Menzel and Pekeris (1935), and Burgess (1958), Eq. (1.230), and the radiative transition probability, A_{mn}. The effective recombination coefficient, $\alpha_{mn}(T_e)$, for hydrogen-like atoms is given by (Baker and Menzel, 1938)

$$\alpha_{mn}(T_e) = \frac{10.394 \times 10^{-14} g_{bb}(m, n)}{mn(m^2 - n^2)} Z \lambda^{3/2} b_m \exp(x_m) \text{ cm}^3 \text{ sec}^{-1} , \quad (2.108)$$

where the bound-bound Kramers-Gaunt factor, $g_{bb}(m, n)$, and the equilibrium departure coefficients, b_m, are tabulated by Baker and Menzel (1938) and Green, Rush, and Chandler (1957), the nuclear charge of the atom is Z,

$$\lambda = \frac{Z^2 I_H}{kT_e} = 157{,}890 \frac{Z^2}{T_e} ,$$

where the ionization potential of hydrogen, $I_H = 13.60$ eV, and

$$x_m = \frac{I_m}{kT_e} = \frac{\lambda}{m^2} ,$$

where the threshold ionization energy of level m is $I_m = Z^2 I_H / m^2$. Because α_{mn}/Z depends only on T_e/Z^2, it is clear from Eq. (2.107) that a measurement of the relative intensities of lines of a given atom or ion will lead to a measure of T_e, whereas a measure of the intensity of one line will give the volume integral of $N_e N_+$ assuming a value for T_e.

Osterbrock (1974, 1989) has provided an overview of the recombination coefficients of gaseous nebulae, with a comparison of theory to observations. Additional references to this and related processes can be found in Section 2.9.

Baker and Menzel (1938) considered two cases, case A and case B, which correspond, respectively, to an optically thin and an optically thick nebula, the latter case being considered first by Zanstra (1927). Case A assumes that the excited states of a hydrogen atom are populated by radiative capture from the continuum, and by cascade from all higher states, and depopulated by cascade to lower levels. Case B assumes that the rate of depopulation of excited states by emission of Lyman lines is exactly equal to the rate of population by absorption of Lyman quanta. Burgess (1958), Seaton (1960), and Pengelly (1964) have considered these two cases when orbital degeneracy is taken into account. Of particular interest are the hydrogen (HI) line of H(β) = H(4,2) at $\lambda = 4861$ Å and the ionized helium (HeII) line of $He^+(4,3)$ at $\lambda = 4686$Å. For these lines we have (Seaton, 1960)

	Case A	*Case B*
$\alpha_{4,2}$(HI)	$1.98 \times 10^{-14} \mathrm{cm}^3 \mathrm{sec}^{-1}$	$2.99 \times 10^{-14} \mathrm{cm}^3 \mathrm{sec}^{-1}$
$\alpha_{4,3}$(HeII)	$9.0 \times 10^{-14} \mathrm{cm}^3 \mathrm{sec}^{-1}$	$20.8 \times 10^{-14} \mathrm{cm}^3 \mathrm{sec}^{-1}$,

for $T_e = 10^4\,°$K. Values of $\alpha_{mn}(T_e)$ may be approximated for other values of T_e by assuming that $\alpha_{mn} \propto T_e^{-1}$. Pengally (1964) has given values of $h\nu_{mn}\alpha_{mn}(T_e)$ for the λ4861 and the λ4686 lines as a function of T_e for case A and case B. He has also tabulated the relative intensities of the hydrogen Balmer and Paschen series relative to $I_{4,2}$(HI) = 100 and the Pickering and Pfund series of ionized helium relative to $I_{4,3}$(HeII) = 100, in both cases for case A and case B at $T_e = 1$ and $2 \times 10^4\,°$K.

The observed intensities must be corrected for reddening, or absorption by interstellar dust particles, before Eqs. (2.107) and (2.108) may be used. If $I_c(\lambda)$ and $I_0(\lambda)$ denote, respectively, the corrected and observed intensities at wavelength λ, then we have (Berman, 1936; Whitford, 1948, 1958; Burgess, 1958)

$$\log I_c(\lambda) = \log I_0(\lambda) + C f(\lambda) \ , \tag{2.109}$$

where the units of $f(\lambda)$ are chosen so that $f(\lambda) = 0$ for the H(β) line at $\lambda = 4861$ Å, and $f(\infty) = -1$. In this way the observed value of the H(β) intensity must be multiplied by 10^C to correct for extinction. The mean value of the constant C is 0.19 (Pengelly, 1964). Values of C for planetary nebulae are tabulated by Lang (1992).

Observations at radio frequencies do not need to be corrected for reddening, and the bremsstrahlung formula (1.222) leads to the relation

$$E = \int_0^L N_e^2 \, ds = \frac{S_\nu T_e^{1/2} c^2}{2k\xi\Omega} \ , \tag{2.110}$$

for an optically thick nebula. Here the emission measure, E, is the line integral of the square of electron density, N_e, across the extent, L, of the nebula, S_ν is the observed flux density at frequency, ν, the apparent solid angle of the source is Ω, and

$$\xi = 9.786 \times 10^{-3} \ln\left(4.9 \times 10^{7} \frac{T_e^{3/2}}{\nu}\right) .$$

Observed flux densities at $\nu = 5000$ MHz are given by Lang (1992). At this frequency, Eq. (2.110) becomes

$$E \approx \frac{S_\nu T_e^{1/2}}{2\theta^2} \mathrm{cm}^{-6}\mathrm{pc} ,$$

where θ is the angular extent in degrees and S_ν is the flux density in flux units. The mass, M, of the nebula is given by

$$M = \frac{4\pi R^3}{3} \varepsilon N_e m_H ,$$

where $m_H = 1.673 \times 10^{-24}$ grams is the mass of the hydrogen atom. It follows from Eqs. (2.107) and (2.109) that the distance, D, to the nebula is given by

$$D^5 = \frac{3}{\varepsilon}\left(\frac{M}{4\pi m_H}\right)^2 \frac{hc}{\lambda\theta^3 F \times 10^C} ,$$

where F is the flux observed at wavelength, λ, the angular radius of the nebula is θ, and C is the extinction correction. Angular sizes and distances of planetary nebulae are also given by Lang (1992).

As it was shown in Sect. 1.31, recombination continuum radiation becomes important at optical frequencies, and the magnitude of its contribution is a sensitive function of frequency, ν, electron density, N_e, and electron temperature, T_e. The total continuum power, $P(\nu)d\nu$, radiated per unit volume per frequency interval between frequencies ν and $\nu + d\nu$ is

$$P(\nu)d\nu = N(\mathrm{HII})N_e\gamma_{\mathrm{eff}}d\nu , \qquad (2.111)$$

where $N(\mathrm{HII})$ is the density of ionized hydrogen, N_e is the electron density, and the effective emission coefficient, γ_{eff}, is given by

$$\gamma_{\mathrm{eff}} = \gamma(\mathrm{HI}) + \gamma(2q) + \frac{N(\mathrm{HeII})}{N(\mathrm{HI})}\gamma(\mathrm{HeI}) + \frac{N(\mathrm{HeIII})}{N(\mathrm{HII})}\gamma(\mathrm{HeII}) ,$$

where $\gamma(\mathrm{HI})$, $\gamma(\mathrm{HeI})$ and $\gamma(\mathrm{HeII})$ denote, respectively, the emission coefficients of neutral hydrogen, neutral helium, and singly ionized helium, $\gamma(2q)$ is the emission coefficient for two photon emission from the $2^2S_{1/2}$ level of hydrogen, and $N(\mathrm{HeII})$ and $N(\mathrm{HeIII})$ denote, respectively, the densities of HeII and HeIII. The emission coefficients for $\gamma(\mathrm{HI})$ and $\gamma(\mathrm{HeI})$ are given by

$$\gamma = \frac{P_r(\nu)}{N_i N_e} + \frac{4\pi\varepsilon_\nu}{N_i N_e} = \gamma_{\mathrm{fb}} + \gamma_{\mathrm{ff}} ,$$

where the recombination power density, $P_r(\nu)$, is given in Eq. (1.236), the volume emissivity for bremsstrahlung, ε_ν, is given in Eq. (1.219), and the subscripts ff and fb denote, respectively, free-free and free-bound emission. Brown and Mathews (1970) have tabulated γ(HI) and γ(HeI) as a function of temperature for optical wavelengths. Their results are illustrated in Fig. 2.1 for a temperature of 10^4 °K. These data include that $\gamma(\mathrm{HI}) \approx \gamma(\mathrm{HeI})$, and that two photon emission becomes important near the Balmer discontinuity at $\lambda = 3647$Å. For normal nebulae, N(HeII)/N(HII) $\approx$ 0.1 and N(HeIII)/ N(HII) $\approx$ 0.01, and the contribution of helium to the continuum radiation is of second order.

The two photon emission process has been discussed by Mayer (1931), Breit and Teller (1940), and Spitzer and Greenstein (1951). The emission coefficient is given by

$$\gamma(2q) = \frac{hA(y)yX\alpha_{\mathrm{B}}(T)}{A_2} = g(v)X\alpha_{\mathrm{B}}(T) \ ,$$

where $y = v/v_{12}$, the hv_{12} is the excitation energy of the $2S_{1/2}$ level, $A_2 = 8.227\mathrm{sec}^{-1}$ is the total probability per second of emitting a photon of energy yhv_{12} as one member of the pair, X is the probability per recombination that two photon decay results, and the total recombination coefficient $\alpha_{\mathrm{B}}(T)$ to excited levels of hydrogen is given by (Hummer and Seaton, 1963)

$$\alpha_{\mathrm{B}}(T) = 1.627 \times 10^{-13} t^{-1/2}[1 - 1.657 \log t + 0.584 t^{1/3}]\,\mathrm{cm}^3\mathrm{sec}^{-1} \ ,$$

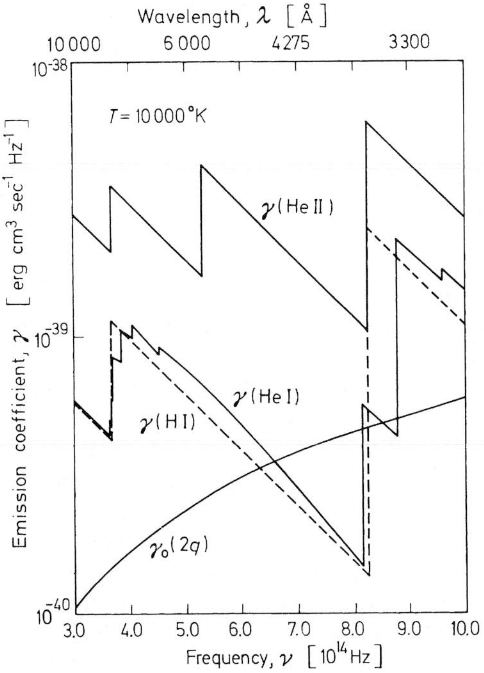

Fig. 2.1. Frequency variation of the continuous emission coefficients γ(HI) (dashed line), γ(He I), γ(He II), and $\gamma_0(2q)$ at $T = 10,000$ °K. (After Brown and Mathews, 1970, by permission of the American Astronomical Society and the University of Chicago Press)

where

$$t = 10^{-4}T \ .$$

The function 0.229 $A(\gamma)$ is tabulated by Spitzer and Greenstein (1951), and the function $g(v)$ is tabulated by Brown and Mathews (1970). Values of X depend upon the population of atoms in the $2S_{1/2}$ and 2^2P states as well as the number density of $L\alpha$ photons (cf. Cox and Mathews, 1969); and approximate values of X at optical wavelengths are tabulated by Brown and Mathews (1970).

When the contributions of recombination radiation, bremsstrahlung, and two photon emission have been taken into account, the Balmer decrement, D_B, may be calculated using Eq. (2.111). It is defined as the logarithm of the intensity ratio on each side of the Balmer discontinuity at $\lambda = 3647$Å, it is illustrated in Fig. 2.2, and is given by

$$D_B = \log\left[\frac{I(3647^-)}{I(3647^+)}\right] \ , \qquad (2.112)$$

where I denotes intensity and $-$ and $+$ denote, respectively, wavelengths lower and higher than 3647 Å. The intensity jump at the head of the Balmer series may be reduced by either raising the temperature or by lowering the electron density (which increases the contribution of the two-photon continuum emission). Consequently, a measurement of D_B leads to a measure of either the electron temperature, T_e, or the electron density, N_e. Boyarchuk, Gershberg, and Godovnikov (1968) have calculated D_B as a function of T_e and N_e by taking into account the recombination radiation, bremsstrahlung, and two photon emission of hydrogen, and their results are also illustrated in Fig. 2.2.

2.12 Ionized Hydrogen (H II) Regions and Atomic Recombination Lines

The first calculation of interstellar ionization by starlight was carried out by Arthur Stanley Eddington, who called attention to the fact that the ultraviolet radiation of hot stars will ionize the atoms between the stars (Eddington, 1926). He predicted that the heating by photoionization is balanced by cooling through radiative recombination of electrons and ions to give a temperature of about 10,000 °K. When it was discovered that hydrogen was by far the most abundant element in stars (Russell, 1929), it followed that hydrogen ought to be the dominant constituent of the interstellar matter from which stars form. Bengt Strömgren next derived the relationship among the density of interstellar hydrogen, the temperature of the exciting star, and the radius of the sphere of ionization (Strömgren, 1939-also see Sect. 3.3.1.5). He found that although in general interstellar hydrogen is neutral, designated H I, very hot stars (particularly O and B stars) can generate enormous but sharply bounded spheres of ionization, now called H II regions. Subsequent detailed analysis has shown that the temperature of the neutral H I regions is about 100 °K, a factor of 100 lower than that of the H II regions (Spitzer and Savedoff, 1950; Spitzer, 1978).

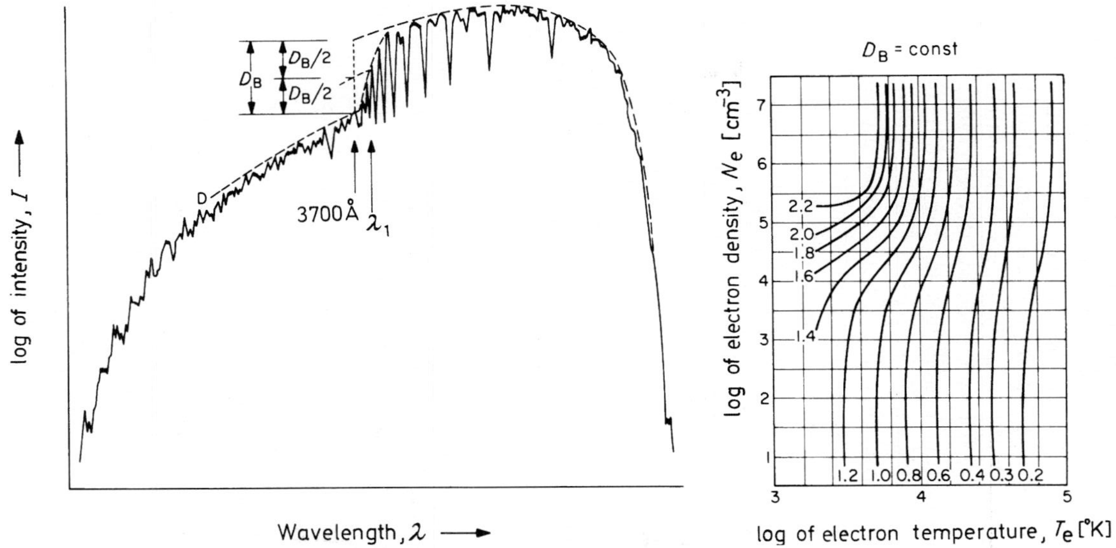

Fig. 2.2. The Balmer decrement, D_B, shown on the continuum spectrum of a star, and given as a function of electron density, N_e, and electron temperature, T_e. [After Boyarchuk, Gershberg, and Godovnikov, 1968]

Photoionization and free–bound recombination radiation of gaseous nebulae were discussed in Sect. 1.31. When the free electrons in H II regions recombine with the protons, they will not always fall into the state of lowest energy but will cascade through a series of intermediate levels, thereby producing the observed optical (Balmer series) emission lines of hydrogen. Van de Hulst (1945) predicted that such recombination lines might also be detected at radio wavelengths, while also arguing that neutral hydrogen would be detected by its 21-cm spin transition (see Sect. 2.14, Ewen and Purcell, 1951 and Muller and Oort, 1951). Early papers on the detailed physical processes in gaseous nebulae have been collected by Menzel (1962). Their photoionization equilibrium and thermal equilibrium are discussed by Osterbrock (1974, 1989), who includes references to fundamental articles. These books also place H II regions within the galactic context, and provide a comparison of theory with observation, including temperature and density measurements of H II regions from both optical emission lines and radio recombination lines.

The formulae for the free-free continuum emission, or bremsstrahlung, from H II regions are given in Sect. 1.30. The continuum radiation of gaseous nebulae at optical wavelengths is discussed by Brown and Mathews (1970) and Osterbrock (1989). The free-free continuum radiation of H II regions at radio wavelengths is delineated by Scheuer (1960) and Brown (1987). Radio recombination lines are also reviewed by Brown (1987) and Gordon (1988). Basic observational data for H II regions are provided by Lang (1992), including radio measurements of electron temperatures, emission measures, and electron densities, as well as positions, angular sizes, distances, radii and velocities.

If the density on both sides of the boundary between the neutral and ionized gases is roughly equal, then the difference in pressure will naturally result in an outward expansion of the ionized gas. Such an evolution of H II regions is discussed in Sect. 3.3.1.5, and has been reviewed by Mathews and O'Dell (1969) and Yorke (1986); detailed formulae for ionization fronts and expanding H II regions are given by Osterbrock (1974, 1989) and Shu (1992).

When a hydrogen-like atom undergoes a transition from an upper level m to a lower level n, it will radiate at the frequency, ν_{mn}, given by (Rydberg, 1890; Ritz, 1908; Planck, 1910; Bohr, 1913)

$$\nu_{mn} = cR_A Z^2 (n^{-2} - m^{-2}) \approx 2cR_A Z^2 n^{-3}(m-n) \ , \tag{2.113}$$

where the velocity of light, $c = 2.997925 \times 10^{10}$ cm sec^{-1}, Z is the "effective" charge of the nucleus (or ionic charge) given in units of the proton charge, and the atomic Rydberg constant, R_A, is given by

$$R_A = R_\infty \left(1 + \frac{m_e}{M_A}\right)^{-1} \approx R_\infty \left(1 - \frac{m_e}{M_A}\right) \ ,$$

where the Rydberg constant for infinite mass is $R_\infty = 2\pi^2 m_e e^4/(ch^3) = 109{,}737.31 \text{cm}^{-1}$, the electron mass $m_e = 5.48597 \times 10^{-4}$ a.m.u., and the atomic mass is M_A. The atomic masses and the Rydberg constants for the more abundant atoms are given in Table 2.1.

Table 2.1. Atomic masses and Rydberg constants for the most abundant atoms

Atom	Atomic mass, M_A (a.m.u.)	Rydberg constant, R_A (cm^{-1})
Hydrogen, H^1	1.007825	109,677.6
Helium, He^4	4.002603	109,722.3
Carbon, C^{12}	12.000000	109,732.3
Nitrogen, N^{14}	14.003074	109,733.0
Oxygen, O^{16}	15.994915	109,733.5
Neon, Ne^{20}	19.99244	109,734.3

The important low n transitions for hydrogen were first observed by Lyman (1906) for $n = 1$, Balmer (1885) for $n = 2$, Paschen (1908) for $n = 3$, Brackett (1922) for $n = 4$, Pfund (1924) for $n = 5$, and Humphreys (1953) for $n = 6$. The wavelengths of these lines are given in Table 2.2 together with the Pickering (1896, 1897) series for singly ionized helium.

The transitions with $m - n = 1, 2, 3, 4, \ldots$ are respectively designated by the lower case letters of the Greek alphabet $\alpha, \beta, \gamma, \delta, \ldots$, and the letters L, H and P used for the Lyman ($n = 1$), Balmer ($n = 2$) and Paschen ($n = 3$) lines. Thus, for example, the wavelengths of the principal $m - n = 1$ transitions are designed $L_\alpha = 1216$ Å, $H_\alpha = 6563$ Å and $P_\alpha = 18751$ Å. The first few lines in

Table 2.2. The wavelengths in Å of the $m \rightarrow n$ transitions of hydrogen for $n = 1$ to 6, $m = 2$ to 21, and $m = \infty$, and for the $n = 4$ Pickering series for ionized helium (HeII)[1]. Here the wavelengths are in Å where 1 Å $= 10^{-8}$ cm

Series m	Lyman ($n = 1$)	Balmer ($n = 2$)	Paschen ($n = 3$)	Brackett ($n = 4$)	Pfund ($n = 5$)	Humphreys ($n = 6$)	Pickering (He^+, $n = 4$)
2	1,215.67						
3	1,025.72	6,562.80					
4	972.537	4,861.32	18,751.0				
5	949.743	4,340.46	12,818.1	40,512.0			10,123.64
6	937.803	4,101.73	10,938.1	26,252.0	74,578		6,560.10
7	930.748	3,970.07	10,049.4	21,655.0	46,525	123,680	5,411.52
8	926.226	3,889.05	9,545.98	19,445.6	37,395	75,005	4,859.32
9	923.150	3,835.38	9,229.02	18,174.1	32,961	59,066	4,541.59
10	920.963	3,797.90	9,014.91	17,362.1	30,384	51,273	4,338.67
11	919.352	3,770.63	8,862.79	16,806.5	28,722	46,712	4,199.83
12	918.129	3,750.15	8,750.47	16,407.2	27,575	43,753	4,100.04
13	917.181	3,734.37	8,665.02	16,109.3	26,744	41,697	4,025.60
14	916.429	3,721.94	8,598.39	15,880.5	26,119	40,198	3,968.43
15	915.824	3,711.97	8,545.39	15,700.7	25,636	39,065	3,923.48
16	915.329	3,703.85	8,502.49	15,556.5	25,254	38,184	3,887.44
17	914.919	3,697.15	8,467.26	15,438.9	24,946	37,484	3,858.07
18	914.576	3,691.55	8,437.96	15,341.8	24,693	36,916	3,833.80
19	914.286	3,686.83	8,413.32	15,260.6	24,483	36,449	3,813.50
20	914.039	3,682.81	8,392.40	15,191.8	24,307	36,060	3,796.33
21	913.826	3,679.35					3,781.68
∞	911.5	3,646.0	8,203.6	14,584	22,788	32,814	3,644.67

[1] Data from Wiese, Smith, and Glennon [1966].

the Balmer series are $H_\alpha = 6563$ Å, $H_\beta = 4861$ Å, $H_\gamma = 4340$ Å, and $H_\delta = 4101$ Å.

In the radio domain where $(m-n) \ll n$ and $\nu_{mn} \approx 2cR_A Z^2 (m-n)/n^3$, the approximate frequency separation, $\Delta\nu$, between adjacent lines is $\Delta\nu \approx 6cR_A Z^2(m-n)/n^4 \approx 3\nu_{mn}/n$. The frequencies and separations of some radio recombination lines for the α, or $m - n = 1$, transition of hydrogen are given in Table 2.3.

When an optically thin gas is observed with an antenna whose beamwidth is narrower than the angular extent of the gas cloud, the antenna temperature due to the line radiation, ΔT_L, is given by

$$\Delta T_L = \eta_A T_e \int_0^L \alpha_\nu \, dl \ , \tag{2.114}$$

where η_A is the antenna efficiency, T_e is the electron temperature of the ionized gas, L is the extent of the gas cloud along the line of sight, the integral is along the line of sight, and the quantum mechanical absorption coefficient, α_ν, is given by Einstein (1917) Eq. (2.66)

$$\alpha_\nu = \frac{c^2 N_n g_m}{8\pi\nu^2 g_n n_\nu^2} A_{mn} \left[1 - \frac{g_n}{g_m}\frac{N_m}{N_n}\right] \varphi_{mn}(\nu) \ , \tag{2.115}$$

where N_n is the population of level n, the A_{mn} is the Einstein coefficient for spontaneous transition from level m to level n, the statistical weight of level n is g_n, and the index of refraction of the gas is n_ν, which will be assumed to be unity in the following. When the gas is in local thermodynamic equilibrium (LTE), the population, N_n, of level n is given by Saha (1921) Eq. (3.129)

$$N_n = \frac{N_e N_i h^3}{(2\pi m_e k T_e)^{3/2}} \frac{g_n}{2} \exp(-\chi_n / k T_e) \ , \tag{2.116}$$

where N_e is the electron density, N_i is the density of the recombining ions, T_e is the electron temperature, and χ_n is the excitation energy of the nth level below the continuum. Under LTE conditions, $\exp(-h\nu_{mn}/kT) = g_n N_m/(g_m N_n)$ and at radio frequencies $h\nu_m \ll kT$. It then follows from Eqs. (2.71) and (2.113) to

Table 2.3. Hydrogen radio recombination lines.

Transition	Frequency, ν (GHz)	Separation between adjacent lines, $\Delta\nu$ (MHz)
H50α	51.071	2918
H75α	15.281	591
H100α	6.478	190
H125α	3.327	78.3
H150α	1.929	38.0
H175α	1.216	20.6
H200α	0.816	12.1
H225α	0.573	7.6
H250α	0.418	5.0

(2.116) that for hydrogen at the frequency ν_{mn} we have from Eq. (2.8) $\chi_n = -hcR_A Z^2/n^2$, and

$$\Delta T_L = 4\pi \left(\frac{\ln 2}{\pi}\right)^{1/2} \frac{e^2 h^4 R_A \eta_A Z^2}{m_e k (2\pi m_e k T_e)^{3/2}} \frac{m-n f_{nm}}{\Delta\nu_D \; n} \exp\left(\frac{-\chi_n}{kT_e}\right) \int_0^L N_e N_i \, dl$$

$$\approx 10^4 Z^2 \eta_A \frac{m-n f_{nm}}{\Delta\nu_D \; n} T_e^{-3/2} \exp\left(\frac{1.58 \times 10^5 Z^2}{n^2 T_e}\right) \int_0^L N_e N_i \, dl \, ^\circ\mathrm{K} \; , \qquad (2.117)$$

where $g_n = 2n^2$, the full width to half maximum of the Doppler broadening is $\Delta\nu_D$, and the numerical approximation is for $\Delta\nu_D$ in kHz, and the line emission measure $E_L = \int_0^L N_e N_i \, dl$ in pc cm^{-6}. The oscillator strength, f_{nm}, is related to the Einstein coefficient for spontaneous emission, A_{mn}, in Eq. (2.40), and the f_{nm} for the hydrogen atom are given by the Kramer's formula (Kramers, 1923; Menzel and Pekeris, 1935)

$$f_{nm} = \frac{-g_m}{g_n} f_{mn} = \frac{2^5}{3\sqrt{3}\pi} \frac{1}{m^2} \frac{1}{\left[\frac{1}{m^2} - \frac{1}{n^2}\right]^3} \left|\frac{1}{n^3} \frac{1}{m^3}\right| g_{bb} \; , \qquad (2.118)$$

where $g_n = 2n^2$ and g_{bb}, is the Kramers–Gaunt factor for the bound–bound transition (Kramers, 1923; Gaunt, 1930; Baker and Menzel, 1938; Burgess, 1958). Values of f_{mn} for low m and n are given by Wiese, Smith, and Glennon (1966).

The oscillator strengths for the Balmer series of hydrogen ($n = 2$) are $f_\alpha = 0.641$ ($m = 3$), $f_\beta = 0.119$ ($m = 4$), $f_\lambda = 0.044$ ($m = 5$), $f_\delta = 0.021$ ($m = 6$) and $f_\varepsilon = 0.012$ ($m = 7$). A comprehensive catalogue of emission nebulae observed at optical wavelengths is given by Lang (1992), together with a listing of bright named H II regions.

For $n \gg 1$ and $\Delta n = m - n \ll n$, we can use the approximation (Brown, 1987)

$$f_{mn} \approx \mathrm{C}(\Delta n)[1 + 1.5(\Delta n/n)] \; ,$$

where the constant $\mathrm{C}(\Delta n) = 0.19077$, 0.026332, 0.0081056 and 0.0034917 for $\Delta n = 1, 2, 3,$ and 4 respectively. Approximation formulae for g_{bb} and numerical evaluations of f_{nm} at large n are given by Goldwire (1968) and Menzel (1969). Corrections to Menzel's expression for the oscillator strength of hydrogen at large n are given by Perrin (1994) and included below. Useful formulae for computing hydrogen transitions between neighboring states for large principal quantum number are given by Malik, Malik and Varma (1991).

$$f_{n+c,\,n} = \frac{(2/c)^{2-2c} n}{3(c!)^2} \left[\frac{(n+c)!}{n! n^c}\right]^2 \left[\frac{(1+c/n)^{2n+2}}{(1+c/2n)^{4n+2c+3}}\right] \{\} \; , \qquad (2.119)$$

where

$$\{\} = {}_2F_1 \times {}_3F_2 \; .$$

The F's, in turn, are hypergeometric functions of four and six variables, respectively:

$$ {}_2F_1\left[-n, -n+1, c+1, -\frac{c^2}{4n(n+c)}\right] $$
$$ = 1 - \frac{n(n-1)}{1!(c+1)}\frac{c^2}{4n(n+c)} + \frac{n(n-1)^2(n-2)}{2!(c+1)(c+2)}\left[\frac{c^2}{4n(n+c)}\right]^2 - \cdots, $$

and

$$ {}_3F_2\left[-n, -n, \theta+1, c+1, \theta, -\frac{c^2}{4n(n+c)}\right] $$
$$ = 1 - \frac{n^2(\theta+1)}{1!(c+1)\theta}\frac{c^2}{4n(n+c)} + \frac{n^2(n-1)^2(\theta+2)}{2!(c+1)(c+2)\theta}\left[\frac{c^2}{4n(n+c)}\right]^2 $$
$$ - \frac{n^2(n-1)^2(n-2)^2(\theta+3)}{3!(c+1))(c+2)(c+3)\theta}\left[\frac{c^2}{4n(n+c)}\right]^3 + \cdots, $$

where $\theta = nc/(2n+c)$ and the transition is the $n+c \rightarrow n$ transition.

As first pointed out by Van de Hulst (1945), Wild (1952) and Kardashev (1959), the $m-n$ transitions of hydrogen and helium may be observed in ionized hydrogen (H II) regions at radio frequencies when n is large. Confirming observations were first made by Dravskikh and Dravskikh (1964). The frequencies of these lines are specified by Eq. (2.113), and the transition is usually designated by the value of n together with a Greek suffix denoting the values of $m-n$. When $m-n=1$, 2, 3, 4, and 5, respectively, the appropriate symbols are $\alpha, \beta, \gamma, \delta$ and ε. Tables of hydrogen α, β, and γ line frequencies and helium α and β line frequencies have been tabulated by Lilley and Palmer (1968) for $n = 51$ to $n = 850$. Observations of radio frequency hydrogen recombination lines result in the specification of the turbulent velocity, the kinematic distance, the electron temperature and the electron density of H II regions. An extensive survey providing this information has been completed using the H 109 α line at 5008.923 MHz (Reifenstein et al., 1970; Wilson et al., 1970). Observations of recombination lines may also be used to obtain the relative cosmic abundance of hydrogen and helium. The radiofrequency data (Churchwell and Mezger, 1970) indicate that the number ratio of singly ionized helium to ionized hydrogen is 0.08 ± 0.03. An example of the recombination lines from the H II region, Orion A, is shown in Fig. 2.3.

Derivation of nebular electron densities from radio recombination lines is discussed by Lockman and Brown (1976). The chemical abundances of a large sample of galactic H II regions is given by Shaver et al. (1983). Model H II regions and elemental abundances within them are given by Mathis (1985), Rubin (1985) and Simpson, Rubin, Erickson and Haas (1986).

In order to illustrate how the parameters of an H II region are obtained from radio observations, we note that a recombination line is observed against the background continuum radiation of the gas cloud. It follows from Eq. (1.222) that for an optically thin gas whose angular extent is larger than that of the

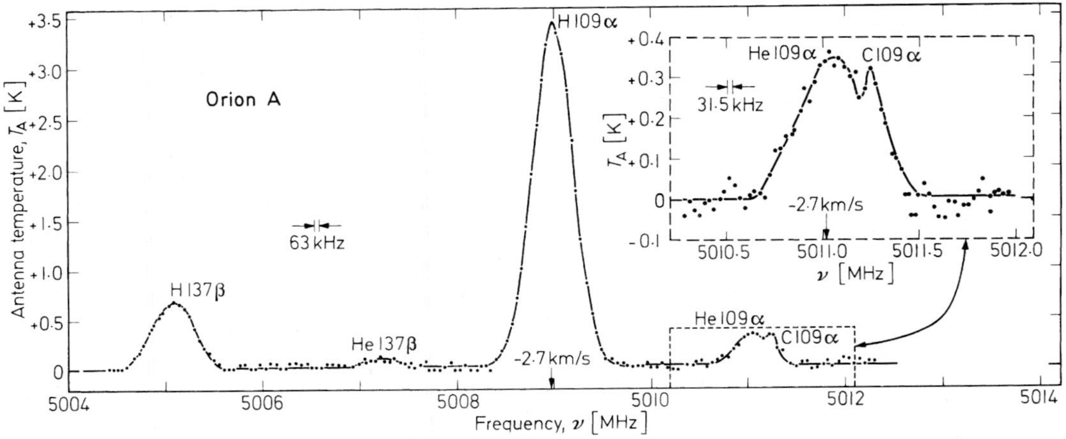

Fig. 2.3. Broadband spectrogram of the 109α region of the spectrum of the Orion Nebula. The frequency resolution is 63 kHz for the broadband spectrogram and 31.5 kHz for the narrow band spectrum centered on the He 109 α line. [After Churchwell and Mezger, 1970, by permission of Gordon & Breach Science Publishers]

antenna beamwidth, the continuum antenna temperature, T_c, is given by

$$T_c = \frac{32\pi^2 e^6}{3\sqrt{2\pi}\, m^3 c\, n_\nu} \frac{\eta_A T_e}{(2\pi\nu)^2} \left(\frac{m}{kT_e}\right)^{3/2} \int_0^L N_e \sum_j Z_j^2 N_j \ln \Lambda\, dl$$

$$\approx 3.01 \times 10^{-2} \eta_A T_e^{-1/2} \nu^{-2} \int_0^L N_e \sum_j Z_j^2 N_j \, ln\, \Lambda\, dl\, {}^\circ\mathrm{K} \ , \qquad (2.120)$$

where η_A is the antenna efficiency, the summation, j, is over all ions, the index of refraction is n_ν, the extent of the gas cloud along the line of sight is L, and the logarithmic free–free Gaunt factor, Λ, is given in Table 1.2. The numerical approximation is for unity n_ν, a frequency ν in GHz, and the extent L in parsecs.

When the radiofrequency approximation for optical depth given in Eq. (1.223) is used with Eq. (2.117), we obtain the relation

$$\frac{\Delta\nu_D \Delta T_L}{T_c} \approx 2.33 \times 10^4 \nu_{mn}^{2.1} T_e^{-1.15} \left(\frac{E_L}{E_c}\right) \quad \mathrm{kHz} \ , \qquad (2.121)$$

where the full width to half maximum of the line, $\Delta\nu_D$, is in kHz, the line frequency, ν_{mn}, is in GHz, the temperatures are in °K, and the ratio of the line emission measure, E_L, to the continuum emission measure, E_c, is given by

$$\frac{E_L}{E_c} \approx \frac{N_i/N_{H^+}}{1 + (N_{He^+} + 4N_{He^{++}}/N_{H^+})} \approx \frac{1}{1 + N_{He^+}/N_{H^+}} \approx 0.93 \ ,$$

where N_i is the number density of the recombining ions, and N_{H^+}, N_{He^+} and $N_{He^{++}}$, are, respectively, the number densities of ionized hydrogen, singly ionized helium and doubly ionized helium. Eq. (2.121) indicates that the electron temperature, T_e, may be obtained by observing $\Delta\nu_D, \Delta T_L$, and T_c. The emission measure $E_L = \int_0^L N_e N_i \, dl$ may then be obtained by using this value of T_e in Eq. (2.117).

These equations apply under conditions of local thermodynamic equilibrium and Doppler broadening with negligible impact broadening.

The full velocity line width $\Delta V = c\Delta\nu/\nu \approx 0.299\ \Delta\nu_D/\nu_{mn}$ km s^{-1}, so that (Brown, 1987)

$$\Delta V \Delta T_L / T_c \approx 6.35 \times 10^3 \nu^{1.1} T_e^{-1.15} \ ,$$

and the electron temperature is given by:

$$T_e \approx [6.35 \times 10^3 \nu^{1.1} \mathrm{T}_c / (\Delta T_L \Delta V)]^{0.87} \ , \qquad (2.122)$$

where ΔV is in units of km s^{-1}. Odegard (1985) discusses a method for determining nebular electron temperatures and local electron densities from observations of two radio recombination lines that are well separated, together with radio continuum measurements. Values of electron temperatures and emission measures using Eqs. (2.117) and (2. 121) or (2.122) are given by Lang (1992) for nearly three hundred H II regions.

In non-LTE conditions, the formulae developed in Sect. 2.10 may be used to relate observed line parameters to the physical parameters of H II regions. However, in most cases the differences between the electron temperatures determined by LTE and non-LTE analysis is within 15% . Shaver (1980) suggested that there is a unique frequency ν that minimizes non-LTE corrections, for which the LTE-derived electron temperatures are reliable and accurate. That frequency is:

$$\nu \approx 0.081\ E_c^{0.36} \quad , \tag{2.123}$$

where the emission measure $E_c = N_e^2 L$, the integral along the line of sight through an extent L of the H II region, can be obtained from continuum observations by determining the optical depth for free-free emission, Eq.(1.223).

2.13 Atomic Fine Structure

For an atom with several electrons, the fine structure is usually described by the quantum numbers based on Russell–Saunders (LS) coupling (Russell and Saunders, 1925). (The LS-coupling is valid for the elements with not very large Z, except for the noble gases (Ne, Ar, ...) and the highly excited states).

$n =$ principal quantum number (cf. Bohr, 1913) and Eqs. (2.4), (2.8).

$L =$ total orbital quantum number in units of $h/2\pi$. The L is the vectoral sum of the orbital angular momentum, l, of all of the electrons. L is always integral and the $L =$ 0, 1, 2, 3, 4, and 5 is designated as S, P, D, F, G, and H, respectively. The l of the electron takes on $n - 1$ values.

$S =$ total spin angular momentum in units of h/2π. The S is the vectoral sum of the spin angular momentum, s, of all the electrons. An electron has $s = 1/2$, and S takes on integral or half integral values for atoms having an even or odd number of electrons (cf. Uhlenbeck and Goudsmit, 1925, 1926; Thomas, 1926, 1927; Pauli, 1925, 1927).

$J =$ the total angular momentum in units of $h/2\pi$. It is the vector sum $L + S$ for LS coupling. The number of possible J values is $2L + 1$ or $2S + 1$ according as $L < S$ or $L \geq S$.

Each set of values of L and S designate a term. As long as $L \geq S$, the term consists of $2S + 1$ energy levels with different J values. The number $2S + 1$ is called the multiplicity of the term (even for $L < S$). The term symbol is given by $^{2S+1}L_J$, where L is designated by its alphabetical symbol $S, P, D, F, \ldots$ for $L = 0, 1, 2, 3, \ldots$ Some terms and J-values for different L and S in Russell–Saunders (LS) coupling are given in Table 2.4.

Some consequences of Russell–Saunders coupling, known as Hund's rules, were first realized from spectroscopic observations. They are: higher S implies lower energy, higher L implies lower energy, and higher J implies higher energy if less than half-filled, lower energy if more than half-filled.

A transition between two energy levels is called a spectral line. The possible transitions between all the levels in two terms produce a neighboring set of lines called a multiplet, first observed by Catalán (1922). The possible transitions are limited to the electron dipole selection rules for Russell–Saunders (LS) coupling: $\Delta S = 0, \Delta L = 0, \pm 1$, and $\Delta J = 0, \pm 1$, except no transitions from $J = 0$ to $J = 0$. Transitions that follow these selection rules are allowed

Table 2.4. Terms and J-values for different quantum numbers L and S in Russell–Saunders coupling

Term	$S = 0$ $2S + 1 = 1$ Singlet	$S = 1/2$ $2S + 1 = 2$ Doublet	$S = 1$ $2S + 1 = 3$ Triplet	$S = 3/2$ $2S + 1 = 4$ Quartet
$L = 0$ S-Term	$J = 0$	$J = 1/2$	$J = 1$	$J = 3/2$
$L = 1$ P-Term	$J = 1$	$J = 1/2\ 3/2$	$J = 0\ 1\ 2$	$J = 1/2\ 3/2\ 5/2$
$L = 2$ D-Term	$J = 2$	$J = 3/2\ 5/2$	$J = 1\ 2\ 3$	$J = 1/2\ 3/2\ 5/2\ 7/2$
$L = 3$ F-Term	$J = 3$	$J = 5/2\ 7/2$	$J = 2\ 3\ 4$	$J = 3/2\ 5/2\ 7/2\ 9/2$

transitions with emission or absorption of electric dipole radiation. The selection rules for forbidden transitions are additionally given in Table 2.5.

The symbol $\nleftrightarrow$ denotes " cannot combine with". For example, the $0 \nleftrightarrow 0$ following $\Delta J = 0, \pm 1$ means that a level with $J = 0$ cannot combine with another level with $J = 0$. A parity change means that terms whose summed electronic l are even can only combine with those whose summed l are odd, and vice versa. Once the appropriate transitions have been determined, line frequencies ν_{mn}, and the appropriate A_{mn}, f_{nm}, or S_{mn} may be computed using the detailed formulae of quantum mechanics, or by measuring them in a laboratory. As an example, the fine structure energy, E_{fs}, arising between the interaction of the L and S of a hydrogen-like atom is given by (Landé, 1923; Thomas, 1926).

$$E_{\text{fs}} = \alpha^2 \frac{hcRZ^4}{n^3} \left[\frac{J(J+1) - L(L+1) - S(S+1)}{L(L+1)(2L+1)} \right], \tag{2.124}$$

where the fine structure constant $\alpha = 2\pi e^2/(hc) \approx 7.2973 \times 10^{-3}$, the Rydberg constant $R = R_{\text{A}} = 2\pi^2 \mu e^4/(ch^3)$ where R_{A} is given just after Eq. (2.113), the reduced mass $\mu = [mM_{\text{A}}/(m + M_{\text{A}})]$, and Z is the ionic charge. The frequency of any transition then follows from Eqs. (2.7) and (2.124). Eq. (2.124) expresses Landé's interval rule which states that the interval between the level of a given J and that of $J - 1$ is proportional to J.

Table 2.5. Selection rules for atomic spectra [cf. Pauli, 1925; Laporte, 1924; Shortley, 1940]

Electric dipole (allowed)	Magnetic dipole (forbidden)	Electric quadrupole (forbidden)
(1) $\Delta J = 0, \pm 1$	$\Delta J = 0, \pm 1$	$\Delta J = 0, \pm 1, \pm 2$
$(0 \nleftrightarrow 0)$	$(0 \nleftrightarrow 0)$	$(0 \nleftrightarrow 0, \frac{1}{2} \nleftrightarrow \frac{1}{2}, 0 \nleftrightarrow 1)$
(2) $\Delta M = 0, \pm 1$	$\Delta M = 0, \pm 1$	$\Delta M = 0, \pm 1, \pm 2$
(3) Parity change	No parity change	No parity change
(4) One electron jump	No electron jump	One or no electron jump
$\Delta l = \pm 1$	$\Delta l = 0$	$\Delta l = 0, \pm 2$
For L – S coupling	$\Delta n = 0$	
(5) $\Delta S = 0$	$\Delta S = 0$	$\Delta S = 0$
(6) $\Delta L = 0, \pm 1$	$\Delta L = 0$	$\Delta L = 0, \pm 1, \pm 2$
$(0 \nleftrightarrow 0)$		$(0 \nleftrightarrow 0, 0 \nleftrightarrow 1)$

The energies and wavelengths are conveniently found from the Grotrian (1930) diagrams which graphically represent the relations of characteristic spectrum lines to the quantum energy levels of an atom or ion. The Grotrian diagrams for atoms and ions of astrophysical interest are given in Fig. 2.4. Multiplets are represented by straight lines: the line is solid for transitions between terms of the same multiplicity, dashed for transitions between terms of different multiplicity (intersystem multiplets) and dot-dashed for transitions whose upper terms are metastable (forbidden transitions between terms of the same parity). Wavelengths greater than 2000 Å are values in air, whereas those less than 2000 Å are values in vacuum. In order to use one diagram for more than one element, corresponding lines in spectra of various elements are given in Table 2.6. Both the diagrams and the table are from data given by Moore and Merrill (1968).

The line strength for the electric dipole radiation in LS coupling is given by

$$S_{mn} = \mathscr{S}(\mathscr{M})\mathscr{S}(\mathscr{L})\sigma_{mn}^2 , \qquad (2.125)$$

where $\mathscr{S}(\mathscr{M})$ depends on the particular multiplet of the transition array, $\mathscr{S}(\mathscr{L})$ depends on the line of the multiplet, and

$$\sigma_{mn}^2 = (4l^2 - 1)^{-1}\left\{\int_0^\infty P_m P_n r' \, dr'\right\}^2 ,$$

where P_m/r' and P_n/r' are the radial wave functions for the upper and lower states of the transition when the optical electron is in configurations with azimuthal quantum numbers l and $l-1$. Here the wave functions are normalized in atomic units (Hartree, 1928) for which $h/2\pi = e = m = 1$. They are solutions to the time independent wave equation (Schrödinger, 1925, 1926)

$$H\psi = E\psi , \qquad (2.126)$$

where the wave function has an amplitude, ψ, and energy state, E, and is operated on by the Hamiltonian operator, H, given by

$$H = \left[-\frac{h^2}{8\pi^2 m}\nabla^2 + V(r)\right] , \qquad (2.127)$$

for a particle of mass, m, in a scalar potential $V(r)$. Eqs. (2.126) and (2.127) follow directly from assuming that a particle of velocity v has a de Broglie wavelength (de Broglie, 1923, 1925) given by $\lambda = h/(mv)$, that its quantized angular momentum is given by Eq. (2.4), and that the wave function satisfies a plane wave equation. The quantum mechanics of multi-electron atoms is discussed by Leighton (1959), Merzbacher (1961), Wu (1986) and Shu (1991).

For a complete solution to the Russell–Saunders case, we must use the Hamiltonian

$$H = \sum_{i=1}^{p}\left[-\frac{\hbar^2}{2m}\nabla_i^2 - \frac{Ze^2}{r_i}\right] + \sum_{ij}\frac{e^2}{r_{ij}} , \qquad (2.128)$$

where Z is the atomic number, p is the number of electrons, the term $\sum Ze^2/r_i$ accounts for the Coulomb interaction of the electrons with the nucleus, and the

Table 2.6. Corresponding lines in spectra of various elements – for use with Grotrian diagrams[1]

Key letter	Multiplet designation	λ	Mult. no.	λ	Mult. no.	λ	Mult. no.
		C I		N II		O III	
A...	$2p\,^{2\,3}P - 2p\,^{2\,1}D$	(9,849.3*P*)	(1F)	6,583.4	(1F)	5,006.9	(1F)
		(9,823.1*P*)		6,548.1		4,958.9	
B...	$2p\,^{2\,1}D - 2p\,^{2\,1}S$	(8,727.6*P*)	(3F)	5,754.6	(3F)	4,363.2	(2F)
C...	$3s\,^{3}P^{\circ} - 3p\,^{3}P$	9,094.9	(3)	4,630.5	(5)	3,047.1	(4)
		9,078.3		4,613.9		3,035.4	
		9,111.8		4,643.1		3,059.3	
		9,088.6		4,621.4		3,043.0	
		9,061.5		4,601.5		3,023.4	
		9,062.5		4,607.2		3,024.6	
D...	$3s\,^{1}P^{\circ} - 3p\,^{1}P$	14,540.2		6,482.1	(8)	5,592.4	(5)
E...	$3s\,^{1}P^{\circ} - 3p\,^{1}D$	9,405.8	(9)	3,995.0	(12)	2,983.8	(6)
		C II		N III		O IV	
A...	$2p\,^{2}P^{\circ} - 3s\,^{2}S$	858.6	(UV4)	452.2	(UV4)	279.9	(UV4)
		858.1		451.9		279.6	
B...	$2p\,^{2}P^{\circ} - 2p\,^{2\,2}D$	1,335.7	(UV1)	991.6	(UV1)	790.2	(UV1)
		1,334.5		989.8		787.7	
				991.5		790.1	
C...	$3s\,^{2}S - 3p\,^{2}P^{\circ}$	6,578.0	(2)	4,097.3	(1)	3,063.5	(1)
		6,582.8		4,103.4		3,071.7	
D...	$3p\,^{2}P^{\circ} - 3d\,^{2}D$	7,236.2	(3)	4,640.6	(2)	3,411.8	(2)
		7,231.1		4,634.2		3,403.6	
				4,641.9		3,413.7	
		C III		N IV		O V	
A...	$2s\,^{2\,1}S - 2p\,^{1}P^{\circ}$	977.0	(UV1)	765.1	(UV1)	629.7	(UV1)
B...	$2s\,^{2\,1}S - 3p\,^{1}P^{\circ}$	386.2	(UV2)	247.2	(UV2)	172.2	(UV2)
C...	$2p\,^{3}P^{\circ} - 3s\,^{3}S$	538.3	(UV5)	322.7	(UV4)	215.2	(UV4)
		538.2		322.6		215.1	
		538.1		322.5		215.0	

D...	$2p\,^1P^\circ - 2p^{2\,1}D$	2,296.9	(UV8)	1,718.5	(UV7)	1,371.3	(UV7)
E...	$2p\,^1P^\circ - 2p^{2\,1}S$	1,247.4	...	955.3	(UV8)	774.5	(UV8)
F...	$2p\,^1P^\circ - 3s\,^1S$	690.5	(UV10)	387.4	(UV9)	248.5	(UV9)
G...	$2p\,^1P^\circ - 3d\,^1D$	574.3	(UV11)	335.0	(UV10)	220.4	(UV10)
H...	$3s\,^3S - 3p\,^3P^\circ$	4,647.4	(1)	3,478.7	(1)	2,781.0	...
		4,650.2		3,483.0		2,787.0	
		4,651.4		3,484.9		2,789.9	
I...	$3p\,^1P^\circ - 3d\,^1D$	5,696.0	(2)	4,057.8	(3)	3,144.7	(2)
		N I		O II		Ne IV	
A...	$2p^{3\,4}S^\circ$–$2p^{3\,2}D^\circ$	5,200.4	(1 F)	3,728.8	(1 F)	(2,441 ± *P*)	...
		5,197.9		3,726.0		(2,438 ± *P*)	
B...	$2p^{3\,2}D^\circ$–$2p^{3\,2}P^\circ$	(10,396.5 *P*)	(3 F)	7,319.9	(2 F)	4,714.2	(1 F)
		(10,406.2 *P*)		7,330.2		4,725.6	
		(10,396.5 *P*)		7,319.9		4,715.6	
		(10,406.2 *P*)		7,330.2		4,724.2	
C...	$2p^{3\,4}S^\circ$–$2p^{3\,2}P^\circ$	(3,466.4 *P*)	(2 F)	(2,470.4 *P*)	...	(1,609 ± *P*)	...
		(3,466.4 *P*)		(2,470.3 *P*)		(1,609 ± *P*)	
D...	$3s\,^4P$–$3p\,^4D^\circ$	8,680.2	(1)	4,649.1	(1)	–	–
		8,683.4		4,641.8			
		8,686.1		4,638.9			
		8,718.8		4,676.2			
		8,711.7		4,661.6			
		8,703.2		4,650.8			
E...	$3s\,^4P$–$3p\,^4P^\circ$	8,216.3	(2)	4,349.4	(2)	(2.203.6*P*)	...
		8,210.6		4,336.9		(2,192.4*P*)	
		8,200.3		4,325.8		(2,188.0*P*)	
		8,242.3		4,367.0		(2,220.3*P*)	
		8,223.1		4,345.6		(2,206.2*P*)	
		8,184.8		4,319.6		(2,176.1*P*)	
		8,188.0		4,317.1		(2,174.4*P*)	
F...	$3s\,^4P$–$3p\,^4S^\circ$	7,468.3	(3)	3,749.5	(3)	(1,875.2*P*)	...
		7,442.3		3,727.3		(1,855.3*P*)	
		7,423.6		3,712.8		(1,842.4*P*)	

[1] After Moore and Merrill (1968).

double sum accounts for the Coulomb interaction of the electrons themselves (hence including each pair only once, omitting $i = j$, and having $p(p-1)/2$ terms). A simpler Hamiltonian is obtained in the central field approximation (Hartree, 1928) for which the Hamiltonian is given by

$$H = \sum_{i=1}^{p} \left[-\frac{\hbar^2}{2m} \nabla_i^2 + V(r_i) \right] , \tag{2.129}$$

where the potential $V(r)$ tends to $-Ze^2/r$ as $r \to 0$ and $(-Z - p + 1)e^2/r$ for large r.

Solutions for the relative strengths of the lines within a multiplet were first computed by Russell (1925) and Dirac (1926). Useful tables for solutions are given by Russell (1936). With reference to Eq. (2.125), Bates and Damgaard (1949) provide tables for evaluating σ_{mn}; whereas the relative multiplet strengths $\mathscr{S}(\mathscr{M})$, compiled by Goldberg (1935) and Menzel and Goldberg (1936), and the relative line strengths $\mathscr{S}(\mathscr{L})$, compiled by White and Eliason (1933) and Russell (1936), are given in the book by Allen (1963). Certain sum rules are useful for checking the relative strengths of lines. The Burger–Dorgelo–Ornstein sum rule (Burger and Dorgelo, 1924; Ornstein and Burger,

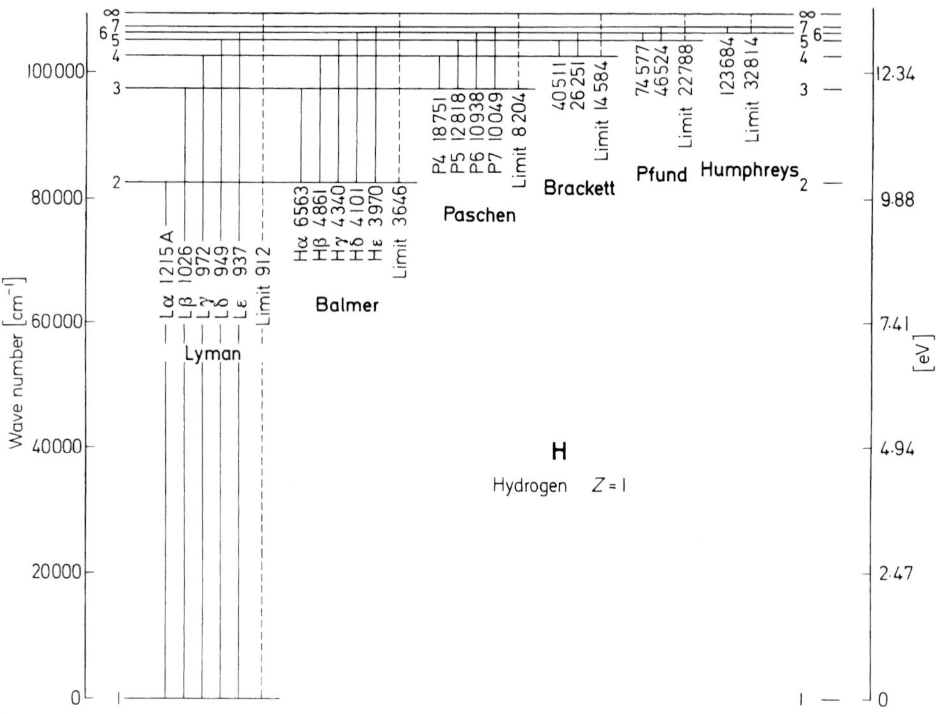

Fig. 2.4. Grotrian diagrams for H, He I, He II, C II, C III, C IV, O I, O II, O III, N II, Mg I, Mg II, Ca I, Ca II, Fe I, and Fe II. [After Moore and Merrill, 1968]

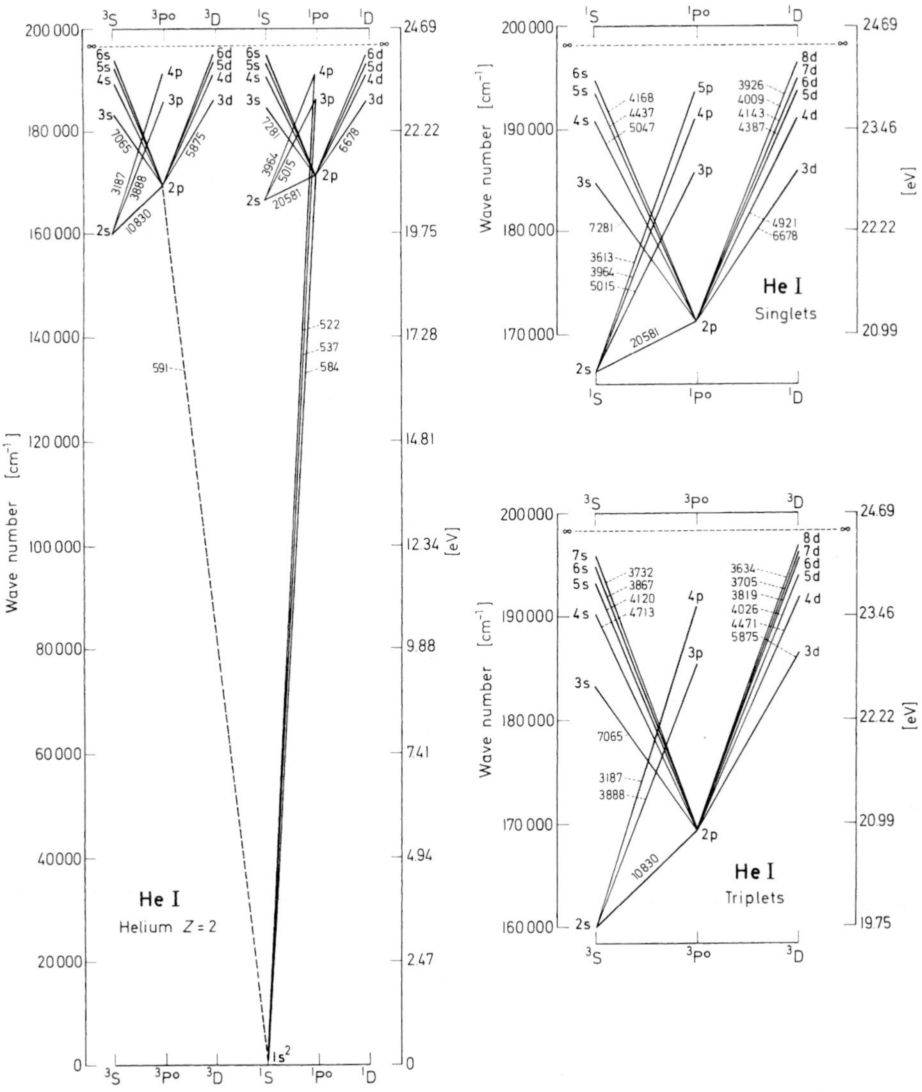

Fig. 2.4. Grotrian diagrams (continued)

1924) states that the sum of the intensities of all lines in a LS multiplet which belong to the same initial or final state is proportional to the statistical weight, $2J + 1$, of the initial or final state, respectively. The Reiche–Thomas–Kuhn f-sum rule (Thomas, 1925; Kuhn, 1925) states that the sum of the f values for all lines arising from a single atomic energy level shall equal the number of optical electrons (electrons which participate in giving the optical line and continuous spectrum). Additional useful sum rules have been derived by Shortley (1935), Menzel and Goldberg (1936), Menzel (1947), and Rohrlich (1959).

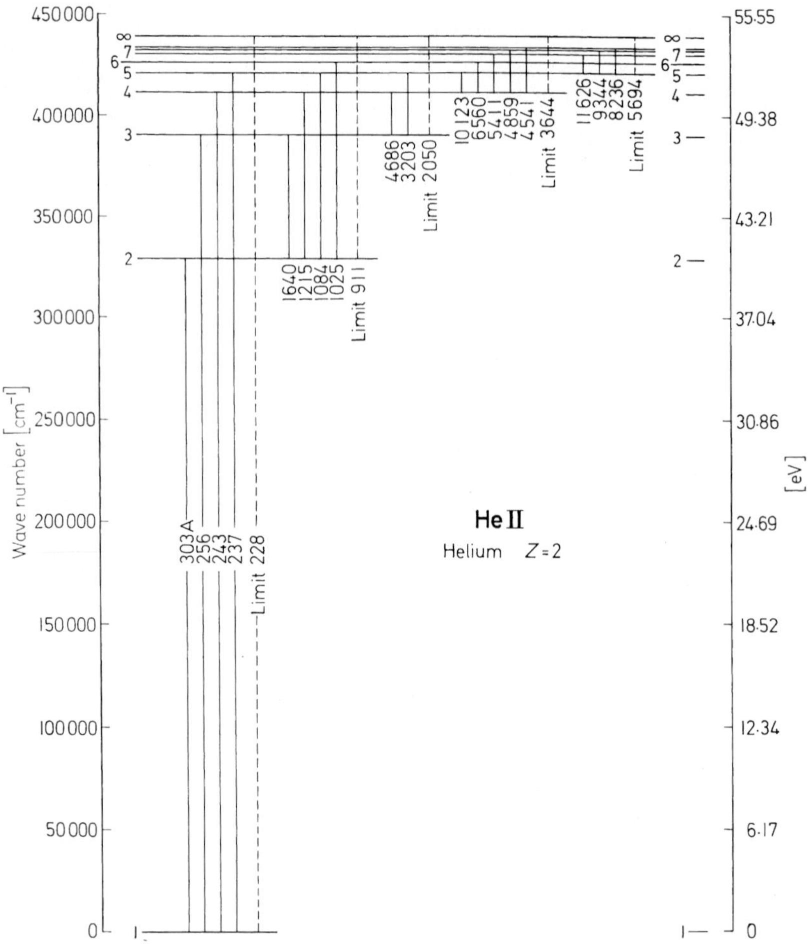

Fig. 2.4. Grotrian diagrams (continued)

Tables of atomic energy levels are given by C. E. Moore (1949, 1952, and 1958). Values of line frequencies are given by C. E. Moore (1950–69) and P. W. Merrill (1958). A bibliography of atomic forbidden line frequencies and strengths is given by Garstang (1962). A bibliography of f values and transition probabilities has been published by Glennon and Wiese (1962). Some line strengths are given by Varsavsky (1961). Ultraviolet and forbidden line strengths are given by Osterbrock (1963), Burbidge and Burbidge (1967), and Tarter (1969). Allen (1963) compiled a complete catalog of allowed and forbidden atomic line frequencies together with their oscillator strengths and transition probabilities. Sources of atomic spectroscopic data were reviewed by Martin (1992). The National Institute of Standards and Technology (NIST) maintains an Atomic Spectroscopic Database with contemporary reports on

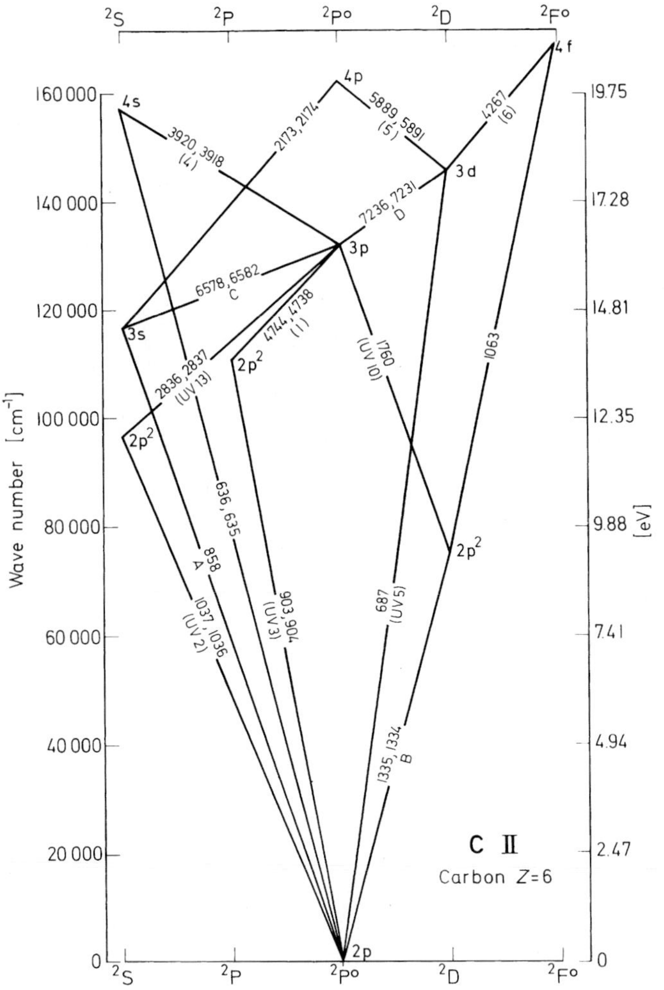

Fig. 2.4. Grotrian diagrams (continued)

the Status of Atomic Spectroscopic Data at their World Wide Web site http://physics.nist.gov/ under Physical Reference Data. Extensive bibliographies for atomic and molecular line strengths are given in the reports of Commission 14 of the International Astronomical Union. The Chianti atomic database provides atomic data needed for calculating astrophysical emission line spectra at wavelengths greater than 50 Angstroms as a function of both density and temperature (Dere et al., 1997). It is available on the World Wide Web at http://wwwsolar.nrl.navy.mil/chianti.html and contains atomic energy levels, wavelengths, radiative data and electron excitation data for ions which are abundant in cosmic plasmas.

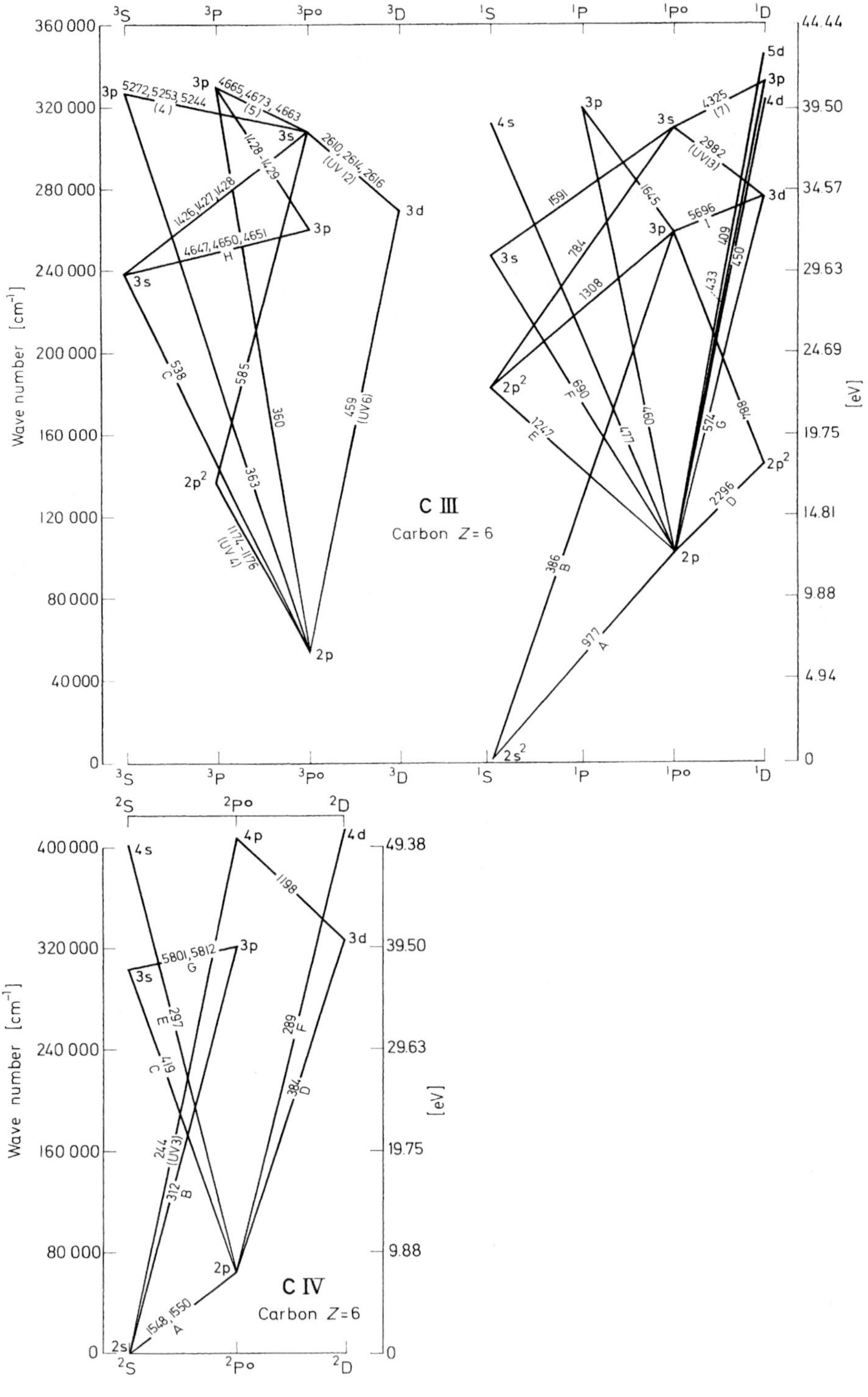

Fig. 2.4. Grotrian diagrams (continued)

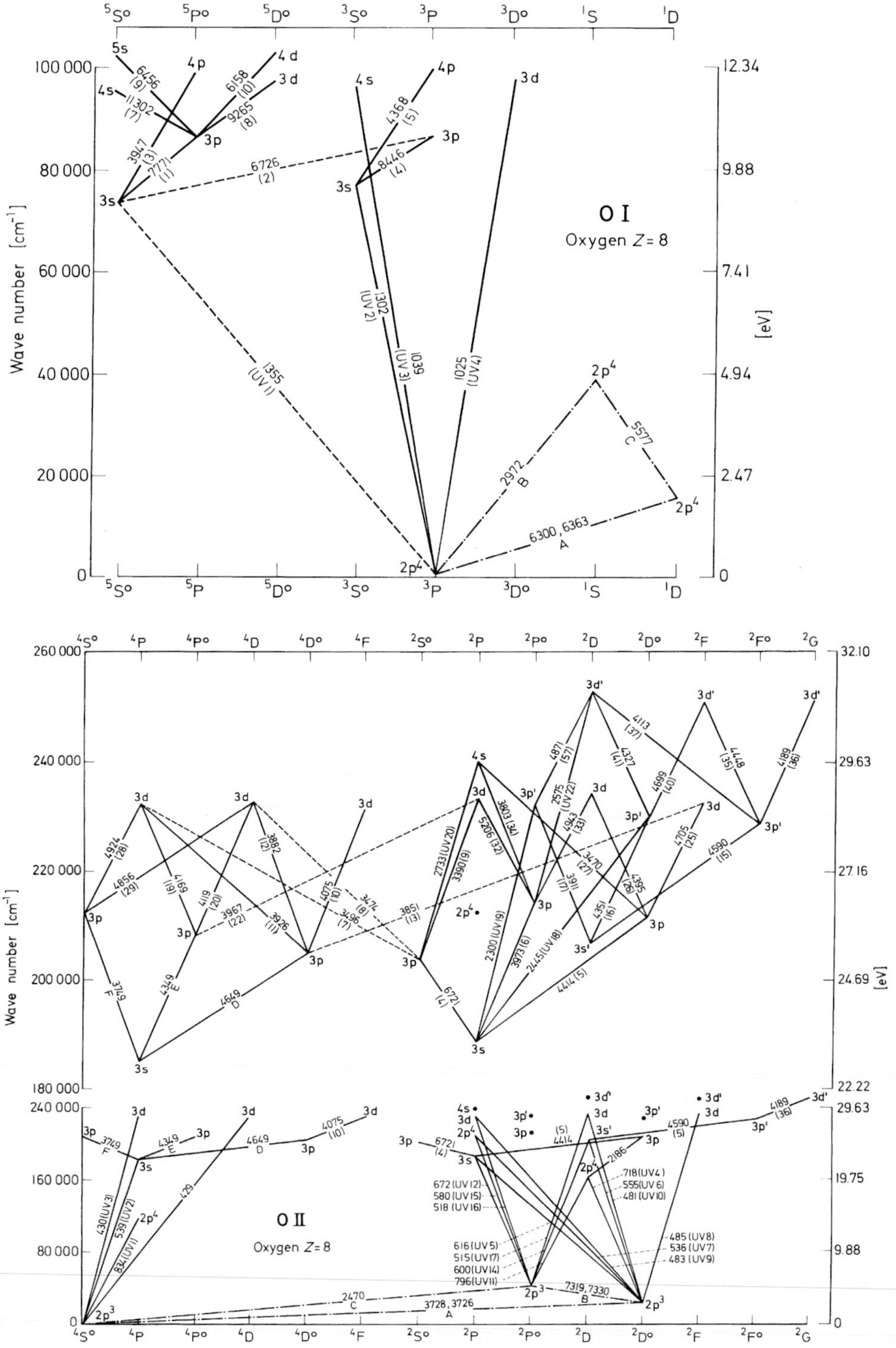

Fig. 2.4. Grotrian diagrams (continued)

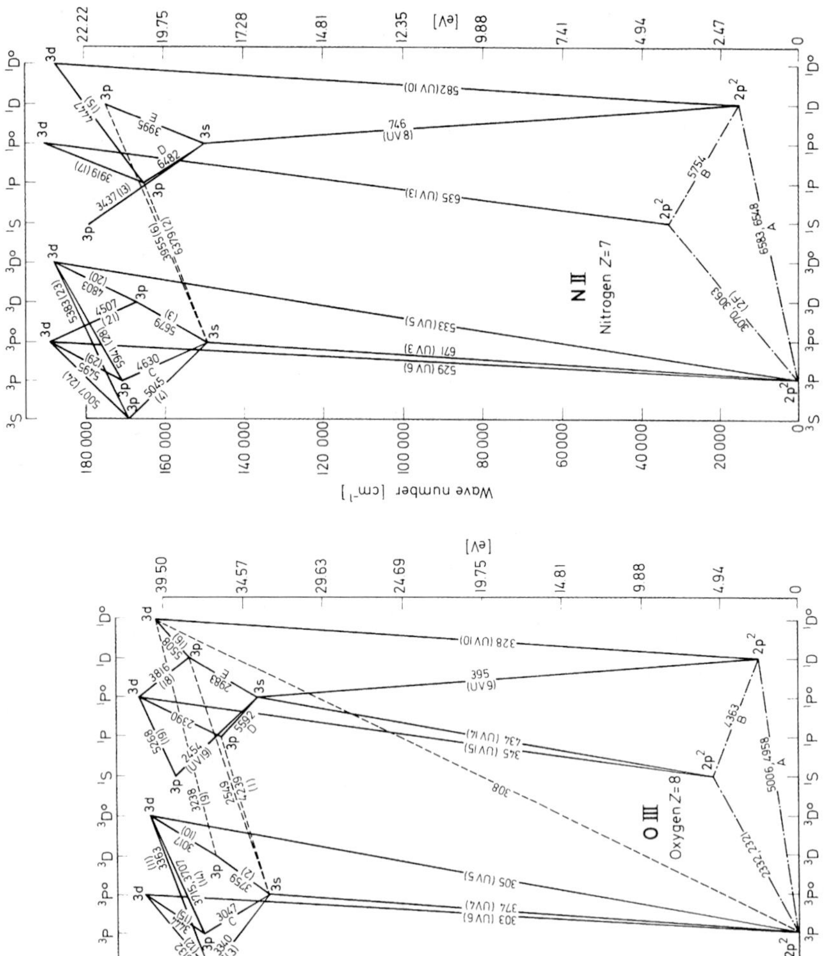

Fig. 2.4. Grotrian diagrams (continued)

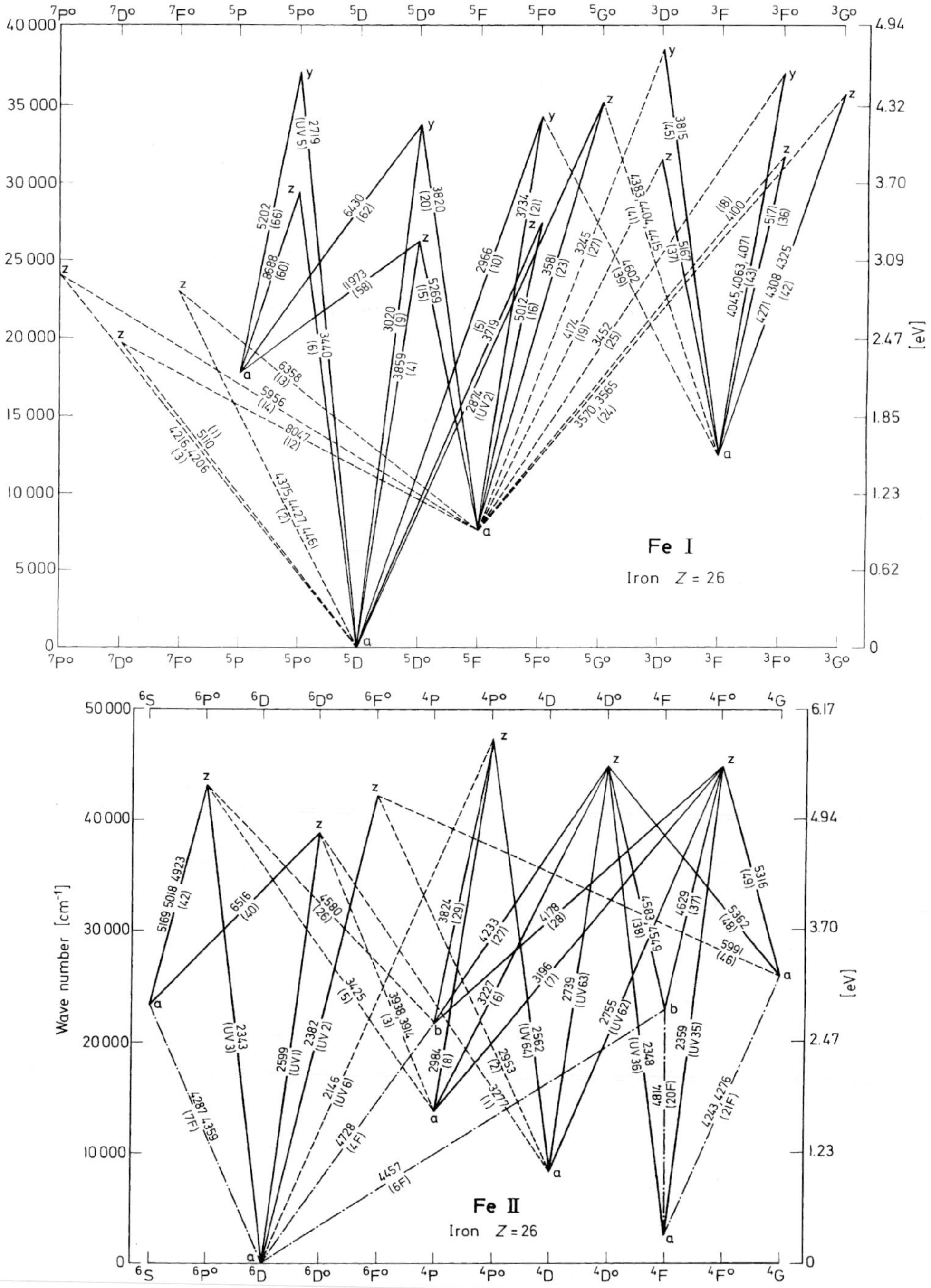

Fig. 2.4. Grotrian diagrams (continued)

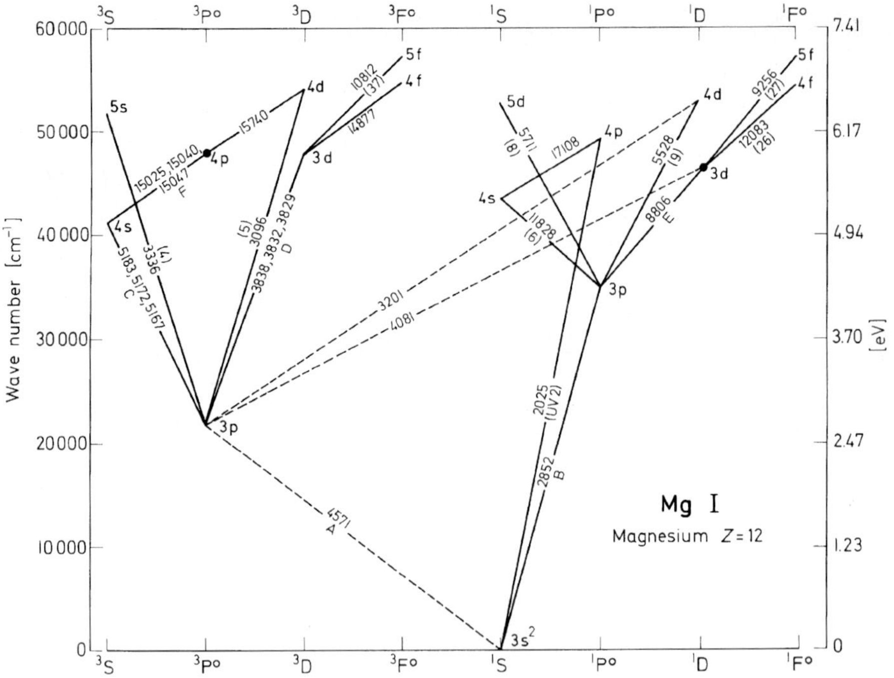

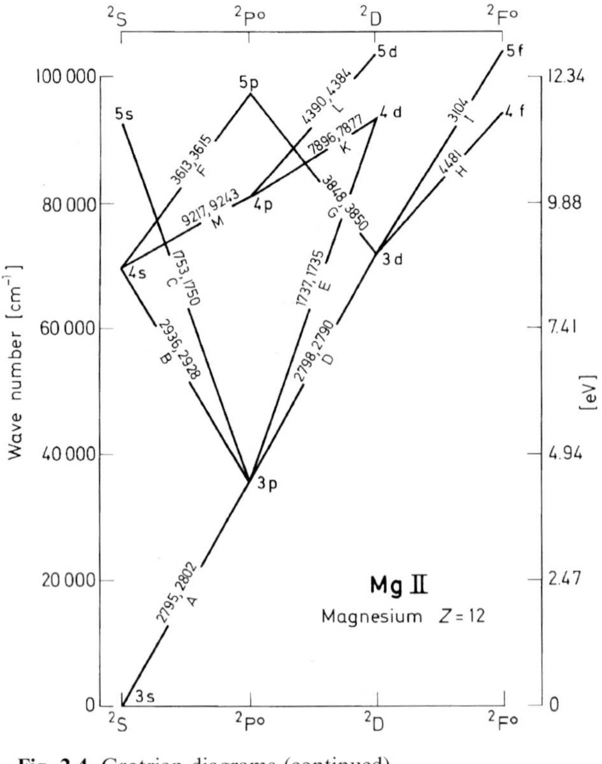

Fig. 2.4. Grotrian diagrams (continued)

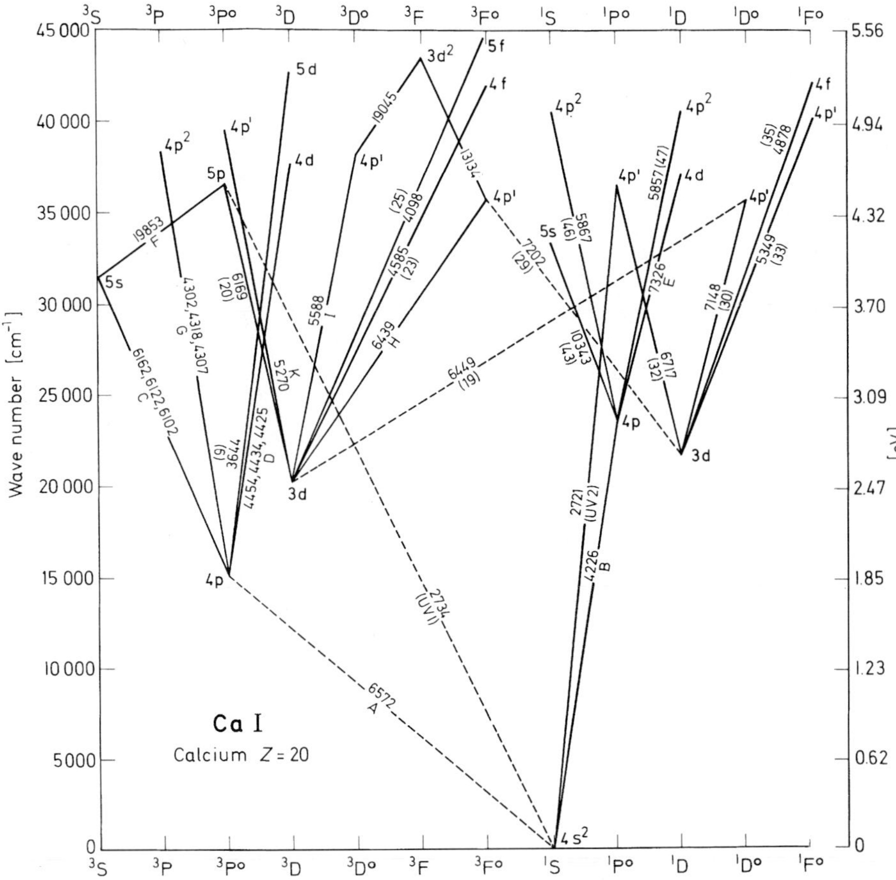

Fig. 2.4. Grotrian diagrams (continued)

The wavelengths of the strongest observed atomic lines from the most abundant elements [hydrogen (H), helium (He), oxygen (O), carbon (C), nitrogen (N), neon (Ne), silicon (Si), magnesium (Mg), sulpher (S), and argon (Ar)] are given in Table 2.7. The symbols I, II, III, ... denote atoms which are, respectively, neutral, singly ionized, doubly ionized.

2.14 Interstellar Hydrogen, Atomic Hyperfine Structure, and the Interstellar Medium

The possibility of detecting a radio frequency spectral line from neutral, or unionized, hydrogen, designated H I, was first suggested by Hendrik van de Hulst (Van de Hulst, 1945). He realized that the space between the stars would be cold and that most of the hydrogen atoms would be in their lowest energy

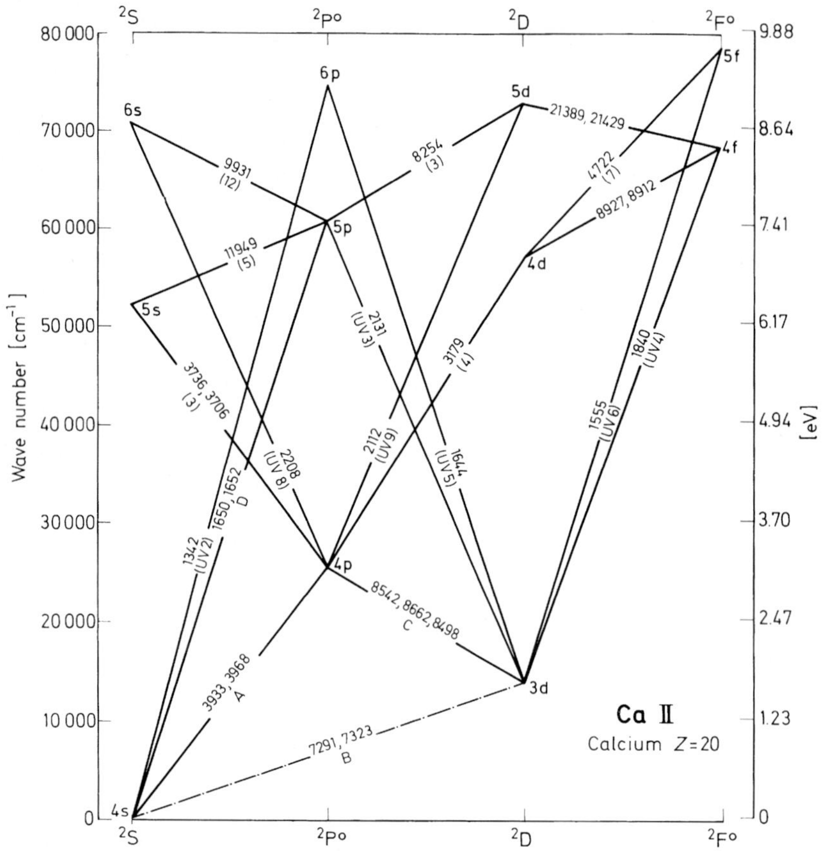

Fig. 2.4. Grotrian diagrams (continued)

ground state. The electron of the hydrogen atom in this state has two possibilities for its spin, and a change from one to the other can give rise to emission or absorption at a wavelength of 21 cm or a frequency of 1,420 MHz $= 1.420 \times 10^9$ Hz; these spectral lines result from a hyperfine transition. Although such changes occur rarely in the tenuous interstellar gas, a radio observer can detect them when looking through the vast extent of interstellar space.

Van de Hulst's prediction was confirmed by Harold Ewen and Edward Purcell who used a novel switched-frequency technique to reduce background noise and detect the hyperfine transition of neutral hydrogen in emission at 1,420 MHz (Ewen and Purcell, 1951). They delayed publication until a coordinated report could be issued with Dutch and Australian radio astronomers (Muller and Oort, 1951).

An interesting aspect of the first observation was that the detected line was seen in emission. The equilibrium populations of the hyperfine components of the hydrogen atom are determined by frequent collisions between the atoms,

Table 2.7. A table of the wavelengths, λ, of the strongest ultraviolet, optical and near-infrared lines from the most abundant elements in their most abundant stages of ionization. Forbidden transitions are denoted by brackets[1]

Ion	λ Å	Ion	λ Å	Ion	λ Å	Ion	λ Å	Ion	λ Å
N II	917	NIV	1,488	OIII	3,047	[SII]	4,069, 4,076	Hα	6,563
Ar II	920, 932	CIV	1,549, 1,551	[NII]	3,063	NIII	4,097	[NII]	6,583
S VI	933	[NeV]	1,575	OIII	3,133	HeII	4,100	HeI	6,678
C III	977	[NeIV]	1,602	HeII	3,203	Hδ	4,102	[SII]	6,717, 6,734
N III	990, 992	HeII	1,640	OIII	3,299, 3,312	HeII	4,200	[ArV]	7,006
He II	993	OIII	1,661, 1,663	OIII	3,341	Hγ	4,340	Hel	7,065
Ar VI	992, 1,002	OIII	1,667	[NeIII]	3,343	[OIII]	4,363	[ArIII]	7,136
Ar VI	1,014, 1,023	NIII	1,750	[NeV]	3,346	HeI	4,388, 4,471	[OII]	7,320, 7,325, 7,330
Ne VI	1,020	SiII	1,808, 1,817	OIV	3,412	HeII	4,541	[ArIII]	7,751
O VI	1,035	[NeIII]	1,815	[NeV]	3,426	MgI	4,571	[CLIV]	8,046
C II	1,037	Hγ, CIII	1,909	OIII	3,429, 3,444	NIII	4,634, 4,641	HeII	8,237
S IV	1,073	NII	2,141	HI	3,691, 3,697	CIII	4,647	P_{13}	8,665
N II	1,084, 1,086	[OIII]	2,321	HI	3,704, 4,712	CIV,[FeIII]	4,658	P_{12}	8,750
He II, N II	1085	CII	2,326	HI, [SIII]	3,722	HeII	4,686	P_{11}	8,862
Si II	1,194	[NeIV]	2,424, 2,426	[OII]	3,727, 3,729	[ArIV], HeI	4,711	p_{10}	9,014
S III	1,201	[OII]	2,470	HI	3,734, 3,750	[NeIV]	4,724, 4,726	[SIII]	9,069
Si III	1,207	HeII, [MgVII]	2,512	OIII	3,760	[ArIV]	4,740	P_9	9,229
Lyα, O V	1,216	[MgVII]	2,632	HI	3,771, 3,798	Hβ	4,861	Pε, [SIII]	9,545
N V	1,240, 1,243	HeII	2,734	HeI	3,820	HeI	4,922	CIII	9,710
S II	1,261	[MgV]	2,786	HI	3,835	[OIII]	4,959, 5,007	NaI	9,961
Si II	1,265	MgII	2,796, 2,799	[NeIII]	3,869	HeII	5,412	P_δ	10,049
C II	1,336	MgII	2,804	Hζ	3,889	[NII]	5,755	HeII	10,120, 10,124
Si IV	1,394, 1,403	[ArIV]	2,855	HeI	3,965	HeI	5,876	[SII]	10,320
O IV	1,402, 1,405	[ArIV]	2,869	[NeIII]	3,968	[OI] [SIII]	6,300	SiI	10,371, 10,603
O IV	1,406	[MgV]	2,931	Hε,[NeIII]	3,970	[OI]	6,364	SiI	10,627, 10,689
O IV	1,410, 1,413	[NeV]	2,973	HeI, HeII	4,026	[NII]	6,549	CI	10,691

[1] From Aller, Bowen, and Wilson (1963), O'Dell (1963), Osterbrock (1963), Burbidge and Burbidge (1967), and Tarter (1969).

which means that the observed excitation temperature of the lines is equal to the kinetic temperature of the interstellar gas (Field, 1959). The excitation temperature of the observed lines confirmed theoretical expectations that the neutral hydrogen in interstellar space has a temperature of about 100 degrees Kelvin (Spitzer and Savedoff, 1950; also see Sect. 3.5.3).

The newly discovered radio wavelength line could also be used to penetrate the curtains of dust hiding most of our Galaxy from view at optical wavelengths. By mapping the intensity and velocities of the 21-cm line in different directions, great circular sections of spiral arms could be delineated (Oort, Kerr and Westerhout, 1958). Early surveys of the galactic distribution of neutral hydrogen were completed by Westerhout (1969), Weaver and Williams (1971) and Heiles and Habing (1974). The structure of our Galaxy, as derived from observations of neutral hydrogen, or H I, has been reviewed by Burton (1988), and many of the relevant formulae are given in Volume II, including kinematic distances and the best estimates for the distance to the center of our Galaxy. The H I emission and absorption in the diffuse interstellar medium are reviewed by Kulkarni and Heiles (1988); they also discuss the thermodynamics and astrophysics of the atomic phase. Dickey and Lockman (1990) also focus on the spatial organization and vertical distribution of interstellar atomic hydrogen in our Galaxy. An overview of the distribution and observational properties of the interstellar medium is given by Burton (1992), including galactic rotation and the linear scale of the Milky Way.

Neutral hydrogen in other galaxies is reviewed by Giovanelli and Haynes (1988). Its role in detecting "invisible", "dark" matter beyond the luminous boundaries of galaxies is given in Volume II. A reference catalogue of neutral hydrogen, H I, observations of galaxies is given by Huchtmeier and Richter (1989), including the H I mass, the H I line width, the radial velocity, distance, name, and celestial position of the galaxy. Tully (1980) also provides the neutral hydrogen radial velocity and line widths together with the neutral hydrogen flux, in units of millions of solar masses per square Megaparsec (Mpc) for nearby galaxies within 40 Mpc.

When the spin of the nucleus is important, two additional quantum numbers are added to the description of the atom.

I = total spin angular momentum of the nucleus in units of $h/2\pi$. The I is positive and takes on integral or half integral values.
F = the total atomic angular momentum in units of $h/2\pi$. It is the vector sum of J and I, and takes on $2J+1$ or $2I+1$ values according as $J < I$ or $J \geq I$.

The nucleus has a magnetic moment, μ, given by

$$\mu = \mu_N g_N I \ , \tag{2.130}$$

where g_N is the Landé nuclear factor ($g_N \approx 1$) and $\mu_N = he/(4\pi m_p c)$ is the Bohr nuclear magneton for the proton of mass m_p. This magnetic moment interacts with the magnetic field produced by the electrons to give the hyperfine splitting. The energy, E_{hfs}, of the hyperfine structure of hydrogen like atoms is (Fermi, 1930; Bethe, 1933)

$$E_{\mathrm{hfs}} = g_{\mathrm{N}}\left(\frac{m}{m_{\mathrm{p}}}\right)\frac{\alpha^2 hcRZ^3}{n^3}\left[\frac{F(F+1) - I(I+1) - J(J+1)}{J(J+1)(2J+1)}\right] , \tag{2.131}$$

where the fine structure constant $\alpha = 2\pi e^2/(hc) = 7.2973 \times 10^{-3}$, the Rydberg constant $R = R_{\mathrm{A}} = 2\pi^2 \mu e^4/(ch^3)$ where R_{A} is given just below Eq. (2.113), the reduced mass $\mu = [mM_{\mathrm{A}}/(m + M_{\mathrm{A}})]$, and Z is the ionic charge. The selection rules for the transitions are $\Delta J = 0, \pm 1$, $\Delta F = 0, \pm 1$, and $F = 0 \nleftrightarrow F = 0$. The radiation is magnetic dipole with a dipole matrix element, μ, of about one Bohr magneton $\mu = eh/(4\pi mc) \approx 0.927 \times 10^{-20}$ erg Gauss^{-1}. It follows from Eq. (2.44) that

$$A_{mn} \approx \frac{\pi^2 e^2 h}{c^5 m^2} \nu_{mn}^3 \approx 10^{-42} \nu_{mn}^3 \text{ sec}^{-1} . \tag{2.132}$$

Detectable radio frequency radiation may arise from the $\Delta F = \pm 1$, $\Delta L = \Delta F = \Delta J = 0$ transitions of the abundant atoms or molecules. Radio frequency hyperfine transitions are given in Table 2.8.

Provided that the angular extent of the emitting gas cloud is larger than that of the antenna beam, and provided the gas is optically thin, measurements of the peak emission line antenna temperature, ΔT_{L}, leads to the column density, N_{H}, using the equation (Milne, 1930; Wild, 1952 or Eqs. (2.69) and (2.75)]

$$N_n = \frac{8\pi \nu_{mn}}{A_{mn} c^2} \frac{g_n}{g_m} \frac{k}{h} \frac{\Delta T_{\mathrm{L}} \Delta \nu_{\mathrm{L}}}{\eta_{\mathrm{A}}} \tag{2.133}$$

or

$$N_{\mathrm{H}} \approx \frac{4N_1}{3} \approx 3.88 \times 10^{14} \Delta T_{\mathrm{L}}\ \Delta \nu_{\mathrm{L}} / \eta_{\mathrm{A}} \text{ cm}^{-2} ,$$

where N_n is the line integrated density of atoms in the nth state, N_{H} is the total line integrated density of neutral hydrogen, ν_{mn} is the frequency of the m—n transition, g_n is the statistical weight of the nth level, A_{mn} is the Einstein coefficient for spontaneous transition between levels m and n, the h and k are, respectively, Planck's and Boltzmann's constants, c is the velocity of light, $\Delta \nu_{\mathrm{L}}$ is the half width of the observed emission line, and η_{A} is the antenna efficiency. The numerical approximation given in Eq. (2.133) is for the hyperfine ground state hydrogen transition for which $\nu_{mn} = 1420.40575$ MHz, $A_{mn} = 2.85 \times 10^{-15}$ sec^{-1}, $g_n = 2n^2$, and it is assumed that ΔT_{L} is given in degrees Kelvin and $\Delta \nu_{\mathrm{L}}$ is in Hz. If the line width, $\Delta \nu_{\mathrm{L}}$, is given in km sec^{-1}, Eq. (2.133) becomes

$$N_{\mathrm{H}} \approx 1.823 \times 10^{18} \Delta T_{\mathrm{L}} \Delta \nu_{L} / \eta_{\mathrm{A}} \text{ cm}^{-2} . \tag{2.134}$$

The emission line measurements provide only a measure of the line integrated density of the neutral hydrogen. When absorption line measurements are combined with the emission line measurement, the temperature of the gas can also be determined. If T_{s} denotes the spin temperature of the gas, and τ denotes the optical depth determined from the absorption line study, then the spin temperature is related to the emission line temperature, ΔT_{L}, by the equation

Table 2.8. A table of hyperfine transitions at radio frequencies[1]

Atom or molecule	Spin	Transition	Frequency (Hz)	A_{mn} (sec^{-1})
HI neutral hydrogen	$\frac{1}{2}$	${}^2S_{1/2},\ F = 0 - 1$	$1.420405751.786 \times 10^9 \pm 0.01$	2.85×10^{-15}
D deuterium	1	${}^2S_{1/2},\ F = \frac{1}{2} - \frac{3}{2}$	$3.27384349 \times 10^8 \pm 5$	4.65×10^{-17}
HeII singly ionized helium	$\frac{1}{2}$	${}^2S_{1/2},\ F = 1 - 0$	$8.66566 \times 10^9 \pm 1.8 \times 10^5$	6.50×10^{-13}
N VII ionized nitrogen	1	${}^2S_{1/2},\ F = \frac{1}{2} - \frac{3}{2}$	5.306×10^7	1.49×10^{-19}
NI neutral nitrogen	1	${}^4S_{3/2},\ F = \frac{3}{2} - \frac{5}{2}$	2.612×10^7	1.78×10^{-20}
		$F = \frac{1}{2} - \frac{3}{2}$	1.567×10^7	3.84×10^{-21}
H_2^+ ionized molecular hydrogen	1	$F_2,\ F\ \frac{3}{2}, \frac{5}{2} - \frac{1}{2}, \frac{3}{2}$	$1.40430 \times 10^9 \pm 10^7$	2.75×10^{-15}
		$F_2,\ F\ \frac{3}{2}, \frac{3}{2} - \frac{1}{2}, \frac{3}{2}$	$1.41224 \times 10^9 \pm 10^7$	2.80×10^{-15}
NaI neutral sodium	$\frac{3}{2}$	${}^2S_{3/2},\ F = 1 - 2$	1.77161×10^9	5.56×10^{-15}

[1] From Townes (1957), Field, Somerville, and Dressler (1966), and Kerr (1968).

$$T_s = \frac{\Delta T_L}{\eta_A[1-\exp(-\tau)]} \approx \frac{\Delta T_L}{\tau \eta_A} \ . \tag{2.135}$$

The column density of interstellar atomic hydrogen, N_H, can be determined from the excitation, or spin, temperature, T_s, and the optical depth, τ, of the 1420 MHz transition. $N_H \approx 3 \times 10^{14} T_s \tau$, and for $\tau \ll 1$ the brightness temperature $T_B = T_s \tau$. If $\tau(v)$ denotes the optical depth at velocity v, the column density $N_H(v)$ at that velocity is given by $N_H(v) = C\, T_s \tau(v)$ where $C = 1.83 \times 10^{18}\ \text{cm}^{-2}\,\text{K}^{-1}\,(\text{km/s})^{-1}$. For the optically thin case, $\tau(v) \ll 1$, the brightness temperature $T_B(v) = T_s \tau(v) = N(v)/C$, and the measured brightness is proportional to the column density per unit velocity. For the optically thick case $\tau(v) \gg 1$, the brightness temperature is the spin temperature, or $T_B(v) = T_s$.

For an optically thin gas seen in absorption against a source of effective brightness temperature, T_{0s}, the optical depth, τ, is given by

$$\tau = \frac{\Delta T_A}{\eta_A T_{0s}} \tag{2.136}$$

where ΔT_A is the absorption line temperature (cf. Eq. (2.75)). Absorption line observations of neutral hydrogen in our Galaxy are given by Clark (1965), Hughes, Thompson, and Colvin (1971), and Radhakrishnan et al. (1972).

The excitation or spin temperature, T_s, is related to the kinetic temperature, T_k, of the gas and the radiation temperature, T_R, of the background radiation by the equation of statistical equilibrium (Eq. (2.82)). If R_{mn} denotes the probability coefficient for transitions induced by collisions, and A_{mn} denotes the Einstein coefficient for a spontaneous transition, then we have the relation

$$T_s = T_k \left[\frac{T_R + T_0}{T_k + T_0}\right] \ , \tag{2.137}$$

where

$$T_0 = \frac{h\nu_{mn} R_{mn}}{k A_{mn}} \ ,$$

and ν_{mn} is the frequency of the transition. The kinetic temperature, T_k, may be measured from the Doppler broadening of the line. For the 21 cm hydrogen line we have $A_{mn} = 2.85 \times 10^{-15}\ \text{sec}^{-1}$, and the probability coefficient for collision induced transitions is

$$R_{mn} = n_H \sigma_F \langle v \rangle = n_H \sigma_F \left(\frac{8kT_k}{\pi m_H}\right)^{1/2} \approx 7 \times 10^{-11} n_H T_k^{1/2}\ \text{sec}^{-1} \ , \tag{2.138}$$

where n_H is the volume density of hydrogen atoms, the average effective cross section, σ_F, for mutual collisions between hydrogen atoms goes from $6.6 \times 10^{-15}\ \text{cm}^2$ to $2.9 \times 10^{-15}\ \text{cm}^2$ when T_k ranges from 2,000 to 10,000° K, and $\langle v \rangle$ is the average thermal velocity. It follows from Eqs. (2.137) and (2.138) that the spin and kinetic temperatures will be equal when

$$n_H > 6 \times 10^{-4} T_k^{1/2}\ \text{cm}^{-3} \ , \tag{2.139}$$

provided that hydrogen collision is the dominant collision process.

Attempts to detect intergalactic neutral hydrogen in emission (Penzias and Wilson, 1969) and in absorption (Allen, 1969) have been unsuccessful. These observations lead to the limits $n_H < 3 \times 10^{-6}$ atoms cm^{-3} and $n_H/T_s < 2.7 \times 10^{-8}\, cm^{-3}\, {}^\circ K^{-1}$ if the Hubble constant is assumed to be 75 km sec^{-1} Mpc^{-1}. The absence of the absorption of the continuum radiation of quasars below the redshifted Lyman α line puts much tighter constraints of $n_H < 5 \times 10^{-12}\, cm^{-3}$ on the number density, n_H, of neutral (unionized), intergalactic hydrogen (Shklovskii, 1964; Scheuer, 1965; Gunn and Peterson, 1965) – also see Volume II. A "forest" of discrete absorption features on the short-wavelength side of the redshifted Lyman alpha radiation of quasars has been attributed to neutral hydrogen clumped into galaxy-sized clouds that are highly photoionized (see Weymann, Carswell and Smith, 1981; Blades, Turnshek and Norman, 1988; and Volume II).

The interstellar matter of our Galaxy does not consist solely of atomic hydrogen, or H I regions. Indeed, the first evidence for gas between the stars was provided by absorption lines of calcium (Dunham, 1937). Interstellar dust was discussed in Sect. 1.40, and the ways that bright stars transform nearby interstellar matter were presented in Sect. 2.11 on planetary nebulae and in Sect. 2.12 on ionized hydrogen, or H II regions. The next Sect. 2.15 reviews line radiation from interstellar molecules, including molecular hydrogen, H_2, that can account for about half the mass of the interstellar medium. The interstellar magnetic field is mentioned in Sect. 2.17, and the relevant gas processes are discussed in Chapter 3, including the Saha–Boltzmann ionization equation (3.3.1.3), the Strömgren radius (3.3.1.5), heating and cooling mechanisms (3.5.3), supernovae explosions (3.5.9) and gravitational collapse that leads to star formation (3.5.11).

Many of these diverse ingredients of the interstellar medium are brought together in comprehensive discussions, including those by Dyson and Williams (1980), Dalgarno and Layzer (1987), Hollenbach and Thronson (1987), Scheffler and Elsässer (1987), and Longair (1994). The local interstellar medium has been reviewed by Cox and Reynolds (1987), focusing on our local bubble within it. High resolution optical and ultraviolet absorption-line studies of interstellar gas are reviewed by Cowie and Songaila (1986), while Turner (1988) provides an overview of radio molecular line investigations of the interstellar medium. The composition of the interstellar medium evolves with time because of an ongoing enrichment with heavy elements forged inside stars that have completed their lives and exploded; aspects of this chemical evolution are reviewed by Wheeler, Sneden and Truran (1989) and Rana (1991).

2.15 Line Radiation from Molecules

The first cosmic molecules were discovered in the optical spectra of bright stars, planetary atmospheres, and comets. Absorption features of diatomic molecules were observed in the light of bright stars, due to absorption by molecules in interstellar space (Swings and Rosenfeld, 1937; McNalley, 1968), and in the photospheres of the Sun and cool, late-type stars (Russell, 1934; Tsuji, 1986; Jaschek and Jaschek, 1988). The optical spectra of carbon dioxide, CO_2, was discovered in the atmospheres of Mars and Venus, and those of ammonia, NH_3, and methane, CH_4, in the outer atmospheres of Jupiter, Saturn, Uranus and Neptune (Dunham, 1933). Radicals such as OH, CO and NH, observed in cometary spectra, were attributed to parent molecules such as water, H_2O, carbon dioxide and ammonia (Bobrovnikoff, 1942; Swings, 1948). The dissociation energies and ionization potentials for diatomic molecules, molecular ions and radicals are given by Singh and Chaturvedi (1987), together with their identification in planetary and cometary atmospheres.

Then, after the discovery of interstellar hydrogen at radio frequencies, astronomers began to speculate about the possibility of detecting molecules at radio wavelengths. Iosif Shklovskii considered the possibility of detecting the microwave emission of the OH radical (Shklovskii, 1953), and Charles Townes provided laboratory measurements of the low-temperature rotational transitions of molecules that might be detected by radio astronomers, including OH, ammonia, water and carbon monoxide (Townes, 1957).

The first observations of interstellar OH at microwave frequencies (Weinreb, Barrett, Meeks and Henry, 1963) was soon followed by the discovery of emission-line sources of OH that are exceptionally compact and have brightness temperatures exceeding a million, million degrees (Gundermann, 1965; Weaver et al., 1965; Weinreb et al., 1965). In these cases, the interstellar OH sources were acting like a gigantic amplifier similar to masers built in terrestrial laboratories (Moran et al., 1968; Litvak, 1969). Maser is an acronym for the microwave amplification by stimulated emission of radiation.

Due to the overwhelming cosmic abundance of atomic hydrogen, the diatomic hydrogen molecule, H_2, was expected to exist in portions of interstellar space (Eddington, 1937; Strömgren, 1939), and detailed studies by Salpeter (1963) and Hollenbach et al. (1971) predicted large fractional abundances of H_2 in interstellar clouds. The symmetric hydrogen molecule does not radiate in the radio range, and remained unobserved in the interstellar medium until its ultraviolet absorption signatures were detected in the light of bright stars using detectors aboard a rocket (Carruthers, 1970) and the Copernicus satellite (Spitzer et al., 1973; Spitzer, 1976). Interstellar molecular hydrogen is discussed by Allen and Robinson (1976, 1977) and Hill and Hollenbach (1976), and reviewed by Shull and Beckwith (1982). The ultraviolet wavelengths of the electronic transitions of molecular hydrogen are between 1013 and 1108 Angstroms, and infrared rotation-vibration transitions have also been observed. Although hydrogen is the most abundant molecule in interstellar space, the ultraviolet absorption studies are limited to relatively unreddened clouds, and cannot probe the dense molecular clouds that are examined through the radio wavelength emission of asymmetric molecules.

The discovery of OH emission did not at first stimulate a search for other, more complex interstellar molecules, since at the time astronomers thought that no more than two atoms could come together at one time in the low density regions of interstellar space. In addition, even if somewhat more complex molecules were formed, astronomers expected that they would be quickly destroyed by energetic ultraviolet starlight and by high-energy cosmic rays. It was therefore something of a surprise when the radio signatures of the polyatomic molecules of ammonia, NH_3, and water, H_2O, were discovered in interstellar space (Cheung et al., 1968). These were soon followed by the detection of formaldehyde, H_2CO, by Snyder et al. (1969), and carbon monoxide, CO, by Wilson, Jefferts and Penzias (1970). These discoveries triggered an avalanche of molecular searches using the latest laboratory measurements, soon resulting in the detection of more than 100 interstellar molecular lines at radio wavelengths. These discoveries had not been anticipated even a decade earlier, because astronomers had overlooked the importance of interstellar dust grains in shielding molecules from ultraviolet starlight and in acting as a catalyst in forming complex molecules. Reviews of these early investigations of the radio radiation from interstellar molecules have been provided by Rank, Townes and Welch (1971), Zuckerman and Palmer (1974) and Townes (1977).

Complex interstellar molecules have now been extensively studied, dramatically changing our view of the interstellar medium. Despite the great abundance of atomic interstellar hydrogen, we now recognize that by mass about one-half of the entire interstellar medium in our Galaxy is in molecular form, and that in some large regions about 90 percent by mass is in molecular form. Using molecules as probes, temperatures between 0.1 and 1,000 °K and densities of up to 10^{10} cm^{-3} can be detected. By way of contrast, 21-cm studies of interstellar atomic hydrogen were, for some time, modeled with a temperature of about 100 °K and densities near 1 cm^{-3}. Interstellar ammonia, reviewed by Ho and Townes (1983), can be used as a sensitive thermometer. Because of its ubiquity, the carbon monoxide molecule, CO, has been used to trace the general distribution of molecular gas in our Galaxy; this distribution has been reviewed by Combes (1991). In the cold interstellar medium, the CO molecule is excited to the first rotational level by collisions with H_2 molecules, so it provides a good tracer of molecular hydrogen in galaxies. Other molecules, such as CS, HCN, HC_3N and HNCO, have been used to infer number densities.

Laboratory measurements of the frequencies of interstellar molecular transitions have played an important role from the beginning. Lovas, Snyder and Johnson (1979) provided a list of recommended rest frequencies of all molecules detected in interstellar clouds at the time. This report was updated by Lovas (1986) and Lovas (1991). The rest frequencies of observed interstellar molecular microwave transitions are provided by the National Institute of Standards and Technology (NIST) on their World Wide Web site http://physics.nist.gov/ under Physical Reference Data. A greatly abbreviated table of molecular lines that are often observed at radio wavelengths is given in Table 2.9.

Turner (1988) reviews the use of molecular radio emission as a probe of the interstellar medium, while Turner and Ziurys (1988) provide an overview of the astrophysics and chemistry of the molecules themselves. Interstellar molecule reactions were also reviewed by Watson (1976). The physics of molecular clouds is discussed by Genzel (1992), and molecular clouds and gas in external

Table 2.9. Some frequently observed interstellar molecules; the transition frequencies are given in MHz = 10^6 Hz or GHz = 10^9 Hz and the transition wavelengths are given in cm or mm = 0.1 cm

Chemical symbol	Name of molecule	Year of discovery	Frequency	Wavelength
OH	Hydroxyl	1963	1665.4 MHz	18.0 cm
CO	Carbon monoxide	1970	115.27 GHz	2.60 mm
CS	Carbon monosulfide	1971	146.97 GHz	2.04 mm
H_2O	Water	1968	22.235 GHz	1.35 cm
HCN	Hydrogen cyanide	1970	88.632 GHz	3.38 mm
NH_3	Ammonia	1968	23.694 GHz	1.27 cm
H_2CO	Formaldehyde	1969	4,829.7 MHz	6.21 cm
HNCO	Isocyanic acid	1971	87.925 GHz	3.41 mm
HC_3N	Cyanoacetylene	1970	9,098.4 MHz	3.30 cm
CH_3CN	Methyl cyanide	1971	110.38 GHz	2.72 mm
CH_3C_2H	Methylacetylene	1971	85.457 GHz	3.51 mm

galaxies have been reviewed by Morris and Rickard (1982) and Young and Scoville (1991). Several articles on molecular clouds in the Milky Way and external galaxies are edited by Dickman, Snell and Young (1988). Molecular masers have been reviewed by Reid and Moran (1981) and Elitzur (1992).

Present-day star formation takes place within giant molecular clouds whose properties are summarized by Goldsmith (1987) and in Table 2.10. They have a typical radius of 10 parsecs and are about 100,000 times as massive as the Sun. By way of comparison, the separation between adjacent stars outside interstellar clouds is about 1 parsec = 3.18×10^{18} cm. The molecular gas is extremely cold, about 10°K, and mainly composed of molecular hydrogen, H_2, with a typical number density of $N_{H2} \approx 200\,cm^{-3}$. Up to 10 percent of the mass of giant molecular clouds exists in dense cores with $N_{H2} \geq 10^9\,cm^{-3}$. Shu, Adams and Lizano (1987) review the observation and theory of star formation in molecular clouds, and Turner (1988) discusses molecular radio emission as probes of star formation. Maps of the CO emission of the molecular clouds in

Table 2.10. Giant molecular clouds

Category	Radius (pc)	Density (cm^{-3})	Mass (solar masses)	Temperature (°K)
Giant Molecular Cloud Complex	10–40	100–300	10^5–10^6	7–15
Giant Molecular Cloud	1–10	10^3–10^4	10^3–10^5	15–40
Giant Molecular Cloud Core	0.2–1.5	10^4–10^6	10–10^3	30–100
Giant Molecular Cloud Clump	< 0.2	> 10^6	30–10^3	30–200

Orion and Monoceros (Maddalena et al., 1980), and of the molecular clouds in Perseus, Taurus and Auriga (Ungerechts and Thaddeus, 1987) are reproduced in Lang (1992). The Orion molecular and star forming region is reviewed by Genzel and Stutzki (1989). Molecular outflows from young stellar objects are reviewed by Lada (1985) and Bachiller (1996). Outflow velocities, mass loss rates and ages of such molecular outflows are provided with their celestial positions and distances by Lang (1992). Formulae related to the gravitational collapse of molecular clouds to form stars are given in Sect. 3.5.11.

2.15.1 Energies and Frequencies of the Molecular Transitions

The interaction of electrons in molecules may be assumed to be similar to that of the atomic Russell–Saunders (*LS*) coupling discussed in Sect. 2.13. The molecular fine structure is usually described by the following quantum numbers.

J = the total angular momentum excluding nuclear spin in units of $h/2\pi$.
N = the total orbital angular momentum including rotation in units of $h/2\pi$.
K = the projection of N on the molecular axis.
O = the orbital angular momentum due to molecular rotation in units of $h/2\pi$.
Λ = the projection of the electronic orbital angular momentum on the molecular axis in units of $h/2\pi$. The molecular state is designated as $\Sigma, \Pi, \Delta, \varphi, \ldots$ according as $\Lambda = 0, 1, 2, 3, \ldots$ The Λ is always positive for $\Lambda = |M_{\mathrm{L}}|$ where $M_{\mathrm{L}} = L, L-1, \ldots, -L$ where L is the electronic angular momentum.
Σ = the projection of the electron spin angular momentum on the molecular axis in units of $h/2\pi$ (not to be confused with the symbol Σ for $\Lambda = 0$). The Σ takes on 2S + 1 values, where S is the electron spin.
Ω = the total electronic angular momentum about the molecular axis in units of $h/2\pi$. The $\Omega = |\Lambda + \Sigma|$, for the Hund coupling case (a).

Each pair of Λ and Σ designate a term which has the symbol ${}^{2S+1}\Lambda_{\Omega}$, where $2S+1$ is the numerical value of the multiplicity, the Λ is designated by its Greek symbol, and Ω is the numerical value of the total electronic angular momentum. A Σ state is called Σ^{+} or Σ^{-} according to whether its electronic eigenfunction remains unchanged or changes sign upon reflection in any plane passing through the internuclear axis. The electronic state is even (g) or odd (u) according to whether the electronic eigenfunction remains unchanged or changes sign for a reflection at the center of symmetry. A rotational level is positive (+) or negative (−) according to whether the total eigenfunction changes sign for reflection at the origin. When the nuclei are identical, the term is symmetric (s) or antisymmetric (a) according to whether the total eigenfunction remains unchanged or changes sign when the nuclei are exchanged. Transitions between levels must obey selection rules which depend on the quantum numbers, the symmetry properties, and the type of coupling. These rules are given in Table 2.11 for diatomic molecules.

The symbol $\nleftrightarrow$ denotes "cannot combine with". The coupling symbols (a) and (b) refer, respectively, to Hund's coupling cases (a) and (b). For case (a), the electronic motion is coupled strongly to the line joining the nuclei so that

Table 2.11. Selection rules for diatomic molecular spectra[1]

Coupling	Electric dipole (allowed)	Magnetic dipole (forbidden)	Electric quadrupole (forbidden)
(1) General	$\Delta J = 0, \pm 1$ $(0 \nleftrightarrow 0)$	$\Delta J = 0, \pm 1$ $(0 \nleftrightarrow 0)$	$\Delta J = 0, \pm 1, \pm 2$ $(0 \nleftrightarrow 0, 0 \nleftrightarrow 1, \frac{1}{2} \nleftrightarrow \frac{1}{2})$
(2) General	$(+ \leftrightarrow -, + \nleftrightarrow +, - \nleftrightarrow -)$	$(+ \leftrightarrow +, - \leftrightarrow -, + \nleftrightarrow -)$	$(+ \leftrightarrow +, - \leftrightarrow -, + \nleftrightarrow -)$
(3) General	$(s \nleftrightarrow a, s \leftrightarrow s, a \leftrightarrow a)$	$(s \leftrightarrow s, a \leftrightarrow a, s \nleftrightarrow a)$	$(s \leftrightarrow s, a \leftrightarrow a, s \nleftrightarrow a)$
(4) General	$(g \leftrightarrow u, g \nleftrightarrow g, u \nleftrightarrow u)$	$(g \leftrightarrow g, u \leftrightarrow u, g \nleftrightarrow u)$	$(g \leftrightarrow g, u \leftrightarrow u, g \nleftrightarrow u)$
(5) (a) and (b)	$\Delta S = 0$	$\Delta S = 0$	$\Delta S = 0$
(6) (a)	$\Delta \Lambda = 0, \pm 1$	$\Delta \Lambda = 0$ if $\Delta \Sigma = \pm 1$ $\Delta \Lambda = \pm 1$ if $\Delta \Sigma = 0$	$\Delta \Lambda = 0, \pm 1, \pm 2$
(b)	$\Delta \Lambda = 0, \pm 1$	$\Delta \Lambda = 0, \pm 1$	
(7) (a)	$\Delta \Sigma = 0$	see (6) (a)	$\Delta \Sigma = 0$
(8) (a)	$\Delta \Omega = 0, \pm 1$	$\Delta \Omega = \pm 1$	$\Delta \Omega = 0, \pm 1, \pm 2$
(9) (a)	$\Delta J \neq 0$ for $\Omega = 0 \leftrightarrow \Omega = 0$		$\Delta J \neq 1$ for $\Omega = 0 \leftrightarrow \Omega = 0$
(10) (b)	$\Delta K = 0, \pm 1$	$\Delta K = 0, \pm 1$	$\Delta K = 0, \pm 1, \pm 2$
(11) (b)	$\Delta K \neq 0$ for $\Sigma \leftrightarrow \Sigma$ transitions	$\Delta K = 0$ for $\Sigma \leftrightarrow \Sigma$ transitions	$\Delta K = 0, \pm 1, \pm 2$ for $\Sigma \leftrightarrow \Sigma$ transitions
(12) (b)	$\Sigma^+ \nleftrightarrow \Sigma^-$	$\Sigma^+ \nleftrightarrow \Sigma^-$	$\Sigma^+ \nleftrightarrow \Sigma^-$

[1] For (3) the nuclei are assumed to be identical, and the term is symmetric, *s*, or antisymmetric, *a*, according to whether the total eigenfunction remains unchanged or changes sign for an exchange of the nuclei. The *g* and *u* in (4) denote, respectively, even and odd electronic states for nuclei of equal charge. An electronic state is even or odd according to whether the electronic eigenfunction remains unchanged or changes sign for a reflection at the center of symmetry. For (5) through (12) the (a) and (b) denote, respectively, Hund's case a and case b. For (12) a Σ state is called Σ^+ or Σ^- according to whether its electronic eigenfunction remains unchanged or changes sign upon reflection in any plane passing through the internuclear axis.

$\Omega = |\Lambda + \Sigma|$. For case (b) only the electronic orbital angular momentum, Λ, is strongly coupled to the molecular axis so that Λ and N couple to form K, and K and S couple to give J.

The total energy, E, of a molecule may be regarded as the superposition of the energy due to the electrons, E_e, the vibrational energy, E_v, and the rotational energy, E_r,

$$E = E_r + E_v + E_e \ . \tag{2.140}$$

An order of magnitude discussion for diatomic molecules indicates that (Shu, 1991):

$$E_e \approx e^2/a_0,$$

$$E_v \approx (m_e/M)^{1/2} E_e \ ,$$

and

$$E_r \approx (m_e/M) E_e \ ,$$

where the Bohr radius $a_o = 0.529 \times 10^{-8}$ cm, the electron charge $e = 4.8 \times 10^{-10}$ e.s.u., the electron mass $m_e = 9.1 \times 10^{-28}$ g, and M is the mass of a typical nucleus. For $m_e/M \approx 10^{-3}$ to 10^{-4}, the vibrational energies, E_v, correspond to radiation in the near- or mid-infrared, and rotational photon energies, E_r, have wavelengths in the millimeter-wave region of the radio spectrum.

Transitions between electronic states occur at optical and ultraviolet frequencies, the vibrational energies usually correspond to transitions in the infrared, and the rotational transitions are at radio frequencies. For end-over-end rotation of a rigid molecule, the rotational energy, E_r, solution to the Schrödinger equation (Schrödinger, 1925, 1926, Eq. (2.126)) is given by

$$E_r = hBJ(J+1) \ , \tag{2.141}$$

where the rotational angular quantum number, J, is zero or an integer, the total angular momentum in units of $h/2\pi$ is $[J(J+1)]^{1/2}$, and the rotational constant, B, is given by

$$B = \frac{h}{8\pi^2 I} \ ,$$

where I is the moment of inertia of the molecule. For a diatomic molecule with atoms of respective masses m_1 and m_2, and separation r_{12},

$$I = \frac{r_{12}^2 m_1 m_2}{m_1 + m_2} = r_{12}^2 \mu \ ,$$

where μ is the reduced mass. Similarly for a triatomic molecule

$$I = \frac{m_1 m_2 r_{12}^2 + m_1 m_3 r_{13}^2 + m_2 m_3 r_{23}^2}{m_1 + m_2 + m_3} \ ,$$

where r_{ij} represents the distance between the i and j atoms with respective masses m_i and m_j.

The frequencies, ν_r, observed when a diatomic molecule makes a transition between a lower rotational state of energy E_1 to an upper state of energy E_2 is given by:

$$\nu_r = (E_2 - E_1)/h = h[J_2(J_2+1) - J_1(J_1+1)]/(8\pi^2 I) \ . \tag{2.142}$$

The transition frequencies, ν_r, and representative line antenna temperatures are listed for observed interstellar molecular microwave transitions by Lovas (1992) and in the NIST World Wide Web site http://nist.physics.gov/ under Physical Reference Data.

For Hund's coupling case (a) Eq. (2.141) becomes

$$E_r = hB[J(J+1) - \Omega^2] \ . \tag{2.143}$$

For the coupling case (b) and the $^2\Sigma$ states, we have (Hund, 1927)

$$E_r = hBK(K+1) + \tfrac{1}{2}h\gamma K \quad \text{for} \quad J = K + \tfrac{1}{2}$$

and

$$E_r = hBK(K+1) - \tfrac{1}{2}h\gamma(K+1) \quad \text{for} \quad J = K - \tfrac{1}{2} \ ,$$

where the splitting constant γ is small compared to B.

When the molecule is not a rigid rotator and centrifugal stretching is taken into account, Eq. (2.141) becomes

$$E_r = hBJ(J+1) - hDJ^2(J+1)^2 \ , \tag{2.144}$$

where for a vibrational radian frequency, ω, the change D in B due to centrifugal stretching is given by

$$D = \frac{4B^3}{\omega^2} \ .$$

Using the selection rule $\Delta J = \pm 1$ with Eq. (2.144), the allowed rotational transition frequencies, ν_r, are given by

$$\nu_r = 2B(J+1) - 4D(J+1)^3 \ . \tag{2.145}$$

A symmetric top molecule has equal moments of inertia along the directions of two of the three principle axes of the molecule. The angular momentum has a fixed component along an internal molecular axis given by $Kh/2\pi$ where K takes on $2J+1$ values from $-J$ to J. The rotational energy, E_r, is given by

$$E_r = h[BJ(J+1) + (A - B)K^2] \ , \tag{2.146}$$

where $A = h/(8\pi^2 I_a)$ and $B = h/(8\pi^2 I_b)$ when I_a and I_b denote the moments of inertia along the a and b axis. The selection rule is $\Delta K = 0$, however, and the frequency of the transition is the same as that resulting from Eq. (2.142). When the molecule is an asymmetric top molecule, the quantum number J is supplemented by the values of K_{-1} for the limiting prolate symmetric top, $I_a < I_b = I_c$, and K_{+1} for the limiting oblate symmetric top, $I_a = I_b < I_c$, where I_i denotes the moment of inertia along the axis i.

The solution to the time independent Schrödinger equation (2.126) for a harmonic oscillator of frequency, ν_0, and force constant k is

$$E_v = h\nu_0\left(v + \tfrac{1}{2}\right) = hc\omega_e\left(v + \tfrac{1}{2}\right) \ , \tag{2.147}$$

where E_v is the vibrational energy for vibrational quantum number v, the $\omega_e = \nu_0/c$ is the term value, and

$$\nu_0 = \frac{1}{2\pi}\left(\frac{k}{m}\right)^{1/2} ,$$

where m is the mass of the oscillator. The quantum number, v, is zero or an integer and the selection rule is $\Delta v = \pm 1$, so that the vibration transition frequencies are equal to ν_0. Born and Oppenheimer (1927) were the first to show that the massive nuclei could be assumed to be fixed in order to calculate the potential energy contribution of the electrons, after which the rotational-vibrational energy levels could be calculated. Morse (1929) suggested a diatomic molecular potential energy, $U(r)$, given by

$$U(r) = D_e\{1 - \exp[-a(r - r_e)]\}^2 , \tag{2.148}$$

where D_e is the dissociation energy of the molecule, r_e is the equilibrium distance between the nuclei, a is a constant, and r is the internuclear distance. The function $U(r)$ is shown in Fig. 2.5 where it is compared with the actual $U(r)$ for the hydrogen molecule together with its vibrational energy levels. When $U(r)$ and the potential energy due to rotation, $J(J+1)/r^2$, are substituted into the Schrödinger equation (2.126), the solution for the rotational-vibrational energy is given by Townes and Schawlow (1955):

$$\begin{aligned}\frac{E_r + E_v}{h} &= \omega_e\left(v + \tfrac{1}{2}\right) - x_e\omega_e\left(v + \tfrac{1}{2}\right)^2 + B_eJ(J+1)\\ &\quad - D_eJ^2(J+1)^2 - \alpha_e\left(v + \tfrac{1}{2}\right)J(J+1) ,\end{aligned} \tag{2.149}$$

where

$$\omega_e = \frac{a}{2\pi}\left(\frac{2D_e}{\mu}\right)^{1/2} ,$$

$$x_e = \frac{h\omega_e}{4D_e} ,$$

$$B_e = \frac{h}{8\pi^2 I_e} ,$$

$$D_e = \frac{4B_e^3}{\omega_e^2} = \frac{h^3}{128\pi^6\mu^3\omega_e^2 r_e^6} ,$$

$$\alpha_e = 6\left(\frac{x_e B_e^3}{\omega_e}\right)^{1/2} - \frac{6B_e^2}{\omega_e} ,$$

J and v are, respectively, the rotational and vibrational quantum numbers, and μ is the reduced mass of the molecule. Values of the rotation-vibration constants of Eq. (2.149) are given for most simple molecules of astrophysical interest in the appendices of the books by Herzberg (1950) and Townes and Schawlow (1955).

Although the rotational quantum number, J, obeys the simple selection rule $\Delta J = \pm 1$, there is no similar simple selection rule for the vibrational quantum number, v. Nevertheless, Franck (1925) and Condon (1926, 1928)

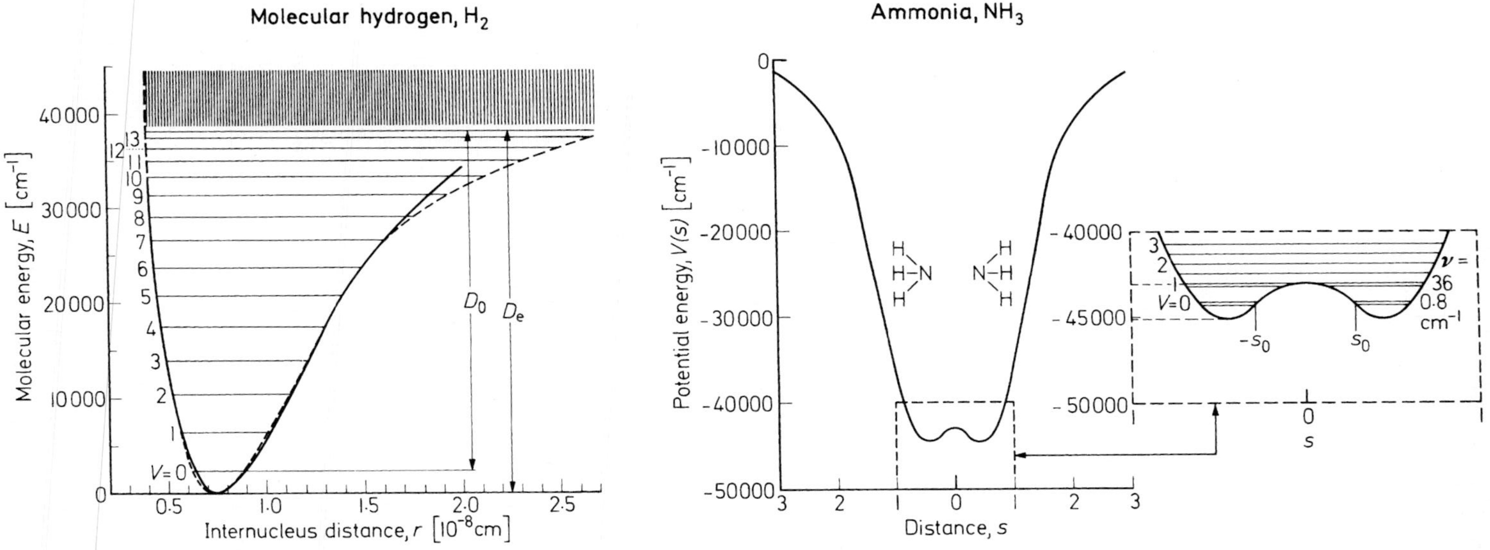

Fig. 2.5. Potential energy curves for the hydrogen molecule, H_2, and the ammonia molecule, NH_3 [after Herzberg, 1950; Townes and Schalow, 1955, by permission, respectively, of the Van Nostrand Reinhold Co. and the McGraw-Hill Book Co.]. The vibrational energy levels are denoted by horizontal lines for different values of the vibrational quantum number, V. For molecular hydrogen, the full curve is experimental data whereas the broken curve is the Morse curve. The variable, r, denotes the internuclear distance, and the continuous term spectrum above $V = 14$ is indicated by vertical hatching. For the ammonia molecule, the variable, s, denotes the distance between the nitrogen and the plane of the hydrogens. The ammonia molecule resonates between the two potential minima, and the resonance slips the virbational levels into the doublets shown in the figure

have developed a principle which relates the favored vibrational transitions to the potential energy curves of the two states under consideration. The favored transition is found to be the one in which there is no instantaneous change of nuclear momentum or position in the transition. This means that $\Delta v = 0$ is favorable for states with similar potential energy curves, whereas for displaced curves a wide band of Δv are possible.

Theoretically there are two equivalent potential energy configurations separated by a barrier as shown in Fig. 2.5 for ammonia. This property reflects the inversion of the eigenfunction about its origin. Quantum mechanically the vibrating molecule can tunnel through the barrier, and the frequency of penetration, ν, is given by (Dennison and Uhlenbeck, 1932)

$$\nu = \frac{\nu_{\mathrm{V}}}{\pi A^2} , \tag{2.150}$$

where ν_{V} is the vibration frequency for one of the potential minima, and the A represents the area under the potential hill. Here

$$A = \exp\left\{\frac{2\pi}{h}\int_0^{s_0}[2\mu(V - E_{\mathrm{v}})]^{1/2}\,ds\right\} ,$$

where μ is the reduced mass, V is the potential energy, ds is an element of the atomic separation, s, the s_0 is shown in Fig. 2.5, and E_{v} is the total vibrational energy. (The factor A^{-2} determines a probability of the tunnel effect.) For ammonia, the inversion barrier is sufficiently low that the vibrational levels have doublets whose frequency of separation lies in the radio frequency range (cf. Table 2.9). Detection of interstellar ammonia was first reported by Cheung et al. (1968). Observations of interstellar ammonia are reviewed by Ho and Townes (1983); they provide values for the frequencies and Einstein probability coefficients for the radio wavelength transitions of ammonia molecules near a frequency of 23 GHz or a wavelength of about 1.27 cm.

Lambda-type doubling, designated by a Λ, is produced by an interaction between the rotational and electronic motions. Because positive and negative values of Λ represent rotation of the orbital electrons in one sense or the other, they have the same energy. Unless $\Lambda = 0$ the energy levels are therefore doubly degenerate. For a rotating molecule with an unpaired electron, however, the electronic-rotational interaction removes this degeneracy and each energy level splits into two components. The most important transitions between the lambda doublet levels are those for the $\Lambda = 1$ or Π state, and the energies, E, of these transitions are given by Van Vleck (1929) and Townes and Schawlow (1955).

Electron spin-nuclear spin interactions lead to further hyperfine splitting of each doublet level. For the ${}^2\Pi$ state, the energy, ΔE, of this splitting is given by Townes (1957)

$$\Delta E = \pm\frac{d(X + 2 - A/B)}{4XJ(J+1)}\left(J + \frac{1}{2}\right)\boldsymbol{I}\cdot\boldsymbol{J} , \tag{2.151}$$

where

$$\boldsymbol{I}\cdot\boldsymbol{J} = \frac{F(F+1) - I(I+1) - J(J+1)}{2}$$

and

$$d = 3\mu_0 \frac{\mu_I}{I} \left(\frac{\sin^2 \theta}{r^3} \right) .$$

Here A is the fine structure constant of energy $A\bar{s} \cdot \bar{\Lambda}$, B is the rotational constant, F is the total molecular angular momentum including nuclear spin in units of $h/2\pi$, the I is the nuclear spin angular momentum in units of $h/2\pi$, the Bohr magneton is μ_0, the nuclear magnetic moment is μ_I, the distance from the nucleus to the electron is r, and θ is the angle between the molecular axis and the radius vector.

Transitions between the hyperfine split, lambda doublet lines of interstellar OH were first detected by Weinreb, Barrett, Meeks, and Henry (1963) for the $^2\Pi_{3/2}$ state. The radio frequencies of transitions of OH masers are given by Reid and Moran (1981); also see Lovas (1986, 1991).

2.15.2 Line Intensities and Molecular Abundances

It follows from Eqs. (2.72) and (2.64) that under conditions of local thermodynamic equilibrium (LTE), the line intensity observed for a line emitted by an optically thin gas is proportional to the line integral of the absorption coefficient per unit length, α_ν(LTE). For a rotational transition at frequency, ν_r, we have from Eq. (2.69)

$$\alpha_{\nu_r}(\mathrm{LTE}) = \frac{c^2}{8\pi\nu_r^2} \frac{N_J}{\Delta\nu_L} \exp\left(\frac{h\nu_r}{kT}\right) \left[1 - \exp\left(\frac{-h\nu_r}{kT}\right)\right] A_J , \tag{2.152}$$

where T is the excitation or spin temperature of the gas, $\Delta\nu_L$ is the full width to half maximum of the line, the Einstein coefficient for the spontaneous electric dipole transition is

$$A_J = \frac{64\pi^4\nu_r^3}{3hc^3} |\mu_J|^2 ,$$

where the electric dipole matrix element, μ_J, for symmetric molecules is given by

$$|\mu_J|^2 = \mu^2 \frac{(J+1)}{(2J+1)} \quad \text{for the } J+1 \leftarrow J \text{ transition}$$

and

$$|\mu_J|^2 = \mu^2 \frac{(J+1)}{(2J+3)} \quad \text{for the } J+1 \rightarrow J \text{ transition} .$$

Transition probability data for molecules of astrophysical interest are given by Nicholls (1977). Measured values of the electric dipole moment, μ, of simple molecules in the gas phase of the most abundant elements lie between 0.1 and 4.5 Debye, with 1 Debye $= 10^{-18}$ e.s.u. (Nelson, Lide and Maryott, 1967). Probabilities for the vibration-rotation transitions of molecular hydrogen are given by Black and Dalgarno (1976) and Turner, Kirby-Docken and Dalgarno (1977).

The number of molecules, N_J, in the J state is given by the Boltzmann equation (3.125) and Eq. (2.141)

$$N_J = \frac{(2J+1)}{U} N \exp\left[\frac{-hBJ(J+1)}{kT}\right] . \tag{2.153}$$

where T is the excitation or spin temperature of the gas, the statistical weight of level J is $(2J + 1)$, N is the total number of molecules, and the partition function, U, is given by

$$U = \sum_{J=0}^{\infty} (2J+1) \exp\left[\frac{-hBJ(J+1)}{kT}\right] \approx \frac{kT}{hB} \quad \text{for} \quad hB \ll kT .$$

Eqs. (2.152) to (2.153) may be combined to give the relation

$$\alpha_{\nu_r}(\mathrm{LTE}) = \frac{8\pi^3 |\mu_J|^2}{3hc} \left(\frac{h\nu_r}{kT}\right)^2 \frac{BN(2J+1)}{\Delta\nu_L} \exp\left[\frac{-hBJ(J+1)}{kT}\right] , \tag{2.154}$$

for $hB \ll kT$ and $h\nu_r \ll kT$. For low values of J, the exponential factor in Eq. (2.154) can be approximated as unity. The line absorption coefficient given by Eq. (2.154) can be related to the observed antenna temperatures of the line and the antenna efficiency by Eqs. (2.74) to (2.77).

For a rotational molecular transition of frequency ν_r and excitation temperature T_s, the excess brightness temperature, ΔT_B, due to the transition is given by Turner (1988) and Turner and Ziurys (1988):

$$\Delta T_B = (T_s - T_{bg})[1 - \exp(-\tau_\nu)] , \tag{2.155}$$

where it is assumed that $h\nu_r \ll kT_s$ and the background temperature is T_{bg}, often equal to the 2.726 °K comic microwave background radiation. The excitation or spin temperature $T_s = T_{mn}$ between two levels m and n is defined by the Boltzmann equation (3.126)

$$\frac{N_m}{N_n} = \frac{g_m}{g_n} \exp\left(\frac{-h\nu_{mn}}{kT_s}\right)$$

where $\nu_{mn} = \nu_r$ is the frequency of the m–n transition, and N_m and g_m are, respectively, the number density and statistical weight of level m. The optical depth or opacity, τ_ν, is related to the molecular column density in the lower state, N_m, by:

$$\tau_\nu = (\phi_\nu/4\pi)\lambda_r^2 A_{mn}(g_m/g_n)N_m[1 - \exp(-h\nu_r/(kT_s))] ,$$

where the Einstein probability coefficient for spontaneous emission of the transition is $A_{mn} = (64\pi^4\nu_r^3/3hc^3)(|\mu_{mn}|^2/g_m)$ and μ_{mn} is the electric dipole matrix element of the transition. For a Gaussian line shape of full width to half maximum $\Delta\nu_L$, the line profile function $\phi_\nu = (\ln 2/\pi)^{1/2}/\Delta\nu_L$. For the optically thin case, with $\tau_\nu \ll 1$, in the Rayleigh-Jeans limit for $\mathrm{h}\nu_r \ll \mathrm{kT_s}$,

$$N_n = \frac{k}{\xi h\nu} \frac{T_s}{T_s - T_{bg}} \int \Delta T_B \, d\nu \tag{2.156}$$

where $\xi = (\phi_\nu/4\pi)\lambda_r^2 \mathrm{A}(\mathrm{g_m}/\mathrm{g_n})$ and the integral of the excess brightness temperature is over frequency ν. For $\tau_\nu \ll 1$ and at millimeter wavelengths, $T_r \gg \mathrm{T_{bg}} = 2.726\,°\mathrm{K}$, and

$$N_n = \frac{8\pi^{3/2}}{2\sqrt{\ln 2}} \frac{g_n}{g_m} \frac{k}{hc\,A} \int \Delta T_{\mathrm{B}}\, dv$$

where the excess brightness temperature is integrated over velocity (km s^{-1}). If $T_{\mathrm{s}} = T_{mn}$ characterizes all rotational transitions, then the total column density N of the molecule can be found from

$$\frac{N_n}{N} = \frac{g_n[\exp(-E_1/kT_{\mathrm{s}})]}{Q} \tag{2.157}$$

where Q is the rotational partition function and E_m is the energy of the lowest state.

When vibrations are taken into account, Eq. (2.156) is modified by multiplication of N with the fraction, f_{v}, of the molecules in the vibrational state of energy $h\omega_{\mathrm{e}}(v + \frac{1}{2})$. This fraction is given by

$$f_{\mathrm{v}} = \exp\left(\frac{-vh\omega_{\mathrm{e}}}{kT}\right)\left[1 - \exp\left(\frac{-h\omega_{\mathrm{e}}}{kT}\right)\right] \ . \tag{2.158}$$

Assuming a Maxwellian distribution of velocities, the gas is characterized by the kinetic temperature, T_{k}, which is related to the Doppler broadening of the line. The transition rate, R_{mn}, for transitions induced by collisions is given by

$$R_{mn} = \frac{1}{\tau_{\mathrm{c}}} = N\langle v\rangle\sigma \ , \tag{2.159}$$

where τ_{c} is the collision lifetime, N is the density of the colliding particles, the average velocity, $\langle v\rangle$, is given by

$$\langle v\rangle = \left[\frac{8kT_{\mathrm{k}}}{\pi\mu}\right]^{1/2} ,$$

where μ is the reduced mass of the molecule and the colliding particle, and the collision cross section, σ, is given by Purcell (1952)

$$\sigma = \frac{16e^2|\mu_{mn}|^2}{3\hbar^2\langle v\rangle^2} \ln\left[\frac{(0.706)3\hbar^2\pi\langle v\rangle^4}{32e^2|\mu_{mn}|^2\omega^2}\right] ,$$

for $\langle v\rangle > 10^6$ cm sec^{-1}. Here μ_{mn} is the dipole matrix element and ω is the transition frequency. Values of R_{mn} for strong collisions and lower values of $\langle v\rangle$ are given by Rogers and Barrett (1968) and Goss and Field (1968).

For transitions induced by radiation of intensity, I_ν, the Einstein probability coefficients, A_{mn} and B_{mn}, are related by the equation

$$A_{mn} = \frac{1}{\tau_{\mathrm{r}}} = I_\nu B_{mn}\left(\frac{h\nu}{kT_{\mathrm{R}}}\right) , \tag{2.160}$$

where the radiation lifetime is τ_{r}, and the radiation temperature, T_{R}, is related to the intensity, I_ν, by the Rayleigh–Jeans approximation

$$I_\nu = \frac{2kT_{\mathrm{R}}\nu^2}{c^2} \ \text{for } h\nu \ll kT_{\mathrm{R}} \ . \tag{2.161}$$

The three temperatures T_{s}, T_{k}, and T_{R} are related by the equation

$$T_{\mathrm{s}} = T_{\mathrm{k}}\left[\frac{T_{\mathrm{R}} + T_0}{T_{\mathrm{k}} + T_0}\right] , \tag{2.162}$$

where

$$T_0 = \frac{h\nu R_{mn}}{kA_{mn}} = \frac{h\nu\tau_R}{k\tau_c} \ ,$$

which follows from the equation of statistical equilibrium (Eq. (2.82)) and from Eqs. (2.159) to (2.161).

When the radiation from the interstellar OH molecule was found to be polarized, to have anomalous relative line intensities, and to imply very high brightness temperatures, it was postulated that the population of the levels had been inverted and that maser amplification occurred. The integrated flux density received from a masing source is given by

$$\int S_\nu \, d\nu = \frac{h\nu R_m}{D^2 \Omega_m} \ , \tag{2.163}$$

where S_ν is the flux density observed at frequency, ν, the distance to the source is D, the solid angle of the maser emission is Ω_m, and the time rate of microwave photons, R_m, is proportional to

$$R_m \propto \exp(\tau_\nu)$$

for an unsaturated maser, and

$$R_m \propto \tau_\nu$$

for a saturated maser. Detailed formulae for the proportionality constants are given by Litvak (1969) for infrared pumping of the OH molecule. Here τ_ν, is given by

$$\tau_\nu = \frac{hB_{mn} g_m \nu_{mn}}{\Delta\nu_{\mathrm{L}}} \int \left(\frac{N_m}{g_m} - \frac{N_n}{g_n} \right) dl \ , \tag{2.164}$$

where B_{mn} is the coefficient for stimulated emission of the line, g_m is the degeneracy of the upper state, ν_{mn} is the line frequency, $\Delta\nu_{\mathrm{L}}$ is the line width, and $\int (N_m/g_m - N_n/g_n) dl$ is the line integral of the population inversion. For an unsaturated maser, the observed Doppler-broadened lines will be narrowed and the line width, $\Delta\nu_L$, is given by

$$\Delta\nu_{\mathrm{L}} = \frac{1}{\sqrt{\tau_\nu}} \left[\frac{1.67}{\lambda} \left(\frac{2kT_k}{M} \right)^{1/2} \right] \ , \tag{2.165}$$

where λ is the wavelength of the line, T_k is the kinetic temperature, M is the molecular mass, and the expression in square brackets is the thermal Doppler width of the line. Saturated masers do not exhibit line narrowing. Molecular masers have been reviewed by Reid and Moran (1981) and Elitzur (1992).

2.15.3 The Formation and Destruction of Molecules

The energetics of the various reactions which create or destroy molecules are determined by the dissociation energy and ionization potential of the molecule together with the energy of other reactants such as photons or charged particles. The dissociation energy of a stable electronic state of a diatomic molecule is that energy required to dissociate it into atoms from the lowest rotation-vibration level. The dissociation energy referred to the ground electronic state is termed D_0^0 when the dissociation products are normal atoms. The ionization potential of a molecule is defined as that energy necessary to remove an electron from the outermost filled molecular orbital of the ground state. Wilkinson (1963) has listed the ionization potentials and dissociation energies for 148 diatomic molecules of astrophysical interest. Stief et al. (1972) and Stief (1973) provide additional data, while Singh and Chaturvedi (1987) give the dissociation energies and ionization potentials of 153 diatomic molecules, molecular ions and radicals of astrophysical interest.

Stellar ultraviolet, or UV, radiation strongly affects the molecular composition of diffuse interstellar clouds Black and Dalgarno (1977). As a result, diffuse cloud chemistry is primarily that of atoms and diatomic molecules; the formation of molecules in diffuse interstellar clouds is discussed by Black and Dalgarno (1977) and Mitchell, Ginsburg and Kuntz (1977). However, molecules are shielded from UV by dust grains in dense molecular clouds (Stief et al., 1972). Photodissociation and photoionization can therefore be negligible in dense molecular clouds where the visual extinction is large. Gas phase reactions are important in dense molecular clouds, whose densities are still very low by terrestrial standards and whose temperatures are far below those associated with most chemical processes. Herbst (1987) has reviewed gas phase chemical processes in molecular clouds, including ion-molecule reactions and radiative association processes, while Watson (1976) provides an overview of chemical reactions on the surfaces of interstellar dust grains as well as gas phase reactions. Ion-molecule reactions were reviewed by Watson (1974) and gas phase reactions were reviewed by Watson (1978). The formation of molecules from ionized atoms is discussed by Langer (1978). The general importance of radiative association reactions in molecular clouds is demonstrated by Smith and Adams (1978) and Huntress and Mitchell (1979).

Photodissociation. The most important destruction mechanism for interstellar molecules is their photodissociation by the interstellar radiation field. The photodestruction rate, or dissociation probability, P, is given by

$$P = \frac{1}{h} \int_{912\,\text{Å}}^{\lambda_{\mathrm{T}}} U_\lambda \sigma_\lambda \Phi \lambda \, d\lambda \ \mathrm{sec}^{-1} \ , \tag{2.166}$$

where h is Planck's constant, U_λ is the energy density of the radiation field at wavelength, λ, the absorption cross section is σ_λ, and Φ is the primary quantum yield for dissociation. The limits to the integration in this equation are $\lambda = 912$ Å where hydrogen is photoionized, and $\lambda_{\mathrm{T}} = 12,396.3/E$ where E is the dissociation energy in eV. Threshold wavelengths, λ_{T}, lie in the ultraviolet range of wavelengths, and the radiation energy density in this range is given by Habing (1968)

$$U_\lambda = 4 \times 10^{-17} \ \mathrm{erg\ cm^{-3}\ Å^{-1}} \quad \text{for } 912\ \text{Å} \leq \lambda \leq 2400\ \text{Å} \ . \tag{2.167}$$

When obscuring clouds are present, the radiation field is attenuated by the factor

$$\left[a_\lambda + (1 - a_\lambda)10^{0.4A_\lambda}\right]^{-1} , \tag{2.168}$$

where the grain albedo $a_\lambda \sim 0.5$, and the extinction, A_λ, is in magnitudes and is given by

$$A_\lambda = 2.3 \times 10^7 Q r_{\mathrm{gr}}^2 N^{2/3} M^{1/3} \mathrm{mag} ,$$

for the center of a cloud of M solar masses and average particle number density, N. Here the extinction efficiency $Q \approx 2.5$, the grain radius $r_{\mathrm{gr}} \approx 0.12 \times 10^{-4}$ cm, and the dust-to-gas ratio $N_{\mathrm{g}}/N \approx 10^{-12}$.

Unattenuated photodestruction rates for various ions and molecules have been compiled by Klemperer (1971), Solomon and Klemperer (1972), Stief et al. (1972), Stief (1973) and Herbst and Leung (1986). Photoionization of molecular hydrogen is discussed by Ford, Docken and Dalgarno (1975).

Gas Exchange Reactions. The astrophysically important gas exchange reactions are of the bimolecular type given by (Watson, 1978; Herbst, 1987)

$$A + BC \rightarrow AB + C ,$$

where A and C are atoms and AB and BC are molecules. The number densities, N, of the reactants are related by the equation

$$-\frac{dN_A}{dt} = -\frac{dN_{BC}}{dt} = \frac{dN_{AB}}{dt} = \frac{dN_B}{dt} = \kappa N_A N_{BC} ,$$

where t is the time variable, and the rate constant, κ, is given by (Polanyi, 1962)

$$\kappa = P r_{AB}^2 \left(\frac{8\pi}{\mu} kT\right)^{1/2} \exp\left[-\frac{E_a}{kT}\right] \mathrm{cm}^3 \mathrm{sec}^{-1}$$

or

$$\kappa = A \exp\left[-\frac{E_a}{kT}\right] \mathrm{cm}^3 \mathrm{sec}^{-1} . \tag{2.169}$$

Here the Arrhenius factor, A increases slowly with temperature, and the activation energy, E_a, is the difference in internal energy between the activated and normal molecule. The steric factor, P, is an orientation parameter close to unity, the reagent molecules are assumed to approach each other to within a distance r_{AB} which is the mean of the gas kinetic collision parameters r_{AA} and r_{BB}, the reduced mass of the atom, A, and the molecule, BC, is μ, and T is the gas kinetic temperature. Most gas exchange reactions of astrophysical interest are exothermic, and therefore not very temperature dependent. The A factors measured at room temperature are summarized by Kaufman (1969), and for most atom molecule reactions of astrophysical interest we have

$$A \sim 4 \times 10^{-11} \mathrm{cm}^3 \mathrm{sec}^{-1} .$$

Specific rate constants are given by Herbst and Klemperer (1973).

Ion-Molecule Reactions, Associative Detachment, and Charge Exchange Reactions. For ion-molecule reactions, associative detachment, and charge exchange reactions we have reactions of the form (Watson, 1974, 1976)

$$A + B \rightarrow C + D \ ,$$

where A is an ion, B is an atom or molecule, C is an atom or molecule, and D is an ion or a charged particle. A rate constant, κ, is defined by

$$-\frac{dN_A}{dt} = -\frac{dN_B}{dt} = \frac{dN_C}{dt} = \frac{dN_D}{dt} = \kappa N_A N_B \ ,$$

where t is the time variable, and N_A, N_B, N_C, and N_D, are, respectively, the number densities of A, B, C, and D. The interaction potential, V, between the ion, A, and the atom or molecule, B, is given by Rapp and Francis (1962)

$$V = -\frac{\alpha e^4}{2r^4} \ ,$$

where α is the dipole polarizability of the atom or molecule, e is the charge of the electron, and r is the internuclear distance. According to Gioumousis and Stevenson (1958), the critical impact parameter, r_{AB}, is given by

$$r_{AB} = \left[\frac{4e^2\alpha}{\mu v^2}\right]^{1/4} \ ,$$

so that the rate constant is given by

$$\kappa = \langle \sigma v \rangle = 2\pi e f \left(\frac{\alpha}{\mu}\right)^{1/2} \ . \tag{2.170}$$

Here σ is the cross section for charge transfer, v is the relative velocity of the reactants, the angular brackets denote averaging over velocities, μ is the reduced mass of the reactants, and f is a statistical factor which takes into account the fact that not all collisions lead to charge transfer. For most ion-molecule, associative detachment, and charge exchange reactions of astrophysical interest we have

$$\kappa \sim 1 \times 10^{-9}\ \mathrm{cm}^3\ \mathrm{sec}^{-1} \ .$$

Specific rate constants are given by Herbst and Klemperer (1973).

Surface Recombination. Van de Hulst (1949) first suggested that molecules might form on the interstellar grains in a three body process where lattice vibrations in the grain absorbed the excess energy liberated when gas atoms combined to form molecules. This idea was developed as a mechanism for the formation of molecular hydrogen by Gould and Salpeter (1963) and Gould, Gold, and Salpeter (1963). The efficiency, γ, at which atoms strike a grain surface and recombine to form a molecule is the product of two factors: the sticking coefficient, S, or probability that an atom hitting the grain surface from the interstellar gas becomes thermalized and sticks to the grain; and the recombination efficiency, γ', or probability that the first adsorbed atom will remain adsorbed and not evaporate before a second atom strikes the grain, becomes adsorbed, and recombines with the first atom. Hollenbach and Salpeter (1970, 1971) show that the sticking coefficient is given by

$$S = \frac{\Gamma^2 + 0.8\Gamma^3}{1 + 2.4\Gamma + \Gamma^2 + 0.8\Gamma^3} \tag{2.171}$$

if

$$kT_{gr} < kT_{gas} < D \ .$$

Here $\Gamma = E_c/(kT_{gas})$ where E_c is the characteristic total energy transferred to the grain surface, T_{gas} is the gas temperature, T_{gr} is the grain temperature, and D is the binding energy for the adsorption ground state. For hydrogen atoms, $\Gamma \approx 1$ and $S \approx 0.3$ at $T_{gas} \approx 100\,^\circ$K. For the heavier atoms C, N, and 0, we have $S \approx 1$ according to Watson and Salpeter (1972). The recombination efficiency, γ', is unity if the time, t_s, for a new atom to strike the grain surface is much less than the time, t_{eV}, for an atom to evaporate from the surface. If N denotes the number density of atoms and V is their thermal velocity, then

$$t_s \approx [SNV\pi r_{gr}^2]^{-1} \ , \tag{2.172}$$

where r_{gr} is the grain radius, usually taken to be $r_{gr} \approx 0.17 \times 10^{-4}$ cm. The characteristic lattice vibration frequency of the grain is $\nu_0 \approx 10^{12}$ sec^{-1} and the evaporation time is

$$t_{eV} \approx \nu_0^{-1} \exp\left(\frac{D}{kT_{gr}}\right) \ , \tag{2.173}$$

where D is the atom adsorption energy and T_{gr} is the grain temperature. Arguments about the efficiency of molecule formation have centered about different estimates of the grain temperature, T_{gr}, and the adsorption binding energy, D (cf. Knapp et al. 1966; Stecher and Williams, 1968; Wentzel, 1967; Hollenbach and Salpeter, 1970, 1971). Current arguments give $T_{gr} \lesssim 25\,^\circ$K and values of D such that $t_s \ll t_{eV}$ for most atoms, especially when dislocations and chemical impurity sites are included in calculating D.

The rate of formation of a molecule, AB, and the rate of depletion of the element, A, are governed by the equations

$$\frac{dN_{AB}}{dt} = \kappa N_A N_{gr}$$

and

$$\frac{dN_A}{dt} = \frac{-\alpha\kappa}{(1-\alpha)} N_A N_{gr} \ ,$$

where N_{AB}, N_A, and N_{gr} denote, respectively, the densities of the molecules, AB, atoms, A, and grains, gr, the time variable is t, the probability that an atom will become permanently locked to the grain is α, and $(1-\alpha)$ is the probability that a molecule will evaporate and return to the gas. The rate constant, κ, is given by

$$\begin{aligned}\kappa &= S(1-\alpha)\pi r_{gr}^2(1+\gamma Z)v \\ &\approx 6 \times 10^{-9}(1-\alpha)(1+2.5Z)r_{gr}^2 T^{1/2}\,\text{cm}^3\,\text{sec}^{-1} \ ,\end{aligned} \tag{2.174}$$

where S is the sticking coefficient, the grain radius $r_{gr} \approx 0.17 \times 10^{-4}$ cm, γ is determined from the equation describing the balance of charge on the grain, Z is the charge of the element, A, and v is its thermal velocity given by

$$v = \left(\frac{8kT_{gas}}{\pi M_A}\right)^{1/2} \ ,$$

where the gas temperature is $T_{\rm gas}$ and the mass of element, A, is M_A. Hollenbach, Werner, and Salpeter (1971) discuss the formation of molecular hydrogen, H_2, on grains. Because H_2 is a light saturated molecule with a low adsorption energy, it is easily evaporated thermally and $\alpha \approx 0$. In calculating molecular abundances, a constant gas to dust density ratio is used. Typical grains have a density $\rho_{\rm gr} \approx 2\,{\rm gm\,cm^{-3}}$, a radius $r_{\rm gr} \approx 0.17 \times 10^{-4}$ cm, a mass $M_{\rm gr} \approx 4 \times 10^{-14}$ gm, and a number density, $N_{\rm gr}$, given by

$$N_{\rm gr} \approx 4 \times 10^{-13} N_A \ ,$$

where $N_A = N_{\rm H} + 2{\rm N}_2$ is the total number density of hydrogen atoms in atomic, $N_{\rm H}$, and molecular, ${\rm N}_2$, form. The formation of the heavier molecules of C, O, N and H is discussed by Watson and Salpeter (1972). In this case $S = 1$, but the mechanism by which the molecules return to the gas is unclear. For the saturated molecules such as CH_4, H_2O, and NH_3, the adsorption energy is low and they may be evaporated thermally. The formation of molecules on interstellar dust grains is reviewed by Watson (1976), and discussed by Allen and Robinson (1975), Barlow and Silk (1978), Hunter and Watson (1978), and Jura (1974).

Radiative Association. At the low gas densities of the interstellar medium, some simple molecules may be formed by the radiative association process (Swings, 1942; Kramers and Ter Haar, 1946; Smith and Adams, 1978; Huntress and Mitchell, 1979).

$$A + B \rightarrow AB + h\nu \ .$$

Here A and B are two ground state atoms which collide, and find themselves in the continuum of an excited molecular state. A molecule, AB, is formed if the excited complex relaxes and emits a photon of energy, $h\nu$, before the atoms separate. Radiative association routes for the diatomic molecules of the more abundant species are given by Lutz (1972).

Bates (1951) first gave the correct theory for calculating the radiative association rate constant, κ. If the number densities of atoms A and B, and the molecule AB are, respectively, N_A, N_B, and N_{AB}; then

$$\frac{dN_{AB}}{dt} = \kappa N_A N_B \ ,$$

where t is the time variable, and the rate constant for two atoms which meet with energy, E, is

$$\kappa = g \int \sigma(E) v(E) W(E) dE \ ,$$

where g is the probability that two atoms find themselves in the continuum of the required excited molecular state, $v(E)$ is the relative velocity of the colliding atoms, $\sigma(E)$ is the cross-section for radiative association, and $W(E)$ is the energy distribution of the atoms. The total cross-section for radiative association is (Lutz, 1972)

$$\sigma(E) = \frac{32 \times 2^{1/2} \pi^4 h^2}{3 \mu^{3/2} E^{1/2} c^3 G' v} \sum_{v''} \sum_{J''} |\langle \psi'' | D | \psi' \rangle|^2 \rho'(E) \nu(E, v'')^3 \ ,$$

where μ is the reduced mass of the two atoms, G' is the electronic degeneracy of the excited state, $|\psi'\rangle$, the density of initial continuum states is $\rho'(E)$, and the frequency $\nu(E, v'')$ is the frequency of the photon emitted for the transition from the continuum energy E to a vibrational level v'' of the lower state, $|\psi''\rangle$. For a Maxwellian velocity distribution, the integration over energy gives a rate constant, κ, of

$$\kappa = g \frac{64 \times 2^{1/2} \pi^{7/2} h^2}{3\mu^{3/2}(kT)^{1/2} c^3 G'} S_e \sum_{v''} |\langle \psi''_{v''} | \psi'_E \rangle|^2 \rho'(E_0) \nu(E_0 v'')^3 \ , \tag{2.175}$$

where E_0 is taken to be the mean energy at temperature, T, the molecular electronic dipole strength is S_e, and the $|\langle \psi''_{v''} | \psi'_E \rangle|^2$ are the Franck–Condon factors.

Bates and Spitzer (1951) first calculated the equilibrium abundance of CH and CH^+ by assuming that they were formed by the radiative association processes

$$C + H \rightarrow CH + h\nu$$

and

$$C^+ + H \rightarrow CH^+ + h\nu \ .$$

They concluded that the observed densities of CH and CH^+ demand hydrogen number densities of several hundred cm^{-3}. Solomon and Klemperer (1972) and Smith, Liszt, and Lutz (1973) have revaluated the rates for these reactions, the more recent values being

$$\kappa = 1.5 \times 10^{-17}\ cm^3\ sec^{-1} \quad \text{for CH formation at } T = 100\ {}^\circ K \ ,$$

and

$$\kappa = 5 \times 10^{-18}\ cm^3 sec^{-1} \quad \text{for } CH^+ \text{ formation at } T = 100\ {}^\circ K \ .$$

Julienne, Krauss, and Donn (1971) showed that OH molecules could be formed by an indirect radiative association through resonance states in the molecular continuum. This process is called inverse predissociation. The reaction rate for indirect radiative association is given by (Julienne and Krauss, 1973)

$$\kappa = \hbar^2 \left(\frac{2\pi}{\mu k T} \right)^{3/2} \frac{1}{g_A g_B} \sum_n (2J_n + 1) \frac{\Gamma_{nr} \Gamma_{np}}{\Gamma_{nr} + \Gamma_{np}} \exp\left(\frac{-E_n}{kT} \right) \ , \tag{2.176}$$

where μ is the reduced mass of the reacting atoms, g_A and g_B are the degeneracies of the ground atomic levels, Γ_{nr} is the natural radiation width, Γ_{np} is the predissociation width, and E_n is the energy of the resonance level n above the lowest atomic fine structure asymptotic energy. Using this equation, Julienne, Krauss, and Donn (1971) obtain

$$\kappa \approx 2 \times 10^{-20} cm^3\ sec^{-1} \quad \text{for OH formation at } T \geq 50\ {}^\circ K \ . \tag{2.177}$$

2.16 Line Radiation from Stellar Atmospheres – The Fraunhofer Spectrum and the Curve of Growth

As illustrated in Fig. 2.2, the spectrum of the Sun contains Fraunhofer (1817) lines of different wavelengths and intensities which are seen in absorption against the bright continuum. The wavelengths of the most notable of these lines are tabulated in Table 2.12, whereas the most intense solar emission lines are tabulated in Table 2.13. Kirchhoff and Bunsen (1861) showed that the wavelengths of Fraunhofer lines correspond to certain transitions of elements observed on the earth; and the wavelengths of the lines for different atomic and ionic states of different elements are tabulated by Abt, Meinel, Morgan, and Tabscott (1969), Moore, Minnaert, and Houtgast (1966), Morgan, Keenan, and Kellman (1943), and Striganov and Sventitskii (1968). Atomic energy levels, wavelengths and transition probabilities are provided by the National Institute of Standards and Technology (NIST) on their World Wide Web site http://physics.nist.gov/ under Physical Reference Data and NIST Atomic. Spectroscopic Database. The Chianti atomic database provides atomic data needed for calculating astrophysical emission line spectra at wavelengths greater than 50 Angstroms as a function of both density and temperature (Dere et al., 1997); it is available on the World Wide Web at http://wwwsolar.nrl.navy.msl/chianti.html.

Observations of line intensities and widths may be compared with theoretical expectations in order to determine the excitation temperature, the turbulent velocity, the electron and gas pressures, the surface gravity, and the abundance of the elements in the stellar atmosphere. The two classical model atmospheres are the Schuster–Schwarzschild (SS) atmosphere (Schuster, 1905; Schwarzschild, 1906), and the Milne–Eddington (ME) atmosphere (Milne, 1921, 1930; Eddington, 1917, 1926). In the SS approximation the continuum spectrum is assumed to be formed in the photosphere and the line spectrum is formed entirely in an overlying "reversing layer". In the ME approximation it is assumed that both the line and continuum spectrum are formed in the same layers in such a way that the ratio of the line and continuum absorption coefficients is a constant. Here we will outline a general approach which includes these classical solutions. A detailed discussion of this approach is given in the book by Mihalas (1970). Bowers and Deeming (1984) provide a clear presentation of radiative transfer in stars, and theoretical computations of the observed line intensities in the Sun's atmosphere.

A model of the average quiet solar atmosphere, including the transition region between the photosphere and chromosphere, is given by Fontenla, Avrett and Loeser (1993), building on the work of Fontenla, Avrett and Loeser (1991) and Vernazza, Avrett and Loeser (1981). Similar empirical models of the photosphere are given by Holweger and Müller (1974) and Maltby et al. (1988). A theoretical line-blanketed, local thermodynamic equilibrium (LTE) photospheric model is provided by Kurucz (1979), and theoretical non-LTE line-blanketed chromospheric models are given by Anderson and Athay (1989). Papers in Stenflo (1990) and Cox, Livingston and Matthews (1991) discuss related studies and include references to earlier work. Important early model atmospheres which best fit the observed continuum and line data for the Sun were given by Gingerich, Noyes, Kalkofen and Cuny (1971) and Gingerich and de Jager (1968). The former is a non-LTE model whereas the latter is LTE.

The equation of transfer for the intensity of radiation, I_ν, at frequency, ν, is given by

Table 2.12. The wavelength, λ, element, equivalent width, W, and letter, L, for the most intense Fraunhofer lines in the solar spectrum. Fraunhofer [1817] labeled some of the most prominent lines with the letters A, B, C, D, E, F, G, H and K, but he did not resolve the components in his D line. The A and B lines are due to molecular oxygen in the terrestrial atmosphere, and the CH molecule produces the G line. All the other spectral lines are due to absorption by atoms or ions in the solar atmosphere. The wavelengths and equivalent widths are in units of Angstroms, or Å, where 1Å = 10^{-8} cm[1]

λ (Å)	Element	W (Å)	L	λ (Å)	Element	W (Å)	L
3581.21	Fe I	2.14	N	4920.51	Fe I	0.43	
3719.95	Fe I	1.66		4957.61	Fe I	0.45	
3734.87	Fe I	3.03	M	5167.33	Mg I	0.65	b_4
3749.50	Fe I	1.91		5172.70	Mg I	1.26	b_2
3758.24	Fe I	1.65		5183.62	Mg I	1.58	b_1
3770.63	H_{11}	1.86		5232.95	Fe I	0.35	
3797.90	H_{10}	3.46		5269.55	Fe I	0.41	
3820.44	Fe I	1.71	L	5324.19	Fe I	0.32	
3825.89	Fe I	1.52		5238.05	Fe I	0.38	
3832.31	Mg I	1.68		5528.42	Mg I	0.29	
3835.39	H_9	2.36		5889.97	Na I	0.63	D_2
3838.30	Mg I	1.92		5895.94	Na I	0.56	D_1
3859.92	Fe I	1.55		6122.23	Ca I	0.22	
3889.05	H_8	2.35		6162.18	Ca I	0.22	
3933.68	Ca II	20.25	K	6562.81	H_α	4.02	C
3968.49	Ca II	15.47	H	6867.19	O_2	Tell	B
4045.82	Fe I	1.17		7593.70	O_2	Tell	A
4101.75	H_δ	3.13	h	8194.84	Na I	0.30	
4226.74	Ca I	1.48	g	8498.06	Ca II	1.46	
4310 ± 10	CH		G	8542.14	Ca II	3.67	
4340.48	H_γ	2.86		8662.17	Ca II	2.60	
4383.56	Fe I	1.01		8688.64	Fe I	0.27	
4361.34	H_β	3.68		8736.04	Mg I	0.29	
4891.50	Fe I	0.31					

[1] Adopted from Moore, Minnaert and Houtgast (1966) and Gray (1992).

Table 2.13. Prominent solar emission lines at wavelengths below 2000 Å[1]

Wavelength (Å)	Element	Wavelength (Å)	Element	Wavelength (Å)	Element
1.8499	Fe XXV	335.41	Fe XVI	1215.67	H I
15.012	Fe XVII	499.41	Si XII	1393.75	Si IV
21.602	O VII	624.94	Mg X	1401.16	O IV
132.83	Fe XXIII	629.73	O V	1548.20	C IV
192.04	Fe XXIV	770.41	Ne VIII	1640.4	He II
284.16	Fe XV	977.02	C III	1892.03	Si III
303.78	He II	1031.91	O VI	1908.73	C III

[1] Courtesy of Kenneth P. Dere.

Table 2.13. (Continued) Chromospheric emission lines observed during solar eclipse[2]

Wavelength (Å)	Integrated intensity at Sun (10^{11} erg sec^{-1} cm^{-1} $ster^{-1}$)	Element	Wavelength (Å)	Integrated intensity at Sun (10^{11} erg sec^{-1} cm^{-1} $ster^{-1}$)	Element	Wavelength (Å)	Integrated intensity at Sun (10^{11} erg sec^{-1} cm^{-1} $ster^{-1}$)	Element
3,685.196	90	Ti II	3,835.39	228	H I (H 9)	4,861.342	1,632	H I (H β)
3,691.56	29	H I (H 18)	3,838.302	60	Mg I	5,015.67	6	He I
3,697.15	35	H I (H 17)	3,889.05	381	H I (H 8)	5,183.619	65	Mg I
3,703.86	43	H I (H 16)	3,933.66	818	Ca II	5,875.65	994	He I (D 3)
3,711.97	53	H I (H 15)	3,968.47	615	Ca II	6,562.808	4,738	H I (H α)
3,721.94	73	H I (H 14)	3,970.076	306	H I (H ε)	7,065.18	138	He I
3,734.37	99	H I (H 13)	4,026.36	24	He I	7,771.954	91	O I
3,750.15	108	H I (H 12)	4,077.724	75	Sr II	7,774.177	75	O I
3,759.299	90	Ti II	4,101.748	459	H I (H δ)	7,775.395	53	O I
3,761.320	82	Ti II	4,215.539	51	Sr II	8,498.02	512	Ca II
3,770.63	116	H I (H 11)	4,226.740	22	Ca I	8,542.09	1,362	Ca II
3,797.90	157	H I (H 10)	4,246.837	18	Sc II	8,545.38	23	H I (P 15)
3,819.61	5	He I	4,340.425	505	H I (Hγ)	8,598.39	26	H I (P 14)
3,820.436	14	Fe I	4,471.69	121	He I	8,662.14	1,181	Ca II
3,829.365	20	Mg I	4,685.68	2	He I	8,665.02	34	H I (P 13)
3,832.310	46	Mg I	4,713.14	9	He I	8,750.47	46	H I (P 12)

[2] After Dunn et. al. [1968] by permission of the American Astronomical Society and the University of Chicago Press.

Table 2.13 (Continued) Coronal emission lines at visible wavelengths[3]

Wavelength (Å)	Equivalent width (m Å)	Element	Wavelength (Å)	Equivalent width (m Å)	Element	Wavelength (Å)	Equivalent width (m Å)	Element
3,329	0.7	Ca XII	4,232.0	1.1	Ni XII	6,374.5	5	Fe X
3,388.0	10	Fe XIII	4,256.4	0.1	K XI	6,701.9	1.2	Ni XV
3,534.0	1	V X	4,351.0	0.1	Co XV	6,740	0.1	K XIV
3,600.9	1.3	Ni XVI	4,412.4	0.3	Ar XIV	7,059.6	0.8	Fe XV
3,642.8	0.4	Ni XIII	4,566.6	0.5	Cr IX	7,891.9	6	Fe XI
3,685	0.2	Mn XII	5,116.0	0.8	Ni XIII	8,024.2	0.3	Ni XV
3,800.7	0.5	Co XII	5,302.9	20	Fe XIV	10,776.8	50	Fe XIII
3,987.1	0.7	Fe XI	5,445.5	0.2	Ca XV	10,797.9	30	Fe XIII
3,998	0.1	Cr XI	5,536	0.3	Ar X			
4,086.5	0.4	Ca XIII	5,094.5	0.3	Ca XV			

[3] After Allen [1963] by permission of the Athlone Press – University of London.

$$\frac{\mu\, dI_\nu}{d\tau_\nu} = I_\nu - S_\nu \ . \tag{2.178}$$

where $\mu = \cos\theta$, the radiation is evaluated at angle θ with respect to the surface normal of the atmosphere, the optical depth $d\tau_\nu = k_\nu\, ds$ for radiation passing through a slab of thickness ds with total opacity k_ν, and the source function $S_\nu = j_\nu / k_\nu$ for an emissivity j_ν. In local thermodynamic equilibrium, $S_\nu = B_\nu$, the Planck function for the radiation brightness of a black body, given by Eq. (1.119) and Eq. (2.189).

The observed radiation is specified by the emergent intensity, $I_\nu(0, \mu)$, at the Sun's surface. The general solution to the equation of transfer is

$$I_\nu(0, \mu) = \int_0^\infty S_\nu(t_\nu) \exp\left(\frac{-t_\nu}{\mu}\right) \frac{dt_\nu}{\mu} \ , \tag{2.179}$$

which is a weighted mean of the source function as a function of depth, t, and gives an outward surface flux of radiation, F_ν, of:

$$F_\nu = 2 \int_0^\infty S_\nu(t_\nu) E_2(t_\nu) dt_\nu \ , \tag{2.180}$$

where the exponential integral $E_n(x)$ is given by

$$E_n(x) = \int_1^\infty \frac{\exp(-xt)}{t^n}\, dt = x^{n-1} \int_x^\infty \frac{\exp(-t)}{t^n}\, dt \ . \tag{2.181}$$

Here $\mu = \cos\theta$ where θ is the angle between the surface normal and the direction of radiation, the optical depth, τ_ν, is given by

$$\tau_\nu = (\kappa + l_\nu + \sigma + \sigma_\nu)\rho\, dz \ , \tag{2.182}$$

where the absorption coefficients per unit mass for the continuum and line radiation are, respectively, κ and l_ν, the coefficients for the coherent scattering of the continuum and line radiation are, respectively, σ and σ_ν, the mass density is ρ, and z is the height measured normal to the plane of stratification of the atmosphere

The mass absorption coefficient, κ_ν, is related to the mass density, ρ, by

$$\kappa_\nu = \frac{\alpha_\nu}{\rho} \ , \tag{2.183}$$

where the absorption coefficient per unit length, α_ν, is given by Eqs. (2.69) and (2.71), (1.237) and (1.221), and (1.219) and (1.221), respectively, for resonance absorption of line radiation, absorption by photoionization, and absorption by free-free transitions of hydrogen like atoms. For hot stars, $T_k > 7000\,°K$, continuous absorption by hydrogen atoms predominates, whereas for cooler stars absorption by the negative hydrogen ion, H^-, predominates. The absorption coefficient for the latter transition is given by Chandrasekhar and Breen (1946), who give $\alpha(H^-) = 5.81 \times 10^{-26}\ cm^{-2}$ at 5600 °K and unit electron pressure. Absorption by He I and He II and Thomson and Rayleigh scattering (Eqs. (1.311) and (1.305)) are also important in some cases. More recent determinations of continuum absorption cross sections are given by Mihalas (1970). When a frequency independent absorption coefficient is desired, the Rosseland mean opacity, κ_R, may be used (Rosseland, 1924)

$$\frac{1}{\kappa_R} = \frac{\pi \int_0^\infty \frac{1}{\kappa_\nu} \frac{dB_\nu(T)}{dT} d\nu}{4\sigma T^3} , \tag{2.184}$$

where the Stefan-Boltzmann constant $\sigma \approx 5.669 \times 10^{-5} \mathrm{erg\,cm^{-2}\,{}^\circ K^{-4}\,sec^{-1}}$, and

$$\frac{dB_\nu(T)}{dT} = \frac{2h^2\nu^4}{c^2 kT} \frac{\exp[h\nu/(kT)]}{\{\exp[h\nu/(kT)] - 1\}^2} . \tag{2.185}$$

Rosseland opacities for population II compositions are given by Cox and Stewart (1970).

The frequency independent absorption coefficient used near the surface of the star is the Planck mean opacity, κ_P, given by

$$\kappa_P = \frac{\pi}{\sigma T^4} \int_0^\infty \kappa_\nu B_\nu(T) d\nu , \tag{2.186}$$

where the Planck function, $B_\nu(T)$, is given by Eq. (2.189). If the continuum absorption coefficient is independent of frequency, then the temperature $T(\tau)$ will vary with the optical depth, τ, according to the gray body relation given by (Milne, 1930)

$$T^4(\tau) = \frac{3}{4}\left(\tau + \frac{2}{3}\right) T_{\mathrm{eff}}^4 , \tag{2.187}$$

where T_{eff} is the effective temperature of a black body radiator which radiates a flux equal to that of the star.

If the continuum emission is thermal, then the source function, S_ν, is given by (Mihalas, 1970)

$$S_\nu = \lambda_\nu B_\nu(T) + (1 - \lambda_\nu) J_\nu , \tag{2.188}$$

where

$$\lambda_\nu = \frac{(1-\rho) + \varepsilon\eta_\nu}{1 + \eta_\nu},$$

$$\eta_\nu = \frac{l_\nu}{\kappa + \sigma},$$

$$\rho = \frac{\sigma}{\kappa + \sigma},$$

the Planck function, $B_\nu(T)$, is given by (Planck, 1901)

$$B_\nu(T) = \frac{2h\nu^3}{c^2}\left[\exp\left(\frac{h\nu}{kT}\right) - 1\right]^{-1} , \tag{2.189}$$

the temperature is T, the fraction of absorbed photons scattered to form line radiation is $(1 - \varepsilon)$, the J_ν is the mean intensity, and κ is the mass absorption coefficient corrected for stimulated emission by multiplication with the factor $[1 - \exp(-h\nu/kT)]$.

When the star is in radiative equilibrium the mean intensity is given by (Schwarzschild, 1906; Milne, 1921)

$$J_\nu(\tau_\nu) = \frac{1}{2}\int_0^\infty S_\nu(t_\nu)E_1(t_\nu - \tau_\nu)dt_\nu = \frac{1}{2}\int_{-1}^{1} I_\nu(\tau_\nu, \mu)d\mu \ , \tag{2.190}$$

where the E_1 is given by Eq. (2.181) with $n = 1$. Using the Eddington (1926) approximation

$$J_\nu(\tau_\nu) = \frac{3}{2}\int_{-1}^{1} I_\nu(\tau_\nu, \mu)\mu^2\, d\mu \ , \tag{2.191}$$

together with the Milne (1930) expansion of the Planck function

$$B_\nu(T) = a + b\tau = a + p_\nu\tau_\nu \ , \tag{2.192}$$

where τ and τ_ν are, respectively, the continuum and line optical depths, and

$$p_\nu = \frac{b}{1 + \eta_\nu} \ ; \tag{2.193}$$

the Milne-Eddington solution to the Schwarzschild–Milne equation (2.190) is obtained

$$J_\nu = a + p_\nu\tau_\nu + \left[\frac{p_\nu - \sqrt{3}a}{\sqrt{3}(1 + \sqrt{\lambda_\nu})}\right]\exp(-\sqrt{3\lambda_\nu}\tau_\nu) \ . \tag{2.194}$$

When studying line radiation it is convenient to use the absorption depths A_ν and a_ν defined by

$$A_\nu = 1 - \frac{F_\nu^L}{F_\nu^C}$$

and

$$a_\nu = 1 - \frac{I_\nu^L(0, \mu)}{I_\nu^C(0, \mu)} \ , \tag{2.195}$$

where the I_ν and F_ν are given, respectively, by Eqs. (2.179) and (2.180), and the superscripts L and C denote, respectively, the line and continuum radiation. When the Milne–Eddington solution is used,

$$A_\nu = 1 - \left[\frac{p_\nu + \sqrt{3\lambda_\nu}a}{1 + \sqrt{\lambda_\nu}}\right]\left[\frac{1 + (1 - \rho)^{1/2}}{b + a[3(1 - \rho)]^{1/2}}\right] \ , \tag{2.196}$$

and a_υ can be obtained from Eq. (2.195), using

$$I_\nu^L(0, \mu) = a + p_\nu\mu + \frac{p_\nu - \sqrt{3}a}{\sqrt{3}(1 + \sqrt{\lambda_\nu})}\frac{(1 - \lambda_\nu)}{(1 + \sqrt{3\lambda_\nu}\mu)} \tag{2.197}$$

and the $I_\nu^C(0, \mu)$ which is given by Eq. (2.197) with $\eta_\nu = 0$.

The equivalent width, W_ν, or W_λ, is given by

$$W_\nu = \int_0^\infty A_\nu d\nu = \frac{\nu_{mn}^2}{c} W_\lambda \ , \tag{2.198}$$

where ν_{mn} is the center frequency of the line. The equivalent width is a useful parameter because it is independent of the instrumental distortion of the profile, and because it provides a measure of the relative abundances of the elements in a stellar atmosphere. The equivalent width of a function is the area of the function divided by its central ordinate. Expressed differently, the equivalent width of a function is the width of the rectangle whose height is equal to that of the function. For a Gaussian function of standard deviation, σ, the equivalent width is 2.5066 σ and the full width to half maximum is 2.355 σ.

If we assume that line formation takes place in an isothermal layer in local thermodynamic equilibrium, then both the (ME) and (SS) approximations lead to the approximate relation (Menzel, 1936)

$$\frac{W_\nu}{2A_0 \Delta\nu_D} = \int_0^\infty \frac{\eta_0 H(a,b)}{1+\eta_0 H(a,b)} db \approx \frac{\eta_0 \pi^{1/2}}{2} \quad \text{for } \eta_0 \ll 1$$
$$\approx (\ln \eta_0)^{1/2} \quad \text{for } 10 \le \eta_0 \le 1000 \approx \frac{(\pi a \eta_0)^{1/2}}{2} \quad \text{for } \eta_0 \ge 1000 \ . \tag{2.199}$$

Here A_0 is the central depth of the line, $\Delta\nu_D$ is the Doppler broadened full width to half maximum, H(a, b) is the Voigt function discussed in Sect. 2.22, the parameter a is given by

$$a = \frac{\Gamma}{4\pi \Delta\nu_D} \ , \tag{2.200}$$

where Γ is the damping constant for the line, and

$$\eta_0 = \frac{\sqrt{\pi}}{\kappa} \frac{e^2}{mc} \frac{f\lambda}{V} N_{r,s} \ , \tag{2.201}$$

where κ is the continuum opacity, the most probable velocity, V, is given by

$$V = \left[\frac{2kT_K}{M} + V_{tur}^2\right]^{1/2} \ , \tag{2.202}$$

the T_K is the gas kinetic temperature, the atomic mass is M, and V_{tur} is the most probable turbulent velocity. The oscillator strengths, f, are tabulated by Corliss and Bozman (1962) and Corliss and Warner (1964), the wavelength is λ, and the number of atoms in the sth state of the rth ionization stage is given by the Saha equation (3.127) (Saha, 1921)

$$N_{r,s} = \frac{N_r g}{U} \exp\left(\frac{-\chi}{kT_{exc}}\right) \ , \tag{2.203}$$

where N_r is the total number of atoms in the rth stage of ionization, g is the statistical weight of the sth level, U is the partition function, and χ is the excitation potential of the sth level. We may then write Eq. (2.201) as

$$\log \eta_0 = \log(gf\lambda) - \frac{5040\chi}{T_{exc}} + \log C \ , \tag{2.204}$$

where $\log C = \log N_r + \log\left[\sqrt{\pi}e^2/(m\,c)\right] - \log V - \log U - \log \kappa$, and $\log\left[\sqrt{\pi}e^2/(m\,c)\right] = -1.826$. Theoretical curves of growth are plots of $\log(W_\lambda/\Delta\lambda_D)$ versus $\log \eta_0$, and an example is shown in Fig. 2.6. The curve of growth method was anticipated by Struve and Elvey (1934). Curves of growth for the (ME) and (SS) approximations are given by Van der Held (1931), Wrubel (1950), Wrubel (1954), and Hunger (1956). Empirical curves of growth are plots of $\log(W_\lambda/\lambda)$ versus $\log(g f \lambda)$, and curves for the Sun are given by Cowley and Cowley (1964) and Pagel (1965). As shown in Fig. 2.6, the ordinates of the empirical and theoretical curves are displaced by the amount

$$\log\left(\frac{V}{c}\right) = \log\left(\frac{W_\lambda}{\lambda}\right) - \log\left(\frac{W_\lambda}{\Delta\lambda_D}\right) , \tag{2.205}$$

from which one can extract V. The excitation temperature, $T_{\rm exc}$, is found by comparing an empirical curve of growth with various theoretical curves, or by comparing the empirical curves for lines of one excitation potential with similar

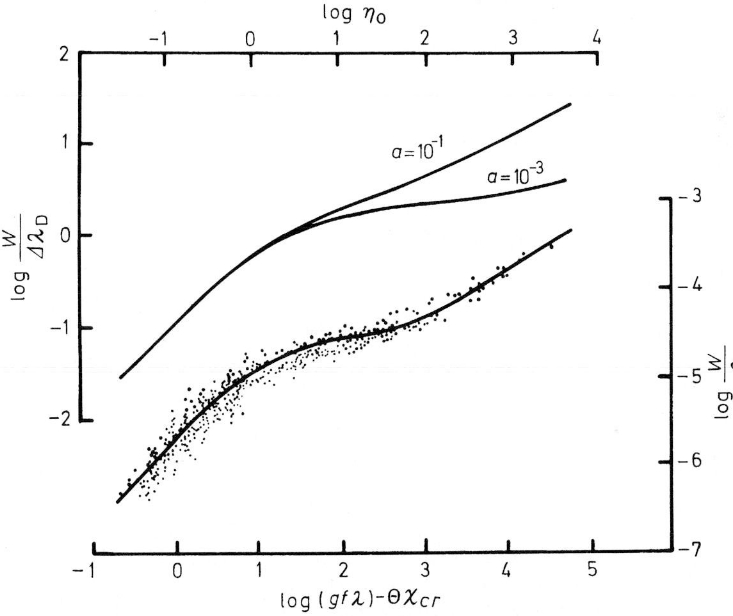

Fig. 2.6. Theoretical curves of growth [cf. Eq. (2.199)] and the solar curve of growth [after Cowley and Cowley, 1964, by permission of the American Astronomical Society and the University of Chicago Press]. For the solar curve of growth, W is the equivalent width of the line, the designation χ_{cr} indicates that the abscissa is given in the chromium scale, and each point represents an individual line. An excitation temperature of 5143 °K was found to best approximate the data; and assuming that the excitation temperature equals the gas kinetic temperature, a turbulent velocity of $1.4 \pm 0.2\,\mathrm{km\,sec^{-1}}$ is obtained. A comparison of the theoretical and empirical curves of growth also yields a value of 1.4 for the logarithm of the ratio of the observed damping constant to the classical damping constant. The parameter $a = \Gamma/(4\pi\Delta\nu_D)$ where Γ is the damping constant and $\Delta\nu_D$ is the full width to half maximum of the Doppler broadened line profile. The variable, η_0, is related to the other parameters of the abscissa by Eq. (2.204)

curves for lines of another excitation potential. Provided that the excitation temperature is equal to the gas kinetic temperature, the turbulent velocity, V_{tur}, can be estimated from V and T_{exc} using Eq. (2.202). For the Sun, Cowley and Cowley (1964) obtain $T_{exc} = 5143\,°K$, and $V_{tur} = 1.4 \pm 0.2\,\mathrm{km\,sec^{-1}}$. The difference in abscissae between the theoretical and empirical curves leads to the $\log C$, which in turn leads to a measure of the relative abundances of the elements.

The solar and meteoritic abundances of the elements have been given by Anders and Grevesse (1989), and reproduced by Lang (1992). An earlier compilation of solar system abundances was given by Cameron (1982); his values are also given in Lang (1992). The relative abundances of the elements in the atmosphere of the Sun and other normal stars have been previously given by Russell (1929), Goldberg et al. (1960), Aller (1961), Lambert and Warner (1968), Unsöld (1969) and Mitler (1970).

The value of the slope of the damping part of the curve of growth may lead to an estimate of the gas pressure, P_g, or electron pressure, P_e, according as the collision damping is dominated by collisions with neutral hydrogen or by the quadratic Stark effect. The gas pressure and electron pressure are related to the gas mass density, ρ, by the equation

$$\rho k T_K = (P_g - P_e)\mu m_H \ , \tag{2.206}$$

where m_H is the mass of hydrogen, μ is the mean molecular weight, and T_K is the gas kinetic temperature.

2.17 Effects Which Alter the Emitted Line Frequency

Normal Zeeman effect. For atoms with singlet lines (spin $S = 0$), a magnetic field will split any term into $2J + 1$ equally spaced levels (where J is the total angular momentum). The energies, E_M, of the levels are given by (Zeeman, 1896, 1897; Lorentz, 1897)

$$E_M = E_n \pm \frac{e h H M}{4\pi m c} = E_n \pm h o M \ , \tag{2.207}$$

where M takes on integral values between 0 and J, E_n is the energy of the term without the magnetic field, H is the magnetic field strength, the factor $eh/(4\pi m c) = 9.2741 \times 10^{-21}$ erg Gauss^{-1} is the magnetic moment of the Bohr magneton, and $o = eH/(4\pi m c) = 1.400 \times 10^6 H$ Hz is the Larmor frequency of precession.

The selection rules for allowed transitions between the levels of different terms are $\Delta M = 0, \pm 1$ and a level with $M = 0$ cannot combine with another level of $M = 0$. Because there is equal splitting for all terms, the frequency of any singlet line, ν_{mn}, will be split into three components; the π component at frequency ν_{mn} and the two σ components at frequencies (Lorentz, 1897)

$$\nu = \nu_{mn} \pm \frac{eH}{4\pi m c} = \nu_{mn} \pm 1.400 \times 10^6 H\,\mathrm{Hz} \ , \tag{2.208}$$

where the magnetic field strength H is given in Gauss. The π component is plane polarized in the plane containing the line of sight and the vector of the magnetic field. The two σ components are elliptically polarized. For observation in the direction of the magnetic field, the central π component is absent and the two σ components are circularly polarized with opposed directions of rotation (the lower frequency having right hand rotation). Observation in a direction transverse to the field shows all three components, the undisplaced one has linear polarization parallel (π) to the field and the others show linear polarization perpendicular (σ), to the field.

In general, the observed intensity, $I(\nu)$, at frequency, ν, will depend on the intensity, $I_0(\nu)$, that would be produced if there were no magnetic field, the inclination, γ, of the magnetic vector to the line of sight, and the polarization reception characteristics of the antenna. The total intensity in a beam with right hand circular polarization is (Seares, 1913)

$$\begin{aligned} I_{\mathrm{R}}(\nu) &= 1/8(1-\cos\gamma)^2 I_0(\nu+\Delta\nu_0) + 1/4\sin^2\gamma I_0(\nu) \\ &\quad + 1/8(1+\cos\gamma)^2 I_0(\nu-\Delta\nu_0) \ , \end{aligned} \tag{2.209}$$

where the Zeeman splitting frequency $\Delta\nu_0 = 1.4 \times 10^6 H$ Hz. The intensity in a beam with left hand polarization, $I_{\mathrm{L}}(\nu)$, is obtained from $I_{\mathrm{R}}(\nu)$ by changing the sign of the $\cos\gamma$ term.

When the magnetic fields are weak, the Zeeman splitting, $\Delta\nu_0 = 1.4 \times 10^6 H$, may be much smaller than the full width to half maximum, $\Delta\nu_{\mathrm{L}}$, of the observed line. In that event, the line of sight component of the magnetic field, $H\cos\gamma$, may be detected by comparing the line profiles observed with right hand, $I_{\mathrm{R}}(\nu)$, and left hand, $I_{\mathrm{L}}(\nu)$, circularly polarized antennae. The two profiles will be displaced in frequency by $\Delta\nu = 1.4 \times 10^6 H\cos\gamma$ Hz. By adding and subtracting the two profiles the sum, $I(\nu)$, and difference, $D(\nu)$, profiles are obtained

$$\begin{aligned} I(\nu) &= I_{\mathrm{L}}(\nu) + I_{\mathrm{R}}(\nu) \\ D(\nu) &= I_{\mathrm{L}}(\nu) - I_{\mathrm{R}}(\nu) = 2.8 \times 10^6 I'(\nu) H\cos\gamma \ , \end{aligned} \tag{2.210}$$

where $I'(\nu)$ denotes the first derivative of $I(\nu)$ with respect to frequency. In terms of the peak antenna temperature, T_{A}, and half width, $\Delta\nu_{\mathrm{L}}$, of $I(\nu)$, and the peak antenna temperature, ΔT_{A}, of $D(\nu)$, we have

$$H\cos\gamma \approx 10^{-6} \frac{\Delta\nu_{\mathrm{L}}\,\Delta T_{\mathrm{A}}}{4T_{\mathrm{A}}} \text{ Gauss} \ . \tag{2. 211}$$

Effective Zeeman frequency displacements as small as $0.005\,\Delta\nu_{\mathrm{L}}$ have been detected using this technique.

George Ellery Hale first used the size of the Zeeman splitting to measure the magnetic field strength of sunspots (Hale, 1908), indicating intense magnetic fields of up to several thousand Gauss. Hale and his colleagues found that the majority of sunspots occur in pairs of opposite magnetic polarity; the magnetic field direction or polarity is determined by the sense of circular polarization of the Zeeman effect. The preceding and following spots of binary groups are of opposite polarity, and the corresponding spots of such groups in the northern and southern hemispheres are also opposite in sign. Furthermore, all of the spots reverse their polarity every 22 years, at twice the 11-year period of the variation in the number of sunspots (Hale, Ellerman,

Nicholson, and Joy, 1919). The global, dipolar magnetic field of the Sun is about 10 Gauss in strength (Babcock and Babcock, 1955).

Nowadays arrays of tiny detectors measure the Zeeman effect across the visible solar disk (Lang, 1995). Two images are produced, one in each polarization, and the difference of these images produces a magnetograph. The brightness at each point portrays the strength of the magnetic field, or the size of the Zeeman splitting, while the circular polarization gives the direction. Magnetic fields directed out along the line of sight have right-hand circular polarization, those pointing in the opposite inward direction have left-hand circular polarization. Elements and patterns in the solar magnetic field were reviewed by Zwaan (1987).

Magnetic fields on other normal, or nondegenerate, stars have been detected using the Zeeman effect in optical absorption lines. Most of these magnetic stars are peculiar stars of A or B type, usually called Ap stars. They have magnetic fields of up to 34,0000 Gauss (Babcock, 1960; Cameron, 1967). The magnetic field strengths, names and celestial positions of magnetic stars have been given by Lang (1992), after the compilation by Didelon (1983). Magnetic stars have been reviewed by Borra, Landstreet and Mestel (1982).

Because magnetic flux is conserved in gravational collapse, the stellar magnetic fields can be amplified to more than a million Gauss when a star contracts to the degenerate, white dwarf stage. These intense magnetic fields are measured by either the Zeeman displacement of their spectral lines or from their wavelength-dependent, circularly-polarized light (Kemp et al., 1970; Landstreet, 1980; Angel et al., 1981). Isolated white dwarf stars with Mega-Gauss magnetic field strengths (strength $H \geq 10^6$ Gauss) are given by Lang (1992), adapted from Schmidt (1988) and McCook and Sion (1987). Intensely- magnetized white dwarf stars that are members of binary stellar systems include the so-called polars whose magnetic field is strong enough to force the white dwarf to rotate synchronously and for the accreting material to be funneled onto the white dwarf's surface near a magnetic pole. The magnetic field strengths of these objects are determined by Zeeman spectroscopy or cyclotron spectral features; they are included in the tables of polars given by Lang (1992) and Cropper (1990). White dwarf stars and their intense magnetic fields have been reviewed by Angel (1978) and Liebert (1980); the magnetic fields of degenerate stars are reviewed by Chanmugam (1992). Observations of magnetic fields in planets, stars, pulsars, masers and protostellar cloudlets are reviewed by Vallée (1998).

The interstellar magnetic field has been measured using the Zeeman effect with the 21-cm line of atomic hydrogen (Verschuur, 1968, 1971), a possibility suggested by Bolton and Wild (1957). The measured strength, H, of the interstellar magnetic field is $H \approx 10^{-6}$ Gauss, which is consistent with the Faraday rotation and rotation measures of pulsars (see Sect. 1.3.3 and Lang (1992)). Observation and theory of interstellar magnetic fields in our Galaxy are described in Beck and Gräve (1987) and reviewed by Beck et al. (1996). The global structure of magnetic fields in spiral galaxies are reviewed by Sofue, Fujimoto and Wielebinski (1988). Observations of the magnetic field of our Galaxy are summarized by Verschuur (1979), and observations of magnetic fields inside and outside the Milky Way are reviewed by Vallée (1997).

Anomalous Zeeman effect. When the spin $S \neq 0$, and there is Russell–Saunders coupling, the magnetic field splits a term into $2J+1$ equally spaced levels. The energies, E_M, of the levels are given by (Preston (1898); Landé (1920, 1923); Shu (1991))

$$E_M = E_n \pm \frac{e\,h\,H\,g\,M}{4\pi\,m\,c} = E_n \pm h\,o\,g\,M = E_n \pm 9.27 \times 10^{-21} HgM \text{ erg} \;, \tag{2.212}$$

where E_n is the energy of the term in the absence of the magnetic field, H is the magnetic field intensity in Gauss, o is the Larmor frequency of precession, M is component of J along H, and the Landé g factor for LS coupling is given by Uhlenbeck and Goudsmit (1925, 1926)

$$g = 1 + \frac{J(J+1) + S(S+1) - L(L+1)}{2J(J+1)} \;, \tag{2.213}$$

where J is the total angular momentum, S is the spin angular momentum, and L is the orbital angular momentum.

Because there is not equal line splitting between different terms, however, the number of line components will depend on the J, L and S of the upper and lower levels. The selection rule for allowed transitions between the levels of different terms is $\Delta M = 0$ (π components) and $\Delta M = \pm 1$(σ components). Observations in a direction transverse to the field will show the π component linearly polarized parallel to the field and the σ components linearly polarized perpendicular to the field. For observations in a direction parallel to the field, only the two σ components are observed and they are circularly polarized in opposed directions of rotation (the lower frequency being right hand).

When the magnetic moment of the atomic nucleus is taken into account, the appropriate Landé g factor is given by (Back and Goudsmit, 1928)

$$g(F) = g\frac{F(F+1) + J(J+1) - I(I+1)}{2F(F+1)} \;, \tag{2.214}$$

where g is the electronic g factor for the atom which is given above, I is the nuclear spin quantum number, J is the total angular momentum excluding nuclear spin, and F is the total angular momentum quantum number.

For molecules, the electronic Landé g factor is given by Townes and Schawlow (1955)

$$g = \frac{(\Lambda + 2.002\Sigma)\Omega}{J(J+1)} = \frac{(\Omega + 1.002\Sigma)\Omega}{J(J+1)} \tag{2.215}$$

for Hund's coupling case (a), and

$$g = \frac{1}{2J(J+1)}\left\{\Lambda^2 \frac{[N(N+1) - S(S+1) + J(J+1)]}{N(N+1)} \right.$$
$$\left. + 2.002[J(J+1) + S(S+1) - N(N+1)]\right\} \tag{2.216}$$

for Hund' coupling case (b). Here Λ is the projection of the electronic orbital angular momentum on the molecular axis, Σ is the projection of the electron spin angular momentum on the molecular axis, Ω is the projection of the total angular momentum excluding nuclear spin on the molecular axis, and N is the total angular momentum including rotation of the molecule.

Intensities of Zeeman components. The relative intensities, I, of the ordinary Zeeman components (weak fields) for observations at right angles to the magnetic field are:

For the $J \rightarrow J$ transition

$$\begin{aligned} &\pi \text{ components } (\Delta M = 0) \quad I = 4AM^2 , \\ &\sigma \text{ components } (\Delta M = \pm 1) \quad I = A[J(J+1) - M(M \pm 1)] \ . \end{aligned} \tag{2.217}$$

For the $J \rightarrow J+1$ transition

$$\begin{aligned} &\pi \text{ components } (\Delta M = 0) \quad I = 4B[(J+1)^2 - M^2] , \\ &\sigma \text{ components } (\Delta M = \pm 1) \quad I = B[J \pm M + 1][J \pm M + 2] \ . \end{aligned} \tag{2.218}$$

Here A and B are constants for a given Zeeman pattern. The σ components will be observed at this same relative intensity when viewed parallel to the magnetic field. These formulae are given by Condon and Shortley (1963), and were first obtained empirically by Ornstein and Burger (1924) and theoretically by Kronig (1925) and Hönl (1925).

Zeeman effect for strong magnetic field – the Paschen–Back effect, the quadratic Zeeman effect, and the circular polarization of thermal radiation. When magnetic splitting becomes greater than the multiplet splitting, for magnetic fields stronger than a few thousand Gauss, the anomalous Zeeman effect changes over to the normal Zeeman effect. This Paschen–Back effect (Paschen and Back, 1912, 1913) is the result of magnetic uncoupling of L and S. To a first approximation the energies, E_M, of the split levels are given by

$$E_M = E_n + \frac{eh}{4\pi mc}(M_L + 2M_S)H \ , \tag{2.219}$$

where E_n is the energy of the term in the absence of the magnetic field, $eh/(4\pi mc) = 9.2741 \times 10^{-21}$ erg Gauss^{-1}, the quantum numbers M_L and M_S take on integral values between $\pm L$ and $\pm S$, respectively, and the selection rules are $\Delta M_L = 0, \pm 1$ and $\Delta M_S = 0$.

If magnetic flux is conserved in gravitational collapse, it is expected that white dwarf and neutron stars will have magnetic fields of $\sim 10^6$ and 10^{12} Gauss, respectively. For these very large field strengths, the quadratic Zeeman effect is expected to dominate over the linear effect (Preston, 1970). The quadratic Zeeman effect produces a displacement, $\Delta\lambda$, of spectral lines towards the short wavelengths. This displacement is given by Van Vleck (1932); Jenkins and Segré (1939)

$$\Delta\lambda = \frac{-e^2 a_0^2}{8mc^3 h}\lambda^2 n^4 (1 + M^2) H^2 \approx -4.98 \times 10^{-23} \lambda^2 n^4 (1 + M^2) H^2 \, \text{Å} \tag{2.220}$$

for hydrogen. Here a_0 is the radius of the first Bohr orbit, n is the principal quantum number, M is the magnetic quantum number, H is the magnetic field intensity, and the numerical approximation is for $\Delta\lambda$ and λ in Angstroms (1Å = 10^{-8} cm). The quadratic Zeeman effect for hydrogen Balmer lines is given by Surmelian and O'Connell (1974). Energy levels of hydrogen in strong magnetic fields are provided by Rau and Spruch (1976), Surmelian and O'Connell (1974) and Wadehra (1978).

Unfortunately, white dwarfs and neutron stars show little spectral structure from which to observe the quadratic Zeeman effect. Any emission from a thermal source may, however, exhibit a diffuse spectral component which is circularly polarized. This may be seen by regarding thermal radiation as a superposition of the emission from a collection of harmonic oscillators which emit circularly polarized emission at their Zeeman split frequencies, $\nu \pm eH/(4\pi mc)$. The net fractional polarization of the thermal radiation is given by Kemp (1970).

$$q(\nu) = \frac{I_+(\nu) - I_-(\nu)}{I_+(\nu) + I_-(\nu)} \approx \frac{-eH}{4\pi m\nu} \approx 4.2 \times 10^{16}\frac{H}{\nu} \ , \tag{2.221}$$

where $I(\nu)$ denotes the observed radition intensity at frequency, ν, the + and − denote, respectively, right and left hand circular polarization, the H is the component of the magnetic field along the line of sight, and $q(\nu)$ is right handed for electrons. The percentage circular polarization will fall off faster than ν^{-1} for frequencies $\nu > \nu_p$, where the plasma frequency $\nu_p \approx 8.9 \times 10^3 N_e^{1/2}$ Hz. Kemp et al. (1970) and Angel and Landstreet (1971) have detected circular polarization in the optical emission of D.C. white dwarfs. They observe a $q(\nu)$ of a few percent indicating magnetic field strengths of 10^6 to 10^7 Gauss.

Stark effect for hydrogen atoms. When a hydrogen atom is subjected to an electric field of strength, E, the alteration, ΔE, of the energy of a level is given by Stark (1913); Schwarzschild (1916); Epstein (1916, 1926); Bethe and Salpeter (1957)

$$\Delta E = AE + BE^2 + CE^3, \tag{2.222}$$

where

$$A = \frac{3h^2}{8\pi^2 me} n(n_2 - n_1),$$

$$B = \frac{-h^6 n^4}{2^{10}\pi^6 m^3 e^6}[17n^2 - 3(n_2 - n_1)^2 - 9m_l^2 + 19]$$

and

$$C = \frac{3h^{10}}{2^{15}\pi^{10} m^5 e^{11}} n^7 (n_2 - n_1)[23n^2 - (n_2 - n_1)^2 + 11m_l^2 - 71] \ .$$

Here n is the principal quantum number, m_l is the component of the orbital quantum number, l, in the direction of the electric field, m_l takes on integral values between $-l$ and $+l$, and the parabolic quantum numbers n_2 and n_1 may assume integer values between 0 and $n - 1$. The coefficients of A, B, and C are 1.27×10^{-20} erg cm volt^{-1}, 1.03×10^{-31} erg cm^2 volt^{-2} and 2.97×10^{-41} erg cm^3 volt^{-3}, respectively. The factor −71 given in brackets for the coefficient C

is from Condon and Shortley (1963), whereas Bethe and Salpeter (1957) give this term as +39.

To the first approximation, each level of principal quantum number, n, is split into $2n-1$ equidistant levels with energies, E_s, given by

$$E_s = E_n + \frac{3h^2}{8\pi^2 me} n(n_2 - n_1)E$$
$$= E_n + 3.81 \times 10^{-18} n(n_2 - n_1)E \text{ erg} , \tag{2.223}$$

where E is given in e.s.u. (1.71×10^7 e.s.u. $= 5.142 \times 10^9$ volt cm^{-1}), E_n is the energy of the nth level in the absence of an electric field, and n_2 and n_1 may assume integer values between 0 and $n-1$. For a transition between an upper level m and a lower level n, the frequency is split into the frequencies $\nu_{m,n}$ given by

$$\nu_{m,n} = \nu_{mn} + \frac{3hE}{8\pi^2 m_e e}\{m(m_2 - m_1) - n(n_2 - n_1)\}$$
$$= \nu_{mn} + 1.91 \times 10^6 E\{m(m_2 - m_1) - n(n_2 - n_1)\} \text{ Hz} , \tag{2.224}$$

where the field strength E is in volt cm^{-1}, n_2 and n_1 may assume integer values between 0 and $n-1$, and m_2 and m_1 may assume integer values between 0 and $m-1$. The change in wavelength, $\Delta\lambda$, corresponding to $\nu_{m,n} - \nu_{mn}$, is

$$\Delta\lambda \approx 0.64 \times 10^{-4} \lambda^2 E\{m(m_2 - m_1) - n(n_2 - n_1)\} \text{ cm} , \tag{2.225}$$

where λ is in cm and E is in volt cm^{-1}.

If we define the new quantum numbers $|kn| = n - 1 - n_1 - n_2$ and $|km| = m - 1 - m_1 - m_2$, then for transverse observation the radiation is polarized parallel (π) to the electric field if $kn = km$, and perpendicular (σ) to the field if $km - kn = \pm 1$.

When considering astrophysical plasmas, the electric field is not uniform and constant; and we are concerned with the probability distribution of the electric field of any moving charged particles. For slowly moving ions, the net effect is to broaden and shift the line to higher frequencies (cf. Sect. 2.20).

Doppler shift. When line radiation of frequency, ν_L, is emitted from an object travelling at a velocity, V, with respect to an observer at rest, the observed frequency, ν_{obs}, is (Doppler, 1842; Lorentz, 1904; Einstein, 1905)

$$\nu_{obs} = \nu_L \left(1 - \frac{V}{c}\cos\theta\right)^{-1} \left(1 - \frac{V^2}{c^2}\right)^{1/2}$$
$$\approx \nu_L \left(1 - \frac{V_r}{c}\right) \text{ for } V \ll c \text{ and } \theta \ll \pi/2 , \tag{2.226}$$

where θ is the angle between the velocity vector and the radition wave vector. The radial velocity $V_r = -V\cos\theta$ is positive when the radiating object is moving away from the observer. Consequently, when receding objects are observed at optical frequencies, the lines are shifted towards the red (towards longer wavelengths). The redshift, z, is defined as

$$z = \frac{\lambda_{obs} - \lambda_L}{\lambda_L} = \frac{\nu_L - \nu_{obs}}{\nu_{obs}} \approx \frac{V_r}{c} , \tag{2.227}$$

where λ_{obs} and λ_L denote, respectively, the wavelengths of the observed and emitted line radiation, υ_{obs} and υ_L denote, respectively, the frequencies of the observed and emitted line radiation, and V_r is the radial velocity. For large velocities the special relativistic Doppler effect is expressed by the equations

$$1 + z = \left[\frac{c + V_r}{c - V_r}\right]^{1/2} \tag{2.228}$$

and

$$\frac{V_r}{c} = \frac{(z+1)^2 - 1}{(z+1)^2 + 1} \ . \tag{2.229}$$

Measured velocities are often corrected for the orbital motion of the earth about the Sun, and the Sun's motion with respect to the local group of stars. The corrections to obtain this velocity with respect to the local standard of rest are discussed in Volume II.

The radial velocities V_r, determined from the absorption line shifts of the 446 nearest and brightest stars are given by Lang (1992). Radial velocities are also included by Lang (1992) in his catalogues of globular star clusters, H II regions (emission nebulae), open star clusters, and planetary nebulae.

Heliocentric radial velocities for nearby galaxies, also determined from the Doppler shifts of optical absorption lines in their spectra, are given by Tully (1980) and included in the Huchtmeier and Richter (1989) catalogue of radio observations of neutral hydrogen in nearby galaxies. The radial velocities of these normal galaxies are also compiled by Hirshfeld and Sinnott (1985). Earlier determinations of the redshifts for the normal optical galaxies are given by Humason, Mayall, and Sandage (1956), Mayall and De Vaucouleurs (1962) and De Vaucouleurs and De Vaucouleurs (1964). A comparison of radio and optical redshifts of normal galaxies is given by Ford, Rubin and Roberts (1971).

The redshifts of intense extragalactic radio sources, called radio galaxies, are also provided by Hirshfeld and Sinnott (1985). These objects are listed in the Third Cambridge, 3C R(revised), and Parkes (PKS) radio surveys, but their Doppler shifts require identifications of optical counterparts that have been provided by Burbidge and Crowne (1979), Kristian (1975), Laing et al. (1978), Longair and Gunn (1975), Riley, Longair and Gunn (1980), Smith, Spinrad and Smith (1976) and Wyndham (1966).

The redshifts of quasi-stellar objects, or quasars, ar given by Hewitt and Burbidge (1980, 1981), and reproduced by Hirshfeld and Sinnot (1985). Radial velocities for quasars must be inferred from the redshifts using the relativistic expression Eq. (2.229).

Gravitational redshift. The energy, E, of a particle of mass, m, and velocity, υ, at the surface of a body of mass, M, and radius, R, is given by

$$E = \frac{m\upsilon^2}{2} - \frac{GMm}{R} \ , \tag{2.230}$$

so that the velocity of escape, for which $E = 0$, is given by

$$V_{esc} = \left(\frac{2GM}{R}\right)^{1/2} . \tag{2.231}$$

For a photon $V_{esc} = c$ and the limiting Schwarzschild radius, R_g, for its escape is given by

$$R_g = \frac{2GM}{c^2} . \tag{2.232}$$

When a photon of energy $h\nu = mc^2$ leaves the surface of the massive body it loses the energy

$$\Delta E = h\Delta\nu = \frac{GMm}{R} = \frac{GMh\nu}{Rc^2} , \tag{2.233}$$

so that the gravitational redshift is given by

$$z_g = \frac{\Delta\nu}{\nu} = \frac{GM}{Rc^2} = \frac{R_g}{2R} = 1.47 \times 10^5 \frac{M}{M_\odot R} , \tag{2.234}$$

where the solar mass $M_\odot \approx 2 \times 10^{33}$ grams. The more exact expression is given by (Einstein, 1911, 1916)

$$z_g = \frac{\nu_L - \nu_{obs}}{\nu_{obs}} = \left(1 - \frac{2GM}{Rc^2}\right)^{-1/2} - 1 . \tag{2.235}$$

For the Sun, $z_g \approx 2 \times 10^{-6}$ which is a value which would be given by the Doppler effect due to a recessional velocity of about 0.64 km sec^{-1}, and as such is very difficult to measure. Nevertheless, such a measurement has been reported (Blamont and Roddier, 1961). White dwarf stars which are members of a binary system provide lines with large gravitational redshifts which can be compared with the unreddened lines of the less dense companion. Measurements of white dwarf spectra indicate a redshift equivalent to 19 km sec^{-1} for Sirius B (Adams, 1925) and 21 ± 4 km sec^{-1} for 40 Eridani B (Popper, 1954). More recent measurements of the gravitational redshift of these two objects give respective values of 89 ± 16 km sec^{-1} for Sirius B (Greenstein, Oke, and Shipman, 1971) and 23 ± 5 km sec^{-1} for 40 Eri B (Greenstein and Trimble, 1972). Wegner (1980) provides a gravitational redshift of 23.9 ± 1.3 km s^{-1} for 40 Eri B; the mass and radius of 40 Eri B are given by Shipman, Provencal, Høg and Thejll (1997). Additional gravitational redshift measurements of white dwarf stars are given by Koester (1987) and Stauffer (1987). The gravitational redshifts, masses and radii of white dwarf stars are given by Lang (1992), and masses and radii of white dwarfs are also given by Shipman (1972, 1977). The rotation and gravitational redshift of white dwarfs are discussed by Greenstein et al. (1977), and the unimportance of pressure shifts in the measurement of gravitational redshifts in white dwarfs is given by Shipman and Mehan (1976). The maximum gravitational redshifts of white dwarfs are discussed by Shapiro and Teukolsky (1976).

Refraction effects. Although the frequency of a wave remains unchanged as it propagates through most media, its wavelenghth, λ, does not. When measuring a wavelength, λ_{air}, in air it must be compared with laboratory measurements in vacuum, λ_{vac}

$$\lambda_{air} = \frac{\lambda_{vac}}{n_{air}} \ , \tag{2.236}$$

where the index of refraction of air at optical frequencies (2000–7000 Å) is given by (Edlén, 1953)

$$n_{air} = 1 + 6432.8 \times 10^{-8} + \frac{2,949,810}{146 \times 10^8 - \tilde{\nu}^2} + \frac{25,540}{41 \times 10^8 - \tilde{\nu}^2} \tag{2.237}$$

and the wave number, $\tilde{\nu}$, is given by

$$\tilde{\nu} = \frac{\nu}{c} = \frac{1}{\lambda_{vac}} \ . \tag{2.238}$$

The measured optical redshift, z_{opt}, is usually calculated from the expression

$$z_{opt} = \frac{\lambda_{air} - \lambda_{vac}}{\lambda_{vac}} \ . \tag{2.239}$$

The true redshift, z, is given by the relation

$$1 + z = (1 + z_{opt})n_{air} \ , \tag{2.240}$$

where n_{air} is the index of refraction of air corresponding to λ_{air}, here λ_{air} is the observed wavelength, and the index of refraction, $n_{air} \approx 1.000297$.

2.18 Doppler Broadening (a Gaussian Profile)

When line broadening is caused by the motion of atoms or molecules in thermodynamic equilibrium, or by the random turbulent motion of the gas, Maxwell's distribution of velocities ((Maxwell, 1860), Eq. (3.114)) may be used together with the Doppler formula which relates frequency shift to radial velocity ((Doppler, 1842), Eq. (2.226)). The resulting spectral line intensity distribution is given by (Rayleigh, 1889)

$$\varphi_{mn}(\nu) = \frac{1}{\sqrt{2\pi}\sigma} \exp\left(-\frac{(\nu - \nu_{mn})^2}{2\sigma^2}\right) \ , \tag{2.241}$$

where ν is the frequency under consideration, ν_{mn} is the frequency of the $m - n$ transition, and

$$2\sigma^2 = \frac{\nu_{mn}^2}{c^2}\left(\frac{2kT_k}{M} + V^2\right),$$

where M is the atomic or molecular mass, T_k is the kinetic temperature of the moving atoms or molecules, and V is the most probable turbulent velocity. The turbulent velocities are assumed to have a Maxwellian distribution so that the r.m.s turbulent velocity, $V_{rms} = (3/2)^{1/2}V$, and the mean turbulent velocity $\langle V \rangle = 2V/\pi^{1/2}$. The profile has a full width to half maximum given by

$$\Delta\nu_D = \frac{2\nu_{mn}}{c}\left[\ln 2\left(\frac{2kT_k}{M} + V^2\right)\right]^{1/2} = 2.3556\sigma \tag{2.242}$$

and the peak value of $2(\ln 2/\pi)^{1/2}(\Delta\nu_D)^{-1}$.

Thus, $2\sigma^2 = \Delta\nu_D{}^2/(4\ln 2)$, and the spectral line intensity distribution is:

$$\varphi_{mn}(\nu) = \left(\frac{4\ln 2}{\pi}\right)^{1/2} \frac{1}{\Delta\nu_D} \exp\left[-4\ln 2\left(\frac{\nu_{mn} - \nu}{\Delta\nu_D}\right)^2\right] . \tag{2.243}$$

The full width at half intensity of the line is also given by

$$\Delta\nu_D = \left(\frac{8kT_D}{mc^2}\ln 2\right)^{1/2} \nu_{mn} , \tag{2.244}$$

with T_D being the effective Doppler temperature that characterizes the line; and

$$T_D = \frac{1}{8\ln 2} \frac{Mc^2}{k\nu_{mn}^2} (\Delta\nu_D)^2 \approx 1.17 \times 10^{36}\, M(\Delta\nu_D/\nu_{mn})^2\ {}^\circ\mathrm{K} . \tag{2.245}$$

When a substantial fraction of the radiation of an atom or molecule is scattered by free electrons, the approximate half width of the observed line will be given by Eq. (2.242), where the mass, M, is the electron mass rather than the mass of the radiating atoms or molecules. A measure of the fraction of radiation which is scattered is the optical depth, τ, given by

$$\tau = \int_0^L \sigma_T N_e\, dl , \tag{2.246}$$

where the integral is a line of sight integral over the extent, L, of the radiating source, N_e is the density of free electrons, and the Thomson scattering cross section (Thomson,1903) is given by

$$\sigma_T = \frac{8\pi}{3}\left(\frac{e^2}{mc^2}\right)^2 \approx 6.65 \times 10^{-25}\ \mathrm{cm}^2 .$$

For optical depths greater than 0.5 almost all of the line radiation is in the wider scattered line as opposed to the narrower radiated line, whereas the fraction of the radiation energy in the wider line is about $\tau/(1-\tau)$ for optical depths less than 0.5. If we assume that the electrons have a Maxwellian velocity distribution, the Compton scattering probability $\sigma(\theta, \nu_1, \nu_2)$ that a photon is scattered by an angle θ, and that its frequency is shifted from ν_1 to ν_2 in this process is (Dirac, 1925)

$$\sigma(\theta, \nu_1, \nu_2) = \frac{3}{4}\sigma_T \frac{(1+\cos^2\theta)}{(1-\cos\theta)^{1/2}} \frac{1}{2\pi\Delta\nu_D} \exp\left[\frac{-(\nu_2-\nu_1)^2}{\Delta\nu_D^2(1-\cos\theta)}\right] , \tag{2.247}$$

where $\Delta\nu_D$ is given by Eq. (2.242) when the mass, M, is the electron mass. A more complicated formula for the $\sigma(\theta, \nu_1, \nu_2)$ of Compton scattering by moving atoms is given by Henyey (1940) and Böhm (1960). The Doppler broadened line profile is illustrated in Fig. 2.7 together with the Lorentz and Voigt profiles.

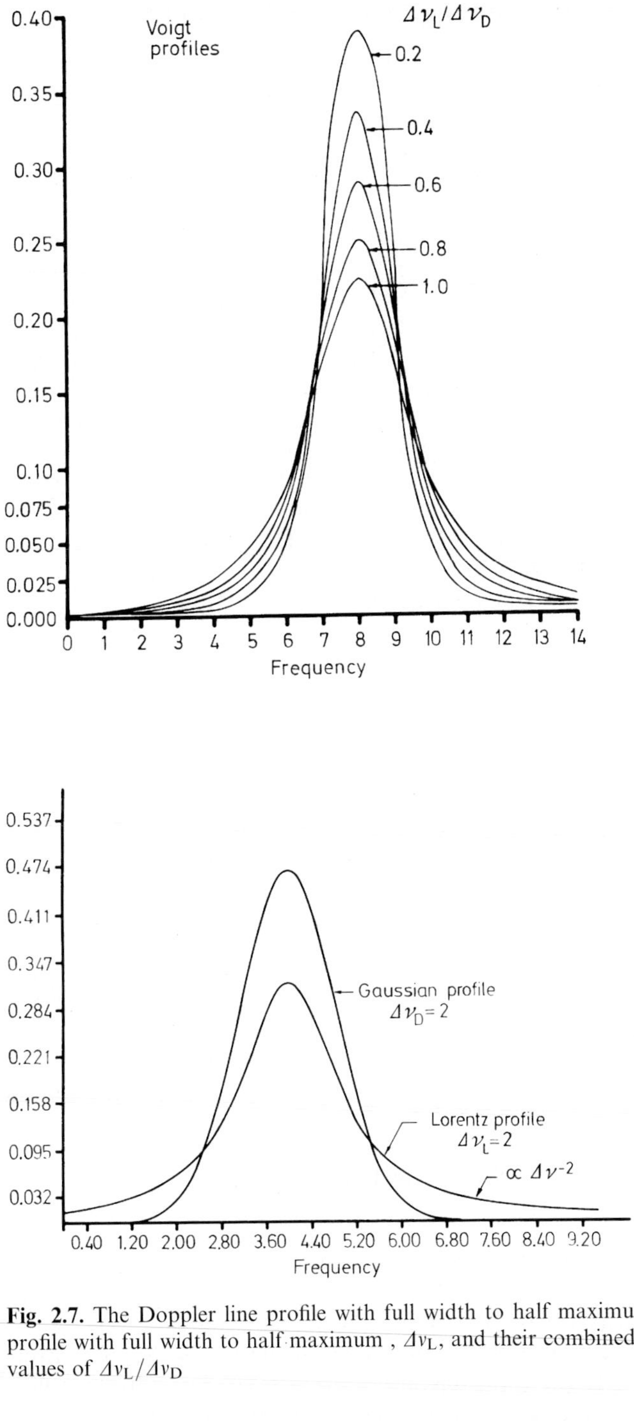

Fig. 2.7. The Doppler line profile with full width to half maximum, $\Delta\nu_D$, the Lorentz line profile with full width to half maximum , $\Delta\nu_L$, and their combined Voigt profile for various values of $\Delta\nu_L/\Delta\nu_D$

2.19 Broadening Due to Rotating or Expanding Sources

When a spherical source is rotating with an equatorial velocity, V_R, then the Doppler equation (2.226) gives a rotational line broadening of

$$\Delta\nu_R = \frac{\nu_{mn} V_R \sin i}{c} , \tag{2.248}$$

where ν_{mn} is the frequency of the $m - n$ transition and i is the inclination of the equator to the celestial plane. The quantity $V_R \sin i$ is called the projected linear equatorial velocity or the observed rotational velocity of the line. The determination of the axial rotation of stars from spectral line widths was first suggested by Abney (1877) and observed by Schlesinger (1909). Theoretical studies of rotational line broadening have been made by Shapley and Nicholson (1919), Shajn and Struve (1929), and Struve (1930), and are summarized by Huang and Struve (1955, 1960). Early type stars, spectral types B, A, and F, have rotational velocities ranging from 200 to 350 km sec^{-1}, whereas the velocities drop rapidly after spectral type F5 ((cf. Slettebak, 1949, 1954, 1955; Slettebak and Howard, 1955; Abt and Hunter, 1962; Kraft, 1965, 1970; Slettebak, 1970) and Fig. 2.8).

A radiating shell of gas which is expanding at a velocity, V_E, will give rise to a double line profile with peaks at

$$\nu = \nu_{mn}\left[1 \pm \frac{V_E}{c}\right] , \tag{2.249}$$

or an asymmetrical line profile centered at

$$\nu = \nu_{mn}\left[1 + \frac{V_E}{c}\right] , \tag{2.250}$$

according as the radiation from the back part of the shell is or is not allowed to pass through the part of the shell nearer the observer. Expansion velocities between 20 and 50 km sec^{-1} have been measured using this effect when observing planetary nebulae (Wilson, 1950, 1958).

The combined effect of rotation and expansion on spectral line shapes is discussed by Duval and Karp (1978). The line shapes of spherically symmetric, radially moving clouds is discussed by Kuiper, Rodriguez, Kuiper and Zuckerman (1978).

2.20 Collision Broadening (Stark or Pressure Broadening)

2.20.1 Ion Broadening – The Quasi-Static Approximation

Ion broadening is calculated by assuming that the slow moving ions have a "quasi-static" electric field distribution, and then calculating the Stark splitting of the line considered. If each ion has an electric field distribution which is independent of the other ions and electrons in the plasma, the probability

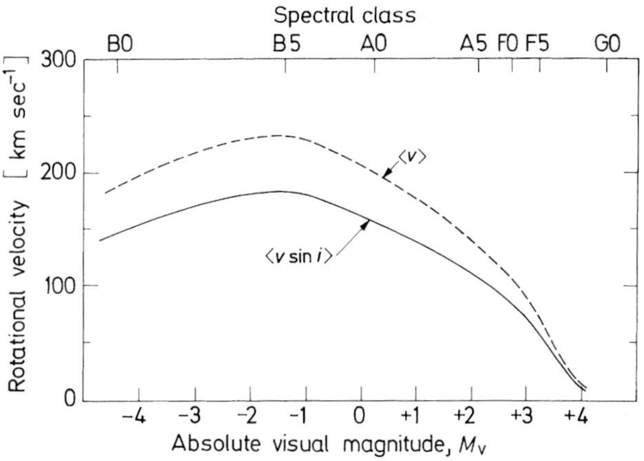

Fig. 2.8. The mean projected rotational velocity, $\langle V \sin i \rangle$, and the mean rotational velocity, $\langle V \rangle$, for stars as a function of spectral class and absolute visual magnitude [after Abt and Hunter, 1962, by permission of the American Astronomical Society and the University of Chicago Press]. The curve for $\langle V \rangle$ is derived from that for $\langle V \sin i \rangle$ on the assumption of a random orientation of rotational axis. The number of stars incorporated in the means for each spectral class was between fifteen and sixty-five

distribution of the ionic electric field is easily calculated. Each ion will have a Coulomb electric field strength given by $E = Ze/r^2$, where r is the distance from the ion of charge Ze. The probability, $P(r)$, that at least one ion will fall into a shell r to $r + dr$ from a radiating atom is given by

$$P(r)dr = \frac{3r^2}{r_0^3} \exp\left[-\left(\frac{r}{r_0}\right)^3\right] dr , \tag{2.251}$$

where the mean ion separation is

$$r_0 = [4\pi N_i/3]^{-1/3} , \tag{2.252}$$

and N_i is the density of ions. Thus, the probability, $P(E)$, of an emitting atom being subjected to the electric field of the nearest ion is (Holtsmark, 1919)

$$P(E) = W_H\left(\frac{E}{E_0}\right) d\left(\frac{E}{E_0}\right) = \frac{3}{2}\left(\frac{E}{E_0}\right)^{-5/2} \exp\left[-\left(\frac{E}{E_0}\right)^{-3/2}\right] d\left(\frac{E}{E_0}\right) , \tag{2.253}$$

where $E = Ze/r^2$ and the mean electric field strength $E_0 = Ze/r_0^2 = 2.61\,eZN_i^{2/3}$. The $W_H(E/E_0)$ is called the Holtsmark ion field strength distribution for $E \gg E_0$ (Holtsmark, 1919). When the electric field, E, is written as the vector sum of the fields of many different ions rather than that of one ion, the Holtsmark distribution for the linear Stark effect is given by Holtsmark (1919); Chandrasekhar (1943)

$$\begin{aligned}W_{\mathrm{H}}(\beta) &= \frac{2}{\pi\beta}\int_0^\infty v \sin v \, \exp\left[-\left(\frac{v}{\beta}\right)^{3/2}\right] dv \\ &= \frac{4}{3\pi}\sum_{n=0}^{\infty}(-1)^n \Gamma\left(\frac{4n+6}{3}\right)\frac{\beta^{2n+2}}{(2n+1)!} \\ &\approx \frac{4\beta^2}{3\pi}(1-0.463\beta^2+0.1227\beta^4 \pm \cdots) \quad \text{for } \beta \ll 1 \\ &= \frac{2}{\pi}\sum_{n=1}^{\infty}(-1)^{n+1}\frac{\Gamma\left(\frac{3n+4}{2}\right)\sin\left(\frac{3\pi n}{4}\right)}{n!\beta^{(3n+2)/2}} \\ &\approx \frac{1.496}{\beta^{5/2}}\left(1+\frac{5.107}{\beta^{3/2}}+\frac{14.43}{\beta^3}+\cdots\right) \quad \text{for } \beta \gg 1 \ , \end{aligned} \tag{2.254}$$

where $\beta = E/E_0$. A plot of $W_{\mathrm{H}}(\beta)$ is shown in Fig. 2.9. For very large β, the $W_{\mathrm{H}}(\beta) \to 1.5\beta^{-5/2}$ just as in Eq. (2.253).

The probability distribution of the ionic field will deviate from the Holtsmark distribution of Eq. (2.254) when ion correlations are important (Ecker and Müller, 1958; Mozer and Baranger, 1960; Hooper, 1968), or when the radiating particle is an ion (Lewis and Margenau, 1958). Ion correlations become important when r_0 is not much less than the Debye radius $r_{\mathrm{D}} = 6.90(T/N_{\mathrm{i}})^{1/2}$ cm, where T is the kinetic temperature of the plasma, and N_{i} is the ion density. The general result is that $W_{\mathrm{H}}(\beta)$ becomes more skewed towards lower β as r_0 approaches r_{D}. When the radiating particle is an ion, the wings of $W_{\mathrm{H}}(\beta)$ are damped by the factor $\exp[-\frac{1}{3}(r_0/r_{\mathrm{D}})^2\beta^{1/2}]$.

For hydrogen-like atoms the frequency displacement, $\Delta\nu$, caused by an electric field of strength, E, is a linear function of E (a relationship first observed by Stark, 1913). For other atoms it is proportional to E^2 (the quadratic Stark

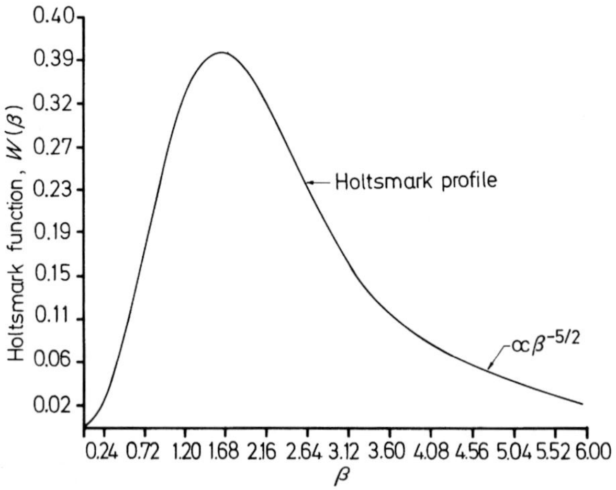

Fig. 2.9. Holtsmark distribution given by Eq. (2-254)

effect) for weak fields and to E for strong fields. Using Eq. (2.253), the spectral line intensity distribution for each Stark component is given by

$$\varphi(\Delta\nu) = \frac{1}{\Delta\nu_0} W_{\mathrm{H}}\left(\frac{\Delta\nu}{\Delta\nu_0}\right) \quad \text{for the linear Stark effect} \tag{2.255}$$

and

$$\varphi(\Delta\nu) = \frac{1}{2}(\Delta\nu_0\,\Delta\nu)^{-1/2}\, W_{\mathrm{H}}[(\Delta\nu/\Delta\nu_0)^{1/2}] \quad \text{for the quadratic Stark effect ,} \tag{2.256}$$

where $\Delta\nu_0$ is the Stark displacement for the mean electric field, E_0.

For hydrogen like atoms, the frequency, ν_{mn}, of a transition from an upper level m to a lower level n is split into the frequencies $\nu_{m,n}$ given by Schwarzschild (1916); Epstein (1916); Bethe and Salpeter (1957)

$$\begin{aligned}\nu_{m,n} &= \nu_{mn} + \frac{3h}{8\pi^2 m_e e Z} E[m(m_2 - m_1) - n(n_2 - n_1)] \\ &\approx \nu_{mn} + 1.92 \times 10^6 \frac{E}{Z}[m(m_2 - m_1) - n(n_2 - n_1)]\ \mathrm{Hz}\ ,\end{aligned} \tag{2.257}$$

where E is in units of volt cm^{-1}, and the frequency of the transition in the absence of the electric field is

$$\nu_{mn} = \frac{2\pi^2 m_e e^4 Z^2}{h^3}(n^{-2} - m^{-2}) \approx 3.288 \times 10^{15} Z^2 (n^{-2} - m^{-2})\ \mathrm{Hz}\ , \tag{2.258}$$

m_2 and m_1 may take on integral values between 0 and $m - 1$, and n_2 and n_1 may take on integral values between 0 and $n - 1$. The frequency shift, $\Delta\nu_{mn}$, and the corresponding wavelength shift, $\Delta\lambda_{mn}$, are given by Struve (1929)

$$\Delta\nu_{mn} = 1.92 \times 10^6 \frac{E}{Z} X_{mn}\ \mathrm{Hz}, \tag{2.259}$$

and

$$\Delta\lambda_{mn} = E\, C_{mn}\ \mathrm{cm}\ ,$$

where

$$X_{mn} = [m(m_2 - m_1) - n(n_2 - n_1)]$$

and

$$\begin{aligned}C_{mn} &= \frac{5.3 \times 10^{-15}}{Z^5}(n^{-2} - m^{-2})^{-2} X_{mn}\ \mathrm{cm}^2\ \mathrm{volt}^{-1} \\ &= 6.4 \times 10^{-5} \frac{\lambda_{mn}^2}{Z} X_{mn}\ \mathrm{cm}^2\ \mathrm{volt}^{-1}\ ,\end{aligned}$$

and E is in units of volt cm^{-1}.

Tables of X_{mn} and C_{mn} are given for optical hydrogen lines by Underhill and Waddell (1959). The total line broadening will be a superposition of all of the Stark components given by Eq. (2.259) and appropriately weighted for the strength of the component. The mean values $\langle X \rangle$ of the weighted X_{mn} for the optical hydrogen lines are

Line	Lα	Lβ	Hα	Hβ	Hγ	Hδ	lim $m \gg n$
$\langle X \rangle$	2.0	4.0	2.24	5.96	11.8	15.9	$\frac{1}{2}m(m-1)$

Tables of the Stark broadening of the first four Lyman lines and the first four Balmer lines of hydrogen are given by Vidal, Cooper and Smith (1973).

Using the mean electric field $E_0 = 2.6\, e\, Z\, N_i^{2/3}$, where N_i is the ion density, together with Eqs. (2.254), (2.255), and (2.259), we obtain the wing spectral line distribution function

$$\varphi(\Delta\nu) \approx N_i(\Delta\nu)^{-5/2}\langle X \rangle^{3/2} , \tag{2.260}$$

where $\langle X \rangle$ is the mean value of X_{mn}.

For a given series of lines (a fixed n), the frequency difference, $\Delta\nu_L$, between adjacent lines of frequencies ν_{m+1} and ν_m is, using Eq. (2.258),

$$\Delta\nu_L = \frac{4\pi^2\, m_e\, e^4\, Z^2}{h^3\, m^3} \approx \frac{6.6 \times 10^{15}}{m^3} Z^2 \text{ Hz} . \tag{2.261}$$

The maximum Stark displacement, $\Delta\nu_{max}$, of a line with quantum number, m, is from Eq. (2.259)

$$\Delta\nu_{max} = \frac{3\, h\, m^2\, E}{8\pi^2\, m_e\, e\, Z} \approx 1.92 \times 10^6 \frac{E\, m^2}{Z} \text{ Hz} . \tag{2.262}$$

Assuming that E has its mean value $E_0 = 2.6\, e\, Z\, N_i^{2/3}$, where N_i is the ion density; and equating $\Delta\nu_{max}$ to $\Delta\nu_L/2$, we obtain the limiting value of the quantum number, m, for which the lines of a series are observed to merge (Inglis and Teller, 1939)

$$N_i^{2/3} = \frac{16\pi^4\, m_e^2\, e^4\, Z^2}{7.8\, h^4\, m^5} = 4.58 \times 10^{15}\, Z^2\, m^{-5} \text{ cm}^{-2}$$

or

$$\log N_i \approx 23 - 7.5 \log m . \tag{2.263}$$

2.20.2 Electron Broadening – The Impact Approximation

In the impact approximation, a radiating atom is assumed to act as an unperturbed harmonic oscillator until it undergoes a collision with a perturbing particle. The effect of such a collision will be to change the phase and possibly the amplitude of the harmonic oscillation. When the time, t, is long enough to include collisions, the dipole moment of the harmonic oscillator is given by

$$d(t) = ex(t) = e\, A(t) \exp\left[i\, \omega_0\, t + i\, \eta(t)\right] , \tag{2.264}$$

where e is the charge of the electron, $x(t)$ denotes the linear displacement of the oscillator, $A(t)$ denotes the time dependent amplitude of the oscillation, ω_0 denotes the unperturbed frequency of the oscillator, and $\eta(t)$ denotes the phase shift induced by collision. The average total energy emitted per unit time by an oscillator in all directions is (Larmor, 1897)

$$\langle I \rangle = \frac{2}{3c^3} \langle \ddot{d}(t)^2 \rangle \propto \langle |x(t)|^2 \rangle \ , \tag{2.265}$$

where $\langle\ \rangle$ denotes a time average, $\ddot{}$ denotes the second differential with respect to time, $|\ |$ denotes the absolute value, and $\propto$ means proportional to. We may express this average intensity in terms of the Fourier transform, $F(\omega)$, of $x(t)$ by using Rayleigh's theorem (Rayleigh, 1889)

$$\langle I(\omega) \rangle \propto \langle |x(t)|^2 \rangle = \langle |F(\omega)|^2 \rangle \ , \tag{2.266}$$

where the Fourier transform, $F(\omega)$, is given by

$$F(\omega) = \int_{-\infty}^{+\infty} x(t) \exp(-i\,\omega\, t) dt \ , \tag{2.267}$$

and $\langle\ \rangle$ denotes an average over time, t, or frequency, ω.

It is also convenient to express the average intensity in terms of the autocorrelation function, $\varphi(s)$, of $x(t)$ by using the Wiener–Khintchine theorem (Wiener, 1930; Khintchine, 1934)

$$I(\omega) \propto \langle |F(\omega)|^2 \rangle = \int_{-\infty}^{+\infty} \langle \varphi(s) \rangle \exp(-i\,\omega s) ds \ , \tag{2.268}$$

where the time averaged autocorrelation function, $\langle \varphi(s) \rangle$, is given by

$$\langle \varphi(s) \rangle = \left\langle \int_{-\infty}^{+\infty} x^*(t) x(t+s) dt \right\rangle \ , \tag{2.269}$$

where $x^*(t)$ denotes the complex conjugate of $x(t)$ and $\langle\rangle$ denotes a time average.

Let us suppose that one collision between an atom and a perturbing particle occurs at a time interval, T. Using Eqs. (2.264) and (2.267)

$$|F(\omega)|^2 = \frac{\sin^2[(\omega_0 - \omega)T/2]}{[(\omega_0 - \omega)/2]^2} = \left| \int_0^T \exp[i(\omega_0 - \omega)t] dt \right|^2 \ . \tag{2.270}$$

If the mean time between collisions is τ, and if the collisions are independent and random, the probability that a collision occurs between time T and $T + dT$ is given by the Poisson distribution

$$P(T) dT = \frac{1}{\tau} \exp\left(-\frac{T}{\tau} \right) dT \ . \tag{2.271}$$

Eqs. (2.266), (2.270), and (2.271) then lead to the Lorentz spectral line intensity distribution (Lorentz, 1906, 1909)

$$\varphi(\nu) = \frac{\Delta \nu_L / 2\pi}{(\nu - \nu_0)^2 + (\Delta \nu_L / 2)^2} \propto \int_0^{\infty} |F(\omega)|^2 P(T) dT \ , \tag{2.272}$$

where $\Delta \nu_L = (\pi \tau)^{-1} = \Gamma_{coll}/(2\pi)$ and we have normalized $\varphi(\nu)$ so that $\int_{-\infty}^{+\infty} \varphi(\nu) d\nu = 1$.

The mean collision time, τ, can be related to the mean density, N, of the perturbing particles by

$$\pi \Delta \nu_{\mathrm{L}} = \pi b^2 v_0 N = (\tau)^{-1} \ , \tag{2.273}$$

where πb^2 is the effective collision cross section, b is the effective impact parameter, and the mean relative speed, v_0, between an atom of mass, m_1, and another of mass, m_2, is

$$v_0 = \left[\frac{8kT}{\pi}\left(\frac{1}{m_1} + \frac{1}{m_2}\right)\right]^{1/2} \ , \tag{2.274}$$

where T is the kinetic temperature of the gas.

Weisskopf (1933) related the critical impact parameter, b, of Eq. (2.273) to a measurable constant by noting that the frequency shift, $\Delta \nu$, of a line caused by a perturber at a distance, r, may be written

$$2\pi\Delta\nu = C_p r^{-p} \ , \tag{2.275}$$

where C_p is a measurable constant, and $p = 2, 3, 4$, and 6, respectively, for the linear Stark effect, resonance broadening, the quadratic Stark effect, and the van der Waals interaction. The total phase shift, $\eta(b)$, induced by a perturbation in which the perturbing particle follows a straight line trajectory is

$$\begin{aligned}\eta(b) &= C_p \int_{-\infty}^{+\infty} \frac{d\tau}{[r(\tau)]^p} = C_p \int_{-\infty}^{+\infty} \frac{d\tau}{(b^2 + v_0^2\tau^2)^{p/2}} \\ &= \frac{\pi^{1/2}\Gamma[(p-1)/2]C_p}{\Gamma(p/2)v_0 b^{p-1}} = \frac{\psi_p C_p}{v_0 b^{p-1}} \ ,\end{aligned} \tag{2.276}$$

where the impact parameter, b, is the perpendicular distance between the radiating atom and straight line trajectory of the perturbing particle, and $\psi_p = \pi$, 2, $\pi/2$, and $3\pi/8$, respectively, for $p = 2, 3, 4$, and 6. The critical value of b to be used in Eq. (2.273) for the line width is the Weisskopf radius

$$b_{\mathrm{w}} = \left(\frac{C_p \psi_p}{\eta_0 v_0}\right)^{1/(p-1)} \ , \tag{2.277}$$

where η_0 is the minimum value of phase shift which contributes to line broadening. The constant η_0 was arbitrarily chosen to be unity by Weisskopf. This impact approximation holds as long as $b_{\mathrm{w}}/v_0 \ll \Delta\nu_{\mathrm{L}}$, and the frequency difference, $\Delta\nu$, satisfies the inequality (Spitzer, 1940)

$$\Delta\nu \ll \frac{v_0^{p/(p-1)}}{C_p^{1/(p-1)}} \ . \tag{2.278}$$

Holstein (1950) has shown that the static theory discussed in the previous section is valid when the inequality is reversed.

Lindholm (1941) and Foley (1946) showed that the impact approximation will generally lead to a line shift as well as a line broadening. They assumed that the amplitude, $A(t)$, of the harmonic oscillator is constant (an adiabatic theory) and noted that the number of collisions per unit time with oscillators whose impact parameters lie between b and $b + db$ is

$$N\, v_0\, 2\pi\, b\, db \;, \tag{2.279}$$

where N is the oscillator number density. Using Eqs. (2.264) and (2.279) in (2.269), the autocorrelation function, $\varphi(s)$, is obtained

$$\varphi(s) = \exp\{-N\, v_0(\sigma_\mathrm{r} - i\sigma_\mathrm{i})s\}\, \exp(-2\pi\, \nu_{mn} s) \;, \tag{2.280}$$

where σ_r and σ_i are the real and imaginary parts of the integral

$$2\pi \int_0^\infty \{\exp[i\eta(b)] - 1\} b\, db \;. \tag{2.281}$$

Thus using Eq. (2.268), the spectral distribution of the line intensity is

$$\varphi_{mn}(\nu) = \frac{\Delta\nu_\mathrm{L}/2\pi}{[\nu - \nu_{mn} - \beta/(2\pi)]^2 + (\Delta\nu_\mathrm{L}/2)^2} \;, \tag{2.282}$$

where $\beta = N\, v_0\, \sigma_\mathrm{i}$ and $\Delta v_\mathrm{L} = \Gamma/(2\pi) = N\, v_0\, \sigma_\mathrm{r}/\pi$. Assuming that $\eta(t)$ is given by Eq. (2.276), we obtain for the linear Stark effect ($p = 2$)

$$\Gamma = 2\pi\Delta\nu_\mathrm{L} = \frac{2N\pi^3 C_2^2}{v_0}\left\{0.923 - \ln\left(\frac{\pi C_2}{v_0 b_\mathrm{max}}\right) + \frac{\pi^2 C_2^2}{24 v_0^2 b_\mathrm{max}^2} + \cdots\right\} \;, \tag{2.283}$$

where the maximum impact parameter, b_max, is often taken to be the Debye radius, and there is no line shift. For other values of p, we have:

p	Γ (sec^{-1})	β (radians)	
3	$2\pi^2 C_3 N$		
4	$11.37 C_4^{2/3} v_0^{1/3} N$	$9.85 C_4^{2/3} v_0^{1/3} N$	(2.284)
6	$8.08 C_6^{2/5} v_0^{3/5} N$	$2.94 C_6^{2/5} v_0^{3/5} N$	

The more exact theory of electron broadening takes into account the effects of inelastic collisions of the electrons (Anderson, 1949; Kolb and Griem, 1958; Baranger, 1958; Margenau and Lewis, 1959; Baranger, 1962; Griem, 1964). This more exact theory is an impact theory in that it assumes the collisions are weak – that it takes many collision times for a light train to lose memory of phase. The interaction between perturbing particles and radiating atoms is, however, specified by a quantum mechanical formalism, and inelastic collisions are taken into account. The general conclusions of this non-adiabatic theory are that the $\Delta\nu_\mathrm{L}$ given by Eq. (2.284) may be underestimated by a factor of two or three, and that $\Delta\nu_\mathrm{L}/\beta$ is generally smaller than the value given. Detailed calculations have been made for hydrogen-like atoms and are given in the next section.

2.20.3 Wing Formulae for Collisional Broadening of Line Radiation from Hydrogen-Like Atoms

At optical frequencies, ion broadening of hydrogen lines by the linear Stark effect is important. For a gas with ion density, N_i, the mean electric field strength is $E_0 \approx eZ(4\pi N_i/3)^{2/3}$, where eZ is the charge of the ion. Using this mean field strength, the wing spectral line distribution is, from Eqs. (2.255) and (2.259), the Holtsmark distribution

$$\varphi(\Delta\lambda) = E_0^{3/2}\langle C\rangle \Delta\lambda^{-5/2} \approx 4.4 \times 10^{-14} N_i \langle C\rangle \Delta\lambda^{-5/2} , \tag{2.285}$$

where $\Delta\lambda$ is the wavelength displacement from the wavelength of the $m-n$ transition, and $\langle C\rangle$ is the average Stark displacement constant which is given in c.g.s. units in Table 2.14. They are from Griem, Kolb, and Shen (1959) for Lyα, Lyβ, Hα, Hβ, Hγ and Hδ; Griem (1962) for He II (3203 Å and 4686 Å); and Griem (1960) and Griem (1967) for the higher order optical hydrogen transitions. More recent corrections to the Stark broadening of hydrogen and helium lines are given by Kepple and Griem (1968) and Kepple (1972).

The electron broadening corrections to the asymptotic wing formulae for the line intensities at optical frequencies have been calculated by Griem, Kolb, and Shen (1959) and Griem (1962). The total broadened intensity, $I(\Delta\lambda)$, is given by Griem (1964)

$$I(\Delta\lambda) = \varphi(\Delta\lambda) \begin{cases} \{1 + [(\Delta\lambda_w)^{-1/2} + R(N,T)](\Delta\lambda)^{1/2}\} \\ \left\{1 + \left[(\Delta\lambda_w)^{-1/2} + R(N,T)\dfrac{\ln(\Delta\lambda_w/\Delta\lambda)}{\ln(\Delta\lambda_w/\Delta\lambda_p)}\right](\Delta\lambda)^{1/2}\right\} \\ \{1+1\} \end{cases} , \tag{2.286}$$

for $\Delta\lambda < \Delta\lambda_p$, $\Delta\lambda_p < \Delta\lambda < \Delta\lambda_w$, and $\Delta\lambda_w < \Delta\lambda$, respectively. Here $\Delta\lambda_w$ and $\Delta\lambda_p$ are defined by

$$\Delta\lambda_w = \lambda^2 kT/(hm^2 c) ,$$

m being the principal quantum number of the upper level, and

$$\Delta\lambda_p = \lambda^2 [N_e e^2/(\pi m c^2)]^{1/2} ,$$

where T is the kinetic temperature of the plasma and N_e is the electron density. The $R(N, T)$ are given in Table 2.14 for some transitions of hydrogen and helium. Tables of the Stark broadening of the first four Lyman lines and the first four Balmer lines of hydrogen are also given by Vidal, Cooper, and Smith (1973).

At radio frequencies, most of the Stark broadening of hydrogen like atoms is due to inelastic collisions with electrons. In the wings of the line, the intensity distribution of the line is Lorentzian and is given by Eq. (2.272). The full width to half maximum of the line, $\Delta\nu_L$, for the hydrogen transition from an upper level with principal quantum number m to a lower level n is (Griem, 1960)

$$\Delta\nu_L = \frac{1}{18\pi}\left(\frac{8\pi m_e}{kT_e}\right)^{1/2} N_e \left(\frac{\hbar}{m_e}\right)^2 \int_{y\,\min}^{\infty} \frac{e^{-y}}{y}(m^5+n^5)^2 dy , \tag{2.287}$$

where

$$y_{\min} = \frac{4\pi N_e}{3 m_e}\left(\frac{e h m^2}{2\pi k T_e}\right)^2 ,$$

Table 2.14. Parameters $\langle C_{mn} \rangle$ (in Å per cgs field strength units)$^{3/2}$ and $R(N_e, T)$ in Å$^{-1/2}$ for the asymptotic wing formulae for hydrogen and ionized helium lines[1]. The electron density, N_e, is in cm^{-3} and the electron temperature, T, is in °K

Lα($\langle C_{mn} \rangle = 3.40 \times 10^{-6}$)					Hγ($\langle C_{mn} \rangle = 6.00 \times 10^{-3}$)				
N_e/T_e	5,000	10,000	20,000	40,000	N_e/T_e	5,000	10,000	20,000	40,000
10^{10}	2.11	1.93	1.45	1.09	10^{10}	1.79	1.37	1.04	0.79
10^{12}	2.01	1.54	1.17	0.89	10^{12}	1.32	1.03	0.80	0.62
10^{14}	1.45	1.14	0.89	0.69	10^{14}	0.84	0.70	0.57	0.45
10^{16}	0.88	0.74	0.61	0.49	10^{16}	0.38	0.36	0.33	0.28
Lβ($\langle C_{mn} \rangle = 1.78 \times 10^{-5}$)					Hδ($\langle C_{mn} \rangle = 9.81 \times 10^{-3}$)				
N_e/T_e	5,000	10,000	20,000	40,000	N_e/T_e	5,000	10,000	20,000	40,000
10^{10}	4.30	3.29	2.47	1.86	10^{10}	2.17	1.66	1.27	0.96
10^{12}	3.31	2.56	1.96	1.50	10^{12}	1.57	1.24	0.97	0.75
10^{14}	2.29	1.83	1.45	1.14	10^{14}	0.97	0.81	0.67	0.54
10^{16}	1.26	1.11	0.94	0.77	10^{16}	0.37	0.39	0.37	0.32
Hα($\langle C_{mn} \rangle = 1.30 \times 10^{-3}$)					HeII 4,686 Å($\langle C_{mn} \rangle = 2.62 \times 10^{-4}$)				
N_e/T_e	5,000	10,000	20,000	40,000	N_e/T_e	5,000	10,000	20,000	40,000
10^{10}	1.50	1.05	0.79	0.60	10^{10}				
10^{12}	1.17	0.82	0.63	0.48	10^{12}	1.65	1.28	0.98	0.75
10^{14}	0.85	0.59	0.46	0.36	10^{14}	1.13	0.91	0.72	0.56
10^{16}	0.52	0.35	0.30	0.25	10^{16}			0.46	0.38
$H\beta$($\langle C_{mn} \rangle = 3.57 \times 10^{-3}$)					HeII 3,203 Å($\langle C_{mn} \rangle = 5.52 \times 10^{-4}$)				
N_e/T_e	5,000	10,000	20,000	40,000	N_e/T_e	5,000	10,000	20,000	40,000
10^{10}	1.39	1.05	0.80	0.60	10^{10}				
10^{12}	1.04	0.81	0.62	0.48	10^{12}	1.28	1.00	0.77	0.59
10^{14}	0.69	0.56	0.45	0.35	10^{14}	0.85	0.69	0.55	0.44
10^{16}	0.34	0.31	0.27	0.23	10^{16}			0.34	0.29

[1] After Griem (1964) by permission of the McGraw-Hill Book Co.

and the nuclear charge is e. Such a process is called impact, or collisional, broadening. The "collision" is the perturbation of the energy levels of an excited atom resulting from the electric field of a free electron.

For the special case of radio frequency hydrogen α lines, $m = n + 1$ and $\nu_{mn} \approx 2cR_H n^{-3} = 6.576 \times 10^{15} n^{-3}$ Hz, where R_H is the Rydberg constant for hydrogen. Using the formulae of Griem (1967) for the Stark broadening of radio frequency Hα lines, the full width to half maximum of the line is

$$\Delta\nu_L = \frac{5}{3(2\pi)^{5/2}} \left(\frac{h}{m_e}\right)^2 \left(\frac{m_e}{kT_e}\right)^{1/2} N_e n^4 \left[\frac{1}{2} + \ln\left(2.09 \frac{kT_e}{h\nu_{mn} n^2}\right)\right]$$
$$\approx 2.4 \times 10^{-6} \frac{N_e n^4}{T_e^{1/2}} \left[\frac{1}{2} + \ln(6.64 \times 10^{-6} T_e n)\right] Hz , \tag{2.288}$$

where N_e and T_e are, respectively, the electron density and temperature, and n is the principal quantum number of lowest level of the α transition.

Using Eqs. (2.242) and (2.288) together with $\nu_{mn} \approx 6.576 \times 10^{15} n^{-3}$ Hz, we obtain

$$\frac{\Delta\nu_L}{\Delta\nu_D} \approx \frac{3\sqrt{\pi}}{1.2012} \left(\frac{\hbar c}{m}\right) \left(\frac{\hbar^2}{m_e e^2}\right)^2 \left(\frac{m_e}{kT_e}\right)^{1/2} \left(\frac{m_H}{kT_D}\right)^{1/2}$$
$$\times N_e n^7 \left[\frac{1}{2} + \ln(6.64 \times 10^{-6} T_e n)\right]$$
$$\approx 9.74 \times 10^{-16} \frac{N_e n^7}{(T_e T_D)^{1/2}} [-11.47 + \ln T_e + \ln(n)] , \tag{2.289}$$

where T_D is the effective Doppler temperature, $\Delta\nu_D$ is the full width to half maximum of the Doppler broadened line, the hydrogen mass, $m_H = 1.6733 \times 10^{-24}$ gm, N_e and T_e are, respectively, the electron density and temperature, and n is the principal quantum number of the lowest level of the α transition. The results obtained by taking accurate collision cross sections and degenerate energy levels into account are in close agreement with those of Griem (Brocklehurst and Leeman, 1971; Peach, 1972). Attempts to detect Stark broadening in H II regions at radio frequencies have failed (Pedlar and Davies, 1971), but this may be explained if the electron density, N_e, is low (Brocklehurst and Seaton, 1971).

For hydrogen α lines, the ratio of impact broadening, $\Delta\nu_L$, to the Doppler broadening, $\Delta\nu_D$, is (Brocklehurst and Seaton, 1972):

$$\Delta\nu_L/\Delta\nu_D \approx 0.14(n/100)^{7.4}(N_e/10^4)(10^4/T_e)^{0.1} ,$$

where n is the principal quantum number, N_e is the electron density and T_e is the electron temperature. As an example, for the H157α line (with $n = 150$ at a wavelength $\lambda = 18$ cm) in a typical H II region with $N_e = 10^4$ cm^{-3} and $T_e = 10^4$ °K, we obtain $\Delta\nu_L/\Delta\nu_D \approx 4$. However, in most nebulae the width of the H 150α is nearly the same as the width of the H100α line ($n = 100$), and not 3 or 4 times greater (Brown, 1987). That is, the observed broadening of the hydrogen recombination lines in H II regions does not show the predicted dependence on principal quantum number, n. The disagreement between observation and expectation may be due to the fact that H II regions are not homogeneous (constant density) objects. Observations of lines at low frequencies (high n), where we expect to observe the effect of impact broadening in dense nebular gas, reveal instead the unbroadened line emission from more tenuous material. The very broadened lines, that being broadened are commensurably diminished in intensity, become indistinguishable from the instrumental spectral baseline.

2.20.4 Van der Waals Broadening Due to Collisions with Neutral Hydrogen Atoms

When line broadening is due to interactions between radiating atoms and neutral atoms, the adiabatic impact theory is appropriate. The spectral line intensity distribution is given by Eq. (2.282) with a full width to half maximum, $\Delta\nu_L$, and a line shift, β, which depend on the power, p, of the interaction potential (Eq. (2.275)). For atoms of neutral hydrogen, $p = 6$ (van der Waals broadening). In this case (Margenau, 1939)

$$\Gamma = 2\pi\Delta\nu_L = 8.08\ C_6^{2/5}\ v_0^{3/5}\ N_H\ =\ 2.650\,\beta\ , \tag{2.290}$$

where N_H is the density of neutral hydrogen, v_0 is given by Eq. (2.274), and the constant C_6 is given by

$$C_6 \approx 4.05\ \times\ 10^{-33}\left[R_m^2 - R_n^2\right]\ ,$$

where

$$R_m^2 \approx \frac{m^{*2}}{2\,Z^2}\left[5m^{*2} + 1 - 3l(l+1)\right]\ ,$$

m^* is the effective principal quantum number of the level, l is the angular momentum quantum number, and

$$m^*\ = Z\left[\frac{\chi_H}{\chi_m - \chi_i}\right]^{1/2}\ ,$$

where $\chi_H = 13.6$ eV, and $\chi_m - \chi_i$ is the energy required to ionize the mth level. The numerical coefficient for C_6 is $\alpha\,e^2\,a_0^2/\hbar$, where α is the polarizability of the atom and a_0 is the Bohr radius. For hydrogen $\alpha = 6.7\ \times\ 10^{-25}\,\mathrm{cm}^2$ and we obtain the coefficient $4.05\ \times\ 10^{-33}$. For helium we have the polarizability $\alpha = 2.07\ \times\ 10^{-25}\,\mathrm{cm}^2$.

2.20.5 Resonance Broadening Due to Interactions of Radiating and Ground State Atoms

Resonance broadening was first calculated by Holtsmark (1925) who assumed that the interaction between a radiating atom and a ground state atom could be represented by the coupling of two dipole fields. Weisskopf (1933) used the impact approximation with an interaction potential given by Eq. (2.275) with $p = 3$. The full width to half maximum of the Lorentzian profile (Eq. (2.272)) is

$$\Delta\nu_L = \frac{N\,e^2}{4m\,\nu_{mn}}f_a \approx 6\ \times\ 10^7\frac{N\,f_a}{\nu_{mn}}\ \mathrm{Hz}\ , \tag{2.291}$$

where N is the density of atoms, ν_{mn} is the frequency of the transition, and $f_a = f_{nm}$ is the absorption oscillator strength. A quantum mechanical calculation gives (Ali and Griem, 1965, 1966)

$$\Delta\nu_L \approx 1.9167\left(\frac{g_a}{g_e}\right)^{1/2}\frac{e^2 f_a}{m\nu_{mn}}N \approx 4.8542 \times 10^7\frac{N f_a}{\nu_{mn}}\left(\frac{g_a}{g_e}\right)^{1/2} \text{ Hz} , \qquad (2.292)$$

where g_a and g_e are, respectively, the statistical weights of the absorbing and emitting states of the atom. For Lyman-α the appropriate constants are $f_a = 0.4162$ and $g_a/g_e = 1/3$.

2.21 Natural Broadening (a Lorentz Dispersion Profile)

The total power radiated by an electric dipole is (Larmor, 1897)

$$P = \frac{2}{3}\frac{e^2}{c^3}[\ddot{x}(t)]^2 , \qquad (2.293)$$

where $x(t)$ describes the linear displacement of a charge e during time t, and $\ddot{}$ denotes the second derivative with respect to time. The effective damping force created by this radiation is

$$F_{\text{rad}} = \frac{2}{3}\frac{e^2}{c^3}\dddot{x}(t) , \qquad (2.294)$$

where $\dddot{}$ denotes the third derivative with respect to time. For a harmonic oscillator $x(t) = x_0\cos(2\pi\nu_0 t)$, where ν_0 is the resonant frequency of the oscillator. The equation of motion of the harmonic oscillator is, taking radiation damping into account,

$$m\ddot{x}(t) + m(2\pi\nu_0)^2 x(t) = \frac{2}{3}\frac{e^2}{c^3}\dddot{x}(t) . \qquad (2.295)$$

Assuming that the damping force is small, this equation has the solution

$$x(t) = x_0\exp[i\,2\pi\nu_0 t - \gamma_{cl}t/2] , \qquad (2.296)$$

where the classical damping constant is

$$\gamma_{cl} = \frac{8\pi^2 e^2}{3mc^3}\nu_0^2 \approx 2.5 \times 10^{-22}\nu_0^2 \text{ sec}^{-1} . \qquad (2.297)$$

Using Rayleigh's theorem ((Rayleigh, 1889], Eq. (2.266)) with Eqs. (2.293) and (2.296), the average power radiated by the harmonic oscillator is

$$I(\omega) \propto \langle|x(t)|^2\rangle = \langle|F(\omega)|^2\rangle , \qquad (2.298)$$

where

$$F(\omega) = \int_{-\infty}^{+\infty} x(t)\exp(-i\omega t)dt = \frac{x_0}{i2\pi(\nu - \nu_0) + \gamma_{cl}/2} .$$

The optical line intensity distribution is therefore Lorentzian (Lorentz, 1906,

1909) and is given by

$$\varphi(\nu) = \frac{\Delta\nu_{\rm L}/2\pi}{(\nu-\nu_0)^2 + (\Delta\nu_{\rm L}/2)^2} , \tag{2.299}$$

where the full width to half-maximum $\Delta\nu_{\rm L} = \gamma_{\rm cl}/2\pi \approx 4 \times 10^{-23}\nu_0^2$ Hz, and we have normalized $\varphi(\nu)$ so that $\int_{-\infty}^{+\infty} \varphi(\nu)d\nu = 1$. The halfwidth in wavelength units is $\Delta\lambda = c\gamma_{\rm cl}/(2\pi\nu_0^2) \approx 1.18 \times 10^{-4}$ Å, which is independent of wavelength. The Lorentz profile given by Eq. (2.299) is illustrated in Fig. 2.7.

Weisskopf and Wigner (1930) and Weisskopf (1933) have shown that the classical Lorentz solution of Eq. (2.299) is also a valid quantum mechanical solution if the transition frequency $\nu_{mn} = \nu_0$, and the half width $\Delta\nu_{\rm L} = \Gamma_R/2\pi$ where the quantum mechanical damping constant, Γ_R, is given by

$$\Gamma_R = \Gamma_m + \Gamma_n = t_m^{-1} + t_n^{-1} , \tag{2.300}$$

where t_m is the mean lifetime of an atom in level m, and, under conditions of local thermodynamic equilibrium,

$$\Gamma_m = \sum_{n<m} A_{mn} + \sum_{n<m} B_{mn} I_\nu(mn) + \sum_{k>m} B_{mk} I_\nu(mk) , \tag{2.301}$$

where

$$\frac{g_m}{g_n} A_{mn} = B_{nm}\frac{2h\nu^3}{c^2} \text{ and } B_{nm} = \frac{g_m}{g_n} B_{mn}$$

relate the Einstein coefficients for spontaneous emission, A_{mn}, and stimulated emission, B_{mn}, and $I_\nu(mn)$ is the black body radiation intensity evaluated at the frequency ν_{mn}, of the $m-n$ transition. Here the first term is due to spontaneous emission to lower levels and the summation is taken over all levels to which a transition can occur, the second term is due to induced or stimulated emission (called negative absorption) to a lower level, and the third term is due to ordinary absorption processes and may include ionizations.

We may use the Wien and Rayleigh–Jeans approximations for the brightness of a black body (Eqs. (1.120) and (1.121)) to obtain the number of induced emissions per cm^3 per sec

$$\begin{aligned} B_{mn} I_\nu(mn) &\approx A_{mn}\exp(-h\nu/kT) \quad \text{for } h\nu \gg kT \\ &\approx A_{mn} kT/h\nu \quad \text{for } h\nu \ll kT . \end{aligned} \tag{2.302}$$

Neglecting ordinary absorptions then,

$$\begin{aligned} \Gamma_m &\approx \sum_{n<m} A_{mn} \quad \text{for } h\nu/kT \gg 1 \\ &\approx \frac{kT}{h\nu}\sum_{n<m} A_{mn} \quad \text{for } h\,\nu/kT \ll 1 . \end{aligned} \tag{2.303}$$

Here $h\nu/kT \ll 1$ means $\nu \ll 10^{10}\,T$.

2.22 Combined Doppler, Lorentz, and Holtsmark Line Broadening (the Voigt Profile)

When the Doppler frequency shift, $\Delta\nu_D$, is taken into account in calculating a Lorentz dispersion profile, Eqs. (2.241) and (2.299) may be combined to give the spectral line intensity distribution function

$$\varphi(\nu) = \int_{-\infty}^{+\infty} \frac{\Delta\nu_L \exp(-\Delta\nu^2/\Delta\nu_D^2)d(\Delta\nu)}{2\pi\sqrt{\pi}\Delta\nu_D\left[(\nu-\nu_0-\Delta\nu)^2+(\Delta\nu_L/2)^2\right]} \tag{2.304}$$

where $\Delta\nu_D$ and $\Delta\nu_L$ are, respectively, the full width to half maximum of the Doppler and Lorentz line profiles. Following the development of Voigt (1913),

$$\varphi(\nu) = \frac{2\sqrt{\ln 2}}{\sqrt{\pi}\Delta\nu_D} H(a,b) \ , \tag{2.305}$$

where

$$a = \sqrt{\ln 2}\Delta\nu_L/(2\Delta\nu_D)$$
$$b = 2\sqrt{\ln 2}(\nu-\nu_0)/\Delta\nu_D$$

and

$$H(a,b) = \frac{a}{\pi}\int_{-\infty}^{+\infty} \frac{\exp(-y^2)dy}{(b-y)^2+a^2} \ .$$

Tables of the Voigt function, $H(a,b)$, are given by Hjerting (1938), Harris (1948), and Finn and Mugglestone (1965). Voigt profiles are illustrated in Fig. 2.7.

When there is a Holtsmark type of broadening as well as the Doppler and Lorentz types, the total spectral line intensity distribution is

$$I(\Delta\nu) = \int_{-\infty}^{+\infty} \varphi(\Delta\nu + b\Delta\nu_D)H(a,b)db \ , \tag{2.306}$$

where $\varphi(\Delta\nu)$ is given by Eqs. (2.255) and (2.256).

The Voigt function, $H(a,b)$, may be approximated by

$$H(a,b) \approx \exp(-b^2) + \frac{a}{\sqrt{\pi}b^2} \ , \tag{2.307}$$

illustrating the fact that in the wings the combined broadening practically coincides with a Lorentz broadening, whereas in the core it is Doppler broadening that dominates.

3. Gas Processes

"This notation may be perhaps further explained by conceiving the air near the earth to be such a heap of little bodies, lying one upon another as may be resembled to a fleece of wool. For this consists of many slender and flexible hairs, each of which may indeed, like a little spring, be easily bent or rolled up; but will also, like a spring, be still endeavouring to stretch itself out again."

R. Boyle (1660)

"O dark dark dark. They all go into the dark. The vacant interstellar spaces, the vacant into the vacant."

T.S. Eliot in East Coker III (1940)

3.1 Microstructure of a Gas

3.1.1 Boltzmann's Equation, the Fokker–Planck Equation, the B.B.G.K.Y. Hierarchy, Maxwell's Distribution Function, and the Vlasov Equation

The one particle probability distribution function, $f(\boldsymbol{r},\boldsymbol{p},t)$, is defined so that

$$f(\boldsymbol{r},\boldsymbol{p},t)dx\,dy\,dz\,dp_x\,dp_y\,dp_z = f(\boldsymbol{r},\boldsymbol{p},t)dV_r dV_p \tag{3.1}$$

is the probability that, at the time, t, a particle has momentum, $\boldsymbol{p}$, in the volume element dV_p at $\boldsymbol{p}$ and position, $\boldsymbol{r}$, in the volume element dV_r at $\boldsymbol{r}$. Similarly, the distribution function $f(\boldsymbol{r},\boldsymbol{v},t)$ is defined so that for an average particle density, N,

$$N\,f(\boldsymbol{r},\boldsymbol{v},t)dx\,dy\,dz\,dv_x\,dv_y\,dv_z = N\,f(\boldsymbol{r},\boldsymbol{v},t)dV_r dV_v \tag{3.2}$$

gives the probable number of particles in the six dimensional phase space $dV_r dV_v$ around position, $\boldsymbol{r}$, and velocity, $\boldsymbol{v}$. Boltzmann's equation for $f(\boldsymbol{r},\boldsymbol{p},t)$ may be written as (Boltzmann, 1872)

$$\frac{\partial f}{\partial t} + \frac{\boldsymbol{p}}{m} \cdot \nabla_r f - \nabla_r \varphi \cdot \nabla_p f = \left(\frac{df}{dt}\right)_{\mathrm{coll}} , \tag{3.3}$$

where $\varphi(r)$ is the potential energy acting on every particle, $\boldsymbol{p}$ is the momentum, m is the particle mass, ∇_r is the gradient in position space, ∇_p is the gradient in momentum space, and $(df/dt)_{\text{coll}}$ is the rate of change in f due to collisions. Noting that $\dot{p} = -\nabla_r \varphi$, we may write Eq. (3.3) in Cartesian coordinates as

$$\frac{\partial f}{\partial t} + \dot{x}\frac{\partial f}{\partial x} + \dot{y}\frac{\partial f}{\partial y} + \dot{z}\frac{\partial f}{\partial z} + \dot{p}_x \frac{\partial f}{\partial p_x} + \dot{p}_y \frac{\partial f}{\partial p_y} + \dot{p}_z \frac{\partial f}{\partial p_z} = \left(\frac{df}{dt}\right)_{\text{coll}} ,$$

where $\cdot$ denotes the first derivative with respect to time. The Boltzmann equation for $f(\boldsymbol{r}, \boldsymbol{v}, t)$ is

$$\frac{\partial f}{\partial t} + \boldsymbol{v} \cdot \nabla_r f + \frac{\boldsymbol{F}}{m} \cdot \nabla_v f = \left(\frac{df}{dt}\right)_{\text{coll}} , \tag{3.4}$$

where $\boldsymbol{v}$ is the velocity, $\boldsymbol{F}$ is the force acting on each particle, m is the particle mass, and ∇_r and ∇_v denote, respectively, gradients in position and velocity space. As an example of astrophysical forces, a particle of charge, q, and mass, m, experiences the force

$$\boldsymbol{F} = q\left(\boldsymbol{E} + \frac{1}{c}\boldsymbol{v} \times \boldsymbol{H}\right) - m\, g\, \boldsymbol{n}_r ,$$

in the presence of an electric field of strength $\boldsymbol{E}$, a magnetic field of strength $\boldsymbol{H}$, and a gravitational field of acceleration g. Here $\boldsymbol{n}_r$ is a unit vector in the radial direction from the mass, m, to another mass, M, and the acceleration due to gravity is GM/r^2, where the gravitational constant $G = 6.67 \times 10^{-8}$ dyn cm^2 g^{-2}, and r is the distance between the mass, M, and the particle of mass, m.

For a two particle system of types i and j, the collision term takes the Fokker-Planck form (Fokker, 1914; Rosenbluth, MacDonald, and Judd, 1957).

$$\frac{1}{\Gamma_i}\left(\frac{df_i}{dt}\right)_{\text{coll}} = -\frac{\partial}{\partial \boldsymbol{v}_i} \cdot \left(f_i \frac{\partial H_i}{\partial \boldsymbol{v}_i}\right) + \frac{1}{2}\frac{\partial^2}{\partial \boldsymbol{v}_i \partial \boldsymbol{v}_i}\left(f_i \frac{\partial^2 G_i}{\partial \boldsymbol{v}_i \partial \boldsymbol{v}_i}\right) , \tag{3.5}$$

where

$$\Gamma_i = -\frac{4\pi e_i^4 \ln\left(\sin \dfrac{\theta_{\min}}{2}\right)}{m_i^2} ,$$

e_i and m_i are, respectively, the charge and mass of the particle of type i, the minimum value of the scattering angle is $\theta_{\min}$,

$$H_i = \sum_j N_j \left(\frac{e_j}{e_i}\right)^2 \left(\frac{m_i + m_j}{m_j}\right) \int d\boldsymbol{v}_j \frac{f_j}{g}$$

and

$$G_i = \sum_j N_j \left(\frac{e_j}{e_i}\right)^2 \int d\boldsymbol{v}_j g\, f_j ,$$

where the density of particle type i is N_i, and

$$g = |\boldsymbol{v}_i - \boldsymbol{v}_j| .$$

For a plasma at temperature, T, consisting of electrons of mass, m, and protons of mass, m_i,

$$g \approx \left(\frac{3kT}{m}\right)^{1/2}$$

and

$$\Gamma_i \approx \frac{4\pi e^4}{m_i^2} \ln \Lambda \ ,$$

where

$$\Lambda = 24\,\pi N\, R_{\mathrm{D}}^3 = \frac{2}{\theta_{\min}} \ ,$$

the average particle density is N, the Debye radius, R_{D}, is given by Debye and Hückel (1923)

$$R_{\mathrm{D}} = \left(\frac{kT}{4\pi N e^2}\right)^{1/2} \approx \left(\frac{1}{Ng}\right)^{1/3}$$

and we have assumed that $NR_{\mathrm{D}}^3 \gg 1$.

For an infinite single particle system the distribution function, f_{s}, for s particles is related to the $f_{\mathrm{s}+1}$ by the B.B.G.K.Y. hierarchy given by Bogolyubov (1946); Born and Green (1949); Green (1952) Kirkwood (1947); Yvon (1935)

$$\frac{\partial f_s}{\partial t} + \left\{f_{\mathrm{s}}; H_{\mathrm{s}}\right\} = N \int \left\{\sum_{i=1}^{\mathrm{s}} \varphi_{i,\mathrm{s}+1}; f_{\mathrm{s}+1}\right\} dX_{\mathrm{s}+1} \ , \tag{3.6}$$

where the Poisson bracket of two quantities, A and B, which depend on $\boldsymbol{r}_i$ and $\boldsymbol{p}_i$, is defined by,

$$\{A; B\} = \sum_{i=1}^{\mathrm{s}} \left[\frac{\partial A}{\partial \boldsymbol{r}_i} \cdot \frac{\partial B}{\partial \boldsymbol{p}_i} - \frac{\partial A}{\partial \boldsymbol{p}_i} \cdot \frac{\partial B}{\partial \boldsymbol{r}_i}\right] \ ,$$

where s is the total number of particles, and H_{s} is the Hamiltonian of s particles.

$$H_{\mathrm{s}} = \sum_{i=1}^{\mathrm{s}} \frac{p_i^2}{2m} + \sum_{i<j=1}^{\mathrm{s}} \varphi_{ij} + \text{constant} \ ,$$

where $\varphi_{ij} = e^2/(|\boldsymbol{r}_i - \boldsymbol{r}_j|)$, for electrons of charge e, and the constant term represents the potential of the electrons moving in the field of ions. For an isolated system in thermal equilibrium, the first member of the hierarchy becomes the Maxwellian distribution function (Maxwell, 1860)

$$f\,dp = 4\,\pi \left(\frac{1}{2\pi mkT}\right)^{3/2} \exp\left[-\frac{p^2}{2mkT}\right] p^2\,dp \ . \tag{3.7}$$

In the limit $g = 1/(NR_{\mathrm{D}}^3) = 0$, the Boltzmann equation becomes collisionless, $(df/dt)_{\mathrm{coll}} = 0$, and the B.B.G.K.Y. hierarchy becomes Vlasov's equation given by Vlasov (1938, 1945)

$$\frac{\partial f}{\partial t} + \boldsymbol{v}_1 \cdot \frac{\partial f}{\partial \boldsymbol{x}_1} - \left[\frac{N}{m} \int d\boldsymbol{x}_2 d\boldsymbol{v}_2 \frac{\partial \varphi_{12}}{\partial \boldsymbol{x}_1} f(\boldsymbol{x}_2, \boldsymbol{v}_2)\right] \cdot \frac{\partial f(\boldsymbol{x}_1, \boldsymbol{v}_1)}{\partial \boldsymbol{v}_1} = 0 \ ,$$

or

$$\frac{\partial f}{\partial t} + \boldsymbol{v}_1 \cdot \frac{\partial f}{\partial \boldsymbol{x}_1} - \frac{e}{m} \boldsymbol{E} \cdot \frac{\partial f}{\partial \boldsymbol{v}_1} = 0 \ , \tag{3.8}$$

where $\boldsymbol{E}$ is the electric field intensity. Linearized perturbation solutions to Vlasov's equation were given by Landau (1946), and such solutions lead to electron plasma oscillations discussed in Sect. 1.32, Eqs. (1.244) to (1.256); and by Tonks and Langmuir (1929) and Bohm and Gross (1949). The general theory for calculating transport coefficients and velocity distribution functions from Boltzmann's equation is discussed by Chapman (1916, 1917), Enskog (1917), and in the book by Chapman and Cowling (1953). The main results of these calculations are given in the following subsections.

The Vlasov equation is a simplified version of the collisionless Boltzmann equation, with $(df/dt)_{\text{coll}} = 0$, derived by Boltzmann in 1872 (see Hénon, 1983). It can be written in vector notation as

$$\frac{\partial f}{\partial t} + \boldsymbol{v} \cdot \nabla f - \nabla \Phi \cdot \frac{\partial f}{\partial v} = 0$$

for a gravitational potential Φ and $\dot{\boldsymbol{v}} = \nabla \Phi$. This collisionless Boltzmann equation is an important relation for stellar and galactic systems and has been described together with the Fokker-Planck description of gravitational systems by Saslaw (1985) and Binney and Tremaine (1987).

3.1.2 Collisions – The Mean Free Path and Mean Free Time Between Collisions

For a gas of neutral atoms with number density, N, and effective collision radius, a_0, the mean free path, l, between collisions is (Clausius, 1858)

$$\begin{aligned} l &= (N\pi a_0^2)^{-1} \\ &\approx 10^{16} N^{-1} \text{cm for hydrogen atoms} \ , \end{aligned} \tag{3.9}$$

where we have taken $a_0 \approx 0.5 \times 10^{-8}$ cm, the radius of the first Bohr orbit. The quantity $\sigma_c = \pi a_0^2$ is called the collision cross section.

When two charged particles collide, a measure of the collision radius is the perpendicular distance, b, between the slower particle and the original path of the faster particle. If $Z_1 e$ and $Z_2 e$ are the two charges, then this impact parameter is (Rutherford, 1911)

$$b = \frac{Z_1 Z_2 e^2}{M_1 V_1^2 \tan(\theta/2)} \ , \tag{3.10}$$

where M_1 and V_1 are, respectively, the mass and velocity of the faster charge, and θ is its angle of deflection. Under conditions of thermal equilibrium at temperature, T, the root-mean-square velocity, V_{rms}, is given by

$$\begin{aligned} V_{\mathrm{rms}} &= (3kT/M)^{1/2} = \langle V^2 \rangle^{1/2} \\ &\approx 6.7 \times 10^5\, T^{1/2}\ \mathrm{cm\ sec^{-1}} \quad \text{for electrons} \\ &\approx 1.57 \times 10^4\, T^{1/2}\ \mathrm{cm\ sec^{-1}} \quad \text{for hydrogen atoms}\ . \end{aligned} \tag{3.11}$$

For a gas composed mostly of neutral particles, Eqs. (3.9), (3.10), and (3.11) give the mean free path

$$l \approx \frac{M^2 V^4}{N_e Z^2 e^4} \approx 3.23 \times 10^6\, T^2\, (Z^2 N_e)^{-1} \mathrm{cm}\ , \tag{3.12}$$

where N_e is the free electron density, and an effective collision is assumed to be one for which the angle of deflection is 90°.

When a gas is composed mostly of charged particles, there is a maximum impact parameter given by

$$b_m = R_D = \left(\frac{kT}{4\pi N_e e^2}\right)^{1/2} \approx 6.9 \left(\frac{T}{N_e}\right)^{1/2} \mathrm{cm}\ , \tag{3.13}$$

where R_D is the Debye radius, T is the temperature, and N_e is the electron density. In this case, the impact parameter must be integrated over its range of values to obtain (Pines and Bohm, 1952; Spitzer, 1962)

$$l \approx \frac{M^2 V_{\mathrm{rms}}^4}{Z^2 N_e e^4 \ln \Lambda} \approx \frac{3.2 \times 10^6\, T^2}{Z^2 N_e \ln \Lambda}\ \mathrm{cm}\ , \tag{3.14}$$

where

$$\Lambda = \frac{R_D}{b} = \frac{3}{2 Z_1 Z_2 e^3} \left(\frac{k^3 T^3}{\pi N_e}\right)^{1/2} \approx 1.3 \times 10^4 \frac{T^{3/2}}{N_e^{1/2}}\ . \tag{3.15}$$

It follows from Eq. (3.14) that the mean free path for collisions of electrons with electrons is the same as that for collisions of protons with protons provided that the electrons and protons have the same kinetic temperature. Using the Poisson distribution, the probability, $P_n(x/l)$, that a particle of a perfect gas will have n collisions after it has travelled the distance, x, is given by

$$P_n(x/l) = [(x/l)^n/n!]\ \exp(-x/l)\ , \tag{3.16}$$

where l is the mean free path.

For a perfect gas in equilibrium at temperature, T, there is a Maxwellian distribution of momentum, p, and the mean free time, τ, between collisions is given by

$$\tau = 4\pi m l \int_0^\infty \frac{p}{(2\pi m k T)^{3/2}} \exp\left[\frac{-p^2}{2mkT}\right] dp = l \left(\frac{2m}{\pi k T}\right)^{1/2}\ ,$$

or

$$\tau \approx l v_{\mathrm{th}}^{-1}\ , \tag{3.17}$$

where l is the mean free path, m is the particle mass, k is Boltzmann's constant, and the thermal velocity, $v_{\mathrm{th}} \approx v_{\mathrm{rms}} = (3kT/m)^{1/2}$. Using Eqs. (3.9), (3.11),

(3.12), (3.14), and (3.17) we obtain

$$\begin{aligned} &\tau \approx 10^{12}\, T^{-1/2}\, N^{-1}\,\mathrm{sec} && \text{for hydrogen atoms} \\ &\tau \approx 4T^{3/2}\, N_{\mathrm{e}}^{-1}\,\mathrm{sec} && \text{for electrons, } N_{\mathrm{e}} \ll N \\ &\tau \approx 4T^{3/2}(N_{\mathrm{e}} \ln \Lambda)^{-1}\,\mathrm{sec} && \text{for electrons, } N_{\mathrm{e}} \gg N \ , \end{aligned} \tag{3.18}$$

where N is the density of neutral atoms, and N_{e} is the electron density. Using the Poisson distribution, we may specify the probability, $P_n(t/\tau)$, that a particle of a perfect gas will have n collisions in the time, t.

$$P_n(t/\tau) = [(t/\tau)^n/n!]\,\exp[-t/\tau] \ . \tag{3.19}$$

For conditions very near equilibrium, we may use the mean free time, τ, to give the approximate Boltzmann equation

$$\frac{\partial f}{\partial t} + \frac{\boldsymbol{p}}{m} \cdot \nabla_r f - \nabla_r \varphi \cdot \nabla_p f \approx \frac{1}{\tau}[f_0(r,p) - f] \ , \tag{3.20}$$

or

$$f + \tau\left(\frac{\partial f}{\partial t}\right) = f_0 - \tau\left[\left(\frac{\boldsymbol{p}}{m}\right) \cdot \nabla_r f_0 + \boldsymbol{F} \cdot \nabla_p f_0\right] \ ,$$

where the equilibrium solution, $f_0(r,p)$, is given by Eq. (3.7).

When a stream of test particles of mass, M_+, charge, Z_+, and velocity, V_+, are injected into a plasma which is in thermal equilibrium, the characteristic time, τ_{D}, in which the particles are deflected through an angle of order $\pi/2$ is given by

$$\tau_{\mathrm{D}} \approx \frac{M_+^2 V_+^3}{16\pi N e^4 Z_+^2 \ln \Lambda} \quad \text{for } V_+ \gg \left(\frac{kT}{m}\right)^{1/2} \ , \tag{3.21}$$

where Λ is given by Eq. (3.15), and N is the plasma number density.

3.1.3 Viscosity and the Reynolds Number

If a part of a gas which is in equilibrium is given a drift velocity, v_x, in the x direction, there will be a restoring force $\boldsymbol{F} = \mu \partial v_x/\partial z$ tending to reduce the rate of shear $\partial v_x/\partial z$. Here μ is called the coefficient of dynamic viscosity. The net rate of increase of momentum within a volume element may be obtained by considering the forces acting on all of the faces of the element. This net force is

$$\boldsymbol{F} = \mu \nabla^2 \boldsymbol{v} \ , \tag{3.22}$$

where $\boldsymbol{v}$ is the velocity, and μ is the coefficient of dynamic viscosity. In the absence of other forces, it follows directly from Eqs. (3.7), (3.20), and (3.22) that (cf. Maxwell, 1866)

$$\mu \approx M\, l\, N\, v_{\mathrm{rms}} \approx M\, l\, N \left(\frac{3kT}{M}\right)^{1/2} \ , \tag{3.23}$$

where M is the atomic or molecular mass, l is the mean free path given by Eqs. (3.9), (3.12) or (3.14), N is the number density of atoms or molecules, v_{rms} is the thermal velocity, and T is the temperature. For hydrogen atoms, for example,

$$\mu \approx 5.7 \times 10^{-5}\, T^{1/2}\, \mathrm{gm\ cm^{-1} sec^{-1}} \ , \tag{3.24}$$

and for a fully ionized gas

$$\mu \approx \frac{M_i^{1/2}(3kT)^{5/2}}{Z^4 e^4 \ln \Lambda} \approx 2 \times 10^{-15} \frac{T^{5/2} A_i^{1/2}}{Z^4 \ln \Lambda} \mathrm{gm\ cm^{-1}\ sec^{-1}} \ , \tag{3.25}$$

where M_i is the ion mass, $e\,Z$ is the ionic charge, A_i is the atomic weight of the positive ion, and the factor Λ is given by Eq. (3.15).

When a magnetic field of strength, B, is present, the effective viscosity in the direction parallel to the direction of B is given by Eq. (3.25). For the direction perpendicular to B, the viscosity is (Simon, 1955)

$$\begin{aligned} \mu_\perp &= \frac{2}{5}\left(\frac{\pi}{M_i k T_i}\right)^{1/2} \frac{Z^4 e^4 N_i^2 M_i^2 c^2 \ln \Lambda}{Z^2 e^2 B^2} \\ &\approx 2.7 \times 10^{-26} \frac{A_i^{3/2} Z^2 N_i^2 \ln \Lambda}{T_i^{1/2} B^2} \mathrm{gm\ cm^{-1} sec^{-1}} \ , \end{aligned} \tag{3.26}$$

where M_i and N_i are, respectively, the ion mass and number density, A_i is the atomic weight of the ion, and T_i is the temperature of the ion.

The ability of a gas to damp out turbulent motion is measured by the Reynolds coefficient (Stokes, 1851; Reynolds, 1883). If a turbule of size, L, has velocity, v, the motion will not be damped out by viscous effects if the Reynolds number

$$\mathrm{Re} = L v \rho / \mu \tag{3.27}$$

is greater than unity. Here μ is the coefficient of dynamic viscosity and ρ is the mass density. Often the coefficient of kinematic viscosity, ν, is used, where

$$\nu = \mu/\rho, \quad \text{and} \quad \mathrm{Re} = L v/\nu \ . \tag{3.28}$$

Using Eqs. (3.24), (3.25), and (3.26) in Eq. (3.27), we obtain

$$\mathrm{Re} \approx 2 \times 10^4 \rho L\, v\, T^{-1/2} \tag{3.29}$$

for an unionized gas,

$$\mathrm{Re} \approx 5 \times 10^{14} \frac{\rho Z^4 L v \ln \Lambda}{T^{5/2} A_i^{1/2}} \ , \tag{3.30}$$

for a fully ionized gas without a magnetic field, and

$$\mathrm{Re} \approx 3 \times 10^{25} \frac{\rho\, L\, v\, T^{1/2} B^2}{A_i^{3/2} Z^2 N_i^2 \ln \Lambda} \ , \tag{3.31}$$

for a fully ionized gas when the motion is perpendicular to the magnetic field.

3.1.4 Electrical Conductivity and Mobility

When an ionized gas is subject to an electric field of strength, E, the free electrons and ions will drift, causing a flow of current. When the number density of free electrons, N_e, is small compared with the total number density of neutral atoms or molecules, N, the net drift velocity, V, will be determined by collisions. In this case, the equation of motion of each free electron or ion is given by

$$Z\,e\,E = mV_{\rm rms}V/l = mV/\tau \ , \tag{3.32}$$

where $e\,Z\,E$ is the Lorentz force on the charge $e\,Z$, the particle mass is m, the r.m.s. particle velocity is $V_{\rm rms}$, the mean free path is l, and τ is the mean time between collisions. If the number density of free electrons is $N_{\rm e}$, then the net current density, J, is given by

$$J = \sigma\,E = N_{\rm e}\,e\,V \ , \tag{3.33}$$

where the electrical conductivity, σ, is given by

$$\sigma = \frac{N_{\rm e}e^2\,l}{mV_{\rm rms}} \ . \tag{3.34}$$

Using Eq. (3.11) for $V_{\rm rms}$ and Eq. (3.12) for l in Eq. (3.34), we obtain

$$\sigma = \frac{(3kT)^{3/2}}{e^2m^{1/2}} \approx 10^9\,T^{3/2}\,\text{e.s.u} \tag{3.35}$$

for a partially ionized gas. For conversion, $1\,\text{mho-cm}^{-1} \approx 10^{-9}$ e.m.u $\approx 10^{12}$ e.s.u. The corresponding expression for a fully ionized gas is (Cowling, 1945)

$$\sigma \approx 6.5 \times 10^6\,T^{3/2}\,\text{e.s.u.} \ , \tag{3.36}$$

where electron-electron encounters and electron shielding have been taken into account. As pointed out by Cowling (1945, 1946), a magnetic field of linear scale, L, has a mean lifetime, τ, of order

$$\tau \approx L^2\sigma/c^2 \ , \tag{3.37}$$

where σ is in e.s.u.

Detailed formulae for the resistivity, η, of an ionized are given by Spitzer and Härm (1953). They obtain

$$\eta = \frac{c^2}{\sigma} = \frac{\pi^{3/2}m^{1/2}Z\,e^2\,c^2\ln\Lambda}{2(2kT)^{3/2}\gamma} \approx 3.80 \times 10^3\,\frac{Z\ln\Lambda}{\gamma\,T^{3/2}}\,\text{ohm-cm} \ , \tag{3.38}$$

where Λ is given by Eq. (3.15) and γ is a Z dependent factor having the values $\gamma = 0.582$ for $Z = 1$, and $\gamma = 1$ for $Z = \infty$. For conversion, 1 ohm-cm $\approx 10^{-12}$ e.s.u $\approx 10^9$ e.m.u.

From Eqs. (3.33) to (3.36), the net drift velocity, V, of the free electrons is

$$V \approx \frac{\sigma}{N_{\rm e}\,e}E = \Omega E \ , \tag{3.39}$$

where $N_{\rm e}$ is the density of the free electrons, and the mobility, Ω, is given by

$$\Omega \approx 10^{19}\,T^{3/2}\,N_{\rm e}^{-1}\,\text{e.s.u.} \ , \tag{3.40}$$

for a partially ionized gas, and

$$\Omega \approx 10^{16}\,T^{3/2}\,N_{\rm e}^{-1}\,\text{e.s.u.} \ ,$$

for a fully ionized gas.

3.1.5 Diffusion and the Magnetic Reynolds Number

A spatial gradient, ∇N_i, in the ion density, N_i will cause a current density, J, given by

$$J = e\, l\, Z\, V_{rms} \nabla N_i \ , \tag{3.41}$$

where the ions are assumed to move with the root mean square velocity, V_{rms}, for the mean free path, l. From the equation of continuity (Eq. (1.38)) for charge conservation, we obtain the diffusion equation

$$\frac{\partial N_i}{\partial t} = \frac{\nabla \cdot J}{eZ} = D\nabla^2 N_i \ , \tag{3.42}$$

where the diffusion constant

$$D \approx l\, V_{rms} = l\left(\frac{3kT}{m}\right)^{1/2} = \frac{kT}{e}\Omega \ , \tag{3.43}$$

where T is the temperature, and Ω is the mobility. The approximate mean free paths, l, are given in Eqs. (3.9), (3.12), and (3.14). A characteristic "deflection time", t_D, can be defined as

$$t_D = D/V_{rms}^2 = l/V_{rms} \ , \tag{3.44}$$

and a characteristic diffusion time for the length, L, is

$$t = L^2/D \ . \tag{3.45}$$

Chandrasekhar (1943) has given formulae for the velocity dispersion of test particles which diffuse into a group of field particles. If the subscripts $\|$ and $\perp$ denote, respectively, directions parallel and perpendicular to the original motion of the test particles, then for a Maxwellian distribution of field particles

$$\langle \Delta v_{\|}^2 \rangle \approx \frac{v^3}{l}\left\{ G\left[\left(\frac{mv^2}{2kT}\right)^{1/2}\right]\right\} \tag{3.46}$$

and

$$\langle \Delta v_{\perp}^2 \rangle \approx \frac{v^3}{l}\left\{ \varphi\left[\left(\frac{mv^2}{2kT}\right)^{1/2}\right] - G\left[\left(\frac{mv^2}{2kT}\right)^{1/2}\right]\right\} ,$$

where $\varphi(x)$ is the error function, $G(x) = [\varphi(x) - x\varphi'(x)]/2x^2$, Δv denotes the dispersion in velocity, the factors in { } are of order unity, the velocity of the field particles is v, and the mean free path, l, of the test particles is given by Eq. (3.14). It follows from Eq. (3.46) that a characteristic deflection time is

$$t_D = \frac{v^2}{\langle \Delta v^2 \rangle} \approx \frac{1}{v} \ . \tag{3.47}$$

Maxwell's equations which relate the electric field intensity, $\boldsymbol{E}$, the magnetic induction, $\boldsymbol{B}$, and the current density, $\boldsymbol{J}$, are (cf. Eqs. (1.39) to (1.43))

$$\nabla \times \boldsymbol{E} + \frac{1}{c}\frac{\partial \boldsymbol{B}}{\partial t} = 0 \tag{3.48}$$

and

$$\nabla \times \boldsymbol{B} = \frac{4\pi}{c} \boldsymbol{J} ,$$

where displacement current has been neglected in this m.h.d. approximation. Using $\boldsymbol{J} = \sigma \boldsymbol{E}$ we obtain the diffusion equation

$$\frac{\partial \boldsymbol{B}}{\partial t} = D_{\mathrm{M}} \nabla^2 \boldsymbol{B} , \tag{3.49}$$

where the material velocity has been assumed to be negligible (cf. Eq. (3.258)), and the magnetic diffusion constant, D_{M}, is given by

$$D_{\mathrm{M}} = \frac{c^2}{4\pi\sigma} = \frac{\eta}{4\pi} . \tag{3.50}$$

The conductivity, σ, may be evaluated using Eq. (3.35) or (3.36). If L is the characteristic size of the spatial variation of $\boldsymbol{B}$, the decay time for the diffusion of the magnetic field is (Cowling, 1945, 1946)

$$\tau_{\mathrm{M}} = \frac{L^2}{D_{\mathrm{M}}} = \frac{4\pi\sigma L^2}{c^2} . \tag{3.51}$$

Magnetic lines of force will move with a fluid of velocity, v, and scale, L, if the magnetic Reynolds number

$$R_{\mathrm{M}} = v\tau_{\mathrm{M}}/L \tag{3.52}$$

is greater than unity. For $R_{\mathrm{M}} > 1$, transport of the lines of force with a fluid or gas dominates over diffusion of the lines of force.

3.1.6 Heat Conductivity and the Prandtl Number

The flow of heat per unit area, Q, in the presence of a temperature gradient, ∇T, is given by

$$Q \approx -\kappa \nabla T , \tag{3.53}$$

where the coefficient of heat conductivity, κ, is given by

$$\kappa = \frac{5}{3} \frac{kTl}{V_{\mathrm{rms}}} N \left(\frac{3}{2} \frac{k}{m} \right) = \frac{5}{3} \mu \left(\frac{3}{2} \frac{k}{m} \right) \tag{3.54}$$

for a gas at temperature, T, with a mean free path, l, root-mean-squared velocity $V_{\mathrm{rms}} = (3kT/m)^{1/2}$, particle number density, N, particle mass, m, and viscosity, μ. The mean free path, l, or viscosity, μ, may be evaluated from Eqs. (3.9), (3.12), (3.14), (3.24), (3.25), or (3.26). The quantity $(3k/2m)$ is given by (C_v/M), where C_v and M are, respectively, the specific heat and the molecular weight of the gas. It follows from Eqs. (3.25), (3.26), and (3.54) that (cf. Spitzer, 1962)

$$\kappa \approx 2 \times 10^{-4} \frac{T^{5/2}}{Z^4 \ln \Lambda} \mathrm{erg\ sec^{-1}\ {}^\circ K^{-1}\ cm^{-1}} \tag{3.55}$$

for a fully ionized gas, and

$$\kappa \approx 1.5 \times 10^{-17} \frac{A_{\mathrm{i}}^{1/2} Z^2 N_{\mathrm{i}}^2 \ln \Lambda}{T^{1/2} B^2} \text{erg sec}^{-1}\,{}^{\circ}\mathrm{K}^{-1}\,\mathrm{cm}^{-1} \tag{3.56}$$

for heat conduction in the direction perpendicular to the magnetic field.

For an incompressible fluid (ρ = constant), the change in temperature, T, with time, t, is given by the equation of thermal conduction:

$$\frac{\partial T}{\partial t} + \boldsymbol{v} \cdot \nabla T = \frac{\kappa}{\rho C_v} \nabla^2 T = \chi \nabla^2 T \ , \tag{3.57}$$

where κ is the coefficient of heat conductivity, ρ is the mass density, C_v is the specific heat at constant volume, and the constant χ is called the thermometric conductivity.

The relative magnitudes of viscosity and heat conductivity are measured by the Prandtl number, P_{r}, given by Prandtl (1905)

$$P_{\mathrm{r}} = \frac{C_p \mu}{\kappa} = \frac{C_p \nu \rho}{\kappa} = \frac{\nu}{\chi} \ , \tag{3.58}$$

where C_p is the specific heat at constant pressure, μ is the coefficient of dynamic viscosity, κ is the coefficient of heat conductivity, ν is the coefficient of kinematic viscosity, ρ is the mass density, and χ is the thermometric conductivity.

3.2 Thermodynamics of a Gas

3.2.1 First Law of Thermodynamics and the Perfect Gas Law

The first law of thermodynamics is the expression of the law of conservation of energy which takes into account the energy due to heat. If dQ is the amount of heat energy which is absorbed by a system from its surroundings, then this law states (Mayer, 1842; Joule, 1847; Helmholtz, 1847)

$$dQ = dU + dW \ , \tag{3.59}$$

where dU is the change in internal energy of the system when going from one equilibrium state to another, and dW is the amount of work done by the system on its surroundings. One of the simpler systems is the hydrostatic system in which the work consists of displacing the system boundaries against a uniform hydrostatic pressure, P. In this case, Eq. (3.59) becomes

$$dQ = dU + P\,dV \ , \tag{3.60}$$

where dV is the change in volume, V. For a perfect gas we have the ideal gas law:

$$PV = nRT \tag{3.61}$$

or equivalently

$$P = NkT \ ,$$

where T is the temperature, n is the number of molecules per mole, N is the number of particles per unit volume, the universal gas constant $R = 8.314510$

(70) $\times 10^7$ erg $°K^{-1}$ mole^{-1} and Boltzmann's constant $k = R/N_A = 1.380658$ (12) $\times 10^{-16}$ erg $°K^{-1}$ where Avogadro's number $N_A = 6.0221367(36) \times 10^{23}$ mole^{-1}. Eq. (3.61) follows from the observations that $P \propto T$ for constant V (Gay-Lussac, 1809), that $PV/T =$ constant (Avogadro, 1811), and that $P \propto V^{-1}$ for constant T (Boyle, 1662; Townely, 1662; Power, 1663).

The pressure is often written in the alternative form:

$$P = NkT = \frac{\rho RT}{\mu} = \frac{\rho kT}{\mu m_u} , \tag{3.62}$$

where ρ is the mass density, the atomic mass unit $m_u = 1.6605402(10) \times 10^{-24}$ grams $= 1/N_A$, and the mean molecular weight, μ, is given by:

$$\mu = \left(\sum \frac{X_i(1 + Z_i)}{\mu_i} \right)^{-1} ,$$

where the summation is for all nuclei of type i of weight fraction X_i, molecular weight μ_i, and charge number Z_i. For a neutral gas all the electrons are in the atom and we replace $(1 + Z_i)$ by 1. For a neutral hydrogen gas, called an H I region, $\mu = 1$, for a fully ionized hydrogen gas, dubbed a H II region, the $\mu = 1/2$, and for a fully ionized helium gas $\mu = 4/3$. If X, Y, and Z represent the concentration by mass of hydrogen, helium and heavier elements, respectively, the number density of particles is:

$$N = \frac{\rho}{m_u} \left[2X + \frac{3}{4} Y + \frac{1}{2} Z \right] = \frac{\rho}{\mu m_H}$$

where $X + Y + Z = 1$ and $\mu = 2/(1 + 3X + 0.5Y)$. The electron density is similarly given by:

$$N_e = \frac{\rho}{\mu_e \, m_u} = \frac{1}{2}(1 + X) \frac{\rho}{m_u} ,$$

where the mean molecular weight per electron is μ_e, each nucleus i contributes Z_i electrons, and since $X + Y + Z = 1$ we have $X + 0.5Y + 0.5Z = 1 + X$.

3.2.2 Thermal (or Heat) Capacity, Molecular Heat, and Specific Heat

By definition the thermal or heat capacity, C, of a system is the amount of heat necessary to raise its temperature one degree under the conditions given. Letting the subscripts V and P denote, respectively, conditions of constant volume and pressure, it follows from Eq. (3.60) that for hydrostatic systems

$$C_v = \left(\frac{dQ}{dT} \right)_V = \left(\frac{\partial U}{\partial T} \right)_V \tag{3.63}$$

and

$$C_p = \left(\frac{dQ}{dT} \right)_P = \left(\frac{\partial U}{\partial T} \right)_P + P \left(\frac{\partial V}{\partial T} \right)_P , \tag{3.64}$$

where $\partial/\partial T$ denotes differentiation with respect to temperature. Noting that the internal energy, U, is only a function of pressure, P, temperature, T, and volume, V, we obtain from Eqs. (3.63) and (3.64) the relation

$$C_p - C_v = \left[\left(\frac{\partial U}{\partial V}\right)_T + P\right]\left(\frac{\partial V}{\partial T}\right)_P . \tag{3.65}$$

For an ideal gas, $(\partial U/\partial V)_T = 0$, and it follows from Eqs. (3.61) and (3.65) that

$$C_p - C_v = nR \ , \tag{3.66}$$

where n is the number of gram-moles and R is the universal gas constant. For a nonrelativistic, nondegenerate, monatomic ideal gas, $U = 3nRT/2$, and from Eqs. (3.63) and (3.64), $C_p = 5nR/2$ and $C_v = 3nR/2$. Similarly, for an ideal diatomic gas, $U = 5nRT/2$, $C_p = 7nR/2$ and $C_v = 5nR/2$. The thermal capacity of one mole of a gas is called the molecular heat and is often given the symbol C, whereas the thermal capacity of one gram of a gas is called the specific heat and is usually given the symbol c.

3.2.3 Adiabatic Processes

An adiabatic process is one in which no heat is exchanged between a system and its surroundings. That is, the gain or loss of heat by conduction or radiation can be ignored in an adiabatic process. A measure of this process is the adiabatic index

$$\gamma = C_p/C_v \ . \tag{3.67}$$

The γ takes on the values 5/3, 7/5, and 4/3, respectively, for an ideal monoatomic, diatomic and polyatomic gas. Using the adiabatic condition $dQ = 0$ in Eq. (3.60), it then follows from Eqs. (3.61), (3.63), and (3.64) that

$$PV^\gamma = \text{constant}, \quad TV^{\gamma-1} = \text{constant, and} \quad P^{(1-\gamma)}T^\gamma = \text{constant}$$

for an adiabatic process involving an ideal gas. These equations show, for example, that a gas is cooled or heated according as it undergoes an adiabatic expansion or contraction.

The requirement $dQ = 0$ imposes, through the first law of thermodynamics, one relation between the pressure, P, temperature, T, and the volume, V. Three adiabatic exponents are then defined by

$$\Gamma_1 = -\left(\frac{d\ln P}{d\ln V}\right)_{ad} = \left(\frac{d\ln P}{d\ln \rho}\right)_{ad} ,$$

$$\frac{\Gamma_2}{(\Gamma_2 - 1)} = \left(\frac{d\ln P}{d\ln T}\right)_{ad} , \tag{3.68}$$

and

$$\Gamma_3 - 1 = -\left(\frac{d\ln T}{d\ln V}\right)_{ad} = \left(\frac{d\ln T}{d\ln \rho}\right)_{ad} ,$$

where the specific volume $V = \rho^{-1}$, the mass density is ρ, and the subscript ad means the differential is along an adiabat for which $dS = dQ/T = 0$. Only two of the three gammas are independent and we have the general identify

$$\frac{\Gamma_1}{\Gamma_3 - 1} = \frac{\Gamma_2}{\Gamma_2 - 1} \ .$$

In the classical limit of a nonrelativistic, monatomic, ideal gas with no internal degrees of freedom, $\Gamma_1 = \Gamma_2 = \Gamma_3 = \gamma = C_p/C_v = 5/3$. In general Γ_1 is important for determining the conditions of dynamical instability of stars whereas Γ_2 and Γ_3 are important, respectively, in determining the conditions of convective and pulsational instability of stars.

3.2.4 Polytropic Processes

A polytropic process is defined as one in which the thermal capacity, C, remains constant during the entire process. The polytropic exponent, γ', is defined as

$$\gamma' = \frac{C_p - C}{C_v - C} \ , \tag{3.69}$$

where the thermal capacity $C = dQ/dT$ and C_p and C_v are defined by Eqs. (3.63) and (3.64). For an ideal gas, it follows from Eqs. (3.60), (3.61), (3.63), (3.64), and (3.68) that

$$C = \left(C_v - \frac{nR}{\gamma' - 1}\right) = \text{constant}$$

and

$$PV^{\gamma'} = \text{constant}, \quad TV^{\gamma'-1} = \text{constant}, \quad \text{and } P^{(1-\gamma')}T^{\gamma'} = \text{constant} \ .$$

If the index $n = 1/(\gamma' - 1)$, then the pressure, P, internal energy, E_{int}, gravitational energy, E_{grav}, total energy, E_{tot}, and central pressure, P_{c}, of a polytropic gas sphere of radius, R, and mass, M, can be obtained from the Lane–Emden equation of index n (Lane, 1870; Emden, 1907):

$$\frac{1}{z^2}\left[\frac{d}{dz}\left(z^2\frac{dw}{dz}\right) + w^n\right] = 0 \ , \tag{3.70}$$

where the radial distribution of mass density is given by $\rho(r) = \rho_c w^n$, with a central value of ρ_c at $r = 0$, the index $n = 1/(\gamma - 1)$, and $z = Ar$ where A depends on n and is described by Chandrasekhar (1939) and more recently by Kippenhahn and Wiegert (1990) who provide a good account of the solutions for the Lane–Emden equation. They include:

$$\begin{aligned}
P &\propto \rho^{1+1/n} \ , \\
E_{\text{int}} &= \frac{GM^2}{R}\frac{n}{5-n} \ , \\
E_{\text{grav}} &= -\frac{GM^2}{R}\frac{3}{5-n} \ , \\
E_{\text{tot}} &= E_{\text{int}} + E_{\text{grav}} = \frac{-GM^2}{R}\frac{3-n}{5-n}
\end{aligned} \tag{3.71}$$

and

$$P_c = KGM^{2/3}\rho_c^{4/3} ,$$

where ρ_c is the central mass density, and the constant $K = 0.488, 0.396$, and 0.364 for $n = 3/2(\gamma = 5/3), n = 5/2(\gamma = 7/5)$, and $n = 3(\gamma = 4/3)$, respectively. For a non-relativistic monatomic gas $n = 3/2$, and for a relativistic gas $n = 4/3$. For $n > 3$ a star has no equilibrium state, and for $n = 3$, an equilibrium exists for any density as the energy is independent of density.

3.2.5 Second Law of Thermodynamics and the Entropy of a Gas

The entropy, S, of a system is defined so that a change in entropy is equal to the integral of dQ/T between the terminal states, where dQ is the amount of heat added to the system from the surroundings, and T is the temperature. The second law of thermodynamics states that, for an isolated system,

$$dS = \frac{dQ}{T} \geq 0 , \tag{3.72}$$

where the equality sign holds for a reversible process, and the greater than sign holds for an irreversible process. As expressed by Clausius (1850, 1857, 1865) this law means that heat cannot pass by itself from a colder to a hotter body. For a reversible process which leaves a system in the same state as it was originally,

$$\int \frac{dQ}{T} = 0 , \tag{3.73}$$

which was first stated by Carnot (1824), and first used to define temperature by Kelvin (1848).

From Eqs. (3.60), (3.61), (3.63), (3.64), (3.66), and (3.72), the change in entropy, dS, for a process involving an ideal gas is

$$\begin{aligned} dS &= C_p \frac{dT}{T} - nR\frac{dP}{P} \\ &= C_v \frac{dT}{T} + nR\frac{dV}{V} \\ &= C_v \frac{dT}{T} - (C_p - C_v)\frac{d\rho}{\rho} , \end{aligned} \tag{3.74}$$

where C_p and C_v are the heat capacities at constant pressure, P, and volume, V, the mass density is ρ, the universal gas constant is R, and n is the number of gram-moles.

For a polytropic change of index $n = 1/(\gamma' - 1)$,

$$dS = C_v(\gamma' - \gamma)\frac{d\rho}{\rho} = C_v \frac{(\gamma' - \gamma)}{\gamma' - 1}\frac{dT}{T} , \tag{3.75}$$

where $\gamma = C_p/C_v$ is the adiabatic index.

In statistical mechanics, criteria are given for assigning to a given thermodynamical state the number, W, of corresponding dynamical states.

The integer, W, is usually called the thermodynamic probability of the given thermodynamic state. Because the most stable state of a system is the state of highest probability consistent with the given total energy of the system, there is a functional relationship between the entropy, S, and the probability, W, given by (Boltzmann, 1868, 1877, 1896; Planck, 1901)

$$S = k \ln W \ , \tag{3.76}$$

where the conversion factor between ergs and degrees is given by Boltzmann's constant $k = 1.380658(12) \times 10^{-16}$ erg °K^{-1}.

For a monatomic nonrelativistic gas, in the absence of ionization and excitation, the entropy (entropy per mole) for temperatures not near absolute zero is given by Sackur (1911, 1913); Tetrode (1912)

$$S = N_{\text{tot}} k \ln \left[\frac{e^{5/2} g_0}{N} \left(\frac{2\pi MkT}{h^2} \right)^{3/2} \right] , \tag{3.77}$$

where N_{tot} is the total number of gas atoms, $e = 2.7182812$ is the base of the natural logarithm, g_0 is the statistical weight of the ground state of the atom, N is the number density of the gas, and M is the mass of the gas atom. Eq. (3.77) may be written in the equivalent form

$$S = \frac{5}{2} k N_{\text{tot}} + k N_{\text{tot}} \ln \left[g_0 \left(\frac{4\pi M U}{3h^2 N_{\text{tot}}} \right)^{3/2} \frac{V}{N_{\text{tot}}} \right] ,$$

where the volume is V, and the internal energy, U, of a monoatomic gas is $3k N_{\text{tot}} T/2$. Because the joint probability is the product of the two separate probabilities, it follows from Eq. (3.75) that the entropy of a composite system is additive. The entropy of complex systems is often conveniently calculated using the Helmholtz free energy, and these calculations are discussed in the following section.

Another alternative form of Eq. (3.77) is (cf. Eq. (3.76))

$$S = -\frac{R}{\mu} \ln \left(\frac{\rho}{\mu M} \right) + \frac{3}{2} \left(\frac{R}{\mu} \right) \ln(kT) + \left(C + \frac{5}{2} \right) \left(\frac{R}{\mu} \right) ,$$

where S is the specific entropy (entropy per gram), R is the gas constant, μ is the mean molecular weight ($\mu = 1$ for hydrogen and $\mu = 4$ for ^{4}He), the gas mass density is ρ, and the constant $C = \ln[g_0 (2\pi M/h^2)^{3/2}]$ has the value $C = 20.204$ for atomic hydrogen and $C = 22.824$ for ^{4}He. For an ionized gas, $\mu = \mu_0/(1 + Z/A)$ where μ_0 is the molecular weight of the neutral gas. At full ionization $\mu = 0.5$ and 4/3 for hydrogen and ^{4}He, respectively.

The specific entropy (per gram) of a photon gas is given by

$$S = \frac{4 a T^3}{3\rho} , \tag{3.78}$$

where the radiation density constant $a = 7.566 \times 10^{-15}$ erg cm^{-3} °K^{-4}, and the gas temperature and mass density are, respectively, T and ρ.

3.2.6 Combined First and Second Laws

The combined first and second law for a reversible process is

$$dU = T\,dS - dW \ , \tag{3.79}$$

where dU is the change in the internal energy of the system, dS is the change in entropy of the system, dW is the amount of work done by the system on its surroundings, and T is the temperature. For a hydrostatic system, this combined law becomes

$$d\,U = T\,dS - P\,dV \ , \tag{3.80}$$

where P is the pressure, and dV is the change in volume.

Useful functions which are combinations of thermodynamic variables include the enthalpy, H, given by

$$H = U + PV \ , \tag{3.81}$$

the Helmholtz function, F, (Helmholtz free energy) given by

$$F = U - TS \ , \tag{3.82}$$

and the Gibbs function, G, (Gibbs free energy) given by

$$G = U - TS + PV = F + PV = H - TS \ , \tag{3.83}$$

where U is the internal energy and S is the entropy.

A useful mnemonic diagram relating the thermodynamic variables is shown below

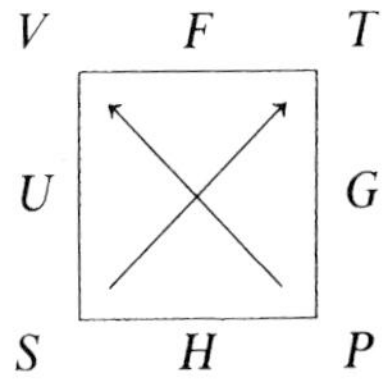

Each of the four thermodynamic potentials U, F, G, and H, is flanked by its natural independent variables. In writing the differential expression for each of the potentials in terms of its independent variables, the coefficient is designated by an arrow. An arrow pointing away from a natural variable implies a positive coefficient, whereas one pointing towards the variable implies a negative coefficient. Hence we have, for example,

$$dF = -P\,d\,V - S\,dT \ ,$$

$$dG = -S\,dT + V\,dP \ ,$$

$$dH = T\,dS + V\,dP \ ,$$

$$dU = T\,dS - P\,dV \ .$$

and

The Maxwell relations

$$\begin{aligned}
\left(\frac{\partial V}{\partial S}\right)_P &= \left(\frac{\partial T}{\partial P}\right)_S , \\
\left(\frac{\partial S}{\partial P}\right)_T &= -\left(\frac{\partial V}{\partial T}\right)_P , \\
\left(\frac{\partial P}{\partial T}\right)_V &= \left(\frac{\partial S}{\partial V}\right)_T , \\
\text{and} \quad \left(\frac{\partial T}{\partial V}\right)_S &= -\left(\frac{\partial P}{\partial S}\right)_V ,
\end{aligned} \tag{3.84}$$

may also be read from the diagram by rotating about the four corners.

The Helmholtz free energy, F, is often called the free energy at constant volume, and the Gibbs free energy, G, is often called the free energy at constant pressure. For thermal equilibrium at constant temperature and volume, F is a minimum, whereas for thermal equilibrium at constant temperature and pressure, G is a minimum. When a system undergoes an isothermal transformation, the work performed by it can never exceed minus the variation, ΔF, of its Helmholtz free energy; and it is equal to $-\Delta F$ if the transformation is reversible. A consequence of this fact is that a thermodynamic system having the temperature of its environment is in a state of stable equilibrium when its Helmholtz free energy is a minimum. A thermodynamic system is equivalently described by the potentials U, H, F, or G depending on the choice of independent variables $U(V, S), H(P, S), F(V, T)$ or $G(P, T)$. The conjugate variables are described by the relations

$$\begin{aligned}
&T = \left(\frac{\partial U}{\partial S}\right)_V , \quad P = -\left(\frac{\partial U}{\partial V}\right)_S , \\
&T = \left(\frac{\partial H}{\partial S}\right)_P , \quad V = \left(\frac{\partial H}{\partial P}\right)_S , \\
&S = -\left(\frac{\partial F}{\partial T}\right)_V , \quad P = -\left(\frac{\partial F}{\partial V}\right)_T , \quad U = -T^2\left(\frac{\partial}{\partial T}\frac{F}{T}\right)_{V,N} ,
\end{aligned} \tag{3.85}$$

and

$$S = -\left(\frac{\partial G}{\partial T}\right)_P , \quad V = \left(\frac{\partial G}{\partial P}\right)_T ,$$

where the subscripts V, T, P, or S denote, respectively, conditions of constant volume, temperature, pressure, or entropy.

According to statistical mechanics, a nongenerate, system consisting of N_{tot} particles has a free energy, F, given by

$$F = -kT \ln Q , \tag{3.86}$$

where the partition function, Q, of an ideal Boltzmann gas is given by (Gibbs, 1902)

$$Q = \sum_n \exp\left(\frac{-E_n}{kT}\right) = \left[Z\left(\frac{e}{N_{\rm tot}}\right)\right]^{N_{\rm tot}} , \tag{3.87}$$

where E_n is the energy of the n th state of the system, Z is the partition function for one particle, and $e = 2.7182812$ is the base of the natural logarithm. If to an energy level, E_n, there belong g_n permissable states, the level is said to be degenerate and g_n is called its statistical weight. In this case,

$$Q = \sum_n g_n \exp\left(\frac{-E_n}{k\,T}\right) . \tag{3.88}$$

The probability, P, that a particle has energy, E_n, is given by

$$P = \frac{1}{Q}\exp\left(\frac{-E_n}{k\,T}\right) . \tag{3.89}$$

The partition function for one molecule, Z, is given by

$$Z = Z_{\rm trans} \cdot Z_{\rm rot} \cdot Z_{\rm vib} \cdot Z_{\rm el} , \tag{3.90}$$

where the subscripts trans, rot, vib, and el denote, respectively, the translational, rotational, vibrational, and electronic partition functions. These partition functions are given by

$$Z_{\rm trans} = \left(\frac{2\pi M\,k\,T}{h^2}\right)^{3/2} V , \tag{3.91}$$

where M is the particle mass and V is the volume occupied by the gas,

$$Z_{\rm rot} = \frac{8\pi^2 Ik\,T}{h^2\sigma} = \frac{kT}{h\,\nu_{\rm rot}\sigma} , \tag{3.92}$$

for a diatomic or linear polyatomic molecule. Here I is the moment of inertia given by $I = h/(8\pi^2\nu_{\rm rot})$ where $\nu_{\rm rot}$ is the rotational frequency, and the symmetry factor, σ is equal to one plus the number of transpositions of identical atoms ($\sigma = 2$ for diatomic molecule of identical atoms and $\sigma = 1$ for one composed of different atoms),

$$Z_{\rm rot} = \frac{8\pi^2}{\sigma}\left(\frac{2\pi IkT}{h^2}\right)^{3/2} , \tag{3.93}$$

for a nonlinear polyatomic molecule,

$$Z_{\rm vib} = \left[1 - \exp\left(\frac{-h\nu_{\rm vib}}{kT}\right)\right]^{-1} , \tag{3.94}$$

where $\nu_{\rm vib}$ is the vibrational frequency, and

$$Z_{\rm el} = \sum_n g_n \exp\left(\frac{-E_n}{kT}\right) = u \exp\left(\frac{-E_0}{kT}\right) , \tag{3.95}$$

where g_n is the statistical weight of level n of energy, E_n, the E_0 is the zero point energy (the energy of the ground state), and the ionic partition function, u, is given by

$$u = \sum_k g_k \exp\left(\frac{-E_k}{kT}\right) = g_0 + g_1 \exp\left(\frac{-E_1}{kT}\right) + \cdots , \tag{3.96}$$

where $E_n = E_k - E_0$ is the excitation energy of the ion in the nth state, E_k is the energy of level k above the ground state energy, and g_k is the statistical weight of level k. An energy level with angular momentum, J, has $g_k = 2J + 1$, and for hydrogen like atoms $g_k = 2k^2$. The statistical weight for the free electron is $g_e = 2$.

It follows from Eqs. (3.86) to (3.96) that the free energy of a nondegenerate, nonrelativistic, monoatomic gas is given by

$$F = -N_{\mathrm{tot}} kT \ln\left[\frac{eVg_0}{N_{\mathrm{tot}}}\left(\frac{2\pi MkT}{h^2}\right)^{3/2}\right] , \tag{3.97}$$

where g_0, the statistical weight of the ground level, is assumed to be equal to Z_{el}. Eq. (3.77) for the entropy of a monoatomic gas follows from Eqs. (3.85) and (3.97). The free energy and entropy for more complex systems may be derived in a similar way using Eqs. (3.85) to (3.95). The internal energy, U, is given by

$$U = \tfrac{3}{2} N_{\mathrm{tot}}\, kT \tag{3.98}$$

for a monoatomic gas,

$$U_{\mathrm{rot}} = N_{\mathrm{tot}}\, kT \quad \text{and} \quad \tfrac{3}{2} N_{\mathrm{tot}}\, kT \ ,$$

for the rotational energy of diatomic or linear polyatomic molecules, and non-linear polyatomic molecules, respectively, and

$$U_{\mathrm{vib}} = N_{\mathrm{tot}} \frac{h\nu_{\mathrm{vib}}}{\left[\exp\left(\frac{h\nu_{\mathrm{vib}}}{kT}\right) - 1\right]} ,$$

for the vibrational energy of N_{tot} identical oscillators (diatomic molecules).

For a system consisting of particles of one kind assumed to be fermions or bosons, we have

$$F = -kT \ln Z + \mu N_{\mathrm{tot}} = -PV + \mu N_{\mathrm{tot}} , \tag{3.99}$$

where the grand partition function, Z, is given by

$$\ln Z = \sum_k g(\varepsilon) \ln\{1 \pm \exp[(\mu - \varepsilon_k)/kT]\}^{\pm 1} , \tag{3.100}$$

where μ is the chemical potential for the kind of particle under consideration, ε_k is the energy of a single representative "particle" in the quantum state k, and $g(\varepsilon)$ is the statistical weight for the state k. The + sign is to be used for fermions, and the − sign for bosons. The total number of particles of the kind under consideration for the whole system is N_{tot}, the pressure is P, and V is the volume. If g_i denotes the statistical weight of the ith discrete state and $4\pi p^2 V\, dp/h^3$ is the statistical weight for the continuous translational states, then

$$g(\varepsilon) = g_i \frac{4\pi p^2 V\, dp}{h^3} ,$$

where p is the momentum and V is the volume.

3.2.7 Nernst Heat Theorem

The Nernst theorem (Nernst, 1906, 1926) states that the entropy of every thermodynamic system at a temperature of absolute zero can always be taken to be zero. In terms of the Boltzmann–Planck relation (3.76), $W = 1$ for $S = 0$ so that the thermodynamic state of a system at absolute zero corresponds to only one dynamical state. This ground state is that dynamical state of lowest energy which is compatible with the system. According to the Nernst theorem, in the limit as the temperature, T, approaches zero

$$S = \Delta S = C_p = C_v = 0 \quad \text{as} \quad T = 0 \ . \tag{3.101}$$

Nernst's theorem can be related to the phenomenological principle (Nernst, 1926) that it is impossible to cool any system down to absolute zero.

3.2.8 Fluctuations in Thermodynamic Quantities

The thermal motion of the particles of a gas produce fluctuations in the temperature, T, pressure, P, volume, V, energy, U, entropy, S, and the total number of particles, N_{tot}, of the gas. In general for any physical quantity, X, these fluctuations are described by the mean square fluctuation $\langle(\Delta X)^2\rangle$ about the mean $\langle X\rangle$, where the brackets $\langle\,\rangle$ denote a time average.

The mean square fluctuations, $\langle(\Delta T)^2\rangle$ and $\langle(\Delta V)^2\rangle$, of temperature, T, and volume, V, respectively, are given by Landau and Lifshitz (1969)

$$\begin{aligned} \langle(\Delta T)^2\rangle &= \frac{kT^2}{C_v} \ , \\ \langle(\Delta V)^2\rangle &= -kT\left(\frac{\partial V}{\partial P}\right)_T \ , \end{aligned} \tag{3.102}$$

where C_v is the specific heat at constant volume and the differential is carried out at constant temperature. The mean square fluctuation, $\langle(\Delta N)^2\rangle$, in the total number of particles, N, is given by

$$\langle(\Delta N)^2\rangle = kT\left(\frac{\partial N}{\partial \mu}\right)_{T,V} \ , \tag{3.103}$$

where the chemical potential, μ, is given by

$$\mu = \left(\frac{\partial U}{\partial N}\right)_{S,V} = -T\left(\frac{\partial S}{\partial N}\right)_{U,V} = \left(\frac{\partial F}{\partial N}\right)_{V,T} = \left(\frac{\partial G}{\partial N}\right)_{P,T} \ . \tag{3.104}$$

For photons in thermodynamic equilibrium (black body radiation) we have $\mu = 0$. For a nonrelativistic, monoatomic gas we have

$$\mu = kT \ln\left[\frac{N_{\mathrm{tot}}}{gV}\left(\frac{2\pi\hbar^2}{mkT}\right)^{3/2}\right] = kT \ln\left[\frac{P}{g(kT)^{5/2}}\left(\frac{2\pi\hbar^2}{m}\right)^{3/2}\right] \ ,$$

where N_{tot} is the total number of particles, m is the mass of one gas particle, and g is the statistical weight (degree of degeneracy) of the ground state. For example, for an electron gas $g = 2$. For an ideal gas, Eq. (3.103) gives $\langle(\Delta N)^2\rangle = N$.

The mean square fluctuations, $\langle(\Delta S)^2\rangle$ and $\langle(\Delta P)^2\rangle$, of entropy, S, and pressure, P, are given by

$$\langle(\Delta S)^2\rangle = C_p$$

and

$$\langle(\Delta P)^2\rangle = -kT\left(\frac{\partial P}{\partial V}\right)_S \ . \tag{3.105}$$

It follows from Eq. (3.103) that for a Boltzmann gas the mean square fluctuation of the number of particles in the kth quantum state is equal to the number of particles in the state. For a Fermi gas, the average number $\langle N_k\rangle$ of particles in the kth quantum state is related to its mean square fluctuation $\langle(\Delta N_k)^2\rangle$ by the relation

$$\langle(\Delta N_k)^2\rangle = \langle N_k\rangle(1 - \langle N_k\rangle) \ , \tag{3.106}$$

whereas for an Einstein-Bose gas (a photon gas)

$$\langle(\Delta N_k)^2\rangle = \langle N_k\rangle(1 + \langle N_k\rangle) \ .$$

For the photon gas, the mean square fluctuation, $\langle(\Delta E)^2\rangle$, for the energy, E, in the frequency range ν to $\nu + d\nu$ is given by Einstein (1925)

$$\langle(\Delta E)^2\rangle = h\nu E + \frac{c^3 E^2}{V 8\pi\nu^2\, d\nu} \ , \tag{3.107}$$

where V is the volume.

3.3 Statistical Properties and Equations of State

3.3.1 The Nondegenerate, Perfect Gas

3.3.1.1 Maxwell Distribution Function for Energy and Velocity

The number density, $n(\varepsilon)$, of perfect gas particles with kinetic energy between ε and $\varepsilon + d\varepsilon$ is given by

$$n(\varepsilon) = \frac{g(\varepsilon)}{V}\exp\left(\frac{\mu}{kT}\right)\exp\left(-\frac{\varepsilon}{kT}\right)d\varepsilon \ , \tag{3.108}$$

where $g(\varepsilon)$ is the number of possible particle states of energy, ε, and is given by

$$g(\varepsilon)\, d\varepsilon = \frac{4\pi g V}{h^3}(2M^3)^{1/2}\varepsilon^{1/2}\, d\varepsilon = \frac{4\pi V g}{h^3}p^2\, dp \ , \tag{3.109}$$

where g is the statistical weight of the discrete energy states, M is the particle rest mass, the temperature is T, the volume is V, and p is the gas momentum. The chemical potential, or Fermi energy, μ, is given in Eq. (3.104). The total number density of particles, N, is given by

$$N = \int_0^\infty n(\varepsilon)\, d\varepsilon = \frac{g}{2\pi^2\hbar^3}\int_0^\infty \exp\left(\frac{\mu}{kT}\right)\exp\left(-\frac{\varepsilon}{kT}\right)p^2\, dp \ , \tag{3.110}$$

where the momentum, p, is related to the kinetic energy, ε, by the equation

$$\varepsilon = (p^2c^2 + M^2c^4)^{1/2} - Mc^2 \ , \tag{3.111}$$

where M is the rest mass of the particle, and c is the velocity of light. For the nonrelativistic, nondegenerate, monatomic gas,

$$\exp\left[\frac{\mu}{kT}\right] = \frac{Nh^3}{g(2\pi MkT)^{3/2}} = \frac{P}{g(kT)^{5/2}}\left(\frac{2\pi\hbar^2}{M}\right)^{3/2} , \tag{3.112}$$

and Eq. (3.108) becomes

$$n(\varepsilon) = \frac{2N}{\pi^{1/2}(kT)^{3/2}}\varepsilon^{1/2}\exp\left(-\frac{\varepsilon}{kT}\right)d\varepsilon . \tag{3.113}$$

Eq. (3.113) was first obtained by Maxwell (1860), who also gave the correct formula for the probability distribution, $f(v)dv$, of gas particles with speeds between v and $v + dv$. If the energy, ε, is the kinetic energy, $Mv^2/2$, it follows from Eq. (3.113) that the number of particles with speeds between v and $v + dv$, for a nonrelativistic, nondegenerate gas, is given by

$$N_{\text{tot}} f(v)\, dv = N_{\text{tot}}\left(\frac{2}{\pi}\right)^{1/2}\left(\frac{M}{kT}\right)^{3/2} v^2 \exp\left[-\frac{Mv^2}{2kT}\right] dv , \tag{3.114}$$

where M is the mass of the atom or molecule, and N_{tot} is the total number of atoms and molecules. Both Eqs. (3.113) and (3.114) follow directly from the Maxwell-Boltzmann solution to Boltzmann's equation (3.4).

From Eq. (3.114) we have the relation

$$N_{\text{tot}}\int_0^\infty f(v)\, dv = N_{\text{tot}} . \tag{3.115}$$

The most probable speed, v_{p}, the mean speed, $\langle v\rangle$, and the root-mean-square speed, v_{rms}, are easily calculated by the relations

$$\begin{aligned} v_{\text{p}} &= (2kT/M)^{1/2} , \\ \langle v\rangle &= (8kT/\pi M)^{1/2} , \\ v_{\text{rms}} &= (3kT/M)^{1/2} = \langle v^2\rangle^{1/2} . \end{aligned} \tag{3.116}$$

The momentum distribution function for the nonrelativistic, nondegenerate gas may be obtained from Eq. (3.114) using $\varepsilon = p^2/(2m) = mv^2/2$. For the relativistic, nondegenerate gas the momentum $p = \varepsilon/c \gg mc$, and the number of particles with momenta between p and $p + dp$ is given by

$$N_{\text{tot}} f(p)\, dp = \frac{N_{\text{tot}}c^3}{8\pi(kT)^3}\exp\left(-\frac{pc}{kT}\right)\cdot 4\pi p^2\, dp . \tag{3.117}$$

3.3.1.2 The Energy Density and Equation of State of a Perfect Gas

It follows directly from Eq. (3.108) that the energy density, U/V, of a nondegenerate perfect gas is given by

$$\frac{U}{V} = \int_0^\infty \varepsilon n(\varepsilon)\, d\varepsilon = \frac{g}{2\pi^2\hbar^3}\int_0^\infty \exp\left[\left(\frac{\mu}{kT} - \frac{\varepsilon}{kT}\right)\right]\varepsilon p^2\, dp , \tag{3.118}$$

where ε is the kinetic energy, the number density, $n(\varepsilon)$, is given by Eqs. (3.108) or (3.113), the Fermi constant, μ, is given by Eq. (3.112), the temperature is T, and the momentum, p, is related to ε by Eq. (3.111). For a nonrelativistic, nondegenerate gas, Eq. (3.118) becomes (Maxwell, 1860)

$$U = \tfrac{3}{2} N_{\mathrm{tot}} kT \ , \tag{3.119}$$

where the total number of particles is N_{tot}. For the relativistic, nondegenerate gas,

$$U = 3N_{\mathrm{tot}} kT \ . \tag{3.120}$$

The entropy, S, of a nonrelativistic, nondegenerate, monoatomic gas is given in Eq. (3.77).

The entropy of a relativistic, nondegenerate gas is given by

$$S = kN_{\mathrm{tot}} \ln\left[\frac{T^4 g_0 8\pi k^4 e^4}{Pc^3 h^3}\right] \tag{3.121}$$

where g_0 is the statistical weight of the ground state of the atom, $e = 2.7182812$ is the base of the natural logarithm, and the pressure, P, of a relativistic, or nonrelativistic, nondegenerate gas is given by

$$PV = N_{\mathrm{tot}} kT \ .$$

For the relativistic, nondegenerate gas, $U = 3PV$. Alternative forms of the equation of state are given in Eq. (3.62), the most useful being

$$P = \frac{R\rho T}{\mu} = 8.314 \times 10^7 \frac{\rho T}{\mu} \ \text{dynes cm}^{-2} \ , \tag{3.122}$$

where here the mean molecular weight per free particle for an ionized gas is given by

$$\mu = \frac{\rho N_{\mathrm{A}}}{N} = \left[\sum_i \frac{X_i(Z_i + 1)}{A_i}\right]^{-1} \approx \frac{2}{1 + 3X + 0.5Y} \ , \tag{3.123}$$

where X_i is the relative mass abundance (mass fraction) of the atom i whose atomic number and atomic weight are, respectively, Z_i and A_i. The weight fraction of hydrogen is X, and the weight fraction of helium is Y. For a hydrogen gas $\mu = 0.5$, whereas for helium $\mu = 4/3$ at full ionization. For population one stars $X \approx 0.60$ and $Y \approx 0.38$. The gas mass density is ρ and the temperature is T. Typical values of the mass density, temperature, and linear extent of various cosmic gases are given in Table 3.1.

3.3.1.3 Boltzmann Equation for the Population Density of Excited States

Under conditions of local thermodynamic equilibrium at a temperature, T, the number density, N_s, of atoms or molecules in the excited state, s, is (Boltzmann, 1872)

$$N_s = \frac{g_s}{g_0} N_0 \exp\left(\frac{-\chi_s}{kT}\right) \ , \tag{3.124}$$

Table 3.1. Values of mass density, ρ, temperature, T, and linear extent, R, of various cosmic gases

Region	ρ (g cm^{-3})	T (°K)	R (cm)
Ionosphere	10^{-20}–10^{-10}	200–1,500	6.4×10^{8}
Magnetosphere	10^{-21}	10^{4}	10^{9}–10^{11}
Sun (stars)	1.4 (mean)	10^{4}–10^{7}	10^{11}
Solar corona	10^{-19}–10^{-16}	10^{6}	10^{12}
Solar system	10^{-23}	10^{5}	10^{15}
Galactic nebulae	10^{-19}–10^{-16}	10^{2}–10^{4}	10^{18}–10^{21}
Galaxy	10^{-24}	10^{2}–10^{4}	10^{23}
Local cluster	10^{-27}	10^{5}?	3×10^{24}
Universe	10^{-29}	10^{5}–10^{6}?	3×10^{28}

where χ_s is the excitation energy (energy above ground level) of the s state, N_0 is the number of molecules in the ground state, and g_s is the statistical weight of the s state. An energy level with total angular quantum numbers, J, has $g_s = 2J + 1$, and for a multiplet term with quantum number L and S, the $g_s = (2S + 1)(2L + 1)$. Eq. (3.124) governs the distribution of particles over different energy states, and is known as the Boltzmann formula.

If the total number density of atoms or molecules is N_{tot},

$$N_s = g_s \frac{N_{tot}}{u} \exp\left(\frac{-\chi_s}{kT}\right) , \tag{3.125}$$

where the partition function $u = \sum g_r \exp(-\chi_r/kT)$. For free particles, u becomes Z_{trans} given by Eq. (3.91).

The population of levels, m and n, with energies, E_m and E_n, are related by

$$\frac{N_n}{g_n} = \frac{N_m}{g_m} \exp\left(-\frac{E_n - E_m}{kT}\right) . \tag{3.126}$$

3.3.1.4 The Saha–Boltzmann Ionization Equation

Under conditions of local thermodynamic equilibrium, the number density, N_r, of atoms in the rth stage of ionization is related to that of the $(r + 1)$th stage, N_{r+1}, by Saha (1920, 1921)

$$\frac{N_{r+1}}{N_r} N_e = \frac{u_{r+1}}{u_r} \frac{2(2\pi mkT)^{3/2}}{h^3} \exp\left(\frac{-\chi_r}{kT}\right) , \tag{3.127}$$

where N_e is the free electron density, u_r is the partition function of the rth stage, and is given by Eq. (3.96), χ_r is the energy required to remove an electron from the ground state of the r-times ionized atom (its ionization potential), T is the temperature of thermal equilibrium, m is the electron mass, and h and k are, respectively, Planck's and Boltzmann's constants.

This relation, known as the Saha equation, relates the degree of ionization of an atom to temperature and pressure; the electron density N_e can be expressed in terms of the electron pressure $P_e = N_e kT$. The Saha equation therefore indicates that the relative intensities of spectral lines are caused, in part, by differences in pressure and temperature in stellar atmospheres – see Kippenhahn and Weigert (1990) for a contemporary treatment.

Table 3.2. Ionization potentials of the elements in electron volts (one eV = 1.6021×10^{-12} erg and $k = 8.617 \times 10^{-5}$ eV °K^{-1})[1]

Z	Element	Stage of Ionization I	II	III	IV	V	VI	VII	VIII	IX	X
1	H	13.598									
2	He	24.587	54.416								
3	Li	5.392	75.638	122.451							
4	Be	9.322	18.211	153.893	217.713						
5	B	8.298	25.154	37.930	259.368	340.217					
6	C	11.260	24.383	47.887	64.492	392.077	489.981				
7	N	14.534	29.601	47.448	77.472	97.888	552.057	667.029			
8	O	13.618	35.116	54.934	77.412	113.896	138.116	739.315	871.387		
9	F	17.422	34.970	62.707	87.138	114.240	157.161	185.182	953.886	1,103.089	
10	Ne	21.564	40.962	63.45	97.11	126.21	157.93	207.27	239.09	1,195.797	1,362.164
11	Na	5.139	47.286	71.64	98.91	138.39	172.15	208.47	264.18	299.87	1,465.091
12	Mg	7.646	15.035	80.143	109.24	141.26	186.50	224.94	265.90	327.95	367.53
13	Al	5.986	18.828	28.447	119.99	153.71	190.47	241.43	284.59	330.21	398.57
14	Si	8.151	16.345	33.492	45.141	166.77	205.05	246.52	303.17	351.10	401.43
15	P	10.486	19.725	30.18	51.37	65.023	220.43	263.22	309.41	371.73	424.50
16	S	10.360	23.33	34.83	47.30	72.68	88.049	280.93	328.23	379.10	447.09
17	Cl	12.967	23.81	39.61	53.46	67.8	97.03	114.193	348.28	400.05	455.62
18	Ar	15.759	27.629	40.74	59.81	75.02	91.007	124.319	143.456	422.44	478.68
19	K	4.341	31.625	45.72	60.91	82.66	100.0	117.56	154.86	175.814	503.44
20	Ca	6.113	11.871	50.908	67.10	84.41	108.78	127.7	147.24	188.54	211.70
21	Sc	6.54	12.80	24.76	73.47	91.66	111.1	138.0	158.7	180.02	225.32
22	Ti	6.82	13.58	27.491	43.266	99.22	119.36	140.8	168.5	193.2	215.91
23	V	6.74	14.65	29.310	46.707	65.23	128.12	150.17	173.7	205.8	230.5
24	Cr	6.766	16.50	30.96	49.1	69.3	90.56	161.1	184.7	209.3	244.4
25	Mn	7.435	15.640	33.667	51.2	72.4	95	119.27	196.46	221.8	243.3
26	Fe	7.870	16.18	30.651	54.8	75.0	99	125	151.06	235.04	262.1
27	Co	7.86	17.06	33.50	51.3	79.5	102	129	157	186.13	276
28	Ni	7.635	18.168	35.17	54.9	75.5	108	133	162	193	224.5
29	Cu	7.726	20.292	36.83	55.2	79.9	103	139	166	199	232
30	Zn	9.394	17.964	39.722	59.4	82.6	108	134	174	203	238
31	Ga	5.999	20.51	30.71	64						
32	Ge	7.899	15.934	34.22	45.71	93.5					
33	As	9.81	18.633	28.351	50.13	62.63	127.6				
34	Se	9.752	21.19	30.820	42.944	68.3	81.70	155.4			
35	Br	11.814	21.8	36	47.3	59.7	88.6	103.0	192.8		
36	Kr	13.999	24.359	36.95	52.5	64.7	78.5	111.0	126	230.9	
37	Rb	4.177	27.28	40	52.6	71.0	84.4	99.2	136	150	277.1
38	Sr	5.695	11.030	43.6	57	71.6	90.8	106	122.3	162	177
39	Y	6.38	12.24	20.52	61.8	77.0	93.0	116	129	146.2	191
40	Zr	6.84	13.13	22.99	34.34	81.5					
41	Nb	6.88	14.32	25.04	38.3	50.55	102.6	125			
42	Mo	7.099	16.15	27.16	46.4	61.2	68	126.8	153		

[1] From C.E. Moore (1970)

Stage of Ionization										
XI	XII	XIII	XIV	XV	XVI	XVII	XVIII	XIX	XX	XXI
1,648.659										
1,761.802	1,962.613									
442.07	2,085.983	2,304.080								
476.06	523.50	2,437.676	2,673.108							
479.57	560.41	611.85	2,816.943	3,069.762						
504.78	564.65	651.63	707.14	3,223.836	3,494.099					
529.26	591.97	656.69	749.74	809.39	3,658.425	3,946.193				
538.95	618.24	686.09	755.73	854.75	918	4,120.778	4,426.114			
564.13	629.09	714.02	787.13	861.77	968	1,034	4,610.955	4,933.931		
591.25	656.39	726.03	816.61	895.12	974	1,087	1,157	5,129.045	5,469.738	
249.832	685.89	755.47	829.79	926.00						
265.23	291.497	787.33	861.33	940.36						
255.04	308.25	336.267	895.58	974.02						
270.8	298.0	355	384.30	1,010.64						
286.0	314.4	343.6	404	435.3	1,136.2					
290.4	330.8	361.0	392.2	457	489.5	1,266.1				
305	336	379	411	444	512	546.8	1,403.0			
321.2	352	384	430	464	499	571	607.2	1,547		
266	368.8	401	435	484	520	557	633	671	1,698	
274	310.8	419.7	454	490	542	579	619	698	738	1,856
324.1										
206	374.0									

Table 3.2 (continued)

Z	Element	Stage of Ionization									
		I	II	III	IV	V	VI	VII	VIII	IX	X
43	Tc	7.28	15.26	29.54							
44	Ru	7.37	16.76	28.47							
45	Rh	7.46	18.08	31.06							
46	Pd	8.34	19.43	32.93							
47	Ag	7.576	21.49	34.83							
48	Cd	8.993	16.908	37.48							
49	In	5.786	18.869	28.03	54						
50	Sn	7.344	14.632	30.502	40.734	72.28					
51	Sb	8.641	16.53	25.3	44.2	56	108				
52	Te	9.009	18.6	27.96	37.41	58.75	70.7	137			
53	I	10.451	19.131	33							
54	Xe	12.130	21.21	32.1							
55	Cs	3.894	25.1								
56	Ba	5.212	10.004								
57	La	5.577	11.06	19.175							
58	Ce	5.47	10.85	20.20	36.72						
59	Pr	5.42	10.55	21.62	38.95	57.45					
60	Nd	5.49	10.72								
61	Pm	5.55	10.90								
62	Sm	5.63	11.07								
63	Eu	5.67	11.25								
64	Gd	6.14	12.1								
65	Tb	5.85	11.52								
66	Dy	5.93	11.67								
67	Ho	6.02	11.80								
68	Er	6.10	11.93								
69	Tm	6.18	12.05	23.71							
70	Yb	6.254	12.17	25.2							
71	Lu	5.426	13.9								
72	Hf	7.0	14.9	23.3	33.3						
73	Ta	7.89									
74	W	7.98									
75	Re	7.88									
76	Os	8.7									
77	Ir	9.1									
78	Pt	9.0	18.563								
79	Au	9.225	20.5								
80	Hg	10.437	18.756	34.2							
81	Tl	6.108	20.428	29.83							
82	Pb	7.416	15.032	31.937	42.32	68.8					
83	Bi	7.289	16.69	25.56	45.3	56.0	88.3				

For computational purposes, it is useful to have Eq. (3.127) expressed in the logarithmic form

$$\log\left(\frac{N_{r+1}N_{\mathrm{e}}}{N_r}\right) = \log\left(\frac{u_{r+1}}{u_r}\right) + 15.6826 + \frac{3}{2}\log T - \frac{5039.95}{T}I \tag{3.128}$$

or

$$\log\left(\frac{N_{r+1}}{N_r}P_{\mathrm{e}}\right) = \log\left(\frac{2u_{r+1}}{u_r}\right) - 0.4772 + \frac{5}{2}\log T - \frac{5039.95}{T}I \ ,$$

where T is in °K, the ionization potential, I, from the r to $r+1$st stage is in electron volts, and the electron pressure $P_{\mathrm{e}} = N_{\mathrm{e}}kT$. The ionization energy is the least energy necessary to remove to infinity one electron from an atom of the element. The ionization potentials, or energies, for elements in different stages of ionization are given in Table 3.2. The ground state configurations and ionization energies of the elements are also given in the CRC Handbook of Chemistry and Physics, in the Review of Particle Properties (Phys. Rev. **D50**, 1244, 1994), and in the Atomic Spectroscopic Database of the National Institute of Standards and Technology (NIST) at their World Wide Web site http://physics.nist.gov/ under Physical Reference Data. The NIST data give ionization energies in units of cm^{-1}, and the other sources give them in units of electron volts, or eV. For conversion, 1 eV $= 1.60217733\ (49) \times 10^{-12}$ erg and the photon energy associated with a wavelength λ is $hc/\lambda = 1.98648 \times 10^{-16}/\lambda$ for λ in cm. Thus, for example, the ionization energy for helium is given as 24.5874 eV or 198310.77227 cm^{-1}.

The population density, N_n, of the nth quantum level is given by (cf. Eq. (3.125) and Eq. (3.127))

$$N_n = N_{\mathrm{e}}N_{\mathrm{i}}\frac{h^3}{(2\pi mkT)^{3/2}}\frac{g_n}{2}\exp\left(\frac{\chi_r - \chi_n}{kT}\right) \ , \tag{3.129}$$

where N_{e} and N_i are, respectively, the free electron and ion densities, g_n is the statistical weight of the nth level, and χ_n is the excitation energy of the nth level above ground level. For hydrogen like atoms, the excitation energy of the nth state is given by

$$\chi_n = I_{\mathrm{H}}Z^2\left(1 - \frac{1}{n^2}\right) \ , \tag{3.130}$$

and the statistical weight is given by

$$g_n = 2n^2 \ . \tag{3.131}$$

Here $I_{\mathrm{H}} \approx 13.5$ eV $= 21.36 \times 10^{-12}$ erg $= \chi_r/Z^2 = 2\pi^2 e^4 m/h^2$ is the ionization potential of hydrogen.

For the first ionization of hydrogen and helium, Eq. (3.127) takes the numerical form

$$\frac{N_1 N_{\mathrm{e}}}{N_0} \approx 2.4 \times 10^{15}\ T^{3/2}\exp(-1.58 \times 10^5 T^{-1})\ \mathrm{cm}^{-3} \tag{3.132}$$

for hydrogen, and

$$\frac{N_1 N_e}{N_0} \approx 9.6 \times 10^{15}\, T^{3/2} \exp(-2.85 \times 10^6 T^{-1})\ \mathrm{cm}^{-3} \ ,$$

for helium. It is sometimes convenient to rewrite Eq. (3.127) in the form relating particle concentrations $\alpha_r = N_r/N_{\mathrm{tot}}$, where N_r and N_{tot} denote the total number of atoms in the rth state and the total number of atoms before ionization, respectively.

$$\frac{\alpha_{r+1}\alpha_e}{\alpha_r} = \frac{V}{N_{\mathrm{tot}}} \frac{N_{r+1}N_e}{N_r} \ , \tag{3.133}$$

where V is the gas volume and $N_{r+1}N_e/N_r$ is given by Eq. (3.127). For a singly ionized region $\alpha = \alpha_1 = \alpha_e$ and the degree of ionization is given by

$$\frac{\alpha^2}{1-\alpha} = 2\frac{u_1}{u_0}\frac{V}{N_{\mathrm{tot}}}\left(\frac{2\pi mkT}{h^2}\right)^{3/2} \exp\left(\frac{-I}{kT}\right) \ , \tag{3.134}$$

where $(2\pi mkT/h^2) = 2.4 \times 10^{15}\ T$, $u_1/u_0 = \frac{1}{2}$ for hydrogen and 2 for helium, I is the ionization potential, N_{tot} is the total number of atoms plus free protons for the case of hydrogen, and $\alpha = N_e/N_{\mathrm{tot}}$ for hydrogen. The total number of particles is $N_{\mathrm{tot}}(1+\alpha)$ so that the gas pressure, P, is given by

$$P = N_{\mathrm{tot}}(1+\alpha)\frac{kT}{V} \ , \tag{3.135}$$

whereas the energy, U, is given by

$$U = \tfrac{3}{2}N_{\mathrm{tot}}(1+\alpha)kT + N_{\mathrm{tot}}\alpha(1+\chi_m) \ , \tag{3.136}$$

where I is the ionization potential, and χ_m is the electron excitation energy of the ion.

3.3.1.5 Strömgren Radius for the Sphere of Ionization

Although most of the hydrogen atoms in interstellar space are neutral, or unionized, the ultraviolet radiation from very hot stars (particularly *O* and *B* stars) can generate enormous but sharply bounded spheres of ionization. For some years these regions were referred to as "Strömgren spheres" after Bengt Strömgren's derivation of the relationship between the number density of interstellar hydrogen, N, the temperature T, of the exciting star, and the radius s_0, of the sphere of ionization (Strömgren, 1939). These regions are now almost invariably called H II regions, although their size is designated as the Strömgren radius, illustrated in Table 3.3. Osterbrock (1974, 1989) and Spitzer (1978) review the photoionization equilibrium and thermal equilibrium of H II regions.

In the immediate neighborhood of a source of ultraviolet radiation, the neutral interstellar hydrogen will become ionized to form an ionized hydrogen region (H II region) with a Strömgren radius, s_0, given by Strömgren (1939); Vandervoort (1963); Osterbrock (1989); Shu, (1991).

$$s_0 = \left(\frac{3S}{4\pi\alpha N^2}\right)^{1/3} \ , \tag{3.137}$$

Table 3.3. Calculated radii, s_0, of Strömgren spheres, or H II regions, for stars of different spectral type and star surface temperature, T_s, assuming an interstellar electron density of $N_e = N_p = 1\ \mathrm{cm}^{-3}$. Here S is the quantity of ultraviolet photons per second, and the Strömgren radius, s_0, is given in parsecs, or pc, with 1 pc = 3.085678×10^{18} cm[1]

Spectral type	T_s (K)	Log S (photons/sec)	s_0 (pc)
O5	48,000	49.67	108
O6	40,000	49.23	74
O7	35,000	48.84	56
O8	33,500	48.60	51
O9	32,000	48.24	34
O9.5	31,000	47.95	29
B0	30,000	47.67	23
B0.5	26,200	46.83	12

[1] Adapted from Osterbrock (1974, 1989)

where α is the recombination coefficient of hydrogen to all states except the ground state, and is given by Kaplan and Pikelner (1970)

$$\alpha \approx 2.6 \times 10^{-13} \left(\frac{10^4}{T}\right)^{0.85} \mathrm{cm}^3\ \mathrm{sec}^{-1} ,$$

the number density, N, of the H II region is $N = \rho/m_{\mathrm{H}}$, where ρ is the gas mass density, m_{H} is the mass of the hydrogen atom, and S is the rate of emission of the ionizing photons from the exciting star

$$S \approx \frac{8\pi^2 R^2 kT}{hc^2} \nu_c^2 \exp\left(\frac{-h\nu_c}{kT}\right) ,$$

where R is the radius of the exciting star, T is the temperature of the star, and $\nu = 3.29 \times 10^{15}$ Hz is the frequency of the limit of the Lyman continuum. Eq. (3.137) indicates that s_0 is a function of N and the spectral class of the exciting star, and this dependence is given by Hershberg and Pronik (1959). These results indicate that $s_0 = UN_e^{-2/3}$ where N_e is the electron density and U varies from 90 to 12 pc cm^{-2} as the spectral class goes from $O5$ to $B1$. Strömgren (1939) first derived a detailed formula for s_0, which is

$$\log s_0 = -6.17 + \frac{1}{3} \log\left[(T_e/T)^{1/2}\left(\frac{2g_1}{g_0}\right)\right] - \frac{1}{3}\log a_u - \frac{1}{3}\theta I + \frac{1}{2}\log T + \frac{2}{3}\log R - \frac{2}{3}\log N , \tag{3.138}$$

where s_0 is in parsecs (1 pc = 3×10^{18} cm), T_e is the electron temperature at s_0, the star's temperature is T, the absorption coefficient for the ionizing radiation per neutral gas atom is a_u, the factor $\theta = 5040/T$, the ionization potential is I, the stellar radius is R solar radii ($R_\odot = 6.96 \times 10^{10}$ cm), and N is the number density of neutral and ionized gas. Using the proper values for hydrogen, $I = 13.5984$ volts and $a_u = 6.3 \times 10^{-18}\ \mathrm{cm}^{-2}$, Eq. (3.138) becomes

$$\log s_0 = -0.44 + \tfrac{1}{3}\log[(T_e/T)^{1/2}] - 4.51\theta + \tfrac{1}{2}\log T + \tfrac{2}{3}\log R - \tfrac{2}{3}\log N . \tag{3.139}$$

Calculations to determine the temperature of a H II region by taking into account energy gains and losses by photoionization, collisions, radiation transfer etc, have been carried out by Spitzer and Savedoff (1950), Osterbrock (1965), Hjellming (1966, 1968), and Rubin (1968) for different exciting stars and hydrogen densities. Strömgren's paper dealt with the ionization balance for nebulae ionized by the ultraviolet radiation of hot massive stars. Later work showed that the H II regions, with temperatures of $T \approx 10,000$ °K, are about 100 times hotter than the temperatures of the surrounding neutral gas, H I regions, with $T \approx 100$ °K (Spitzer and Savedoff, 1950). The ionized nebulae will expand as the result of the pressure difference between the ionized and neutral material.

The ionized gas of a H II region is then separated from the surrounding neutral H I gas by an ionization front and its associated shock which move outward. The conditions ahead and behind of the ionization boundary are discussed in the same terminology as shock fronts, given in Section 3.5.9. The critical parameters are the gas pressure, $P = \rho kT/(\mu m_{\mathrm{H}})$ for mass density ρ, temperature T, and mean molecular weight μ, and the speed of sound s. For the H II region, $T \approx 10^4$ °K and $\mu = 1/2$, for the H I gas $T \approx 10^2$ °K and $\mu = 1$. The speed of sound $s = (\gamma P/\rho)^{1/2}$ for an adiabatic index γ, where $\gamma = 5/3$ for a monoatomic gas.

Ionization fronts and shocks of expanding H II regions are reviewed by Mathews and O'Dell (1969), Spitzer (1978), Yorke (1986), Osterbrock (1989) and Shu (1992). Earlier papers dealing with the expansion of H II regions into the surrounding neutral hydrogen are given by Vandervoort (1963), Mathews (1965) and Lasker (1966, 1967). The results indicate that the ionization front obtains the radius s_0 in about $(10^4/N)$ years and subsequently has a radius, r, given by Eq. (3.137) for s_0 when N is the density of the H II region. This means that $N = N_0 s_0^{3/2} r^{-3/2}$, where s_0 and N_0 are, respectively, the Strömgren radius and the density at time $t = 0$. At a subsequent time, t, the radius r, is given by (Spitzer, 1978)

$$r = s_0 \left(1 + \frac{7C_1 t}{4s_0}\right)^{4/7} , \tag{3.140}$$

and the density, N, is given by

$$N = N_0 \left(1 + \frac{7C_1 t}{4s_0}\right)^{-6/7} ,$$

where the $C_1 = 2s/\sqrt{3}$ and s is the velocity of sound in the H II region. The kinetic energy, E_{k}, transferred to the surrounding neutral region is given by

$$E_{\mathrm{k}} = \frac{4\pi}{3} N_0 m_{\mathrm{H}} s^2 s_0^3 \left[\left(\frac{r}{s_0}\right)^{3/2} - 1\right] = \frac{2}{3}\left[1 - \left(\frac{r}{s_0}\right)^{-3/2}\right] E_{\mathrm{T}} , \tag{3.141}$$

where E_{T} is the thermal energy in the H II region. Some of this energy goes into ionizing the neutral hydrogen. The resultant loss of energy, E_{L}, is given by

$$E_{\mathrm{L}} = \frac{-4\pi}{7} N_0 m_{\mathrm{H}} s^2 s_0^3 \ln\left[\left(\frac{7}{4} s s_0^{3/4} t + s_0^{7/4}\right) s_0^{-7/4}\right] . \tag{3.142}$$

The thickness, Δr, of the H I shell is given by

$$\Delta r = \frac{1}{6}\left[\left(\frac{7}{4}ss_0^{3/4}t + s_0^{7/4}\right)^{6/7} - s_0^{3/2}\right]\left(\frac{7}{4}ss_0^{3/4}t + s_0^{7/4}\right)^{-2/7} \approx \frac{r}{6} \quad \text{for } t \rightarrow \infty \ . \tag{3.143}$$

The variation of velocity, number density, temperature, and pressure of an H II region as a function of distance from the exciting star are illustrated in Fig. 3.1 for an 0 star of $T = 42,000\,^\circ$K and after 3×10^4 and 6×10^4 years.

After a massive, early-type star is born and begins to emit ultraviolet radiation, it takes a few thousand years to ionize the surrounding hydrogen out to the initial Strömgren radius, s_{oi}, given by Eq. (3.137). For an *O5* star born into a region of density $N \approx 10^2$ cm^{-3}, for a typical dense cloud, the $s_{\text{oi}} \approx 6$ pc. The nascent H II region then grows in size as an ionization front and its associated shock expand against the surrounding neutral medium, with a

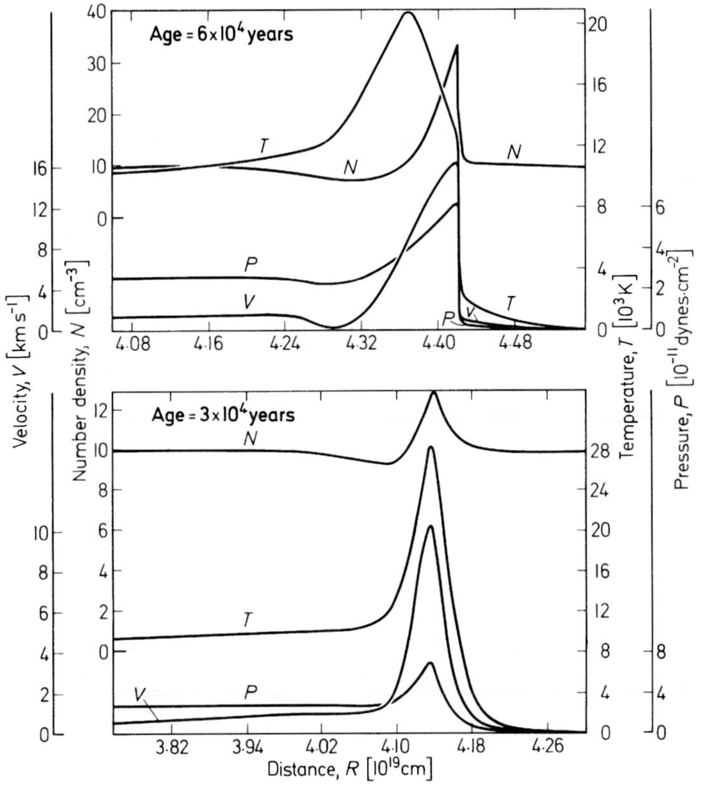

Fig. 3.1. Variation of velocity, V, number density, N, temperature, T, and pressure, P, in the vicinity of an ionization front as a function of distance, R, from an 0 star with temperature $T = 41,958\,^\circ$K after 3.07×10^4 and 6.16×10^4 years. The 30 $M_\odot$ star is assumed to be formed in an infinite medium of neutral hydrogen at rest at 100 °K, and the star's photon luminosity $L = 8.75 \times 10^{48}$ ultraviolet photons per second while on the main sequence [after Mathews, 1965, by permission of the American Astronomical Society and the University of Chicago Press]

radius given by Eq. (3.140) when the sound speed $s = C_1 \approx 10$ km s^{-1}. The expansion lowers the density of the ionized sphere, thus reducing the recombination rate and making more photons available to ionize the gas. Eventually the pressure difference between the ionized and neutral regions will drop to zero with a final radius, s_{of}, given by Eq. (3.137) for $N = 1$ to $10\,\mathrm{cm}^{-3}$ or $s_{of} \approx (5$ to $30)$ s_{oi}. The time to expand to this size at a velocity of about 10 km s^{-1} is comparable to the lifetime of massive stars, or around a few million years, so the H II region might not reach its "final" radius and may eventually be disrupted by the supernova explosion of its progenitor star.

3.3.2 The Degenerate Gas – Number Density, Energy Density, Entropy Density, and the Equation of State

3.3.2.1 Fermi–Dirac Statistics and Functions

In a degenerate gas, the particles are packed together with the maximum density allowed by quantum mechanics for a given pressure. The packing is so tight that the Pauli exclusion principle, rather than Coulomb repulsion, keeps the particles apart. This principle states that one system can never have two elements with exactly the same set of quantum numbers (Pauli, 1927). When the electrons in a degenerate electron gas, or the neutrons in a degenerate neutron gas, are squeezed into too small a volume, they dart away from each other at high speeds to avoid invading each others territory and occupying the same quantum state. This exerts an outward push, called degeneracy pressure, that resists further compression and prevents the electrons or neutrons from being packed too closely together.

An overview of dense astrophysical plasmas is provided by Van Horn (1991); they are present in stars formed at the endpoints of stellar evolution, the white dwarfs and neutron stars, as well as in the brown dwarfs, very cool objects with masses too low for nuclear reactions to be ignited in their cores. A detailed discussion of the properties of dense plasmas is given by Ichimaru (1982). The physics of white dwarfs and neutron stars is discussed by Shapiro and Teukolsky (1983) and Kippenhahn and Weigert (1990). The cooling of white dwarf stars and neutron stars are respectively reviewed by D'Antona and Mazzitelli (1990) and Pethick (1992). The astrophysical theory of the hypothetical brown dwarfs is given in Kafatos, Harrington and Maran (1986); the current status of both brown dwarf theory and searches for these elusive substellar objects are reviewed by Burrows and Liebert (1993).

Elementary particles whose spin is an odd multiple of $\hbar/2$ (electrons, protons, and nuclei with odd mass numbers) obey Fermi–Dirac statistics (Fermi, 1926; Dirac, 1926). Using Pauli's exclusion principle property, Fermi derived the following formula for the number density, $n(\varepsilon)$, of particles with kinetic energy between ε and $\varepsilon + d\varepsilon$ in a volume V

$$Vn(\varepsilon) = \frac{g(\varepsilon)\,d\varepsilon}{\exp\left[\frac{-\mu}{kT} + \frac{\varepsilon}{kT}\right] + 1}\,, \tag{3.144}$$

where $g(\varepsilon)$ is the number of possible particle states of energy, ε, the temperature is T, and the Fermi constant or chemical potential, μ, is given by Eq. (3.104).

For a non-relativistic particle of mass, m, Eq. (3.144) becomes

$$n(\varepsilon) = \frac{2\pi(2m)^{3/2}}{h^3} \frac{g\varepsilon^{1/2}\,d\varepsilon}{\exp\left[\frac{-\mu}{kT} + \frac{\varepsilon}{kT}\right] + 1} \ . \tag{3.145}$$

The total number of particles per unit volume, N, is given by

$$N = \int_0^\infty n(\varepsilon)\,d\varepsilon = \frac{g}{2\pi^2\hbar^3} \int_0^\infty \frac{p^2\,dp}{\exp\left[\frac{-\mu}{kT} + \frac{\varepsilon}{kT}\right] + 1} \ . \tag{3.146}$$

where g is the statistical weight for discrete energy levels and ε is related to the momentum, p, by Eq. (3.111). For the nonrelativistic, degenerate gas ($\varepsilon = p^2/(2m)$, $p \ll mc$, $kT \ll mc^2$)

$$N = \frac{2\pi(2mkT)^{3/2} g}{h^3} \int_0^\infty \frac{x^{1/2}\,dx}{\exp\left[x + \frac{-\mu}{kT}\right] + 1} \ . \tag{3.147}$$

The total energy density, U/V, is given by

$$\frac{U}{V} = \int_0^\infty \varepsilon n(\varepsilon)\,d\varepsilon = \frac{g}{2\pi^2\hbar^3} \int_0^\infty \frac{\varepsilon p^2\,dp}{\exp\left[\frac{-\mu}{kT} + \frac{\varepsilon}{kT}\right] + 1} \ . \tag{3.148}$$

For the nonrelativistic, degenerate gas,

$$\frac{U}{V} = \frac{4\pi}{h^3}(2m)^{3/2}(kT)^{5/2} F_{3/2}\left(\frac{\mu}{kT}\right) = \frac{3}{2} P_e \ , \tag{3.149}$$

where P_e is the electron pressure, and the Fermi–Dirac function

$$F_{3/2}\left(\frac{\mu}{kT}\right) = \int_0^\infty \frac{x^{3/2}\,dx}{\exp\left[x + \frac{-\mu}{kT}\right] + 1} \ .$$

Tables of $F_{3/2}(\mu/kT)$ are given by McDougal and Stoner (1938) and reproduced by Kippenhahn and Weigert (1990). Even at zero temperature, the energy is finite and is given by

$$U_0 = \left(\frac{6\pi^2}{g}\right)^{2/3} \frac{\hbar^2}{2m} N_e^{2/3} = \mu,$$
$$\approx 5.0 \times 10^{-27}\, N_e^{2/3} \text{ erg} \ . \tag{3.150}$$

The chemical potential of a gas at absolute zero is the same as the energy of the electrons. Here N_e is the electron density, and we may write in general

$$U_0 = E_F = \left\{ \left[1.018 \left(\frac{\rho}{\mu_e}\right)^{2/3} \times 10^{-4} + 1 \right]^{1/2} - 1 \right\} mc^2$$

$$= 0.509 \left(\frac{\rho}{\mu_e}\right)^{2/3} \times 10^{-4} mc^2 \quad \text{for nonrelativistic electrons}$$

$$= 1.009 \left(\frac{\rho}{\mu_e}\right)^{1/3} \times 10^{-2} mc^2 \quad \text{for relativistic electrons} \ .$$

Here ρ is the gas mass density, μ_e is the mean electron molecular weight, and the symbol E_F is used to indicate that this quantity is often called the Fermi energy of a completely degenerate gas. The maximum electron kinetic energy is E_F, whereas the average kinetic energy is $3E_F/5$. The electron density, N_e, may be written as

$$N_e = \frac{8\pi}{3h^3}(mc)^3\left[\left(\frac{E_F}{mc^2}+1\right)^2-1\right]^{3/2}$$

or

$$N_e = \frac{\rho N_A}{\mu_e} = \rho N_A \sum_i \frac{X_i Z_i}{A_i} \ , \tag{3.151}$$

where the mass density $\rho = N_e \mu_e m_u$, the atomic mass unit $m_u = 1.6605402(10) \times 10^{-24}$ grams, the reciprocal of Avogadro's number $= N_A = 6.0221367(36) \times 10^{23}$ mole^{-1}. The X_i, Z_i, and A_i are, respectively, the mass fraction, atomic number, and atomic weight of element i. The transition from Fermi–Dirac to Boltzmann statistics occurs at the degeneracy temperature given by

$$T_0 = \frac{E_F}{k} = \frac{1}{8}\left(\frac{3}{\pi}\right)^{2/3}\frac{h^2}{mk}N_e^{2/3} = 4.35 \times 10^{-11} N_e^{2/3}\ {}^\circ\mathrm{K} \ . \tag{3.152}$$

The entropy, S, per unit volume, V, for the nonrelativistic, degenerate gas is given by

$$\frac{S}{V} = \frac{5P_e}{2T} - \frac{\mu}{kT}kN_e = kN_e\left[\frac{5}{2}\frac{\frac{2}{3}F_{3/2}\left(\frac{\mu}{kT}\right)}{F_{1/2}\left(\frac{\mu}{kT}\right)} - \frac{\mu}{kT}\right] \ . \tag{3.153}$$

For a relativistic degenerate gas we have for $\varepsilon = pc$, $p \gg mc$, and $kT \gg mc^2$, with

$$N_e = \frac{8\pi}{c^3h^3}(kT)^3 F_2\left(\frac{\mu}{kT}\right) \ ,$$

$$\frac{U}{V} = \frac{8\pi}{c^3h^3}(kT)^4 F_3\left(\frac{\mu}{kT}\right) = 3P_e \ ,$$

and

$$\frac{S}{V} = 4\frac{P_e}{T} - kN_e\left(\frac{\mu}{kT}\right) = kN_e\left[4 \times \frac{\frac{1}{3}F_3\left(\frac{\mu}{kT}\right)}{F_2\left(\frac{\mu}{kT}\right)} - \left(\frac{\mu}{kT}\right)\right] \ . \tag{3.154}$$

where the Fermi–Dirac integral $F_n(\psi) = \int_0^\infty x^n[\exp(x-\psi)+1]^{-1}\,dx$, and numerical values of F_2 and F_3 are given by Kippenhahn and Weigert (1990).

3.3.2.2 Equation of State of a Degenerate Electron Gas – White Dwarf Stars

The high temperatures and low luminosities of white dwarf stars, discovered more than half a century ago (Adams, 1914, 1915), imply that they are small, a direct consequence of the Stefan–Boltzmann law for thermal radiation (Eq. 1.127). The mean radius of white dwarfs is one hundredth that of the Sun, so they are dwarf-like stars, comparable to the size of the Earth. Their rather ordinary masses of a little more than half that of the Sun, imply an enormously high mass density of $\rho \approx 10^5$ g cm^{-3}. Since the magnetic flux, HR^2, is conserved during stellar collapse, where H is the magnetic field strength and R is the radius, the magnetic field strength of white dwarf stars should be amplified to about a million Gauss. The intense, million-Gauss magnetic fields observed on white dwarf stars (Kemp et al., 1970; Sect. 2.17) provide additional evidence that white dwarfs are gravitationally collapsed normal stars with the small radii and large densities inferred under the assumption that they emit thermal radiation.

Lang (1992) provides the astrometric masses and gravitational redshifts (also see Sect. 2.17) for some white dwarfs, together with a catalogue of the positions, visual magnitudes, proper motions and distances of more than 500 of them. Shipman (1972, 1977) also provides masses and radii for white dwarf stars, and Shipman, Provencal, Høg and Thejll (1997) give accurate values for 40 Eri B.

Because the matter in the hot interiors of stars is ionized, there is nothing to prevent the free electrons from being closely packed with the bare nuclei to form a very dense gas in white dwarf stars. Their enormous gravity is supported by the electrons, which exert an outward degeneracy pressure that keeps the electrons from being pushed too closely together and violating the Pauli exclusion principle. In 1926 Ralph Fowler showed that the individual electrons at absolute zero still have a kinetic energy comparable with the thermal energy of particles in an expanded gas whose temperature is as large as 10 million degrees. Fowler also showed that the pressure of the white dwarf gas is unaffected by its temperature, thereby resolving the paradoxical fate of white dwarf stars (Eddington, 1926; Chandrasekhar, 1984). The light of white dwarfs comes from the slow leakage of the heat contained in the nondegenerate nuclei, and is not sustained by thermonuclear reactions. These stars therefore slowly cool off and fade into darkness as they age; eventually they will become a gigantic black molecule in which all the nuclei and electrons are in their lowest quantum state. The cooling of white dwarfs is reviewed by D'Antona and Mazzitelli (1990); and the numbers of white dwarfs of different luminosities can be used to infer their age ((Winget et al., 1987; Hernanz et al., 1990; Wood, 1992) and Volume II)

By showing that the pressure of a nonrelativistic degenerate gas of mass density ρ is proportional to $\rho^{5/3}$, Fowler (1926) set the stage for subsequent theoretical speculations on the internal constitution of white dwarfs. By 1929 Wilhelm Anderson had demonstrated that the electrons in the centers of degenerate masses begin to attain velocities on the order of the velocity of light, and that in this case the variation of the electron mass with velocity must be taken into account by using the equations of special relativity. For such a relativistic degenerate electron gas, the pressure at mass density ρ is proportional to $\rho^{4/3}$ (Anderson, 1929). In the following sections we provide the detailed formulae for these conditions; they have also been delineated by Chandrasekhar (1939), Shapiro and Teukolsky (1983) and Kippenhahn and Weigert (1990).

For densities higher than the critical density, ρ_{cd}, given by

$$\rho_{\mathrm{cd}} = 2.4 \times 10^{-8}\, T^{3/2} \mu_{\mathrm{e}} \text{ gm cm}^{-3} \quad , \tag{3.155}$$

a gas is completely degenerate. Here T is the temperature, and the mean molecular weight per electron, μ_{e}, has the value 1 for hydrogen and 2 for the other elements. The gas pressure, P, for a nonrelativistic, completely degenerate gas, with $\rho_{cd} \leq \rho \leq \rho_{cr}$ and $\mu \gg kT$, is given by Fowler (1926)

$$P = \frac{h^2}{20m}\left(\frac{3}{\pi}\right)^{2/3} N_e^{5/3} = 3.12 \times 10^{12}\left(\frac{2}{g}\right)^{2/3}\left(\frac{2\rho}{\mu_e}\right)^{5/3} \text{ dynes cm}^{-2}$$
$$= 1.004 \times 10^{13}\left(\frac{\rho}{\mu_e}\right)^{5/3} \text{ dynes cm}^{-2} = \frac{2}{3}\frac{U}{V} \ , \qquad (3.156)$$

where the electron density $N_e = \rho/(\mu_e m_u)$ for mean electron molecular weight $\mu_e \approx 2$ and $m_u \approx 1.66 \times 10^{-24}$ grams, the atomic mass unit, g is the statistical weight, ρ is the mass density, and the energy density is U/V. A gas is said to be relativistic when the density, ρ, is larger than the critical value, ρ_{cr} given by

$$\rho_{cr} = 7.3 \times 10^6 \mu_e \text{ gm cm}^{-3} \ . \qquad (3.157)$$

For the relativistic, completely degenerate gas, the pressure, P, is given by

$$P = \frac{c}{4}\left(\frac{3h^3}{8\pi}\right)^{1/3} N_e^{4/3} = 4.56 \times 10^{14}\left(\frac{g}{2}\right)^{1/3}\left(\frac{2\rho}{\mu_e}\right)^{4/3} \text{ dynes cm}^{-2}$$
$$\approx 1.244 \times 10^{15}\left(\frac{\rho}{\mu_e}\right)^{4/3} \text{ dynes cm}^{-2} = \frac{1}{3}\frac{U}{V} \ , \qquad (3.158)$$

where $N_A \rho = \mu_e N_e$, and N_A is Avogadro's number. Relativistic effects were first considered by Anderson (1929) and Stoner (1930).

The equations of state for the degenerate gas can be written in polytropic form:

$$P = K\rho^{1+1/n} \qquad (3.159)$$

for an Emden polytrope of index n. For a nonrelativistic degenerate electron gas:

$$P = K_1 \rho^{5/3} \quad \text{for } \rho \ll 10^6 \text{ g cm}^{-3} \ , \qquad (3.160)$$

for a polytrope of index $n = 3/2$ and

$$K_1 = \frac{1}{20}\left(\frac{3}{\pi}\right)^{2/3} \frac{h^2}{m(\mu_e m_u)^{5/3}} = \frac{1.0036 \times 10^{13}}{\mu_e^{5/3}}$$

which according to the theory of polytropes is also given by (Chandrasekhar, 1984)

$$K_1 = 0.4242(GM^{1/3}R) \ ,$$

which means that the radius, R, decreases with increasing mass, M, with $R \propto M^{-1/3}$ or

$$\log\left(\frac{R}{R_\odot}\right) = -\frac{1}{3}\log\left(\frac{M}{M_\odot}\right) - \frac{5}{3}\log(\mu_e) - 1.397 \ , \qquad (3.161)$$

where the Sun's radius $R_\odot = 6.96 \times 10^{10}$ cm and the Sun's mass $M_\odot = 1.989 \times 10^{33}$ grams, and the mean molecular weight per electron is μ_e. For a mass equal to the solar mass and $\mu_e = 2$, we obtain $R = 0.0126$ solar radii and a mean density of 7.0×10^5 g cm^{-3}.

For the relativistic degenerate electron gas,

$$P = K_2 \rho^{4/3} \quad \text{for } \rho \gg 10^6 \text{ g cm}^{-3} , \tag{3.162}$$

for an Emden polytrope of index $n = 3$ and

$$K_2 = \frac{1}{8}\left(\frac{3}{\pi}\right)^{1/3} \frac{hc}{(\mu_e m_u)^{4/3}} = \frac{1.2435 \times 10^{15}}{\mu_e^{4/3}} .$$

When a self gravitating sphere of mass, M, is supported by nonrelativistic electron degeneracy pressure, its radius, R, decreases with increasing mass, M, as $R \propto M^{-1/3}$. However, the radius cannot shrink to zero with the mass increasing without limit, for a degenerate gas eventually becomes relativistic and provides less pressure than the nonrelativistic formula would indicate. As shown by Anderson (1929) and Stoner (1930), this means that a white dwarf cannot be more massive than a critical mass, M_c, often called the Chandrasekhar limit, given by (Chandrasekhar 1931, 1935, 1938, 1984; Kippenhahn and Weigert, 1990)

$$M_c = 4\pi \left[\frac{K_2}{\pi G}\right]^{3/2} (2.01824) = 0.197 \left(\frac{hc}{G}\right)^{3/2} \frac{1}{(\mu_e m_u)^2} = 5.836 \mu_e^{-2} M_\odot , \tag{3.163}$$

where the solar mass $M_\odot = 1.989 \times 10^{33}$ grams (also see Section 3.4.3).

For a partially degenerate gas in which the mass density, ρ, is near the degeneracy limit, ρ_{cd}, the pressure, P, is given by

$$P = \frac{g}{6\pi^2 \hbar^3} \int_0^\infty \frac{p^3 \frac{\partial \varepsilon}{\partial p} \, dp}{\exp\left[\frac{-\mu}{kT} + \frac{\varepsilon}{kT}\right] + 1} , \tag{3.164}$$

where g is the statistical weight, the chemical potential, μ, is given by Eq. (3.104), and the momentum, p, is related to the kinetic energy, ε, by Eq. (3.111). Eq. (3.159) may be rewritten as Eq. (3.149)

$$P = \frac{8\pi kT}{3h^3} (2mkT)^{3/2} F_{3/2}\left(\frac{\mu}{kT}\right) = \frac{2}{3} \frac{U}{V} .$$

where the Fermi–Dirac function $F_{3/2}(\mu/kT)$ is given by Eq. (3.149). Similarly, the electron density, N_e, is given by

$$N_e = \frac{4\pi}{h^3} (2mkT)^{3/2} F_{1/2}\left(\frac{\mu}{kT}\right) ,$$

where the Fermi–Dirac integral $F_n(\psi) = \int_0^\infty x^n [\exp(x - \psi) + 1]^{-1} dx$; numerical values of $F_{3/2}(\psi)$ and $F_{1/2}(\psi)$ are tabulated by McDougal and Stoner (1938) and reproduced by Kippenhahn and Weigert (1990).

In the standard treatment of the degenerate electron gas (given above), the mean molecular weight is assumed to be fixed and the electrons are assumed to be noninteracting and to form an ideal Fermi gas. For the range of densities considered here, $\rho_{cd} \leq \rho \leq \rho_{cr}$ or 500 g cm$^{-3} \leq \rho \leq 10^{11}$ g cm^{-3}, the gas is in fact a plasma of electrons and nuclei. Salpeter (1961) has discussed the equation of state for a zero temperature plasma of electrons and nuclei of atomic weight, A, and charge, Z. The corrections include the classical Coulomb energy of an ion lattice with uniformly distributed electrons, the Thomas–Fermi deviations from a uniform charge distribution of electrons, the exchange energy between the electrons, and the spin–spin interactions between the electrons. The total energy per electron, E, is

$$E = E_0 + E_c + E_{TF} + E_{ex} + E_{cor} \ , \tag{3.165}$$

where the energy per electron for a Fermi gas of non-interacting electrons, E_0, is given by

$$E_0 = \frac{mc^2 g(x)}{8x^3} \ ,$$

where

$$g(x) \equiv 8x^3 \left[\sqrt{(1+x^2)} - 1\right] - x(2x^2 - 3)\sqrt{(1+x^2)} - 3 \sinh^{-1} x \ ,$$

and x is related to the gas mass density, ρ, by

$$\rho = 9.738 \times 10^5 \mu x^3 \text{ gm cm}^{-3} \ ,$$

where $\mu = A/Z$. The Coulomb correction, E_c, is given by

$$E_c = -\frac{9}{5}\frac{Z^{2/3}}{r_e} ry \ ,$$

where one Rydberg $= ry = 2\pi^2 me^4/h^2 \approx 2.179 \times 10^{-11}$ erg ≈ 13.605 eV, and r_e is related to the mass density, ρ, by

$$\rho = 2.6787 \, \mu r_e^{-3} \text{ gm cm}^{-3}.$$

The Thomas–Fermi correction, E_{TF}, is given by

$$E_{TF} = -\frac{324}{175}\left(\frac{4}{9\pi}\right)^{2/3} \sqrt{(1+x^2)} Z^{4/3} ry \ .$$

The exchange energy correction, E_{ex}, is given by

$$E_{ex} = -\frac{3}{2}\left(\frac{9}{4\pi^2}\right)^{1/3} \frac{ry}{r_e} \quad \text{for } x \ll 1 \ ,$$

$$E_{ex} = -\left(\frac{3}{4\pi}\right) \alpha mc^2 x \varphi(x) \quad \text{for any } x \ ,$$

where (Fristrom, 1987)

$$\varphi(x) = \frac{1}{4x^4}\left[\frac{9}{4} + 3\left(\beta^2 - \frac{1}{\beta^2}\right) \ln \beta - 6(\ln \beta)^2 - \left(\beta^2 + \frac{1}{\beta^2}\right) - \frac{1}{8}\left(\beta^4 + \frac{1}{\beta^4}\right)\right] \ ,$$

and $\beta \equiv x + \sqrt{(1+x^2)}$.

The correlation energy correction, E_{cor}, is given by

$$E_{cor} = (0.062 \ln r_e - 0.096) ry \ .$$

Similarly, the gas pressure can be written as the sum of five terms

$$P = P_0 + P_c + P_{TF} + P_{ex} + P_{cor} \ , \tag{3.166}$$

where the non-interacting Fermi electron pressure, P_0, is given by

$$P_0 = \frac{1}{24\pi^2} mc^2 \left(\frac{mc}{\hbar}\right)^3 f(x) \ ,$$

where $f(x) = x(2x^2 - 3)(x^2 + 1)^{1/2} + 3\sinh^{-1} x$, the Coulomb and Thomas–Fermi terms are given by

$$P_C + P_{TF}$$
$$= -mc^2\left(\frac{mc}{\hbar}\right)^3\left[\frac{\alpha Z^{2/3}}{10\pi^2}\left(\frac{4}{9\pi}\right)^{1/3} x^4 + \frac{162}{175}\frac{(\alpha Z^{2/3})^2}{9\pi^2}\left(\frac{4}{9\pi}\right)^{2/3}\frac{x^5}{\sqrt{(1+x^2)}}\right] ,$$

the exchange term is given by

$$P_{ex} = -\frac{\alpha}{4\pi^3} mc^2 \left(\frac{mc}{\hbar}\right)^3 \chi(x) \ ,$$

where

$$\chi(x) = \frac{1}{32}\left(\beta^4 + \frac{1}{\beta^4}\right) + \frac{1}{4}\left(\beta^2 + \frac{1}{\beta^2}\right) - \frac{9}{16} - \frac{3}{4}\left(\beta^2 - \frac{1}{\beta^2}\right)\ln\beta + \frac{3}{2}(\ln\beta)^2$$
$$-\frac{x}{3}\left[1 + \frac{x}{\sqrt{(1+x^2)}}\right]\left[\frac{1}{8}\left(\beta^3 - \frac{1}{\beta^5}\right)\right.$$
$$\left. -\frac{1}{4}\left(\beta - \frac{1}{\beta^3}\right) - \frac{3}{2}\left(\beta + \frac{1}{\beta^3}\right)\ln\beta + \frac{3}{\beta}\ln\beta\right] ,$$

and the correlation term is

$$P_{cor} = -\frac{0.0311}{9\pi^2} mc^2 \left(\frac{mc}{\hbar}\right)^3 \alpha^2 x^3 \ .$$

For extremely relativistic electrons, $x \gg 1$, these equations become

$$\frac{P}{P_0} = 1 + \frac{\alpha}{2\pi} - \frac{6}{5}(\alpha Z^{2/3})\left(\frac{4}{9\pi}\right)^{1/3} - \frac{216}{175}\left(\frac{4}{9\pi}\right)^{2/3}(\alpha Z^{2/3})^2$$
$$= 1.00116 - 4.56 \times 10^{-3} Z^{2/3} - 1.78 \times 10^{-5} Z^{4/3} \ .$$

Hamada and Salpeter (1961) have used the equation of state given immediately above to obtain a maximum radius of

$$R_{max} = 0.021\, R_\odot \tag{3.167}$$

with a mass of about $1.02\, M_\odot$. Here the solar radius, $R_\odot \approx 6.96 \times 10^{10}$ cm and the solar mass, $M_\odot \approx 1.989 \times 10^{33}$ gm. This value of R_{max} is near that first estimated by Kothari (1938). The value of the Chandrasekhar limiting mass is given by Eq. (3.162) with a μ depending on chemical composition. Hamada and Salpeter obtain a lowest possible value of M_{cre}, given by

$$M_{cre} = 1.015\, M_\odot \ .$$

Baym, Pethick, and Sutherland (1971) have included a lattice term in calculating the equilibrium species of nuclei present in the white dwarf density range of 10^4 g cm$^{-3} \lesssim \rho \lesssim 10^{11}$ g cm^{-3}. They obtain

$$M_{\mathrm{cre}} = 1.00\,M_\odot$$

where

$$R_{\max} = 2,140\ \mathrm{km}\ .$$

3.3.2.3 Equation of State of a Degenerate Neutron Gas – Neutron Stars

In a remarkably prescient speculation, Walter Baade and Fritz Zwicky predicted that neutron stars would be formed at the center of massive stars that have ended their thermonuclear lives in a supernova explosion (Baade and Zwicky, 1934). If sufficiently massive, the imploding core could collapse right through the white dwarf stage, forcing the electrons to combine with protons to make neutrons and ultimately forming a neutron star with a radius, R, of $R \approx 10$ km and a mass density, ρ, of $\rho \approx 10^{15}$ g cm^{-3}, comparable to that of the atomic nucleus. Such a neutron star might be created when a star of between 8 and 30 times the Sun's mass has exhausted the nuclear fuel in its central regions, the core collapsing to form a neutron star and the outer regions exploding into a supernova remnant.

While observing interplanetary scintillations of distant radio sources, Antony Hewish and his student Joceyln Bell found the pulsars (Hewish et al., 1968; Hewish, 1975) that turned out to be neutron stars. Due to conservation of magnetic flux, HR^2, neutron stars will have magnetic field strengths, H, of up to 10^{12} Gauss (Pacini, 1967), and the radio pulsars were interpreted as rotating neutron stars with an intense dipole magnetic field that is misaligned with the rotation axis ((Gold, 1968) Volume II). These pulsar magnetospheres have been reviewed by Michel (1982, 1991) and Mészáros (1992), and basic observational data of the radio pulsars are given by Lang (1992). Reviews of our understanding of radio pulsars have been provided by Taylor and Stinebring (1986), Lyne and Graham-Smith (1990) and Bailes and Johnston (1993).

So, the neutron stars were observed as isolated radio pulsars, and subsequently as the accreting neutron star member of binary X-ray sources (also see Volume II). This led to renewed interest in the internal structure of neutron stars, reviewed by Baym and Pethick (1975, 1979) and Shapiro and Teukolsky (1983). We next provide an introduction to this theory, followed by a brief overview of changes in the pulse period that may provide information about the internal structure of neutron stars.

The density of matter in a neutron star increases with depth from about 10^4 g cm^{-3} in the outermost layers to greater than 10^{15} g cm^{-3} in the inner core. To the first approximation, the matter is in its ground state and can be approximated by zero temperature equations of state, for the thermal energy of 10^{-4} to 10^{-2} MeV ($10^6\,{}^\circ$K $\lesssim T \lesssim 10^8\,{}^\circ K$) is much less than the typical excitation energies. In the outermost layers, 10^4 g cm$^{-3} \lesssim \rho \lesssim 4.3 \times 10^{11}$ g cm^{-3}, the constituents are electrons and nuclei and the equation of state may be calculated according to the procedures given in the previous Sect. 3.3.2.2 (cf. Salpeter, 1961 and Baym, Pethick, and Sutherland, 1971). For densities in the range 4.3×10^{11} g cm$^{-3} \lesssim \rho \lesssim 3 \times 10^{14}$g cm^{-3}, the neutron rich nuclei begin to "drip" neutrons and the matter consists of nuclei immersed in a neutron gas as well as an electron gas. Oppenheimer and Volkoff (1939) first considered the equation of state of an ideal Fermi gas of nonrelativistic neutrons. For a nonrelativistic, completely degenerate neutron gas of density, ρ_n, and a neutron mass, M_n, the pressure is given by (Eq. (3.156))

$$P = \frac{(3\pi^2)^{2/3}}{5}\frac{\hbar^2}{M_n^{8/3}}\rho_n^{5/3} \approx 10^{10}\rho_n^{5/3}\ \text{dynes cm}^{-2}\ , \tag{3.168}$$

which corresponds to a kinetic energy per particle of

$$W(k,0) = \frac{3\hbar^2(2^{1/3}k)^2}{10M_n}\ ,$$

where the neutron mass $M_n = 1.675 \times 10^{-24}$ grams, and the number density, N_n, of the neutron gas is

$$N_n = \frac{k^3}{1.5\pi^2}\ .$$

When more refined calculations for pure neutron matter are considered, the energy per particle is given by (Siemans and Panharipande, 1971)

$$W(k,0) \approx 19.74\,k^2 - k^3\frac{(40.4 - 1.088\,k^3)}{(1 + 2.545\,k)}\ \text{MeV}$$

for $k \leq 1.5\,\text{fm}^{-1}$, and for $W(k,0)$ in MeV and k in fm^{-1}.

The equation of state of the ideal (non-interacting), fully degenerate neutron gas can be expressed in polytropic form $P = K\rho^{1+1/n}$ = Eq. (3.159). For the nonrelativistic degenerate neutron gas:

$$P = K_1\rho^{5/3} \quad \text{for } \rho \ll 6 \times 10^{15}\ \text{g cm}^{-3}\ , \tag{3.169}$$

for an Emden polytrope of index $n = 3/2$ and

$$K_1 = \frac{1}{20}\left(\frac{3}{\pi}\right)^{2/3}\frac{h^2}{m_n^{8/3}} = 5.3802 \times 10^9\ .$$

For the relativistic degenerate neutron gas:

$$P = K_2\rho^{4/3} \quad \text{for } \rho \gg 6 \times 10^{15}\ \text{g cm}^{-3}\ , \tag{3.170}$$

for an Emden polytrope of index $n = 3$ and

$$K_2 = \frac{1}{8}\left(\frac{3}{\pi}\right)^{1/3}\frac{hc}{m_n^{4/3}} = 1.2293 \times 10^{15}\ ,$$

where the neutron mass $m_n = 1.6749286(10) \times 10^{-24}$ grams, which is roughly equal to m_u the atomic mass unit, and the mass density $\rho = N_n m_n$ for a number density N_n of neutrons.

When actual neutron star matter is concerned, the energy and pressure must be a superposition of terms corresponding to the energy of neutrons, electrons, and nuclei as well as the lattice binding energy for the nuclei. In this case, the total energy per unit volume is given by (Baym, Bethe, and Pethick, 1971)

$$E_{tot} = N_N(W_N + W_L) + (1 - V_N N_N)E_N(N_n) + E_e(N_e)\ , \tag{3.171}$$

where N_N, N_n and N_e are, respectively, the densities of nuclei, neutrons, and electrons, and $1 - V_N N_N$ is the fraction of the volume occupied by the neutron gas. The energy, W_N, of a nucleus is the sum of the surface energy, W_s, and Coulomb energy terms, W_c, as well as a bulk energy term.

$$W_N = [(1-x)M_n c^2 + xM_p c^2 + W(k,x)]A + W_s + W_c \ ,$$

where x is the fractional concentration of protons, M_p is the proton mass, and $W(k,x)$ is the energy per particle of bulk nuclear matter of density $N_N = k^3/(1.5\pi^2)$. Baym, Bethe, and Pethick give detailed formulae for $W(k,x)$, W_s, W_c and W_N. The energy density $E_N(N_n)/N_n$ of the neutron gas can, to first approximation, be given by $W(k,0) + M_n c^2$ where $W(k,0)$ is given by Eq. (3.168), and the energy density, $E_e(N_e)$, of the electrons is the value for a free electron gas given in the previous section. The lattice energy per nucleus is given by

$$W_L = -\frac{1.82 Z^2 e^2}{a} \ ,$$

where the lattice constant $a = (2/N_N)^{1/3}$, and the Coulomb energy, W_c, is of order

$$W_c \approx \frac{3}{5}\frac{Z^2 e^2}{r_N} \ ,$$

where $4\pi N_N r^3 \approx 3$. Baym, Bethe, and Pethick (1971) give the formulae and values of neutron star matter for densities which range between 4.6×10^{11} and 2.4×10^{14} g cm^{-3}. Barkat, Buchler, and Ingber (1972) use an extended nuclear Thomas–Fermi model to calculate the pressure as a function of density for neutron star matter in the range $10^8\ \mathrm{g\,cm^{-3}} \lesssim \rho \lesssim 10^{14}\ \mathrm{g\,cm^{-3}}$. Additional work using the Fermi–Thomas model for nuclei is given by Buchler and Barkat (1971) and Buchler and Ingber (1971). Ravenhall, Bennet, and Pethick (1972) employ a more reliable theory for the nuclear surface than did Baym, Bethe, and Pethick (1971), and find results similar to Barkat, Buchler, and Ingber (1972). Pedagogic notes on the Thomas–Fermi theory of the atom (Thomas, 1927; Fermi, 1927) are provided by Spruch (1991) with applications to neutron stars and white dwarfs.

As with white dwarf stars, there is a maximum mass for a neutron star. The analysis given in the previous Sect. 3.3.2.2 can be applied for a neutron star mass to give an upper limit of 5.73 solar masses, but the extreme gravity inside neutron stars requires Einstein's General Theory of Relativity for an exact treatment (also see Volume II). The precise solution for the upper mass limit to a neutron star was pioneered by Oppenheimer and Volkoff (1939). Subsequent developments using computers and an improved understanding of the nuclear force, has resulted in an upper limit of about 3 solar masses (Rhoades and Ruffini, 1974; Hartle, 1978; Baym and Pethick, 1979; Shapiro and Teukolsky, 1983), in accordance with a recent estimated upper limit to a neutron star mass of $M_{ns} \leq 2.9$ solar masses, including any possible contribution due to rapid uniform rotation (Kalogera and Baym, 1996).

Information on presupernova stars is given by Woosley and Weaver (1995) and the neutron star masses expected after massive star evolution and supernova explosion are given by Timmes, Woosley and Weaver (1996). For stars that explode as Type II supernovae, the neutron star masses average $M_{\rm ns} = 1.28$ or 1.73 solar masses; the average for those arising from Type Ib supernovae is $M_{\rm ns} = 1.32$ solar masses. This compares favorably with the determination of a neutron star mass of $M_{\rm ns} = 1.35 \pm 0.27$ solar masses for 17 systems (Thorsett et al., 1993). Radio observations for four neutron binary star systems give $M_{\rm ns} = 1.01$ to 1.64 solar masses (Finn, 1994), while neutron star masses inferred from X-ray binaries lie in the range $M_{\rm ns} = 1$ to 2 solar masses (Bahcall, 1978; Joss and Rappaport, 1984; Lang, 1992).

At densities higher than 2.4×10^{14} g cm^{-3}, the nuclei begin to touch each other, and the matter becomes a degenerate sea of neutrons in a sea of protons and electrons, each of which has about 4% of the neutron abundance. When the density surpasses 10^{15} g cm^{-3}, hyperons become important as well. Details of the equation of state in the density range 2–10 $\times 10^{14}$ g cm^{-3} are given by Wang, Rose, and Schlenker (1970), and as shown in Ruderman's (1969) review, the matter probably consists of superfluid neutrons together with a few per cent of superfluid protons and normal electrons. For a degenerate system of Fermi particles, superfluidity can exist when the particles are correlated in pairs with the same center of mass momentum. A new ground state of energy is established, and an energy gap, ε, exists in the energy spectrum at zero temperature. According to Bardeen, Cooper, and Schrieffer (1957),

$$\varepsilon = kT_{\rm c} = \frac{4N_p}{N_0} = 4\hbar\omega \exp\left[-\frac{1}{N_0 V}\right] , \tag{3.172}$$

where $N_{\rm p}$ is the number of electrons in pairs virtually excited above the Fermi surface, $N_0 \approx m\sqrt{2mE_{\rm f}}/(2\pi^2\hbar^3)$ is the number of particle states of one spin per unit energy range at the Fermi surface, $E_{\rm f}$ is the Fermi energy, ω is the average lattice frequency and $-V$ is the interaction energy between pairs. At the transition temperature, $T_{\rm c}$, thermal excitation has reduced the number of pairs to zero and the system is in a normal degenerate state. As emphasized by Ginzburg (1969) and Ruderman (1969), neutron star matter with $\rho \approx 10^{14} - 10^{15}$ g cm^{-3} may be superfluid. Details of the physics of the superfluidity of neutron star matter are given by Yang and Clark (1971), Chao, Clark, and Yang (1972), and Krotscheck (1972). Assuming that changes in pulsar periods reflect changes in the rotation rate of a neutron star, the changes may be related to the physics of the superfluid core in a way which has been reviewed by Pines (1970).

The equation of state of matter at densities larger than 10^{15} g cm^{-3} has been discussed by Harrison, Thorne, Wakano, and Wheeler (1964), Buchler and Ingber (1971), Panharipande (1971), Frautschi, Bahcall, Steigman, and Wheeler (1971), and Leung and Wang (1971).

Shapiro and Teukolsky (1983) discuss equations of state for neutron stars of different mass density, ρ, including those of Baym, Bethe and Pethick (1971) for mass densities in the range 4.3×10^{11} g cm$^{-3} \le \rho \le 5 \times 10^{14}$ g cm^{-3}, Bethe and Johnson (1974) for 1.7×10^{14} g cm$^{-3} \le \rho \le 3.2 \times 10^{16}$ g cm^{-3}, and

Pandharipande (1971), Friedman and Pandharipande (1981) and Walecka (1974) for mass densities $\rho \geq 10^{14}\,\mathrm{g\,cm^{-3}}$. Nevertheless, in his review of the cooling of neutron stars, Pethick (1992) argues that it is impossible to predict with confidence the interior constitution of neutron stars on the basis of current physical understanding.

So far the only observable consequences of the structure within a neutron star are the small effect on radio pulsar spin periods. Some isolated radio pulsars, such as the Crab and Vela pulsars, exhibit period decreases, or spinups, dubbed "glitches" that last weeks or months, superposed on a systematic period lengthening or spin down. They have been interpreted as a possible consequence of neutron crust cracking, or neutron starquakes, due to internal changes such as the coupling between the superfluid interior and the crust (Baym and Pines, 1971; Anderson and Itoh, 1975; Greenstein, 1976; Ruderman, 1976; Harding, Guyer and Greenstein, 1978; Alpar, Chau, Chen and Pines, 1993; Lorenz, Ravenhall and Pethick, 1993). Properties of the neutron star crust are summarized by Ruderman (1991); the properties of radio pulsar glitches are reviewed by Ruderman (1997) within the context of neutron star theory.

3.3.2.4 The Neutrino Gas – Number Density, Energy Density, Entropy Density, and the Equation of State

The number density, N, of a neutrino gas with zero rest mass is given by (Kuchowicz, 1963 and private communication)

$$N = 8\pi g \left(\frac{kT}{hc}\right)^3 F_2\left(\frac{\mu}{k\,T}\right) , \tag{3.173}$$

where $g \sim 1$ is the statistical weight of the neutrino, T is the gas temperature, and the Fermi–Dirac integral, $F_{\mathrm{n}}(\mu/k\,T)$, is given by

$$F_{\mathrm{n}}\left(\frac{\mu}{kT}\right) = \frac{1}{n!}\int_0^\infty \frac{x^n\,dx}{\exp\left(x - \frac{\mu}{kT}\right) + 1} , \tag{3.174}$$

where the neutrino chemical potential, μ, is given by Eq. (3.104).

The neutrino energy density, U/V, is given by

$$\frac{U}{V} = 3P = 24\pi gkT\left(\frac{kT}{hc}\right)^3 F_3\left(\frac{\mu}{kT}\right) , \tag{3.175}$$

where P is the neutrino pressure, U is the energy, and V is the volume. The neutrino entropy density, S/V, is given by

$$\frac{S}{V} = 8\pi gk\left(\frac{kT}{hc}\right)^3 \left[4F_3\left(\frac{\mu}{kT}\right) - \left(\frac{\mu}{kT}\right)F_2\left(\frac{\mu}{kT}\right)\right] , \tag{3.176}$$

where S is the entropy.

It follows from Eq. (3.175) that the pressure, P, of a neutrino gas with zero rest mass is given by

$$P = 8\pi gkT\left(\frac{kT}{hc}\right)^3 F_3\left(\frac{\mu}{kT}\right) ,$$

where $g \sim 1$ is the statistical weight of the neutrino, T is the gas temperature, μ is the chemical potential given by Eq. (3.104), and the Fermi–Dirac integral, $F_n(\mu/kT)$, is given in Eq. (3.174).

The asymptotic expressions for the equation of state are (Kuchowicz, 1963 and private communication):

For extreme degeneracy ($\mu \gg kT$):

$$P = \frac{\pi g k T}{3}\left(\frac{kT}{hc}\right)^3 \left\{ \left[\frac{1}{2}y^4 + \frac{1}{3}\pi^6 y^2 + \frac{1}{27}\pi^{12} + D^{1/2}\right]^{1/3} + \left[\frac{1}{2}y^4 + \frac{1}{3}\pi^6 y^2 + \frac{1}{27}\pi^{12} - D^{1/2}\right]^{1/3} - \frac{\pi^4}{5}\right\}$$

$$\approx \frac{\pi g}{3}\frac{(kTy^{1/3})^4}{(hc)^3} , \tag{3.177}$$

where

$$D = \frac{y^2}{4} + \frac{\pi^6}{27} ,$$

and

$$y = \frac{3N}{4\pi g}\left(\frac{hc}{kT}\right)^3 .$$

Here N is the neutrino number density given in Eq. (3.173).

For weak degeneracy ($\mu \ll -0.4\, kT$):

$$P = \frac{16\pi g k T}{3}\left(\frac{kT}{hc}\right)^3 \left[1 + \frac{y}{24} - \left(1 - \frac{y}{12}\right)^{1/2}\right]$$

$$\approx \frac{NkT}{3}\left[1 + \frac{N}{128\pi g}\left(\frac{hc}{kT}\right)^3\right] \quad \text{for } N < T^3 . \tag{3.178}$$

For zero chemical potential we obtain the neutrino analogue to the Stefan-Boltzmann law, the difference in statistics between photons and neutrinos being a $7g/16$ factor. This factor doubles for a neutrino-antineutrino gas.

3.3.3 The Photon Gas

3.3.3.1 Einstein–Bose Statistics

Elementary particles whose spin is an even multiple of $\hbar/2$ (photons and nuclei with even mass numbers) obey Einstein–Bose statistics (Einstein, 1924; Bose, 1924). These particles are indistinguishable from each other, and all values for the number of particles in a particular quantum state are equally likely. The number density, $n(\varepsilon)$, of particles with kinetic energy between ε and $\varepsilon + d\varepsilon$ is given by

$$Vn(\varepsilon) = \frac{g(\varepsilon)\, d\varepsilon}{\exp\left[\frac{-\mu}{kT} + \frac{\varepsilon}{kT}\right] - 1} , \tag{3.179}$$

where $g(\varepsilon)$ is the number of possible particle states of energy, ε, and is given by Eq. (3.109), the volume is V, and the chemical potential, μ, is given by Eq. (3.104). For the special case of photons, $\mu = 0$ and

$$Vn(\varepsilon) = \frac{g(\varepsilon)\,d\varepsilon}{\exp\left(\frac{\varepsilon}{kT}\right) - 1} .$$

The total number of particles per unit volume, N, is given by

$$N = \int_0^\infty n(\varepsilon)\,d\varepsilon = \frac{g}{2\pi^2\hbar^3}\int_0^\infty \frac{p^2\,dp}{\exp\left[\frac{-\mu}{kT} + \frac{\varepsilon}{kT}\right] - 1} , \tag{3.180}$$

where g is the statistical weight, and ε is related to the momentum, p, by Eq. (3.111). For the nonrelativistic gas ($\varepsilon = p^2/(2m)$, $p \ll mc$, $kT \ll mc^2$)

$$N = \frac{4\pi}{h^3}(2mkT)^{3/2}\int_0^\infty \frac{x^{1/2}\,dx}{\exp\left[\frac{-\mu}{kT} + x\right] - 1} .$$

The total energy density, U/V, is given by

$$\frac{U}{V} = \frac{g}{2\pi^2\hbar^3}\int_0^\infty \frac{\varepsilon p^2\,dp}{\exp\left[\frac{-\mu}{kT} + \frac{\varepsilon}{kT}\right] - 1} ,$$

which reduces to

$$\frac{U}{V} = \frac{4\pi}{h^3}(2m)^{3/2}(kT)^{5/2}\int_0^\infty \frac{x^{3/2}\,dx}{\exp\left[\frac{-\mu}{kT} + x\right] - 1} , \tag{3.181}$$

for a nonrelativistic gas.

3.3.3.2 Equation of State, Energy Density, and Entropy of a Photon Gas

For densities lower than the critical density, $\rho_{\rm cp}$, given by

$$\rho_{\rm cp} = 3.0 \times 10^{-23}\mu T^3 \text{ gm cm}^{-3} \quad \text{or } T = 3.20 \times 10^7\left(\frac{\rho_{\rm cp}}{\mu}\right)^{1/3} {}^\circ\text{K} , \tag{3.182}$$

the pressure of a photon gas predominates over the pressure of the non-degenerate perfect gas. Here the constant μ is the mean molecular weight. As the photon energy is $h\nu$, it follows from Eq. (3.181) that the radiation pressure of the photon gas is

$$P = \frac{1}{3}\int_0^\infty h\nu N(\nu)\,d\nu = \frac{1}{3}\frac{U}{V} , \tag{3.183}$$

where U/V is the energy density and the number density of photons, $N(\nu)$, with frequencies between ν and $\nu + d\nu$ is

$$N(\nu) = \frac{8\pi\nu^2}{c^3}\frac{1}{\left[\exp\left(\frac{h\nu}{kT}\right) - 1\right]}\,d\nu . \tag{3.184}$$

Substituting Eq. (3.184) into Eq. (3.183) we obtain

$$P = \frac{1}{3}\frac{U}{V} = \frac{1}{3}aT^4 \ , \tag{3.185}$$

where the radiation constant, a, is given by

$$a = \frac{8\pi^5}{15}\frac{k^4}{h^3c^3} \approx 7.56591(19) \times 10^{-15} \text{ erg cm}^{-3} \, {}^\circ\text{K}^{-4} \ .$$

The radiation constant, a, is related to the Stefan–Boltzmann constant, σ, by the relation (Stefan, 1879; Boltzmann, 1884)

$$\sigma = \frac{ac}{4} = 5.67051(19) \times 10^{-5} \text{ erg cm}^{-3} \, {}^\circ\text{K}^{-4} \, \text{sec}^{-1} \ .$$

Further relations for a photon gas are given in Sect. 1.23 on thermal emission from a black body. For example, the entropy per unit volume is $S/V = 4aT^3/3$, when the total energy per unit volume is $U/V = aT^4$ and the pressure is $P = aT^4/3$.

3.4 Macrostructure of a Gas – The Virial Theorem

3.4.1 The Virial Theorem of Clausius

Consider the motion of a single molecule of mass, m. If $\boldsymbol{r}$ is its position vector measured from an arbitrary origin and $\boldsymbol{F}$ is the force acting on the molecule, then

$$m\frac{d^2\boldsymbol{r}}{dt^2} = \boldsymbol{F} \ .$$

Forming the scalar product with **r** and transforming the resulting equation we obtain

$$\frac{1}{2}m\frac{d^2r^2}{dt^2} = m\frac{d}{dt}\left(\boldsymbol{r}\cdot\frac{d\boldsymbol{r}}{dt}\right) = m\left(\frac{dr}{dt}\right)^2 + \boldsymbol{F}\cdot\boldsymbol{r} \ . \tag{3.186}$$

Letting $\sum$ denote an ensemble average over all molecules, and noting that a macroscopic system which is in equilibrium is stationary, Eq. (3.186) becomes the virial theorem (Clausius, 1870)

$$\sum m\left(\frac{dr}{dt}\right)^2 + \sum \boldsymbol{F}\cdot\boldsymbol{r} = 0 \ , \tag{3.187}$$

where $\sum \boldsymbol{F}\cdot\boldsymbol{r}$ is called the virial. Eq. (3.187) is also known as Poincaré's theorem, as Poincaré derived it by assuming that the only force acting on a system of molecules is that of their mutual gravitational attraction. In this case, Eq. (3.187) becomes (Poincaré, 1811)

$$2T + \Omega = 0 \ , \tag{3.188}$$

where T is the total kinetic energy of the particles, and Ω is the total gravitational potential energy of the system.

As Eddington (1916) noted, Eqs. (3.187) and (3.188) may be extended to include nonequilibrium situations by including the moment of inertia, I. In this nonstationary case,

$$\frac{1}{2}\frac{d^2 I}{dt^2} = \sum m\left(\frac{dr}{dt}\right)^2 + \sum \boldsymbol{F}\cdot\boldsymbol{r} = 2T + \Omega \ , \tag{3.189}$$

where $I = \sum mr^2$.

Binney and Tremaine (1987) and Saslaw (1985) provide both the tensor and scalar versions of the virial theorem, applying them to stellar and galactic systems. They separate the kinetic energy into two components – one due to ordered or bulk motions and the other due to random motions or pressure, and note that in many cases the system is in a steady state with $d^2I/dt^2 = 0$. A simple example of such applications is in the determination of the masses of galaxies in clusters of galaxies (Volume II), while a more complex one is the virial theorem for astrophysical blast waves (Ostriker and Mc Kee, 1988). The application of the tensor form of the virial theorem to rotating masses is discussed in Sect. 3.4.6 and by Chandrasekhar (1969).

When the magnetic energy, $\mathscr{M}$, and the kinetic energy of mass motion, T_{m}, are taken into account, the equilibrium virial theorem becomes (Chandrasekhar and Fermi, 1953)

$$2[T_{\mathrm{m}} + T_{\mathrm{k}}] + \Omega + \mathscr{M} = 0 \ , \tag{3.190}$$

where T_{k} is the kinetic energy of molecular motion, and Ω is the total gravitational potential energy of the system. For a perfect gas sphere of radius, R, mass, M, pressure, P, and temperature, T,

$$2T_{\mathrm{k}} = \frac{3MkT}{\mu m_{\mathrm{H}}} = 4\pi PR^3 = 3(\gamma - 1)U \ , \tag{3.191}$$

where μ is the mean molecular weight, m_{H} is the mass of hydrogen, γ is the adiabatic index, and U is the internal energy due to the molecular motion (the heat energy). For a gaseous mass rotating with angular velocity, ω, the kinetic energy of mass motion, T_{m}, is given by

$$T_{\mathrm{m}} = \tfrac{1}{2} I\omega^2 \ , \tag{3.192}$$

where I is the moment of inertia. For a sphere of mass, M, and radius, R,

$$T_{\mathrm{m}} = \tfrac{1}{2} I\omega^2 = \tfrac{1}{5} MR^2\omega^2 \ . \tag{3.193}$$

The gravitational potential energy, Ω, of such a sphere is given by

$$\Omega = -\frac{3}{5}\frac{GM^2}{R} \ , \tag{3.194}$$

where the gravitational constant, $G = 6.668 \times 10^{-8}$ dyn cm^2 g^{-2}, and the magnetic energy, $\mathscr{M}$, is given by

$$\mathscr{M} = \tfrac{1}{6} H^2 R^3 \ , \tag{3.195}$$

where H is the average magnetic field strength in the sphere. The magnetic virial theorem has been discussed by Chandrasekhar (1961), Mestel (1985), and Shu (1992).

3.4.2 Ritter's Relation

For an adiabatic process in a gas which is in hydrostatic equilibrium, it follows from the first law of thermodynamics that the change, dU, in the internal energy, U, of the system is given by

$$dU = -P\,dV \ , \tag{3.196}$$

where P is the gas pressure, and dV is the change in volume, V. For an adiabatic process involving a perfect gas, the change, dT, in temperature, T, is given by

$$\frac{dT}{T} = -(\gamma - 1)\frac{dV}{V} \ , \tag{3.197}$$

where γ is the adiabatic index. Using Eqs. (3.196) and (3.197) together with the ideal gas law, we obtain

$$dU = \frac{N_{\mathrm{tot}} k\, dT}{(\gamma - 1)} \ ,$$

and

$$U = \frac{N_{\mathrm{tot}} kT}{(\gamma - 1)} = \frac{2T_{\mathrm{k}}}{3(\gamma - 1)} \ ,$$

where k is Boltzmann's constant, N_{tot} is the total number of molecules, and T_{k} is the total internal energy of molecular motion. Using the virial theorem in Poincaré's form, it follows that

$$U = -\frac{\Omega}{3(\gamma - 1)} \ , \tag{3.198}$$

which is Ritter's relation (Ritter, 1880). The total energy, E, is then given by

$$E = U + \Omega = \frac{3\gamma - 4}{3(\gamma - 1)}\Omega \ . \tag{3.199}$$

The energy, E, becomes negative for $\gamma < 4/3$, and therefore a gas mass is unstable against adiabatic pulsations for $\gamma < 4/3$. When the magnetic term, $\mathcal{M}$, given in Eq. (3.195) is included, the criterion for dynamical stability becomes $(3\gamma - 4)(|\Omega| - \mathcal{M}) > 0$. Furthermore, Chandrasekhar (1964) has shown that the Newtonian lower limit of 4/3 for the ratio of specific heats is increased by effects arising from general relativity. As long as γ is finite, dynamical instability occurs if the mass contracts and the radius falls below the critical value

$$R_{\mathrm{c}} = \frac{K}{\gamma - 4/3}\frac{2GM}{c^2} \approx 4.7 \times 10^{17}\mathrm{cm} \ , \tag{3.200}$$

where the constant K ranges from 0.45 to 1.12 for polytropes whose index ranges from 0 to 4. The approximate value is for a polytrope of index 3.

3.4.3 Chandrasekhar Limiting Mass for Degenerate Matter

Both Wilhelm Anderson and Edmund C. Stoner showed that a star can contract only until the change in gravitational potential energy becomes insufficient to balance the increase of the kinetic energy of the electrons; accordingly, for stellar masses larger than about 1 solar mass, there can be no equilibrium white dwarf configurations (Anderson, 1929; Stoner, 1930). Subramanyan Chandrasekhar then derived the detailed equilibrium configurations in which degenerate electron gases support their own gravity, finding the exact upper mass limit, M_c, for white dwarfs. Although Anderson and Stoner had previously called attention to the existence of this upper mass limit, it is often called the Chandrasekhar limit, and is given by Chandrasekhar (1931, 1935, 1938, 1984; Kippenhahn and Weigert, 1990):

$$M_c \approx \frac{5.836}{\mu_e^2} M_\odot \approx 1.459\, M_\odot \ , \qquad (3.201)$$

where the solar mass $M_\odot = 1.989 \times 10^{33}$ grams. The numerical constant in this expression is derived from the explicit solution of the Lane–Emden equation for a polytropic index of $n = 3$. That is, Chandrasekhar noted the relationship of pressure, P, and mass density, ρ, for a relativistic degenerate gas, with $P = K_2 \rho^{4/3}$, is that of a polytrope of index $n = 3$, and obtained the limiting mass (see Sect. 3.3.2.2).

This equation can be inferred, in approximate form, from the virial theorem.

When only the kinetic energy of molecular motion and the gravitational potential energy are taken into account, the virial theorem for a spherical object of radius, R, becomes

$$3PV = 4\pi P R^3 = \frac{3}{5}\frac{GM^2}{R} \ , \qquad (3.202)$$

where P is the gas pressure, V is the volume, M is the mass, and the gravitational constant $G = 6.668 \times 10^{-8}$ dyne cm^2 g^{-2}. For a relativistic completely degenerate electron gas, P, is given by Eq. (3.162) or by $P = K_2 \rho^{4/3} \approx (hc/2\pi)(\rho/m_u)^{4/3}$. Using this expression for P together with $V \approx R^3$ and $\rho V = M$ in Eq. (3.202) one obtains:

$$M \approx \frac{1}{m_u^2}\left(\frac{hc}{2\pi G}\right)^{3/2} \approx 2 M_\odot \ , \qquad (3.203)$$

where $M_\odot$ is the Sun's mass.

3.4.4 Conditions for Gravitational Contraction in the Presence of a Magnetic Field or an External Pressure

Consider a perfect gas sphere of radius, R, and mass, M, with an internal magnetic field of average strength, H. If the gas is in equilibrium, it follows from the virial theorem (Eq. (3.190)) that

$$3(\gamma - 1)U + \Omega + \mathcal{M} = 0 \ , \qquad (3.204)$$

where the total energy

$$E = U + \Omega + \mathcal{M} \ , \qquad (3.205)$$

and U, Ω, and $\mathcal{M}$ are, respectively, the heat energy of molecular motion, the total gravitational potential energy, and the magnetic energy. Eliminating U, we obtain

$$E = -\frac{3\gamma - 4}{3(\gamma - 1)}[|\Omega| - \mathcal{M}] \ , \tag{3.206}$$

which must be greater than zero for dynamical stability. Thus, in the absence of a magnetic field, $\gamma > 4/3$ for dynamical stability. In the presence of a magnetic field. Eqs. (3.194) and (3.195) for Ω and $\mathcal{M}$ show that for masses larger than

$$M_c = \left[\frac{5}{18}\frac{H^2R^4}{G}\right]^{1/2}$$

the magnetic field cannot support gravity alone. In terms of the solar mass, $M_\odot$, and radius, $R_\odot$, the condition that, $\mathcal{M} >| \Omega |$ gives

$$H > 2 \times 10^8 \frac{M}{M_\odot}\left(\frac{R_\odot}{R}\right)^2 \quad \text{Gauss} \ , \tag{3.207}$$

providing an upper limit to the magnetic field for dynamical stability.

When the surface of a perfect gas sphere of radius, R, mass, M, and temperature, T, is subjected to a uniform external pressure, P_e, the equilibrium virial equation (3.190) is

$$2T_k - 4\pi P_e R^3 + \Omega = 0 \tag{3.208}$$

or

$$P_e = \frac{3MkT}{4\pi R^3 \mu m_H} - \frac{3GM^2}{20\pi R^4} \ , \tag{3.209}$$

where μ is the molecular weight, and m_H is the mass of the hydrogen atom. At small R, the second self-gravitation term on the right side of Eq. (3.209) comes into play, and there is less external pressure needed to produce a given decrease in cloud size R. For some critical value, the curve of P_e versus R reaches a critical maximum, P_{ec}. Beyond this critical point a reduction in external pressure, P_e, is required to maintain the cloud equilibrium (Ebert, 1955; Bonner, 1956). Differentiating Eq. (3.209) with respect to radius, R, we obtain (McCrea, 1957):

$$P_{ec} = \frac{3}{16\pi G^3 M^2}\left(\frac{45}{12}\right)^3\left(\frac{kT}{\mu m_H}\right)^4 \ . \tag{3.210}$$

All masses larger than the critical mass,

$$M_c = \frac{\sqrt{3}}{4\sqrt{\pi}G^{3/2}P_{ec}^{1/2}}\left(\frac{45}{12}\right)^{3/2}\left(\frac{kT}{\mu m_H}\right)^2 \ , \tag{3.211}$$

will collapse. The corresponding critical radius below which collapse occurs is

$$R_c = \frac{12}{45}\frac{GM\mu m_H}{kT} \approx 6 \times 10^6 \frac{\mu}{T}\frac{M}{M_\odot} R_\odot \ \text{cm} \ ,$$

and the critical mass density above which collapse occurs is

$$\rho_c = \left(\frac{45}{12}\right)^3 \frac{3}{4\pi G^3 M^2} \left(\frac{kT}{m_H \mu}\right)^3 \approx 10^{-20} \left(\frac{T}{\mu}\right)^3 \left(\frac{M_\odot}{M}\right)^2 \text{ gm cm}^{-3} .$$

Eqs. (3.209) and (3.210) ignore magnetic and self-gravitational effects. The sign of the combined contribution of these effects to the external pressure, P_e, needed to contain a cloud of mass, M, changes at the magnetic critical mass, M_Φ, given by (Shu, 1992):

$$M_\Phi = (\beta/\alpha)^{1/2} \, G^{-1/2} \, \Phi \approx 10^3 (H/30\mu G)(R/2\text{pc})^2 , \tag{3.212}$$

the gravitational potential energy $\Omega = -\alpha G M^2/R$, the magnetic energy is $M = \beta\Phi^2/R$ for a magnetic flux $\Phi \approx \pi \text{ HR}^2$, so the form factors $\alpha \approx 3/5$ and $\beta \approx 1/(6\pi^2)$; detailed calculations give similar values with $(\beta/\alpha)^{1/2} \approx 0.13$ (Mouschovias and Spitzer, 1974; Tomisaka, Ikeuchi and Nakamura, 1988). In the numerical approximation the magnetic field strength is in units of $30\,\mu\text{G} = 30 \times 10^{-6}$ Gauss and the radius is in units of 2 pc, where one parsec $= 1 \text{ pc} = 3.0856 \times 10^{18}$ cm.

The external pressure can then be expressed as (Shu, 1992):

$$P_e = \frac{1}{4\pi} \left[\frac{3\text{a}^2 M}{R^3} + \frac{\alpha \text{G}}{R^4} (M_\Phi^2 - M^2) \right] , \tag{3.213}$$

where $\text{a}^2 = \text{kT/m} = \text{kT}/(\mu \text{m}_\text{u})$. When $M \gg M_\Phi$, we can ignore magnetic effects and Eq. (3.213) is the same as Eq. (3.209). The contraction and equilibria of magnetic interstellar clouds has been considered by Mouschovias (1974, 1976) and Mouschovias and Spitzer (1976, 1978).

3.4.5 Gravitational Contraction, Hydrodynamic Time Scale, Free-Fall Time, and the Kelvin–Helmholtz Contraction Time

Assuming free-fall and equating the kinetic energy of infall to the gravitational potential energy, the time for collapse, τ_H, is obtained. This hydrodynamic time scale, τ_H, or free-fall time, τ_F, is given by (Spitzer, 1978)

$$\tau_H = \tau_F = \left[\frac{3\pi}{32 G \rho_0}\right]^{1/2} \approx 1.92 \times 10^3 \rho_0^{-1/2} \text{seconds} \approx 4.3 \times 10^7 \text{n}_{\text{H}0}^{-1/2} \text{years} , \tag{3.214}$$

where the initial mass density is ρ_0 in units of g cm^{-3} , for a cloud of initial radius R_0 and mass M we have $\rho_0 = 3M/4\pi R_0^3$, the initial number density of hydrogen atoms is $\text{n}_{\text{H}0}$ in units of cm^{-3}, the number density is $\text{n}_{\text{H}0} = \rho_0/\text{m}_\text{H}$ where the hydrogen atom's mass is $\text{m}_\text{H} \approx 1.67 \times 10^{-24}$ grams, and one year $\approx 3.16 \times 10^7$ seconds. If there is an initial uniform density, ρ_0, and radius, R_0, then the radius, R, after a time, t, from the beginning of free-fall is given by the relation

$$\left(\frac{8\pi G}{3}\rho_0\right)^{1/2} t = \left(1 - \frac{R}{R_0}\right)^{1/2} \left(\frac{R}{R_0}\right)^{1/2} + \sin^{-1}\left(1 - \frac{R}{R_0}\right)^{1/2} . \tag{3.215}$$

If λ_J is the critical Jeans length (see Eq. (3.388) in Sect. 3.5.11), then the e-folding time for the gravitational condensation of a mass of diameter $D > \lambda_J$ is

$$\tau \approx \left\{4\pi G\rho\left[1-(\lambda_J/D)^2\right]\right\}^{-1/2} . \tag{3.216}$$

When the dissipation of energy by radiation is taken into account, the appropriate time scale is

$$\tau_{K-H} \approx \frac{GM^2}{RL} , \tag{3.217}$$

where L is the luminosity of the sphere. This time is called the Kelvin-Helmholtz contraction time after Helmholtz (1854) and Kelvin (1861, 1863) who first hypothesized that a star's lifetime might be determined by equating the energy of gravitational contraction to the radiated energy. Such calculations led to the conclusion that a nuclear energy source, as opposed to that of gravity, must keep a star shining.

3.4.6 Stable Equilibrium Ellipsoids of Rotating Liquid Masses

The theory of rotating fluid masses is provided in the books by Lyttleton (1953), Chandrasekhar (1969) and Tassoul (1978). Some of the history of this subject can be found in Todhunter (1962). This section begins by delineating conditions for the instability of uniformly rotating spherical masses, called Maclaurin spheroids (Maclaurin, 1742). The secular instabilities of Maclaurin spheroids were discussed by Lindblom and Detweiler (1977), and the instabilities of rotating gaseous masses discussed by Hunter (1977). Unstable conditions for uniformly rotating disks, called Maclaurin disks, as well as spheroids, are discussed by Binney and Tremaine (1987). Conditions for the secular instability and dynamical instability of Maclaurin spheroids are also given by Shapiro and Teukolsky (1983) who note that these conditions apply for a wide range of angular momentum distributions and stellar equations of state. The stability of rotating white dwarf stars is discussed by Ostriker and Tassoul (1969), Durisen (1975) and Durisen and Imamura (1981). Work on the stability of differentially rotating stars includes that of Lynden-Bell and Ostriker (1967), Tassoul and Ostriker (1968), Ostriker and Bodenheimer (1973), Chandrasekhar (1974) and Bardeen, Friedman, Schutz and Sorkin (1977).

The equilibrium figures which arise when a homogeneous fluid rotates with a uniform angular velocity are derived from the virial equation. As discussed in detail by Chandrasekhar (1969), and in the papers referenced therein, the virial equation is written in tensor form and includes additional terms corresponding to the distribution of pressure, the centrifugal potential, and the Coriolis acceleration. For a homogeneous liquid mass of mass density, ρ, and uniform rotational velocity, ω, the virial theorem becomes

$$\frac{d}{dt}\int\limits_{v} \rho u_i x_j \, dx = 2T_{ij} + \Omega_{ij} + \omega^2 I_{ij} - \omega_i \omega_k I_{kj} + 2\varepsilon_{ilm}\omega_m \int\limits_{v} \rho u_i x_j \, dx + \delta_{ij}\Pi , \tag{3.218}$$

where the integrals are volume integrals,

$$\frac{d}{dt}\int_v \rho u_i x_j \, dx = \frac{1}{2}\frac{d^2 I_{ij}}{dt^2} , \tag{3.219}$$

and becomes zero in the steady state, u_i and x_j denote the velocity and spatial components as measured from the center of mass, the moment of inertia tensor, I_{ij}, is given by

$$I_{ij} = \int_v \rho x_i x_j \, dx , \tag{3.220}$$

the kinetic energy tensor, T_{ij}, is given by

$$T_{ij} = \tfrac{1}{2}\int_v \rho u_i u_j \, dx , \tag{3.221}$$

the gravitational potential energy tensor, Ω_{ij}, is given by

$$\Omega_{ij} = -\frac{1}{2}\int_v \rho U_{ij} \, dx = \int_v \rho x_i \frac{\partial U_{ij}}{\partial x_j} \, dx , \tag{3.222}$$

where U_{ij} is the gravitational potential, the Krönecker delta function $\delta_{i,j} = 1$ if $i = j$ and is zero otherwise, the function $\varepsilon_{ilm} = +1$ if i, l, m are in cyclic order, -1 if i, l, m are in noncyclic order, and 0 if any two subscripts are repeated.

The term, Π, resulting from the distribution of pressure, P, is given by

$$\Pi = \int_v P \, dx . \tag{3.223}$$

If the x_3 axis is chosen to coincide with the direction of rotation, Eq. (3.218) becomes

$$\Omega_{ij} + \omega^2 (I_{ij} - \delta_{i3} I_{3j}) = -\delta_{ij}\Pi , \tag{3.224}$$

under static conditions.

For a homogeneous ellipsoid with semi-axes a_1, a_2, and a_3, the potential energy tensor, Ω_{ij}, is given by

$$\Omega_{ij} = -2\pi G \rho A_i I_{ij} , \tag{3.225}$$

where the moment of inertia tensor, I_{ij}, is given by

$$I_{ij} = \tfrac{1}{5} M a_i^2 \delta_{ij} , \tag{3.226}$$

the mass, M, is given by

$$M = \tfrac{4}{3}\pi a_1 a_2 a_3 \rho , \tag{3.227}$$

and

$$A_i = a_1 a_2 a_3 \int_0^{\infty} \frac{du}{\Delta (a_i^2 + u)} , \tag{3.228}$$

where

$$\Delta^2 = (a_1^2 + u)(a_2^2 + u)(a_3^2 + u) \ . \tag{3.229}$$

The potential, U, at an internal point, x_i, of a solid homogeneous ellipsoid is given by

$$U = \pi G \rho \left(I - \sum_{l=1}^{3} A_l x_l^2 \right) \ , \tag{3.230}$$

where

$$I = a_1 a_2 a_3 \int_0^\infty \frac{du}{\Delta} = \sum_{i=1}^{3} a_i^2 A_i \ . \tag{3.231}$$

The potential, U, of a homogeneous ellipsoid at an external point, x_i, is given by

$$U = \pi G \rho a_1 a_2 a_3 \int_\lambda^\infty \frac{du}{\Delta} \left(1 - \sum_{i=1}^{3} \frac{x_i^2}{a_i^2 + u} \right) \ , \tag{3.232}$$

where the ellipsoidal coordinate of the point considered is λ, and is given by the positive root of the equation

$$\sum_{i=1}^{3} \frac{x_i^2}{a_i^2 + \lambda} = 1 \ . \tag{3.233}$$

It follows from Eqs. (3.224) and (3.225) that the angular rotation velocity, ω, for a Maclaurin spheroid for which $a_1 = a_2$ is given by (Maclaurin, 1742)

$$\omega = \left\{ 2\pi G \rho \left[\frac{(1 - e^2)^{1/2}}{e^3} (3 - 2e^2) \sin^{-1} e - \frac{3}{e^2} (1 - e^2) \right] \right\}^{1/2} \ , \tag{3.234}$$

where the eccentricity, e, is given by

$$e = \left(1 - \frac{a_3^2}{a_1^2} \right)^{1/2} \ . \tag{3.235}$$

The angular momentum, L, of the Maclaurin spheroid is given by

$$L = \left(G M^3 \langle a \rangle \right)^{1/2} \frac{\sqrt{3}}{5} \left(\frac{a_1}{\langle a \rangle} \right)^2 \frac{\omega}{\sqrt{\pi G \rho}} \ , \tag{3.236}$$

where

$$\langle a \rangle = (a_1^2 a_3)^{1/3}$$

is the radius of the sphere of the same mass, M, as the spheroid, and the mass M is given by Eq. (3.227). The angular velocity increases from zero at $e = 0$ to a maximum, ω_{max}, given by

$$\omega_{\mathrm{max}}^2 = 0.449331 (\pi G \rho) \ \text{at} \ e = 0.92995 \ , \tag{3.237}$$

and then decreases again to zero as e approaches unity (cf. Fig. 3.2). The angular momentum, L, increases monotonically from zero to infinity as e varies from 0 to 1. Maclaurin spheroids become, however, secularly unstable via viscous processes for

$$0.81267 < e \leq 0.95289 \ ,$$

and they become dynamically unstable for

$$0.95289 \leq e \leq 1 \ . \tag{3.238}$$

Shapiro and Teukolsky (1983) provide formulae for the kinetic energy, T, and the gravitational potential energy, W, of a Maclaurin spheroid, which rotates with a uniform angular velocity $\omega = \Omega$. They obtain the useful parameter:

$$\frac{T}{|W|} = \frac{3}{2e^2}\left(1 - \frac{e(1-e^2)^{1/2}}{\sin^{-1} e}\right) - 1 \ , \tag{3.239}$$

noting that $T/|W| \geq 0.14$ for secular instability and $T/|W| \geq 0.26$ for dynamical instability for both the Maclaurin spheroids and for a wide range of angular momentum distributions and equations of state.

Jacobi (1834) first realized that equilibrium configurations are possible for ellipsoidal ($a_1 \neq a_2$) as well as for spheroidal ($a_1 = a_2$) configurations. It follows from Eqs. (3.224) and (3.225) that the angular rotation velocity, ω, for the Jacobi ellipsoid is given by (Jacobi, 1834)

$$\omega = \left[2\pi G\rho a_1 a_2 a_3 \int\limits_0^\infty \frac{u\,du}{(a_1^2+u)(a_2^2+u)\Delta}\right]^{1/2} \ , \tag{3.240}$$

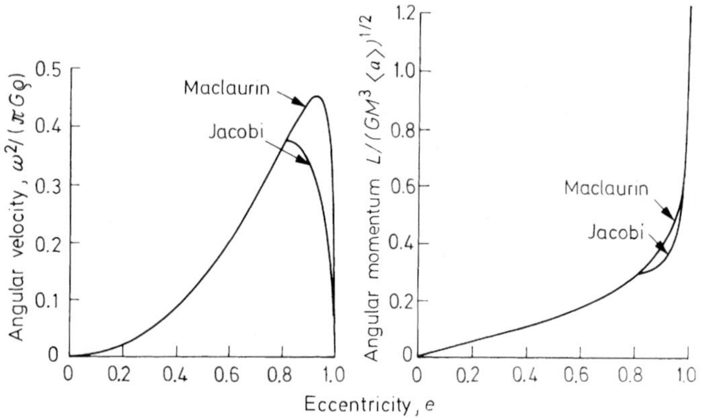

Fig. 3.2. The square of the angular velocity, ω, and the angular momentum, L, for the Maclaurin and Jacobi spheroids of different eccentricity, e. (After Chandrasekhar, 1969, by permission of the Yale University Press). Here G is the Newtonian gravitational constant, M is the mass, ρ is the mass density, and $< a >$ is the mean radius

provided that the geometrical restriction

$$a_1^2 a_2^2 \int_0^\infty \frac{du}{(a_1^2+u)(a_2^2+u)\Delta} = a_3^2 \int_0^\infty \frac{du}{(a_3^2+u)\Delta}$$

is satisfied. This restriction allows a solution for a_3 which satisfies the inequality

$$\frac{1}{a_3^2} > \frac{1}{a_1^2} + \frac{1}{a_2^2} \ .$$

The angular momentum, L, of the Jacobi ellipsoid is given by

$$L = \left(GM^3 \langle a \rangle\right)^{1/2} \frac{\sqrt{3}}{10} \frac{a_1^2 + a_2^2}{\langle a \rangle^2} \frac{\omega}{\sqrt{\pi G \rho}} \ ,$$

where

$$\langle a \rangle = (a_1 a_2 a_3)^{1/3} \ . \tag{3.241}$$

The Maclaurin spheroid is one solution of the Jacobi ellipsoidal equations for eccentricities, e, and angular velocities, ω, up to the bifurcation point given by

$$\omega = [0.37423\pi G\rho]^{1/2} \text{ and } e = 0.81297 \ . \tag{3.242}$$

Here $a_1 = a_2, a_3 = 0.582724 a_1$ and $L = 0.303751(GM^3\langle a \rangle)^{1/2}$. For $\omega \le [0.37423\pi G\rho]^{1/2}$ and $e \ge 0.81297$, there are three equilibrium configurations possible, two Maclaurin spheroids and one Jacobi ellipsoid, whereas for $0.3742 \le \omega^2/(\pi G\rho) \le 0.4493$ only the Maclaurin figures are possible. Along the Jacobi sequence $a_2/a_1, a_3/a_1$, and ω decrease monotonically to zero as e increases, whereas the angular momentum increases monotonically to infinity (cf. Fig. 3.2). Poincaré (1855) showed that the Jacobian sequence bifurcates into a new sequence of pear-shaped configurations when

$$\frac{a_2}{a_1} = 0.432232, \ \frac{a_3}{a_1} = 0.345069, \text{ and } \omega = [0.284030\pi G\rho]^{1/2} \ . \tag{3.243}$$

At this point the Jacobi ellipsoid becomes unstable and in this respect is different from the Maclaurin sequence which remains stable on either side of the bifurcation point (Cartan, 1924, Lyttleton, 1953). The bifurcation from the Maclaurin to the Jacobi sequences is discussed as a second-order phase transition by Bertin and Radicati (1976).

For an infinitesimal, homogeneous satellite of density, ρ, rotation about a rigid spherical planet of mass, M, in circular Keplerian orbit of radius, r, no equilibrium configurations are possible for rotational angular velocities, ω, which exceed the limit given by (Roche, 1847)

$$\frac{\omega^2}{\pi G \rho} = \frac{M}{\pi \rho r^3} \le 0.090093 \ . \tag{3.244}$$

A satellite or moon of a planet will be tidally disrupted when the differential force of the planet's gravitational attraction exceeds the self-gravitation of the satellite. From Eq. (3.244), this occurs when the orbital radius, r, of the satellite passes within the critical Roche limit, r_{cr}, given by

$$r_{cr} = 2.456 \left(\frac{\rho_P}{\rho_S} \right)^{1/3} R_P \ , \tag{3.245}$$

where r_{cr} is the critical distance from the planet center, R_P is the radius of the planet, the average mass density of the satellite is ρ_S, the average mass density, ρ_P, of the planet is $\rho_P = 3M_P/(4\pi R_P^3)$. A similar expression can be derived by equating the satellite's self-gravitation, GM_S/R_S^2, to the differential gravitational attraction of the planet between the near and far sides of the satellite, $2GM_P R_S/r_{cr}^3$. In this case the constant in Eq. (3.245) is 1.3 (Jeans, 1917, 1919).

Roche derived his result for a fluid, prolate satellite. The Roche limit for a solid body, derived by Aggarwal and Oberbeck (1974) is 1.38 R_P from the planet center or at an altitude of 0.38 R_P. The relativistic Roche problem is discussed by Fishbone (1975).

It is no accident that most planetary rings lie within the Roche limit of a planet. Any large satellite that passes within the Roche limit of a planet will be torn into pieces, creating a ring of innumerable small particles in orbit around the planet. The dynamics of planetary rings are reviewed by Goldreich and Tremaine (1982). Small satellites passing within the Roche limit may have enough internal cohesion to withstand the planet's differential gravitational forces. These smaller bodies can form the sharp edges of planetary rings (Borderies, Goldreich and Tremaine, 1982), and confine ring particles within narrow boundaries such as those observed for the thin rings of Uranus and Neptune (Goldreich and Tremaine, 1979).

A large portion of the stars in our Galaxy ($\sim 50\%$) are binaries, and a substantial fraction of the binaries revolve at a distance which is about the same order as the sum of their radii. When the more massive star has depleted its core of hydrogen, it will expand, but its expansion is limited by Roche's criterion (Eq. (3.244)). Such a limit may be related to the tendency for close binary systems to favor more outbursts (cf. Hack, 1963; Paczynski, 1971). If a star of mass, M_1, is in circular orbit about another star of mass, M_2, matter will flow outwards from one of the stars if its radius, R, becomes larger than the critical radius, R_{cr}, given by (Paczński, 1971)

$$\begin{aligned} R_{cr} &= \left[0.38 + 0.2 \log \left(\frac{M_1}{M_2} \right) \right] A \quad \text{for } 0.3 < \frac{M_1}{M_2} < 20, \\ R_{cr} &= 0.46224\, A \left(\frac{M_1}{M_1 + M_2} \right)^{1/3} \quad \text{for } 0 < \frac{M_1}{M_2} < 0.8 \ . \end{aligned} \tag{3.246}$$

Here A is the separation between the star centers and is related to the orbital period, P, by Kepler's law

$$\left(\frac{2\pi}{P} \right)^2 A^3 = G(M_1 + M_2) \ ,$$

where P is the orbital period and the gravitational constant $G = 6.668 \times 10^{-8}$ dyn cm^2 g^{-2}. Once a star has expanded beyond the critical radius, R_{cr}, the rate of mass outflow, dM/dt, is given by

$$\frac{dM}{dt} \approx \left(\frac{R - R_{\mathrm{cr}}}{R}\right)^{n+1.5} ,$$

where R is the stellar radius and n is the polytropic index.

The radii, R_{cr}, of Roche lobes, in units of the star separation, A, can be approximated to better than 1% by the equation (Eggleton, 1983)

$$\frac{R_{\mathrm{cr}}}{A} = \frac{0.49\, q^{2/3}}{0.6\, q^{2/3} + \ln(1 + q^{1/3})} \tag{3.247}$$

where the mass ratio $q = M_1/M_2$ and the approximation is valid for $0 < q < \infty$. If a star just fills its Roche lobe, its mean density, ρ, is related to its orbital period, P, by

$$\rho^{1/2}P = 0.1375\left(\frac{q}{1+q}\right)^{1/2}\left(\frac{R_{\mathrm{cr}}}{A}\right)^{-3/2}$$

for a separation A between star centers. In this numerical approximation, ρ is in units of g cm^{-3} and P is in days. The geometry of the eclipse of a pointlike star by its Roche-lobe-filling companion is given by Chanan, Middleditch and Nelson (1976).

3.5 Gas Macrostructure – Hydrodynamics

3.5.1 The Continuity Equation for Mass Conservation

In many circumstances a gas may be regarded as a continuous fluid. The continuity equation for the conservation of gas mass for a fluid element having volume, V, is, in this case, expressed by

$$\frac{1}{\rho V}\frac{D}{Dt}(\rho V) = \frac{1}{\rho}\frac{D\rho}{Dt} + \frac{1}{V}\frac{DV}{Dt} = 0 , \tag{3.248}$$

where ρ is the mass density, and D/Dt denotes a differentiation following the fluid element. For a scalar, φ, and a vector, $\boldsymbol{F}$,

$$\frac{D\varphi}{Dt} = \frac{\partial\varphi}{\partial t} + \boldsymbol{v}\cdot\nabla\varphi \tag{3.249}$$

and

$$\frac{D\boldsymbol{F}}{Dt} = \frac{\partial\boldsymbol{F}}{\partial t} + (\boldsymbol{v}\cdot\nabla)\boldsymbol{F} ,$$

where $\boldsymbol{v}$ is the velocity of the moving fluid. Using Eq. (3.249) in Eq. (3.248), we obtain the continuity equation

$$\frac{\partial\rho}{\partial t} + \nabla\cdot(\rho\boldsymbol{v}) = \frac{\partial\rho}{\partial t} + \boldsymbol{v}\cdot\nabla\rho + \rho\nabla\cdot\boldsymbol{v} = 0 . \tag{3.250}$$

For an incompressible fluid, Eq. (3.250) becomes $\nabla\cdot\boldsymbol{v} = 0$.

3.5.2 Euler's Equation (the Navier-Stokes and Bernoulli's Equations)

The equation of motion of a fluid element is obtained from the law of conservation of momentum. Taking the time derivative of momentum we obtain Euler's (1755) force equation

$$\rho \frac{D\boldsymbol{v}}{Dt} = \rho \left[\frac{\partial \boldsymbol{v}}{\partial t} + (\boldsymbol{v} \cdot \nabla)\boldsymbol{v} \right] = \boldsymbol{F} - \nabla P \ , \tag{3.251}$$

where ρ is the mass density, $\boldsymbol{v}$ is the velocity, $\boldsymbol{F}$ is the external force (other than that due to gas pressure) acting on a unit volume, and P is the pressure. The contribution of gravity to the external force is

$$\boldsymbol{F} = -\rho \nabla \varphi \ , \tag{3.252}$$

where, from Poisson's equation, (Poisson, 1813), the gravitational potential, φ, is given by

$$\nabla^2 \varphi = 4\pi G \rho \ , \tag{3.253}$$

and the constant of gravity, $G = 6.668 \times 10^{-8}$dyn cm^2 g^{-2}. The viscous force of an incompressible fluid, one in which the mass density $\rho =$ constant, is given by

$$\boldsymbol{F} = \mu \nabla^2 \boldsymbol{v} \ , \tag{3.254}$$

where the coefficient of dynamic viscosity, μ, is given by Eqs. (3.23), (3.25), or (3.26). When Eq. (3.251) includes only the viscous force term, the resulting equation is called the Navier-Stokes equation (Navier, 1822; Stokes, 1845).

The force contribution due to the electric and magnetic fields of respective strengths $\boldsymbol{E}$ and $\boldsymbol{B}$ is

$$\boldsymbol{F} = \frac{1}{c} \boldsymbol{J} \times \boldsymbol{B} \ , \tag{3.255}$$

where the current $\boldsymbol{J}$ is given by

$$\boldsymbol{J} = \sigma \left[\boldsymbol{E} + \frac{\boldsymbol{v}}{c} \times \boldsymbol{B} \right] \ , \tag{3.256}$$

and the electrical conductivity, σ, is given by Eqs. (3.35) or (3.36). For a highly conducting medium, the electromagnetic force term is

$$\boldsymbol{F} = \frac{-1}{4\pi} \boldsymbol{B} \times (\nabla \times \boldsymbol{B}) \ , \tag{3.257}$$

where $\boldsymbol{B}$ and $\boldsymbol{v}$ are related by the diffusion equation

$$\frac{\partial \boldsymbol{B}}{\partial t} = \nabla \times (\boldsymbol{v} \times \boldsymbol{B}) + \frac{c^2}{4\pi\sigma} \nabla^2 B \approx \nabla \times (\boldsymbol{v} \times \boldsymbol{B}) \ . \tag{3.258}$$

For the steady flow of a gas, $\partial \boldsymbol{v}/\partial t = 0$, and Euler's equation (3.251), in the absence of external forces, becomes

$$\rho \boldsymbol{v} \cdot \nabla \boldsymbol{v} = -\nabla P \ , \tag{3.259}$$

which is Bernoulli's equation for compressible flow (Bernoulli, 1738). Here, ρ is the mass density, $\boldsymbol{v}$ is the velocity, and P is the pressure. Bernoulli's equation for compressible steady flow has the integral form

$$\frac{v^2}{2} + \int \frac{dP}{\rho} = \text{constant} \ . \tag{3.260}$$

The equation for incompressible steady flow is

$$\tfrac{1}{2}\rho v^2 + P = \text{constant} \ . \tag{3.261}$$

3.5.3 The Energy Equation

From the law of conservation of energy, the change per unit time in the total energy of the gas in any volume must equal the total flux of energy through the surface bounding that volume. When the effects of thermal conductivity and viscosity are unimportant, and in the absence of body forces like gravity, this relation may be expressed as

$$\frac{\partial}{\partial t}\int \left[\frac{1}{2}\rho v^2 + \rho u\right] dV = -\oint \rho v\left(\frac{1}{2}v^2 + h\right) \cdot \boldsymbol{n}\, ds \ , \tag{3.262}$$

where t is the time variable, ρ is the mass density, v is the velocity, u is the internal energy per unit mass, $h = u + P/\rho$ is the enthalpy per unit mass, $\oint$ denotes a closed surface integral, $\boldsymbol{n}$ is a unit vector normal to the surface, ds is an element of area, and dV is an element of volume. In vector form, the energy equation (3.262) becomes

$$\frac{\partial}{\partial t}\left[\frac{1}{2}\rho \boldsymbol{v}^2 + \rho u\right] = -\nabla \cdot \left[\rho \boldsymbol{v}\left(\frac{1}{2}\boldsymbol{v}^2 + h\right)\right] \ , \tag{3.263}$$

and the quantity $\rho\boldsymbol{v}[(\boldsymbol{v}^2/2) + h]$ is called the energy flux density.

When thermal conductivity and viscous effects are important, the equation of energy balance can be written

$$\rho\frac{\partial}{\partial t}(c_v T) + \rho v_j \frac{\partial}{\partial x_j}(c_v T) = \frac{\partial}{\partial x_j}\left(\kappa \frac{\partial T}{\partial x_j}\right) - \rho P \frac{\partial v_j}{\partial x_j} + \varphi \ , \tag{3.264}$$

where c_v is the specific heat at constant volume, κ is the coefficient of heat conduction, and the rate at which energy is dissipated by viscosity is given by

$$\varphi = \frac{\mu}{2}\left[\left(\frac{\partial v_i}{\partial x_j} + \frac{\partial v_j}{\partial x_i}\right)^2 - \frac{4}{3}\left(\frac{\partial v_j}{\partial x_j}\right)^2\right] \ ,$$

where μ is the coefficient of dynamic viscosity, and $\partial v_i/\partial x_i = 0$ for an incompressible fluid. Under the Boussinesq approximation to be discussed later, Eq. (3.264) becomes Eq. (3.57):

$$\frac{\partial T}{\partial t} + \boldsymbol{v} \cdot \nabla T = \frac{\kappa}{\rho c_v}\nabla^2 T \ .$$

Eq. (3.263) is the energy balance equation which states that the rate of energy change per unit volume is equal to the amount of energy flowing out of this volume in unit time. If conduction, viscosity, and the kinetic energy term are ignored, and if all particles are at the same kinetic temperature, T, then Eq. (3.263) becomes, under conditions of constant pressure, (Field, 1965)

$$\rho \frac{d}{dt}\left(\frac{3}{2}NkT\right) - \frac{5}{2}\rho kT \frac{dN}{dt} = \rho(\Gamma - \Lambda) \ , \tag{3.265}$$

where N is the number density of free particles, the left hand side of Eq. (3.265) is the rate of increase of thermal energy plus the work done by the gas, $\rho(\Gamma - \Lambda)$ is the energy input per gram per second, and Γ and Λ denote, respectively, the rate of energy gain or loss per unit volume. When $\Gamma = \Lambda$ the gas is in thermal equilibrium and has the equilibrium temperature, T_E. When T is different from T_E, then an effective cooling time, t_T, may be defined by the relation

$$\frac{d}{dt}\left(\frac{3}{2}NkT\right) = -\frac{3N\,k(T - T_E)}{2t_T} \ . \tag{3.266}$$

Equilibrium temperatures of ionized hydrogen (H II) regions ($T_E \approx 10^{4\circ}$K), and neutral hydrogen (H I) regions ($T_E \approx 10^{2\circ}$K) are determined by equating the heat gained by photoionization of hydrogen, helium, or carbon to the cooling effects of the excitation of ions by electrons and the excitation of ions and hydrogen molecules by neutral atoms (cf. Spitzer, 1948, 1949, 1954; Spitzer and Savedoff, 1950; Seaton, 1951, 1954, 1955; Axford, 1961, 1964; Osterbrock, 1965). The detailed formulae for determining T_E in these cases are given in Spitzer (1962). As an example, the energy balance equation for the exchange of energy between neutral hydrogen atoms, H, and electrons, e, is given by (Spitzer and Savedoff, 1950)

$$\frac{d}{dt}\left(\frac{3}{2}N_H kT_H\right) = -\gamma N_e N_H \quad \mathrm{erg\,cm^{-3}\,sec^{-1}} \ , \tag{3.267}$$

where

$$\gamma = 8Q\frac{m_e}{M_H}\left(\frac{2kT_e}{\pi m_e}\right)^{1/2} k(T_H - T_e)$$
$$\approx 1.16 \times 10^{-27}\, T_e^{1/2}(T_H - T_e) \quad \mathrm{cm^3\,erg\,sec^{-1}} \ ,$$

and the elastic collision cross section $Q \approx 6.3 \times 10^{-15}\,\mathrm{cm}^2$ (Seaton, 1955). Eqs. (3.266) and (3.267) lead to a cooling time of

$$t_{TH} \approx 1.8 \times 10^{11}\, N_e^{-1}\, T_e^{-1/2} \quad \mathrm{sec} \ , \tag{3.268}$$

or using $N_e \approx 2 \times 10^{-4}\, N_H$,

$$t_{TH} \approx 2.8 \times 10^{7}\, N_H^{-1}\, T_e^{-1/2} \quad \mathrm{years} \ . \tag{3.269}$$

If the cooling time defined in Eq. (3.266) is negative, then the gas is thermally unstable. In this case, T is greater than T_E and the kinetic energy grows until the cooling time changes. Field (1965) uses Eq. (3.265) to show that the gas will remain thermally stable if

$$T\frac{\partial}{\partial T}(\Gamma - \Lambda) - \rho\frac{\partial}{\partial \rho}(\Gamma - \Lambda) < 0$$

or

$$\left[\frac{\partial}{\partial T}(\Gamma - \Lambda)\right]\left[1 + \frac{\rho}{T_E}\frac{dT_E}{d\rho}\right] < 0 \ .$$

Osterbrock (1974, 1989) reviews the thermal equilibrium of gaseous nebulae such as planetary nebulae and ionized hydrogen, H II, regions. Heating and cooling processes for the interstellar medium, denoted by the gain Γ and loss Λ of thermal energy, have been delineated by Spitzer (1978). Representative temperatures, T, of $T \approx 10^2$ °K and $T \approx 10^4$ °K have been obtained for neutral hydrogen, H I regions, and for H II regions, respectively. Typical calculations of the equilibrium temperatures are illustrated in Fig. 3.3. Since the time of those computations, considerations of powerful stellar explosions, known as supernovae, have led to a new picture of a more violent interstellar medium (Mc Cray and Snow, 1979; Tenorio-Tagle and Bodenheimer, 1988), and to research on a hot component of the interstellar gas with temperatures of $T \approx 10^6$ °K (Spitzer, 1990), including a bubble of hot gas in the local interstellar medium (Cox and Reynolds, 1987). We now know that dense molecular clouds, with temperatures as low as $T \approx 10$ °K are also found in the interstellar medium (see Sections 2.14 and 2.15).

3.5.4 Atmospheres – Hydrostatic Equilibrium, the Barometric Equation, Scale Height, Escape Velocity, Stellar Winds, and the Solar Corona

In the absence of mass flow, Euler's equation (3.251) becomes

$$\boldsymbol{F} - \nabla P = 0 \ , \tag{3.270}$$

where $\boldsymbol{F}$ is the external force and $\boldsymbol{P}$ is the pressure. When the outward force of the gas pressure is just balanced by the inward force of gravity, Eq. (3.270) becomes

$$\frac{dP}{dr} = -\rho\frac{GM(r)}{r^2} \ , \tag{3.271}$$

which is the equation of hydrostatic equilibrium for a spherical mass of gas. Here r is the distance from the center of the sphere, P is the pressure, ρ is the mass density, the gravitational constant $G = 6.67 \times 10^{-8}$dyn cm^{2}g^{-2}, and the mass, $M(r)$, within a sphere of radius, r, is given by

$$M(r) = \int_0^r 4\pi r^2 \rho \, dr \ . \tag{3.272}$$

For a perfect gas,

$$P = \frac{k\rho T}{\mu m_H} \ , \tag{3.273}$$

where T is the temperature, μ is the mean molecular weight, Boltzmann's constant $k \approx 1.380658(12) \times 10^{-16}$ erg °K^{-1}, and the mass of the hydrogen atom is $m_H \approx 1.67 \times 10^{-24}$ gm. For a perfect gas atmosphere surrounding a

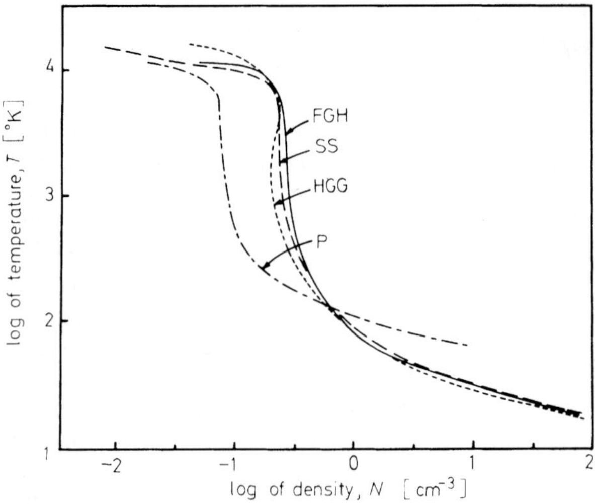

Fig. 3.3. The equilibrium temperature, T, of the interstellar medium as a function of the total number density, N, of nuclei (after Pikel'ner, 1968, (P), Field, Goldsmith, and Habing, 1969 (FGH), Spitzer and Scott, 1969, (SS), and Hjellming, Gordon, and Gordon, 1969 (HGG). The figure shows two thermally stable gas phases which coexist in pressure equilibrium; one for $T = 10^4$ °K and one for $T < 300$ °K. It is assumed that there is an equilibrium between heating by cosmic rays and cooling by inelastic thermal collisions. A cosmic ray energy density of $W = 6 \times 10^{-14}$ erg cm^{-3} in 2 MeV protons and an ionization rate of 4×10^{-16} sec^{-1} are assumed. The observed ionization rate is 1.5×10^{-15} sec^{-1}.

spherical mass, M, of radius, R, Eqs. (3.271) and (3.273) give the barometric equation

$$P(r) = P(R) \exp\left[-\int_R^r \frac{dr}{H}\right] , \tag{3.274}$$

$$= P(R)\exp\left[-\frac{(r-R)}{H}\right] , \tag{3.275}$$

where $P(R)$ is the pressure at the surface of the mass, M, the scale height, H, is given by

$$H = \frac{kT}{\mu m_{\rm H} g} ,$$

$$\approx \frac{kTR^2}{\mu m_{\rm H} GM} , \tag{3.276}$$

and the g is the local acceleration of gravity given by

$$g = \frac{GM(r)}{r^2} \approx \frac{GM}{R^2} . \tag{3.277}$$

In order for a particle of mass, m, to escape from a larger mass, M, the particles must have kinetic energy, $mv^2/2$, greater than the gravitational potential energy mMG/R. This means the velocities, v, must be greater than the escape velocity, v_{esc}, given by

$$v_{esc} = \left(\frac{2MG}{R}\right)^{1/2} . \tag{3.278}$$

The escape velocity for the Sun, for example, is 617.7 km sec^{-1}.

Observations of the optical continuum emission of the Sun during a solar eclipse show a corona of light extending beyond the visual extent of the uneclipsed Sun for several solar radii. Part of this light, the K corona, is thought to be due to Thomson scattering of solar photons by free electrons in the corona. Measurements of the intensity of the coronal light lead to measurements of the coronal electron density, $N_e(r)$, as a function of distance, r, from the Sun. Observations by Baumbach (1937) and Allen (1947), give a coronal model near the Sun ($r \approx R_\odot$)

$$N_e(r) \approx 1.55 \times 10^8 \left(\frac{r}{R_\odot}\right)^{-6} \left[1 + 1.93 \left(\frac{r}{R_\odot}\right)^{-10}\right] \mathrm{cm}^{-3} , \tag{3.279}$$

where the solar radius $R_\odot \approx 6.96 \times 10^{10}$cm. More recent measurements using the scintillations of radio sources give (Erickson, 1964)

$$N_e(r) \approx 7.2 \times 10^5 \left(\frac{r}{R_\odot}\right)^{-2} \mathrm{cm}^{-3} , \tag{3.280}$$

for $r \gg R_\odot$.

Measurements of N_e are given as a function of r for the Sun in Fig. 3.4.

If the corona is isothermal and is in hydrostatic equilibrium, then the barometric equation (3.274) gives

$$N_e(r) = N_e(r_0) \exp\left[\frac{GM_\odot \mu m_H}{R_\odot kT}\left(\frac{1}{r} - \frac{1}{r_0}\right)\right] , \tag{3.281}$$

whereas the equations of hydrostatic equilibrium and thermal conduction give

$$N_e(r) = N_e(r_0)\left(\frac{r}{r_0}\right)^{2/7} \exp\left[-\frac{7r_0}{5H}\left\{1 - \left(\frac{r}{r_0}\right)^{-5/7}\right\}\right] .$$

Here $N_e(r_0)$ is the electron density at some reference distance, r_0. When Eq. (3.281) is used together with measurements of $N_e(r)$ near the Sun, a coronal temperature of $T \approx 1.5 \times 10^6$ °K is obtained (De Jager, 1959). This temperature agrees with that obtained from radio observations of the Sun (Martyn, 1946, 1948; Pawsey, 1946; Bracewell and Preston, 1956) and from the intensity and width of optical lines observed during solar eclipse (Edlén, 1942; Burgess, 1964). Mass and energy flow in the solar chromosphere and corona were reviewed by Withbroe and Noyes (1977). The basic energy source for the outer solar atmosphere is believed to be acoustic waves (chromosphere) or magnetohydrodynamic waves or magnetic reconnection (corona), see Sections 3.5.6 and 3.5.12, but the exact mechanisms remain controversial.

Chapman (1957, 1959) first showed that energy transport by thermal conduction in the outer solar atmosphere exceeds radiative energy loss, with the result that the 10^6 °K observed near the Sun must extend beyond the orbit of the Earth with little decline. The stationary heat flow equation is

$$\nabla \cdot (\kappa \nabla T) = 0 \ , \tag{3.282}$$

where the coefficient of heat conductivity for a fully ionized gas is (Eq. (3.55))

$$\kappa \approx 6 \times 10^{-6} T^{5/2} \quad \mathrm{erg\,cm^{-1}\,sec^{-1}\,°K^{-1}} \ . \tag{3.283}$$

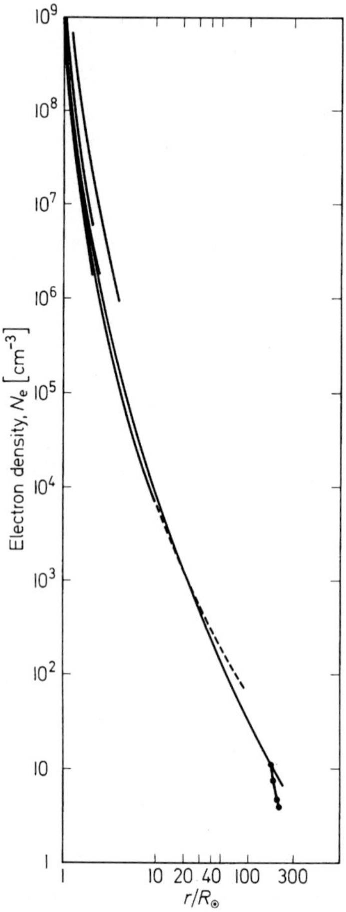

Fig. 3.4. Observed electron densities of the equatorial solar corona as a function of distance, r, from the Sun (after Newkirk, 1967, by permission of Annual Reviews Inc.). No attempt has been made to rectify data taken during different portions of the sunspot cycle. The long solid line corresponds to the theoretical model of Whang, Liu, and Chang (1966). The solar radius, $R_\odot \approx 6.96 \times 10^{10}$ cm

It follows from Eqs. (3.282) and (3.283) that the temperature, $T(r)$, at a distance, r, from the Sun in a static solar atmosphere is given by

$$T(r) = \left(\frac{R_\odot}{r}\right)^{2/7} T(R_\odot) \ , \tag{3.284}$$

where $T(R_\odot) \approx 10^6$ °K and the solar radius, $R_\odot \approx 6.96 \times 10^{10}$ cm. The rate of thermal energy loss, Q, is given by (Eq. (3.53))

$$Q = -4\pi r^2 \kappa \nabla T = -\frac{8\pi}{7} R_\odot \kappa(T_\odot) T(R_\odot) \ . \tag{3.285}$$

From Eq. (3.284), the temperature at the Earth is $T(r \approx 215 R_\odot) \approx 4 \times 10^5$ °K, and from Eq. (3.285) the thermal energy loss rate is $Q \approx 2 \times 10^{27}$ erg sec^{-1}. Measurements of solar wind protons and electrons near the Earth indicate a proton temperature of 1.2×10^5 °K and an electron temperature of 1.4×10^5 °K (Lang, 1992).

Birkeland (1908) and Chapman (1918, 1919) first suggested that particles could escape from the Sun and flow to the Earth causing geomagnetic storms and the aurorae. Then Biermann (1951, 1957) noted that comet tails never fail to point away from the Sun by an amount over and above that caused by the radiation pressure of the Sun, and that this must be due to a continuous radial flow of corpuscles with velocities of a few hundred km sec^{-1}. If all of the heat flux, Q, given by Eq. (3.285) went into an expanding spherical corona, then

$$Q \approx 4\pi r^2 M \frac{V}{2} N(r)(V^2 + V_{esc}^2) \ , \tag{3.286}$$

where $N(r)$ is the coronal density at distance, r, from the Sun, M is the particle mass, V is the constant solar wind velocity, and the escape velocity for the Sun is $V_{esc} \approx 617$ km sec^{-1}. Assuming $V \ll V_{esc}$, and using $Q = 2.5 \times 10^{27}$ erg sec^{-1}, we obtain a proton flux density of $N(r)V \approx 2.4 \times 10^8$ cm^{-2} sec^{-1} at the Earth orbit for which $r = 1.5 \times 10^{13}$ cm. The observed proton density near the Earth is 5 cm^{-3} implying $V \approx 500$ km sec^{-1}, near the 400 km sec^{-1} measured by satellites (Lüst, 1970; Lang, 1992).

Parker (1958) first suggested that the dynamical origin of the solar corpuscular radiation must be sought in the hydrodynamic equations describing an expanding solar atmosphere. For steady flow, Euler's equation (3.251) can be written

$$\rho \boldsymbol{v} \cdot \nabla \boldsymbol{v} = -\nabla P - \rho g = -\nabla P - \frac{\rho M_\odot G}{r^2} \ , \tag{3.287}$$

where ρ is the mass density, v is the velocity, P is the pressure, g is the acceleration due to gravity, the solar mass, $M_\odot \approx 1.989 \times 10^{33}$ g, the gravitational constant, $G \approx 6.668 \times 10^{-8}$ dyn cm^2 g^{-2}, and r is the radial distance from the center of the Sun. For a compressible gas flowing down a tube of varying cross sectional area, A,

$$A\rho v = \text{ constant} \ , \tag{3.288}$$

and if the solar atmosphere expands with spherical symmetry, $A \propto r^2$. Hence using Eqs. (3.287) and (3.288) we obtain

$$\left[\frac{v^2}{s^2} - 1\right]\frac{dv}{v} = \left[2 - \frac{M_\odot G}{s^2 r}\right]\frac{dr}{r} , \tag{3.289}$$

where the speed of sound

$$s = \left(\frac{dP}{d\rho}\right)^{1/2} \approx \left(\frac{\gamma P}{\rho}\right)^{1/2} = \left(\frac{\gamma k T}{\mu m_{\mathrm{H}}}\right)^{1/2} \approx \left(8.5 \times 10^7 \frac{\gamma T}{\mu}\right)^{1/2}$$
$$\approx 1.7 \times 10^7 \mathrm{cm\ sec}^{-1} . \tag{3.290}$$

Here γ is the adiabatic constant, T is the temperature, μ is the mean molecular weight, and m_{H} is the mass of the hydrogen atom. Thus we see that the solar wind expands supersonically, $v > s$, for values of r greater than the critical value

$$r_{\mathrm{c}} = \frac{M_\odot G}{2s^2} \approx 3.5 R_\odot , \tag{3.291}$$

where the solar radius $R_\odot \approx 6.96 \times 10^{10}$ cm. The sound velocity may be compared with the thermal velocity of protons in the solar corona.

$$v_{\mathrm{th}} = \left(\frac{3kT}{m_{\mathrm{H}}}\right)^{1/2} \approx 1.6 \times 10^7 \mathrm{\ cm\ sec}^{-1} , \tag{3.292}$$

where m_{H} is the proton mass and the corona temperature is 10^6 °K.

Since the isothermal sound speed, $s = c_{\mathrm{s}} = \left(\frac{\gamma P}{\rho}\right)^{\frac{1}{2}} = \left(\frac{5kT}{3\mu m_{\mathrm{H}}}\right)^{\frac{1}{2}}$ is roughly equal to the thermal velocity $v_{\mathrm{th}} = \left(\frac{kT}{\mu m_{\mathrm{H}}}\right)^{\frac{1}{2}}$, we can rewrite Parker's basic wind formula as:

$$\frac{1}{v}\frac{dv}{dr}(v^2 - v_{\mathrm{th}}^2) = \frac{2v_{\mathrm{th}}^2}{r} - \frac{GM_\odot}{r^2} \tag{3.293}$$

for a wind speed v and a solar mass of $M_\odot$.

As outlined above, Parker (1958, 1959, 1960) showed that when the temperature of any corona declines less rapidly than r^{-1} the only steady equilibrium state is one of expansion to supersonic velocities at large radial distances from the surface of the parent body. A detailed summary of the velocity, density, and temperature of the stellar wind is given by Parker (1965). Parker's model has been successfully applied to the modulation of cosmic rays, and reviews on this subject are given by Forbush (1966).

Spacecraft measurements of the physical parameters of the solar wind near the Earth, at one astronomical unit from the Sun, have been summarized by Feldman et al. (1977) and Lang (1992). Typical values of some of them are given in Table 3.4.

The solar wind has a fast and a slow component (Lang 1995, 1996, 1997). The slow one moves at about 400 kilometers per second; the fast one travels at twice that speed. The slow component is associated with equatorial regions of the Sun, and may be a thermally-driven wind of the type described by Parker, an expected consequence of the corona's million-degree temperature. The high-speed wind pours forth from the polar coronal holes (Krieger, Timothy and Roelof, 1973; Nolte, 1976; Rickett and Coles, 1991; Phillips et al., 1995), but no one knows what gives this component its additional push.

The solar wind moves past all the planets, modulating their magnetic environments, called magnetospheres, and creating the heliosphere, a vast

Table 3.4. Mean values of solar-wind parameters at one astronomical unit from the Sun[1]

Particle density, N	$N \approx 10\,\mathrm{cm}^{-3}$
Velocity, V	$V \approx 400\,\mathrm{km\,s}^{-1}$ and $V \approx 800\,\mathrm{km\,s}^{-1}$
Temperature, T	$T \approx (1.2$ to $1.4) \times 10^5\,°\mathrm{K}$
Particle energy, kT	$kT \approx 10^{-11}\,\mathrm{erg} \approx 500\,\mathrm{eV}$
Particle energy density, NkT	$NkT \approx 10^{-10}\,\mathrm{erg\,cm}^{-3}$
Magnetic field strength, H	$H \approx 6 \times 10^{-6}$ Gauss

[1]Adapted from Lang (1992).

region in which physical conditions are dominated by the Sun (Lang, 1995). The heliosphere is centered on the Sun and extends to somewhere between 116 and 177 astronomical units from the Sun. The interaction of the solar wind with the Earth's magnetosphere is reviewed by Akasofu (1980); its interaction with the interstellar medium is reviewed by Holzer (1989). The influence of the solar wind upon the local flux of cosmic rays is discussed by Longair (1992).

Equations for thermally driven winds from stars are discussed by Parker (1963); rotation and/or magnetic fields have been included by Weber and Davis (1967), Mestel (1968) and Goldreich and Julian (1970).

We now know that virtually all single nearby main-sequence (dwarf) stars of late spectral type (F, G, K, M) emit detectable, quiescent (non- flaring) X-ray radiation with an absolute X-ray luminosity, L_X, of up to $L_X = 10^{30}\,\mathrm{erg\,s}^{-1}$, compared with the Sun's value of $L_{X\odot} = 2 \times 10^{27}\,\mathrm{erg\,s}^{-1}$ (Rosner, Golub and Vaiana, 1985). The X-ray emission is attributed to the thermal radiation of stellar coronae with temperatures of 1 to 10 million degrees. The thermal X-ray emission increases with stellar rotation speed (Pallavicini et al., 1981; Hempelman et al., 1995), presumably due to enhanced magnetism by internal dynamo action that leads to greater coronal heating.

Radio evidence for quiescent (non-flaring) nonthermal activity on main-sequence stars of late spectral type is reviewed by Lang (1994). The absolute thermal X-ray luminosity, L_X, and the absolute, nonthermal radio luminosity, L_R, are correlated for a wide variety of late-type, main-sequence stars (Benz and Güdel, 1994; Güdel et al., 1993, Güdel, Schmidt and Benz, 1994; Lang, 1996), with $\log(L_X/L_R) = 15.5 \pm 0.5\,\mathrm{Hz}$, where the units of L_X are erg s^{-1} and those of L_R are erg $\mathrm{s}^{-1}\,\mathrm{Hz}^{-1}$. Thus, the nonthermal particle acceleration process is probably related to coronal heating in these stars, and to magnetic activity resulting from internal rotation and convection.

3.5.5 Convection – Schwarzschild Condition, Prandtl Mixing Length Theory, Rayleigh and Nusselt Numbers, Boussinesq Equations

When convective mass motion is absent, and when the temperature of a gas is not constant, the gas will remain stable and in mechanical equilibrium. When convective motion is present, currents appear which tend to mix the gas and equalize its temperature. When a gas is in mechanical equilibrium, Euler's equation (3–251) becomes

$$\nabla P = -g\rho \ , \tag{3.294}$$

where P is the pressure, g is the acceleration due to gravity, and ρ is the mass density. Following an idea first suggested by Kelvin (1862), let us suppose that the surface of constant pressure, P, is not changed by a small perturbation. For a perfect gas,

$$P = \frac{k\rho T}{\mu m_u}, \tag{3.295}$$

where T is the temperature, k is Boltzmann's constant, μ is the mean molecular weight, and m_u is the atomic mass unit. For an adiabatic perturbation in a perfect gas, which is in mechanical equilibrium, we have the temperature gradient

$$\left(\frac{dT}{dr}\right)_{ad} = \frac{T}{P}\left(1 - \frac{1}{\gamma}\right)\frac{dP}{dr} = -\frac{g}{C_p}, \tag{3.296}$$

where it has been assumed that the gas is a sphere of radius, r. Here, γ is the adiabatic index and C_p is the specific heat at constant pressure. As first pointed out by Schwarzschild (1906), convection will occur, in the absence of damping, if

$$\left(\frac{dT}{dr}\right)_{str} > \left(\frac{dT}{dr}\right)_{ad},$$

where $(dT/dr)_{str}$ is the actual structural temperature gradient in the gas. The more general condition for convection in any kind of gas is

$$\left(\frac{dT}{dr}\right)_{str} > -\frac{gT}{C_pV}\left(\frac{\partial V}{\partial T}\right)_p, \tag{3.297}$$

where V is the volume and $(\partial V/\partial T)_p$ denotes a differential at constant pressure.

Sampson (1894) and Schwarzschild (1906) first suggested that the transfer of heat by radiation might be the predominant form of energy transfer in a star. For a star which is in radiative equilibrium, the equation for the temperature gradient is (Eddington, 1917)

$$\left(\frac{dT}{dr}\right)_{str} = \frac{-3\kappa\rho L(r)}{4acT^3 4\pi r^2}, \tag{3.298}$$

where the Rosseland mean opacity is κ, the radiation constant $a = 7.566 \times 10^{-15}$ erg cm^{-3} °K^{-4}, and $L(r)$ is the luminosity at the radius, r. The net radiative energy flux, F_{rad}, is given by

$$F_{rad} = \frac{16\sigma T^3}{3\kappa}\frac{dT}{dr}, \tag{3.299}$$

where $\sigma \approx 5.670 \times 10^{-5}$erg cm^{-2} °K^{-4}sec^{-1} is the Stefan–Boltzmann constant.

When convection occurs, the tendency is to reduce the structural temperature gradient until

$$\left(\frac{dT}{dr}\right)_{str} = \left(\frac{dT}{dr}\right)_{ad} = \frac{T}{P}\left(1 - \frac{1}{\gamma}\right)\frac{dP}{dr} = -\frac{g}{C_p}. \tag{3.300}$$

The equations which specify the convective motion of a gas are the continuity equation (3.250), the Navier–Stokes equation (3.251) to (3.254), and, for an incompressible fluid, the heat transfer equation (3.57)

$$\frac{\partial T}{\partial t} + \boldsymbol{v} \cdot \nabla T = \frac{\kappa}{\rho C_p} \nabla^2 T \ . \tag{3.301}$$

The solution to this equation involves a characteristic length, l, a velocity, v, and a temperature difference, ΔT, between the convective bubbles and their surroundings. When the convective bubble merges with its surroundings after travelling the distance, l, the length is called a "mixing-length". A typical value of l for a perfect gas in hydrostatic equilibrium is the scale height, H, given by

$$l \approx H = \frac{kT}{\mu m_{\mathrm{H}} g} \ , \tag{3.302}$$

where T is the temperature, μ is the mean molecular weight, m_{H} is the mass of the hydrogen atom, and g is the acceleration due to gravity. The temperature difference, ΔT, is given by

$$\Delta T = l\left(\left|\frac{dT}{dr}\right|_{\mathrm{str}} - \left|\frac{dT}{dr}\right|_{\mathrm{ad}}\right) \ . \tag{3.303}$$

Assuming that the pressure remains constant, the equation of state for a perfect gas and Eq. (3.303) give

$$\Delta\rho = \frac{\rho \Delta T}{T} = l\left(\left|\frac{d\rho}{dr}\right|_{\mathrm{str}} - \left|\frac{d\rho}{dr}\right|_{\mathrm{ad}}\right) \ , \tag{3.304}$$

where $\Delta\rho$ is the change in mass density, ρ, and $d\rho/dr$ denotes the radial gradient in ρ. For a bubble of volume, V, the buoyant force will be

$$F_{\mathrm{b}} = V g \Delta\rho = V \rho g \frac{\Delta T}{T} = \frac{V \rho g l}{T}\left(\left|\frac{dT}{dr}\right|_{\mathrm{str}} - \left|\frac{dT}{dr}\right|_{\mathrm{ad}}\right) \ . \tag{3.305}$$

Assuming that all of the work done by this force goes into the kinetic energy of the bubble, the bubble velocity, v, is given by

$$\begin{aligned} v &= \frac{l}{2}\left(\frac{g}{T}\right)^{1/2}\left(\left|\frac{dT}{dr}\right|_{\mathrm{str}} - \left|\frac{dT}{dr}\right|_{\mathrm{ad}}\right)^{1/2} \\ &= \frac{l}{2}\left(\frac{g}{\rho}\right)^{1/2}\left(\left|\frac{d\rho}{dr}\right|_{\mathrm{str}} - \left|\frac{d\rho}{dr}\right|_{\mathrm{ad}}\right)^{1/2} . \end{aligned} \tag{3.306}$$

It then follows that the energy flux, F_{conv}, of the convective flow is

$$\begin{aligned} F_{\mathrm{conv}} &= C_p \rho v \frac{l}{2}\left(\left|\frac{dT}{dr}\right|_{\mathrm{str}} - \left|\frac{dT}{dr}\right|_{\mathrm{ad}}\right) \\ &= \frac{C_p \rho l^2}{4}\left(\frac{g}{T}\right)^{1/2}\left(\left|\frac{dT}{dr}\right|_{\mathrm{str}} - \left|\frac{dT}{dr}\right|_{\mathrm{ad}}\right)^{3/2} . \end{aligned} \tag{3.307}$$

Eqs. (3.302) to (3.307) follow from the mixing-length theory first developed by Ludwig Prandtl (Prandtl, 1925, 1952). The mixing length theory of convective energy transport is discussed by Schwarzschild (1958), Cox and Giuli (1968) and Böhm–Vitense (1992).

The equations are put in a convenient form by using the relation

$$\frac{dT}{dr} = \frac{T}{H}\frac{d \ln T}{d \ln P} = \frac{T}{H}\nabla \ , \tag{3.308}$$

to obtain from Eqs. (3.299), (3.306), and (3.307) the relations (Böhm–Vitense, 1953, 1958; Henyey, Vardya, and Bodenheimer, 1965)

$$F_{\text{rad}} = \frac{16\sigma T^4}{3\kappa H f}\nabla,$$

$$v = \left[\frac{gl^2}{4H}(\nabla - \nabla')\right]^{1/2}, \tag{3.309}$$

$$F_{\text{conv}} = \frac{1}{2}C_p \rho v \frac{l}{H} T(\nabla - \nabla'),$$

and

$$F = \frac{16\sigma T^4}{3\kappa H} \cdot \nabla_{\text{rad}} = \sigma T_{\text{e}}^4 = F_{\text{conv}} + F_{\text{rad}},$$

where ∇' is the logarithmic gradient for individual turbulent elements, ∇_{rad} is the gradient which would be required if the total flux, F, were carried away by radiation, f is a diffusion correction which is near unity, C_p is the specific heat per unit mass at constant pressure, and T_{e} is the effective temperature.

When a layer of gas is considered, convective instability occurs when the Rayleigh number, R, is greater than some critical value. For a layer of thickness, d, under an adverse temperature gradient, $\Delta T/d$, we have (Rayleigh, 1916)

$$R = \frac{g\alpha}{\chi \nu}\frac{\Delta T}{d} d^4 \ , \tag{3.310}$$

where g is the acceleration due to gravity, χ and ν are, respectively, the coefficients of thermometric conductivity and kinematic viscosity, and the coefficient of volume expansion, α, is given by the equation of state

$$\rho = \rho_0[1 - \alpha(T - T_0)] \ , \tag{3.311}$$

where T_0 is the temperature for which the mass density $\rho = \rho_0$. For some gases and fluids, $\alpha \approx 10^{-3}$ to 10^{-4}.

In the "Boussinesq" approximation, the Euler's equation and the equation of heat conduction become (Boussinesq, 1903)

$$\frac{\partial}{\partial r}(\langle P \rangle + \rho \langle v^2 \rangle) = -g\rho$$

and

$$\frac{\partial}{\partial t}\langle T \rangle + \frac{\partial}{\partial r}(\langle v\Delta T \rangle) = \chi \frac{\partial^2 \langle T \rangle}{\partial r^2} \ , \tag{3.312}$$

where ΔT is the difference in temperature from its mean $\langle T \rangle$ and $\langle \ \rangle$ is taken to denote a horizontal mean. These equations can be made to give the mixing length equations (3.302) to (3.307) by replacing the spatial derivative of fluctuating quantities by l^{-1} and dropping the pressure and time derivatives.

The two characteristic numbers, the Peclet number, P_e, and the Reynolds number Re, are then given by

$$P_e = \frac{vl}{\chi} \approx (P_r R)^{1/2} ,$$
$$\mathrm{Re} = \frac{vl}{\nu} = \left(\frac{R}{P_r}\right)^{1/2} , \tag{3.313}$$

where R and P_r denote, respectively, the Rayleigh and Prandtl numbers. These numbers measure the ratio of turbulent motion to the damping effects of thermal conductivity and viscosity. The Nusselt number is a convenient way of expressing the sum, Q, of the convective and conductive heat flux.

$$N = \frac{Qd}{\kappa \Delta T} \tag{3.314}$$

which becomes unity for conduction without convection. From the mixing length theory we have (Spiegel, 1971)

$$N = \left[\frac{\sqrt{1+4P_r} - 1}{2P_r}\right]^{2/3} \left(\frac{P_r R}{R_c}\right)^{1/3} ,$$

where the Prandtl number $P_r = \nu/\chi$ and is given by Eq. (3.58) and R_c is the critical Rayleigh number given below.

Chandrasekhar (1961) has written an excellent text which includes a complete discussion of the convective instability of a layer heated from below. As first suggested by Rayleigh (1916), a layer of thickness, d, becomes unstable for

$$R > R_c , \tag{3.315}$$

where:

Surface	R_c	a	$2\pi/a$
Both free	657.511	2.2214	2.828
Both rigid	1707.762	3.117	2.016
One rigid and one free	1100.65	2.682	2.342

Here a disturbance of wavelength, λ, has wave number

$$a = \frac{2\pi d}{\lambda} .$$

For example, for two free boundaries

$$R = \frac{(\pi^2 + a^2)^3}{a^2} , \tag{3.316}$$

and the critical Rayleigh number for the onset of instability is set by the condition $\partial R/(\partial a^2) = 0$ for which $a = 2.214$ and $R = R_c = 657.511$. The stability criteria for both incompressible and compressible fluids are also discussed by Jeffreys (1926) and Jeffreys (1930), respectively.

If rotation is introduced, convection is inhibited; and an inviscid, ideal fluid becomes stable with respect to the onset of convection for all adverse temperature gradients. This is a consequence of the Taylor–Proudman theorem

(Taylor, 1921; Proudman, 1916) which states that all steady slow motions in a rotating inviscid fluid are necessarily two dimensional for they cannot vary in the direction of rotation. For a viscous, rotating fluid, convection is possible and is characterized by the Taylor number

$$T = \frac{4\omega^2 d^4}{\nu^2} , \tag{3.317}$$

where ω is the angular velocity, ν is the coefficient of kinematic viscosity, and d is the layer thickness. For the case of two free boundaries we have (Chandrasekhar, 1953; Chandrasekhar and Elbert, 1955)

$$R = \frac{1}{a^2}[(\pi^2 + a^2)^3 + \pi^2 T] , \tag{3.318}$$

and for large T,

$$R_c = 8.6956 T^{2/3} ,$$

for which

$$a = 1.3048 T^{1/6} .$$

Magnetic fields also inhibit thermal convection; and the effect of a magnetic field of strength, H, is characterized by the parameter (Thompson, 1951; Chandrasekhar, 1952, 1961)

$$Q = \frac{\mu^2 H^2 \sigma}{\rho \nu} d^2 , \tag{3.319}$$

where μ is the magnetic permeability, ρ is the mass density, ν is the coefficient of kinematic viscosity, σ is the electrical conductivity, and d is the layer thickness. For the case of two free boundaries we have

$$R = \frac{\pi^2 + a^2}{a^2}[(\pi^2 + a^2)^2 + \pi^2 Q] , \tag{3.320}$$

and for large Q,

$$R_c = \pi^2 Q,$$

for which

$$a = [\pi^4 Q/2]^{1/6} .$$

3.5.6 Sound Waves

Assume that a gas is in a static, uniform equilibrium condition in which the velocity, v_0, is zero and the density, ρ_0, and pressure, P_0, are constant. Next assume a perturbation such that the density $\rho = \rho_0 + \rho_1$ and the velocity $v = v_1$. Ignoring all external forces including that due to gravity, the continuity equation (3.250) and Euler's equation (3.251) become

$$\frac{\partial \rho_1}{\partial t} + \rho_0 \nabla \cdot v_1 = 0 \tag{3.321}$$

and

$$\frac{\partial v_1}{\partial t} = -\frac{1}{\rho_0}\nabla P_1 \ . \tag{3.322}$$

For an adiabatic or polytropic process involving an ideal gas

$$\nabla P_1 = \left(\frac{\gamma P_0}{\rho_0}\right)\nabla\rho_1 = \left(\frac{\partial P}{\partial\rho}\right)\nabla\rho_1 \ , \tag{3.323}$$

where γ is the adiabatic index or the polytropic exponent. Differentiating Eq. (3.321) with respect to time and using Eqs. (3.322) and (3.323), we obtain the wave equation

$$\frac{\partial^2\rho_1}{\partial t^2} = \left(\frac{\partial P}{\partial\rho}\right)\nabla^2\rho_1 \ , \tag{3.324}$$

which has the plane wave solution

$$\rho_1 \propto \exp\left[i\left(\frac{2\pi x}{\lambda} - \omega t\right)\right] \ , \tag{3.325}$$

where the frequency, ω, is related to the wavelength, λ, by the formula

$$\omega^2 = \left(\frac{2\pi}{\lambda}\right)^2\left(\frac{\partial P}{\partial\rho}\right) \ . \tag{3.326}$$

Both the pressure, P_1, and the velocity, v_1, also satisfy the wave equation (3.324). The "sound" velocity, s, of the perturbation waves is given by

$$s = \left(\frac{\omega\lambda}{2\pi}\right) = \left(\frac{\partial P}{\partial\rho}\right)^{1/2} = \left(\frac{\gamma P_0}{\rho_0}\right)^{1/2} = \left(\frac{\gamma k T_0}{\mu m_{\mathrm{H}}}\right)^{1/2} \ , \tag{3.327}$$

where the differentiation $\partial P/\partial\rho$ is carried out under conditions of constant entropy. Eq. (3.327) was first derived for the adiabatic case by Laplace (1816). For an adiabatic process, the pressure, P, and mass density, ρ, are related by $P = K\rho^{\gamma}$, where K is a constant. For adiabatic perturbations of a monatomic gas the index $\gamma = 5/3$ (Sect. 3.2.3) and for isothermal perturbations $\gamma = 1$. For adiabatic and isothermal perturbations:

$$s = c_{\mathrm{s}}^{\mathrm{ad}} = \left(\frac{5P}{3\rho}\right)^{1/2} = \left(\frac{5kT}{3\mu m_{\mathrm{H}}}\right)^{1/2} \propto \rho^{1/3} \text{ for adiabatic perturbations}$$

and

$$s = c_{\mathrm{s}}^{\mathrm{iso}} = \left(\frac{P}{\rho}\right)^{1/2} = \left(\frac{kT}{\mu m_{\mathrm{H}}}\right)^{1/2} \text{ for isothermal perturbations.} \tag{3.328}$$

The sound speed, s, is often designated c_{s}. Both $c_{\mathrm{s}}^{\mathrm{ad}}$ and $c_{\mathrm{s}}^{\mathrm{iso}}$ are of the order of the mean thermal speed of the ions of the gas, or numerically,

$$s = c_{\mathrm{s}} \approx 10(T/10^4\,{}^{\circ}\mathrm{K})^{1/2}\ \mathrm{km\ s^{-1}}.$$

As the velocity, v_1, is in the direction of propagation, these sound waves are called longitudinal waves. The Mach number, M, of the wave is the ratio

$$M = v_1/s \ . \tag{3.329}$$

The energy density, U, of the plane wave of sound is

$$U = \rho_0 v_1^2/2 \ , \tag{3.330}$$

and the energy flux density, S, is given by

$$S = s\rho_0 v_1^2/2 \ . \tag{3.331}$$

The boundary conditions for the reflection and transmission of sound waves at a boundary between two gases are determined by holding the pressures and normal velocity components equal at the interface between the two gases.

As a sound wave propagates in a gas, its intensity falls off with distance, x, as $\exp[-2\gamma x]$, where the absorption coefficient, γ, is given by (Landau and Lifshitz, 1959)

$$\gamma = \frac{\omega^2}{2\rho c^3}\left[\left(\frac{4}{3}\mu + \xi\right) + \kappa\left(\frac{1}{C_v} - \frac{1}{C_p}\right)\right] \ . \tag{3.332}$$

Here ω is the frequency of the sound wave, ρ is the mass density, the coefficient of dynamic viscosity, μ, is given by Eqs. (3.23), (3.24), or (3.25), the second coefficient of viscosity, ξ, is usually of the same order of magnitude as μ, the coefficient of heat conductivity, κ, is given by Eqs. (3.55) or (3.56), and C_v and C_p are, respectively, the specific heats at constant volume and pressure. The cross section for scattering of sound waves are the same as those given in Sects. 1.36 to 1.40.

Lighthill (1952, 1954) developed the theory for the generation of sound in a stratified atmosphere by fluid motions, and Proudman (1952) used a Hiesenberg turbulence spectrum to give a sound wave emissivity of

$$\varepsilon_{\mathrm{s}} = \frac{38\rho v^8}{l s^5} \ \mathrm{erg\ cm^{-3}\ sec^{-1}} \ , \tag{3.333}$$

where the constant 38 is a result of the choice of turbulence spectrum, ρ is the gas mass density, v is the turbulent velocity, l is the turbulent length scale, and s is the speed of sound. Osterbrock (1961) used the Böhm-Vitense model for turbulent motion (cf. Sect. 3.5.5) to obtain a solar flux of

$$F_{\mathrm{s}\odot} \approx 10^7 - 10^8 \ \mathrm{erg\ cm^{-2}\ sec^{-1}} \ . \tag{3.334}$$

Biermann (1946, 1947), Schwarzschild (1948) and Schatzman (1949) first suggested that turbulent motion in the convective zone of a star might create sound, or compressional, waves whose energy could heat the star's corona. However, spacecraft observations indicate that sound waves cannot heat the Sun's corona, since the sound energy is dissipated in the underlying parts of the solar atmosphere (Athay and White, 1978, 1979). Sound waves may nevertheless play a role in heating the chromosphere. Mechanisms of chromospheric and coronal heating are discussed in Ulmschneider, Priest and Rosner (1991).

Sound waves cannot propagate at frequencies below the critical frequency, ω_{s}, given by (Lamb, 1909)

$$\omega_{\mathrm{s}} = \frac{\gamma g}{2s} = \frac{s}{2H} \ , \tag{3.335}$$

where H is the scale height.

3.5.7 Helioseismology

We can illuminate the hidden depths of the Sun by observing the in-and-out motions of the photosphere. These oscillations are caused by sounds that echo and resonate inside the Sun. The name helioseismology is used to describe such studies of the solar interior; it is a hybrid name combining the Greek words *helios* for the Sun and *seismos* for quake or tremor.

The relatively new science of helioseismology has been reviewed by Deubner and Gough (1984), Gough and Toomre (1991), Däppen (1995), Gough (1995), Harvey (1995) and Gough, Leibacher, Scherrer and Toomre (1996). The technique is described for the general educated reader in the book by Lang (1995).

The heaving motions of the Sun's photosphere can travel at a few hundred meters per second and reach tens of kilometers in height, moving locally up to one ten thousandth (0.0001) of the Sun's radius. The throbbing motions that these sounds create are imperceptible to the naked eye, but sensitive instruments detect them from tiny, periodic Doppler shifts in a well-defined spectral line, or from minuscule but regular variations in the Sun's total light output.

The solar oscillations were discovered by Doppler-shift observations of spectral lines, indicating that the solar photosphere is moving in and out with a period of about five minutes (Leighton, 1961; Leighton, Noyes and Simon, 1962). We now know that these surface oscillations are the combined effect of about 10 million separate notes – each of which has a unique path of propagation and samples a well defined section inside the Sun (Fig. 3.5).

A major obstacle to obtaining precise solar-oscillation data is the Earth's rotation, which keeps us from observing the Sun around the clock. The nightly gaps in the data create background noise that hides all but the strongest oscillations, especially at low frequencies. This noisy confusion in the observed oscillations is reduced by continuous, uninterrupted observations from space and from the ground.

Three helioseismology instruments aboard the Solar and Heliospheric Observatory (SOHO) observe the Sun 24 hours a day, every day, from their strategic vantage point at the inner Lagrangian point (Lang, 1997). One of them, the Michelson Doppler Imager (MDI), detects averaged oscillation speeds with remarkable precision – to better than one millimeter per second, at a million points across the Sun every minute. The initial results from the MDI instrument aboard SOHO have been reviewed in a general way by Lang (1996, 1997), and in more technical terms by Duvall et al. (1997) and Kosovichev et al. (1997).

The Sun is also observed around the clock by a worldwide network of observatories, known by the acronym GONG for the Global Oscillation Network Group (Harvey et al., 1996). They form an unbroken chain that follows the Sun as the Earth rotates. At each site, imaging spectrometers measure very precise radial velocities at more than 40,000 points on the Sun's surface. (Important continuous data sets have been made by early world-wide networks, such as those of the British BISON group and France's IRIS collaboration, but they were restricted to measurements of the Sun's total light output.) The early results from the ground-based GONG observations have been described by Gough et al. (1996) and Thompson et al. (1996).

The five-minute vertical motions have been attributed to standing sound, or acoustic, waves that are trapped within the Sun and repeatedly strike the photosphere to produce the observed oscillations (Kahn, 1961; Leibacher and Stein, 1971; Wolff, 1972). An inward-traveling sound wave is refracted back up toward the surface because of the hotter gas and greater sound speed at increasing depths within the Sun. The inner turning point, or cavity bottom, depends on the trajectory of the wave (Fig. 3.5). Sound waves with longer trajectories penetrate deep into the Sun before they return, while waves with shorter trajectories travel through shallower and cooler layers and bounce off the visible photosphere more frequently. Most of the upward-traveling sound waves are turned around and deflected back inside the Sun just beneath the photosphere due to a sharp drop in density there; the waves become evanescent, or non-propagating, in the photosphere and above.

The sounds are therefore trapped inside the Sun; they cannot propagate through the near vacuum of space. Even if they could reach the Earth, they are too low for human hearing. The loudest sounds in the Sun are extremely low-

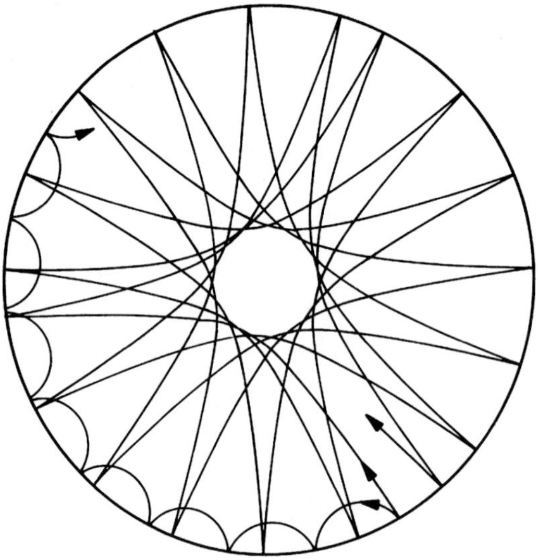

Fig. 3.5. As illustrated by these ray paths, sound waves in the Sun do not travel in straight lines. Instead, the Sun traps them within a spherical shell. The increasing speed of sound toward the center refracts inward-travelling waves outwards. The sharp drop in density at the Sun's visible surface reflects outward-travelling waves back in. How deep a wave penetrates and how far around the Sun it goes before it hits the surface depends on its trajectory. Some waves interfere constructively with themselves as they circle the Sun, creating resonances that are detectable as ripples, or oscillations, on the solar surface. [Adapted from Gough and Toomre (1991)]

pitched; the dominant frequencies cluster around one vibration per five minutes (0.003 Hertz). This is 12.5 octaves below the lowest note audible to humans, 20 vibrations per second (20 Hertz). Many of the Sun's other notes are even lower.

For spherical symmetry, the oscillation eigenfunctions can be specified as a function of spherical coordinates (r, θ, ϕ) and the displacement ξ written as (Gough and Toomre, 1991):

$$\xi = \left(\Xi_{nl} P_l^m, \frac{H_{nl}}{L} \frac{dP_l^m}{d\theta}, \frac{H_{nl}}{L \sin\theta} P_l^m \frac{\partial}{\partial\phi} \right) \cos(m\phi - \omega_{nl} t) \tag{3.336}$$

where the eigenfunctions Ξ_{nl}, H_{nl} are functions of radius, or distance from the Sun center, r, alone, $P_l^m(\cos\theta)$ is the associated Legendre function of the first kind of degree l and order $m, L^2 = l(l+1)$, and t is time. We refer to the integers l and m as the *degree* and *azimuthal order* of the mode, respectively. For zonal modes $(m = 0)$, all the nodal lines of the spherical harmonics are lines of latitude, θ; for sectoral modes $(m = l)$, they are lines of longitude, ϕ.

The eigenfunctions are also labeled with the integer n, called the *order* of the mode, in such a way that ω_{nl} increases with n at fixed l. The frequencies ω_{nl} also increase with l at fixed n; they are degenerate with respect to m. The group of degenerate, or as is actually the case, nearly degenerate (because the Sun is not precisely spherical) modes of like n and l are called a *multiplet*, each member of which is a *singlet*.

When either n or l is sufficiently large, asymptotic methods may be employed; all of the solar modes that have been observed and unambiguously identified fall into this category. The equation determining the eigenfrequencies can then be written using integrals along the ray path between the lower and upper turning points, r_1 and r_2; they are the radii between which the waves are trapped. In this case, we have (Gough and Toomre, 1991):

$$\frac{(n + \tilde{\alpha})\pi}{\omega} = \int_{r_1}^{r_2} \left[1 - \frac{\omega_L^2}{\omega^2} \left(1 - \frac{N^2}{\omega^2} \right) - \frac{\omega_c^2}{\omega^2} \right]^{1/2} \frac{d \ln r}{a} ,$$

where $\tilde{\alpha}$ is a constant of order unity, $a = c/r$, where c is the local adiabatic sound speed, and $\omega_L = La$ is the Lamb frequency, with $L^2 = l(l+1)$ or $(l + 1/2)^2$ depending on whether the spherical harmonic decomposition or ray theory is used,

$$N^2 = g \left(\frac{1}{H} - \frac{g}{c^2} \right) \equiv \frac{gv}{r}$$

is the square of the buoyancy frequency, g being the acceleration due to gravity, and ω_c is the acoustical cutoff frequency:

$$\omega_c^2 = \frac{c^2}{4H^2} \left(1 - 2\frac{dH}{dr} \right) ,$$

where H is the density scale height. Except in the convection zone (where generally N^2 is small and negative), N^2 and ω_c^2 are similar in magnitude; except near the surface, both are much less than the Lamb frequency ω_L, which is an approximate lower bound to the frequencies at which acoustic propagation can occur in the deep interior.

In the case of polytropes (Cowling, 1941; Cowling and Newing, 1949), the nonradial oscillation modes are either high-frequency acoustic modes (*p*-modes), for which pressure perturbations provide the main restoring force, or low-frequency, internal gravity modes (*g* modes) for which buoyancy dominates the restoring force. The sound, or acoustic, waves are dominant within the convection zone, while gravity waves are located within the solar core. Gravity waves become evanescent in regions where the stratified gas is not stable, such as the turbulent convection zone, so they are largely confined within the Sun's deep interior and will have low amplitudes and long periods of an hour or more when reaching the visible solar surface.

Early spatial and temporal analysis of the five-minute oscillations (Frazier, 1968) suggested that most, if not all, of the oscillatory power is indeed in the form of acoustic waves (*p*-modes). The observed oscillations are primarily the result of standing, resonant sound waves in the convection zone; and the gravitational modes of oscillation could not be detected in the available photospheric data. The sound waves are probably generated and continually excited by vigorous turbulence in the convection zone (Goldreich and Kumar, 1990).

For the five-minute oscillations, where ω is high and N^2/ω^2 and ω_c^2/ω^2 can be neglected compared with unity, we have (Duvall, 1982; Christensen-Dalsgaard et al., 1985):

$$\frac{(n+\alpha)\pi}{\omega} \sim \int_{r_1}^{r_2} \left(1 - \frac{a^2}{w^2}\right)^{1/2} \frac{d\ln r}{a} \equiv F(w) \equiv \int_{r_1}^{R_\odot} \left(\frac{r^2}{c^2} - \frac{1}{w^2}\right)^{1/2} \frac{dr}{r} , \tag{3.337}$$

where $w = \omega/L$, L is related to the degree l of the mode by $L^2 = l(l+1) = R_\odot^2 k_h^2$, the radius of the Sun is $R_\odot$, and k_h is the horizontal component of the wave number vector in the photosphere. Although the neglect of ω_c^2/ω^2 is a poor approximation near $r = r_2$ (where $\omega_c \simeq \omega$), this occurs in a relatively thin layer, and its influence can be absorbed into $\bar{\alpha}$, which is replaced by α. The integrand depends only weakly on l near the surface of the star, where $\omega_L^2/\omega^2 \ll 1$, and therefore α and the eigenfunctions can depend only weakly on l and one can ignore such dependence.

For a mode of degree l, the $L^2 = l(l+1)$, the horizontal wavenumber $k_h = L/r$ at radius r, and the vertical component, k_v, of the local wavenumber obeys the acoustic dispersion relation (Christensen-Dalsgaard et al., 1985):

$$k_v^2 = \frac{\omega^2 - \omega_c^2}{c^2} - \frac{L^2}{r^2} = k^2 - k_h^2 , \tag{3.338}$$

where $\omega > \omega_c$ inside the Sun, and k is the local wavenumber. The waves propagate only where the vertical wavenumber is real; elsewhere the mode is evanescent. Near the solar surface $\omega^2/c^2 \gg L^2/r^2$, unless L is very large, $k_v^2 \gg k_h^2$, and waves propagate almost vertically.

The inner turning point, or cavity bottom, depends on the increase of sound speed, or wave velocity, with depth. Because the speed of sound, c, is greater in a hotter gas, it increases in the deeper, hotter layers of the Sun. The deeper part of a wave front traveling obliquely into the Sun moves faster than the shallower part, and pulls ahead of it. Gradually, the advancing wave front is refracted, or curved and bent, until the wave is once again headed toward the Sun's surface.

Because c increases with depth, the mode propagates down to the lower turning point, at which $k_{\mathrm{v}} = 0$, where the waves undergo total internal reflection. In the bulk of the interior, the frequency ω_{c} is negligible, so the radius r_1 at which the wave is turned back is given by:

$$a = c/r_1 = \omega/L = w \ ,$$

where propagation is locally horizontal.

In other words, the dispersion relation for horizontally propagating acoustic waves in spherical geometry with a lower boundary at radius r_1 depends on the spherical harmonic degree, l, the local sound speed, c, and the frequency of oscillation, ω, with (Duvall, 1982; Duvall and Harvey, 1983)

$$\omega^2 = 4\pi^2\nu^2 = l(l+1)\frac{c^2}{r_1^2} \ , \tag{3.339}$$

or equivalently

$$l = -\frac{1}{2} + \left(\frac{1}{4} + \frac{4\pi^2\nu^2 r_1^2}{c^2}\right)^{1/2} .$$

Thus, r_1 is a function only of w/L, or ω/ℓ, or equivalently of ν/L, where $\nu \equiv \omega/2\pi$.

Beneath $r = r_1$ the mode's energy density decays exponentially with distance and hence the frequency is not very sensitive to conditions in that region. Amongst the observed p modes, ν varies by only a factor of about three, but L varies by several thousand. Thus modes of different degree can be used to probe the variation with depth of physical parameters within the solar interior. The modes also have an upper turning point, close to the surface (say at $r = r_2$), at which k_{v} is again zero. This is caused by the decrease in density scale height and consequent increase in ω_{c} in the outer layers. Provided l is not too large, this occurs approximately at the point at which $\omega = \omega_{\mathrm{c}}$.

Theoretical considerations indicate that about 10^7 vibrations are simultaneously excited in the solar interior, and that some of these trapped sound waves combine and reinforce each other to shake the photosphere in a regular way. These resonant sound waves are amplified by constructive interference (also see Fig. 3.5). Each individual vibration moves the surface of the Sun in and out by only a few tens of meters or less, at speeds of less than ten centimeters per second. When millions of these vibrations are superimposed, they produce surface oscillations with peak values as large as a few hundred meters per second, or thousands of times bigger than those of individual vibrations. These large-amplitude combinations are the well-known, five-minute oscillations detected as vertical motions in the photosphere. They grow and decay as individual vibrations go in and out of phase to combine and disperse and then combine again.

The trapped sound waves vary in space and time in a regular way, with the resultant oscillatory power concentrated at specific frequencies and sizes (Fig. 3.6). Theory predicts that the oscillations will be enhanced along narrow bands, or ridges, in a two-dimensional display of temporal frequency, or period, and horizontal wavelength, or size (Ulrich, 1970; Ando and Osaki, 1975). Observations of the predicted pattern have fully confirmed that the five-minute oscillations are due to standing acoustic waves trapped in the convection zone (Frazier, 1968; Deubner, 1975; Rhodes, Ulrich and Simon, 1977; Duvall and Harvey, 1983; Libbrecht, Woodard and Kaufman, 1990).

When integrating between the lower turning point and the upper one:

$$\pi(n+\varepsilon) = \int_{r_1}^{r_2} k_{\rm v}\, dr = \int_{r_1}^{r_2} \left(\frac{\omega^2}{c^2} - \frac{L^2}{r^2} - \frac{\omega_{\rm c}^2}{c^2}\right)^{\frac{1}{2}} dr.$$

Here n is an integer (in fact we can take it to be the order of the mode for the observed solar p modes) and ε is a phase which accounts for the asymptotic behavior near the turning points. By taking certain portions of the integral on the right over to the left-hand side and incorporating them and ε into a new

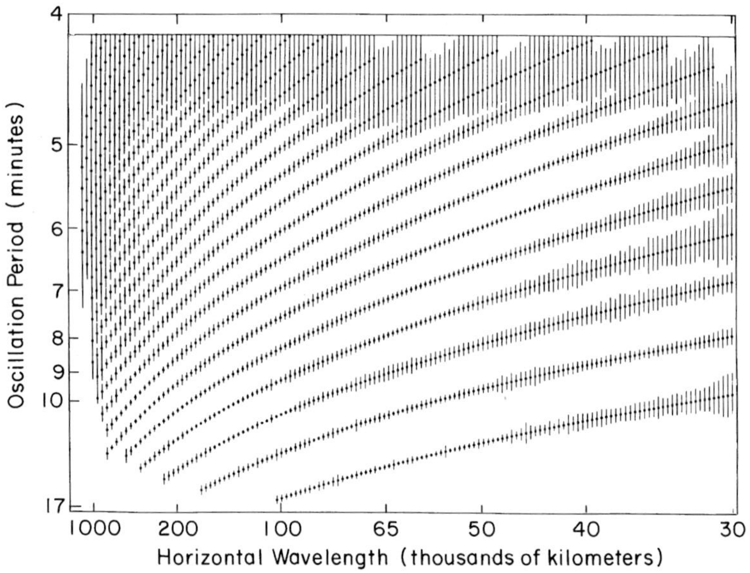

Fig. 3.6. The frequencies, or periods, of sound waves in the Sun are determined with such great precision that the vertical lines shown here represent a thousand times the standard error. Such oscillations are sometimes called p-modes because pressure, or p, perturbations provide the main restoring force. [Adapted from Libbrecht and Woodard (1991), and Lang (1995)]

quantity α, the equation may be rewritten as

$$\pi(n+\alpha) = \int_{r_1}^{r_2} \left(\frac{\omega^2}{c^2} - \frac{L^2}{r^2}\right)^{\frac{1}{2}} dr.$$

It may be shown that, provided l is not too large, α is a function only of the mode frequency ω. At higher l, the dependence of α on the degree as well as on frequency must be taken into account, but for the present purposes, it is sufficient to treat α as a function of ω alone. This equation holds equally for the Sun and for a model; thus we can write down the equation for each, take differences and linearize in (presumed) small quantities to obtain (Däppen, 1996):

$$S(w)\frac{\delta\omega}{\omega} = \int_{r_1}^{R_\odot} \left(1 - \frac{c^2}{w^2 r^2}\right)^{-\frac{1}{2}} \frac{\delta c}{c}\, dr + \pi\frac{\delta\alpha}{\omega} \equiv H_1(w) + H_2(w) \ . \tag{3.340}$$

Here $\delta\omega$ is the difference in frequency between the Sun and model for a particular mode, δc is the difference in sound speed at fixed radius, $\delta\alpha$ is the difference in surface phase function α, the radius of the photosphere is $R_\odot$, and $w \equiv \omega/L$. The function S, which is related to the mode inertia, is defined by

$$S(w) = \int_{r_1}^{R_\odot} \left(1 - \frac{c^2}{w^2 r^2}\right)^{-1/2} dr \ .$$

(A numerically small term involving $d\alpha/d\omega$ has been omitted.) The right-hand side is the sum of two functions, say $H_1(w)$ and $H_2(w)$, which can be distinguished by their different dependencies on frequency, ω, and degree l. The function $H_1(w)$ depends on the sound speed difference δc, while $H_2(w)$ is a function of near surface terms. The equation can be used to analyze the differences $\delta\omega$ in frequency between the Sun and the reference model.

A particular mode depends on a restricted range of depths and on a certain range of the colatitude θ. The spherical harmonic functions $Y_l^m \equiv P_l^m(\cos\theta)\exp(im\phi)$ determine the dependence of the mode eigenfunctions on the angular variables θ and ϕ. The horizontal component of the wavenumber can itself be expressed as the sum of latitudinal and longitudinal components: $k_\mathrm{h}^2 = k_\theta^2 + k_\phi^2$. From the $\exp(im\phi)$ dependence of the eigenfunction it follows that $k_\phi = m/(r\sin\theta)$; hence

$$k_\theta^2 = \frac{L^2 - m^2/\sin^2\theta}{r^2} \ .$$

The mode propagates only where k_θ is real, i.e., in the region $\sin^2\theta \geq m^2/L^2$. The frequency is sensitive essentially only to conditions inside this band centered on the equator. Note that the latitudinal extent of the mode depends only on m/L.

For low-degree modes with $n/l \gg 1$, the lower turning point r_1 is much less than the photosphere's radius $R_\odot$, and the eigenfrequency equation simplifies to (Gough and Toomre, 1991):

$$\omega \sim \left(n + \frac{1}{2}l + \frac{1}{2}\mu + \frac{1}{4}\right)\omega_0 - (AL^2 - B)\omega_0^2/\omega \ , \tag{3.341}$$

where μ is the effective polytropic index near $r = r_2$,

$$\omega_0 = \pi\left(\int_0^{R_\odot} c^{-1}\, dr\right)^{-1}$$

$$A = \frac{1}{2\pi\omega_0}\left[\frac{c(R)}{R} - \int_{r_1}^{r_2} \frac{dc}{dr}\, d\ \ln\ \mathrm{r}\right]$$

and

$$B = \frac{\omega^2}{\pi\omega_0}\int_{r_1}^{r_2}\left(1 - \frac{a^2}{w^2}\right)^{-1/2}\left[1 - \left(1 - \frac{\omega_c^2}{\omega^2}\right)^{1/2}\right]\frac{dr}{c} \ .$$

The characteristic frequency ω_0 is a global quantity, and is a measure of the harmonic mean of the sound speed through the star; A has a comparatively sensitive dependence on conditions near the center of the star and B depends predominantly on conditions near the upper turning point r_2. The second of the two terms on the right-hand side of the equation for ω is much smaller than the first, so for low-degree modes,

$$\omega \simeq \pi\left(n + \frac{1}{2}L + \alpha\right)\left(\int_0^{R_\odot}\frac{dr}{c}\right)^{-1} \equiv \left(n + \frac{1}{2}L + \alpha\right)\omega_0 \ .$$

For gravity modes trapped beneath the convection zone, the approximation is:

$$\frac{(n + \alpha_g)\pi}{L} \sim \int_{r_1}^{r_2}\left(\frac{N^2}{\omega^2} - 1\right)^{1/2} d\ \ln\ r \ ,$$

for $r_1 \ll R_\odot$ and $r_2 \simeq r_c$, where r_c is the radius of the base of the convection zone. The phase constant α_g depends particularly on the stratification near $r = r_c$. Only modes with low l are of practical interest, high-degree modes connot tunnel through the convection zone to be detectable in the photosphere.

When the Doppler-shift observations of photospheric absorption lines, obtained along an equatorial strip, are analyzed in space and time, the five-minute oscillations are resolved into frequencies which have the character of nonradial p-mode eigenfrequencies, typically with horizontal wavelengths $k_h \approx 10^4$ km and spherical harmonics of $l \approx 200$ (Frazier, 1968; Rhodes, Ulrich and Simon, 1977). Longer-duration observations of the five-minute oscillations probe the Sun to greater depths. Early results were obtained from Doppler-shift observations of a single absorption line in light integrated over the entire disk of the Sun, either from the South Pole during the austral summer (Grec, Fossat and Pomerantz, 1980, 1983; Fossat, Grec and Pomerantz, 1981; Duvall and Harvey, 1983), or by combining observations from widely separated telescopes (Claverie et al., 1979). Continuous observations for hundreds of hours show that the entire Sun is oscillating with hundreds of normal modes at low spherical harmonics $l = 0$ to 3, large horizontal wavelengths $k_h > 10^5$ km, and amplitudes as low as 2 cm s^{-1}. Moreover, the oscillations are not confined to periods near 5 minutes, but extend from 3 minutes to 1 hour (also see Fig. 3.6), and can last for days or longer. The low-degree, $l = 0, 1$ and 2 p-mode oscillations in the five-minute band were also detected as modulations in the total solar irradiance, with amplitudes of a few parts per million, using the ACRIM radiometer aboard the SMM spacecraft (Woodard and Hudson, 1983). As discussed during the beginning of this section, modern helioseismological instruments observe the Sun round the clock from the SOHO spacecraft (Duvall et al. 1997; Kosovichev et al. 1997) and with the GONG network (Gough et al. 1996; Thompson et al. 1996).

Pioneering detailed observations of the five-minute oscillations have been used to infer the internal constitution of the Sun, primarily in the convection zone or in the outer third of the Sun containing about 2 percent of its mass. Small but systematic discrepancies between the theoretical and observed frequencies suggested, for example, that the convection zone extends deeper than had previously been expected, to about 30 percent of the Sun's radius (Gough, 1977). A direct determination of the convection-zone depth is derived from measurements of the sound speed, c, as a function of radius, r, using the relation between r and c/r (Christensen-Dalsgaard et al. 1985):

$$r = R_\odot \exp\left\{ \frac{-2}{\pi} \int_{a_s}^{a} \left(w^{-2} - a^{-2}\right)^{-1/2} \frac{dF}{dw}\, dw \right\} \tag{3.342}$$

where $a_s = c/R_\odot$, $a = c/r$, $w = \omega/L$ for $L^2 = l(l+1) = R_\odot^2 k_h^2$ and $F(w)$ is given in a previous equation. This direct determination of the sound speed c as a function of radius r, illustrated in Figure 3.7, shows a small but definite change in sound speed that marks the lower boundary of the convection zone, located at a depth of 28.7 ± 0.3 percent of the radius of the photosphere, or at a radius of 71.3 percent of $R_\odot$ (Christensen-Dalsgaard, Gough and Thompson, 1991).

The dominant factor affecting each sound is its speed, which in turn depends on the temperature and composition of the solar regions through which it passes. For a temperature, T, and mean molecular weight, μ, the square of the sound speed is $c^2 \propto T/\mu$. Helioseismologists compute the expected sound speed using a numerical model. They then use relatively small discrepancies between their computer calculations and the observed sound speed to establish the Sun's radial variations in temperature and composition. Data from the Global Oscillation Network Group (GONG) indicate that very small discrepancies between the predicted and measured sound speed occur immediately beneath the convection zone and at the edge of the energy-generating core (Gough et al. 1996). Theoretical sound velocities and observations made with the MDI instrument aboard the Solar and Heliospheric Observatory (SOHO) are in close agreement, showing a maximum difference of only 0.2 percent. Where these discrepancies occur is, in fact, significant (Kosovichev et al. 1997). They suggest that turbulent material is moving in and out just below the convection zone, and hint that such mixing motions might occur at the boundary of the energy-generating core – which could be very important for studies of stellar evolution.

To a first approximation, the internal structure of the Sun has spherical symmetry, and the frequencies of the five-minute oscillations depend only on radial variations within the spherical shell in which the sound waves propagate. However, internal rotation breaks this symmetry. Waves moving with the rotation will appear to move faster, so their measured periods will be shorter. Waves propagating against the rotation will be slowed and show longer periods. These opposite effects split an oscillation into a pair of close, but distinct, observed frequencies. Since different oscillations probe distinct regions inside the Sun, this splitting sheds light not only on rotation at different depths, but also on how this internal rotation varies with latitude.

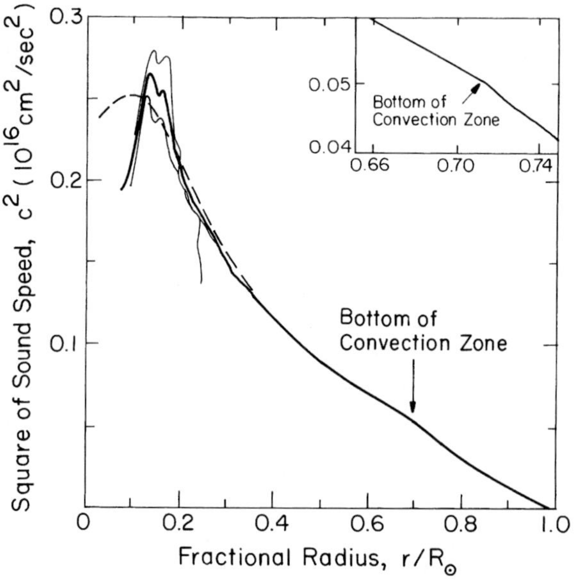

Fig. 3.7. Observations of the Sun's oscillating surface provide a window into the invisible interior of the Sun, including this determination of the square of the sound speed as a function of fractional radius (zero at the center of the Sun and one at the photosphere). The speed of sound increases at greater depths inside the Sun, and for this reason inward-moving sound waves are refracted outward. The bump at a fractional radius of 0.713 is located at the bottom of the convection zone. Here the thick curve describes observational data, flanked by thin lines representing one standard error. The dashed line is from a theoretical solar model. The large discrepancy between observation and theory at small fractional radius results mainly from the fact that the observed five-minute oscillations do not probe the deep interior very well. [Adapted from Libbrecht and Woodard (1991)]

For more than one century, astronomers have known from watching sunspots that the photosphere rotates faster at the equator than at higher latitudes, and that the speed decreases evenly toward each pole (Newton and Nunn, 1951; Snodgrass, 1983). The Sun spins from east to west, with respect to the stars, with a period of about 25 days at the equator; because the Earth orbits the Sun we observe a synodic period of about 27 days at the equator. A precise value for the synodic equatorial rotation period is 26.75 ± 0.05 days, with a differential synodic rotation given by (Sheeley, Wang and Nash, 1992):

$$\omega(\theta) = 13.46 - 2.7\cos^2\theta + 1.2\cos^4\theta - 3.2\cos^6\theta \ \mathrm{deg\,day}^{-1} \ , \tag{3.343}$$

where θ is the colatitude.

The rotational frequency splitting of the solar five-minute oscillations depends on both depth and latitude within the Sun. In the simplest approximation, a p-mode oscillation of frequency v, radial order n, spherical harmonic or angular degree l, and azimuthal order m has the form (Brown and Morrow, 1987)

$$v(n,l,m) - v(n,l,0) = -m\frac{\int_{-l}^{l}\bar{\Omega}(\theta)[P_l^m(\cos\theta)]^2\,d\cos\theta}{\int_{-l}^{l}[P_l^m(\cos\theta)]^2\,d\cos\theta} \ , \tag{3.344}$$

where θ is the colatitude, $\bar{\Omega}(\theta)$ is an appropriate average of the stellar rotation over the depth range sampled by the p-mode in question, and P_l^m is the associated Legendre function. One may represent a general frequency variation with m as a series (Duvall, Harvey and Pomerantz, 1986):

$$v(n,1\ m) - \bar{v}(n,1) = L\sum_{i=0}^{i=N} a_i\,P_i(-m/L) \ ,$$

where $\bar{v}$ is the frequency averaged over m, P_i are the Legendre polynomials of degree i and $L = \sqrt{l(l+1)}$. The argument m/L minimizes the innate l dependence of the coefficients; some analysts evaluate the coefficients a_i for $P_i(-m/l)$ instead of L to facilitate comparison with theoretical results. The p-mode rotational frequency splitting can be used to determine the solar rotation rate as a function of r and θ using the expressions given by Kosovichev (1996).

Rotational p-mode frequency splittings of solar oscillations were first reported by Duvall and Harvey (1984), and given in greater detail by Libbrecht (1989). The depth and latitude dependence of solar rotation has been discussed by Duvall, Harvey and Pomerantz (1986); Brown and Morrow (1987); Brown et al. (1989); Thompson (1990); and Goode et al. (1991). All of the results are in substantial agreement; the differential effect detected at the surface, in which the equator spins faster than the poles, is preserved throughout the convection zone (see Fig. 3.8). Within this zone, there is little variation of rotation with depth, and the inside of the Sun does not rotate any faster than the outside. The first results for the internal rotation of the Sun also indicated that the solar gravitational quadrupole moment, J_2, is small, with a value of

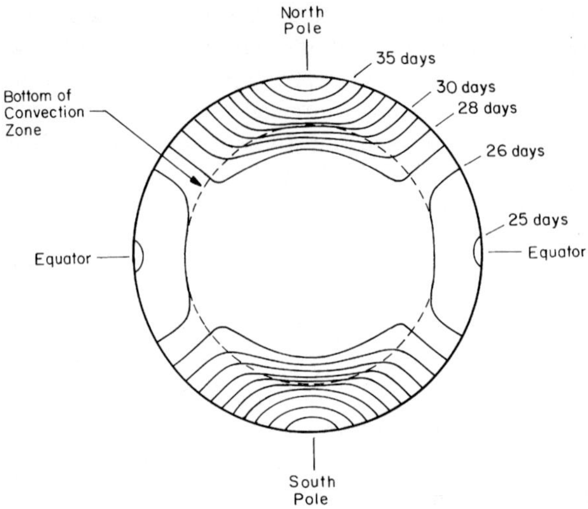

Fig. 3.8. Unlike a solid body such as the Earth, the Sun's visible surface (solid circle) does not rotate at the same rate at all latitudes, or distances from the equator. This differential rotation persists throughout most of the convection zone, the outer 28.7 percent of the Sun by radius. Here the equatorial region completes one rotation in about 25 days while it takes 35 days near the poles, but the rotation rate is roughly independent of depth. The rotation speed becomes uniform from pole to pole roughly one third of the way down at about 220,000 kilometers below the photosphere. Within the underlying radiative zone the rotation rate is uniform and independent of latitude, as it is for a solid body like the Earth – but of course the Sun is not solid. The rotation of the Sun's energy-generating core is not well known. [Adapted from Lang (1995)]

$J_2 = (1.7 \pm 0.4) \times 10^{-7}$, and provides a negligible contribution to current planetary tests of Einstein's General Theory of Relativity (Duvall et al. 1984, also see Volume II).

Helioseismological measurements of the rotation rate deeper down, beneath the convection zone, are discussed by Elsworth et al. (1995), Kosovichev (1996), Kosovichev et al. (1997), Thompson et al. (1996), and Tomczyk, Schou and Thompson (1995). Although differential rotation extends right through the convection zone, the rotation speed becomes uniform from pole to pole roughly one third of the way down, at about 220,000 kilometers beneath the photosphere. Within the radiative zone, the rotation rate is uniform and independent of latitude, as it is for a solid body like the Earth – but of course the Sun is not solid. Thus, the rotation velocity changes sharply at the base of the convection zone. There the outer parts of the radiative interior, which rotates at one speed, meet the overlying convection zone, which spins faster in its equatorial middle.

The MDI instrument aboard SOHO has additionally discovered rivers, or currents, of hot gas that might circle the Sun's polar regions just beneath the photosphere. It has also found subsurface flows of gas from the equator to the poles (Giles, Duvall, Scherrer and Bogart, 1997). Internal rivers of gas additionally move in bands near the equator at different speeds relative to each other; these are not the dominant global rotational motions, but rather the ones found when rotation is removed from the data.

The Sun's magnetism originates from an unseen dynamo in the interior (Cowling, 1981). Hot, circulating gases moving through the Sun's magnetic field generate electric currents that in turn amplify the magnetism, just as in a power-plant dynamo. The same thing happens in the fluid interiors of many planets, including the Earth (Inglis, 1981, Stern and Ness, 1982).

The solar dynamo does not seem to be located in the convection zone, where the electrified gas tends to move up and down in deep, large-scale currents (Rabin et al., 1991; Levy, 1992). The Sun's all-pervasive magnetic fields may instead originate deeper down in a thin base layer of rotational shear. There are two signs of dynamo action at this location – a high level of turbulence and sharply defined sideways flows associated with depth-related changes in rotation rate. This boundary layer, located just below the convection zone, has also been found to be hotter than expected. At deeper levels there is too little variation in solar rotation rate, with either depth or latitude, to play a significant role in the solar dynamo.

The Sun's intense magnetism is concentrated into so-called sunspots, dark cool islands located on the visible solar surface, or photosphere, that are as large as the Earth and thousands of times more magnetic. Sunspots and surrounding regions consume sound waves, absorbing as much as 30 percent of the waves' energy as they move through them. Moreover, the intense magnetism in sunspots must extend deep down inside the Sun, where the magnetism modifies the propagation of sound waves (Bogdan and Braun, 1995). Indeed, the frequencies of the solar oscillations apparently vary with the 11-year cycle of solar magnetic activity (Pallé, Regulo and Roca-Cortés, 1989; Elsworth et al., 1990; Libbrecht and Woodard, 1990). The new technique of travel-time helioseismology (Duvall et al., 1993; Hill, 1995; Kosovichev, 1996) has been used to detect strong downflows beneath sunspots (Duvall et al., 1996).

SOHO's MDI has provided clues to the origin of sunspots by measuring subsurface motions, as well as surface magnetism. Horizontal motions at a depth of about 1,400 kilometers were compared with an overlying magnetic image, also taken by the MDI instrument, indicating that strong magnetic concentrations tend to lie in regions where the subsurface gas flow converges (Duvall et al., 1997). Thus, the churning gas probably forces magnetic fields together and concentrates them, thereby overcoming the outward magnetic pressure that ought to make such localized concentrations expand and disperse.

3.5.8 Isentropic Flow – The Adiabatic Efflux of Gas

For the adiabatic nonconducting flow of a perfect gas, it follows directly from the first law of thermodynamics, Eq. (3.59), Eq. (3.61) for the perfect gas law, and Eq. (3.63) for specific heat, that

$$C_p \, dT + v \, dv = 0 \ , \tag{3.345}$$

where C_p is the specific heat at constant pressure, dT is the change in temperature, T, and dv is the change in velocity, v. It also follows from Euler's equation (3.251) and from Eq. (3.345) that the change in entropy, dS, is

$$dS = C_p \frac{dT}{T} - \frac{R}{\mu}\frac{dP}{P} = 0 \ , \tag{3.346}$$

where R is the universal gas constant. That is, the entropy is constant (isentropic) during the adiabatic flow of a perfect gas. If the subscript 0 denotes initial conditions, and no subscript denotes the condition at a later time, it follows from Eq. (3.346) that

$$\frac{P}{P_0} = \left(\frac{\rho}{\rho_0}\right)^{\gamma} = \left(\frac{T}{T_0}\right)^{\gamma/(\gamma-1)} \ , \tag{3.347}$$

where γ is the adiabatic index (see Section 3.2.3). For a monatomic gas $\gamma = 5/3$. Since the adiabatic flows are isentropic, they are sometimes called isentropic flows.

When a gas is initially static ($v_0 = 0$), it follows from Eqs. (3.347) and (3.345) that the following equations hold

$$\begin{aligned} &\frac{v^2}{2} + C_p T = C_p T_0 \ , \\ &\frac{v^2}{2} + \frac{s^2}{\gamma - 1} = \frac{s_0^2}{\gamma - 1} \ , \\ &\frac{s_0^2}{s^2} = \frac{T_0}{T} = 1 + \frac{\gamma - 1}{2} M^2 \ , \\ &\frac{P_0}{P} = \left(1 + \frac{\gamma - 1}{2} M^2\right)^{\gamma/(\gamma-1)} \ , \end{aligned} \tag{3.348}$$

and

$$\frac{\rho_0}{\rho} = \left(1 + \frac{\gamma - 1}{2} M^2\right)^{1/(\gamma-1)} \ ,$$

where the speed of sound, s, is

$$s = \left(\frac{\partial P}{\partial \rho}\right)^{1/2} = \left(\frac{\gamma P}{\rho}\right)^{1/2} ,$$

and the Mach number, M, is

$$M = v/s \ .$$

A particularly useful point is that for which $M = 1$ where

$$\begin{aligned} &T = 2T_0/(\gamma + 1), \\ &P = P_0 \left(\frac{2}{\gamma + 1}\right)^{\gamma/(\gamma-1)} \ , \end{aligned} \tag{3.349}$$

and

$$\rho = \rho_0 \left(\frac{2}{\gamma + 1}\right)^{1/(\gamma-1)} \ .$$

As first pointed out by Reynolds (1876) equations (3.349) apply for the efflux of a gas into vacuum. The velocity of efflux in this case is

$$v = s_0 \left(\frac{2}{\gamma + 1} \right)^{1/2} . \tag{3.350}$$

For the adiabatic expansion of a radiation dominated gas, we have the relations (c.f. Eq. (3.78))

$$\frac{dT}{T} = \frac{1}{3} \frac{d\rho}{\rho} ,$$

$$\rho \propto T^3 \propto R^{-3} \propto t^{-3} ,$$

and

$$U = aT^4 \propto R^{-4} \propto t^{-4} , \tag{3.351}$$

where T is the temperature, ρ is the gas mass density, R is the radius of the gas cloud, U is the radiation energy density, the radiation constant $a = 7.566 \times 10^{-15}$erg cm^{-3} °K^{-4}, and it is assumed that the gas cloud is in its early stages of expansion so that $R = vt$ where v is the constant velocity of expansion, and t is the time variable.

Similarly, the energy, E, of a relativistic particle in a gas cloud of radius, R, will go as

$$E \propto R^{-1} \propto t^{-1} , \tag{3.352}$$

and the conservation of magnetic flux means that

$$H \propto R^{-2} \propto t^{-2},$$

where H is the magnetic field intensity. The formulae of synchrotron radiation then lead to the conclusion that the synchrotron flux density, S, goes as (Shklovskii, 1960)

$$S \propto R^{-2\gamma} \propto t^{-2\gamma} , \tag{3.353}$$

where γ is the power law index of the relativistic electron energy spectrum. A decrease in the radio flux of Cassiopeia A with ongoing time has been observed (cf. Shklovskii, 1968). However, for a power law index of $p = \gamma = 2.5$, Eq. (3.353) implies a decline at the rate of 1.5% per year, which is more than double the observed rate. The time evolution of supernova remnants, such as Cassiopeia A, is discussed by Reynolds (1988), and compared to their distribution of radio flux per solid angle, Σ, as a function of their diameter, D. Shklovskii's prediction of an adiabatic decrease in mean particle density also predicts paths in the Σ–D plane that are steeper than those observed. Thus, some non-adiabatic input of particle or magnetic field energy seems to occur in young supernova remnants.

In the later stages of expansion, a gas cloud will be slowed down by the interstellar medium. If the medium density is ρ, and the gas cloud has radius, R, mass, M, and initial velocity, V_0, then its velocity, V, is given by (Oort, 1946; Shklovskii, 1968)

$$V = \frac{3MV_0}{4\pi\rho R^3} , \tag{3.354}$$

and

$$R = 4Vt = \left(\frac{3MV_0t}{\pi\rho}\right)^{1/4}$$

provided that $\rho R^3 \gg M$.

For adiabatic compression or expansion the specific entropy (per gram), S, is constant, and we consequently have the relations

$$T = \left(\frac{3S\rho}{4a}\right)^{1/3} \tag{3.355}$$

and

$$P = \frac{3^{1/3}S^{4/3}\rho^{4/3}}{4^{4/3}a^{1/3}} \tag{3.356}$$

for a photon (radiation dominated) gas. Here T and P are, respectively, the temperature and pressure, and the radiation constant $a = 7.566 \times 10^{-15}$ erg cm^{-3} °K^{-4}.

3.5.9 Shock Waves

Energetic phenomena in which material is ejected at supersonic velocities are common in the Universe. Many stars continuously inject material into the interstellar medium by high-velocity winds, such as the solar wind, during their lifetime. Massive stars explode as supernovae when they die. In our Galaxy, shocks from supernovae are thought to be important in the acceleration and collapse of interstellar clouds. Shocks also play a role in the acceleration of cosmic rays. The nuclei of active galaxies can generate shock waves when they eject radio-emitting jets of material at nearly the velocity of light.

In most cases, the sound speed of the ambient medium is less than the velocity of the gas expanding into it, and a shock is produced. This shock wave accelerates, heats and compresses the surrounding ambient gas, causing it to radiate.

In mathematical terms, a shock wave is a discontinuity across which the fluxes of mass, momentum and energy are conserved. There is a discontinuity or jump in the mass density, ρ, and temperature, T, across the shock front, designated in this volume by subscript 1 for the pre-shock values and subscript 2 for the post-shock ones. The flow speed, v, changes from supersonic ahead of the shock (region 1) to subsonic behind the shock (region 2), so $v_1 > s_1 = c_{s1}$, where s or c_s are used to denote the speed of sound. Discontinuity, or jump, conditions give the post shock quantities in terms of the preshock values; they are often called the Rankine–Hugoniot relations after their initial derivation by Rankine (1870) and Hugoniot (1889).

A detailed discussion of shocks can be found in the monographs by Zeldovich and Raizer (1966) and Tidman and Krall (1971), and in the review by McKee and Hollenbach (1980). Astrophysical shocks in diffuse gas are reviewed by McKee (1987). The Rankine–Hugoniot jump conditions are placed in an astrophysical setting, including the luminous shock and ionization fronts of gaseous nebulae, in the texts by Spitzer (1978), Bowers and Deeming (1984), Osterbrock (1989), Longair (1992) and Shu (1992). They are also specified by Frank, King and Raine (1992) who apply them to accretion. Blast waves from supernovae and other cosmic explosions are reviewed by Ostriker and McKee (1988).

A first shown by Rankine (1870), a surface of discontinuity in density, ρ, pressure, P, and temperature, T, may propagate as a "shock" wave in a gas. If u denotes the velocity of the shock front, and if the gas ahead of the shock is at rest, the gas behind the shock travels at the velocity, v_b, given by

$$v_b = u - v_2 = \frac{\rho_2 - \rho_1}{\rho_2} u \ , \tag{3.357}$$

where

$$\begin{aligned} u = v_1 &= \left[\left(\frac{\rho_2}{\rho_1}\right)\frac{P_2 - P_1}{\rho_2 - \rho_1}\right]^{1/2} = \frac{\rho_1}{\rho_2} v_2 \ , \\ v_1 &= M_1 s_1 = M_1(\gamma P_1/\rho_1)^{1/2} \ , \\ v_2 &= M_2 s_2 = M_2(\gamma P_2/\rho_2)^{1/2} \ , \end{aligned} \tag{3.358}$$

and

$$M_2^2 = \frac{1 + \frac{\gamma-1}{2} M_1^2}{\gamma M_1^2 - \frac{\gamma-1}{2}} \ .$$

Here M is the Mach number, s is the velocity of sound, γ is the adiabatic index, and subscripts 1 and 2 denote, respectively, the regions in front and behind of the shock. Eqs. (3.358) indicate that $M_1 > 1$ or < 1 according as $M_2 < 1$ or > 1, which means that either region 1 or 2 must be supersonic whereas the other region is subsonic.

The pressures, densities, and temperatures of the two regions in front (1) and behind (2) of the shock front are related by equations first derived by Rankine (1870) and Hugoniot (1889). These Rankine–Hugoniot relations, which follow directly from Euler's equation, the law of conservation of energy, and the perfect gas law, are

$$\begin{aligned} \frac{P_2}{P_1} &= 1 + \frac{2\gamma}{\gamma+1}[M_1^2 - 1], \\ \frac{\rho_2}{\rho_1} &= \frac{v_1}{v_2} = \frac{(\gamma+1)M_1^2}{(\gamma-1)M_1^2 + 2} = \frac{1 + \frac{\gamma+1}{\gamma-1}\frac{P_2}{P_1}}{\frac{\gamma+1}{\gamma-1} + \frac{P_2}{P_1}} \ , \end{aligned}$$

and

$$\frac{T_2}{T_1} = \frac{\{2\gamma M_1^2 - (\gamma-1)\}\{(\gamma-1)M_1^2 + 2\}}{(\gamma+1)^2 M_1^2} = \frac{P_2}{P_1}\frac{\frac{\gamma+1}{\gamma-1} + \frac{P_2}{P_1}}{1 + \frac{\gamma+1}{\gamma-1}\frac{P_2}{P_1}} \ . \tag{3.359}$$

It follows from Eqs. (3.357) to (3.359) that:

$$v_b = \frac{s_1}{\gamma}\left(\frac{P_2}{P_1} - 1\right)\left[\frac{\frac{2\gamma}{\gamma+1}}{\frac{P_2}{P_1} + \frac{\gamma-1}{\gamma+1}}\right]^{1/2} \tag{3.360}$$

and

$$\frac{P_2 - P_1}{P_1} = \frac{2\gamma}{\gamma+1}(M_1^2 - 1) = \frac{\Delta P}{P_1} .$$

For a very weak shock, $\Delta P/P_1 \ll 1$, we have

$$\frac{\Delta\rho}{\rho_1} = \frac{1}{\gamma}\frac{\Delta P}{P_1} = \frac{v_b}{s_1} , \tag{3.361}$$

$$\frac{\Delta T}{T_1} = \frac{\gamma - 1}{\gamma}\frac{\Delta P}{P_1} ,$$

and

$$v_1 \approx s_1 .$$

For a very strong shock, $\Delta P/P_1 \gg 1$, we have

$$\frac{\rho_2}{\rho_1} = \frac{\gamma+1}{\gamma-1} ,$$

$$\frac{T_2}{T_1} = \frac{\gamma-1}{\gamma+1}\frac{P_2}{P_1} , \tag{3.362}$$

$$v_1 = s_1\left[\frac{\gamma+1}{2\gamma}\frac{P_2}{P_1}\right]^{1/2} ,$$

and

$$v_b = s_1\left[\frac{2}{\gamma(\gamma+1)}\frac{P_2}{P_1}\right]^{1/2} .$$

for a strong shock, with $P_2 \gg P_1$, we also have:

$$P_2 = \frac{2\rho_1 v_1^2}{(\gamma+1)} , \tag{3.363}$$

where v_1 is the velocity of the shock wave. For a monatomic gas the adiabatic index $\gamma = 5/3$, while for an isothermal shock front one uses $\gamma = 1$. The Mach number $M = v/s$, the ratio of velocity to the sound speed. The speed of sound, s, is often designated c_s, and is given by $s = (\gamma P/\rho)^{1/2}$ and the gas pressure is given by the ideal gas law $P = \rho kT/(\mu m_H)$ for a gas of mass density ρ, temperature T, and mean molecular weight μ; Boltzmann's constant $k \approx 1.38\times 10^{-16}$ erg °K^{-1}, and the mass of the hydrogen is $m_H \approx 1.67 \times 10^{-24}$ grams $\approx m_U$ the atomic mass unit.

For a strong shock in a monatomic gas, the adiabatic index or specific heat ratio $\gamma = 5/3$,

$$\rho_2 = 4\rho_1 ,$$

$$v_2 = 0.25 v_1 , \tag{3.364}$$

$$P_2 = 0.75\rho_1 v_1^2 ,$$

and

$$T_2 = \frac{3}{16}\frac{\mu m_H v_1^2}{k} \ .$$

The gas is compressed by a factor of 4 by a strong shock and the velocity drops to the subsonic value behind the shock, since the velocity of sound there is:

$$s_2 = c_{s2} = \left(\frac{\gamma P_2}{\rho_2}\right)^{1/2} = \frac{\sqrt{5}}{4} v_1 \ . \tag{3.365}$$

The thermal pressure, P_2, behind the shock is:

$$P_2 = \frac{3}{4}\rho_1 v_1^2 \ , \tag{3.366}$$

and the temperature, T_2, of the post-shock gas is found using the perfect gas law $P = \rho kT/(\mu m_H)$ for a gas of mass density ρ and mean molecular weight μ:

$$T_2 = \frac{3}{16}\frac{\mu m_H}{k} v_1^2 \ . \tag{3.367}$$

For gas of cosmic abundances with 10% helium by number, the post-shock temperature is (McKee, 1987):

$$T_2 = \text{constant} \times 10^5 \left(\frac{v_1}{10^7 \ \text{cm s}^{-1}}\right) \text{K} \ , \tag{3.368}$$

where the constant is 1.38 for a fully ionized gas and 2.9 for a neutral atomic gas. The constant takes on the value 5.3 for a molecular gas.

For a strong shock, the shock front thickness, Δx, is of the order of the molecular mean free path, $\Delta x \approx l$. For a weak shock the front thickness is inversely proportional to the wave strength $\Delta x \approx v_1 l/(v_2 - v_1)$.

Powerful explosions, such as supernovae, generate expanding blast waves that will shock, heat and accelerate the surrounding ambient medium. The theoretical paradigm for the simple blast wave is the Sedov–Taylor (ST) self-similar solution (Sedov, 1959; Taylor, 1950) for a point explosion in a homogeneous medium with zero pressure and a fixed ratio of specific heats. This solution depends only on the blast wave energy, E, and the ambient density, ρ_1.

For a strong explosion of initial energy, E, in a perfect gas of mass density, ρ_1, the dimensionless variable

$$\xi = r\left(\frac{\rho_1}{Et^2}\right)^{1/5}$$

can serve as a similarity variable (cf. Sedov, 1959). The shock front radius, R_s, and velocity, V_s, at a time, t, are given by

$$R_s = \xi_1 \left(\frac{E}{\rho_1}\right)^{1/5} t^{2/5} \text{ and } V_s = \frac{2}{5}\xi_1^{5/2}\left(\frac{E}{\rho_1}\right)^{1/2} R_s^{-3/2} \ , \tag{3.369}$$

where the parameter, ξ_1, is of order unity. For this strong shock case, the parameters for the shock front are given by

$$\rho_2 = \frac{\gamma+1}{\gamma-1}\rho_1, \quad V_2 = \frac{2}{\gamma+1}V_s \ ,$$

$$P_2 = \frac{2}{\gamma+1}\rho_2 V_s^2 \approx \rho_1 \left(\frac{E}{\rho_1}\right)^{2/5} t^{-6/5} \ ,$$

and

$$T_2 = \frac{2(\gamma-1)}{(\gamma+1)^2}\frac{\mu m_{\rm H}}{k}V_s^2 \approx \frac{\gamma-1}{\gamma+1}\frac{\mu m_{\rm H}}{k}\rho_1\left(\frac{E}{\rho_1}\right)^{2/5} t^{-6/5} \ . \tag{3.370}$$

Similarity solutions for the spherical and cylindrical shock waves produced by supernovae explosions are given in Sedov's (1959) book. Ostriker and McKee (1988) present a general discussion of spherical, nonrelativistic blast waves in an astrophysical context, focusing on situations where self-similarity is assumed. Blast waves are also discussed by Shu (1992).

Shocks with jump discontinuities described by the Rankine–Hugoniot relations have been called J-shocks. The high temperature behind J-shocks leads to atomic line emission at optical and ultraviolet wavelengths (see McKee, 1987). In weakly-ionized, magnetized molecular gas, a different sort of C-shocks (Draine, 1980) are found, in which the hydrodynamic quantities vary continuously through the shock. Such C-shocks are relatively cool, radiate primarily infrared molecular lines, and move with velocities $v \le 50$ km s^{-1}. Physical conditions in C-shocks are reviewed by McKee (1987).

It is now believed that shocks can efficiently accelerate relativistic particles including cosmic rays in our Galaxy and those in extragalactic radio sources (Bell, 1978; Blandford and Ostriker, 1978, 1980; Blandford and Königl, 1979; Axford, 1981; Blandford and Eichler, 1981). The effect of cosmic rays on the shock jump conditions, for a fast shock with velocity ≥ 100 km s^{-1} are discussed by Chevalier (1983) and reviewed by McKee (1987). In the presence of cosmic-ray acceleration, the post-shock temperature, is, for example, reduced by a factor of between 0.4 and 0.8 depending on the fraction of the post-shock pressure in relativistic particles.

3.5.10 Hydrodynamic Gravity Waves

The motion of an inviscid fluid executing small oscillations is irrotational. That is

$$\nabla \times \boldsymbol{v} = 0 \ , \tag{3.371}$$

where $\boldsymbol{v}$ is the velocity. Such flow is called potential flow. It follows from Eq. (3.371) that we may let

$$\boldsymbol{v} = \nabla\varphi \ , \tag{3.372}$$

where the velocity potential, φ, satisfies Poisson's equation

$$\nabla^2\varphi = 0 \ . \tag{3.373}$$

If the fluid is incompressible (ρ = constant), and the prevalent force is that of gravity whose direction is along the z axis, then it follows from Euler's equation (3.251) that

$$\left(\frac{\partial\varphi}{\partial z}+\frac{1}{g}\frac{\partial^2\varphi}{\partial t^2}\right)_{z=0}=0 \ , \tag{3.374}$$

where the wave equation is evaluated at the surface $z = 0$, and g is the acceleration due to gravity. The solution to Eqs. (3.373) and (3.374) is

$$\varphi = A\cos(kz-\omega t)\exp[kz] \ , \tag{3.375}$$

where the gas occupies the region $z < 0$, and the frequency, ω, is related to the wave vector, k, by the relation

$$\omega = (kg)^{1/2} \ . \tag{3.376}$$

The velocity of propagation, u, of this gravity wave is

$$u=\frac{\partial\omega}{\partial k}=\frac{1}{2}\left(\frac{g}{k}\right)^{1/2}=\frac{1}{2}\left(\frac{g\lambda}{2\pi}\right)^{1/2} \ , \tag{3.377}$$

where the wave vector $k = 2\pi/\lambda$, and λ is the wavelength. It follows from Eqs. (3.375) and (3.377) that the velocity distribution in the moving fluid is given by

$$\begin{aligned} v_z &= -Ak\cos(kx-\omega t)\exp(kz) \\ v_x &= Ak\sin(kx-\omega t)\exp(kz) \ . \end{aligned} \tag{3.378}$$

For a viscous gas, the energy of the wave dies off as $\exp(-2\gamma t)$, where t is the time variable, and the damping coefficient, γ, is given by (Landau and Lifshitz, 1959)

$$\gamma = 2\nu\omega^4/(g^2) \ , \tag{3.379}$$

where ν is the coefficient of kinematic viscosity, ω is the wave frequency, and g is the acceleration of gravity.

The dispersion relation for the gravity waves in a viscous medium is (Chandrasekhar, 1961)

$$Y^4+2Y^2-4Y+1+Q+Q^{1/3}S=0 \ , \tag{3.380}$$

where

$$Y=\left[1+\frac{\omega}{k^2\nu}\right]^{1/2},$$

$$Q=\frac{g}{k^3\nu^2} \ ,$$

and

$$S=\frac{T}{\rho(g\nu^4)^{1/3}} \ .$$

Here ω is the wave frequency, k is the wave number, ν is the coefficient of kinematic viscosity, T is the surface tension, g is the acceleration due to gravity, and ρ is the mass density.

3.5.11 Jeans Condition for Gravitational Instability, Star Formation

Assume that a gas is in a static, uniform, equilibrium condition in which the velocity, v_0, is zero, and the density, ρ_0, and the pressure, P_0, are constant. Next assume a perturbation such that the density, $\rho = \rho_0 + \rho_1$, and the velocity $v = v_1$. Assuming that the only external force is that of the gravitational potential, φ, the continuity equation (3.250) and Euler's equation (3.251) become

$$\frac{\partial \rho_1}{\partial t} + \rho_0 \nabla \cdot v_1 = 0$$

and

$$\frac{\partial v_1}{\partial t} = -\nabla \varphi_1 - \frac{\nabla P_1}{\rho_0} \ . \tag{3.381}$$

For an isothermal process in an ideal gas,

$$P_1 = \frac{kT}{\mu m_{\mathrm{H}}} \rho_1 \ , \tag{3.382}$$

where T is the temperature, μ is the mean molecular weight, and m_{H} is the mass of the hydrogen atom. From Poisson's equation

$$\nabla^2 \varphi_1 = 4\pi G \rho_1 \ . \tag{3.383}$$

Eqs. (3.381) to (3.383) may be combined to form the wave equation

$$\frac{\partial^2 \rho_1}{\partial t^2} = 4\pi G \rho_0 \rho_1 + \frac{kT}{\mu m_{\mathrm{H}}} \nabla^2 \rho_1 \ , \tag{3.384}$$

which has the plane wave solution

$$\rho_1 = \exp\left[i\left(\frac{2\pi x}{\lambda} - \omega t\right)\right] \ , \tag{3.385}$$

where the frequency, ω, is related to the wavelength, λ, by the formula

$$\omega^2 = \left(\frac{2\pi}{\lambda}\right)^2 \left(\frac{kT}{\mu m_{\mathrm{H}}}\right) - 4\pi G \rho_0 \ . \tag{3.386}$$

The velocity of propagation, V_{J}, of small density fluctuations is therefore given by (Jeans, 1902)

$$V_{\mathrm{J}} = \frac{\lambda \omega}{2\pi} = s\left[1 - \frac{G\rho_0 \lambda^2}{\pi s^2}\right]^{1/2} \ , \tag{3.387}$$

where the velocity of sound, s, is given by

$$s = \left(\frac{P_0}{\rho_0}\right)^{1/2} = \left(\frac{kT}{\mu m_{\mathrm{H}}}\right)^{1/2} .$$

A fluctuation with wavelength, λ, greater than the critical wavelength (Jeans, 1902)

$$\lambda_{\mathrm{J}} = s\left(\frac{\pi}{G\rho_0}\right)^{1/2} = \left(\frac{\pi k T}{\mu m_{\mathrm{H}} G \rho_0}\right)^{1/2} \approx 6 \times 10^7 \left(\frac{T}{\mu \rho_0}\right)^{1/2} \ \mathrm{cm} \tag{3.388}$$

will grow exponentially in time and the waves are unstable. The velocity, V_{J}, is called the Jeans velocity, and length, λ_{J}, is the Jeans length. For dimensions larger than λ_{J}, a mass will be gravitationally unstable and will contract continuously. The corresponding critical mass of a sphere whose diameter is λ_{J}

is the Jeans mass:

$$M_{\mathrm{J}} = \frac{\pi}{6}\rho_0 \lambda_{\mathrm{J}}^3 \approx 10^{23} \frac{(T/\mu)^{3/2}}{\rho^{1/2}} \text{ gm} ,$$

or

$$M_{\mathrm{J}} = 1.2 \times 10^5 \left(\frac{T}{100 \text{ K}}\right)^{3/2} \left(\frac{\rho}{10^{-24} \text{ g cm}^{-3}}\right) \mu^{-3/2} \text{ solar masses.} \tag{3.389}$$

Using typical conditions for interstellar clouds of neutral hydrogen, $\rho = 10^{-24}$ g cm^{-3}, $T = 100$ K and $\mu = 1$, we obtains a Jeans mass of $M_{\mathrm{J}} \approx 10^5$ solar masses. Thus, only masses large compared to those of stars seem able to collapse because of the Jeans instability. However, massive clouds that exceed the Jeans mass undergo fragmentation during collapse, so smaller submasses can condense into stars. Estimates by Rees (1976) indicate that the Jeans mass at the end of fragmentation is comparable to the Sun's mass, and not of the order of smaller planetary masses or of larger star clusters.

Gravitational collapse and star formation are considered in the books by Spitzer (1978) and Kippenhahn and Weigert (1990). Theoretical models for star formation by gravitational collapse are reviewed by Larson (1973), Woodward (1978) and Shu, Adams and Lizano (1987).

The extention of the Jeans criteria to adiabatic processes is discussed by Chandrasekhar (1951) and Parker (1952). The Jeans criteria for the gravitational instability of an infinite homogeneous medium is unaffected by the presence, separately, or simultaneously, of uniform rotation or a uniform magnetic field. The effects of rotation and a magnetic field are discussed by Chandrasekhar and Fermi (1953), Chandrasekhar (1953, 1954), and Pacholczyk and Stódólkiewicz (1960). Conditions for gravitational contraction in the presence of a magnetic field or an external pressure were given in Section 3.4.4. Mouschovias (1974, 1976), Mouschovias and Spitzer (1976) and Mestel (1985) have discussed the equilibrium and collapse of magnetic molecular clouds.

When the direction of wave propagation is at right angles to the direction of rotation, the dispersion relation is

$$\omega^2 = k^2 s^2 + 4\Omega^2 - 4\pi G\rho ,$$

where Ω is the rotation velocity, and the minimum unstable wavelength is

$$\lambda_{\mathrm{J}} = \frac{\pi s}{(\pi G\rho - \Omega^2)^{1/2}} . \tag{3.390}$$

In this special case, gravitational instability cannot occur if $\Omega^2 > \pi G\rho$. For wave propagation in every other direction, gravitational instability will occur if Jean's condition $s^2 k^2 - 4\pi G\rho < 0$ is satisfied. For an infinite magnetic field strength, the Jeans disturbance can only propagate along the magnetic field and not perpendicular to it. The general dispersion relation when a magnetic field of strength, H, is present is given by

$$\omega^4 - \left(\frac{\pi H^2}{\lambda^2 \rho} + \frac{4\pi^2 s^2}{\lambda^2} - 4\pi G\rho\right)\omega^2 + \frac{\pi H^2}{\rho\lambda^2}\left(\frac{4\pi^2 s^2}{\lambda^2} - 4\pi G\rho\right)\cos^2\theta = 0 ,$$

where θ is the inclination of the direction of the magnetic field to the direction of wave propagation.

A classic numerical study of the collapse of a gaseous mass slightly in excess of the Jeans mass or wavelength was provided by Larson (1969). He showed that dense central regions fall in well before the extended envelope does, and that the temperature of a collapsing molecular cloud remains constant over many orders of magnitude in density. Self-similar solutions for the gravitational collapse of such an isothermal sphere have been provided by Larson (1969), Penston (1969) and Shu (1977); and investigated using numerical hydrodynamics by Foster and Chevalier (1993). The similarity solution for the gravitational collapse of an isothermal sphere yields a central object whose mass increases linearly with time, t, with a mass infall rate, dM/dt, given by (Shu, 1992):

$$dM/dt = m_{\mathrm{o}} a^3/G \ , \tag{3.391}$$

where $m_{\mathrm{o}} = 0.975$, and for an ideal gas $a^2 = P/\rho = kT/(\mu m_{\mathrm{H}})$. An accreting protostar at the center of the cloud has a mass $M = t dM/dt$ after time t. For a molecular cloud with a temperature of $T = 10°\mathrm{K}$ and $a = 0.19$ km s^{-1}, we obtain $dM/dt = 2 \times 10^{-6}$ solar masses per year, and it would take 5×10^5 years to build up a star with the Sun's mass. This is much longer than previous studies of star formation might have suggested (Walker, 1956; Hayashi, 1961, 1966).

Shu, Adams and Lizano (1987) have presented a comprehensive picture of star formation in molecular clouds. Present-day star formation begins with the collapse of cold density enhancements in giant molecular clouds. Such clouds have been mapped by their CO emission (Maddalena et al. 1986; Ungerechts and Thaddeus, 1987; Lang, 1992). The energy source of the protostar is the accretion of matter onto the core, and a rotating disc forms perpendicular to the rotation axis. A stellar wind also breaks out along the rotation axis of the system, creating a bipolar outflow of cold molecular gas, with kinetic energies of 10^{43} to 10^{47} erg – see Lada (1985), Bachiller (1996), and Lang (1992). Overviews of molecular clouds and star formation are provided by Elmegreen (1985) and Evans (1985). Observations of circumstellar discs have been discussed by Aumann et al. (1984), Smith and Terrile (1984), Aumann (1985), Sargent and Beckwith (1987), Beckwith and Sargent (1996) and Burrows et al. (1996). The detection of large planets in orbit about Sun-like stars is reviewed in the books by Boss (1998), Croswell (1997), Goldsmith (1997), and Halpern (1997), and in the references contained therein.

Although collapse might naturally occur in the cold, dense places within molecular clouds, the dust and gas in interstellar space is normally moving too fast for stars to be formed by spontaneous gravitational collapse. Some agent has to compress this interstellar matter and thereby provide the extra pressure needed to start gravitational collapse. Ernst Öpik reasoned that the expanding remnants of supernova explosions can trigger collapse (Öpik, 1953), while Jan Oort suggested that the expanding H II regions found around hot, massive stars both trigger collapse and give rise to expanding stellar associations (Oort, 1954). Since massive stars burn their thermonuclear fuel quickly, they can complete their evolution and explode as supernovae while other stars are still forming in molecular clouds. Shocks associated with the supernova explosion of massive stars (Woodward, 1976), would therefore compress the molecular cloud, leading to sequential star formation of successive generations of massive stars (Blaauw, 1964; Elmegreen and Lada, 1971; Elmegreen, 1992).

3.5.12 Magnetohydrodynamics, Alfvén Waves

Magnetohydrodynamics, commonly known as MHD, concerns the dynamics of electrically conducting fluids in the presence of magnetic fields. This area of astrophysics was pioneered by Hannes Alfvén, who received the Nobel Prize for physics in 1970 for fundamental MHD work with fruitful application in plasma physics. Here we will discuss one of these applications, now called Alfvén waves. Thorough discussions of the field of magnetohydrodynamics are found in the monographs by Alfvén (1950), Alfvén and Fälthammer (1963) and Cowling (1957, 1976). Various aspects of solar MHD are provided by Priest (1982), as well as Benz (1993) and Sturrock (1994). The MHD instabilities are discussed in Section 3.5.16 and in the references contained therein.

Astrophysical gases often have a large electrical conductivity, σ, and low electrical resistivity, $\eta = c^2/(4\pi\sigma)$. For example, in a plasma of electron density, N_e, the electrical conductivity is $\sigma = e^2 N_e/(m\nu_c)$ where ν_c is the collision frequency and $\eta = c^2\nu_c/\omega_p^2$ for a plasma frequency $\omega_p = (4\pi e^2\ N_e/m)^{1/2}$, and the charge and mass of the electron are respectively denoted by e and m.

In the case of zero resistivity or "infinite conductivity" we have a perfectly conducting medium and the magnetic field satisfies the relation:

$$\frac{\partial \boldsymbol{B}}{\partial t} + \nabla \times (\boldsymbol{B} \times \boldsymbol{v}) = 0 \quad , \tag{3.392}$$

where $\boldsymbol{v}$ is the bulk speed of the fluid. This equation expresses a condition in which the magnetic field is tied to matter (field freezing); and the matter (free electrons and ions) is tied to the field. The time dependence of the magnetic field for finite conductivity is more generally written:

$$\frac{\partial \boldsymbol{B}}{\partial t} = \nabla \times (\boldsymbol{v} \times \boldsymbol{B}) + \frac{c^2}{4\pi\sigma}\nabla^2 \boldsymbol{B} \ , \tag{3.393}$$

which reduces to the magnetic diffusion Eq. (3.49) for a fluid at rest with $v = 0$. The magnetic diffusion time, τ_m, over a linear scale L is $\tau_m = 4\pi\sigma L^2/c^2$, which was given in Eq. (3.51). For times short compared to the diffusion time τ_m, which is usually very large in astrophysical situations, the temporal behavior of the magnetic field is given by Eq. (3.392). The other equations of magnetohydrodynamics are those of hydrodynamics, such as the continuity equation (3.250) and Euler's equation (3.251) together with the ideal gas law and the expression for electromagnetic force.

Assume that a gas is in a static, uniform, equilibrium condition in which the velocity, v_0, is zero and the density, ρ_0, and pressure, P_0, are constant. Next assume a perturbation such that the density $\rho = \rho_0 + \rho_1$ and the velocity

$v = v_1$. Ignoring the external force due to gravity, but assuming a uniform magnetic field of strength, $\boldsymbol{B}_0$, in a perfectly conducting medium, the continuity equation (3.250) and Euler's equation (3.251) become:

$$\frac{\partial \rho_1}{\partial t} + \rho_0 \nabla \cdot \boldsymbol{v}_1 = 0$$

and

$$\frac{\partial \boldsymbol{v}_1}{\partial t} = -\frac{\nabla P_1}{\rho_0} - \frac{\boldsymbol{B}_0}{4\pi\rho_0} \times (\nabla \times \boldsymbol{B}_1) \ , \tag{3.394}$$

where $\boldsymbol{B}_1$ is the perturbation in the magnetic field, and from Maxwell's equations

$$\frac{\partial B_1}{\partial t} = \nabla \times (\boldsymbol{v}_1 \times \boldsymbol{B}_0) \ .$$

For an ideal gas,

$$\nabla P_1 = s^2 \nabla \rho_1 \ ,$$

where $s^2 = (\partial P / \partial \rho)$ is the square of the velocity of sound. Introducing the Alfvén velocity (Alfvén, 1942)

$$\boldsymbol{v}_{\mathrm{A}} = \boldsymbol{B}_0 (4\pi\rho_0)^{-1/2} \ , \tag{3.395}$$

Eqs. (3.394) and (3.395) may be combined to give the equation

$$\frac{\partial^2 \boldsymbol{v}_1}{\partial t^2} - s^2 \nabla(\nabla \cdot \boldsymbol{v}_1) + \boldsymbol{v}_{\mathrm{A}} \times \nabla \times [\nabla \times (\boldsymbol{v}_1 \times \boldsymbol{v}_{\mathrm{A}})] = 0 \ . \tag{3.396}$$

As first pointed out by Alfvén (1942) and Aström (1950), Eq. (3.396) has the plane wave solution

$$v_1 = \exp[i(\boldsymbol{k} \cdot \boldsymbol{x} - \omega t)] \ , \tag{3.397}$$

where ω is the wave frequency, and the wave vector

$$\boldsymbol{k} = \frac{2\pi}{\lambda} \boldsymbol{n} \ ,$$

where λ is the wavelength and $\boldsymbol{n}$ is a unit vector in the direction of propagation. When the wave vector $\boldsymbol{k}$ is perpendicular to $\boldsymbol{v}_{\mathrm{A}}$, Eq. (3.396) has a plane wave solution with phase velocity

$$v_p = \frac{\lambda\omega}{2\pi} = (s^2 + v_{\mathrm{A}}^2)^{1/2} = \frac{\omega}{k} \ . \tag{3.398}$$

This wave is a longitudinal wave, in which the particles oscillate in the direction of propagation. The dispersion relation represents alternating compressions and rarefactions of the gas and field; they are called a fast magnetoacoustic wave since it is faster than both the sound and Alfvén waves. When the velocity of sound, s, is $s \ll v_{\mathrm{A}}$, we have a compressional Alfvén wave with the dispersion relation $\omega/k = v_{\mathrm{A}}$.

Expression (3.398) is a special case for perpendicular propagation derived from a more general dispersion relation that describes both fast, slow and Alfvén magnetohydrodynamic (MHD) waves (Shu, 1992):

$$\frac{\omega^2}{\mathrm{k}^2} = \frac{1}{2}\left\{(v_\mathrm{A}^2 + s^2) \pm \left[(v_\mathrm{A}^2 + s^2)^2 - 4v_\mathrm{A}^2 s^2 \cos^2\Psi\right]^{1/2}\right\} . \qquad (3.399)$$

The upper sign, which gives a large wave speed, yields the fast MHD wave; the lower sign, the slow MHD wave. Here Ψ is the angle between the direction of wave propagation and the magnetic field. For $\cos^2\Psi = 0$, we have $\boldsymbol{k}$ perpendicular to the magnetic field $\boldsymbol{B}$ and Eq. (3.399) becomes $\omega^2/k^2 = 0$ and $\omega^2/k^2 = \mathrm{s}^2 + v_\mathrm{A}^2$, or Eq. (3.398).

For parallel propagation, with $\boldsymbol{k}$ parallel to the magnetic field $\boldsymbol{B}$, we have $\cos^2\Psi = 1$, so that:

$$\frac{\omega^2}{k^2} = \frac{1}{2}\left[(v_\mathrm{A}^2 + s^2) \pm |v_\mathrm{A}^2 - s^2|\right] . \qquad (3.400)$$

In this case the phase velocity ω/k can either be that of a longitudinal sound wave, where the particles oscillate in the direction of propagation, with (Benz, 1993).

$$\frac{\omega^2}{k^2} = s^2 = \frac{\gamma P_0}{\rho_0}$$

where $\gamma = 5/3$ for a monatomic gas, or a shear Alfvén wave with:

$$\frac{\omega^2}{k^2} = v_\mathrm{A}^2 ,$$

where the particles oscillate in transverse motion to both the magnetic field and the direction of propagation.

Magnetohydrodynamic waves are of particular interest for the solar corona and solar wind, where the plasma beta parameter $\beta = 8\pi P/B^2 \ll 1$ for a gas pressure P and magnetic field intensity B. Heating of the solar atmosphere by acoustic processes, or sound waves (Section 3.5.6), only becomes important when $\beta \gg 1$, perhaps within "nonmagnetic" regions of the chromosphere. Early papers on coronal heating by magnetohydrodynamic waves include those by Alfvén (1947), Giovanelli (1949), Piddington (1956) and Osterbrock (1961). Since these early speculations, Alfvén waves have been seen in the solar equatorial regions over a vast range of distances from the Sun by in-ecliptic spacecraft (Belcher, Davis and Smith, 1969; Belcher and Davis, 1971; see Tu and Marsch, 1995 for a review). Such waves may contribute to the heating of the solar corona (Hollweg, 1972, 1978; Wentzel, 1974, 1976; Chashei and Shishov, 1986, 1987). Alfvén waves have also been detected by the Ulysses spacecraft far above the Sun's poles (Tsurutani et al., 1994; Smith et al., 1995), where they might help accelerate the fast component of the solar wind. These Alfvén waves may also block the incoming flow of cosmic rays in the Sun's polar regions (Jokipii and Kóta, 1989; Simpson et al., 1995).

Coronal heating may alternatively be attributed to the dissipation of currents (Tucker, 1973; Rosner et al., 1978; Hinata, 1979, 1980). The currents could release energy by current dissipation, magnetic reconnection or double layers (see Kuperus, Ionson and Spicer, 1981 for a review). Reconnection or merging of oppositely directed magnetic fields could, for example, dissipate some magnetic energy as heat (Levine, 1974; Heyvaerts and Priest, 1984; Browning and Priest, 1986; Browning, Sakurai and Priest, 1986). As first suggested by Gold and Hoyle (1960), stored magnetic energy is the only plausible source for powering explosive solar activity. Continued low-level explosive magnetic activity, dubbed nanoflares or microflares, may play a role in coronal heating (Gold, 1964; Parker, 1988).

The ultimate source of coronal heating will most likely be magnetic energy. The tearing-mode instability and magnetic reconnection are discussed in greater detail in Sect. 3.5.16. The theory of coronal heating mechanisms is reviewed by Kuperus, Ionson and Spicer (1981), and mechanisms of chromospheric and coronal heating are discussed in the collection of articles edited by Ulmschneider, Priest and Rosner (1990).

3.5.13 Turbulence

When studying the flow of fluids in a pipe, Reynolds (1883) found that turbulent motion is damped by viscosity when the Reynolds number

$$\mathrm{Re} = \frac{L\boldsymbol{v}}{\nu} \tag{3.401}$$

falls below a critical value of approximately 2000. Here L is the pipe diameter, $\boldsymbol{v}$ is the average flow velocity, and ν is the coefficient of kinematic viscosity. Formulae for the dynamic viscosity $= \mu = \nu/\rho$ and Reynolds number of a gas are given in Eqs. (3.23) to (3.31). As first pointed out by Rosseland (1929), the distance scale involved with astronomical objects is so large that the Reynolds number must often exceed the critical value for the onset of turbulence.

Turbulent motion is fluctuating and turbulent velocity must be treated as a random continuous function of position and time. The appropriate statistical theory assumes its simplest form when the medium is assumed to be homogeneous and isotropic. In this case the average properties of the motion are independent of position or the direction of the axis of reference. The turbulent velocity, v is then described by the correlation function (Taylor, 1935)

$$R_{ij}(\boldsymbol{r}) = \langle v_i(\boldsymbol{x})v_j(\boldsymbol{x}+\boldsymbol{r})\rangle \ , \tag{3.402}$$

and its Fourier transform (Taylor, 1938)

$$\Phi_{ij}(\boldsymbol{k}) = \frac{1}{8\pi^3}\int_{-\infty}^{+\infty} \mathrm{R}_{ij}(\boldsymbol{r})\exp[-i\boldsymbol{k}\cdot\boldsymbol{r}]\,d\boldsymbol{r} \ .$$

Here the correlation is between two points, i, j, seperated by the space vector, $\boldsymbol{r}$, the $\langle\ \rangle$ denotes a spatial average, and $\boldsymbol{k}$ is the wave number vector. When the turbulence is isotropic and homogeneous, the correlation function is of the form (von Kármán, 1937; von Kármán and Howarth, 1938)

$$R_{ij}(\boldsymbol{r}) = F(r)r_ir_j + G(r)\delta_{ij} \ , \tag{3.403}$$

where F and G are scalar functions of r, and δ_{ij} is the Dirac delta function. When the velocity correlation function is further divided into its longitudinal,

$f(r)$, and lateral, $g(r)$, components, we have the relations

$$G = v^2 g$$

and

$$F = \frac{v^2}{r^2}(f - g) \ ,$$

where v^2 is the mean square turbulent velocity, and

$$g(r) = \frac{\langle v_n(\boldsymbol{x}) v_n(\boldsymbol{x}+\boldsymbol{r})\rangle}{v^2}$$

$$f(r) = \frac{\langle v_p(\boldsymbol{x}) v_p(\boldsymbol{x}+\boldsymbol{r})\rangle}{v^2} \ ,$$

where the subscripts p and n denote, respectively, the components of velocity parallel (longitudinal) and perpendicular (lateral) to $\boldsymbol{r}$.

Of particular physical interest is the energy spectrum function

$$E(k) = 2\pi k^2 \Phi_{ii}(k) \ . \tag{3.404}$$

The contribution of the total energy from that part of the wave number space between spheres of radii k and $k + dk$ is $E(k)dk$.

For the case in which turbulent velocity elements have a characteristic size $L = \alpha^{-1}$, we might take

$$v(x) = \alpha^2 x \exp\left[-\frac{\alpha^2 x^2}{2}\right] \tag{3.405}$$

to obtain

$$E(k) \propto k^4 \exp\left[-\frac{k^2}{\alpha^2}\right] \ .$$

This special case shows that $E(k)$ has a maximum at 1.4α, a variance of $\sigma_k = 0.4\alpha$ and a wave number $k \approx L^{-1}$. That is, larger turbulent eddies have smaller wave numbers and vice versa.

The total turbulent kinetic energy density is given by

$$\frac{\rho}{2} R_{ii}(0) = \frac{\rho}{2}\langle v_i v_i\rangle = \frac{\rho}{2}\int\int_{-\infty}^{\infty}\int \Phi_{ii}(\boldsymbol{k})\, d\boldsymbol{k} = \rho\int_0^{\infty} E(k)\, dk \ .$$

It follows from the Navier–Stokes equation that the time rate of change of kinetic energy is given by (cf. Batchelor, 1967)

$$\frac{\partial}{\partial t}\left[\frac{\rho R_{ii}(0)}{2}\right] = -2\rho\nu \int_0^{\infty} k^2 E(k)\, dk = -\rho\varepsilon \ , \tag{3.406}$$

where ν is the coefficient of kinematic viscosity, and ε is called the dissipation integral. It follows from Eq. (3.406) that kinetic energy is dissipated by viscosity into heat, and that most of this dissipation is done by the smaller eddies for which k is large.

When inertial forces are considered, it is seen that they do not change the total kinetic energy; but they do serve to transfer kinetic energy from larger turbulent elements to smaller ones. Kolmogoroff (1941) postulated that when the Reynolds number is large, the smaller scale components of turbulence are in statistical equilibrium. In this case, the size distribution of the smaller eddies depends only on the viscosity and the characteristics of the larger eddies. Kolmogoroff (1941) obtained the energy spectrum

$$E(k) \propto \varepsilon^{2/3} k^{-5/3} \ , \tag{3.407}$$

where ε is the rate of transfer of energy from the bigger eddies to the smaller eddies and is given by Eq. (3.406). A simple way of interpreting the Kolmogoroff spectrum is to assume that a large eddy of scale, L_0, will exist for time

$$t_{L0} = L_0 / v_{L0} \ ,$$

where v_{L0} is its velocity with respect to neighboring eddies. If there is a constant flow of energy to smaller eddies, we will have

$$v_L \propto L^{1/3}$$

and

$$t_L \propto L^{2/3} \ , \tag{3.408}$$

where v_L and t_L denote, respectively, the velocity and lifetime of an eddy of size, L. When the velocities, v_L, are less than the Alfvén velocity, v_{A}, the magnetic field will damp them, whereas when v_L is larger than the velocity of sound, s, part of the kinetic energy may be dissipated as shock waves. In these cases we have (Kaplan, 1954)

$$\begin{aligned} v_L &\propto L \quad \text{for } v_L > s \\ v_L &\propto L^{1/2} \quad \text{for } v_L < v_{\mathrm{A}} \ . \end{aligned} \tag{3.409}$$

Heisenberg (1948, 1949) calculated the distribution of energy over wavenumbers within the equilibrium range and found that (cf. Chandrasekhar, 1949)

$$\begin{aligned} E(k) &= \frac{c\varepsilon^{2/3} k^{-5/3}}{[1 + (8/3)(k/k_d)^4]^{4/3}} \\ &\propto \varepsilon^{2/3} k^{-5/3} \quad \text{for} \quad k \ll k_d \\ &\approx k^{-7} \quad \text{for} \quad k \gg k_d \ . \end{aligned} \tag{3.410}$$

Here c is a constant, and the limiting wave number $k_d = (\varepsilon/v^3)^{1/4}$.

The suggestion that turbulence might play a role in astrophysics was put forth by Rosseland (1929) and Chandrasekhar (1949). Turbulence theories were first applied to the motion of solar granules by Richardson and Schwarzschild (1950); to the determination of the strength of the interstellar magnetic field by Chandrasekhar and Fermi (1953); and to the formation of galaxies by Weizsäcker (1951). The Kolmogoroff $k^{-5/3}$ energy spectrum is found to be valid for atmospheric turbulence (cf. Lumley and Panofsky, 1964).

Spectra of the interplanetary magnetic field, velocity, and density all indicate that a Kolmogoroff spectrum is appropriate for the solar wind in the frequency range 10^{-5} Hz $\lesssim f \lesssim 10$ Hz (cf. Jokipii, 1974). Here $f \approx v_W k/2\pi$, where k is the wave number and the solar wind velocity $v_W \approx 350$ km sec^{-1}. Furthermore, Ozernoi and Chibisov (1971) argue that a Kolmogoroff spectrum is valid for the scale sizes of galaxies.

3.5.14 Accretion

In the following we discuss steady accretion from a single body, such as a grain of dust or a planet, star or galaxy. A compact star can also accrete matter from a close stellar companion, giving rise to X-ray radiation as the accreting mass falls into a neutron star or black hole. This topic is discussed in detail in Volume II, and reviewed by Pringle (1981), Longair (1994), Lin (1995) and Lin and Papaloizou (1996). Details were provided by Shakura and Sunyaev (1973) and in the monograph by Frank, King and Raine (1992). The enormous energy released by active galactic nuclei, radio galaxies and quasars may be attributed to accretion onto supermassive black holes located at their centers. This topic is discussed in Volume II, and reviewed by Begelman, Blandford and Rees (1984), Rees (1984) and Begelman and Rees (1996).

When a stationary seed object is imbedded in a gas which is in thermal equilibrium at temperature, T, the collision time, τ_c, between collisions with gas atoms is

$$\tau_c \approx (\pi r^2 N v_{th})^{-1} \approx \left[\pi r^2 N \left(\frac{kT}{m}\right)^{1/2}\right]^{-1} ,$$

where r is the radius of the seed object, v_{th} is the thermal velocity of the gas atoms, and N and m are, respectively, the number density and mass of the atoms. If every atom which hits a seed object stays with it, the rate of increase, A, of the mass, M, of the seed object is

$$A = \frac{dM}{dt} \approx \frac{m}{\tau_c} \approx \pi r^2 N (kTm)^{1/2} , \tag{3.411}$$

and the rate of increase of the radius is

$$\frac{dr}{dt} \approx \frac{N}{\rho}(kTm)^{1/2} \approx \frac{A}{\pi r^2 \rho} ,$$

where ρ is the gas mass density. This type of accretion in a gas cloud is thought to account for the formation of grains in interstellar matter (Lindblad, 1935; Oort and van de Hulst, 1946).

If a planet, star, or galaxy has a relative velocity, v, with respect to a gas cloud, the duration of their encounter is roughly

$$\tau \approx r/v , \tag{3.412}$$

where r is the smallest distance between the two objects. If the planet, star, or galaxy has mass, M, and the cloud has mass density, ρ, the gravitational force, F, exerted by the object on a unit volume of gas is

$$F = GM\rho/r^2 \ , \tag{3.413}$$

and the momentum, p, transferred by this force is approximately

$$p = \tau F = \frac{GM\rho}{rv} \ . \tag{3.414}$$

The original momentum, ρv, becomes equal to p at the critical distance

$$R_A = \frac{GM}{v^2} = 1.3 \times 10^{14} \left(\frac{10 \text{ km sec}^{-1}}{v}\right)^2 \left(\frac{M}{M_\odot}\right) \text{cm} \ . \tag{3.415}$$

As a seed object moves through a gas, it will capture or accrete a gas within the critical radius, R_A. Provided that R_A is larger than the mean free path of the gas, the captured gas will be maintained as a turbulent cloud surrounding the seed object. Turbulent friction will cause a loss of the gas kinetic energy, and most of the gas will eventually be united with the planet, star, or galaxy. As the total mass within this gaseous sphere is ρR_A^3, and as it is replenished in the time $\tau_A = R_A/v$, the rate of increase of mass, A, is (Hoyle and Lyttleton, 1939; Bondi and Hoyle, 1944)

$$A = \frac{dM}{dt} \approx \pi \rho R_A^2 v = \pi R_A^2 \mu m_{\rm H} N v \ , \tag{3.416}$$

where μ is the molecular weight, $m_{\rm H}$ is the mass of the hydrogen atom, and N is the density of the gas cloud. The corresponding formula for the rate of increase of the radius is

$$\frac{dr}{dt} = \frac{v}{4}\frac{\rho}{\rho_s} \ , \tag{3.417}$$

where $\rho_s = 3M/(4\pi R^3)$ is the mass density of the planet, star, or galaxy. The terrestrial mass accretion rate of cosmic dust is, for example, given by Love and Brownlee (1993) as $(40 \pm 20) \times 10^6$ kilograms per year for meteoroids in the mass range 10^{-9} to 10^{-4} grams, a mass input comparable with or greater than the average input from extraterrestrial bodies in the 1-centimeter to 10-kilometer size range.

If the average kinetic energy of the incoming gas is less than the gravitational energy of the planet, star, or galaxy, all of the gas will be pulled to the massive object by gravitational attraction. Eq. (3.416) is then written as

$$A = \pi \rho [R(R + R_A)] v \ , \tag{3.418}$$

where R is the radius of the planet, star, or galaxy. With $\rho = 10^{-24}$ g cm^{-3}, the value for the interstellar medium, $R = R_\odot$ and $v = 10$ km sec^{-1}, Eq. (3.418) gives $A \approx 10^{-18}$ $M_\odot$ per year for $M = M_\odot$.

Accretion into a planet, star, or galaxy of mass, M, becomes supersonic when the object's radius, R, is less than the critical radius, R_s, given by

$$R_s = \frac{5 - 3\gamma}{4}\frac{GM}{s^2} \ , \tag{3.419}$$

where γ is the adiabatic index, and the velocity of sound of the gas at infinity, s, is given by

$$s = \left(\frac{\gamma P}{\rho}\right)^{1/2} \approx \left(\frac{kT}{m_{\mathrm{H}}}\right)^{1/2} , \tag{3.420}$$

where P and T are, respectively, the gas pressure and temperature, and m_{H} is the mass of the hydrogen atom. When $v < s$ and for accretion, then $R < R_s$. In this case the critical accretion radius, R_A, is given by (Bondi, 1952; Zeldovich and Novikov, 1971)

$$R_A = \delta(\gamma)\frac{GM}{s^2} , \tag{3.421}$$

where

$$\delta(\gamma) = \frac{1}{2}\left[\frac{2}{(5-3\gamma)}\right]^{(5-3\gamma)/(3\gamma-3)} ,$$

and $\delta(\gamma) = 0.5$ and 1 for $\gamma = 5/3$ and $4/3$, respectively. The accretion rate, A, is for the case of $v < s$,

$$A = 4\pi R_s^2 U_s \rho_s = \alpha(\gamma)\left(\frac{2GM}{c^2}\right)^2 c\rho\left(\frac{m_{\mathrm{H}}c^2}{kT}\right)^{3/2} \approx \pi R_A^2 s\rho , \tag{3.422}$$

where the velocity of sound, U_s, at the critical radius, R_s, is given by

$$U_s = s\left[\frac{2}{(5-3\gamma)}\right]^{1/2} ,$$

and

$$\alpha(\gamma) = \frac{\pi}{4\gamma^{3/2}}\left[\frac{2}{(5-3\gamma)}\right]^{(5-3\gamma)/[2(\gamma-1)]} ,$$

which has the values of 1.5, 0.3 and 1.4 for $\gamma = 1$, $5/3$ and $4/3$, respectively. Eqs. (3.421) and (3.422) are roughly the same as Eqs. (3.415) and (3.416) if v is replaced by s when v is less than s. For $\rho = 10^{-24}$ g cm^{-3}, $M \approx M_\odot$ and $s = 1$ km sec^{-1}, Eq. (3.422) gives $A \approx 5 \times 10^{-12}\ M_\odot$ per year.

Equation (3.422) can be written in the equivalent form (Frank, King and Raine, 1992):

$$\begin{aligned} A = \frac{dM}{dt} &= \alpha(\gamma)G^2M^2\frac{\rho(\infty)}{[s(\infty)]^3} \\ &\approx 1.4 \times 10^{11}\left(\frac{M}{M_\odot}\right)^2\left(\frac{\rho(\infty)}{10^{-24}\ \mathrm{g\ cm^{-3}}}\right) \\ &\left(\frac{s(\infty)}{10\ \mathrm{km\ s^{-1}}}\right)^{-3} \text{ grams per second} \end{aligned} \tag{3.423}$$

where $\rho(\infty)$ and $s(\infty)$ respectively denote the mass density and sound velocity at large distances from the accreting body, and for the numerical approximation $\gamma = 1.4$ and $M_\odot$ denotes the solar mass.

The luminosity of a star due to accretion is given by (Zeldovich and Novikov, 1971)

$$L \approx \varphi \frac{dM}{dt} \approx 2 \times 10^{31} \left(\frac{\varphi}{0.1c^2}\right) \left(\frac{M}{M_\odot}\right)^2 \left(\frac{10^4}{T}\right)^{3/2} N \,\text{erg sec}^{-1} \ , \qquad (3.424)$$

where φ is the gravitational potential near the surface of the star, T and N are, respectively, the temperature and number density of the gas near R_A, and γ has been taken to be 4/3. Although the spectrum is difficult to estimate, one approach is to assume that the radiation emerges from the surface as a black body radiator of luminosity, L, and temperature, T. The Wien displacement law then gives the wavelength of maximum radiation once T is found from L and the star's radius, R.

A detailed discussion of the density distribution about an accreting object is given by Danby and Camm (1957). Danby and Bray (1967) show that when a gas is not cold, accretion may not operate as efficiently as indicated above. Spiegel (1970) gives formulae for the drag force, F_{D}, due to accretion shocks, and shows that the dissipation of gravitational energy by supersonically moving galaxies may be sufficient to heat a galactic cluster to around 10^8 °K. Spiegel gives

$$F_{\mathrm{D}} = \pi R_A^2 \rho v \ln \left\{ \frac{\pi}{R k_{\mathrm{J}}} \left[\frac{M}{(M^2 - 1)^{1/2}} \right] \right\} , \qquad (3.425)$$

where

$$k_{\mathrm{J}} = \left[\frac{4\pi G \rho}{s^2} \right]^{1/2} , \qquad (3.426)$$

and the Mach number $M = v/s$ is greater than one. The rate of work by the supersonic object is vF_{D}.

3.5.15 Stellar Variability and Oscillation Theory

Fabricus (1594) first found that the light from the star, Mira, is variable; and Goodricke (1784) and Bailey (1899), respectively found that the light of δ Cephei and RR Lyrae is periodically variable. Vogel (1889) and Belopolski (1897) found that the radial velocities of classical Cepheid variables change with the same period as the light variations, and that the time of maximum brightness is very near the time of their maximum velocity of recession. This situation led Henry Plummer and Harlow Shapley to suggest that the periodic displacement of spectral lines is caused by the radial pulsations of isolated individual stars, rather than by the orbital motion of binary stars as had been previously imagined (Plummer, 1913; Shapley, 1914). Meanwhile, the absolute luminosities of the Cepheid variable stars had been shown to increase with their periods of light variation (Leavitt, 1912), which now provides a basic technique for determining the distances of stars and of nearby galaxies (see Volume II).

Catalogues of thousands of variable stars have subsequently been prepared by Kukarin and Parenago (1969) and by Hirshfeld and Sinnott (1985). The latter authors also define variable star types for stellar luminosity variations due to eruptions as well as pulsations. A description of the different kinds of stellar variability is also given by Hoffmeister, Richter and Wenzel (1985) and Petit (1987). The variable subclass of pulsating stars has periods ranging from 0.3 to 1000 days. The RR Lyrae type variables have periods ranging from 0.3 to 0.9 days, whereas the classical Cepheid variables have periods ranging from 1 to 50 days.

The idea that certain kinds of periodic stellar variability could be due to the periodic expansion and contraction of the star was put on a firm mathematical foundation by Eddington (1918), following Ritter's (1880, 1881) early investigations of the adiabatic, radial oscillations of a star. Eddington applied his theory of radiative equilibrium to the Cepheid variables, predicting that the period of oscillations should vary inversely with the square root of the mass density of the star. His calculations also suggested that the pulsations must be determined primarily by conditions in the envelope of the star, rather than by those in the central regions where most of the mass is located. He argued that the material in the outer atmosphere must act as a heat engine by absorbing heat when it is hottest and most compressed and releasing heat when it is coolest and most expanded. Eddington later argued that pulsation originates in an outer, convective zone where hydrogen is alternatively ionized and neutral during the course of pulsation (Eddington, 1941, 1942), and Epstein (1950) showed that the source of the instability was primarily in the outermost nonadiabatic regions of a star. S. A. Zhevakin showed that Eddington's convective ionized-hydrogen zone cannot maintain pulsations because it does not absorb sufficient energy during the contraction of the star, but that an outer region of doubly ionized helium can provide the necessary periodic valve for the radiant flux of the star (Zhevakin, 1953, 1954). A similar idea was suggested by Cox and Whitney (1958). The ionized helium zone absorbs the outward flow of energy from the center of the star during contraction and returns it during expansion. The effectiveness of helium ionization as a driving mechanism for pulsating stars was first conclusively demonstrated by the linear nonadiabatic calculations of Baker and Kippenhahn (1962) and Cox (1963). Summaries of some of the early history of pulsation theory are given in the monographs by Eddington (1926) and Rosseland (1949), as well as by Ledoux and Walraven (1958) and Cox (1974, 1980).

Hertzsprung (1905) showed that the pulsating stars occupy an "instability" strip in the Hertzsprung-Russell (HR) diagram, and that the Cepheid variable stars are "giant" stars in a particular stage of evolution. During almost a century of subsequent investigation of stellar evolution, it has been realized that stellar pulsations, either as radial or nonradial dynamical variabilities, are found in many phases of stellar evolution and vastly differing regions on the HR diagram. A review of stellar pulsations across the H-R diagram is given by Gautshy and Saio (1995, 1996).

Reviews of the pulsation theory of variable stars have been given by Ledoux and Walraven (1958), Zhevakin (1963), Christy (1966) and Cox (1974). Both radial and nonradial oscillations are reviewed in the monograph by Cox (1980) and in the stellar text by Kippenhahn and Weigert (1990). Nonradial oscillations are discussed in the book by Uno et al. (1989). Basic equations for radial and nonradial adiabatic pulsations, as well as pulsations in rotating stars, are given by Gautschy and Saio (1995). Both rotation and a magnetic field have been added to the interpretation of certain rapidly oscillating stars, designated roAp stars, interpreted in terms of an oblique pulsator model (Dziembowski and Goode, 1985; Kurtz, 1990). Other periodic magnetic variations of the Ap stars had been explained in terms of an oblique rotator model (Stibbs, 1950).

Ritter (1880, 1881) first advanced the hypothesis that adiabatic, radial oscillations of a star might account for their observed periodic variability. Eddington (1918, 1919, 1926) used the linearized forms of the continuity equation (3.250) and Euler's equation (3.251) together with the adiabatic relations

$$\begin{aligned}
\frac{\Delta P}{P} &= \Gamma_1 \frac{\Delta \rho}{\rho} \ , \\
\frac{\Delta T}{T} &= (\Gamma_3 - 1)\frac{\Delta \rho}{\rho} \ , \\
\Gamma_1 &= \beta + \frac{(4-3\beta)^2(\gamma-1)}{\beta + 12(\gamma-1)(1-\beta)} \ , \\
\Gamma_3 &= 1 + \frac{\Gamma_1 - \beta}{4 - 3\beta} \ ,
\end{aligned} \tag{3.427}$$

and

$$\beta = \frac{P_G}{P} = \frac{Nk\rho T}{Nk\rho T + aT^4/3} \ ,$$

to obtain the wave equation for adiabatic radial oscillations (Ledoux and Walraven, 1958; Cox, 1980; Gautshy and Saio, 1995, 1996)

$$-\frac{1}{\rho r^4}\frac{d}{dr}\left(\Gamma_1 P r^4 \frac{d\xi}{dr}\right) - \frac{1}{r\rho}\left\{\frac{d}{dr}[(3\Gamma_1 - 4)P]\right\}\xi = \sigma^2 \xi$$

or

$$\frac{d^2\xi}{dx^2} + [4 - V(x)]\frac{1}{x}\frac{d\xi}{dx} + \frac{V(x)}{x^2}\left[\frac{4-3\gamma}{\gamma} + \frac{x^3 R^3 \sigma^2}{\gamma G M(x)}\right]\xi = 0 \ . \tag{3.428}$$

Here a star has total pressure, P, gas pressure $P_G = Nk\rho T$, where N is the total number of free electrons and ions per unit mass, radiation pressure $P_R = aT^4/3$, where $ac/4$ is the Stefan–Boltzmann constant, mass density, ρ, temperature, T, and adiabatic coefficient, γ. Linear radial oscillations result in a change, Δr, in radius, r, given by

$$\Delta r = r\xi \exp[i\sigma t] \ , \tag{3.429}$$

where ξ and σ are, respectively, the amplitude and frequency of the oscillation. These oscillations are regarded as being superposed on a state of complete hydrostatic and thermal equilibrium. For an equilibrium radius, R, the variable $x = r/R$, and

$$V(x) = \frac{G\rho(x)M(x)}{rP(x)} \quad , \tag{3.430}$$

where $M(x)$ is the mass interior to x, and is given by

$$M(x) = \int_0^x 4\pi r^2 \rho(r) dr \ .$$

The wave equation (3.428) is subject to the boundary conditions

$$\Delta r = 0 \quad \text{at} \quad r = 0$$

and

$$\frac{\Delta P}{P} = -\left[4 + \frac{\sigma^2 R^3}{GM}\right]\xi \quad \text{at} \quad r = R \ ,$$

or

$$\frac{d\xi}{dx} = \left[4 - 3\gamma + \frac{\sigma^2 R^3}{GM}\right]\frac{\xi}{\gamma} \quad \text{at} \quad x = 1 \ . \tag{3.431}$$

Eq. (3.428) admits eigenfunction solutions, ξ_i, for certain eigenvalues, σ_i^2, of the parameter σ^2, which increase as the integer i increases. Schwarzschild (1941) has computed amplitudes, ξ, for the first four modes, $i = 0, 1, 2, 3$, for $\Gamma_1 = 5/3$, and Ledoux and Walraven (1958) have summarized the properties of the first two modes of radial oscillation for different stellar models with $\Gamma_1 = 5/3$. In general, the fundamental mode, $i = 0$, has the solution (Ledoux and Pekeris, 1941; Ledoux, 1945; Gautshy and Saio, 1995)

$$(3\Gamma_1 - 4)\frac{4\pi G\rho}{3} \leq \sigma_0^2 \leq -(3\Gamma_1 - 4)\frac{\Omega}{I} \quad , \tag{3.432}$$

where $\Omega \approx -GM^2/R$ is the gravitational potential energy and $I \approx MR^2$ is the moment of inertia. If a magnetic field of strength, H, is present, Ω must be replaced by $\Omega + \mathscr{M}$ (Chandrasekhar and Limber, 1954) where the magnetic energy $\mathscr{M} = R^3H^2/6$. For a homogeneous star we have (Ritter, 1881)

$$\sigma_0^2 = (3\Gamma_1 - 4)\frac{GM}{R^3} \quad , \tag{3.433}$$

so that the period, Π_0, is given by

$$\Pi_0 = \frac{2\pi}{\sigma_0} \approx \frac{2R}{s} \propto \rho^{-1/2} \quad , \tag{3.434}$$

where s is the velocity of sound, and the pulsation constant

$$Q_0 = \Pi_0 \left(\frac{\rho}{\rho_\odot}\right)^{1/2} \approx 0.116\,\text{days} \quad \text{for} \quad \Gamma_1 = 5/3 \ , \tag{3.435}$$

where the solar density $\rho_\odot \approx 1.41\,\mathrm{gm\,cm^{-3}}$. Epstein (1950) showed that the fundamental mode solution to Eq. (3.428) is determined by a weighting function which peaks at

$$x = \frac{r}{R} \approx 0.75 \ . \tag{3.436}$$

This means that the period of the fundamental mode is determined primarily by conditions in the envelope of the star, and is almost independent of conditions in the central regions where most of the mass is located. Period ratios, Π_i/Π_0, are given as a function of Q_0 by Schwarzschild (1941), Baker and Kippenhahn (1965), and Christy (1966). Observed data (Christy 1966) indicate that $\Pi_i/\Pi_0 \approx 0.77$ and $Q_0 = 0.033$ days. The homogeneous model gives $Q_0 \approx 0.12$ whereas the standard model for a polytrope of index, 3, has $Q_0 \approx 0.038$. Christy (1966) gives

$$Q_0 \approx 0.022 \left(\frac{R}{R_\odot}\right)^{1/4} \left(\frac{M_\odot}{M}\right)^{1/4} \text{days} \ , \tag{3.437}$$

for a star of mass, M, and radius, R. Here the solar radius, $R_\odot \approx 6.96 \times 10^{10}$ cm and the solar mass $M_\odot \approx 2 \times 10^{33}$ gm.

Although the periods of variable stars are relatively insensitive to non-linear or non-adiabatic effects, adiabatic oscillations quickly decay and some non-linear or non-adiabatic effect is needed to explain their continual pulsation. Such effects are also needed to explain the observed phase shift between the oscillation of brightness and the stellar radius. Eddington (1941) suggested that a possible cause of instability might be the non-adiabatic hydrogen ionization zone near the surface of the star; and Zhevakin (1953, 1963) and Cox and Whitney (1958) showed that stellar oscillations might be excited by the region of ionization of helium (He^+). Later linear, non-adiabatic calculations by Baker and Kippenhahn (1962, 1965) and Cox (1963) verified Zhevakin's conclusion. These calculations utilize the linear forms of the equation of motion

$$\frac{\partial^2 r}{\partial t^2} = -\frac{GM(r)}{r^2} - \frac{1}{\rho}\frac{\partial P}{\partial r} \ , \tag{3.438}$$

and the heat flow equation

$$\frac{\partial Q}{\partial t} = \frac{\partial E}{\partial t} - \frac{P}{\rho^2}\frac{\partial \rho}{\partial t} = \varepsilon - \frac{1}{4\pi r^2 \rho}\frac{\partial L}{\partial r} \ . \tag{3.439}$$

Here $M(r)$ is the mass interior to radius, r, $\partial Q/\partial t$ is the net rate of gain of heat per unit mass, E is the total internal energy per unit mass, ε is the rate per unit mass of thermonuclear energy generation

$$\varepsilon = \varepsilon_0 \rho^\lambda T^\nu$$

or

$$\frac{\Delta\varepsilon}{\varepsilon} = \lambda\frac{\Delta\rho}{\rho} + \nu\frac{\Delta T}{T} \approx \frac{\Delta\rho}{\rho}[\lambda + \nu(\Gamma_3 - 1)] \ , \tag{3.440}$$

and the "interior" luminosity, $L(r)$, is given by

$$L(r) = -4\pi r^2 \frac{4ac}{3}\frac{T^3}{\kappa\rho}\frac{\partial T}{\partial r} \ , \tag{3.441}$$

where the Rosseland mean opacity, κ, is tabulated by Cox, Stewart, and Eilers (1965). Often the formula

$$\kappa = \kappa_0 \, \rho^n \, T^{-s} \ , \tag{3.442}$$

is employed to give

$$\frac{\Delta\kappa}{\kappa} = n\frac{\Delta\rho}{\rho} - s\frac{\Delta T}{T}$$

and

$$\frac{\Delta L}{L} = 4\frac{\Delta r}{r} + (4+s)\frac{\Delta T}{T} - n\frac{\Delta\rho}{\rho} + \frac{1}{d\ln T/dx}\frac{\partial}{\partial x}\left(\frac{\Delta T}{T}\right) . \tag{3.443}$$

Alternative forms of the heat flow equation (3.439) are

$$\frac{\partial P}{\partial t} = \frac{\Gamma_1 P}{\rho}\frac{\partial\rho}{\partial t} + \rho(\Gamma_3 - 1)\left[\varepsilon - \frac{1}{4\pi r^2\rho}\frac{\partial L}{\partial r}\right] ,$$

and

$$\frac{\partial \ln T}{\partial t} = (\Gamma_3 - 1)\frac{\partial \ln\rho}{\partial t} + \frac{1}{c_v T}\left[\varepsilon - \frac{1}{4\pi r^2\rho}\frac{\partial L}{\partial r}\right] . \tag{3.444}$$

When $\xi = \Delta r/r$ is used with the linear forms of Eqs. (3.438) and (3.439) or (3.444), we obtain the wave equation for non-adiabatic radial oscillations.

$$\dddot{\xi} - \frac{\dot{\xi}}{r\rho}\frac{\partial}{\partial r}\left[(3\Gamma_1 - 4)P\right] - \frac{1}{\rho r^4}\frac{\partial}{\partial r}\left[\Gamma_1 P r^4\frac{\partial\dot{\xi}}{\partial r}\right] = -\frac{1}{r\rho}\frac{\partial}{\partial r}\left[\rho(\Gamma_3 - 1)\Delta\left(\varepsilon - \frac{1}{4\pi r^2\rho}\frac{\partial L}{\partial r}\right)\right] , \tag{3.445}$$

where the dot over a symbol means the Stokes derivative $\partial/\partial t$. Baker and Kippenhahn (1962, 1965). Cox (1963), and Zhevakin (1963) have given computer solutions to Eqs. (3.438), (3.439), (3.444), and (3.445), whereas Ledoux (1963, 1965) has given integral solutions for the eigenvalues.

When general departures from an equilibrium spherical shape are considered, the radius, r, of the deformed surface of a fluid is given by

$$r = R + \varepsilon\, Y_l^m(\varphi, \theta) \ , \tag{3.446}$$

where R is the equilibrium radius, Y_l^m is a spherical harmonic, and $\varepsilon = \varepsilon_0 \exp[-\sigma t]$ where ε_0 is a constant and the oscillatory period is $2\pi/\sigma$. Both radial and nonradial oscillations of an inviscid fluid have the Kelvin modes

$$\sigma^2 = -\frac{2l(l-1)}{(2l+1)}\frac{GM}{R^3} = -\frac{8}{3}\pi G\rho\frac{l(l-1)}{2l+1} . \tag{3.447}$$

When viscosity is introduced, the Kelvin modes are damped with a mean lifetime of

$$\tau = \frac{R^2}{\nu(l-1)(2l+1)} ,$$

where the kinematic viscosity, ν, is assumed to be vanishingly small. Lamb (1881) and Chandrasekhar (1959) have considered the oscillations of a viscous globe.

Nonradial oscillations are characterized by the dispersion relation (Cowling, 1941; Cox, 1980; Gautshy and Saio, 1995):

$$k_r^2 \approx \frac{(\sigma^2 - L_l^2)(\sigma^2 - N^2)}{\sigma^2 c_s^2} \tag{3.448}$$

describing two types of oscillations, sound waves (p-mode) that are restored by the compressibility of the gas, which propagate with $\sigma > L_l$ and $\sigma > N$ and a gravity (g-mode), restored by buoyancy force, which propagates with $\sigma < L_l$ and $\sigma < N$. Here the Lamb frequency, L_l, and the Brunt-Väisäla frequency, N, are defined by

$$L_l = \sqrt{\frac{l(l+1)\,c_s^2}{r^2}}$$

and

$$N = \sqrt{g\frac{\delta}{c_p}\frac{dS}{dr}}\ ,$$

where $c_s = s$ is the adiabatic sound speed, g is the local gravitational acceleration, c_p is the specific heat, and $\delta = \partial \log \rho / \partial \log T$, both at constant pressure, S is the specific entropy, and r is the radial coordinate.

3.5.16 Instabilities in Fluids and Plasmas

Two common instabilities occur when two distinct fluids are in pressure equilibrium across a common interface. The Rayleigh–Taylor instability occurs when a heavy fluid rests on top of a light fluid, and the Kelvin–Helmholtz instability occurs at the interface of two fluids in relative motion. Our discussion begins with the dispersion relations for these instabilities, together with some of their magnetohydrodynamic and plasma counterparts. Many of the details of these two types of instabilities can be found in the monograph by Chandrasekhar (1961). The Kelvin–Helmholtz instability is additionally reviewed by Gerwin (1968). Other fluid instabilities, not included in this section but presented in other parts of this book, are the convective, gravitational, rotational and thermal instabilities.

We also consider plasma instabilities, which were reviewed by Hasegawa (1971) for the Earth's magnetosphere. Plasma instabilities can arise because of anisotropies in the velocity distribution. Examples include the "two-stream" instability due to a difference in the velocity of plasma beams (Sturrock, 1994) and the "fire-hose" instability (Benz, 1993). Other plasma instabilities, such as the "pinch" instability, can be studied within the context of magnetohydrodynamic (MHD) theory (Boyd and Sanderson, 1969; Bateman, 1980).

Some plasma instabilities differ from MHD ones by permitting nonzero resistivity, and allowing magnetic fields to move independently of the plasma.

One of these resistive instabilities is the "tearing-mode" instability that can develop a current sheet (Furth, Killeen and Rosenbluth, 1963; White, 1983; Sturrock, 1994). Since the magnetic structure can change, the magnetic field lines can reconnect (Dungey, 1953; Parker, 1963). Magnetic reconnection in astrophysics was reviewed by Syrovatskii (1981), while critical problems in plasma astrophysics including reconnection were discussed by Sagdeev (1979).

A uniform plasma in thermodynamic equilibrium is stable; whereas the introduction of nonuniformities in plasma density, temperature, velocity, or magnetic field might cause instabilities. The instabilities are usually analyzed by assuming an initial equilibrium state, φ_0, subject to an infinitely small perturbation, φ_1, of the form

$$\varphi_1 \propto \exp i(\boldsymbol{k} \cdot \boldsymbol{x} - \omega t) \ , \tag{3.449}$$

where $\boldsymbol{k}$ is the wave number vector and ω is the frequency. The equation of motion and the conservation equations are then linearized and assumed to have a solution $\varphi = \varphi_0 + \varphi_1$. As a consequence, a dispersion relation between ω and $\boldsymbol{k}$ is found. The plasma is said to be linearly unstable if ω has a positive imaginary part for any real value of k. If all possible ω are real or have negative imaginary parts, then the plasma is said to be stable.

3.5.16.1 Rayleigh–Taylor Instability

When an incompressible fluid has a nonuniform distribution of mass density, ρ, it may exhibit the Rayleigh (1883) – Taylor (1950) instability. For two vertically adjacent, hydrostatic, inviscid fluids, the lower fluid having density ρ_1 and the upper layer having density ρ_2, the dispersion relation is (Rayleigh, 1883)

$$\omega^2 = -gk\left(\frac{\rho_2 - \rho_1}{\rho_2 + \rho_1}\right) \ , \tag{3.450}$$

where g is the gravitational acceleration. If $\rho_2 < \rho_1$ the situation is stable, whereas it is unstable for all wave numbers if $\rho_2 > \rho_1$. If there is a surface tension, T, however, the arrangement is stabilized for sufficiently short wavelengths. In this case, the fluid is unstable for all wave numbers, k, in the range $0 < k < k_c$, where

$$k_c = \left[\frac{(\rho_2 - \rho_1)}{T}g\right]^{1/2} \ . \tag{3.451}$$

When a horizontal magnetic field of strength, B, is introduced, the arrangement is also stabilized with an effective surface tension, T_{eff}, given by (Kruskal and Schwarzschild, 1954)

$$T_{\mathrm{eff}} = \frac{\mu B^2}{2\pi k}\cos^2\theta \ , \tag{3.452}$$

where μ is the magnetic permeability, and θ is the angle between the wave vector and the direction of B. If the fluid is rotating with angular velocity, Ω, the dispersion relation becomes (Hide, 1956; Chandrasekhar, 1961)

$$\omega^2 \left[1 - \frac{4\Omega^2}{\omega^2}\right]^{1/2} = \omega_0^2 \ , \tag{3.453}$$

where ω_0^2 is the ω^2 given in Eq. (3.450). Rotation, therefore, puts a lower limit to $\omega = 2\Omega$.

For a stratified medium of density $\rho = \rho_0 \exp(-z/H)$, where H is the scale height, the Rayleigh-Taylor instability occurs for negative H, whereas stable gravity waves occur for positive H. When an inviscid fluid is confined between two right planes at $z = 0$ and $z = d$, the dispersion relation is (Rayleigh, 1883; Chandrasekhar, 1955)

$$\frac{g}{H\omega^2} = 1 + \frac{[d^2/(4H^2)] + m^2\pi^2}{k^2d^2} \ , \tag{3.454}$$

where m is an integer. Hence, for a gradient, ∇N, in density, N, we have instability for

$$g\nabla N < 0 \ , \tag{3.455}$$

for which

$$\omega = \pm i(gH^{-1})^{1/2} \ , \tag{3.456}$$

where

$$H^{-1} = -\frac{d}{dz}[\ln N] \ . \tag{3.457}$$

Rotation stabilizes this arrangement for all wave numbers, k, less than a minimum wave number, $k_{\min}$, given by

$$k_{\min}^2 = -\frac{4\Omega^2 H}{gd^2}\left[\frac{d^2}{4H^2} + \pi^2\right] \ , \tag{3.458}$$

where the angular rotational velocity is Ω, and the layer thickness is d.

When plasma particles move along curved magnetic field lines of radius, R, the flow creates a centrifugal acceleration

$$g \approx \frac{P}{\rho R} \ , \tag{3.459}$$

where the field lines are assumed to pass through an atmosphere of density, ρ, and pressure, P. For the ideal "flute mode" instability, we have

$$\omega^2 \approx gH^{-1} \approx \frac{P}{\rho RH} \ , \tag{3.460}$$

which is seen to be the same dispersion relation as that of the Rayleigh-Taylor instability. Kulsrud (1967) gives an approximate dispersion relation for the interchange instability

$$\omega^2 = -\left[\frac{\gamma P}{U}\frac{dU}{dP} + 1\right]\frac{P^2}{\rho U}\frac{dU}{dP} \ , \tag{3.461}$$

where it is assumed that the gas pressure, P, is much less than the magnetic pressure, $B^2/(8\pi)$, and the quantity

$$U = \int \frac{dl}{B} ,$$

where the integral is along a magnetic field line for a magnetic field of strength B. Instability occurs if

$$\frac{dP}{dU} > 0 ,$$

and stability results if

$$\frac{d}{dP}[P U^{\gamma}] < 0 . \tag{3.462}$$

Here γ is the ratio of specific heats. The system is stable if the interchange of two equal flux tubes requires more compression of the plasma than expansion. Hasegawa (1971) gives a general interchange dispersion relation

$$\frac{\omega_{\mathrm{ci}}}{Hk_{\perp}} \left[\frac{1}{\omega + k_{\perp} g/\omega_{\mathrm{ci}}} - \frac{1}{\omega} \right] = 1 - \frac{M_{\mathrm{i}}}{M_{\mathrm{e}}} \left[\frac{k_{\parallel} \omega_{\mathrm{ci}}}{k_{\perp} \omega} \right]^2 , \tag{3.463}$$

where the gravitational acceleration, g, is taken to simulate the centrifugal force due to particle motion parallel to the curved field lines, $k_{\perp}$ and $k_{\parallel}$ denote, respectively, the two orthogonal wave vectors which are perpendicular to the field direction, the subscripts i and e denote, respectively, the ions and electrons, and the cyclotron frequency, $\omega_{\mathrm{c}j}$, for the jth particle type of charge eZ_j is

$$\omega_{\mathrm{c}j} = \frac{Z_j eB}{M_j c} . \tag{3.464}$$

For the flute mode, $k_{\parallel} = 0$ and Eq. (3.463) becomes Eq. (3.460). For $k_{\parallel} \neq 0$, however, the instability is stabilized for a perturbation of any size if

$$Rk_{\parallel} > 2 \left(\frac{M_{\mathrm{e}}}{M_{\mathrm{i}}} \right)^{1/2} \tag{3.465},$$

where R is the radius of curvature of the field line.

Even without the action of a gravitational field, a plasma with a density gradient becomes unstable for waves whose parallel phase velocity is between the thermal velocities of the electrons and ions. This drift wave instability has frequency, ω, given by the drift wave frequency (Moiseev and Sagdeev, 1963; Hasegawa, 1971)

$$\omega = \frac{v_{\mathrm{Te}}^2 k_{\parallel}}{H \omega_{\mathrm{ce}}} \tag{3.466}$$

when

$$v_{\mathrm{Te}} > \frac{\omega}{k_{\parallel}} > v_{\mathrm{Ti}} ,$$

where H is the scale height given by Eq. (3.457) and v_{T} is the thermal velocity.

3.5.16.2 Kelvin–Helmholtz Instability

When both a nonuniformity in density and shear flow are considered, instability between two fluids is found to be possible even when $\rho_2 < \rho_1$. If $v_1 - v_2 = v$ denotes the relative velocity between the two fluids, then this Kelvin (1871)-Helmholtz (1868) instability occurs for all wavenumbers

$$k > \frac{g(\rho_2 - \rho_1)(\rho_1 + \rho_2)}{\rho_1 \rho_2 (v_1 - v_2)^2 \cos^2 \varphi} , \tag{3.467}$$

where φ is the angle between the directions of $\boldsymbol{k}$ and $\boldsymbol{v}$. A surface tension, T, will suppress the instability if

$$(v_1 - v_2)^2 < \frac{2(\rho_1 + \rho_2)}{\rho_1 \rho_2} [Tg(\rho_1 - \rho_2)]^{1/2} . \tag{3.468}$$

A magnetic field of strength B will also suppress the instability if the inequality in Eq. (3.468) is found to hold with T replaced by the T_{eff} given in Eq. (3.452).

An electrostatic Kelvin-Helmholtz instability exists in a plasma due to the shear flow caused by an E cross H drift. For an electron sheet of thickness $2a$ which is parallel to a uniform magnetic field of strength B, the dispersion relation becomes

$$\frac{4\omega^2}{\omega_0^2} = \left[1 - \frac{2kv_0}{\omega_0}\right]^2 - \exp[-4ka] , \tag{3.469}$$

where

$$\omega_0 = 4\pi ecN/B = \omega_{\mathrm{pe}}^2/\omega_{\mathrm{ce}} ,$$

and the shear velocity, v_0, is given by

$$v_0 = E(a)/B ,$$

where ω_{ce} and ω_{pe} are, respectively, given in Eqs. (3.464) and (3.471), and E(a) is the electric field intensity at a. The wave becomes unstable for all wave numbers, k, satisfying the inequality

$$ka \lesssim 0.7 , \tag{3.470}$$

which is sufficient to make the right side of Eq. (3.469) negative. The Kelvin-Helmholtz instability at the interface between two fluids in relative motion has been studied by Lamb (1949), where earlier references may be found, and reviewed by Gerwin (1968) where references are given to related MHD and plasma situations including the stability of the Earth's magnetosphere in the solar wind, the stability of charged particles emanating from the Sun, and the stability of jets, plasma jets and laboratory plasmas.

3.5.16.3 Instabilities Due to Velocity Anisotropies

As was shown in Sect. 1.32, the basic frequency of oscillation, $\omega_{\mathrm{p}j}$, of a thermal equilibrium plasma of one species of particles, j, is given by (Tonks and Langmuir, 1929)

$$\omega_{pj} = \left[\frac{4\pi e^2 Z_j^2 N_j}{M_j}\right]^{1/2} , \tag{3.471}$$

where the particle charge is eZ_j, the mass is M_j, and the volume density is N_j. In the absence of collisions, these oscillations are damped with a damping constant given by (Landau, 1946)

$$\mathscr{d}(\omega) \approx -\omega_{pj}\sqrt{\frac{\pi}{8}}\left(\frac{\omega_{pj}}{k\,v_T}\right)^3 \exp\left(-\frac{\omega^2}{2\,k^2\,v_T^2}\right) , \tag{3.472}$$

where $\mathscr{d}$ denotes the imaginary part of the term in parenthesis, k is the wave number of the disturbance, and v_T is the root mean square thermal velocity. Bohm and Gross (1949) pointed out that the physical mechanism of the damping described above is the trapping by the electric field of particles moving at approximately the phase velocity of the wave, with a consequent exchange of energy between particles and plasma oscillations.

It follows from the linearized Boltzmann equation that the general dispersion relation for a plasma is (Vlasov, 1945; Landau, 1946; Bohm and Gross, 1949)

$$1 - \frac{\omega_{pj}^2}{k^2}\int \frac{\partial f_0/\partial v}{v-(\omega/k)}\,dv = 0 , \tag{3.473}$$

where $f_0(v)$ is the equilibrium (unperturbed) distribution function of the velocity, v. If $f_0(v)$ has a single hump, it can be shown (Jackson, 1960) that there can only be damped oscillations. In the case of a Maxwellian distribution, the dispersion relation is

$$\omega^2 \simeq \omega_{pj}^2\left[1 + \frac{3k^2\,v_T^2}{\omega_{pj}^2}\right] , \tag{3.474}$$

with a damping constant given by Eq. (3.472). However, if the velocity distribution is two humped, damped, steady state, or growing oscillations are possible. For example, for a plasma of j beams of velocity, $\boldsymbol{v}_j$, the dispersion relation for longitudinal oscillations becomes (Bailey, 1948; Pierce, 1948)

$$1 = \sum_j \frac{\omega_{pj}^2}{(\omega - \boldsymbol{k}\cdot\boldsymbol{v}_j)^2} , \tag{3.475}$$

where the plasma frequency, ω_{pj}, is given by Eq. (3.471). For one beam of velocity, v_s, in a stationary plasma, the dispersion relation becomes

$$1 - \frac{\omega_p^2}{\omega^2} - \frac{\omega_{ps}^2}{(\omega - k\,v_s)^2} = 0 , \tag{3.476}$$

ω_p and ω_{ps} denote, respectively, the plasma frequencies of the plasma and the stream. A "two-stream" instability can occur when the negative-energy wave of the stream ($\omega < kv_s$) is coupled to the positive energy wave of the plasma. For weak coupling, four waves are possible with dispersion relations given by $\omega = kv_s \pm \omega_{ps}$ and $\omega = \pm\omega_p$; and the details of the stable and unstable oscillations for these cases are given by Sturrock (1958, 1994). One has an instability and growing oscillations for all wave numbers, k, such that

$$\frac{\omega}{v_s} \approx k \lesssim \frac{\omega_p}{v_s} \quad . \tag{3.477}$$

The growth rate for this instability is given by

$$\mathscr{d}(\omega) = \frac{\omega_{ps} k\, v_s}{(\omega_p^2 - k^2 v_s^2)^{1/2}} \ , \tag{3.478}$$

if kv_s is not close to ω_p. For an electronic plasma consisting of two streams of electrons of the same density but equal and opposite velocities, v_s, growing oscillations occur for

$$k \leq 0.5\frac{\omega_p}{v_T} \approx \sqrt{2}\frac{\omega_p}{v_s} \ , \tag{3.479}$$

where v_T is the thermal velocity spread of the stream, and the wave frequency $\omega = kv_s$. For a relativistic stream, the appropriate mass to use in the equation for the plasma frequency is

$$M = \gamma^3 M_0 \ , \tag{3.480}$$

where M_0 is the rest mass and $\gamma = (1 - v_s^2/c^2)^{-1/2}$. If the plasma is not cold, the two stream instability discussed above still occurs provided that the thermal velocity of the plasma, v_T, is much smaller than the stream velocity. In this case, however, the growth of the plasma wave is accompanied by a flattening of the velocity distribution and the stream loses its identity in the time (Zheleznyakov and Zaitsev, 1970)

$$\tau = \frac{N_p}{N_s}\left(\frac{\Delta v_s}{v_s}\right)^2 \frac{1}{\omega_p} \ , \tag{3.481}$$

where N_s, v_s, and Δv_s denote, respectively, the particle density, velocity, and velocity dispersion of the stream, and N_p and ω_p denote the density and plasma frequency of the plasma.

Buneman (1958, 1959) and Jackson (1960) have discussed the dispersion equation for an electron-ion plasma, which takes the general form

$$1 - \frac{\omega_{pe}^2}{k^2}\int_{-\infty}^{+\infty} \frac{\partial f_{0e}/\partial v\, dv}{v - (\omega/k)} - \frac{\omega_{pi}^2}{k^2}\int_{-\infty}^{+\infty} \frac{\partial f_{0i}/\partial v\, dv}{v - (\omega/k)} = 0 \ , \tag{3.482}$$

where the subscripts e and i denote, respectively, the electrons and ions. For a Maxwellian distribution of electrons and ions with respective thermal velocities v_{Te} and v_{Ti}, growing oscillations occur for wave numbers, k, such that

$$k \lesssim \sqrt{2}\left[1 + \left(\frac{M_e}{M_i}\right)^{1/2}\right]\frac{\omega_{pe}}{v} \ , \tag{3.483}$$

where the velocity difference $v = v_e - v_i$, is the difference between the mean velocities of the electrons and ions and is assumed to be large. For ions streaming through cold electrons, or for electrons streaming through electrons,

Eq. (3.483) becomes, respectively, Eqs. (3.477) or (3.479). For electrons moving thermally with respect to cold ions, however, the real part of the dispersion relation becomes

$$\omega = ks_{\mathrm{i}} \quad \text{for} \quad \omega \ll \omega_{\mathrm{pi}} \ , \tag{3.484}$$

where the ion sound speed, s_{i}, is given by

$$s_{\mathrm{i}} = v_{\mathrm{Te}} \left(\frac{M_{\mathrm{e}}}{M_{\mathrm{i}}} \right)^{1/2} \ .$$

The wave represented by the dispersion relation given in Eq. (3.484) is called the ion acoustic wave, and unstable oscillations occur for (Hasegawa, 1971)

$$v_{\mathrm{D}} > s_{\mathrm{i}} \quad \text{for} \quad \omega \lesssim \omega_{\mathrm{pi}},$$

and

$$\frac{\partial f_{0\mathrm{e}}}{\partial v} > 0 \quad \text{for} \quad \omega \sim \omega_{\mathrm{pe}} \ , \tag{3.485}$$

where v_{D} is the electron drift velocity, and $f_{0\mathrm{e}}$ is the unperturbed electron velocity distribution function.

When a magnetic field of strength, B, is present, velocity anisotropies lead to electromagnetic instabilities as well as the electrostatic instabilities discussed above. For a cold plasma in which j beams of velocity v_j occur, the dispersion relation is

$$k^2 = \frac{\omega^2}{c^2} + \frac{1}{c^2} \sum_j \frac{\omega_{\mathrm{p}j}^2 \omega}{\omega + kv_j + \omega_{\mathrm{c}j}} \ , \tag{3.486}$$

where the cyclotron frequency, $\omega_{\mathrm{c}j}$, for the jth particle is given by Eq. (3.464), and the plasma frequency, $\omega_{\mathrm{p}j}$, is given in Eq. (3.471). For cold ions and drifting electrons, the dispersion relation becomes (Hasegawa, 1971)

$$k^2c^2 - \omega^2 + \omega_{\mathrm{pe}}^2 \frac{\omega - kv_{\mathrm{D}}}{\omega - kv_{\mathrm{D}} + \omega_{\mathrm{ce}}} + \omega_{\mathrm{ci}}^2 \frac{\omega}{\omega - \omega_{\mathrm{ci}}} = 0 \ , \tag{3.487}$$

where v_{D} is the electron drift velocity, and the subscripts e and i denote, respectively, electrons and ions. Waves which satisfy the dispersion relation are called ion cyclotron waves, and they become unstable for $\omega \approx \omega_{\mathrm{ci}}$ or $\omega \approx 0$.

The fire-hose (or garden-hose) instability is driven by beam pressure parallel to a magnetic field (Benz, 1993). When a magnetic field is present and collision effects are negligible, the particle pressures along, $P_{\|}$, and perpendicular, $P_{\perp}$, to the magnetic field become decoupled and instabilities are possible. If $f_j(v)$ is the velocity distribution function of species, j,

$$P_{\|} = \sum M_j \int v_{\|}^2 f_j(v) d^3v \ ,$$

and

$$P_{\perp} = \sum M_j \int v_{\perp}^2 f_j(v) d^3v \ , \tag{3.488}$$

where the summation is over the species of particles. If we let $\beta = 8\pi P/B^2$ denote the ratio of the pressure of a plasma species to the pressure of the magnetic field, then the dispersion relation at low frequencies ($\omega \ll \omega_{ci}$) and large wavelengths ($kv_{Ti} \ll \omega_{ci}$) becomes (Kutsenko and Stepanov, 1960)

$$\frac{\omega^2}{k_\parallel^2 v_A^2} = 1 - \sum_{\text{species}} \frac{1}{2}(\beta_\parallel - \beta_\perp) \quad \text{for} \quad \langle v_\parallel \rangle = 0 \ , \tag{3.489}$$

and

$$k_\parallel^2\left[1 + \sum_{\text{species}} \frac{\beta_\perp - \beta_\parallel}{2}\right] + k_\perp^2\left[1 + \sum_{\text{species}} \beta_\perp\left(1 - \frac{\beta_\perp}{\beta_\parallel}\right) - i\frac{\beta_{\perp i}^2}{\beta_{\parallel i}}\frac{\omega}{k_\parallel \langle v_{\parallel i}\rangle}\left(\frac{\pi}{2}\right)^{1/2}\right] = 0$$

for $\omega \ll kv_A$, when again the subscripts $\perp$ and $\parallel$ denote, respectively, directions perpendicular and parallel to the magnetic field, the subscript i denotes the ion component, and v_A is the Alfvén velocity. For $k \gg k_\perp$, an Alfvén wave propagates along (parallel) the lines of force, and the wave becomes unstable for

$$1 + \sum_j \frac{\beta_{\perp j} - \beta_{\parallel j}}{2} < 0$$

or

$$P_\parallel > P_\perp + \frac{B^2}{4\pi} \ . \tag{3.490}$$

This "fire hose" instability occurs for both the shear mode (no variation in the parallel component of the magnetic field) and for the compressional mode (variation in the parallel component ($k_\parallel > k_\perp$) of the magnetic field. For almost perpendicular propagation ($k_\parallel \ll k_\perp$), the compressional mode allows the "mirror" instability which occurs for

$$1 + \sum_j \beta_{\perp j}\left(1 - \frac{\beta_{\perp j}}{\beta_{\parallel j}}\right) < 0$$

or

$$\frac{P_\perp^2}{P_\parallel} > P_\perp + \frac{B^2}{8\pi} \ . \tag{3.491}$$

3.5.16.4 Pinch Instability

A current flowing through a cylindrical plasma column with an axial magnetic field will produce an external magnetic pressure that can pinch the plasma and confine it by balancing the gas pressure (Tayler, 1957; Kruskal and Tuck, 1958; Chandrasekhar, Kaufman and Watson, 1958). The total current, I, flowing in the z direction is given by Bennett's relation (Bennett, 1934)

$$I = 2(NkT)^{1/2} \ , \tag{3.492}$$

where N is the number of electrons per unit length, L, of the cylinder

$$N = \int_0^L 2\pi n r \, dr$$

for a plasma electron density n. There is a minimum value of the strength, B_z, of the internal, axial magnetic field that is required to stabilize the pinch. It depends on the cylindrically symmetric mode, m, of instability; for example, $m = 0$ for the sausage instability and $m = 1$ designates the kink instability. If B_ϕ denotes the strength of the external magnetic field that is generated, then the plasma column is unstable to sausage distortions ($m = 0$) for (Jackson, 1962)

$$B_z^2 \leq B_\phi^2/2 \ . \tag{3.493}$$

Sturrock (1994) derives a more exact expression. A column of radius, R, is kinked ($m = 1$) or bent over a length, L, when

$$B_z^2 \leq B_\phi^2 \ \ln\left(\frac{L}{R}\right) \ . \tag{3.494}$$

3.5.16.5 Tearing-Mode Instability and Magnetic Reconnection

Of particular interest in the theory of solar flares is the "tearing" instability which occurs when two magnetic fields of opposite sign move against each other and reconnect. If a magnetic field of strength, B, is brought into a region of thickness, δ, and extension, L, along the field line, the outward diffusion of the field is balanced by the inward flow of the plasma. If v denotes the plasma velocity, then we have (Sweet, 1958, 1969; Parker, 1963; Furth, Killeen, and Rosenbluth, 1963)

$$v = (4\pi\delta\sigma/c^2)^{-1} \ , \tag{3.495}$$

where σ is the electrical conductivity, and matter is squeezed out of the region at the Alfvén velocity

$$v_\mathrm{A} = \frac{B}{(4\pi\rho)^{1/2}} \ .$$

Continuity of matter flow then requires that

$$vL = v_\mathrm{A}\delta \ , \tag{3.496}$$

to give an instability lifetime of

$$\tau \approx \frac{L}{v} = \frac{L R_\mathrm{m}^{1/2}}{v_\mathrm{A}} \ , \tag{3.497}$$

where the magnetic Reynolds number, R_m, is given by

$$R_\mathrm{m} = \left(\frac{L}{\delta}\right)^2 = \left(\frac{v_\mathrm{A}}{v}\right)^2 = \frac{4\pi L\sigma v_\mathrm{A}}{c^2} \ . \tag{3.498}$$

Petschek (1964) matched the reconnection geometry to the external surroundings and found that the reconnection rate Mach number, M, is given by

$$M = \frac{v}{v_A} = \frac{\pi}{4} \frac{1}{\ln[2M^2 R_m]} , \tag{3.499}$$

and in general

$$R_m^{-1/2} \lesssim \frac{v}{v_A} \lesssim 1 .$$

In this case, the timescale, τ, of the flux annihilation is given by

$$\tau = \frac{L}{v} = \frac{4L}{\pi v_A} \ln\left[2R_m \left(\frac{L}{v_A \tau}\right)^2\right] . \tag{3.500}$$

As emphasized by Gold and Hoyle (1960), the energy of the solar flare must come from the release of magnetic energy, $(\Delta B)^2/8\pi$, which would occur in the field annihilation-reconnection mechanism discussed above. Sturrock (1966) has proposed a flare mechanism which incorporates the Petscheck annihilation mode in the open field region of a magnetic loop.

Magnetic reconnection in astrophysics has been reviewed by Syrovatskii (1981). It has a particularly rich history in the study of solar flares. In one theoretical model, large scale, oppositely-directed magnetic fields come together in the corona, annihilate each other, and reconnect, resulting in high-energy particle acceleration. It has been dubbed the CSHKP model after the first letters of the last names of various researchers who have developed it (Carmichael, 1964; Sturrock, 1968; Hirayama, 1974; and Kopp and Pneuman, 1976). Observations at radio, X-ray and ultraviolet wavelengths, respectively using the Very Large Array, and the Yohkoh and SOHO spacecraft, indicate that the violent release of energy is often due to forced magnetic reconnection within the low corona and/or chromosphere (Kundu and Lang, 1985; Tsuneta et al., 1992; Masuda et al., 1994; Manoharan et al., 1996; Innes et al., 1997).

4. High Energy Astrophysics

"Certain physical investigations in the past year, make it probable to my mind that some portion of sub-atomic energy is actually being set free in the stars. F.W. Aston's experiments seem to leave no room for doubt that all the elements are constituted out of hydrogen atoms bound together with negative electrons. The nucleus of the helium atom, for example, consists of four hydrogen atoms bound with two electrons. But Aston has further shown that the mass of the helium atom is less than the sum of the masses of the four hydrogen atoms which enter into it. ... Now mass cannot be annihilated, and the deficit can only represent the energy set free in the transmutation. ... If only five per cent of a star's mass consists initially of hydrogen atoms, which are gradually being combined to form more complex elements, the total heat liberated will more than suffice for our demands, and we need look no further for the source of a star's energy. ... If, indeed, the sub-atomic energy in the star is being freely used to maintain their great furnaces, it seems to bring a little nearer to fulfillment our dream of controlling this latent power for the well being of the human race – or for its suicide."

A.S. Eddington 1920

"We therefore feel justified in advancing tentatively the hypothesis that cosmic rays are produced in the super-nova process. ... With all reserve we advance the view that a super-nova represents the transition of an ordinary star into a neutron star, consisting mainly of neutrons."

W. Baade and F. Zwicky 1934

"When the conditions depart widely from being static, there is no necessary tendency towards equipartition, but the energy may instead become enormously concentrated into certain small parts of the system. Thus in the crack of a whip the tip of the lash is moving faster than the speed of sound, though the coachman's wrist never moves fast at all. Again, when a large sea-wave strikes the wall of a lighthouse, spray is thrown up to a great height, and this in spite of its later rise being much slowed by air resistance.... It is suggested that cosmic rays may originate from some mechanism of this kind, and though there may be other possibilities, the most obvious source is from the stormy seas that must cover the surface of many of the stars."

C. Darwin 1949

4.1 Early Fundamental Particles, Symbols, and Definitions

4.1.1 The Electron, Proton, Neutron, and Photon and Their Antiparticles

At about the same time that Thomson (1897) discovered that all atoms emit electrons, photons with energy in the range 1–500 keV, called X-rays, were observed by Röntgen (1896). Photons with energy greater than 500 keV, called gamma (γ) rays, were subsequently observed by Villard (1900). Einstein (1905) then suggested that a photon particle of energy, $h\nu$, and zero mass is an electromagnetic wave of frequency, ν, and vice versa. The nuclear theory of matter was then introduced by Rutherford (1911, 1914) who proposed that an atom, which has a radius of approximately 10^{-8} cm, actually consists of a swarm of electrons surrounding a positively charged nucleus whose radius is less than 10^{-12} cm. The subsequent discovery of the proton by Rutherford and Chadwick (1921) further confirmed the speculation that the nucleus contains positively charged particles. The neutron was then discovered (Chadwick, 1932; Curie and Joliot, 1932), and Heisenberg (1932) proposed that the atomic nucleus contains the neutral neutrons as well as the protons.

The mass, charge, magnetic moment and mean life of the electron, proton and neutron are given in Table 4.1.

In 1932 Carl D. Anderson discovered a new particle in cosmic-ray cloud chamber tracks that differs from the electron only in that its charge is positive, and in the following year he proposed the name positron (Anderson, 1932, 1933). The positron is a form of antimatter, first predicted by Paul A.M. Dirac (Dirac, 1930, 1931). He noticed that equations describing the electron have two solutions. One of them characterizes the electron and the other its anti-particle, now called the positron. However, Dirac's theory apparently played no role whatsoever in the discovery of the positron (Kragh, 1990). Although similar antiparticles for the proton and neutron were also expected on theoretical grounds, they were not observed until the advent of large particle accelerators (Chamberlain et al., 1950).

Table 4.1. Properties of the electron, proton and neutron[1]

Particle	Electron	Proton	Neutron
Rest Mass (grams)	$9.109\ 389\ 7(54) \times 10^{-28}$	$1.672\ 623\ 1(10) \times 10^{-24}$	$1.674\ 928\ 6(10) \times 10^{-24}$
Rest Mass (a.m.u.)	$5.485\ 799\ 03(13) \times 10^{-4}$	1.007 276 470(12)	1.008 664 904(14)
Rest Mass (MeV/c^2)	0.510 999 06(15)	938.272 31(28)	939.565 63(28)
Charge (e.s.u.)	$-4.803\ 242(14) \times 10^{-10}$	$+4.803\ 242(14) \times 10^{-10}$	0
Magnetic Moment[2] (erg Gauss^{-1})	$9.28\ 477\ 01(31) \times 10^{-21}$	$1.410\ 607\ 61(47) \times 10^{-21}$	$0.966\ 237\ 1(4) \times 10^{-21}$
Mean Life[3]	$> 2.7 \times 10^{23}$ years	$> 1.6 \times 10^{26}$ years	887.0 ± 2.0 seconds

[1] Adapted from Lang (1992), the Review of Particle Properties, Physical Review **D50**, 1173 (1994), and Physical Reference Data at http://physics.nist.gov/. The numbers in parenthesis after the value give the one standard-deviation uncertainties in the last digits.

[2] The Bohr magneton is $\mu_B = 9.274\,015\,4(31) \times 10^{-21}$ erg Gauss^{-1} so the ratio of the electron magnetic moment, μ_e, and the Bohr magneton, μ_B, is $\mu_e/\mu_B = 1.001\,159\,652\,193(10)$. The nuclear magneton is $\mu_N = 5.050\,786\,6(17) \times 10^{-24}$ erg Gauss^{-1}, so the ratio $\mu_e/\mu_N = 1\,838.282\,000(37)$.

[3] The mean life of the proton, τ_p, given here is independent of mode. Values of $\tau_p > 10^{31}$ to 5×10^{32} years are mode dependent. The half life, $\tau_{1/2}$, is given by $\tau_{1/2} = 0.693\,\tau$, where τ is the mean life. Thus, the half life of the neutron, $\tau_{n1/2}$, for the mean life given here, $\tau_n = 887.0$ seconds, is $\tau_{n1/2} = 614.69$ seconds $= 10.25$ minutes. Mampe et al. (1993) obtain a mean life for the neutron of $\tau_n = 882.6 \pm 2.7$ seconds.

4.1.2 Symbols, Nomenclature, and Units

A nucleus is defined by the numbers:

$$\begin{aligned} &\text{Atomic number} = Z = \text{number of protons} \\ &\text{Neutron number} = N = \text{number of neutrons} \\ &\text{Mass number} = A = N + Z = \text{number of nucleons} \\ &\text{Isotopic number} = N - Z = A - 2Z \ . \end{aligned} \tag{4.1}$$

The nuclear mass, $M_{\rm nucl}$, can be calculated from the atomic mass, $M_{A,Z}$, using the relation (Fermi, 1928; Thomas, 1927)

$$M_{\rm nucl} = M_{A,Z} - 5.48593 \times 10^{-4}\, Z + 1.67475 \times 10^{-8}\, Z^{7/3} \text{a.m.u.} \ , \tag{4.2}$$

where Z is the atomic number of the nucleus. The second term on the righthand side of Eq. (4.2) corrects for the electron mass, and the last term represents the Fermi-Thomas binding energy of 15.6 $Z^{7/3}$ eV. The mass number, A, is the integer nearest in value to the exact mass, M, expressed in atomic mass units.

Special names are given to nuclei having the same values of some of the numbers Z, N, and A.

$$\begin{aligned} &\text{Isotope} = \text{same } Z, \text{different } N \\ &\text{Isotone} = \text{same } N, \text{different } Z \\ &\text{Isobar} \ = \text{same } A, \text{different } N, Z \\ &\text{Isomer} = \text{same } A, \text{same } Z \ . \end{aligned} \tag{4.3}$$

Nuclei are given the symbols (Z, A) for unexcited nuclei, and $(Z, A)^*$ for excited nuclei.

A reaction in which a particle, a, interacts with a nucleus, X, to produce a nucleus, Y, and a new particle, b, is designated by

$$a + X \rightarrow Y + b + Q \quad \text{or} \quad X(a, b)Y \ , \tag{4.4}$$

where Q is the energy released in the reaction. An element, B, is given the symbol

$${}_ZB^A_N \quad \text{or} \quad {}^A_NB \quad \text{or} \quad {}^AB \quad \text{or} \quad B^A, \tag{4.5}$$

where A is the mass number, Z is the atomic number, and N is the neutron number. Elements appearing inside the parentheses of a reaction are given the symbols:

$$p \text{ for } \mathrm{H}^1,\ D \text{ for } \mathrm{H}^2,\ T \text{ for } \mathrm{H}^3,\ \tau \text{ for } \mathrm{He}^3 \ \text{ and } \alpha \text{ for } \mathrm{He}^4 \ . \tag{4.6}$$

A fundamental particle such as the pion, π, which has positive, +, negative, –, or neutral, 0, charge is given the symbols

$$\pi^+, \pi^-, or\, \pi^0 \ . \tag{4.7}$$

An antiparticle is denoted by a raised bar. For example, the antiproton is denoted by $\bar{p}$.

Typical units used in nuclear astrophysics are:

$$\begin{aligned}
&\text{One Fermi} = 10^{-13}\ \text{cm} \\
&\text{One barn} = 10^{-24}\ \text{cm}^2 \\
&\text{One MeV} = 1.602\ 177\ 33(49) \times 10^{-6}\ \text{erg} \\
&\text{One a.m.u.} = u = 1.660\ 540\ 2(10) \times 10^{-24}\ \text{grams} \\
&\qquad = 931.494\ 32(28) \times \text{MeV}/c^2 \\
&\text{Boltzmann constant, } k = 1.380\ 658(12) \times 10^{-16}\ \text{erg}\ {}^\circ\text{K}^{-1} \\
&\qquad = 8.617\ 385(73) \times 10^{-11}\ \text{MeV K}^{-1} \\
&\text{Planck constant, } h = 6.626\ 075\ 5(40) \times 10^{-27}\ \text{erg sec} \\
&\qquad = 4.135\ 669\ 2(12) \times 10^{-21}\ \text{MeV sec}
\end{aligned} \tag{4.8}$$

The values of the physical constants are from Lang (1992), the Review of Particle Properties, Physical Review **D50**, 1 (1994), and Physical Reference Data at http://physics.nist.gov/.

4.1.3 Binding Energy, Mass Defect, Mass Excess, Atomic Mass, Mass Fraction, Packing Fraction, Energy Release, Magic Numbers, and Mass Laws

The total mass of the nucleus is less than the sum of the masses of the constituent protons and neutrons, or nucleons. The difference, called the nuclear mass defect ΔM_{nucl}, is given by:

$$\Delta M_{\text{nucl}} = M_{\text{nucl}} - ZM_{\text{P}} - NM_{\text{n}} \ , \tag{4.9}$$

where the atomic number Z is the number of protons, N is the number of neutrons, the mass number $A = Z + N$ is the number of nucleons, and M_{P} and M_{n} respectively denote the mass of the proton and neutron (see Table 4.1). An energy ΔE is released in assembling a nucleus from its constituent nucleons and is gained at the expense of the mass of the nucleus. This energy is referred to as the binding energy required to separate the nucleus into its constituent nucleons. The binding energy, ΔE, is given by

$$\Delta E = \Delta M_{\text{nucl}} c^2 \ , \tag{4.10}$$

where the velocity of light $c = 2.99\ 792\ 458 \times 10^{10}$ cm s^{-1}. Values of the binding energy, ΔE and binding energy per nucleon, $\Delta E/A$, are given in Table 4.2 for representative nuclei.

The separation energy, S_{N}, required to remove a neutron to infinity follows from Einstein's (1905, 1906, 1907) energy-mass equivalence and is given by

$$S_{\text{N}} = [M_{A-1,Z} + M_{\text{N}} - M_{A,Z}]c^2 \ . \tag{4.11}$$

The atomic mass, M_{AZ}, of a given isotope is the quantity that is usually measured and tabulated. The atomic mass excess, ΔM_{AZ}, in units of energy is given by:

$$\Delta M_{AZ} = (M_{AZ} - AM_u)c^2 \ , \tag{4.12}$$

where A is the atomic number, $M_u = u$ is the atomic mass unit (a.m.u.), and $M_u c^2 = 931.494$ MeV. Table 4.3 gives the atomic mass excesses ΔM_{AZ} for the

Table 4.2. Binding energies, ΔE, and binding energy per nucleon, $\Delta E/A$, for several nuclei[1]

Nucleus	Binding Energy, ΔE (MeV)	Binding Energy per Nucleon, $\Delta E/A$ (MeV)
^{2}D	2.22	1.11
^{4}He	28.30	7.07
^{12}C	92.16	7.68
^{16}O	127.62	7.98
^{40}Ca	342.05	8.55
^{56}Fe	492.26	8.79
^{238}U	1801.70	7.57

[1] Adapted from Wapstra and Audi (1985).

lightest isotopes. Atomic mass evaluations as well as nuclear-reaction and separation energies are given by Wapstra and Audi (1985).

Atomic masses may be calculated using these mass excesses and Eq. (4.12). Some frequently used atomic masses are:

$$\begin{aligned} A_{\mathrm{N}} &= 1.008665 \text{ a.m.u.} \\ A_{\mathrm{H}} &= 1.007825 \text{ a.m.u.} \\ A_{\mathrm{D}} &= 2.014102 \text{ a.m.u.} \\ A_{\mathrm{T}} &= 3.016050 \text{ a.m.u.} \\ A_{\mathrm{He}^3} &= 3.016030 \text{ a.m.u.} \\ A_{\mathrm{He}^4} &= 4.002603 \text{ a.m.u.} \ , \end{aligned} \tag{4.13}$$

where the symbol A_i is used to denote the atomic mass of element, i.

Table 4.3. Atomic mass excess ΔM_{AZ} of the lightest isotopes[1]

Z	Element	A	Mass Excess (keV)	Z	Element	A	Mass Excess (keV)
0	n	1	8071.37				
1	H	1	7289.03	4	Be	6	18374.00
		2	13135.82			7	15768.70
		3	14949.91			8	4941.73
		4	25840.00			9	11347.70
2	He	3	14931.32	4	Be	10	12607.00
		4	2424.92			11	20174.00
		5	11390.00				
		6	17592.30	5	B	7	27870.00
		7	26110.00			8	22920.30
						9	12415.80
3	Li	4	25120.00			10	12050.78
		5	11680.00			11	8668.00
		6	14085.60			12	13369.50
		7	14906.90			13	16562.30
		8	20945.40				
		9	24953.90	12	C	12	0.0

[1] Adapted from Wapstra and Audi (1985)

In describing the abundance of a given element, i, in a gas of mass density, ρ, the mass fraction, X_i, is often used

$$X_i = \frac{A_i N_i}{\rho N_{\mathrm{A}}} \tag{4.14}$$

where A_i is the mass in a.m.u., the number density of element i is N_i, and Avogadro's number, $N_A = 6.022\,136\,7(36) \times 10^{-23}$ mole^{-1}.

The packing fraction, f, is given by

$$f = (M_{AZ} - A)/A \ , \tag{4.15}$$

where M_{AZ} is the atomic mass of the atom with mass number, A, and atomic number, Z.

The energy release, Q, in the reaction $a + X \rightarrow b + Y + d + Q$ follows from Einstein's (1905, 1906, 1907) relation, $E = Mc^2$, between energy, E, and mass, M, and is given by

$$\begin{aligned} Q &= E_{byd} - E_{ax} \\ &= c^2(M_a + M_x - M_b - M_y - M_d) \\ &= 931.481(A_a + A_x - A_b - A_y - A_d) \text{ MeV} \ . \end{aligned} \tag{4.16}$$

Here E_{ax} and E_{byd} are the center-of-mass kinetic energies of the incident and outgoing particles, M_i is the mass of the particle i, the velocity of light is c, and A_i is the atomic mass of the particle, i, in atomic mass units. Provided that the number of nucleons is conserved in the reaction, we may also write

$$Q = \Delta M_b + \Delta M_y + \Delta M_d - \Delta M_a - \Delta M_x \ , \tag{4.17}$$

where ΔM_i denotes the mass excess of i in energy units. When a nuclear reaction includes the emission of a positron, it is customary to add the annihilation energy $2mc^2 = 1.022$ MeV $= 1.637 \times 10^{-6}$ erg to the value of Q given by Eqs. (4.16) or (4.17).

The Q-values based on atomic and nuclear masses differ only by the differences, ΔB_{e}, of electron binding energies involved in the entrance and exit channels:

$$Q_{\mathrm{atomic}} = Q_{\mathrm{nucl}} + \Delta B_{\mathrm{e}} \ ,$$

where the binding energy correction can usually be neglected unless it is comparable to Q_{nucl} (Yokoi, Takahashi and Arnould, 1983). The Q-value of a nuclear reaction is the energy liberated for each event, and is known from the atomic mass tables. Such nuclear reaction energies are given by Wapstra and Audi (1985).

Nuclear binding energies, E_{B}, have a narrow range of values per nucleon, 7.4 MeV $\leq E_{\mathrm{B}}/A \leq 8.8$ MeV for $A > 10$ (Aston, 1927). Nevertheless, some nuclei are extremely stable when compared with others (Elsasser, 1933), and especially stable nuclei are those with "magic number" values of Z or N. These numbers are (Mayer, 1948)

$$2, 8, 14, 20, 28, 50, 82, \text{ or } 126 \ .$$

A semi-empirical formula which gives the atomic mass, M_{AZ}, for a given value of A and Z was first derived by Weizsäcker (1935). His mass law is

$$M_{AZ} = M_{\rm N}A - (M_{\rm N} - M_{\rm H})Z - E_{\rm B}(A,Z)/c^2 \ , \tag{4.18}$$

where the neutron mass, $M_{\rm N}$ = 1.008665 a.m.u., the mass of the hydrogen atom, $M_{\rm H}$ = 1.007825 a.m.u., the atomic mass unit is 1 a.m.u. = 931.494 MeV, and the nuclear binding energy, $E_{\rm B}(A,Z)$, is given by

$$-E_{\rm B}(A,Z) = -a_1 A + a_2 A^{2/3} + a_3(Z^2 A^{-1/3}) + 0.25 a_4 (A - 2Z)^2 A^{-1} \ .$$

Green (1954) gives numerical values for the constants a_1 = 16.9177 MeV , a_2 = 19.120 MeV, a_3 = 0.76278 MeV, and a_4 = 101.777 MeV. An additional term of $\pm 132 A^{-1}$ MeV is often added to the binding energy expression, where the + and – of the ± sign correspond, respectively, to the cases where $N = A - Z$ and Z are both odd or both even. Modern attempts at deriving mass laws involve extrapolating from known nuclear masses to predict the masses and binding energies of yet unmeasured nuclei. Considerable effort has also gone into estimating theoretical values for atomic and nuclear masses (cf. Myers and Swiatecki, 1966; Garvey et al., 1969; Truran et al., 1970; Kodama, 1971). Using the semi-empirical mass formula and taking shell effects into account, the nuclear binding energy, $E_{\rm B}(A,Z)$, to be used in Eq. (4.18) is (Myers and Swiatecki, 1966)

$$\begin{aligned} -E_{\rm B}(A,Z) = & - c_1 A + c_2 A^{2/3} + c^3(Z^2 A^{-1/3}) - c_4 Z^2 A^{-1} \\ & + [4E^3/(9F^2)] - [8E^3/(27F^2)] \ , \end{aligned} \tag{4.19}$$

where

$$\begin{aligned} c_1 &= 15.677\left[1 - 1.79\left(\frac{N-Z}{A}\right)^2\right] \mathrm{MeV}, \\ c_2 &= 18.56 \ \left[1 - 1.79\left(\frac{N-Z}{A}\right)^2\right] \mathrm{MeV}, \\ c_3 &= 0.717 \ \mathrm{MeV}, \\ c_4 &= 1.21129 \ \mathrm{MeV}, \\ E &= \frac{2}{5} c_2 A^{2/3}(1-x)\alpha_0^2, \\ F &= \frac{4}{105} c_2 A^{2/3}(1+2x)\alpha_0^3 \ , \\ x &= c_3 Z^2/(2c_2 A), \end{aligned}$$

and

$$\alpha_0^2 = 0.3645 \ A^{-2/3}.$$

A discussion of some of the physics behind Eq. (4.19) is given by Myers (1970), and individual mass values may be found from Garvey's law and from his mass table (cf. Garvey et al. (1969)).

4.1.4 Alpha Decay and Other Natural Nuclear Reactions

Following Becquerel's (1896) discovery that uranium salts emit particles, Rutherford (1899) showed that there were two such radioactive particles, the β

and α particles which were, respectively, more or less penetrating. Subsequently, Rutherford and Soddy (1902, 1903) showed that in the emission of an α particle the mass number, A, decreased by four and the atomic number, Z, decreased by two. It follows that two protons and two neutrons come together within a nucleus to form an α particle which is a He^4 nucleus with a charge $Z_\alpha = 2$ and a mass $M_\alpha = 4.002603$ a.m.u. Rutherford and Soddy (1902, 1903) also showed that the mass number, A, stays the same when a nucleus emits a β particle, and that the charge, Z, changes by one. This meant that β particles were probably electrons and positrons, which was subsequently shown to be true. In fact, under different thermonuclear conditions a nucleus may emit an α particle, positrons, β^+, electrons, β^-, neutrons, n, protons, p, deuterons, d, tritons, t, and helium, He^3. The nucleus may also capture many of these particles. The theory of beta decay and electron capture is discussed in Sect. 4.3 on weak interactions, whereas the theory of alpha decay follows.

The α disintegration energies and half-lives of many elements are given on the chart of nuclides, which is useful in interpreting the flow of patterns in the transmutation of the elements. The relative locations on the chart of nuclides of the products of various nuclear processes are illustrated in Fig. 4.1.

The α disintegration energy, E, which is defined as the sum of the kinetic energies of the α particle and the recoil nucleus, is given by

$$E = [M(Z,A) - M(Z-2,A-4) - M_\alpha]c^2 = E_\alpha[1 + M_\alpha/M(Z-2,A-4)] \,, \quad (4.20)$$

where $M(Z,A)$ denotes the mass of the nucleus (Z,A), the kinetic energy and mass of the α particle are, respectively, E_α and M_α, and the mass equivalents of the electron binding energies have been ignored. Eq. (4.20) gives the maximum value of E, for the nucleus may sometimes α decay to an excited state and then radiate a gamma ray photon. Typical values of E range from 4 to 9 MeV.

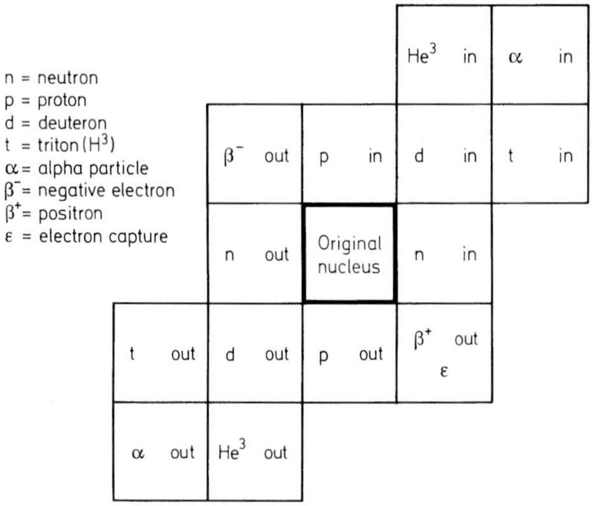

Fig. 4.1. Relative location on the chart of nuclides of the products of various nuclear processes

Although the disintegration energy, E, is less than the energy of the Coulomb barrier of the nucleus, there will be a finite probability per second, λ_α, that the α particle will escape the nucleus. Often the half-life, $\tau_{1/2}$, or mean lifetime, τ, are measured.

$$\lambda_\alpha = \frac{1}{\tau} = \frac{\ln 2}{\tau_{1/2}} \ , \tag{4.21}$$

where $\ln 2 \approx 0.69315$. If the number of nuclei which can undergo α decay is N_0 at time zero, then the number, N, at time, t, is given by

$$N(t) = N_0 \exp[-\lambda_\alpha t] \ ,$$

or

$$\frac{dN(t)}{dt} = -\lambda_\alpha N(t) \ . \tag{4.22}$$

The nuclide chart shows that there are a total of thirty nuclides which α decay naturally and are found on the Earth. These nuclides are elements in three chains which begin with the elements ${}^{90}\mathrm{Th}^{232}$, ${}^{92}\mathrm{U}^{238}$, and ${}^{92}\mathrm{U}^{235}$, terminate on elements with $Z \geq 82$, and have half-lives, $\tau_{1/2}$, of 14.05 billion years, 4.47 billion years and 704 million years, respectively. The use of radioactive isotopes, and their decay products, in determining the ages of the Earth, Moon and meteorites is given in Volume II together with the half life, $\tau_{1/2}$, of radioactive isotopes commonly used for dating. Since these elements were synthesized inside former stars, they can also be used to infer the age of our Galaxy (see Volume II).

A theoretical formula for λ_α was first derived by Gamow (1928) and Gurney and Condon (1928). They assumed that the potential energy, $V(r)$, of the α particle and the nucleus is given by $V(r) = -V_0$ for $r < R$ and $V(r) = 2(Z-2)e^2/r$ for $r > R$. Here r is the separation of the center of the nucleus and the α particle, and the nuclear radius, R, is defined as the greatest distance for which nuclear forces are significant. Experiments involving electron and neutron scattering indicate that

$$R \approx 1.2 \times 10^{-13} A^{1/3}\mathrm{cm} \ . \tag{4.23}$$

When $V(r)$ is used in the Schrödinger wave equation (Eq. (2.126)), and the continuity conditions are satisfied at $r = R$, it can be shown (Preston, 1962) that

$$\begin{aligned} \lambda_\alpha &= \frac{2V}{R}\frac{\tan\alpha_0}{f}\exp\left[\frac{-8(Z-2)e^2}{\hbar}\left(\frac{M}{2E}\right)^{1/2}(\alpha_0 - \sin\alpha_0\cos\alpha_0)\right] \\ &\approx \frac{V}{2R}\{\exp[2.97(Z-2)^{1/2}R^{1/2} - 3.95(Z-2)E^{-1/2}]\} \ , \end{aligned} \tag{4.24}$$

where, in the numerical approximation the disintegration energy, E, is in MeV and R is in Fermis (1 Fermi $= 10^{-13}$ cm), and the reduced mass, M, is given by

$$M = \frac{M_\alpha M_{Z-2}}{M_\alpha + M_{Z-2}} \ ,$$

where M_{Z-2} denotes the mass of the product nucleus, the final relative velocity of the α particle and the nucleus is given by

$$V = V_\alpha - V_{Z-2} \approx V_\alpha[1 + M_\alpha/M_{Z-2}] \ , \tag{4.25}$$

where V_α and V_{Z-2} are the respective velocities of the α particle and the recoil nucleus,

$$\alpha_0 = \arccos\left[\frac{4(Z-2)e^2}{VR}\left(\frac{1}{2ME}\right)^{1/2}\right]^{-1/2} \approx \arccos\left[\frac{2(Z-2)e^2}{ER}\right]^{-1/2}$$

and

$$f = \mathrm{cosec}^2\,(KR) - \cot\,(KR)/KR \ ,$$

where

$$K\cot\,(KR) = -\frac{MV}{\hbar}\tan\alpha_0 \ ,$$

and the energy of the state $E \approx 2(Z-2)e^2/R$. Experiments show that Eq. (4.24) holds quite well for nuclei with even Z and A with $R = 1.57 \pm 0.015 \times 10^{-13}$ $A^{1/3}$ cm, $E + V_0 = 0.52 \pm 0.01$ MeV, and $KR = 2.986 \pm 0.005$.

4.2 Thermonuclear Reaction Rates

The fundamental formulae for thermonuclear reactions in stars are given in the excellent texts by Clayton (1968, 1983) and Rolfs and Rodney (1988). Nuclear energy generation in the solar interior is reviewed by Parker and Rolfs (1991) and an evaluation of the fundamental proton-proton reaction is provided by Kamionkowski and Bahcall (1994). Stellar reaction rates, determined from laboratory measurements, are reviewed by Fowler (1984), and rate constants are provided by Fowler, Caughlan and Zimmerman (1967, 1975), Harris, Fowler, Caughlan and Zimmerman (1983), Caughlan, Fowler, Harris and Zimmerman (1985), and Caughlan and Fowler (1988). Thermonuclear reaction rates are also discussed by Langanke (1991).

4.2.1 Definition and Reciprocity Theorem for Cross Sections

The cross section, σ, for any event is defined as the number of desired events per second divided by the number of particles incident per unit area per second. The cross section, σ_{12}, for the reaction, $1 + 2 \rightarrow 3 + 4$, is related to the cross section, σ_{34}, for the reverse reaction, $3 + 4 \rightarrow 1 + 2$, by the relation (Blatt and Weisskopf, 1952 as modified by Fowler, Caughlan, and Zimmerman, 1967)

$$\frac{\sigma_{34}}{\sigma_{12}} = \frac{(1+\delta_{34})g_1 g_2 A_1 A_2 E_{12}}{(1+\delta_{12})g_3 g_4 A_3 A_4 E_{34}} \ , \tag{4.26}$$

where the Kronecker delta function, δ_{12}, is one if $1 = 2$ and zero if $1 \neq 2$, the statistical weight, g_i, of nucleus, i, is given by $g_i = 2I_i + 1$, where I_i is the spin of the nucleus, the mass number is A_i, and E_{12} and E_{34} are the kinetic energies, in the center of mass system, of the two sides of the nuclear reaction equation. At very high temperatures the nuclei may be in excited states which are in thermal equilibrium with their ground states, and in this case the g_i are replaced by the nuclear partition functions $G_i = \sum_j g_{ij}\exp(-E_j/kT)$ where the E_j is the excitation energy of the jth state. Schematic plots of typical charged particle and neutron cross sections are shown in Fig. 4.2 together with the Maxwell-Boltzmann distribution function and the experimental cross section for the $^{13}C(p,\gamma)^{14}N$ reaction. Since the cross sections for nuclear reactions are of the order of 10^{-24} cm^2, this value, called the barn, has been adopted as a unit of area.

4.2.2 Nonresonant Neutron Capture Cross Section

When a nucleus (Z,A) captures a neutron, n, it becomes the isotope $(Z,A+1)$ of the same element, and a photon, γ, can be radiated according to the reaction

$$(Z,A) + n \rightarrow (Z,A+1) + \gamma \ . \qquad (4.27)$$

Under normal conditions in stellar interiors, the relative velocity, v, between a neutron and a nucleus is determined by the Maxwell–Boltzmann distribution. The effective cross section, $\langle\sigma\rangle$, for a Maxwell–Boltzmann distribution of particle velocities is given by

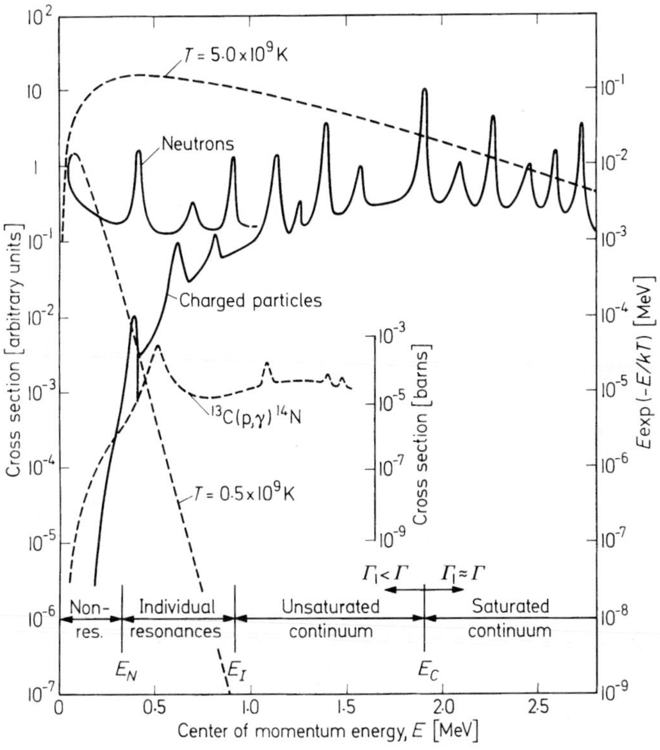

Fig. 4.2. Schematic plot of typical charged particle and neutron cross sections as a function of center-of-momentum energy, E, together with the Maxwell–Boltzmann distribution function (dashed lines). Also shown is the measured cross section in barns of the $^{13}C(p,\gamma)^{14}N$ reaction, where in this case E is the proton laboratory energy. The former curves are after Wagoner (1969, by permission of the American Astronomical Society and the University of Chicago Press), whereas the latter is after Seagrave (1952)

$$\langle\sigma\rangle = \frac{\langle\sigma v\rangle}{v_T} = \frac{\int_0^\infty \sigma v \varphi(v) dv}{v_T} , \tag{4.28}$$

where the Maxwell weighting function, $\varphi(v)$, is given by

$$\varphi(v) = \frac{4}{\pi^{1/2}} \left(\frac{v}{v_T}\right)^2 \exp\left[-\left(\frac{v}{v_T}\right)^2\right] \frac{dv}{v_T} , \tag{4.29}$$

the most probable velocity, v_T, is given by

$$v_T = \left(\frac{2kT}{M}\right)^{1/2} \approx 1.284 \times 10^4 T^{1/2} \text{ cm sec}^{-1} ,$$

the gas temperature is T, and the reduced neutron mass, $M = M_N M_A/(M_N + M_A) \approx M_N = 1.6749\,286(10) \times 10^{-24}$ grams. Here M_N and M_A are, respectively, the mass of the neutron and the nucleus.

For heavy elements, nuclei more massive than the iron group nuclei, the effective neutron capture cross section, $\langle\sigma\rangle$, at the most probable energy $kT = 8.617\,385(73) \times 10^{-5} T$ eV, is given by

$$\langle\sigma\rangle = \sigma_T \text{ measured at } v_T = (1.648 \times 10^8 T)^{1/2} \text{ cm sec}^{-1} . \tag{4.30}$$

4.2.3 Nonresonant Charged Particle Cross Section

At low energies, nonresonant charged particle interactions are dominated by the Coulomb-barrier penetration factors first discussed by Gamow (1928) and Gurney and Condon (1928, 1929). It is therefore convenient to factor out this energy dependence and express the cross section, $\sigma(E)$, by

$$\sigma(E) = \frac{S(E)}{E} \exp[-(E_G/E)^{1/2}] , \tag{4.31}$$

where the kinetic energy, E, of the center-of-mass system is given by

$$E = \frac{Mv^2}{2} = \frac{M_1 E_{L2}}{M_1 + M_2} , \tag{4.32}$$

the reduced mass, M, of interacting particles 1 and 2, is given by $M = M_1 M_2/(M_1 + M_2)$, the particles have masses, M_1 and M_2, and relative velocity, v, and E_{L2} is the laboratory energy of particle 2 – i.e., the energy of particle 2 in the frame where particle 1 is at rest.

The Gamow energy, E_G, is given by

$$E_G = (2\pi\alpha Z_1 Z_2)^2 (Mc^2/2) = [0.98948 Z_1 Z_2 A^{1/2}]^2 \quad \text{MeV} . \tag{4.33}$$

Here the fine structure constant $\alpha = e^2/(\hbar c) = [137.0359895(61)]^{-1}$, the particle charges in units of the proton charge are Z_1 and Z_2, and the reduced nuclear mass $A = A_1 A_2/(A_1 + A_2)$ which differs little from the reduced atomic mass.

Far from a nuclear resonance, the cross section factor, $S(E)$, is a slowly varying function of the center of momentum energy, E, and it may be conveniently expressed in terms of the power series expansion

$$S(E) = S(0)\left[1 + \frac{S'(0)}{S(0)} E + \frac{1}{2} \frac{S''(0)}{S(0)} E^2\right] , \tag{4.34}$$

where the prime denotes differentiation with respect to E. Experimental measurements of the constants in this expression are given by Fowler, Caughlan, and Zimmerman (1967, 1975), Harris, Fowler, Caughlan and Zimmerman (1983), Caughlan, Fowler, Harris and Zimmerman (1985), and Caughlan and Fowler (1988). When the reaction proceeds through the wings of a resonance, $S(E)$ may be evaluated using measurements of the resonance at higher energies (cf. Eq. (4.51)).

4.2.4 Resonant Cross Sections for Neutrons and Charged Particles – Breit–Wigner Shapes

For a single resonance of energy, E_R, the cross section, $\sigma(E)$, of the nuclear reaction, $1 + 2 \rightarrow 3 + 4 + Q$, is given by (Breit and Wigner, 1936; Rolfs and Rodney, 1988)

$$\sigma(E) = \pi\lambda\!\!\!^{-2} \frac{\omega\Gamma_{12}\Gamma_{34}}{(E - E_R)^2 + (\Gamma^2/4)} = \frac{0.657}{AE} \frac{\omega\Gamma_{12}\Gamma_{34}}{(E - E_R)^2 + (\Gamma^2/4)} \text{ barn} , \quad (4.35)$$

where the center of momentum energy, E, of 1 and 2 is in MeV in the numerical approximation, and the reduced de Broglie wavelength $\lambda\!\!\!^{-} = \hbar/(Mv)$, where M is the reduced mass of 1 and 2 and v is their relative velocity. The statistical factor $\omega = (2J + 1)/[(2J_1 + 1)(2J_2 + 1)]$, where J is the angular momentum of the resonant state, and J_1 and J_2 are, respectively, the angular momenta of particles 1 and 2. The total width, Γ, of the resonant state is given by $\Gamma = \hbar/\tau = \Gamma_{12} + \Gamma_{34} + \cdots$, where τ is the effective lifetime of the state. The partial width Γ_{12} is the width for reemission of 1 with 2, and Γ_{34} is the width for emission of 3 and 4. Detailed formulae for evaluation of partial widths are given in Sect. 4.2.7.

4.2.5 Reaction Rate, Mean Lifetime, and Energy Generation

The reaction rate, r_{12}, between two nuclei, 1 and 2, with a relative velocity, v, is given by

$$r_{12} = \frac{N_1 N_2}{(1 + \delta_{12})} \int_0^\infty v\sigma(v)\varphi(v)dv = \frac{N_1 N_2 \langle\sigma v\rangle}{(1 + \delta_{12})} \quad \text{reactions cm}^{-3}\text{ sec}^{-1}$$

$$= \frac{N_1 N_2}{(1 + \delta_{12})} \int_0^\infty \left(\frac{2E}{M}\right)^{1/2} \sigma(E)\psi(E)dE \quad \text{reactions cm}^{-3}\text{ sec}^{-1} , \quad (4.36)$$

where the kinetic energy in the center of mass system is $E = Mv^2/2$, the reduced mass $M = M_1 M_2/(M_1 + M_2)$ where the nuclei have masses M_1 and M_2, the reaction cross section is $\sigma(v)$ or $\sigma(E)$, N_1 and N_2 are the number densities of 1 and 2, the relative velocity spectrum is $\varphi(v)dv = \psi(E)dE$, and the Kronecker delta function, δ_{12}, is one if $1 = 2$ and zero if $1 \neq 2$. For a gas of mass density, ρ, the number density, N_i, of the nuclide, i, is often expressed in terms of its mass fraction, X_i, by the relation

$$N_i = \rho N_A \frac{X_i}{A_i} \text{cm}^{-3} , \tag{4.37}$$

where Avogadro's number $N_A = 6.0221367(36) \times 10^{23} (\text{mole})^{-1}$, and A_i is the atomic mass of i in atomic mass units.

For a nondegenerate, nonrelativistic gas, the relative velocity spectrum is Maxwellian and is given by

$$\varphi(v)d^3v = \left(\frac{M}{2\pi kT}\right)^{3/2} \exp\left(-\frac{Mv^2}{2kT}\right) 4\pi v^2\, dv \tag{4.38}$$

or

$$\psi(E)dE = \frac{2E}{\sqrt{\pi}kT} \exp\left(-\frac{E}{kT}\right) \frac{dE}{(kTE)^{1/2}} ,$$

where the gas temperature is T, the reduced mass is M, the relative velocity is v, and the kinetic energy in the center-of-mass system is E. Combining Eq. (4.37) and Eq. (4.38) we obtain an expression for $\langle \sigma v \rangle$, the reaction rate per interacting pair:

$$\langle \sigma v \rangle = \left(\frac{8}{\pi \mu}\right)^{1/2} \frac{1}{(kT)^{3/2}} \int_0^\infty \sigma(E) \exp\left(-\frac{E}{kT}\right) dE , \tag{4.39}$$

where $\mu = M = M_1 M_2 / (M_1 + M_2)$ is the reduced mass of the interacting nuclei of mass M_1 and M_2.

The mean lifetime, $\tau_2(1)$, of nucleus 1 for destruction by nucleus 2 is given by the relation

$$\lambda_2(1) = \frac{1}{\tau_2(1)} = N_2 \langle \sigma v \rangle = \rho N_A \frac{X_2}{A_2} \langle \sigma v \rangle \ \text{sec}^{-1} , \tag{4.40}$$

where $\lambda_2(1)$ is the decay rate of 1 for interactions with 2. The energy generation, ε_{12}, for the forward reaction $1 + 2 \rightarrow 3 + 4 + Q$ is given by

$$\varepsilon_{12} = r_{12} Q / \rho \ \text{erg g}^{-1} \text{sec}^{-1} , \tag{4.41}$$

where r_{12} is the reaction rate, Q is the energy release, and ρ is the gas mass density.

4.2.6 Nonresonant Reaction Rates

If follows from Eqs. (4.28), (4.29), (4.30) and (4.36) that the reaction rate, r_{1n}, for the nonresonant neutron capture reaction is given by

$$r_{1n} = N_1 N_n \sigma_T \left(\frac{2kT}{M}\right)^{1/2} = N_1 N_n \langle \sigma v \rangle , \tag{4.42}$$

where N_1 and N_n are, respectively, the number densities of the nuclei and the neutrons, σ_T is the neutron capture cross section at temperature, T, and the reduced mass $M = M_1 M_N / (M_N + M_1) \approx M_N = 1.6749286(10) \times 10^{-24}$ grams. The effective energy and energy spread of the neutron capture reaction is on the order of $kT \approx 0.08617 T_9$MeV, where T_9 is $T/10^9$ when T is in °K.

It follows from Eqs. (4.31), (4.32), (4.33), (4.34), and (4.36) that the reaction rate, r_{12}, for a nonresonant charged particle reaction in a nondegenerate, non-relativistic gas is given by (Burbidge, Burbidge, Fowler, and Hoyle (B^2FH), 1957; Fowler, Caughlan, and Zimmerman, 1975; Clayton, 1963, 1983; Rolfs and Rodney, 1988)

$$\begin{aligned} r_{12} &= \frac{N_1 N_2 \langle \sigma v \rangle}{(1+\delta_{12})} \\ &= \frac{N_1 N_2}{(1+\delta_{12})} \left(\frac{8}{\pi M}\right)^{1/2} \frac{1}{(kT)^{3/2}} \int S(E) \exp\left[-\left(\frac{E_G}{E}\right)^{1/2} - \frac{E}{kT}\right] dE \qquad (4.43) \\ &= \frac{N_1 N_2}{(1+\delta_{12})} \left(\frac{2}{M}\right)^{1/2} \frac{\Delta E_0}{(kT)^{3/2}} S_{\mathrm{eff}} \exp\left[-\frac{3E_0}{kT} - \left(\frac{T}{T_0}\right)^2\right] , \end{aligned}$$

where N_1 and N_2 are the number densities of 1 and 2, the Kronecker delta function, δ_{12}, is one if $1 = 2$ and zero if $1 \neq 2$, the reduced mass $M = M_1 M_2/(M_1 + M_2)$, the temperature is T, the cutoff temperature, T_0, occurs when the nuclear cross section no longer varies according to the Gamow penetration factor given in Eq. (4.33) (Fowler, Caughlan, and Zimmerman, 1974). This condition may occur when the effective energy, E_0, is larger than E_G (given in Eq. (4.33)), or when resonance or the continuum set in.

A closed-form evaluation of the reaction-rate integral in Eq. (4.43) is provided by Haubold and John (1978, 1981) and Haubold and Mathai (1986).

The integrand in Eq. (4.43) has its peak value at the effective thermal energy, E_0, (Gamow peak) given by (Fowler and Hoyle, 1964)

$$E_0 = [\pi \alpha Z_1 Z_2 kT (Mc^2/2)^{1/2}]^{2/3} = 0.1220 (Z_1^2 Z_2^2 A)^{1/3} T_9^{2/3} \ \text{MeV} ,$$

or

$$E_0 = 1.22 (Z_1^2 Z_2^2 \mu T_6^2)^{1/3} \ \text{keV} , \qquad (4.44)$$

where $T_6 = T/10^6$ and μ is the reduced mass in a.m.u. Here the fine structure constant $\alpha = e^2/\hbar c = (137.03602)^{-1}$, M is the reduced mass, Z_1 and Z_2 are, respectively, the charges of 1 and 2 in units of the proton charge, $A = A_1 A_2/(A_1 + A_2)$ is the reduced atomic mass, and $T_9 = T/10^9$, where T is the temperature. The quantity E_0 is the effective mean energy for thermonuclear fusion reactions at a given temperature T. For example, at a stellar core temperature of $T_6 = T_9/10^3 = 15$, the solar value, the proton-proton reaction, $p+p$, has a value of $E_0 = 5.9$ keV. The peak of the integrand in Eq. (4.43) has a full width at $1/e$ of the maximum value given by

$$\Delta E_0 = 4(E_0 kT/3)^{1/2} = 0.2368 (Z_1^2 Z_2^2 A)^{1/6} T_9^{5/6} \ \text{MeV} , \qquad (4.45)$$

which is the width of the effective range of thermal energy.[1] The effective value, S_{eff}, of the cross section factor, $S(E)$, is given by

[1] Nuclei with the energies close to E_0 and spread over the energy range ΔE_0 contribute mainly to the total rate of the thermonuclear reaction.

$$S_{\text{eff}} = S(0)\left[1 + \frac{5kT}{36E_0} + \frac{S'(0)}{S(0)}\left(E_0 + \frac{35}{36}kT\right) + \frac{1}{2}\frac{S''(0)}{S(0)}\left(E_0^2 + \frac{89}{36}E_0 kT\right)\right] \text{ MeV-barn} , \tag{4.46}$$

where the prime denotes differentiation with respect to the kinetic energy, E, in the center-of-mass system. Eqs. (4.37), (4.43), (4.44), and (4.46) may be combined to give the useful result

$$N_A\langle\sigma v\rangle = C_1 T_9^{-2/3} \exp[-C_2 T_9^{-1/3} - (T_9/T_0)^2]\{1 + C_3 T_9^{1/3} + C_4 T_9^{2/3} + C_5 T_9 + C_6 T_9^{4/3} + C_7 T_9^{5/3}\}\text{cm}^3\,\text{sec}^{-1}(\text{mole})^{-1} , \tag{4.47}$$

where

$$C_1 = 7.8324 \times 10^9 (Z_1^2 Z_2^2 A)^{1/6} \frac{S(0)}{A^{1/2}} ,$$

$$C_2 = 4.2475(Z_1^2 Z_2^2 A)^{1/3} ,$$

$$C_3 = 9.810 \times 10^{-2}(Z_1^2 Z_2^2 A)^{-1/3} ,$$

$$C_4 = 0.1220 \frac{S'(0)}{S(0)}(Z_1^2 Z_2^2 A)^{1/3} ,$$

$$C_5 = 8.377 \times 10^{-2} \frac{S'(0)}{S(0)} ,$$

$$C_6 = 7.442 \times 10^{-3} \frac{S''0)}{S(0)}(Z_1^2 Z_2^2 A)^{2/3} ,$$

and

$$C_7 = 1.299 \times 10^{-2} \frac{S''(0)}{S(0)}(Z_1^2 Z_2^2 A)^{1/3} .$$

Here the temperature $T_9 = T/10^9$, where T is the temperature in °K, the effective cutoff temperature is T_0, the charges of 1 and 2 are Z_1 and Z_2 in units of the proton charge, the reduced atomic mass $A = A_1 A_2/(A_1 + A_2)$ is given in atomic mass units, and the quantities $S(0), S'(0)$ and $S''(0)$ have respective units of MeV-barns, barns, and barns MeV^{-1}. Laboratory measurements of $S(0)$, $S'0)$, and $S''(0)$ for many stellar reactions are given by Fowler, Caughlan and Zimmerman (1967, 1974), Harris, Fowler, Caughlan and Zimmerman (1973), Caughlan, Fowler, Harris and Zimmerman (1985), and Caughlan and Fowler (1988).

The reaction rate, $r_{\ln}$, for the neutron induced, nonresonant reactions, can be alternatively expressed by (Rolfs and Rodney, 1988):

$$\langle\sigma v\rangle = \left(\frac{2}{\mu}\right)^{1/2} \frac{\Delta}{(kT)^{3/2}} S(E_0) \exp\left(-\frac{3E_0}{kT}\right)$$

or

$$\langle\sigma v\rangle = 7.20 \times 10^{-19} \frac{1}{\mu Z_1 Z_2} \tau^2 \exp(-\tau) S(E_0) \ \ \text{cm}^3\,\text{s}^{-1} , \tag{4.48}$$

where $\langle \sigma v \rangle$ is the reaction rate per interaction particle pair, the factor $S(E_0)$ is in units of keV barn in the numerical approximation, the reduced mass μ is in a.m.u., and

$$\Delta = \frac{4}{3^{1/2}}(E_0 kT)^{1/2} = 0.749(Z_1^2 Z_2^2 \mu T_6^5)^{1/6} \text{ keV} ,$$

where the dimensionless parameter τ is given by

$$\tau = \frac{3E_0}{kT} = 42.46(Z_1^2 Z_2^2 \mu / T_6)^{1/3} .$$

The temperature dependence of the reaction is inferred from:

$$\langle \sigma v \rangle \propto T^{\tau/3 - 2/3} . \tag{4.49}$$

As an example, for the proton-proton reaction, p+p, the temperature dependence near $T_6 = 15$ is $\langle \sigma v \rangle \propto T^{3.9}$, while for oxygen burning $^{16}O + ^{16}O$, the $\langle \sigma v \rangle \propto T^{182}$, showing the dramatic sensitivity of thermonuclear reaction rates on temperature.

4.2.7 Resonant Reaction Rates

It follows from Eqs. (4.35), (4.36), and (4.38) that the reaction rate, r_{12}, for a resonant reaction in a nonrelativistic, nondegenerate gas is given by

$$r_{12} = \frac{N_1 N_2}{(1+\delta_{12})} \langle \sigma v \rangle = \frac{N_1 N_2}{(1+\delta_{12})} \int_0^\infty \frac{0.657}{AE} \frac{\omega \Gamma_{12} \Gamma_{34}}{(E - E_R)^2 + (\Gamma^2/4)} \times \left[\frac{2E}{M}\right]^{1/2} \left[\frac{2E}{\sqrt{\pi} kT} \exp\left(-\frac{E}{kT}\right) \frac{1}{(kTE)^{1/2}}\right] dE , \tag{4.50}$$

where N_1 and N_2 are the number densities of 1 and 2, the kinetic energy in the center-of-mass system is E, the reduced atomic mass $A = A_1 A_2/(A_1 + A_2)$, the reduced mass $M = M_1 M_2/(M_1 + M_2)$, the resonant energy is E_R, the gas temperature is T, the statistical factor $\omega = (2J+1)/[(2J_1+1)(2J_2+1)]$ where J is the angular momentum of the resonant state and J_1 and J_2 are, respectively, the angular momenta of 1 and 2, the width of the resonant state is Γ, the partial width Γ_{12} is the width for reemission of 1 with 2, and Γ_{34} is the width for emission of 3 and 4. The numerical constant of 0.657 is for a cross section in barns when the energy, E, is in MeV. A closed-form evaluation of the resonant reaction rate integral given in Eq. (4.50) is given by Haubold and John (1979) and Haubold and Mathai (1984, 1986).

When the effective thermal energy, E_0, given in Eq. (4.44), is much smaller than the resonant energy, E_r, then the Breit-Wigner resonant cross section may be evaluated in its wings to give the cross section factor

$$S(E) = \frac{0.657}{A} \frac{\omega \Gamma_{12}(E) \Gamma_{34}}{(E - E_r)^2 + (\Gamma^2/4)} \exp[0.98948 Z_1 Z_2 A^{1/2} E^{-1/2}] , \tag{4.51}$$

where $S(E)$ is in MeV barns, and E is in MeV. Eq. (4.51) may be evaluated using measurements of a resonance at high energies, and then used with the nonresonant formalism of Eq. (4.34), (4.43), and (4.46) to obtain reaction rates at low energies.

When the range of effective steller energies [2] includes the resonance energy, then Eq. (4.50) must be used. Provided that the width of the resonance, Γ, is much less then the effective spread in energy of the interacting particles (less than kT when neutrons are involved and less than ΔE_0 for two charged particles). Eq. (4.50) may be integrated to give the useful result

$$N_A\langle\sigma v\rangle = C_8 T_9^{-3/2}\exp(-C_9/T_9)\mathrm{cm}^3\,\mathrm{sec}^{-1}(\mathrm{g\ mole})^{-1} , \tag{4.52}$$

where

$$C_8 = 1.53986\times 10^{11}A^{-3/2}(\omega\gamma) ,$$
$$C_9 = 11.605\ E_\mathrm{R} ,$$

the temperature $T_9 = T/10^9$, where the temperature T is in °K, the reduced atomic mass, A, is in atomic mass units, the factor $(\omega\gamma) = \omega\Gamma_{12}\Gamma_{34}/\Gamma$ is in MeV, and the resonant energy E_R is in MeV. When several resonances occur, the net reaction rate is determined by the superposition

$$N_A\langle\sigma v\rangle = \sum_n C_{8n} T_9^{-3/2}\exp(-C_{9n}/T_9) , \tag{4.53}$$

where the constants C_{8n} and C_{9n} are the appropriate constants given by Eq. (4.52) for the nth resonance.

The quantity $\gamma = \Gamma_{12}\Gamma_{34}/\Gamma$ is referred to as the strength of the resonance. Laboratory measurements of $(\omega\gamma)$ and E_R, or equivalently C_8 and C_9 are given by Fowler, Caughlan and Zimmerman (1967, 1974), Harris, Fowler, Caughlan and Zimmerman (1983), Caughlan, Fowler, Harris and Zimmerman (1985), and Caughlan and Zimmerman (1988).

Rolfs and Rodney (1988) express Eq. (4.52) in the form:

$$\langle\sigma v\rangle = \left(\frac{2\pi}{\mu kT}\right)^{3/2}\hbar^2(\omega\gamma)_\mathrm{R}\exp\left(-\frac{E_\mathrm{R}}{kT}\right)f_{12} , \tag{4.54}$$

which is the stellar reaction rate per particle pair for a narrow resonance, and the electron screening factor, f_{12}, has been added (see Sect. 4.2.9). The number of nuclei N_{12} in a resonance state is given by:

$$N_{12} = \frac{N_1 N_2}{1+\delta_{12}}\left(\frac{2\pi}{\mu kT}\right)^{3/2}\hbar^3\omega\exp\left(-\frac{E_\mathrm{R}}{kT}\right) . \tag{4.55}$$

When the density of resonances lying within the Gamow peak becomes sufficiently large, we can replace the summation in Eq. (4.53) by the integration dE/D, where D is the average level distance within the Gamow peak. This occurs mainly in elements heavier than about Mg, and for energies in excess of a few MeV. Thus we have (Hayashi, Hoshi, and Sugimoto, 1962)

[2] The effective energy range is of the order kT for neutron interactions and is given by ΔE_0 from (4.45) centered at E_0 from (4.44) for charged particle interactions.

$$r_{12} = \rho \frac{X_1}{A_1} N_2 N_A \langle \sigma v \rangle$$
$$= \frac{(2\pi)^{3/2} \hbar^2 N_1 N_2}{(MkT)^{3/2}} \frac{\omega}{D} \int_0^\infty \frac{\langle \Gamma_1(E) \rangle \langle \Gamma_2(E) \rangle}{\langle \Gamma(E) \rangle} \exp\left(-\frac{E}{kT}\right) dE \ , \tag{4.56}$$

where $M = A$ (in a.m.u.) is the reduced mass of the interacting nuclei, the reduced atomic mass number $A = A_1 A_2/(A_1 + A_2)$, the average spin factor is ω (the average of $(2J+1)/[(2I_1+1)(2I_2+1)]$ where J is the resonance value of angular momentum and I_1 and I_2 are the nuclear spin, Γ_1, Γ_2, and Γ denote, respectively, the absorption width, the emission width, and the total width, and the $\langle\ \rangle$ denotes an average value. Considering that the emission width is normally much larger than the absorption width in the relevant range of temperature, Eq. (4.56) becomes

$$\langle \sigma v \rangle = \frac{\sqrt{\pi}}{2} \left(\frac{2\pi\hbar^2}{MkT}\right)^{3/2} \frac{\Delta E_0}{\hbar} \frac{\omega \Gamma_1(E_0)}{D} \exp\left(-\frac{E_0}{kT}\right)$$
$$\approx \alpha T_9^{-2/3} \exp(-\beta T_9^{-1/3}) \ \mathrm{cm^3\ sec^{-1}} \ , \tag{4.57}$$

where M is the reduced mass of the interacting nuclei, the effective thermal energy, E_0, is given in Eq. (4.44), the effective range of thermal energy, ΔE_0, is given in Eq. (4.45), and

$$\alpha = 1.78 \times 10^{-12} (Z_1^5 Z_2^5 A^{-11})^{1/6} \frac{\theta^2}{DR^{3/2}} \exp\left[1.05 (Z_1 Z_2 A R)^{1/2}\right] \ ,$$
$$\beta = 4.25 (Z_1^2 Z_2^2 A)^{1/3} \ ,$$

and the average level distance, D, is measured in MeV. For $A \gtrsim 20$, we have $\omega \approx 1$ and $D \approx 0.1$ MeV. The reduced width, θ, is related to D by the equation

$$\theta^2 = \frac{2\lambda_0 M}{3\pi\hbar^2} RD \ ,$$

where $\lambda_0 \approx 10^{-13}$ cm is the characteristic wavelength of nucleons inside the nucleus, and the interaction radius, R, is given by

$$R \approx 1.4 \times 10^{-13} [A_1^{1/3} + A_2^{1/3}] \ \mathrm{cm} \ .$$

Partial widths, Γ, may be evaluated using the expression

$$\Gamma = \frac{3\hbar v}{R} \theta_l^2 P_l \ , \tag{4.58}$$

where R is the interaction radius, $v = (2E/M)^{1/2}$ is the relative velocity of the particles, the reduced width, θ_l^2, for particles with relative angular momentum, l, usually lies in the range $0.01 \lesssim \theta_l^2 \lesssim 1.0$, and the penetration factor, P_l, is given by

$$P_l = [F_l^2(R) + G_l^2(R)]^{-1} \ , \tag{4.59}$$

where $F_l(R)$ and $G_l(R)$ are the regular and irregular solutions of the Schrödinger wave equation for a charged particle in a Coulomb field evaluated at the interaction radius, R. The Coulomb wave functions $F_l(R)$ and $G_l(R)$ are discussed and tabulated by Fröberg (1955) and Hull and Breit (1959). Approximate values of the penetration factor are derived by using the WKB

(Wentzel, 1926; Kramers, 1926; Brillouin, 1926) solution to the Schrödinger equation (cf. Bethe, 1937, and Van Horn and Salpeter, 1967). If the Coulomb barrier energy, E_c, is given by

$$E_c = \frac{Z_1 Z_2 e^2}{R} \approx 1.4 \frac{Z_1 Z_2}{R} \quad \text{MeV} \ , \tag{4.60}$$

where R is in Fermis ($R \approx 1.4\, A^{1/3}$ fm), then the WKB solution for the penetration factor is

$$P_l \approx \left(\frac{E_c}{E}\right)^{1/2} \exp[-W_l] = \left(\frac{E_c}{E}\right)^{1/2} \exp\left[\frac{-2\pi Z_1 Z_2 e^2}{\hbar v} + 4\left(\frac{E_c}{\hbar^2/2MR^2}\right)^{1/2} - 2l(l+1)\left(\frac{\hbar^2/2MR^2}{E_c}\right)^{1/2}\right]$$

$$\approx \left(\frac{E_c}{E}\right)^{1/2} \exp[-(\mathrm{E_G/E})^{1/2} + 1.05(\mathrm{ARZ_1Z_2})^{1/2} - 7.62\left(l+\frac{1}{2}\right)^2 (ARZ_1Z_2)^{-1/2}] \ , \tag{4.61}$$

where the centre of momentum energy, E, is $\ll E_c$, the Gamow energy , E_G, is given in Eq. (4.33); M is the reduced mass of the interacting nuclei, all energies are in MeV and R is in Fermis. Substituting Eq. (4.61) into Eq. (4.58), we obtain

$$\Gamma \approx 6.0\ \theta_l^2 \left(\frac{\hbar^2}{2MR^2} E_c\right)^{1/2} \exp(-W_l)$$

$$\approx 33.0\ \theta_l^2 \left(\frac{Z_1 Z_2}{AR^3}\right)^{1/2} \exp(-W_l) \ \text{MeV} \ , \tag{4.62}$$

where again R is in Fermis.

At high temperatures, where the energies are high enough to be on the order of resonance energies, the low lying nuclear states may be excited. For the special case of a thermal distribution of excited states of initial nuclei, and for reactions which proceed through resonances in the compound nucleus, we have the relation (Bahcall and Fowler, 1969)

$$N_A\langle\sigma v\rangle_e = \sum_n \frac{G_1^g}{G_1} N_A\langle\sigma v\rangle_g \left[1 + \sum_{e \neq 0} \frac{\Gamma_{12}^e}{\Gamma_{12}^g}\right] \ , \tag{4.63}$$

where the summation is over n resonances, $N_A\langle\sigma v\rangle_g$ is the rate of the reaction when all target nuclei are assumed to be in their ground state, g, and is given by Eq. (4.52), the Γ_{12}^e is the partial width for the decay of the resonance by reemission of the projectile 2 leaving nucleus 1 in the excited state, e, Γ_{12}^g is the decay width to the ground state of the target nucleus and is larger than Γ_{12}^e (for the same resonance) because of barrier-penetration factors, and the nuclear-partition functions, G_1, and G_1^g, are given by

$$G_1 = \sum_{e=0}^{n} G_1^e \ ,$$

where

$$G_1^e = (2J_1^e + 1)\exp\left(-\frac{E_1^e}{kT}\right) ,$$

and

$$G_1^g = 2J_1^g + 1 ,$$

the summation is over n excited states $e = 0, 1, 2, \ldots, n$, the spin of the target nucleus, 1, is J_1^e, and the excitation energy for the excited state in question is E_1^e. Nuclear partition functions for nuclei with mass numbers $A \leq 40$ that have low lying excited states have been calculated as a function of temperature by Bahcall and Fowler (1970). The influence of excited states on thermonuclear reactions has been further discussed by Michaud and Fowler (1970) and Arnould (1972). When the effective thermal energy, E_o, is larger than the Coulomb barrier energy, E_c, and the outgoing particle is a photon, γ, or if the penetration factor does not vary greatly over the range of interest, then the additional reaction rate term is given by (Wagoner, 1969)

$$N_A\langle\sigma v\rangle = C_{10}\exp(-C_{11}/T_9) , \tag{4.64}$$

where the constant C_{10} may be measured experimentally, and $C_{11} = 11.605E_c$, where E_c is given by Eq. (4.60).

4.2.8 Inverse Reaction Rates and Photodisintegration

For the reaction $1 + 2 \rightarrow 3 + 4 + Q$, we have the reciprocity relation

$$N_A\langle 34\rangle = C_{12}[\exp(-C_{13}/T_9)]N_A\langle 12\rangle , \tag{4.65}$$

where $\langle 12\rangle$ and $\langle 34\rangle$ denote, respectively, the $\langle\sigma v\rangle$ for the forward and reverse reactions, and the constants C_{12} and C_{13} are given by (Blatt and Weisskopf, 1952; Fowler, Caughlan, and Zimmerman, 1967)

$$C_{12} = \frac{(1+\delta_{34})\,g_1 g_2}{(1+\delta_{12})\,g_3 g_4}\left[\frac{A_1 A_2}{A_3 A_4}\right]^{3/2} , \tag{4.66}$$

and $C_{13} = 11.605Q_6$, where the Kronecker delta function, δ_{12}, is one if $1 = 2$ and zero if $1 \neq 2$, the statistical weight, g_i, of the nucleus i is $g_i = (2I_i + 1)$, where I_i is the nuclear spin, A_i is the atomic mass of i in atomic mass units, and Q_6 is the energy release of the forward reaction in MeV. Similar reciprocity formulae for reactions involving five particles are given by Fowler, Caughlan, and Zimmerman (1967) and Bahcall and Fowler (1969). At high temperatures, $T > 10^9$ °K, some nuclear states may be excited and Eq. (4.66) may be generalized by replacing $g_1 g_2/(g_3 g_4)$by $G_1 G_2/(G_3 G_4)$, where G_i is the partition function for the ith nucleus (Bahcall and Fowler, 1969). Nuclear partition functions for nuclei with mass numbers $A \leq 40$, which have low lying excited states, have been calculated as a function of temperature by Bahcall and Fowler (1970).

For the special case of the radiative capture reaction $1 + 2 \rightarrow 3 + \gamma + Q$, the lifetime, $\tau_\gamma(3)$, for the reverse photodisintegration is given by (Fowler, Caughlan, and Zimmerman, 1967)

$$\lambda_\gamma(3) = \frac{1}{\tau_\gamma(3)} = C_{14} T_9^{3/2} [\exp(-C_{13}/T_9] N_A \langle 12 \rangle \quad \sec^{-1} , \tag{4.67}$$

where $\langle 12 \rangle$ is the $\langle \sigma v \rangle$ for the forward reaction, the constant C_{13} is given following Eq. (4.66), and the constant C_{14} is given by

$$C_{14} = 0.987 \times 10^{10} \frac{g_1 g_2}{(1 + \delta_{12}) g_3} \left(\frac{A_1 A_2}{A_3} \right)^{3/2} . \tag{4.68}$$

4.2.9 Electron Shielding – Weak and Strong Screening

The Coulomb repulsion between nuclei plays an important role in controlling reactions between them. In hot stellar interiors, the nuclei are immersed in a sea of electrons that tend to cluster near the nuclei and reduce their Coulomb potential. The shielding effect of the electrons therefore increases the nonresonant reaction rate by an electron screening factor, f_{12}, and the nonresonant reaction rate $\langle \sigma v \rangle$ per particle pair (1,2) is enhanced by

$$\langle \sigma v \rangle_{\text{screening}} = f_{12} \langle \sigma v \rangle_{\text{no screening}} \cdot \tag{4.69}$$

The screening factor, f_{12}, is given by (Salpeter, 1954, Ichimaru, 1982):

$$f_{12} = \exp\left[\frac{E_D}{kT}\right] = \exp\left[\frac{Z_1 Z_2 e^2}{R_D kT}\right] \approx 1 - \frac{Z_1 Z_2 e^2}{R_D kT} , \tag{4.70}$$

where the Debye-Hückel radius, R_D, is given by (Debye and Hückel, 1923)

$$R_D = \left(\frac{kT}{4\pi e^2 \rho N_A \xi^2} \right)^{1/2} , \tag{4.71}$$

Avogadro's number is N_A and the quantity ξ^2 is given by:

$$\xi^2 = \sum_i (Z_i^2 + Z_i) \frac{X_i}{A_i} . \tag{4.72}$$

For normal stars with temperatures $T \approx 10^7$K and mass densities $\rho = 1$ to 10^2 g cm^{-3}, the Debye-Hückel radius, R_D, has typical values of $R_D = 10^{-8}$ to 10^{-9} cm, which is far from the nucleus with a radius of about 10^{-13} cm, and electron screening has negligible effect on reaction rates. However, as the mass density increases and the temperature decreases, R_D and the screening electrons move closer to the nucleus; the reaction rates can then be significantly increased above their values without the electrons. The enhancement is important for a degenerate gas. At very high densities, both electron screening and quantum mechanical effects become important, and this case is discussed in the next Sect. 4.2.10 on pycnonuclear reactions.

Salpeter (1954), Salpeter and Van Horn (1959), and Ichimaru and Utsumi (1984) have computed the electron screening enhancement factors, f_{12}, for nonresonant thermal reactions, and a review has been given by Ichimaru (1982). Here we provide formulae for the weak and strong screening conditions given by:

$$\begin{aligned} &E_c \ll kT \text{ and } E_0 \gg E_c \quad \text{(weak screening)} \\ &E_c \gg kT \text{ and } E_0 \gg E_c \quad \text{(strong screening)} \end{aligned} \tag{4.73}$$

where $E_0 = 0.122(Z_1^2 Z_2^2 \mu)^{1/3} T_9^{2/3}$ MeV, see Eq. (4.44), and the electrostatic interaction energy, E_c, for a Wigner–Seitz (1934) sphere of radius a_s is given by:

$$E_c = \frac{9}{10}\frac{Z^2 e^2}{a_s} = 1.8278\,\lambda\,E^* \ , \tag{4.74}$$

where N_e is the electron density, and

$$\lambda = r^* \left(\frac{N_e}{2Z_1}\right)^{1/3} = \frac{A_1 + A_2}{2A_1A_2Z_1Z_2}\left[\frac{1}{Z_1\mu_e}\frac{\rho}{1.3574 \times 10^{11}\text{g cm}^{-3}}\right]^{1/3} . \tag{4.75}$$

Here the gas mass density is ρ, and the mean electron molecular weight, μ_e, is given by

$$\mu_e^{-1} = \sum_i X_i \frac{Z_i}{A_i}\left[1 + \frac{Z_i m}{A_i m_H}\right]^{-1} , \tag{4.76}$$

where X_i, Z_i, and A_i are, respectively, the mass fraction, the number of protons, and the atomic weight of the nuclei of species, i, the electron mass is m, and $m_H \approx 1.6735 \times 10^{-24}$ g.

The weak screening multiplicative correction factor for nonresonant reaction rates is given by (Salpeter, 1954; Schatzman, 1948, 1958; Salpeter and Van Horn, 1969)

$$f_{12} = \exp\left[\sqrt{3}\frac{Z_2}{Z_1}\left(\frac{\mu_e}{Z_1}\right)^{1/2} \xi \Gamma_{Z_1}^{3/2}\right] \approx \exp[0.188 Z_1 Z_2 \xi \rho^{1/2} T_6^{-3/2}] \ , \tag{4.77}$$

where the gas mass density is ρ, the $T_6 = T/10^6$, the factor, ξ , is given by

$$\xi^2 = \sum_i X_i \frac{Z_i^2}{A_i}\left[1 + \frac{Z_i\ m}{A_i\ m_H}\right]^{-1} + \frac{f'}{f}\frac{1}{\mu_e} \tag{4.78}$$

$$\approx \sum_i \frac{X_i}{A_i}[Z_i^2 + Z_i] \ , \tag{4.79}$$

where f'/f is the logarithmic derivative of the Fermi-Dirac distribution function, f, with respect to its argument, and is unity for a non-degenerate gas and zero for a fully degenerate gas of electrons. The dimensionless parameter, Γ_Z, which compares the Coulomb and thermal energies, is given by

$$\begin{aligned} \Gamma_Z &= Z^{5/3}\frac{e^2}{kT}\left(\frac{4\pi N_e}{3}\right)^{1/3} \\ &\approx Z^{5/3}\frac{5.7562 \times 10^8\ {}^\circ\text{K}}{T}\left(\frac{1}{\mu_e}\frac{\rho}{6.203 \times 10^{13}\text{ g cm}^{-3}}\right)^{1/3} , \end{aligned} \tag{4.80}$$

for a nucleus of charge Z.

For strong screening, the multiplicative correction factor for nonresonant reaction rates is given by (Salpeter, 1954; Schatzman, 1948, 1958; Wolf, 1965; Salpeter and Van Horn, 1969)

$$f_{12} = \exp[U_{s0} + U_{s1}] \tag{4.81}$$

$$\approx \exp\left\{0.205\left[(Z_1 + Z_2)^{5/3} - Z_1^{5/3} - Z_2^{5/3}\right]\left(\frac{\rho}{\mu_e}\right)^{1/3} T_6^{-1}\right\} , \tag{4.82}$$

where $T_6 = T/10^6$, the

$$U_{s0} = 0.9[\Gamma_{Z_1+Z_2} - \Gamma_{Z_1} - \Gamma_{Z_2}] , \tag{4.83}$$

where $\Gamma_{Z_1+Z_2}$, Γ_{Z_1}, and Γ_{Z_2} are given by Eq. (4.80) with Z replaced, respectively, by $(Z_1 + Z_2), Z_1$, and Z_2, and

$$\exp[U_{s1}] = \frac{(1 + 0.3\Gamma_{Z_1+Z_2} + 0.266\Gamma_{Z_1+Z_2}^{3/2})}{(1 + 0.3\Gamma_{Z_1} + 0.266\Gamma_{Z_1}^{3/2})(1 + 0.3\Gamma_{Z_2} + 0.266\Gamma_{Z_2}^{3/2})} . \tag{4.84}$$

Van Horn and Salpeter (1969) have also discussed the screening corrections for resonant reactions, and give explicit correction formulae for the triple alpha reaction. They also give interpolation formulae connecting the weak and strong screening cases. Generalized screening functions for arbitrary charge conditions in a plasma screening nuclear reactions are given by De Witt, Graboske, and Cooper (1973) and Graboske, De Witt, Groosman, and Cooper (1973).

4.2.10 Pycnonuclear Reactions

At very high densities, two effects combine to enhance thermonuclear reaction rates, even at low temperatures. First, electron screening, which is important at high density and low temperature, reduces the Coulomb barrier (see previous Sect. 4.2.9). Second, as the density increases the nuclei are packed into a smaller volume, eventually becoming so close that the uncertainty principle becomes important. It can result in enough energy to initiate nuclear reactions between neighboring nuclei even at zero temperature. Such pycnonuclear reactions are most effective when both the background nuclei and the reacting nuclei are bound in a Coulomb lattice, in which case the zero point energy, at zero temperature, contributes to their vibrational energy about the equilibrium position. A pycnonuclear reaction is relatively insensitive to temperature, but very sensitive to density. Pycnonuclear conditions hold when the effective thermal energy, E_0, given by Eq. (4.44) is much less than the electrostatic energy per nucleus, E_c, given by Eq. (4.74).

That is

$$E_0 \ll E_c$$

and

$$\beta = 0.032234\lambda\left[AZ_1^2Z_2^2\,\frac{7.6696 \times 10^{10}\,{}^\circ\mathrm{K}}{T}\right]^{2/3} \gg 1 \tag{4.85}$$

for pycnonuclear reactions. Here the density parameter, λ, is given by

$$\lambda = \frac{1}{2AZ_1Z_2}\left(\frac{1}{Z_1\mu_e}\frac{\rho}{1.3574 \times 10^{11}\mathrm{g\ cm^{-3}}}\right)^{1/3} , \tag{4.86}$$

the reduced atomic weight $A = A_1A_2/(A_1 + A_2)$, the mass density is ρ, the μ_e is the mean electron molecular weight given in Eq. (4.76), and Z_1 and Z_2 are the respective number of protons in the two reacting nuclei.

Pycnonuclear reactions have been considered by Wildhack (1940), Zeldovich (1958), Cameron (1959), Wolf (1965), and Salpeter and Van Horn (1969). The latter authors give a "zero temperature" reaction rate of

$$r = \begin{pmatrix} 3.90 \\ 4.76 \end{pmatrix} 10^{46} \frac{\rho}{2\mu_A} A^2Z^4S\lambda^{7/4} \exp\left[-\begin{pmatrix} 2.638 \\ 2.516 \end{pmatrix}\lambda^{-1/2}\right]$$

$$\text{reactions cm}^{-3}\ \text{sec}^{-1} . \tag{4.87}$$

Here the upper and lower numbers in parentheses correspond to two different approximations to the potential function, the mass density, ρ, is in g cm^{-3}, and the cross section factor, S, is in units of MeV barns. The mean molecular weight, μ_A, is given by

$$\mu_A^{-1} = \sum_i \frac{X_i}{A_i}\left[1 + \frac{Z_i}{A_i}\frac{m}{m_H}\right]^{-1} , \tag{4.88}$$

where X_i, Z_i, and A_i are, respectively, the mass fraction, number of protons, and atomic weight of nuclei of species i, the electron mass is m, and $m_H = 1.6735 \times 10^{-24}$ g. The cross section factor, S, is defined in Eqs. (4.34) and (4.51), and the density factor, λ, is given in Eq. (4.86). It follows from Eqs. (4.86) and (4.87) that the reaction rate is a sensitive function of mass density, ρ, and goes as $r \propto \rho^{19/12} \exp[-C\rho^{-1/6}]$, where C is a constant. It follows from Eq. (4.41) that the energy generation, ε, is given by

$$\varepsilon = rQ/(\rho) \ \ \mathrm{erg\ g^{-1}\ sec^{-1}} , \tag{4.89}$$

where r is given by Eq. (4.87) and Q is the energy release per reaction.

Cameron (1959) has given approximate numerical expressions for the pycnonuclear reaction rates as functions of both temperature and density for the reactions

$$\begin{aligned} &3\,\mathrm{He}^4 \rightarrow && \mathrm{C}^{12} \\ &\mathrm{C}^{12} + \mathrm{He}^4 \rightarrow && \mathrm{O}^{16} + \gamma \\ &\mathrm{N}^{14} + \mathrm{He}^4 \rightarrow && \mathrm{F}^{18} + \gamma \\ & && \downarrow \\ & && \mathrm{O}^{18} + \beta^+ + \nu_e \end{aligned} \tag{4.90}$$

and

$$\mathrm{O}^{16} + \mathrm{He}^4 \rightarrow \mathrm{Ne}^{20} + \gamma ,$$

as well as for the reactions of the heavy ions C^{12}, O^{16}, Ne^{24}, Mg^{32}, Si^{38}, S^{44}, A^{50} and Ca^{56} with themselves. He concludes that helium pycnonuclear reactions might ignite stellar explosions in the advanced stages of stellar evolution.

Salpeter and Van Horn (1969) also give pycnonuclear corrections to the strong screening case when $E_0 \approx E_c$, and give a temperature corrected pycnonuclear reaction rate, $r(T)$.

$$\frac{r(T)}{r} - 1 = \left(\begin{matrix}0.0430\\0.0485\end{matrix}\right)\lambda^{-1/2}\left[1 + \left(\begin{matrix}1.2624\\2.9314\end{matrix}\right)\exp\left(-8.7833\beta^{3/2}\right)\right]^{-1/2}$$
$$\times \exp\left\{-7.272\beta^{3/2} + \lambda^{-1/2}\left(\begin{matrix}1.2231\\1.4331\end{matrix}\right)\exp\left(-8.7833\beta^{3/2}\right)\right.$$
$$\left.\times\left[1 - \left(\begin{matrix}0.6310\\1.4654\end{matrix}\right)\exp\left(-8.7833\beta^{3/2}\right)\right]\right\} , \qquad (4.91)$$

where r is given in Eq. (4.87) and λ and β are given, respectively, in Eqs. (4.86) and (4.85).

4.3 Weak Interaction Processes

4.3.1 Electron Neutrino, Mu Neutrino, Muons, Pions, and Weak Interaction Theory

In order to satisfy the law of conservation of energy during the beta decay of radioactive elements, Wolfgang Pauli suggested that a hitherto-unknown, electrically neutral particle, later named the neutrino by Enrico Fermi, was carrying away energy that could not be observed (Pauli, 1930, 1933). Fermi developed the theory of beta decay in analogy with the process of photon emission from atoms, incorporating the neutrino hypothesis of Pauli (Fermi, 1934). Because of its tiny cross section for interaction with matter, detection of the neutrino requires the production of large quantities of them, as well as a massive detector. They were first observed in 1956 by Clyde Cowan, Frederick Reines and collaborators using a large liquid scintillation detector in the vicinity of a powerful fission reactor. Antineutrinos associated with the negative beta decay of neutron-rich fission fragments interacted with protons in the liquid scintillator to produce neutrons and positrons. These reactions were signaled by the delayed coincidence associated with the positron and neutron-capture pulses (Reines and Cowan, 1953).

Pauli (1930) first showed that the radioactive beta decay could only be explained if the proton, p, and the neutron, n, have the weak interactions

$$n \to p + \beta^- + \bar{\nu}_e , \qquad (4.92)$$

and

$$p \to n + \beta^+ + \nu_e ,$$

where the negative electron, β^-, is called the negatron, β^+ is called the positron, ν_e is the electron neutrino, and $\bar{\nu}_e$ is the antielectron neutrino. Properties of the negatron are given in Table 4.1, whereas properties of the electron neutrino and the other particles introduced in this section are given in Table 4.4.

The strong interaction, which accounts for the interaction nuclear force which holds protons and neutrons together, was thought to be due to another fundamental particle called the pion or the π-meson (Yukawa, 1935, 1937; Oppenheimer and Serber, 1937). Just as a photon results from the quantization of the electromagnetic field, the pion results from the quantization of the field which explains the nuclear force. The strong interactions are given by

$$\begin{aligned} &p \rightarrow n + \pi^+ \\ &n \rightarrow p + \pi^- \\ &p \rightarrow p + \pi^0 \\ &n \rightarrow n + \pi^0 \ , \end{aligned} \tag{4.93}$$

where π^+ and π^- are the charged pions, and the neutral pion, π^0, was introduced (Yukawa, 1938) to explain the charge independence of the nuclear force.

Before Yukawa's theoretical arguments, extraterrestrial high energy particles, called cosmic rays, had been observed (Hess, 1911, 1912; Kolhörster, 1913). It was thought that the massive π meson of Yukawa's theory might be a component of the cosmic rays, and such a component was soon found (Anderson and Neddermeyer, 1936). Detailed examinations of the cross section and life-time of the cosmic ray component (Neddermeyer and Anderson, 1937; Street and Stevenson, 1937; Nishina, Takeuchi, and Ichimiya, 1937; Rossi, 1939) indicated that this particle was different from the nuclear π meson. The cosmic ray meson was therefore called a muon or μ-meson, and a two meson theory was introduced (Sakata and Inoue, 1946; Marshak and Bethe, 1947). According to this theory, the nuclear π meson and the cosmic ray μ meson were interrelated by the decay reactions

Table 4.4. Properties of leptons: the muon, pion, tau particle, and the electron neutrino[1]

Particle	Symbol	Rest Mass (MeV/c^2)	Mean Life (seconds)	Decade Mode (≥99%)
Muon[2] (μ-meson)	μ^+, μ^-	105.658 389(34)	$2.197\ 03(4) \times 10^{-6}$	$\mu \rightarrow \nu_e, \nu_\mu$, e (see Eq. 4.96)
Pion (π-meson)	π^+, π^-	139.569 95(35)	$2.603\ 0(24) \times 10^{-8}$	$\pi^+ \rightarrow \mu^+ + \nu_\mu$
Pion (π-meson)	π^0	134.976 4(6)	$(8.4 \pm 0.6) \times 10^{-17}$	$\pi^0 \rightarrow \gamma + \gamma$
Tau	τ^+, τ^-	1777.1(4)	$2.956(3) \times 10^{-13}$	(see Eq. 4.97)
Electron Neutrino	ν_e	<0.000 016[3]		

[1] Adapted from the Review of Particle Properties, Physical Review **D50**, 1173 (1994). The numbers in parenthesis after the value give the one standard-deviation uncertainties in the last digits.

[2] The magnetic moment of the muon is $4.490\ 451\ 4(15) \times 10^{-21}$ erg Gauss^{-1} and the mass of the muon is $1.883\ 532\ 7(11) \times 10^{-25}$ grams.

[3] See Sect. 4.3.8.

$$\begin{aligned} \pi^+ &\rightarrow \mu^+ + \nu_\mu \\ \pi^- &\rightarrow \mu^- + \bar{\nu}_\mu \\ \pi^0 &\rightarrow \gamma + \gamma \ , \end{aligned} \tag{4.94}$$

where ν_μ is the mu-neutrino, $\bar{\nu}_\mu$ is the anti-mu-neutrino, and γ denotes a gamma ray photon. The π-meson was then also found to be a cosmic ray component (Lattes, Occhialini, and Powell, 1947) as well as a product of particle accelerator bombardments (Gardner and Lattes, 1948). Substituting Eq. (4.94) into Eq. (4.93) we obtain the reactions

$$n \rightarrow p + \mu^- + \bar{\nu}_\mu \ , \tag{4.95}$$

and

$$p \rightarrow n + \mu^+ + \nu_\mu \ ,$$

which are similar to the electron neutrino reactions of Eq. (4.92). Apart from the interactions given in Eq (4.95), the muon decays according to the reactions

$$\begin{aligned} \mu^- &\rightarrow \beta^- + \bar{\nu}_e + \nu_\mu \\ \mu^+ &\rightarrow \beta^+ + \nu_e + \bar{\nu}_\mu \ , \end{aligned} \tag{4.96}$$

with a mean lifetime of $\tau = 2.19703(4) \times 10^{-6}$ seconds.

The shape of the muon decay spectrum, as well as its decay rate, was understood in terms of the emission of two neutrinos according to Eq. (4.96). However, the second type of neutrino, the mu neutrino, ν_μ, associated with the muon, μ, was not detected until the early 1960s using the Brookhaven particle accelerator and neutrino beams (Danby et al., 1962, also see Pontecorvo, 1960, and Schwartz, 1960).

Electrons, muons and tau particles, along with their corresponding neutrinos, are collectively known as leptons, the Greek word for "slender", as they are all significantly less massive than most other elementary particles. The unanticipated discovery of the production and decay of the tau lepton, designated τ, by Martin Perl and his colleagues using the Stanford Linear Accelerator (Perl et al., 1975, 1976), is interpreted to imply the existence of yet another type of neutrino, the tau neutrino, ν_τ, and its antiparticle, $\bar{\nu}_\tau$. Because of its greater mass, given in Table 4.4, the tau particle has many decay modes such as:

$$\begin{aligned} \tau^- &\rightarrow \nu_\tau + e^- + \bar{\nu}_e \ , \\ \tau^- &\rightarrow \nu_\tau + \mu^- + \bar{\nu}_\mu \ , \\ \tau^- &\rightarrow \nu_\tau + \pi^- \ , \\ \tau^- &\rightarrow \nu_\tau + \pi^- + \pi^0 \ , \\ \tau^- &\rightarrow \nu_\tau + \pi^- + \pi^+ + \pi^- , \end{aligned} \tag{4.97}$$

with corresponding modes for τ^+.

The mu-neutrino, ν_μ, has the same physical properties as the electron neutrino, ν_e, except that it reacts with the muon instead of the electron. In fact, both neutrinos have been detected with interaction cross sections given by (Reines and Cowan, 1953)

$$\sigma \approx 10^{-44} \quad \text{cm}^2 \text{ at } 1\,\text{MeV} \ , \tag{4.98}$$

for the reaction $\bar{\nu}_e + p \rightarrow e^+ + n$, and (Danby *et al.*, 1962)

$$\sigma \approx 10^{-38} \quad \text{cm}^2 \text{ at } 1\text{GeV} \ ,$$

for the reactions $\bar{\nu}_\mu + p \rightarrow n + \mu^+$ and $\nu_\mu + n \rightarrow p + \mu^-$. These data are consistent with a universal coupling constant, g, applicable to both neutrino interactions. The interaction cross section, σ, is given by

$$\sigma \approx \frac{g^2 E_\nu^2}{\hbar^4 c^4} \approx 10^{-44} \left(\frac{E_\nu}{mc^2} \right)^2 \quad \text{cm}^2 \ , \tag{4.99}$$

where E_ν is the center of mass neutrino energy, and the Fermi constant, g, is given by

$$g = 1.4102 \pm 0.0012 \times 10^{-49} \quad \text{erg cm}^3 \ . \tag{4.100}$$

Fermi (1934) first realized that the β decay (Eq. (4.92)) was caused by a weak interaction which was different from the nuclear interaction; and postulated an interaction matrix element, v_{fi}, given by

$$v_{\text{fi}} = (\bar{\psi}_{\text{n}} O \psi_{\text{p}})(\bar{\psi}_{\text{e}} O_{\text{L}} \psi_{\nu_{\text{e}}}) \ , \tag{4.101}$$

where the wave function ψ has subscripts n, p, e or v_e denoting the neutron, n, the proton, p, the electron, β^- or e^- or e, and the neutrino, ν_e, and O and O_L denote, respectively, operators which operate on the nucleon (proton and neutron) and the lepton (electron and electron neutrino) wave functions. The transition probability, W, for a transition from an initial state, whose wave function is ψ_i, to a final state, whose wave function is ψ_f, is given by the golden rule of time dependent perturbation theory

$$\begin{aligned} W &= \frac{2\pi}{\hbar} v_{\text{fi}}^* v_{\text{fi}} \rho(E_{\text{f}}) = \frac{2\pi}{\hbar} \left| \int \psi_{\text{f}}^* H_{\text{int}} \psi_i \, d\tau \right|^2 \rho(E_{\text{f}}) \\ &= \frac{2\pi}{\hbar} |\langle \psi_{\text{f}} | H_{\text{int}} | \psi_i \rangle|^2 \rho(E_{\text{f}}) \ , \end{aligned} \tag{4.102}$$

where v_{fi} is the interaction matrix element, H_{int} is the interaction Hamiltonian capable of matrix elements between the initial and final states, and $\rho(E_f)$ is the total number of final states in the energy range E_f to $E_f + dE_f$.

During Fermi's time, it was unknown what the exact form of the operator, O, was; but it was known that it must be some mixture of coupling constants which measure the strength of the interactions and the five possible sets of combinations of Dirac matrices (cf. Preston, 1962). It was also known that in nature the protons, neutrons, electrons, muons, and neutrinos interact according to the combinations given by (Wheeler, 1947; Tiomno and Wheeler, 1949)

$$(\text{p}, \text{n}) \leftrightarrow (\mu, \nu_\mu) \leftrightarrow (\text{e}, \nu_\text{e}) \leftrightarrow (\text{p,n}) \ . \tag{4.103}$$

Feynman and Gell-Mann (1958) postulated the existence of a universal weak interaction in which any of the pairs given in Eq. (4.103) can interact with each other, as well as with themselves. Such self-weak interactions have a weak interaction Hamiltonian given by

$$H_{\mathrm{int}} = \sqrt{2} g_V J^* J \ , \tag{4.104}$$

where the weak interaction current, J, is given by

$$J = \bar{\psi}_{\mathrm{n}}(V-A)\psi_{\mathrm{p}} + \bar{\psi}_{\mathrm{e}}(V-A)\psi_{\nu_{\mathrm{e}}} + \bar{\psi}_{\mu}(V-A)\psi_{\nu_{\mu}} \ , \tag{4.105}$$

the wave function, ψ, has a subscript denoting the appropriate particle, and the vector J^* is the Hermitian conjugate of J. Here the operator $(V-A)$ reflects the experimental conclusion that only the vector, V, and axial vector, A, operators exist in the weak interaction; as well as Feynman and Gell-Mann's hypothesis that they appear in the mixture $(V-A)$. The vector coupling constant, C_V, is measured from the beta decay of ^{26}Al, and has the value (Freeman et al., 1972)

$$g_v = C_v = 1.4102 \pm 0.0012 \times 10^{-49} \text{ erg cm}^3 \ . \tag{4.106}$$

The ratio of the axial vector coupling constant, C_A, to C_V is measured from the neutron half life. Christensen *et al.* (1967) measure a neutron half life of $\tau_{\mathrm{n}1/2} = 10.61 \pm 0.16$ minutes. The more recent value of $\tau_{\mathrm{n}1/2} = 614.29$ seconds $= 10.25$ minutes, with a mean life of $\tau = 887.0 \pm 2.0$ seconds, results in (Review of Particle Properties, Physical Review **D50**, 1, 1994)

$$\frac{C_A}{C_V} = \frac{g_A}{g_V} = -1.2573 \pm 0.0028 \ . \tag{4.107}$$

This ratio follows from the neutron $(ft)_{\mathrm{n}}$ value given by

$$(\mathrm{ft})_{\mathrm{n}} = \frac{2\pi^3(\ln 2)\hbar^7}{m^5c^4[C_V^2|\mathcal{M}_V|^2 + C_A^2|\mathcal{M}_A|^2]} = \frac{1.230627 \times 10^{-94}}{[C_V^2 + 3C_A^2]} \ . \tag{4.108}$$

(For neutron $|\mathcal{M}_V|^2 = 1$ and $|\mathcal{M}_A|^2 = 3$.)

The current-current interaction results in a β decay interaction Hamiltonian, H_β, and an electron-neutrino self interaction Hamiltonian, H_W, given by

$$H_\beta = \frac{g_V}{\sqrt{2}}[\bar{\psi}_\nu \gamma_\alpha(1+\gamma_5)\psi_{\mathrm{e}}][\bar{\psi}_{\mathrm{n}}\gamma_\alpha(C_V - C_A\gamma_5)\psi_{\mathrm{p}}] \quad \text{for (p, n)(e, } \nu), \tag{4.109}$$

and

$$H_W = \sqrt{2} g_V \left[\bar{\psi}_{\mathrm{e}}\gamma_\mu \frac{(1+\gamma_5)}{\sqrt{2}}\psi_\nu\right]\left[\bar{\psi}_\nu \gamma^\mu \frac{(1+\gamma_5)}{\sqrt{2}}\psi_{\mathrm{e}}\right] \quad \text{for (e, } \nu)(\text{e, } \nu) \ . \tag{4.110}$$

Here the operator $(V-A)$ has been expressed in terms of combinations of Dirac matrices, γ, and the symbols have the meanings described in the texts of Konopinski (1966) and Preston (1962). Eq. (4.110) is often rearranged to give

$$H_W = g_V[\bar{\psi}_{\mathrm{e}}\gamma_\mu\psi_{\mathrm{e}}]\left[\bar{\psi}_\nu\gamma^\mu\frac{(1+\gamma_5)}{\sqrt{2}}\psi_\nu\right] - g_V[\bar{\psi}_{\mathrm{e}}\gamma_5\gamma_\mu\psi_{\mathrm{e}}]\left[\bar{\psi}_\nu\gamma^\mu\frac{(1+\gamma_5)}{\sqrt{2}}\psi_\nu\right] \ . \tag{4.111}$$

4.3.2 Beta Decay

As suggested by Pauli (1930), the particle called the electron neutrino, ν_e, must be present if reactions involving β decay satisfy the laws of conservation of energy and momentum. The nuclear reaction for negative beta, β^-, decay is

$$(Z-1, A) \rightarrow (Z, A) + \beta^- + \bar{\nu}_e \ , \tag{4.112}$$

where (Z, A) denotes a nucleus of charge, Z, and mass number, A, the β^- particle is an electron, and $\bar{\nu}_e$ denotes the antielectron neutrino. Negative beta decay will only occur if

$$M_{at}(Z-1, A) > M_{at}(Z, A) \ , \tag{4.113}$$

or equivalently

$$M(Z-1, A) > M(Z, A) + m_e \ ,$$

where $M_{at}(Z, A)$ denotes the mass of the atom, $M(Z, A)$ denotes the mass of the nucleus, and the electron mass $m_e = 5.48579903(13) \times 10^{-4}$ a.m.u. $= 0.5109906(15)\text{MeV/c}^2$. Negative beta particles are usually ejected with energies ranging from a few keV to 15 MeV.

The nuclear reaction for positive beta, β^+, decay is

$$(Z+1, A) \rightarrow (Z, A) + \beta^+ + \nu_e \ , \tag{4.114}$$

where the positive beta particle, β^+, is the positron, and ν_e is the electron neutrino. Positive beta decay will only occur if

$$M_{at}(Z+1, A) > M_{at}(Z, A) + 2m_e \ , \tag{4.115}$$

or equivalently

$$M(Z+1, A) > M(Z, A) + m_e \ .$$

In general, it is found that for a given odd A only one nucleus is stable against beta decay, and for a given even A there are only two nuclei stable against beta decay.

If follows from Eqs. (4.102), (4.104), and (4.105) that the transition probability per unit time, W, for a single negative beta decay of a nucleus (Z, A) is given by

$$W = \frac{2\pi g_V^2}{\hbar} |\mathcal{M}|^2 \rho(E_f) \ , \tag{4.116}$$

where the nuclear matrix element, $\mathcal{M}$, is given by

$$\begin{aligned} \mathcal{M} &= \int (\bar{\psi}_{Z,A}(V-A)\psi_{Z+1,A})(\bar{\psi}_e(V-A)\psi_{\nu_e}) d\tau \\ &\approx \int \bar{\psi}_{Z,A}(V-A)\psi_{Z+1,A} \, d\tau \ . \end{aligned} \tag{4.117}$$

Here the integrals are over the volume of the nucleus, and the electron and neutrino wave functions have wave solutions with arguments so small that the wave functions contribute a constant factor near unity to the integral of Eq. (4.117). For the allowed transitions of nuclear β decay, the nuclear matrix element, $\mathcal{M}$, satisfies the equation

$$|\mathscr{M}|^2 = |\mathscr{M}_{\mathrm{F}}|^2 + \frac{g_A^2}{g_V^2}|\mathscr{M}_{\mathrm{GT}}|^2 \ , \tag{4.118}$$

here, $\mathscr{M}_{\mathrm{F}}$ and $\mathscr{M}_{\mathrm{GT}}$ denote, respectively, the matrix elements for the Fermi and the Gamow-Teller transitions. For neutron decay, for example, $|\mathscr{M}_{\mathrm{F}}| = |\mathscr{M}_V| = 1$ and $|\mathscr{M}_{\mathrm{GT}}| = |\mathscr{M}_A| = \sqrt{3}$ where we have included subscripts V and A for the vector and axial vector nuclear matrix elements. The Fermi transition corresponds to the vector interaction and has the selection rules $\Delta J = 0$, no change of parity; whereas the Gamow–Teller transition corresponds to the axial-vector interaction and has the selection rules $\Delta J = \pm 1, 0$, no $0 \to 0$ transition, and no change of parity. Here J is the total angular momentum of the nucleus.

As the product nucleus $(Z+1, A)$ may be assumed to be in a single state, only the electron and the neutrino contribute to the energy density of states, $\rho(E_{\mathrm{f}})$. If p_{e} and $p_{\nu_{\mathrm{e}}}$ denote, respectively, the momentum of the electron and the electron neutrino, and W_0 denotes the maximum energy release of the β decay, then it follows from Eq. (4.116) that

$$W = \frac{g_V^2|\mathscr{M}|^2}{2\pi^3\hbar^7c^3}(W_0 - E)^2 p_{\mathrm{e}}^2\, dp_{\mathrm{e}} \ , \tag{4.119}$$

where the number of final states, $\rho(E)$, with energy between E and $E + dE$ is

$$\rho(E) = \frac{(W_0 - E)^2 p_{\mathrm{e}}^2\, dp_{\mathrm{e}}}{4\pi^4\hbar^6c^3} \ .$$

Integration of Eq. (4.119) over all momenta up to the maximum corresponding to the maximum electron energy, W_0, available for beta decay gives the total transition probability per unit time, λ_β. This beta decay rate is given by

$$\lambda_\beta = \frac{\ln 2}{t} = \frac{f \ln 2}{ft} = \frac{1}{2\pi^3}\frac{m^5c^4g_V^2}{\hbar^7}|\mathscr{M}|^2 \int_1^{W_0} pE(W_0 - E)^2 F(Z, E)dE \ , \tag{4.120}$$

where the factor $m^5c^4g_V^2|\mathscr{M}|^2/\hbar^7$ is of order unity. This factor assumes the value of one if the wave functions are normalized so that $2\pi|\mathscr{M}|^2$ is the probability of a transition per unit time per unit of energy and no state density factor is needed. The $\ln 2 = 0.693$, the half lifetime for beta decay is t, and the Fermi function, $F(Z, E)$ is given in detail by Konopinski (1966) and Preston (1962) and has the approximate value

$$F(Z, E) \approx 2\pi\eta[1 - \exp(-2\pi\eta)]^{-1} \ , \tag{4.121}$$

where

$$\eta = \frac{Ze^2}{\hbar v} \ ,$$

and the velocity of the emitted electron is v. The quantity f is given by

$$f = f(Z, W_0) = \int_1^{W_0} F(Z, E)pE(W_0 - E)^2\, dE \ , \tag{4.122}$$

where E and W_0 are in terms of mc^2, and p is in terms of mc, and both $F(Z,E)$ and $f(Z,W_0)$ have been extensively tabulated by Feenberg and Trigg (1950), Feingold (1951), Dismuke et al. (1952), Major and Biedenharn (1954), and Ajzenberg-Selove (1960). For an allowed transition (no parity change and $\Delta J = 0, \pm 1$), the product ft is well defined by the relation

$$ft = \frac{2\pi^3 \hbar^7 \ln 2}{m^5 c^4} \left[g_V^2 |\mathcal{M}_V|^2 + g_A^2 |\mathcal{M}_A|^2\right]^{-1}$$
$$\approx 1.230627 \times 10^{-94} \left[g_V^2 |\mathcal{M}_V|^2 + g_A^2 |\mathcal{M}_A|^2\right]^{-1} \text{ sec} . \quad (4.123)$$

Measured values of ft range from $10^{4.5}$ to $10^{6.0}$ sec, and therefore λ_β is often estimated using tabulated values of f together with an assumed value of $ft = 10^{5.5}$ sec.

As first suggested by Hoyle (1946), beta decay in stellar interiors might be inhibited due to the prepopulation of momentum states in the electron phase space into which the final electron would normally decay (called "exclusion principle inhibition"). For allowed decays (no parity change and $\Delta J = 0, \pm 1$) and for most first forbidden decays (parity change and $\Delta J = 0, \pm 1$), the stellar beta decay rate, λ_s, is related to the terrestrial value, λ_β, by the equation (Peterson and Bahcall, 1963)

$$\lambda_s = \lambda_\beta (1 - \delta) , \quad (4.124)$$

where the exclusion principle inhibition factor, δ, is given by

$$\delta = \frac{\int S \, d\lambda}{\int d\lambda} , \quad (4.125)$$

where (cf. Eq.(4.120))

$$d\lambda = \frac{p}{2\pi^3}(E + mc^2)(W_0 - E)^2 F(Z,E) \, dE \propto (E + mc^2)^2 (W_0 - E)^2 \, dE ,$$

the relativistic energy of the electron is E, the end point energy for decay is W_0, and the probability, S, that the electron states corresponding to an energy between E and $E + dE$ are occupied, is given by

$$S = \left[1 + \exp\left(\frac{E}{kT} - \frac{E_F}{kT}\right)\right]^{-1} ,$$

where E_F is the Fermi energy. We have the approximate relations

$$\delta \approx 0 \quad \text{for a nondegenerate gas} ,$$

and

$$\delta \approx \left[1 - \exp\left(-\frac{E_F}{kT}\right)\right] \approx 1 \quad \text{for a degenerate gas and} \quad W_0 \ll E_F . \quad (4.126)$$

Peterson and Bahcall (1963) give a lengthy approximate relation for δ in the degenerate case of $W_0 \gg E_F$, where $0 \lesssim \delta \lesssim 1$.

As pointed out by Hoyle (1946) and Cameron (1959), beta decay lifetimes are considerably shortened by decay from thermally populated excited states of the nuclear species in high temperature stellar interiors. The decay rate for these excited state reactions, λ_e, is given by (Cameron, 1959)

$$\lambda_e = \frac{1}{P(T)} \sum_i \sum_j \lambda_{sij}(2J_i + 1)\exp\left(-\frac{E_i}{kT}\right) , \tag{4.127}$$

where λ_s is the stellar beta decay rate between parent states, i, and daughter states, j, the $E_i(J_i)$ is the excitation energy (spin) of the parent state, i, and the nuclear spin of state i is J_i. It is assumed that states i are populated according to a Boltzmann distribution, and the nuclear partition function, $P(T)$, is given by

$$P(T) = \sum_i (2J_i + 1)\exp\left(-\frac{E_i}{kT}\right) . \tag{4.128}$$

The partition function is often taken to be equal to the statistical weight of the ground state, $2I + 1$, where I is the nuclear spin of the ground state. For example, for a two level nucleus with ground state spin J_1 and decay rate $\lambda_\beta(1)$, and an excited state of spin J_2 and decay rate $\lambda_\beta(2)$, the total beta decay rate, λ_β, is given by

$$\lambda_\beta = \frac{2J_1 + 1}{G}\lambda_\beta(1) + \frac{2J_2 + 1}{G}\lambda_\beta(2)\exp\left(-\frac{E^*}{kT}\right) ,$$

where

$$G = (2J_1 + 1) + (2J_2 + 1)\exp\left(-\frac{E^*}{kT}\right) , \tag{4.129}$$

and E^* is the excitation energy of the excited state. The energy criterion for the occurence of this beta decay is

$$M(Z - 1, A) > M(Z, A) + m_e + E^*/c^2 , \tag{4.130}$$

where $M(Z, A)$ is the mass of the nucleus in the ground state. Cameron (1959) has called the beta decay from thermally excited states "photobeta" decay because the excited state is caused by the exchange of energy between matter and radiation (photons) under conditions of statistical equilibrium. To calculate the excited state beta decay rates, level distributions and parameters must be known for the nuclei in question, and much of this data can be obtained from the nuclear energy levels given in the Nuclear Data Sheets of the Oak Ridge National Laboratory or from the Isotope Tables (Lederer et al., 1967). Otherwise, level parameters may be estimated using the nuclear level density formula of Gilbert and Cameron (1965). Beta decay rates for both excited and unexcited nuclei are given by Hansen (1968) and Wagoner (1969).

When excited state beta decay is unfavored due to the large excitation energies of all favorable excited states, another "photobeta" process may be important. In this process the photon can be considered to decay virtually into an electron-positron pair, with the positron being absorbed by the nucleus, giving rise to an antineutrino according to the reaction

$$\gamma + (Z, A) \to (Z + 1, A) + e^- + \bar{\nu}_e , \tag{4.131}$$

subject to the energy criterion

$$\Delta = M(Z + 1, A) + m_e - M(Z, A) . \tag{4.132}$$

Here Δ is the atomic mass difference, $M(Z,A)$ is the nuclear mass of the nucleus (Z,A), and m_e is the electron mass. The decay rate, λ_p, of this photon induced beta decay is given by (Shaw, Clayton, and Michel, 1965)

$$\lambda_p = \frac{2\ln 2}{\pi}\frac{\alpha}{ft}\left(\frac{kT}{mc^2}\right)^{9/2}\exp\left[-\frac{\Delta}{kT}\right] \quad \text{for } kT < 0.3mc^2 \text{ and } \Delta < 2mc^2 \ ,$$

and

$$\lambda_p = \frac{\ln 2}{ft}\frac{\alpha}{\pi}\left[5.78\left(\frac{kT}{mc^2}\right)^5 \ln\left(\frac{kT}{mc^2}\right) + 3.10\left(\frac{kT}{mc^2}\right)^5 + \cdots\right] \quad \text{for } kT \gg mc^2, \Delta \ , \tag{4.133}$$

where $\alpha = 2\pi e^2/(hc) \approx (137.036)^{-1}$ is the fine structure constant, the ft value is that which characterizes the nuclear matrix element connecting the states (Z,A) and $(Z+1,A)$, the gas temperature is T, and the mass difference is Δ.

When the thermal energy of a star becomes greater than the electron rest mass energy, $T \gtrsim 10^9\,^\circ$K, the production of electron-positron pairs sets in. If there is a significant density of free positrons, even terrestrially stable nuclei might undergo beta transformation. Although this process may not be astrophysically important, we give here the probability per unit time, λ, that a nucleus (Z,A) will capture a positron with kinetic energy $W = (E-1)mc^2$ to form a nucleus $(Z+1,A)$. According to Reeves and Stewart (1965)[1]

$$\lambda = \sum_i A_i\lambda_i \ , \tag{4.134}$$

where the fractional population, A_j, of level j with spin J_j is

$$A_j = \frac{(2J_j+1)\exp[-E_j/kT]}{\sum_i (2J_i+1)\exp[-E_i/kT]} \ ,$$

the excitation energy of the ith level is E_i, the $\sum$ is over all levels of the nucleus, and

$$\lambda_i = \frac{N^+ \ln 2}{4\pi ft}\left(\frac{h}{mc}\right)^3 \int_{E_m}^{\infty}(W-W_m)^2 F(Z,W)N(E)dE \Big/ \int_1^{\infty} N(E)dE \ ,$$

where W_m is the threshold energy of the reaction, $W = (E-1)mc^2$ is the kinetic energy of the positron, the positron density, N^+, is given by

$$N^+ = \{[(N_0^2/4) + 4I^2/h^6]^{1/2} - N_0/2\} \ .$$

the residual electron density $N_0 = \rho/(2M)$ where ρ is the gas mass density, M is the nuclear mass, and $I = 4\pi(mc)^3 Z^{-1}K_2(Z)$, where $Z = M_e c^2/(kT)$ and K_2 is the modified Bessel function of the second kind. The energy spectrum of the positrons is given by

$$N(E)dE = \frac{8\pi(mc/h)^3 E(E^2-1)^{1/2}dE}{(2I/N^+h^3)[\exp(mc^2E/kT)+1]} \ . \tag{4.135}$$

[1] The value of λ_i is overestimated by a factor by a factor of 2 (see footnote two on page 370)

4.3.3 Electron Capture

A nucleus (Z, A) which is unstable to positive beta, β^+, decay can instead capture an electron from the surrounding electrons (Yukawa, 1935; Möller, 1937). Moreover, every nucleus can capture an electron provided that the energy of electrons in stellar material is sufficiently high. The process of successive electron captures is called neutronization and it plays a fundamental role in the theory of stellar evolution. Under normal terrestrial conditions electron capture only occurs when

$$(Z+1, A) + e^- \rightarrow (Z, A) + \nu_e \ , \tag{4.136}$$

where the symbol e^- is used to denote the electron in the electron capture reaction. Electron capture will only occur if

$$M_{at}(Z+1, A) > M_{at}(Z, A) \tag{4.137}$$

or equivalently

$$M(Z+1, A) + m_e > M(Z, A) \ ,$$

where $M_{at}(Z, A)$ denotes the atomic mass of atom (Z, A), and $M(Z, A)$ denotes the mass of the nucleus (Z, A). It can be shown (Feenberg and Trigg , 1950) that the terrestrial decay rates for positron emission, λ_{β^+}, and the electron capture to the K shell, λ_K, are related by a function of the atomic number, Z, of the product nucleus and the maximum energy release, Q. For large Z and small Q electron capture becomes more probable then β^+ emission, whereas for small Z and large Q, positron emission is more probable. Precise values of the ratio $\lambda_K/\lambda_{\beta^+}$ are given in Feenberg and Trigg (1950) and Preston (1962). The f values for electron capture are also tabulated by Major and Biedenharn (1954). As Bahcall (1962) has pointed out, the atoms in stellar interiors are highly ionized and cannot easily capture electrons from bound orbits. Nuclei which decay terrestrially by electron capture from bound orbits can, however, decay in stars by the capture of free electrons from continuum orbits.

The stellar decay rate, λ_{ce}, for continuum electron capture on nuclei, which decay terrestrially by allowed and first-forbidden positron emission, is given by (Bahcall, 1964)

$$\lambda_{ce} = \frac{K \ln 2}{ft} \ , \tag{4.138}$$

where ft is the terrestrially measured value, $\ln 2 = 0.693$, and the stellar phase-space function, K, is given by

$$K = \int_{p_0}^{\infty} p^2 q^2 F(Z, W) \left[1 - \exp\left(-\frac{\mu}{kT} + \frac{W}{kT}\right)\right]^{-1} dp \ , \tag{4.139}$$

where W is the total relativistic energy of the electron, μ is the chemical potential, $\mu \approx E_F$, where E_F is the Fermi energy, p is the electron momentum, $F(Z, W)$ is the Fermi function and $q = (W + W_0)/(mc^2)$ is the energy in mc^2 units (0.511 MeV) of the neutrino emitted in the electron capture process. If the difference between the initial and final nuclear masses is W_0, then the threshold value of momentum, p_0, is given by

$$p_0 = 0 \text{ if } W_0 \gtrsim -mc^2 \quad \text{(exoergic capture)}$$

and

$$p_0 = \frac{1}{mc^2}(W_0^2 - m^2c^4)^{1/2} \quad \text{if } W_0 \lesssim -mc^2 \quad \text{(endoergic capture)} . \tag{4.140}$$

For nonrelativistic electron energies, the Fermi function, $F(Z, W)$, is given by

$$F(Z, W) = 2\pi\eta[1 - \exp(-2\pi\eta)]^{-1} ,$$

where

$$\eta = \frac{\alpha Z W}{p} = \frac{\alpha Z c}{v} \approx \alpha Z\left(\frac{mc^2}{3kT}\right)^{1/2} ,$$

$\alpha = 2\pi e^2/(hc) \approx (137.036)^{-1}$, the velocity of the captured electron is v, and we assume that $\alpha^2 Z^2 \ll 1$. As Fowler and Hoyle (1964) have pointed out, $F(Z, W)$ is a slowly varying function of p and can be factored out of the integral by assigning it a suitable average value $\langle F \rangle$, which they give. Hansen (1968) has evaluated the resultant expression for K in the computer and gives values of λ_{ce} for several nuclei.

For both a degenerate and nondegenerate gas, and a beta transition threshold energy of W_0,

$$K \approx \pi^2\left(\frac{\hbar}{mc}\right)^3 N_e \langle F \rangle (W_0 + 1)^2 . \tag{4.141}$$

Here N_e is the electron density, and $\langle F \rangle$ is the Fermi function evaluated at $W = 1 + \langle E \rangle$, where the average energy, $\langle E \rangle \approx kT$, for a nondegenerate gas, and $\langle E \rangle \approx E_F$, the Fermi energy, for a degenerate gas. For nonrelativistic and extremely relativistic electrons,

$$E_F = (3\pi^2)^{2/3} \frac{\hbar^2}{2m} N_e^{2/3} \quad \text{(nonrelativistic)}$$

and

$$E_F = (3\pi^2)^{1/3} \hbar c N_e^{1/3} \quad \text{(relativistic)} , \tag{4.142}$$

but note (4.141) is not applicable for a relativistic degenerate gas. For a nondegenerate gas for which Boltzmann statistics apply, Bahcall (1962) gives

$$K \approx \pi^2 W_0^2 N_e \quad \text{if } \eta \ll 1, \alpha^2 Z^2 \ll 1, kT \ll mc^2 \text{ and } W_0 \ll kT . \tag{4.143}$$

For a degenerate gas, Bahcall (1962) and Tsuruta and Cameron (1965) give

$$\begin{aligned} K \approx \langle F \rangle \big[& \tfrac{1}{5} P_F^5 + \tfrac{1}{3} P_F^3 (1 + W_0^2) \\ & + \tfrac{1}{4} W_0 \{2 P_F E_F^3 - P_F E_F - \ln(P_F + E_F)\}\big] \\ & \text{if } W_0 > -mc^2 \quad \text{(exoergic) and } \eta_F \ll 1 , \end{aligned} \tag{4.144}$$

$$K \approx \langle F \rangle \Big\{ \tfrac{1}{5}(P_{\mathrm{F}}^5 - P_0^5) + \tfrac{1}{3}(P_{\mathrm{F}}^3 - P_0^3)(1 + W_0^2) - \tfrac{1}{4}|W_o|[2(P_{\mathrm{F}}E_{\mathrm{F}}^3 - P_0|W_0|^3)$$
$$- (P_{\mathrm{F}}E_{\mathrm{F}} - P_0|W_0|) - \ln\,(P_{\mathrm{F}} + E_{\mathrm{F}}) + \ln\,(P_0 + |W_0|)] \Big\}$$
$$\text{if } W_0 < -mc^2 \quad \text{(endoergic) and } \eta_F \ll 1 \ , \tag{4.145}$$

and for extreme degeneracy we obtain the simpler expressions

$$K \approx \frac{2\pi\alpha Z}{5} \left[\left(E_{\mathrm{F}}^5 + \frac{5}{2} W_0 E_{\mathrm{F}}^4 + \frac{5}{3} W_0^2 E_{\mathrm{F}}^3 \right) - \left(1 + \frac{5}{2} W_0 + \frac{5}{3} W_0^2 \right) \right]$$
$$\text{if } W_0 > -mc^2 \quad \text{(exoergic) and } \eta_{\mathrm{F}} \gg 1 \ ,$$
$$K \approx \frac{2\pi\alpha Z}{5} \left[(E_{\mathrm{F}} - |W_0|)^5 + \frac{5}{2}|W_0|(E_{\mathrm{F}} - |W_0|)^4 + \frac{5}{3}|W_0^2|^2 (E_{\mathrm{F}} - |W_0|)^3 \right]$$
$$\text{if } W_0 < -mc^2 \quad \text{(endoergic) and } \eta_{\mathrm{F}} \ll 1 \ . \tag{4.146}$$

Here the energies E_{F} and W_0 are expressed in units of mc^2, and

$$\langle F \rangle = 2\pi\alpha Z \quad \text{for } Z > (2\pi\alpha)^{-1} \approx 23$$

and

$$\langle F \rangle = 1 \quad \text{if } Z < (2\pi\alpha)^{-1} \approx 23 \ ,$$

the fine structure constant is α, the Fermi momentum $P_{\mathrm{F}} = c^{-1}(E_{\mathrm{F}}^2 - m^2c^4)^{1/2}$ where E_{F} is the Fermi energy, and $P_0 = c^{-1}(W_0^2 - m^2c^4)^{1/2}$, where W_0 is the threshold energy for the beta transition.

Finzi and Wolf (1967) have shown that contracting white dwarfs with magnesium rich cores or with calcium rich shells may lead to Type I supernovae. They show that the contractions are caused by the electron capture reactions of Mg^{24} and Ca^{40} and give formulae for the reaction lifetimes.

4.3.4 The URCA Processes

A nucleus (Z, A) may produce neutrinos, ν_e, through the beta decay reactions

$$(Z, A) \rightarrow (Z + 1, A) + e^- + \bar{\nu}_e$$

and

$$(Z, A) \rightarrow (Z - 1, A) + e^+ + \nu_e \ , \tag{4.147}$$

or by the electron capture reaction

$$(Z + 1, A) + e^- \rightarrow (Z, A) + \nu_e \ . \tag{4.148}$$

In the ordinary URCA process, a nucleus alternately captures an electron, e^-, and undergoes negative beta decay, meanwhile emitting a neutrino, ν_e, and an antineutrino, $\bar{\nu}_e$, according to the reaction (Gamow and Schönberg, 1941)

$$e^- + (Z, A) \rightarrow e^- + (Z, A) + \nu_e + \bar{\nu}_e \ . \tag{4.149}$$

Only nuclei of odd mass number can participate in the ordinary URCA process in a degenerate gas. This is because a nucleus with even mass number has a lower threshold for electron capture after capturing one electron. Hence only even-mass-number nuclei of even charge number can exist in a degenerate gas, and these will, in general, be stable against both electron capture and beta decay. As pointed out by Gamow and Schönberg (1941), neutrino pair emission by the URCA process may affect the rates of stellar evolution, and what is of interest in this case is the rate of energy loss by the neutrinos. By assuming that the nuclei are in statistical equilibrium, and that neutrinos and antineutrinos are produced continuously through negative beta decay and electron capture, neutrino luminosities can be calculated using the decay rates given in the two previous sections (cf. Bahcall, 1962, 1964; Peterson and Bahcall, 1963; Tsuruta and Cameron, 1965).

Tsuruta and Cameron (1965) consider conditions of nuclear statistical equilibrium in a degenerate gas and obtain an average neutrino energy production rate per nucleus in the excited level, i, given by

$$P_\nu = \lambda_i \omega_i = \frac{\ln 2 f_i}{(ft)_i} \omega_i \ , \tag{4.150}$$

where for electron capture

$$\omega_i = \frac{5}{6}(E_\mathrm{F} + W_0)\left[\frac{(1-y^6) - \frac{12}{5}x(1-y^5) + \frac{3}{2}x^2(1-y^4)}{(1-y^5) - \frac{5}{2}x(1-y^4) + \frac{5}{3}x^2(1-y^3)}\right] \ , \tag{4.151}$$

and

$$\begin{aligned} x &= W_0/(E_\mathrm{F} + W_0) \\ y &= (1 + W_0)/(E_\mathrm{F} + W_0) \end{aligned} \qquad \text{if } W_0 \gg -mc^2 \ ,$$

and

$$\omega_i = \frac{5}{6}(E_\mathrm{F} - |W_0|)\left[\frac{1 + \frac{12}{5}x + \frac{3}{2}x^2}{1 + \frac{5}{2}x + \frac{5}{3}x^2}\right] \ , \tag{4.152}$$

where $x = |W_0|/(E_\mathrm{F} - |W_0|)$ if $W_0 \ll -mc^2$.

For electron emission (beta decay)

$$\omega_i = \frac{D_2(W_0, W_0)}{D_1(W_0, W_0)}\left[1 - \frac{D_2(E_\mathrm{F}, W_0)}{E_2(W_0, W_0)}\right] \Big/ \left[1 - \frac{D_1(E_\mathrm{F}, W_0)}{D_1(W_0, W_0)}\right] \ , \tag{4.153}$$

where

$$D_1(x, y) = \tfrac{1}{5}(x^5 - 1) - \tfrac{1}{2}(x^4 - 1)y + \tfrac{1}{3}(x^3 - 1)y^2$$

and

$$D_2(x, y) = -\tfrac{1}{6}(x^6 - 1) + \tfrac{3}{5}(x^5 - 1)y - \tfrac{3}{4}(x^4 - 1)y^2 + \tfrac{1}{3}(x^3 - 1)y^3 \ .$$

Here W_0 is the threshold energy for the beta transition, E_F is the electron Fermi energy given in Eqs. (4.142), and W_0 and E_F are assumed to be in units of mc^2.

Under electron-degenerate conditions, the rate of the ordinary URCA process is only significant if the electron Fermi energy is near the electron capture threshold of the nucleus $(Z + 1, A)$, and then only if the phase space is made available by thermal rounding of the Fermi surface, by a vibration of it, or by a macroscopic transport of nuclei via convection. Because of this restriction, the URCA process of a given pair of nuclei proceeds significantly only in a shell within a stellar core. A complete discussion of the thermal and vibrational energy losses due to URCA shells in stellar interiors is given by Tsuruta and Cameron (1970). The rate of neutrino energy loss at finite temperature due to the presence of a URCA shell is given by

$$L_\nu = 4\pi r_s^2 X \left[\left|\frac{dE_F}{dr}\right|\right]^{-1} (F_1 \xi_0^5 + F_2 T^5) \ , \qquad (4.154)$$

where r_s is the radius of the URCA shell, X is the abundance by mass fraction of the nuclear pair under consideration, E_F is the Fermi energy,

$$\left|\frac{dE_F}{dr}\right| \approx 4 \times 10^{-8} \text{ MeV cm}^{-1} \ ,$$

T is the temperature, the relative amplitude of radial oscillation is $\xi_0 = \Delta r / r_0$, where Δr is the displacement from the equilibrium radius, r_0, and F_1 and F_2 are tabulated by Tsuruta and Cameron (1970) for 132 pairs of nuclei. The F_1 and F_2 are characteristic of a given pair of nuclei and are determined by the ft value, threshold energy, and charge. Tsuruta and Cameron show that in a stellar interior where URCA shells are present, the URCA neutrino energy losses dominate those of other neutrino energy losses in the temperature region up to about 2×10^9 °K. The effects of URCA shells on carbon ignition in degenerate stellar cores are discussed by Bruenn (1972), Paczynski (1972), Couch and Arnett (1974, 1975), Couch and Loumos (1975), Iben (1978).

Pinaev (1964) first pointed out that when the thermal energy, kT, becomes greater than the electron rest mass energy, mc^2, electron-positron pairs are produced, and the positron capture reaction

$$(Z-1, A) + e^+ \to (Z, A) + \bar{\nu}_e \qquad (4.155)$$

becomes important as well. Under equilibrium conditions, the neutrino energy loss when the reactions given in Eqs. (4.147), (4.148), and (4.155) are included, is given by (Pinaev, 1964)[2]

$$P_\nu \approx \frac{4 \times 10^{18}}{Aft} \rho \left[\frac{Q + mc^2}{mc^2}\right]^2 \left(\frac{kT}{mc^2}\right)^4 \exp\left(-\frac{Q + mc^2}{kT}\right) \text{erg cm}^{-3} \text{ sec}^{-1} \ , \qquad (4.156)$$

for $kT \ll mc^2, kT \ll Q$, and

$$P_\nu \approx \frac{0.8 \times 10^{20}}{Aft} \rho \left(\frac{kT}{mc^2}\right)^6 \text{erg cm}^{-3} \text{ sec}^{-1} \ , \qquad (4.157)$$

---

[2] In Pinaev's (1964) work and also in a number of works of that time, the overestimated cross sections of electron and positron captures were used. This error has been revealed by V. S. Imshennik, D. K. Nadyozhin and V. S. Pinaev (Astron. Zh., **44**, 768, 1977). Here we give the values for P_ν which are a factor of two larger than the correct values.

for $kT \gg mc^2, kT \gg Q$. Here ft is the experimental value, and Q is the threshold energy. Beaudet, Salpeter, and Silvestro (1972) have given more detailed formulae which include the effects of capture of both positrons and electrons by both beta stable and beta unstable nuclei, as well as normal beta decay of electrons and positrons. They also give simple fitting formulae for all of these transitions as a function of mass density and temperature; and calculate the overall URCA energy loss rates for a number of common stable even-even nuclei under equilibrium conditions.

When there is an abundance of free protons, p, and neutrons, n, the appropriate URCA reactions are

$$\mathrm{p} + \mathrm{e}^- \rightarrow \mathrm{n} + \nu_e$$

$$\mathrm{n} + \mathrm{e}^+ \rightarrow \mathrm{p} + \bar{\nu}_e$$

and

$$\mathrm{n} \rightarrow \mathrm{p} + \mathrm{e}^- + \bar{\nu}_e \ . \tag{4.158}$$

Hansen (1968) has extended Pinaev's (1964) results to take into account electron degeneracy, partial electron degeneracy, and electron-positron pair formation. For electron capture on protons, Hansen obtains a capture rate

$$\lambda = 4.7 \times 10^{-4}\left(\frac{u^5}{5} + 8 \times 10^{-3}u^3T_9^2 + 3.2 \times 10^{-3}T_9^5\right) \text{ sec}^{-1} \text{ per proton} , \tag{4.159}$$

and a neutrino luminosity of

$$L_\nu = 3.1 \times 10^{-10}\left(\frac{u^6}{6} + 6 \times 10^{-2}u^4T_9^2 + 3.2 \times 10^{-3}T_9^6\right) \text{ erg sec}^{-1} \text{ per proton} \tag{4.160}$$

for $T_9 \gtrsim 10$ and $\mu \gtrsim kT$. Here $T_9 = T/10^9$, the chemical potential is μ, and

$$u = \frac{\mu}{mc^2} = \text{lesser of } 10^{-2}\left(\frac{\rho}{\mu_e}\right)^{1/3}, 6 \times 10^{-6}\left(\frac{\rho}{\mu_e}\right)T_9^{-2} \ . \tag{4.161}$$

For positron capture on neutrons,

$$\lambda = 10^{-6}T_9^5 \exp\left[-5.8\frac{(u+1)}{T_9}\right] \quad \text{sec}^{-1} \text{ per neutron} , \tag{4.162}$$

and

$$L_\nu = 7 \times 10^{-13}T_9^6 \exp\left[-5.8\frac{(u+1)}{T_9}\right] \quad \text{erg sec}^{-1} \text{ per neutron} \tag{4.163}$$

for $T_9 \gtrsim 5$. For the noninteracting gas model, a zero temperature neutron star would (Bahcall and Wolf, 1965) have

$$N_\mathrm{n} \approx 2 \times 10^{38}\left(\frac{\rho}{\rho_\mathrm{nucl}}\right) \text{ cm}^{-3} , \tag{4.164}$$

and

$$N_\mathrm{e} = N_\mathrm{p} \approx 2 \times 10^{36}\left(\frac{\rho}{\rho_\mathrm{nucl}}\right)^2 \text{ cm}^{-3} , \tag{4.165}$$

for $\rho \lesssim 2\rho_{\rm nucl}$. Here $N_{\rm n}$, $N_{\rm e}$, and $N_{\rm p}$ denote, respectively, the number densities of neutrons, electrons, and protons, the gas mass density is ρ, and the nuclear matter density $\rho_{\rm nucl} \approx 3.7 \times 10^{14}$ g cm^{-3}. Arnett (1967) applies Pinaev's analysis to the reactions given in Eq. (4.158) to obtain a total neutrino emissivity, Q_ν, given by[3]

$$Q_\nu = 8.1 \times 10^{16} X_n \left(\frac{T_9}{6}\right)^6 \text{ erg g}^{-1} \text{ sec}^{-1} \ , \tag{4.166}$$

where $T_9 = T/10^9$, and X_n is related to the number density of protons, $N_{\rm p}$, and neutrons, $N_{\rm n}$, by

$$X_{\rm n} = \frac{(N_{\rm p} + N_{\rm n})}{6 \times 10^{23} \rho} \ . \tag{4.167}$$

Fassio-Canuto (1969) has given the beta decay rate in a strong magnetic field, and Canuto and Chou (1971) have discussed the URCA energy loss rates in an intense magnetic field of strength, H. The results are given in terms of the parameter

$$\theta = \frac{H}{H_q} = \frac{e\hbar H}{m^2 c^3} = \frac{H}{4.414 \times 10^{13} \text{ Gauss}} \ . \tag{4.168}$$

They give detailed formulae for the neutrino luminosity, L_H, and the beta decay and electron capture reaction rates under the conditions of statistical equilibrium and charge neutrality. Analytic approximations to these formulae show that when a neutron-proton gas is considered and $\theta \gg 1$, the neutrino luminosity L_H, from a nondegenerate gas may be up to 100 times the neutrino luminosity, L_0, when the field is absent. For a degenerate neutron-proton gas, however, $L_H \approx 10^{-2} L_0$ for $\theta \gtrsim 1$. When a degenerate gas of nuclei is considered, it is found that the URCA energy loss rates are reduced up to a factor of 10^{-3} for $\theta \approx 1$ and $T \lesssim 10^9$ °K.

The interior of a neutron star has a large amount of free neutrons in equilibrium with a small amount of protons and electrons. Under these conditions, the neutrons can scatter and undergo the modified URCA reactions (Chiu and Salpeter, 1964)

$$\rm n + n \rightarrow n + p + e^- + \bar{\nu}_e \tag{4.169}$$

and

$$\rm n + p + e^- \rightarrow n + n + \nu_e \ . \tag{4.170}$$

If the neutron Fermi energy is greater than the muon rest energy, $m_\mu c^2$, muon neutrinos will also be formed according to the modified URCA reactions

$$\rm n + n \rightarrow n + p + \mu^- + \bar{\nu}_\mu \tag{4.171}$$

and

$$\rm n + p + \mu^- \rightarrow n + n + \nu_\mu \ . \tag{4.172}$$

If quasi-free pions are also present in neutron matter, there are the additional URCA type reactions

[3] The numerical coefficient in (4.166) is overestimated (see footnote 2). The correct value is equal to 3.6×10^{16} (Imshennik et al., 1977).

$$\begin{aligned} &\pi^- + \mathrm{n} \rightarrow \mathrm{n} + \mathrm{e}^- + \bar{\nu}_\mathrm{e} \\ &\mathrm{n} + \mathrm{e}^- \rightarrow \mathrm{n} + \pi^- + \nu_\mathrm{e} \end{aligned} \tag{4.173}$$

and

$$\begin{aligned} &\pi^- + \mathrm{n} \rightarrow \mathrm{n} + \mu^- + \bar{\nu}_\mu \\ &\mathrm{n} + \mu^- \rightarrow \mathrm{n} + \pi^- + \nu_\mu \ . \end{aligned} \tag{4.174}$$

Bahcall and Wolf (1965) assume that the neutron gas is not superfluid and that the temperature and density are constant over the star. The independent particle model was then used to obtain the neutrino luminosities

$$L_\nu = 6 \times 10^{38} \left(\frac{M}{M_\odot}\right) \left(\frac{\rho_{\mathrm{nucl}}}{\rho}\right)^{1/3} T_9^8 \quad \mathrm{erg\ sec}^{-1} \ , \tag{4.175}$$

for the modified URCA reactions given in Eqs. (4.169) and (4.170),

$$\begin{aligned} L_\nu &= 6 \times 10^{38} \left[1 - 2.25\left(\frac{\rho_{\mathrm{nucl}}}{\rho}\right)^{4/3}\right]^{1/2} \left(\frac{M}{M_\odot}\right) \left(\frac{\rho_{\mathrm{nucl}}}{\rho}\right)^{1/3} T_9^8 \quad \mathrm{erg\ sec}^{-1} \\ &\qquad\qquad\qquad\qquad\qquad\qquad\qquad\qquad \text{for } \rho > 1.8\ \rho_{\mathrm{nucl}} \ , \\ &= 0 \qquad \text{for } \rho < 1.8\ \rho_{\mathrm{nucl}} \ , \end{aligned} \tag{4.176}$$

for the modified URCA reactions given in Eqs. (4.171) and (4.172), and

$$L_\nu = 10^{46}\ T_9^6 \left(\frac{N_\pi}{N_b}\right) \left(\frac{M}{M_\odot}\right) \quad \mathrm{erg\ sec}^{-1} \ , \tag{4.177}$$

for the pion reactions given in Eqs. (4.173) and (4.174). Here M is the neutron star mass, the solar mass is $M_\odot$, the neutron star mass density is ρ, the density of nuclear matter $\rho_{\mathrm{nucl}} \approx 3.7 \times 10^{14}$ g cm^{-3}, T_9 is the temperature of the stellar core in units of 10^9 °K, and N_π/N_b is the ratio of the number density of quasi-free pions to the number density of baryons. For comparison, for a neutron star of effective surface temperature, T_e, the photon luminosity, L_γ, is given by

$$L_\gamma = 7 \times 10^{36}\ T_{\mathrm{e}7}^4 R_{10}^2 \quad \mathrm{erg\ sec}^{-1} \ , \tag{4.178}$$

where $T_{\mathrm{e}7} = T_\mathrm{e}/10^7$, and R_{10} is the radius of the star in units of 10 km.

Bahcall and Wolf (1965) also consider the cooling times for the loss of thermal energy from a neutron star in statistical equilibrium. Models for neutron star cooling are often based upon neutrino emission from the interior that is dominated by the modified URCA reactions given in Eqs. (4.169, 4.170). Lattimer *et al.* (1991) showed that the direct URCA process, given by

$$\begin{aligned} &\mathrm{n} \rightarrow \mathrm{p} + \mathrm{e}^- + \bar{\nu}_\mathrm{e} \\ &\mathrm{p} + \mathrm{e}^- \rightarrow \mathrm{n} + \nu_\mathrm{e} \end{aligned} \tag{4.179}$$

can occur in neutron stars if the proton concentration exceeds a critical value of 11–15%, enhancing neutrino emission and neutron star cooling rates. Neutron star cooling has been reviewed by Pethick (1992).

Finzi (1965) pointed out that the exponential light curve of Type I supernovae might be due to the dissipation of the vibrational energy of a neutron star by the URCA process. When the neutron star is vibrating, equilibrium conditions are not satisfied, and the rate of dissipation of vibrational energy per unit mass, dw/dt, is given by (Finzi and Wolf, 1968)

$$\frac{dw}{dt} = K - Q_\nu \ , \tag{4.180}$$

where the neutrino emissivity due to the modified URCA process given in Eqs. (4.169) and (4.170) is

$$Q_\nu = 3 \times 10^5 \left(\frac{\rho_{\text{nucl}}}{\rho}\right) T_9^8 \quad \text{erg g}^{-1} \text{ sec}^{-1} \ ,$$

and the constant K, ranges from 10^{-1} to 10^3 as the ratio of the chemical potential, μ, to the thermal energy, kT, ranges from one to sixteen.

4.3.5 Neutrino Pair Emission

As first suggested by Pontecorvo (1959), neutrino pair emission may rapidly accelerate the evolution of a star in its later stages. The major neutrino pair emission processes were outlined by Fowler and Hoyle (1964). Neutrino processes in stellar interiors have been reviewed by Barkat (1975). The neutrinos usually escape directly from a star, robbing it of thermal energy and thus cooling it. The cooling of neutron stars was reviewed by Pethick (1992).

Three important neutrino processes occur during advanced stellar evolution, at high mass densities ($\rho = 10^3$ to 10^{10} g cm^{-3}) and temperatures ($T = 10^8$ to 10^{10} °K). They are the production of neutrinos during electron-positron annihilation (pair annihilation), during photon-electron collisions (photoneutrino process) and during photon-plasma interaction (plasma process). Neutrinos can also be produced at very high mass densities during bremsstrahlung (electron-ion interaction). The mass densities, ρ, and temperatures, T, at which different types of neutrino energy losses dominate are given in Fig. 4.3.

In very hot environments ($T > 10^9$ °K), there are enough energetic photons to create large numbers of electron-positron (e^-, e^+) pairs. These pairs are usually annihilated to produce photons, γ, via the reaction $e^- + e^+ \rightarrow \gamma + \gamma$, but the pair annihilation can sometimes result in a significant number of neutrino-antineutrino pairs at very high temperatures $T \geq 10^9$ °K and relatively low mass densities $\rho \leq 10^6$ g cm^{-3}. When the temperature is too low for significant e^-, e^+ pair production, but the density is still relatively low, neutrinos can be produced during the interaction of a photon and an electron; this photoneutrino process is the analogue of the normal Compton scattering in which a photon is scattered by an electron. When the stellar core density becomes large, $\rho \approx 10^8$ g cm^{-3}, the photons interact with the plasma producing a plasmon that can decay to form neutrinos. At very high mass densities $\rho \approx 10^{10}$ g cm^{-3} and relatively low temperatures $T \approx 10^8$ °K, neutrinos are also produced during bremsstrahlung, the braking radiation emitted by electrons interacting with ions.

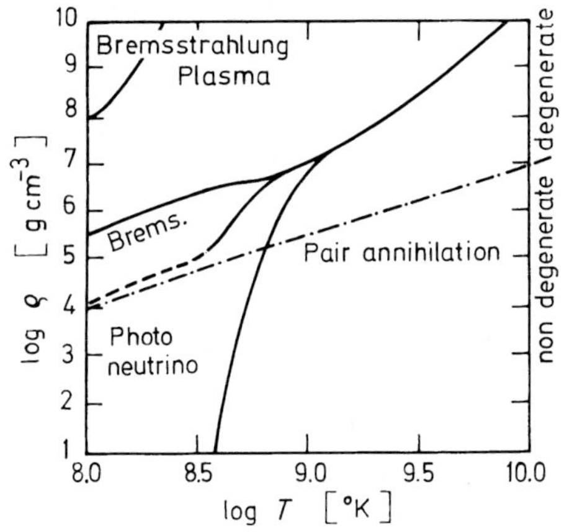

Fig. 4.3. Dominant regimes for various neutrino pair emission processes. The gas density and temperature are denoted, respectively, by ρ and T. The neutrino bremsstrahlung is calculated assuming the nuclear charge Z = 26, and lattice corrections are neglected after Festa and Ruderman (1969)

4.3.5.1 Neutrino Bremsstrahlung and Neutrino Synchrotron Radiation

As first suggested by Pontecorvo (1959), neutrinos will be emitted by bremsstrahlung radiation according to the reaction

$$e^- + (Z, A) \rightarrow e^- + (Z, A) + \nu_e + \bar{\nu}_e \ . \tag{4.181}$$

Here electrons, e^-, collide with nuclei, (Z, A), to emit an electron, e^-, together with a neutrino, ν_e, and antineutrino, $\bar{\nu}_e$, pair. This process is analogous to ordinary photon bremsstrahlung with the neutrino pair replacing the usual photon emitted in inelastic electron scattering. Gandelman and Pinaev (1960) have considered the free–free neutrino bremsstrahlung of a nondegenerate gas, and obtain an effective cross section

$$\sigma = \sigma_0 Z^2 \left(\frac{E_e}{mc^2}\right)^3 , \tag{4.182}$$

where E_e is the kinetic energy of the incident electron, and

$$\sigma_0 = \frac{8r_0^2}{525\pi^3}\left[\frac{g}{mc^2}\left(\frac{\hbar}{mc}\right)^3\right]^2 = 3.52 \times 10^{-52}\ \mathrm{cm}^2 \ . \tag{4.183}$$

Here r_0 is the classical electron radius, and the weak interaction coupling constant $g \approx 1.41 \times 10^{-49}$ erg cm^{-3}. If the electrons are assumed to have a Maxwellian distribution, and the nondegenerate gas is fully ionized, the electron neutrino luminosity density, P_ν, is given by

$$P_\nu = 2.75 \times 10^{-10} \frac{\rho^2}{\mu_e \nu} T^{4.5}\ \mathrm{erg\ cm^{-3}\ sec^{-1}} , \tag{4.184}$$

where ρ is the gas mass density, the temperature, T, is in keV (1 keV = 1.1605 × 10^7 °K), and the mean electron molecular weight, μ_e, and the factor, ν, are given by

$$\mu_e^{-1} = \sum_i \frac{X_i Z_i}{A_i} \text{ and } \nu^{-1} = \sum_i X_i \frac{Z_i^2}{A_i} \ , \tag{4.185}$$

where X_i, Z_i, and A_i are, respectively, the mass fraction, nuclear charge, and atomic weight of the nucleus of species, i. When the electrons have a Fermi–Dirac distribution and are nearly degenerate,

$$P_\nu \approx 0.82 \times 10^{-7} \frac{\rho}{\nu} T^6 \ln\left(0.89 \frac{E_0}{T}\right) \quad \text{erg cm}^{-3} \text{ sec}^{-1} \ , \tag{4.186}$$

where T is in keV, and

$$E_0 = 5.07 \times 10^{-5} mc^2 \left(\frac{\rho}{\mu_e}\right)^{2/3} \ ,$$

where $mc^2 = 511$ keV should be used if E_0 is to be expressed in keV.

The neutrino bremsstrahlung of a relativistic degenerate gas of electrons Coulomb scattering on nuclei has been considered by Festa and Ruderman (1969). For a random gas of nuclei, the neutrino emissivity, Q_ν, is given by

$$\begin{aligned} Q_\nu &= 20 g^2 \frac{Z^2 e^4}{\pi^5} (kT)^6 \left\{ B_1 B_2 - (\beta^2 - 1) B_3 \left[\frac{2}{3} + \frac{1}{2} \right.\right. \\ &\qquad \left.\left. \times \left(\beta - \beta \frac{(\beta^2+1)}{2} \ln \frac{\beta+1}{\beta-1} \right) \right] \right\} \quad \text{erg nucleus}^{-1} \text{ sec}^{-1} \\ &\approx 0.76 \frac{Z^2}{A} \left(\frac{T}{10^8 \, {}^\circ\mathrm{K}} \right)^6 \quad \text{erg g}^{-1} \text{ sec}^{-1} \quad \text{for } \rho \to \infty \ . \end{aligned} \tag{4.187}$$

Here $\beta = E_F/P_F$, where E_F and P_F denote, respectively, the Fermi energy and momentum, and the factors B_1, B_2, and B_3 are given by

$$B_1 = (1 - \beta^2) \left[\frac{1}{2} \beta \ln\left(\frac{\beta+1}{\beta-1} \right) - 1 \right] + \frac{1}{3} \ ,$$

$$B_2 = -4 + 2(1 + \alpha^2) \ln\left(\frac{2+\alpha^2}{\alpha^2} \right) \ ,$$

and

$$B_3 = \ln\left(\frac{2+\alpha^2}{\alpha^2} \right) - \frac{2}{2+\alpha^2} \ ,$$

where $\alpha^2 = \beta/215$. When the nuclei are arranged in a zero temperature, rigid lattice, Q_ν is reduced at low Z going to zero at $Z = 1$ and to $\gtrsim 0.4 \ Q_\nu$ for $Z \gtrsim 25$. This suppression is less, however, when the temperature is finite. As it was illustrated in Fig. 4.3, neutrino pair emission by neutrino bremsstrahlung dominates over other pair emission processes at high density and moderate temperatures. The other processes are suppressed at high densities because of the absence of accessible unoccupied electron states, and because of the increase in plasma frequency. At low temperatures, the normally dominant photoneutrino emissivity decreases as T^9, whereas the neutrino bremsstrahlung emissivity decreases as T^6, making the bremsstrahlung dominant in the moderate density-temperature region shown in Fig. 4.3.

Pinaev (1964) has calculated the neutrino bremsstrahlung for recombination on the K shell of an atom. The total cross section, σ, for recombination on the K shell is given by

$$\sigma = \sigma_0 Z^5 \frac{c}{v}(E_e + I) \ ,$$

where

$$\sigma_0 = \frac{4\alpha^5}{15\pi^2}\frac{g^2 m^2}{\hbar^4} = 0.76 \times 10^{-56}\ \mathrm{cm}^2 \ , \qquad (4.188)$$

the kinetic energy of the electron, E_e, and I, are in units of mc^2, the velocity of the initial electron is v, and the ionization potential of the K electron is $I = \alpha^2 Z^2 mc^2/2$, where α is the fine structure constant. For a nondegenerate gas with a Maxwellian electron distribution, the free–bound neutrino luminosity density is

$$P_\nu = 1.54 \times 10^{-8} Z^6 A^{-2} \rho^2 T^2 (1 - f) \quad \mathrm{erg\ cm^{-3}\ sec^{-1}} \ , \qquad (4.189)$$

where Z is the nuclear charge, A is the atomic weight, ρ is the gas mass density, T is the temperature in keV, and

$$f = \left[1 + 320\frac{AT^{3/2}}{Z\rho}\right]^{-1} \ .$$

Equation (4.189) holds for $kT \gg I$.

For a degenerate electron gas, the free–bound neutrino luminosity density is

$$P_\nu = 1.45 \times 10^{-13} Z^4 \left(\frac{Z\rho}{A}\right)^{10/3} \exp\left(-\frac{E_F}{T}\right) \quad \mathrm{erg\ \ cm^{-3}\ sec^{-1}} \ . \qquad (4.190)$$

Equation (4.190) holds for $E_F \gg kT \gg I$.

Beaudet, Petrosian and Salpeter (1967) noted the possible importance of neutrino energy loss in stellar interiors by the recombination neutrino process. Kohyama, Itoh, Obama and Mutoh (1993) showed that the recombination neutrino process is dominant for relatively large Z-values and relatively low densities and low temperatures. However, the neutrino energy loss rates are more than 1.5 orders of magnitude lower than previously supposed, so the influence of the neutrino recombination process on stellar evolution is limited.

Neutrino bremsstrahlung of an electron accelerated in an intense magnetic field was first calculated by Landstreet (1967). For a magnetic field of strength, H, the neutrino synchrotron luminosity density is given by (Landstreet, 1967; Canuto, Chiu, Chou, and Fassio–Canuto, 1970)

$$P_\nu = 3 \times 10^{-44} H_8^6 T_7 \rho^4 \quad \mathrm{erg\ cm^{-3}\ sec^{-1}} \quad \text{if } H_8 \rho^{2/3} \lesssim 8 \times 10^6 T_7$$

and

$$P_\nu = 4 \times 10^{-7} T_7^{19/3} \rho^{4/9} H_8^{2/3} \quad \mathrm{erg\ cm^{-3}\ sec^{-1}} \quad \text{if } H_8 \rho^{2/3} \gtrsim 8 \times 10^6 T_7 \ , \qquad (4.191)$$

for relativistic electrons in a degenerate gas of mass density, ρ, and temperature, T. Here $H_8 = H/10^8$ and $T_7 = T/10^7$. For a nondegenerate, nonrelativistic gas

$$P_\nu = 2 \times 10^{-46} H_8^6 N_e \quad \text{erg cm}^{-3}\ \text{sec}^{-1} \ , \tag{4.192}$$

whereas for a degenerate, nonrelativistic gas

$$P_\nu = 1 \times 10^{-28} T_7 H_8^6 N_e^{1/3} \quad \text{erg cm}^{-3}\ \text{sec}^{-1} \ , \tag{4.193}$$

where N_e is the electron number density. Canuto, Chiu, Chou, and Fassio-Canuto (1970) derive general expressions for the neutrino synchrotron radiation of a relativistic electron gas, and also give numerical approximations. These data agree with Landstreet's formulae at high densities, but disagree at lower densities.

4.3.5.2 Electron Positron (Pair) Annihilation Neutrinos

The emission of neutrinos by the annihilation of electrons, e^-, and positrons, e^+, according to the reaction

$$e^- + e^+ \rightarrow \nu_e + \bar{\nu}_e \ , \tag{4.194}$$

was considered by Chiu and Morrison (1960), Chiu (1961), Chiu and Stabler (1961) and Chiu (1968). The neutrino luminosity density, P_ν, is given by

$$P_\nu = 4.90 \times 10^{18} T_9^3 \exp\left(-\frac{11.86}{T_9}\right) \quad \text{erg cm}^{-3}\ \text{sec}^{-1}$$

(nondegenerate, nonrelativistic)

$$P_\nu = 4.22 \times 10^{15} T_9^9 \quad \text{erg cm}^{-3}\ \text{sec}^{-1} \quad \text{(nondegenerate, relativistic)}$$

$$P_\nu = 1.93 \times 10^{13} \left(\frac{\rho}{\mu_e}\right) T_9^{3/2} \quad \exp\left[-\frac{(E_F + mc^2)}{kT}\right] \quad \text{erg cm}^{-3}\ \text{sec}^{-1}$$

(degenerate, nonrelativistic) (4.195)

and

$$P_\nu = 1.44 \times 10^{11} \left(\frac{\rho}{\mu_e}\right) \left(\frac{E_F}{mc^2}\right)^2 T_9^4 \left[1 + \frac{5kT}{E_F}\right] \quad \exp\left(-\frac{E_F}{kT}\right) \quad \text{erg cm}^{-3}\ \text{sec}^{-1} \ ,$$

(degenerate, relativistic)

where the gas mass density is ρ, the gas temperature, T, is in °K, the factor $T_9 = T/10^9$, the gas is degenerate if

$$\rho > 2.4 \times 10^{-8} T^{3/2} \mu_e \quad \text{g cm}^{-3} \quad \text{(degenerate)} \ , \tag{4.196}$$

and it is relativistic if

$$\rho > 7.3 \times 10^6 \mu_e \quad \text{g cm}^{-3} \quad \text{(relativistic)} \ , \tag{4.197}$$

the mean molecular weight per electron is

$$\mu_e^{-1} = \frac{N_e}{\rho N_A} = \sum \frac{X_Z Z}{A_Z} \ , \tag{4.198}$$

where N_e is the electron density, Avogadro's number $N_A \approx 6.022 \times 10^{23}\,(\text{mole})^{-1}$, the mass fraction is X_Z for an element whose atomic number is Z, the element mass number is A, and $E_F = \varepsilon_F + mc^2$ where the Fermi energy, ε_F, of a completely degenerate gas is given by

$$\varepsilon_F = mc^2 \left\{ \left[1.018 \times 10^{-4} \left(\frac{\rho}{\mu_e} \right)^{2/3} + 1 \right]^{1/2} - 1 \right\}$$

$$= 0.509 \times 10^{-4} \left(\frac{\rho}{\mu_e} \right)^{2/3} m_e c^2 \quad \text{(nonrelativistic)}$$

$$= 1.009 \times 10^{-2} \left(\frac{\rho}{\mu_e} \right)^{1/3} m_e c^2 \quad \text{(relativistic)} \ . \qquad (4.199)$$

These equations can be used to give a pair annihilation neutrino emissivity $Q_\nu = P_\nu / \rho$, or

$$Q_\nu = \frac{4.8}{\rho} \left(\frac{T}{10^3} \right)^3 \exp\left(-\frac{2mc^2}{kT} \right) \quad \text{erg g}^{-1}\,\text{sec}^{-1} \quad \text{for } kT \ll mc^2$$

$$\text{(nondegenerate)} \qquad (4.200)$$

$$Q_\nu = \frac{4.3 \times 10^{24}}{\rho} \left(\frac{T}{10^{10}} \right)^9 \quad \text{erg g}^{-1}\,\text{sec}^{-1} \quad \text{for } kT \gg mc^2$$

$$\text{(nondegenerate)}$$

$$Q_\nu = \frac{4.5}{\mu_e} \left(\frac{T}{10^4} \right)^{3/2} \exp\left[-\frac{(E_F + mc^2)}{kT} \right] \quad \text{erg g}^{-1}\,\text{sec}^{-1} \text{ for } kT \ll mc^2; E_F \ll mc^2$$

$$\text{(degenerate)}$$

and

$$Q_\nu = \frac{0.14}{\mu_e} \left(\frac{T}{10^6} \right)^4 \left(\frac{E_F}{mc^2} \right)^2 \exp\left[-\frac{E_F}{kT} \right] \quad \text{erg g}^{-1}\,\text{sec}^{-1} \quad \text{for } kT \gg mc^2$$

$$\text{(degenerate)}$$

A comparison of the neutrino energy loss rates of the pair annihilation process and the following photoneutrino and plasma neutrino processes is given by Beaudet, Petrosian, and Salpeter (1967). Stellar electron and positron emission rates and continuum electron and positron capture rates, as well as associated neutrino energy loss rates were considered by Fuller, Fowler and Newman (1980, 1982).

4.3.5.3 Photoneutrino Process

The emission of neutrinos by the collision of a photon, γ, and an electron, e^-, according to the photoneutrino reaction

$$\gamma + e^- \rightarrow e^- + \nu_e + \bar{\nu}_e \qquad (4.201)$$

was first considered by Chiu and Stabler (1961) and Ritus (1962). Petrosian, Beaudet, and Salpeter (1967) consider the neutrino energy loss rate due to photoneutrino processes in a hot plasma, including the contribution of positrons present in the black body radiation. They obtain the photoneutrino luminosity densities given by

$$P_\nu = 0.976 \times 10^8 T_9^8 \left(\frac{\rho}{\mu_e}\right) \quad \text{erg cm}^{-3}\,\text{sec}^{-1}$$

(nondegenerate, nonrelativistic)

$$P_\nu = 1.477 \times 10^{13} T_9^9 [\log T_9 - 0.536] \quad \text{erg cm}^{-3}\,\text{sec}^{-1}$$

(nondegenerate, relativistic) (4.202)

$$P_\nu = 0.976 \times 10^8 T_9^9 \left(\frac{\rho}{\mu_e}\right) \left[\frac{(\rho/\mu_e)}{3.504 \times 10^5}\right]^{-2/3} \quad \text{erg cm}^{-3}\,\text{sec}^{-1}$$

(degenerate, nonrelativistic)

and

$$P_\nu = 1.514 \times 10^{13} T_9^9 \quad \text{erg cm}^{-3}\,\text{sec}^{-1} \text{ (degenerate, relativistic)} ,$$

where the conditions for degeneracy and the relativistic criterion are given in the previous section, $T_9 = T/10^9$, ρ is the gas mass density, and μ_e is the mean molecular weight per electron. For most applications, the photoneutrino luminosity density can be calculated from the expression (Petrosian, Beaudet, and Salpeter, 1967)

$$P_\nu = 1.103 \times 10^{13} T_9^9 \exp\left(-\frac{5.93}{T_9}\right) + 0.976 \times 10^8 T_9^8 (1 + 4.2 T_9)^{-1} \left(\frac{\rho}{\mu_e}\right)$$
$$\times \left[1 + \frac{6.446 \times 10^{-6} \rho}{\mu_e T_9 (1 + 4.2 T_9)}\right]^{-1} \quad \text{erg cm}^{-3}\,\text{sec}^{-1} . \tag{4.203}$$

A comparison of energy losses due to the pair annihilation, photoneutrino, and plasma neutrino processes is given by Beaudet, Petrosian, and Salpeter (1967).

These equations can be used to give a photoneutrino emissivity $Q_\nu = P_\nu/\rho$, or

$$Q_\nu \approx \mu_e^{-1} \left(\frac{T}{10^8}\right)^8 \quad \text{erg g}^{-1}\,\text{sec}^{-1} \quad \text{for } kT \ll mc^2$$

(nondegenerate, nonrelativistic)

$$Q_\nu \approx \frac{2.5 \times 10^{14}}{\mu_e} \left(\frac{T}{10^{10}}\right)^8 \left[\log\left(\frac{T}{10^{10}}\right) + 1.6\right] \quad \text{erg g}^{-1}\,\text{sec}^{-1} \quad \text{for } kT \gg mc^2$$

(nondegenerate, relativistic)

$$Q_\nu \approx 1.5 \times 10^2 \left(\frac{T}{10^8}\right)^9 (\mu_e \rho^2)^{-1/3} \quad \text{erg g}^{-1}\,\text{sec}^{-1} \quad \text{for } kT \ll mc^2$$

(degenerate, nonrelativistic) (4.204)

and

$$Q_\nu \approx \frac{6.3 \times 10^6}{\mu_e} \left[1 + 5\left(\frac{T}{10^9}\right)^2\right] \left(\frac{T}{10^9}\right)^7 \left(\frac{mc^2}{E_F}\right)^3 \quad \text{erg g}^{-1}\,\text{sec}^{-1}$$

for $kT \gg mc^2$ (degenerate, relativistic) .

4.3.5.4 Plasma Neutrino Process

When a photon propagates in an ionized gas, it creates virtual electron-hole pairs and behaves as if it had a rest mass, M, given by

$$M = \frac{\hbar\omega_p}{c^2} \ , \tag{4.205}$$

where the plasma frequency, ω_p, is given by

$$\omega_p^2 = \frac{4\pi N_e e^2}{m} \quad \text{(nondegenerate)}$$

$$\omega_p^2 = \frac{4\pi N_e e^2}{m}\left[1+\left(\frac{\hbar}{mc}\right)^2 (3\pi^2 N_e)^{2/3}\right]^{-1/2} \quad \text{(degenerate)} \ . \tag{4.206}$$

Such a particle, called a plasmon, may decay and emit neutrinos according to the reaction $\Gamma \rightarrow \nu_e + \bar{\nu}_e$. The plasmon, Γ, may propagate in both the longitudinal and transverse modes giving rise to respective neutrino luminosity densities, $P_{l\nu}$ and $P_{t\nu}$ given by (Adams, Ruderman, and Woo, 1963; Inman and Ruderman, 1964)

$$P_{l\nu} = 1.224 \times 10^{13} T_9^9 x^9 (e^x - 1)^{-1} \quad \text{erg cm}^{-3}\,\text{sec}^{-1} \ , \tag{4.207}$$

$$\begin{aligned} P_{t\nu} &= 3.214 \times 10^{14} T_9^9 x^9 F(x) \quad \text{erg cm}^{-3}\,\text{sec}^{-1} \\ &\approx 7.4 \times 10^{21}\left(\frac{\hbar\omega_p}{mc^2}\right)^6 \left(\frac{mc^2}{kT}\right)^{-3} \quad \text{erg cm}^{-3}\,\text{sec}^{-1} \quad \text{for } x \ll 1 \\ &\approx 3.85 \times 10^{21}\left(\frac{\hbar\omega_p}{mc^2}\right)^{7.5}\left(\frac{mc^2}{kT}\right)^{-1.5} \exp\left(-\frac{\hbar\omega_p}{kT}\right) \quad \text{erg cm}^{-3}\,\text{sec}^{-1} \\ &\qquad \text{for } x \gg 1 \ , \end{aligned} \tag{4.208}$$

where

$$\begin{aligned} x = \frac{\hbar\omega_p}{kT} &= \left\{\frac{4\pi e^2 \rho \hbar^2}{m_p m \mu_e (kT)^2}\left[1+\left(\frac{\hbar}{mc}\right)^2\left(\frac{3\pi^3\rho}{m_p\mu_e}\right)^{2/3}\right]^{1/2}\right\}^{1/2} \\ &\approx \frac{3.345 \times 10^{-4}(\rho/\mu_e)^{1/2}}{T_9[1+1.0177\times 10^{-4}(\rho/\mu_e)^{2/3}]^{1/4}} \ , \end{aligned}$$

and $F(x) = \sum_{n=1}^{\infty}[K_2(nx)/nx]$ where $K_2(nx)$ is the modified Bessel function of the second kind. Here we have incorporated the corrections given by Zaidi (1965), who showed that the constants of Adams, Ruderman, and Woo (1963) for the longitudinal and transverse luminosity density were off by the respective factors of 3/8 and 1/4. Beaudet, Petrosian, and Salpeter (1967) compare the neutrino luminosity due to pair annihilation, photoneutrinos, and the plasma neutrinos. They obtain the relations

$$\begin{aligned} \frac{P_{l\nu}}{P_{t\nu}} &= 0.0158\left(\frac{\hbar\omega_p}{kT}\right)^2 \quad \text{for } \hbar\omega_p \ll kT \\ &= 1.078 \quad \text{for } \hbar\omega_p \gg kT \ . \end{aligned} \tag{4.209}$$

The emission of plasmon neutrinos in a strong magnetic field has been given by Canuto, Chiuderi, and Chou (1970). The results are normalized in terms of the magnetic field parameter

$$H_q = \frac{m^2c^3}{e\hbar} = 4.414 \times 10^{13} \text{ Gauss} . \tag{4.210}$$

Although the transverse neutrino emission is not seriously altered unless the magnetic field strength, H, is $\gtrsim 10^{13}$ Gauss; the longitudinal plasmons propagate in a new mode and give rise to a neutrino luminosity density, $P_{l\nu}$, given by

$$P_{l\nu} = 14.8T\rho_6^{-2}\left(\frac{H}{H_q}\right)^6 \quad \text{erg cm}^{-3}\text{ sec}^{-1} , \tag{4.211}$$

where $T \geq 10^9$ °K is the temperature, and $\rho_6 = \rho/10^6$ where ρ is the gas mass density. If $P_{l\nu H}$ and $P_{l\nu 0}$ indicate the luminosity densities in the presence and absence of a magnetic field, respectively, $P_{l\nu H} \gg 10^{10} P_{l\nu 0}$ for $\rho > 10^{11}$ g cm^{-3}; whereas $P_{l\nu 0} \gg 10^5 P_{l\nu H}$ for $\rho < 10^{11}$ g cm^{-3} if $H = 10^{11}$ Gauss and $T \gtrsim 10^9$ °K.

4.3.5.5 Photocoulomb and Photon–Photon Neutrinos

Matinyan and Tsilosani (1962) and Rosenberg (1963) have discussed the neutrino-pair production by photons, γ, in the Coulomb field of a nucleus (Z, A) according to the reaction

$$\gamma + (Z, A) \rightarrow (Z, A) + \nu_e + \bar{\nu}_e , \tag{4.212}$$

where ν_e and $\bar{\nu}_e$ denote, respectively, the electron neutrino and antielectron neutrino. Rosenberg (1963) obtained the photocoulomb neutrino luminosity density, P_ν, given by

$$P_\nu \approx \frac{4.6 \times 10^8 \rho}{\nu}\left(\frac{kT}{mc^2}\right)^{10} \quad \text{erg cm}^{-3}\text{ sec}^{-1} \tag{4.213}$$

where $\nu^{-1} = \sum_i X_i Z_i^2/A_i = 5\text{ g}^{-1}$.

Chiu and Morrison (1960), Rosenberg (1963), and Van Hieu and Shabalin (1963) have discussed the conversion of a photon, γ, into a neutrino pair upon collision between two photons according to the reaction

$$\gamma + \gamma \rightarrow \gamma + \nu_e + \bar{\nu}_e . \tag{4.214}$$

Van Hieu and Shabalin obtained by means of detailed calculations a photon-photon neutrino luminosity density, P_ν, given by

$$P_\nu = 1.7 \times 10^{-28} T^{17} \quad \text{erg cm}^{-3}\text{ sec}^{-1} , \tag{4.215}$$

while according to a rough estimate by Rosenberg (1963) we have

$$P_\nu \approx 1.6 \times 10^{-20} T^{13} \quad \text{erg cm}^{-3}\text{ sec}^{-1} , \tag{4.216}$$

where T is in keV (1 keV $\approx 1.16 \times 10^7$ °K).

4.3.5.6 The Muon and Pion Neutrino Processes

At temperatures exceeding 10^{11} °K (or $kT \approx 50$ MeV), the radiation field of a star can create muon, μ, or pion, π, pairs which can subsequently annihilate to form neutrino-antineutrino pairs according to the reactions

$$\begin{aligned} \mu^- &\rightarrow e^- + \bar{\nu}_e + \nu_\mu \\ \mu^+ &\rightarrow e^+ + \nu_e + \bar{\nu}_\mu \ , \end{aligned} \tag{4.217}$$

and

$$\begin{aligned} \pi^- &\rightarrow \mu^- + \bar{\nu}_\mu \\ \pi^+ &\rightarrow \mu^+ + \nu_\mu \ . \end{aligned} \tag{4.218}$$

According to Arnett (1967), these processes may be important cooling mechanisms for highly evolved massive stars. The energy loss rate, Q_μ, for muon decay is given by

$$\begin{aligned} Q_\mu &= -\frac{2N_{\text{pair}}E_{\text{av}}}{\rho\tau_\mu} \\ &\approx -9.9 \times 10^{38} \frac{\exp(-\beta)}{\rho\beta^{3/2}} \quad \text{erg g}^{-1}\ \text{sec}^{-1} \ , \end{aligned} \tag{4.219}$$

where ρ is the mass density, the average decay energy, $E_{\text{av}} \approx 35$ MeV, the mean lifetime, $\tau_\mu = 2.197 \times 10^{-6}$ sec,

$$\beta \approx \frac{m_\mu c^2}{kT} \ ,$$

where $m_\mu c^2 \approx 105.658$ MeV is the muon neutrino rest mass energy. The number density of particle-antiparticle pairs, N_{pair}, is obtained by assuming equilibrium with the radiation field with $kT \ll mc^2$, and an equal concentration of particles and antiparticles

$$N_{\text{pair}} \approx \frac{1}{\sqrt{2}\pi^{3/2}} \left(\frac{mc}{\hbar}\right)^3 \frac{\exp(-\beta)}{\beta^{3/2}} \ . \tag{4.220}$$

The energy loss for pion decay, Q_π, is given by Eq. (4.219) with $E_{\text{av}} \approx 34$ MeV and $\tau_\pi \approx 2.60 \times 10^{-8}$ sec to obtain[4]

$$Q_\pi \approx -9.6 \times 10^{40} \frac{\exp(-\beta)}{\rho\beta^{3/2}} \quad \text{erg g}^{-1}\ \text{sec}^{-1} \ ,$$

where

$$\beta = \frac{m_\pi c^2}{kT} \ , \tag{4.221}$$

and $m_\pi c^2 \approx 139.570$ MeV is the rest mass energy for pions. Hansen (1968) has estimated the muon neutrino luminosity, L_μ, due to process $\mu^+ + \mu^- \rightarrow \nu_\mu + \bar{\nu}_\mu$, by assuming equal concentrations of μ^+ and μ^-. He obtains

$$\begin{aligned} L_\mu &\approx 4 \times 10^{32} T_9^3 \left[1 + 4.75 \times 10^{-3} T_9 + 6.5 \times 10^{-6} T_9^2\right] \\ &\quad \times \exp\left[-\frac{2.45 \times 10^3}{T_9}\right] \quad \text{erg cm}^{-3}\ \text{sec}^{-1} \ , \end{aligned} \tag{4.222}$$

[4] This equation is obtained by the change of muon momentum, as compared with pion momentum in reactions (4.218) being neglected. With due regard for this effect, the numerical coefficient in (4.221) becomes 8.4×10^{40} (G. V. Domogatsky, Preprint No. 96, Institute for Nuclear Research, Moscow, 1973).

for $50 \ll T_9 \ll 500$, where $T_9 = T/10^9$. Hansen (1968) also gives a lower limit to the muon neutrino luminosity due to the reaction $e^- + \mu^+ \to \nu_e + \bar{\nu}_\mu$.

4.3.6 Neutrino Opacities

Measured cross sections for the interaction of electron and mu neutrinos with matter are on the order of $10^{-44}\,\mathrm{cm}^2$ and were given in Eqs. (4.98) and (4.99). Neutrino absorption cross sections for all targets of interest in solar neutrino experiments have been developed by John Bahcall and his colleagues (Bahcall, 1989); the resulting computations are compared with the observed solar neutrino flux in the next Sect. 4.3.7. In the following we provide cross sections for neutrino–electron and neutrino–nucleon scattering. For the neutrino–electron and neutrino–muon scattering reactions given by

$$(\nu_e \text{ or } \bar{\nu}_e) + e^- \to (\nu_e' \text{ or } \bar{\nu}_e') + e^{-\prime}$$

and

$$(\nu_\mu \text{ or } \bar{\nu}_\mu) + \mu^- \to (\nu_\mu' \text{ or } \bar{\nu}_\mu') + \mu^{-\prime} \ . \tag{4.223}$$

Bahcall (1964) finds cross sections which are proportional to

$$\sigma_{0e} = \frac{4}{\pi}\left(\frac{\hbar}{mc}\right)^{-4}\left(\frac{g^2}{m^2c^4}\right) \approx 1.7 \times 10^{-44}\,\mathrm{cm}^2$$

and

$$\sigma_{0\mu} = \left(\frac{m_\mu}{m}\right)^2 \sigma_{0e} \approx 7.3 \times 10^{-40}\,\mathrm{cm}^2 \ , \tag{4.224}$$

where $g \approx 1.41 \times 10^{-49}$ erg cm^3 is the weak interaction coupling constant, and the subscripts e and μ denote, respectively, the scattering of electron neutrinos and mu neutrinos. For a neutrino of energy, E_ν, the cross sections for electron scattering in a nondegenerate gas are (Bahcall, 1964)

$$\begin{aligned}
\sigma &= \frac{\sigma_0}{2}\left(\frac{E_\nu}{mc^2}\right) \quad \text{for } E_\nu \gg mc^2 \quad \text{and} \quad kT \ll mc^2 \ , \\
\sigma &= 1.6\left(\frac{kT}{mc^2}\right)\sigma_0\left(\frac{E_\nu}{mc^2}\right) \quad \text{for } E_\nu \gg mc^2 \quad \text{and} \quad kT \gg mc^2 \ , \\
\sigma &= \sigma_0\left(\frac{E_\nu}{mc^2}\right)^2 \quad \text{for } E_\nu \ll mc^2 \quad \text{and} \quad kT \ll mc^2 \ ,
\end{aligned} \tag{4.225}$$

and

$$\sigma = 17\left(\frac{kT}{mc^2}\right)^2 \sigma_0\left(\frac{E_\nu}{mc^2}\right)^2 \quad \text{for } E_\nu \ll mc^2 \quad \text{and} \quad kT \gg mc^2 \ ,$$

where the rest mass energy of the electron, $mc^2 \approx 8.2 \times 10^{-7}$ erg ≈ 0.51 MeV, and the thermal energy $kT \approx 0.86 \times 10^{-4}\,T$ eV. The first two Eqs. (4.225) should be divided by three for antineutrino–electron scattering, whereas the last two Eqs. (4.225) are valid for both neutrino–electron and antineutrino–electron scattering.

For a degenerate gas, the neutrino–electron scattering cross sections are given by (Bahcall, 1964 as corrected by Bahcall and Wolf, 1965)

$$\sigma = \sigma_0 \left(\frac{E_\nu}{mc^2}\right)^2 \left(\frac{E_\nu}{E_F}\right) \quad \text{for } E_\nu \ll E_F$$

and

$$\sigma = \sigma_0 \left(\frac{E_\nu}{mc^2}\right) \left(\frac{E_F}{mc^2}\right) \quad \text{for } E_\nu \gg E_F \ . \tag{4.226}$$

Here E_F is the Fermi energy for the electron, and the equations should be multiplied by one third for antineutrino–electron scattering. The $E_\nu \approx 3kT$, and the Fermi energy is given by

$$E_F = mc^2 \left\{ \left[\left(\frac{\rho}{10^6 \mu_e}\right)^{2/3} + 1 \right]^{1/2} - 1 \right\}$$

$$\approx mc^2 \left(\frac{\rho}{10^6 \mu_e}\right)^{1/3} = (3\pi^2)^{1/3} \hbar c N_e^{1/3} \quad \text{for } \rho \gg 10^6 \mu_e \ ,$$

and the mass density, ρ, is related to the electron density, N_e, by the equation

$$\rho = N_A^{-1} \mu_e N_e = 10^6 \mu_e x^3 \ ,$$

where

$$x^3 = 3\pi^2 N_e \left(\frac{\hbar}{mc}\right)^3 \ ,$$

$\mu_e \approx 2$ is the electron molecular weight, and $N_A \approx 6.02 \times 10^{23}$ mole^{-1} is Avogadro's number. For nonrelativistic conditions, $\rho \ll 10^6 \mu_e$, we have

$$E_F \approx \frac{mc^2}{2} \left(\frac{\rho}{10^6 \mu_e}\right)^{2/3} \approx (3\pi^2)^{2/3} \frac{\hbar^2}{2m} N_e^{2/3} \ .$$

Hansen (1968) defines the chemical potential, μ, as the lesser of

$$\mu = 10^{-2} mc^2 \left(\frac{\rho}{\mu_e}\right)^{1/3}$$

or

$$\mu = 6 \times 10^{-6} mc^2 T_9^{-2} \left(\frac{\rho}{\mu_e}\right) \ ,$$

and obtains the electron–neutrino scattering cross sections

$$\sigma = \tfrac{2}{3} \sigma_0 E_\nu T_9 \quad \text{for } \mu \ll kT \ ,$$

$$\sigma = \frac{\sigma_0}{2} E_\nu \left(\frac{\mu}{mc^2}\right) \quad \text{for } \mu \gg kT \quad \text{and} \quad E_\nu \gg \mu \ , \tag{4.227}$$

and

$$\sigma = \frac{\sigma_0}{4} \frac{E_\nu}{\mu} mc^2 T_9 \left[1 + \frac{3E_\nu^2}{T_9}\right] \left[1 + \frac{T_9}{E_\nu}\right] \quad \text{for } \mu \gg kT \quad \text{and} \quad E_\nu \ll \mu \ ,$$

where E_ν is the neutrino energy in MeV, and $T_9 = T/10^9$. These results are quite close to those given above if we let $\mu = E_F + mc^2 \approx E_F$ and note that E_ν in MeV $\approx E_\nu/(mc^2)$ when $mc^2 = 0.51$ MeV. The exact expressions for the corresponding cross sections for $\mu \gg kT, E_\nu \gg kT$ are given by

$$\sigma = \sigma_0 E_\nu^2 \frac{2}{5} x \left\{ 1 + \frac{2}{3} x + \frac{1}{7} x^2 \right\} \quad \text{for } x = \frac{E_\nu}{\mu} < 1 \ ,$$

$$\sigma = 2\sigma_0 E_\nu \mu \left\{ 1 - \frac{2}{5} x - \frac{1}{5} x^2 + \frac{4}{105} x^3 \right\} \quad \text{for } x = \frac{\mu}{E_\nu} < 1 \ ,$$

where E_ν and μ are in MeV. For $E_\nu < \mu$ the asymptotics practically coincides with that of Hansen, but for $E_\nu > \mu$ it is twice as large.

The opacity due to neutrino–nucleon scattering has been given by Bahcall and Frautschi (1964), who give

$$\sigma \lesssim 10^{-2} \sigma_{0e} \left(\frac{E_\nu}{mc^2} \right)^2 \ ,$$

for the scattering reactions

$$\nu_e + p \rightarrow \nu_e' + p' \ ,$$

$$\nu_e + n \rightarrow \nu_e' + n' \ ,$$

$$\bar{\nu}_e + p \rightarrow \bar{\nu}_e' + p' \ ,$$

$$\bar{\nu}_e + n \rightarrow \bar{\nu}_e' + n' \ ,$$

where the n and p are, respectively, free neutrons and protons, E_ν is the neutrino energy, and σ_{0e} is given in Eqs. (4.225). Bahcall and Frautschi (1964) also give the formulae for neutrino absorption by bound nucleons.

Hansen (1968) has derived expressions for the total scattering rate, Q, by assuming that the incident neutrinos are distributed as a black body spectrum at the same temperature as the local electrons or nucleons. For neutrino-electron scattering, he obtains

$$Q \approx 9.1 \times 10^{-35} N_e N_\nu T_9^2 \quad \text{sec}^{-1}\,\text{cm}^{-3} \quad \text{for } \mu \leq 4kT \ , \tag{4.228}$$

and

$$Q \approx 3.64 \times 10^{-34} \frac{N_e N_\nu}{(\mu/kT)} T_9^2 \quad \text{sec}^{-1}\,\text{cm}^{-3} \quad \text{for } \mu \geq 4kT \ .$$

Here N_e is the electron number density, $T_9 = T/10^9$, the chemical potential, μ, is given before Eqs. (4.227), and the neutrino number density, N_ν, is given by

$$N_\nu = 7.65 \times 10^{27} T_9^3 \text{cm}^{-3} \ .$$

Similarly, for absorption by protons, p, or neutrons, n,

$$Q \approx 1.5 \times 10^{-6} N_p T_9^5 \quad \text{sec}^{-1}\text{cm}^{-3} \quad \text{for } T_9 \gtrsim 10 \ ,$$

and

$$Q \approx 1.5 \times 10^{-6} N_n T_9^5 \exp\left(-\frac{\mu}{kT} \right) \quad \text{sec}^{-1}\,\text{cm}^{-3} \quad \text{for } T_9 \geq 5 \ ,$$

where N_p and N_n are, respectively, the number densities of free protons and neutrons.

4.3.7 Solar Neutrinos

The nuclear reactions in the Sun's central furnace create prodigious quantities of neutrinos. Small amounts of them have been observed using massive, subterranean detectors, confirming that the Sun shines by thermonuclear reactions. However, these experiments have been finding only one-third to one-half the number of neutrinos that theory says they should, a discrepancy known as the solar neutrino problem. This paradox has been reviewed by Bahcall et al. (1995), Haxton (1995) and Bahcall (1996, 1997), together with its implications for solar models and elementary particle physics.

The flux of solar neutrinos expected at the Earth is calculated using large computers that produce theoretical models, culminating in the Standard Solar Model that best describes the Sun's luminous output, size and mass at its present age. The models use appropriate equations describing nuclear energy generation by hydrogen-burning reactions in the central core of the Sun, hydrostatic equilibrium that balances the outward force of gas pressure and the inward force of gravity, energy transport by radiative diffusion and convection, and an opacity determined from atomic physics calculations. Since the Standard Solar Model determines a helium abundance that agrees with other astronomical measurements, and can also be used to calculate the observed oscillations of the Sun (Bahcall et al., 1997), most astronomers are convinced that it correctly predicts the amount of solar neutrinos expected at Earth.

The nuclear reactions that emit most solar neutrinos comprise the hydrogen burning proton–proton (pp) chain, which is thought to provide most of the Sun's thermonuclear energy. The overall reaction can be written symbolically as

$$4\mathrm{p} \rightarrow \alpha + 2\mathrm{e}^+ + 2\nu_\mathrm{e} + 26\ \mathrm{MeV} \ , \tag{4.229}$$

where hydrogen nuclei, ^{1}H, or protons, p, are converted into helium nuclei, ^{4}He, or alpha particles, α, positrons, e^+, and electron neutrinos, ν_e, with a release of about 26 MeV of thermal energy for every four protons burned.

About 600 million tons of hydrogen are consumed every second to supply the solar luminosity, and in the process the Sun emits 2×10^{38} neutrinos. Each second about 100 billion solar neutrinos pass through every square centimeter of the surface of the Earth facing the Sun, and out through the opposite surface unimpeded.

The most important neutrino-producing reactions in the proton–proton chain are given in Table 4.5.

The fundamental reaction in the solar energy-generating process is the proton–proton (pp) reaction. This reaction produces the great majority of solar neutrinos; but these pp neutrinos have energies below the detection thresholds of some experiments (see Fig. 4.4). Most of the time, 86% in the Standard Solar Model, the proton–proton chain is terminated by two ^{3}He nuclei fusing to form an alpha particle and two protons. About 14% of the time, a ^{3}He nucleus will capture an already existing alpha particle to form beryllium ^{7}Be plus a gamma ray. The ^{7}Be nucleus will usually undergo electron, e^-, capture, producing neutrinos with an energy of $q = 0.862$ MeV (89.7% of the time).

Table 4.5. Neutrino producing reactions[1]

Name	Reaction	Termination (%)	Neutrino Energy, q
Proton–proton (pp) chain			
pp	$p + p \rightarrow {}^2H + e^+ + \nu_e$	100	$(q \leq 0.420$ MeV),
pep	$p + e^- + p \rightarrow {}^2H + \nu_e$	0.4	$(q = 1.442$ MeV),
hep	${}^3He + p \rightarrow {}^4He + \nu_e$	0.00002	$(q \leq 18.773$ MeV),
^{7}Be	${}^7Be + e^- \rightarrow {}^7Li + \nu_e$	15	$(q = 0.862$ MeV, 89.7%), $(q = 0.384$ MeV, 10.3%)
^{8}B	${}^8B \rightarrow {}^7Be^* + e^+ + \nu_e$	0.02	$(q \leq 15$ MeV),

[1]Adapted from Bahcall (1989). The termination percentage is the fraction of terminations of the pp chain, $4p \rightarrow \alpha + 2e^+ + 2\nu_e$, in which each reaction occurs. Since in essentially all terminations at least one pp neutrino is produced and in a few terminations one pp and one pep neutrino are created, the total of pp and pep terminations exceeds 100%.

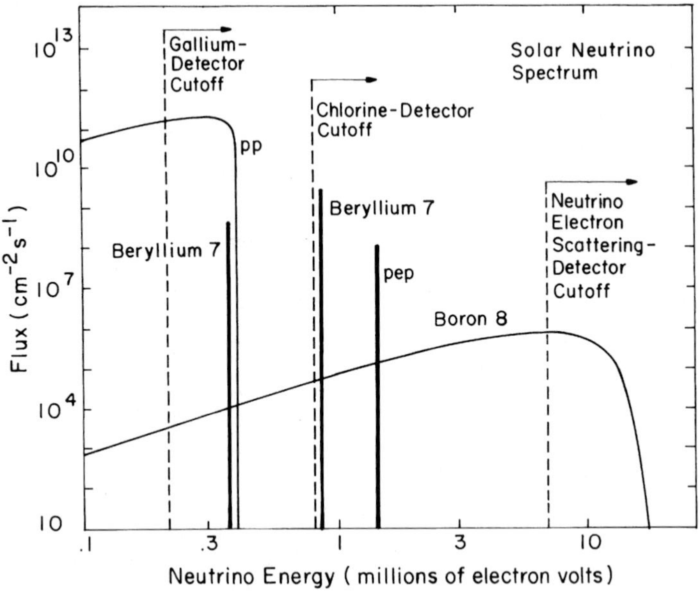

Fig. 4.4. Neutrinos are produced inside the Sun as a byproduct of fusion reactions in its core, but both the amounts and energies of the neutrinos depend on the element fused and the detailed model of the solar interior. Here we show the neutrino flux predicted by the Standard Solar Model. The largest flux of solar neutrinos is found at low energies; they are produced by the proton–proton (pp) reaction in the Sun's core. Less abundant, high-energy neutrinos are produced by a rare side reaction involving boron 8. Broken vertical lines mark the detection thresholds of the gallium, chlorine and neutrino-electron scattering (water) experiments; their detectors are sensitive to neutrinos with energies at the right side of the broken lines. The gallium experiment can detect the low-energy pp neutrinos, as well as those of higher energy; both the chlorine and water detectors are sensitive to the high-energy boron-8 neutrinos. The neutrino fluxes from continuum sources (pp and boron 8) are given in the units of number per square centimeter per second per million electron volts (MeV) at one astronomical unit. Neutrinos should also be generated at two specific energies when beryllium-7 captures an electron, and also during a relatively rare proton–electron–proton (pep) reaction; their fluxes are given in number per square centimeter per second. [Adapted from Bahcall (1989)].

Neutrinos with energy $q \leq 15$ MeV are produced from the rare termination in which ^7Be captures a proton to form radioactive boron, ^8B, ultimately producing neutrinos. Reactions involving ^8B occur only once in every 5,000 terminations of the pp chain, or 0.02% of the time, but the associated neutrinos dominate many detection experiments. The hep reaction (Table 4.5) is extremely rare, occurring about twice in every 10 million terminations.

The energy spectrum of neutrinos predicted by the Standard Solar Model is shown in Fig. 4.4, together with the detection thresholds of some experiments. The pp and ^8B reactions produce neutrinos with all energies from zero up to a maximum energy, resulting in a continuous spectrum. The easiest neutrinos to detect are the relatively high-energy neutrinos from the ^8B reaction. The pep and ^7Be neutrinos have well-defined energies and are known as neutrino lines; the shape of these lines may provide detailed physical information about the Sun's energy-generating core, but current experiments cannot distinguish these low-energy neutrino lines.

The solar neutrino fluxes expected at the Earth's surface for different neutrino sources are given in Table 4.6; they have been computed using the Standard Solar Model.

Solar neutrino astronomy involves massive detectors placed beneath mountains or deep inside mines; the intervening rocks shield them from deceptive signals caused by cosmic rays. The neutrino detectors consist of large amounts of material, literally tons of it, because of the very small interaction cross section between neutrinos and matter. The more energetic neutrinos have larger interaction cross sections, but there are fewer of them. The neutrino absorption cross sections for different material averaged over energy spectra are given in Table 4.7.

The neutrino reaction rate within even massive, underground detectors is so slow that a special unit has been invented to specify the experiment-specific flux. This solar neutrino unit, or SNU, is equal to one neutrino interaction per second for every 10^{36} atoms, or 1 SNU $= 10^{-36}$ neutrino absorptions per target atom per second.

Each experiment sets a limit to the summation $\sum$ of the products of the absorption cross section, σ_i, in cm^2 and the incident neutrino flux, ϕ_i, in cm^{-2} sec^{-1} for the i neutrino sources. These measurements of $\sum_i \phi_i \sigma_i$ may then be compared with theoretical calculations.

By finding solar neutrinos in roughly the predicted numbers, four pioneering neutrino detectors have now demonstrated that the Sun is indeed

Table 4.6. Calculated solar neutrino fluxes[1]

Neutrino Source	Flux (10^{10} cm^{-2} s^{-1})
pp	6.0 (1.00 ± 0.02)
pep	0.014 (1.00 ± 0.05)
hep	8×10^{-7}
^7Be	0.47 (1.00 ± 0.15)
^8B	5.8×10^{-4} (1.00 ± 0.37)

[1]Adapted from Bahcall (1989), where the estimated uncertainties are given in parenthesis.

Table 4.7. The neutrino absorption cross sections for various target materials and different neutrino sources. The units are 10^{-46} cm^2 for all sources except hep and ^{8}B, for which the unit is 10^{-42} cm^2. Contributions from excited states and from forbidden effects are included[1]

Target	pp	pep	hep†	^{7}Be	^{8}B†	^{13}N	^{15}O	^{17}F
^{7}Li	0.0	655	8.4	9.6	3.9	42.4	246	249
^{37}Cl	0.0	16	3.9	2.4	1.06	1.7	6.8	6.9
^{71}Ga	11.8	215	7.3	73.2	2.43	61.8	116	117
^{81}Br	0.0	75	9.0	18.3	2.7	14.5	36.7	37.0
^{98}Mo	0.0	0.0	10	0.0	3.0	0.0	0.0	0.0
^{115}In	78.0	576	6.1	248	2.5	224	355	356

[1]Adapted from Bahcall (1989).
†Unit is 10^{-42} cm^2.

energized by hydrogen fusion – see Table 4.8. However, these experiments have been finding only one-third to one-half the number of predicted neutrinos. This discord between measurements and predictions is known as the solar neutrino problem.

The first solar neutrino detection experiment is based on the reaction (Davis, Harmer, and Hoffman, 1968; Bahcall et al., 1995))

$$\nu_e + {}^{37}\mathrm{Cl} \rightarrow e^- + {}^{37}\mathrm{Ar} \ , \tag{4.230}$$

which has a threshold energy of 0.814 MeV, permitting the detection of all the major solar neutrino sources except those from the fundamental pp reaction. The underground detector, constructed by Raymond Davis, Jr. in 1967, is a 615-ton tank containing 10^5 gallons of C_2Cl_4(perchloroethylene, a cleaning fluid), located 1.5 kilometers underground in the Homestake Gold Mine near Lead, South Dakota. On the average, about one neutrino-induced nuclear conversion is detected every 2.17 days, producing one argon atom, , ^{37}Ar, out of the total of 2×10^{30} atoms of chlorine, ^{37}Cl. This corresponds to an average capture rate in the tank of (Bahcall and Krastev, 1996, 1997; Bahcall, 1994):

$$\text{Observed Homestake Capture Rate} = 2.55 \pm 0.16 \ (\text{stat.}) \pm 0.14 \ (\text{syst.}) \ \text{SNU}$$

which can be compared with theoretical predictions by the Standard Solar Model for a ^{37}Cl detector (Bahcall and Krastev, 1996, 1997; Bahcall, 1994; Bahcall and Pinsonneault, 1992):

$$\text{Theoretical } {}^{37}\mathrm{Cl} : \sum_i \phi_i \sigma_i = 9.5 \ ^{+1.2}_{-1.4} \ \text{SNU} \ ,$$

where the quoted errors are at the one sigma level. Thus, the chlorine experiment is detecting about one third of the expected number of neutrinos, or to be exact 0.27 ± 0.022 times the predicted value. The chlorine experiment has been run for more than a quarter of a century, always observing fewer solar neutrinos than expected theoretically (see Fig. 4.5). Approximately 77% (7.3 SNU) of the predicted event rate is contributed by ^{8}B neutrinos; the next largest contribution is the 14% (1.3 SNU) from the ^{7}Be neutrinos; the remaining 9% is due to a combination of all the other neutrino-producing reactions in the pp chain.

Table 4.8. Solar Neutrino Experiments[1]

Target	Experiment	Threshold Energy (MeV)	Measured Neutrino Flux (SNU)*	Predicted Neutrino Flux (SNU)*	Ratio: Measured/ Predicted
Chlorine 37	HOMESTAKE	0.814	$2.56 \pm 0.16 \pm 0.14$	$9.5^{+1.2}_{-1.4}$	0.27 ± 0.022
Water	KAMIOKANDE*	7.5	$2.80 \pm 0.19 \pm 0.35$	$6.62^{+0.93}_{-1.12}$	0.42 ± 0.060
Gallium 71	GALLEX	0.2	$69.7 \pm 6.67^{+3.9}_{-4.5}$	136.8^{+8}_{-7}	0.51 ± 0.058
Gallium 71	SAGE	0.2	$72^{+12}_{-10}{}^{+5}_{-7}$	136.8^{+8}_{-7}	0.53 ± 0.095

[1] Adapted from Bahcall and Krastev (1996, 1997). Also see Bahcall (1994) and Bahcall and Pinsonneault (1992). Here the uncertainties are one standard deviation, or 1σ, and the first and second measurement uncertainties correspond, respectively, to the statistical and systematic uncertainties.
* The units of the measured and predicted values for the Kamiokande experiment are 10^6 cm^{-2} s^{-1}, while the SNU or solar neutrino unit, is used for the other three experiments.

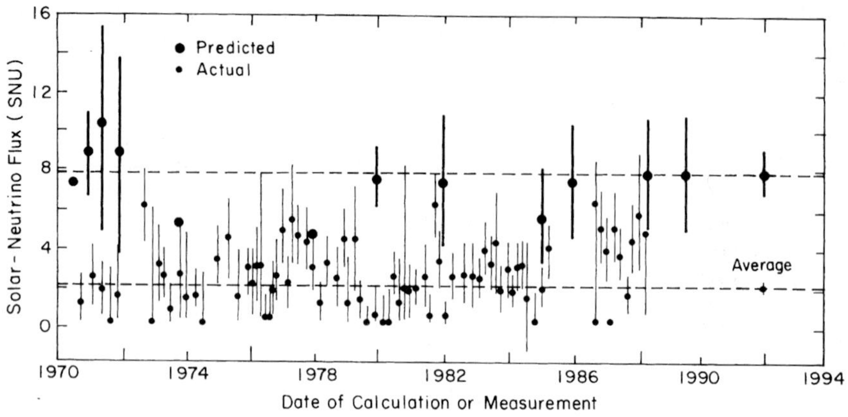

Fig. 4.5. Calculated and measured solar neutrino fluxes have consistently disagreed over the past three decades. Fluxes are measured in solar neutrino units (SNU), defined as one neutrino interaction per 10^{36} atoms per second. Measurements from the chlorine neutrino detector (small dots) give an average solar neutrino flux of 2.56 SNU (lower broken line). Theoretical calculations (large dots) predict at least 8.0 SNU (upper broken line) for the flux of solar neutrinos using the Standard Solar Model. Other experiments have also observed a deficit of solar neutrinos, suggesting that our current understanding of neutrinos is incomplete. [Adapted from Lang (1995)].

In 1987, a second experiment, located in a mine at Kamioka, Japan and known as Kamiokande II, independently confirmed that the neutrino flux is less than that predicted by the Standard Solar Model and showed that neutrinos are really coming from the Sun (Hirata et al., 1989, Totsuka, 1991). It consists of a 3,000-ton tank of pure water (680-ton fiducial volume). This neutrino-electron scattering experiment is based upon the reaction

$$\nu + e \rightarrow \nu' + e' \ , \tag{4.231}$$

which has an energy threshold of 7.5 MeV and is sensitive only to the ^{8}B neutrinos. Here the unprimed and primed sides respectively denote the incident and scattered neutrinos, ν, and electrons, e. The recoil electrons, e$'$, are primarily scattered in the forward direction in which the neutrinos are arriving.

As indicated by this reaction, a passing electron neutrino occasionally knocks a high-speed electron from a water molecule. The electron moves through the water faster than light travels in water, generating an electromagnetic shock wave and a light cone of Čerenkov radiation (see Section 1.34). Thousands of light detectors lining the water tank measure the axis of the light cone, which tells the direction of the incoming neutrino. The observed signals indicate that the electrons are preferentially scattered along the direction of an imaginary line joining the Earth to the Sun, confirming that the neutrinos are indeed produced by nuclear reactions in the Sun's core (Totsuka, 1991). The number of scattered electrons detected by Kamiokande II indicates that the flux of high-energy neutrinos is just under half the neutrino flux expected from the Standard Solar Model, or to be exact 0.42 ± 0.06 times the predicted value.

The numerous low-energy pp neutrinos, as well as fewer high-energy neutrinos, are detected in two gallium, ^{71}Ga, experiments based upon the reaction

$$\nu_e + {}^{71}\mathrm{Ga} \rightarrow \mathrm{e}^- + {}^{71}\mathrm{Ge} \ , \tag{4.232}$$

which has a low threshold of 0.233 MeV. The radioactive germanium, ^{71}Ge, produced in this way decays by electron capture with a half-life of 11.43 days. In 1990, the Soviet-American Gallium Experiment, or SAGE, began operation in a long tunnel, some 2 kilometers below the summit of Mount Andyrchi in the northern Caucasus (Abasov et al., 1991). A second multinational experiment, dubbed GALLEX for gallium experiment, started operating in 1991, located in the Gran Sasso Underground Laboratory some 1.4 kilometers below a peak in the Appenine Mountains of Italy. SAGE and GALLEX, which respectively use 60 tons and 30 tons of gallium, obtain (Bahcall and Kratsev, 1996, 1997; Bahcall, 1994):

$$\text{Observed GALLEX Capture Rate} = 69.7 \pm 6.7\ (\text{stat.}) \pm 4\ (\text{syst.})\ \ \text{SNU}$$

$$\text{Observed SAGE Capture Rate} = 72\ ^{+12}_{-10}\ (\text{stat.})\ ^{+5}_{-7}\ (\text{syst.})\ \ \text{SNU}$$

which can be compared to theoretical predictions by the Standard Solar Model for a ^{71}Ga detector (Bahcall and Krastev, 1996, 1997; Bahcall, 1994; Bahcall and Pinsonneault, 1992):

$$\text{Theoretical } {}^{71}\text{Ga} : \sum_{\mathrm{i}} \phi_{\mathrm{i}}\sigma_{\mathrm{i}} = 136.8\ ^{+8}_{-7}\ \text{SNU} \ .$$

So, the two gallium experiments observe about half the number of expected neutrinos. To be exact, the GALLEX and SAGE experiments have respectively measured 0.51 ± 0.058 and 0.53 ± 0.095 times the predicted value. Moreover, if you believe the chlorine and neutrino-electron scattering results, then the gallium experiments have apparently detected the low-energy pp neutrinos, confirming that the Sun shines by hydrogen fusion. (According to the Standard Solar Model, the low-energy pp neutrinos should contribute about 71 SNU, or slightly more than half of the predicted 132 SNU for the gallium experiments; ^{7}Be and ^{8}B respectively contribute about 34 SNU and 14 SNU.)

There are two possible explanations for the solar neutrino problem. Either we don't really know how the Sun and stars create their energy, or we don't understand neutrinos.

One method of solving the solar neutrino problem would be the creation of a nonstandard solar model that modifies our astrophysical description of the Sun and produces the observed number of neutrinos. This hypothetical solution seems to be inconsistent with the observed surface oscillations of the Sun (Bahcall, Pinsonneault, Basu and Christensen-Dalsgaard, 1997). The temperatures and composition assumed in calculating the expected flux of solar neutrinos, using the Standard Solar Model, have been compared with helioseismological measurements of the internal sound velocity that depends on these two parameters. Since the deep solar interior behaves essentially as a fully ionized perfect gas, the velocity of sound, c, varies as $c^2 \propto T/\mu$ for a temperature, T, and a mean molecular weight, μ. Thus even tiny fractional errors in the model values of T or μ would produce measurable discrepancies in the precisely determined helioseismological sound speed. However, the measured and predicted velocities do not differ from each other by more than 0.2 percent, from 0.95 solar radii down to 0.05 solar radii from the center (Bahcall, Pinsonneault, Basu and Christensen-Dalsgaard, 1997; Guenther and Demarque, 1997). This plausibly pins down the central temperature to be very close to the calculated value of 15.6 million degrees Kelvin, and strongly disfavors astrophysical explanations of the solar neutrino problem.

Indeed, a number of authors have shown that you cannot solve the solar neutrino problem by any astrophysical solution as long as you accept the four experimental results (Bahcall and Bethe, 1990; Heeger and Robertson, 1996; Bahcall, 1996, 1997). This is essentially because the shapes of individual neutrino energy spectra, such as those of the pp reaction or from ^{8}B decay, are independent of solar physics to an accuracy of 1 part in 10^5 (Bahcall, 1991). Thus, the solar neutrino problem cannot be solved with plausible variations in solar models, and particle physics solutions involving changes in our current understanding of neutrinos seem to be required.

4.3.8 Neutrino Oscillations

In the other solution to the solar neutrino problem, the neutrinos are produced at the Sun's center in the quantity predicted by the Standard Solar Model, but the neutrinos change form as they propagate out from the Sun. The neutrino comes in three types, or flavors, each named after the particle with which it is most likely to interact. All of the neutrinos generated inside the Sun are electron neutrinos, ν_e; this is the kind that interacts with electrons, e. The other two flavors, the muon neutrino, ν_μ, and the tau neutrino, ν_τ, interact with muons, μ, and tau particles, τ, respectively. The solar neutrino problem might be explained if the electron neutrinos produced inside the Sun are converted to neutrinos of a different flavor, say muon neutrinos, in transit to the Earth where they would remain invisible to current solar neutrino detectors. The electron neutrino is the type generated in the Sun's core. It is the only kind that the pioneering solar-neutrino detectors could respond to and the only flavor they could taste.

However, neutrinos can only transform, or oscillate, between states if the neutrino, long thought to be a massless particle, possesses a very small rest mass. Neutrinos are assumed to be completely massless in the present theory uniting the electromagnetic force with the "weak" force that governs neutrino interactions with other particles, so the proposed explanation of the solar neutrino problem involves a profound change in the theory of fundamental particles. As discussed in greater detail later in this section, there is evidence that muon neutrinos, generated by cosmic rays in the terrestrial atmosphere, may be undergoing oscillations, and that some neutrinos may therefore possess a very small mass (Fukuda et al., 1998).

The possibility that neutrinos may change form was first suggested by Bruno Pontecorvo and his colleagues (Pontecorvo, 1958, 1968; Gribov and Pontecorvo, 1969); they have considered neutrino oscillations in a vacuum (see Bilenky and

Pontecorvo, 1978; Bilenky and Petcov, 1987 for reviews). In the two-flavor problem with two interacting neutrinos, the flavor eigen-states, $|f\rangle$, can be expressed in terms of the mass eigenstates, $|m\rangle$, denoted by subscripts 1 and 2, with the aid of a two-dimensional orthogonal matrix, $|f\rangle = U_\mathrm{v}|m\rangle$, where,

$$U_\mathrm{v} = \begin{pmatrix} \cos\theta_\mathrm{v} & \sin\theta_\mathrm{v} \\ -\sin\theta_\mathrm{v} & \cos\theta_\mathrm{v} \end{pmatrix} . \tag{4.233}$$

Here θ_v denotes the vacuum mixing angle. Without loss of generality, we can choose $0 \le \theta_\mathrm{v} < \pi/4$ so that $|\nu_\mathrm{e}\rangle$ is "mostly" $|\nu_1\rangle$. The second flavor eigenstate, $|\nu_\mathrm{x}\rangle$, may be identified in practice with ν_μ, ν_τ, or with a linear combination of ν_μ and ν_τ.

The time evolution of an electron-neutrino state may be expressed by

$$|\nu_\mathrm{e}\rangle_\mathrm{t} = \cos\theta_\mathrm{v}\exp(-iE_1t)|\nu_1\rangle + \sin\theta_\mathrm{v}\exp(-iE_2t)|\nu_2\rangle , \tag{4.234}$$

where E_1 and E_2 are the energies of the two mass eigenstates with the same momentum p. The probability amplitude for an electron neutrino remaining an electron neutrino after travelling for a time t is (Bahcall, 1989)

$$\langle\nu_\mathrm{e}|\nu_\mathrm{e}\rangle_\mathrm{t} = \cos^2\theta_\mathrm{v}\exp(-i\mathrm{E}_1t) + \sin^2\theta_\mathrm{v}\exp(-i\mathrm{E}_2t) , \tag{4.235}$$

which corresponds to a probability of a ν_e remaining a ν_e of

$$|\langle\nu_\mathrm{e}|\nu_\mathrm{e}\rangle_\mathrm{t}|^2 = 1 - \sin^2 2\theta_\mathrm{v}\sin^2\left[\tfrac{1}{2}(\mathrm{E}_2 - \mathrm{E}_1)t\right] .$$

The two mass eigenstates are assumed to have the same momentum which implies that they have slightly different energies if they have finite masses. The energy difference for relativistic neutrinos is

$$E_2 - E_1 = \frac{m_2^2 - m_1^2}{2E} \equiv \pm\frac{\Delta m^2}{2E} ,$$

where the appearance of the plus and minus signs reflects the introduction of a positive definite quantity, Δm^2,

$$\Delta m^2 \equiv |m_1^2 - m_2^2| .$$

The plus sign applies if $m_2 > m_1$ and the minus sign in the opposite case.

The probability that an electron neutrino remains an electron neutrino may be written therefore in the following form that is often used.

$$|\langle\nu_\mathrm{e}|\nu_\mathrm{e}\rangle_\mathrm{t}|^2 = 1 - \sin^2 2\theta_\mathrm{v}\sin^2\left(\frac{\pi R}{L_\mathrm{v}}\right) , \tag{4.236}$$

where R is the distance travelled in a time t and the vacuum oscillation length $L_\mathrm{v} \equiv 4\pi E/\Delta m^2$. When the event rates from each neutrino source are averaged over the region of emission in the Sun, the region of absorption in transit to the Earth, and the energy spectrum of the neutrino source, one obtains (Bahcall and Frautschi, 1969):

$$|\langle\nu_\mathrm{e}|\nu_\mathrm{e}\rangle_\mathrm{t}|^2_\mathrm{average} = 1 - \tfrac{1}{2}\sin^2 2\theta_\mathrm{v} .$$

The probability of observing a neutrino of a different flavor, $|\nu_\mathrm{x}\rangle$, is

$$|\langle\nu_\mathrm{x}|\nu_\mathrm{e}\rangle_\mathrm{t}|^2 = \sin^2 2\theta_\mathrm{v}\sin^2\left(\frac{\pi R}{L_\mathrm{v}}\right) .$$

The vacuum oscillation length, $L_V = 2\pi/|E_1 - E_2|$, can be written in the form

$$L_V = \frac{4\pi E\hbar}{\Delta m^2 c^3} = 2.48\left(\frac{E}{\text{MeV}}\right)\left(\frac{\text{eV}^2}{\Delta m^2}\right) \quad \text{meters} \ , \tag{4.237}$$

which is often used in discussing terrestrial oscillation experiments that employ beams from reactors or accelerators. For discussing solar neutrinos, it is more convenient to rewrite the distance-dependent dimensionless argument of the sine function as follows:

$$\frac{\pi R}{L_V} \approx 10^{11}\left(\frac{1\text{MeV}}{E}\right)\left(\frac{\Delta m^2}{1\text{eV}^2}\right)\left(\frac{R}{1\text{AU}}\right) \ ,$$

where 1 AU $= 1.496 \times 10^{13}$ cm is the average distance between the Earth and the Sun.

Solar neutrino experiments are sensitive to mass differences for which the argument $\pi R/L_V$ can significantly affect the probability of observing an electron neutrino at Earth. The minimum mass difference squared, $\Delta m^2_{\text{solar}}$, that can be studied with solar neutrinos is

$$\Delta m^2_{\text{solar}} = 1.6 \times 10^{-12}\left(\frac{E}{1\text{MeV}}\right)\text{eV}^2, \ \left(\frac{\pi R}{L_V}\right) = 0.3 \ . \tag{4.238}$$

Vacuum oscillations provide an important theoretical background, but they are an unlikely solution to the solar neutrino problem because the amount of mixing required is large, and because the maximum reduction achievable is only 1/2 with two neutrino flavors. In contrast, neutrino oscillations in matter can cause the almost complete conversion of solar neutrinos of the electron type to neutrinos of a different flavor.

The transformation of electron neutrinos in propagating through the dense solar material is known as the MSW effect after Wolfenstein (1978, 1979) who originated the theory, and Mikheyev and Smirnov (1986, 1987) who further developed it and showed that a nontrivial suppression of the Sun's neutrino flux can occur for a wide range of mass differences $\Delta m^2 = 10^{-4}$ to10^{-8} $(\text{eV})^2$ and mixing angles $\sin^2\theta_m > 10^{-4}$. The MSW effect was suggested as a possible explanation for the solar neutrino problem by Bethe (1986, 1989) and Bahcall and Bethe (1990). Neutrino oscillations in matter have been reviewed by Kuo and Pantaleone (1989).

When neutrinos propagate through matter, the forward scattering of neutrinos off the background matter will induce an index of refraction for the neutrinos, just as light is refracted when travelling through matter. However, the neutrino index of refraction depends on the flavor; electron neutrinos will scatter off electrons in the Sun and thereby have a different index of refraction than muon neutrinos that do not scatter off electrons. Moreover, if the neutrinos are massive, then neutrino flavors will mix during propagation.

In the two-flavor solution to neutrino oscillations in matter, the flavor eigenstates are related to the mass eigenstates by a unitary transformation U_m and a mixing angle θ_m that are analogous to the vacuum oscillation parameters. The mixing angle in matter is determined by the relation (Bahcall, 1989):

$$\tan 2\theta_m = \frac{\tan 2\theta_v}{[1 \pm (L_v/L_e)\sec 2\theta_v]} \;, \tag{4.239}$$

where the plus sign applies for $m_2 < m_1$ and the minus sign for $m_2 > m_1$. The neutrino-electron interaction length, L_e, which appears in the formula for $\tan 2\theta_m$, is given by

$$L_e = \frac{\sqrt{2}\pi\hbar c}{G_F n_e} = 1.64 \times 10^5 \text{ meter } (100 \text{ g cm}^{-3}/\mu_e\rho) \;,$$

for an electron density n_e. The electron mean molecular weight, μ_e, can be calculated with the aid of the hydrogen mass fraction, X, using

$$\mu_e = \frac{(1+X)}{2} \;.$$

The Fermi constant $G_F = 1.436 \times 10^{-49}$ erg cm^3 [or $10^{-5}/m_p^2$ in units ($\hbar = c = 1$) preferred by particle physicists]. Useful relations for numerical calculations are

$$\frac{G_F N_A R_\odot}{\hbar c} = 1.94 \times 10^2, \quad \frac{n_e(0)}{N_A} = 98.6 \text{ cm}^{-3} \;,$$

where $R_\odot$ is the Sun's radius, $n_e(0)$ is the electron number density at the center of the Standard Solar Model and N_A is Avogadro's number ($N_A = 6.02 \times 10^{23}$). The interaction length, L_e, is independent of energy, unlike the oscillation lengths in vacuum or in matter.

The resonant character of matter oscillations is shown by the equation for $\tan\theta_M$. If $m_2 \geq m_1$, the mixing angle in matter is maximal ($\theta_m = \pi/4$) for an electron density determined by

$$\left(\frac{L_v}{L_e}\right)_{res} = \cos 2\theta_v \;, \tag{4.240}$$

which is often referred to as the "resonance condition." Even for a very small vacuum mixing angle, the matter mixing angle is $\theta_m = \pi/4$ if the electron density satisfies the resonance condition. Mixing between an electron neutrino, ν_e, and other neutrinos, ν_x, of type x can become very large at resonance and the smallness of the vacuum mixing angle is no longer important.

The MSW resonance density is

$$n_{e,res} = \frac{|\Delta m^2|\cos 2\theta_v}{2\sqrt{2}G_F E}$$

or

$$\frac{n_{e,res}}{N_A} \cong 66\cos 2\theta_v \left(\frac{|\Delta m^2|}{10^{-4}\text{eV}^2}\right)\left(\frac{10\text{MeV}}{E}\right) \;. \tag{4.241}$$

An electron neutrino will always pass through resonance if its energy exceeds the resonance value for an electron density corresponding to the solar center. Therefore, the minimum neutrino energy that can be resonant in the Sun is

$$E_{min} = 6.6\cos 2\theta_v \left(\frac{\Delta m^2}{10^{-4}\text{eV}^2}\right) \text{MeV} \;. \tag{4.242}$$

A significant fraction of the ^{8}B neutrinos created in the solar core will pass through a resonance density for mass differences of order 10^{-4} eV^2 or smaller.

Matter oscillations are particularly pronounced when the electron density is variable; in this case, electron neutrinos can be almost completely converted to neutrinos of another flavor. The probability for an electron neutrino, ν_e, that is created at a density larger than the resonance density to remain a ν_e is (Parke, 1986),

$$|\langle \nu_e | \nu_e \rangle_t|^2 = \frac{1}{2} + \left(\frac{1}{2} - P_{\text{jump}}\right) \cos 2\theta_m \cos 2\theta_v \ , \tag{4.243}$$

where the value of θ_m, the mixing angle in matter, is given by the relation:

$$\sin 2\theta_m = \frac{\sin 2\theta_v}{D_m} ,$$

with

$$D_m = \left[1 - 2\left(\frac{L_v}{L_e}\right) \cos 2\theta_v + \left(\frac{L_v}{L_e}\right)^2\right]^{1/2}$$

using the density at which ν_e was formed. The quantity P_{jump} represents the probability of jumping from one adiabatic mass eigenstate to another. Under the assumption that the density varies linearly with radius,

$$P_{\text{jump}} = \exp\left[\frac{-\pi \Delta m^2 \sin^2 2\theta_v}{4E \cos 2\theta_v} \left(\frac{n_e}{|dn_e/dr|}\right)_{\text{res}}\right] , n_e > n_{\text{res}} \ . \tag{4.244}$$

For ν_e produced at high densities, $n_{e,\text{prod}} \gg n_{e,\text{res}}$,

$$|\langle \nu_e | \nu_e \rangle_t|^2 \approx \sin^2 \theta_v + P_{\text{jump}}(E) \cos 2\theta_v \ . \tag{4.245}$$

For small vacuum mixing angles, the survival probability is approximately equal to P_{jump}.

Oscillation parameters in matter are given in Table 4.9 for different electron densities.

The results of different solar neutrino experiments can be used to provide constraints on the mass difference Δm and the mixing angle θ_v. They are often plotted in a MS diagram, named after Mikheyev and Smirnov (1986), whose orthogonal coordinates are $\sin^2 2\theta_v / \cos 2\theta_v$ and Δm^2. Examples of MS diagrams for the ^{37}Cl and ^{71}Ga experiments are given in Bahcall's (1989) book; also see Hirata et al. (1990) for the Kamiokande II experiment and Gelb, Kwong and Rosen (1992) for the GALLEX experiment.

Table 4.9. Parameters for matter oscillations[1]

n_e	0	$n_{e,\text{res}}$	$2n_{e,\text{res}}$	$\to \infty$
$E_2 - E_1$	$\Delta m^2/2E$	$\Delta m^2/2E \sin 2\theta_v$	$\Delta m^2/2E$	$\to \infty$
θ_m	θ_v	$\pi/4$	$\pi/2 - \theta_v$	$\to \pi/2$
L	L_v	$L_v/\sin 2\theta_v$	L_v	$\to 0$

[1]Adapted from Parke (1986)

Experiments currently or soon to be under way should settle the question of whether some neutrinos switch identities en route to Earth. These underground detectors include the Japanese Super-Kamiokande and the Sudbury Neutrino Observatory (Bahcall et al., 1995; Haxton, 1995; Lang, 1996; Bahcall, 1996, 1997).

The Super-Kamiokande experiment, located 1,000 kilometers down in a mine under the Japanese Alps to shield it from cosmic rays, uses 50,000 tons of ultrapure ordinary water. Like its 3,000-ton Kamiokande II predecessor, light detectors placed along the walls of the water tank detect the Čerenkov radiation of the electrons knocked free of the water molecules by passing neutrinos. Because of its larger volume, Super-Kamiokande provides a 30-fold increase in the observed rate of neutrino-electron collisions. It has confirmed the long-standing deficit in solar electron neutrinos, and should measure the neutrinos' energy spectrum via the energies of the scattered electrons. A distortion in the energy spectrum could reveal the way neutrinos transform from one type to another. Super-Kamiokande could therefore eventually provide evidence for solar neutrino oscillations.

Recent measurements with the Super-Kamiokande detector indicate that muon neutrinos, produced when energetic cosmic-ray protons and nuclei hit the Earth's atmosphere, exhibit a zenith angle dependent deficit that can be explained by the metamorphosis of muon neutrinos into tau neutrinos (Fukuda et al., 1998). It follows from Eqs. (4.236) and (4.237) that the probability, P, for such a two-neutrino oscillation while travelling a distance L through a vacuum is $P = \sin^2 2\theta_v \sin^2 (1.27\, \Delta m^2\, L/E_\nu)$, where θ_v is the vacuum mixing angle between the muon and tau flavor eigenstates, Δm^2 is the mass squared difference of the neutrino eigenstates in eV^2, the distance L is in kilometers, and E_ν is the neutrino energy in GeV. The Super-Kamiokande results indicate that $\sin^2 2\theta_\nu > 0.82$, that $0.0005\ eV^2 < \Delta m^2 < 0.006\ eV^2$, and that if one neutrino is much lighter than the other then the heavier neutrino has a mass of about 0.05 eV. This atmospheric neutrino data has nothing to do with electron neutrinos, the only kind made by nuclear reactions in the Sun, and therefore has no direct bearing on the solar neutrino problem. Although solar electron neutrinos are less energetic than the atmospheric ones, far more of the Sun's neutrinos rain down on Super-Kamiokande, so it may soon contribute to this solar neutrino problem as well.

The Sudbury Neutrino Observatory, or SNO, is also poised to search for a change in solar neutrino properties before reaching Earth. This novel detector is buried 2 kilometers underground in a working nickel mine near Sudbury, Ontario. The detector is a huge, spherical vat holding 1,000 tons of heavy water, a form of H_2O that contains deuterium, or heavy hydrogen, ^2H, with a neutron and a proton in its nucleus. The vat of heavy water is surrounded by a 7,000-ton jacket of ordinary water that shields it from weak natural radiation in the underground environment. As with the other neutrino detectors, the overlying rock blocks cosmic rays. The SNO should detect about 10 solar neutrinos per day, about 50 times the rate of previous experiments other than Super-Kamiokande.

Unlike the ordinary-water detectors, which are sensitive only to electron neutrinos, Sudbury's heavy water will be sensitive to all three types of neutrinos, because they can knock the neutron from a deuterium nucleus to produce a different light signal. When a neutrino of type x, denoted ν_X, interacts with deuterium, ^{2}H, two reactions are possible (Bahcall, et al., 1995):

$$\nu_e + {}^2\mathrm{H} \rightarrow \mathrm{p} + \mathrm{p} + \mathrm{e}^- \ , \tag{4.246}$$

and

$$\nu_X + {}^2\mathrm{H} \rightarrow \mathrm{p} + \mathrm{n} + \nu_X \ ,$$

where ν_e is the electron neutrino, and p, n and e^- respectively denote the proton, the neutron and the electron. The first reaction can only be induced by electron-type neutrinos, whereas the cross-section for the second reaction is the same for neutrinos of all three types. If more neutrinos are detected via the first reaction than by the second one, that would be direct evidence that some electron-type neutrinos have changed, or oscillated, into neutrinos of some other type.

Calculations of the expected results of future solar neutrino experiments (Super-Kamiokande, SNO, Borexino, Icarus, Hellaz and Heron) are provided by Bahcall and Krastev (1996); they show how these experiments will restrict the range of the allowed neutrino mixing parameters.

Future and ongoing experiments should tell us whether solar neutrinos change identity or not, and therefore determine if they have some mass however tiny. Because of the long proper time that is available for a neutrino to transform its state, solar neutrino experiments are sensitive to very small neutrino masses, m_ν, that can cause neutrino oscillations in vacuum. Quantitatively (Bahcall, 1996)

$$m_\nu(\text{solar level of sensitivity}) \approx 10^{-6}\ \text{to} 10^{-5}\text{eV (vacuum oscillations)} \ ,$$

provided the electron-type neutrino that is created by beta decay contains appreciable portions of at least two different neutrino mass eigenstates (i.e., the neutrino mixing angle is relatively large). For matter-induced oscillations, the MSW effect, the planned or operating experiments are sensitive to neutrino masses, m_ν, in the range

$$m_\nu(\text{solar level of sensitivity}) \approx 10^{-4}\ \text{to} 10^{-2}\text{eV (matter-induced oscillations)} \ . \tag{4.247}$$

Neutrino burst events detected from the supernova 1987 have placed upper limits on the neutrino's rest mass. The fact that high-energy neutrinos from SN 1987A arrived within just a few seconds of each other, after travelling 160,000 light-years from the Large Magellanic Cloud, sets the upper limit to the neutrino mass at about 16 electron volts, or 16 eV (Burrows, 1988). Terrestrial laboratory experiments have a sensitivity to electron-type neutrino masses of order 1 eV. As previously mentioned, muon or tau neutrinos, generated by cosmic rays in our atmosphere, may have a mass of about 0.05 eV. By comparison the lightest known particle, the electron, has a rest mass of 511,000 eV.

4.3.9 Neutrino Emission from Stellar Collapse and Supernovae

When an old massive star expends all its nuclear fuel, and can no longer support itself, the star's core will collapse under its own weight in less than a second. If the core mass reaches the Chandrasekhar mass limit of approximately 1.459 solar masses, the electrons are forced into the nuclei, converting protons into neutrons with the emission of electron neutrinos. A neutron star can then be formed at the center of collapse with nuclear densities of $\approx 10^{14}$ g cm^{-3} and a radius of about 10 km, while the stellar envelope is hurled outward in a supernova explosion. About 10^{57} neutrons are formed when the stellar core collapses to create a neutron star, resulting in the release of about 10^{57} electron neutrinos.

Most of the binding energy, E_b, that is released when a neutron star is formed, and a supernova occurs, is believed to be emitted in the form of neutrinos. That energy is about 10% of the rest mass energy of the star, or $E_b \approx 0.1 M_{\odot} c^2 \approx 2 \times 10^{53}$ ergs. The largest part of this energy, about 99% , is emitted by neutrinos during a cooling period of about 10 seconds, when the newly-formed neutron star rids itself of excess heat; only about 1% of the binding energy is released by neutrinos during the previous collapse, lasting around 0.01 seconds (Mayle, Wilson and Schramm, 1987; Woosley, Wilson and Mayle, 1986).

The number of neutrinos per unit area incident at the Earth, called the fluence F, depends on the distance, D, and the temperature, T, of the collapsing star. For a neutrino or antineutrino of type j, we have

$$F(\nu_j, T_j) = \frac{f_j (l - \epsilon) E_b}{12 \pi D^2 k T_j}, \quad j = 1, \ldots 6 , \tag{4.248}$$

where f_j is the fraction of the energy $(1 - \epsilon) E_b$ that is carried off during thermal cooling, the energy ϵE_b is the fraction of the binding energy released during the collapse, T_j is the temperature, and k is Boltzmann's constant. For a galactic supernova (Bahcall, 1989):

$$F_{\mathrm{cool}}(\nu_j, T_j) = 1.5 \times 10^{11} \left(\frac{10\mathrm{kpc}}{D}\right)^2 \left(\frac{5\mathrm{MeV}}{T}\right) (6f) \quad \mathrm{cm}^{-2} ,$$

where $10\,\mathrm{kpc} = 3.086 \times 10^{22}$ cm is the average expected distance of a supernova in our Galaxy, and the cooling temperature of neutrinos and antineutrinos of the electron type is $T_j = 5\mathrm{MeV} = 8 \times 10^{-6}\mathrm{erg} \approx 6 \times 10^{10}\mathrm{K}$ (Schramm, 1987). Muon neutrinos and muon antineutrinos have a temperature of $T_j = 10\mathrm{MeV}$, but the electron antineutrinos, $\bar{\nu}_e$, are most easily observed with existing water detectors. The time scale, Δt, over which the neutrinos are emitted, as determined by the diffusion time out of the interior of the collapsed star, is $\Delta t \approx 1$ to 10 seconds (Burrows and Lattimer, 1986).

Within our Galaxy, stellar collapse and supernovae, that are like likely to give rise to a detectable pulse of neutrinos, are expected to occur every 10 to 100 years (Type II supernovae, with progenitor star mass $M \geq 8$ solar masses; Narayan, 1987; Van Den Bergh, Mc Clure and Evans, 1987); but only seven or eight have been recorded in our Galaxy during the past millennium (Lang, 1992), in part because interstellar dust obscures the light of all but the nearest ones.

On 23 February 1987 the explosion of a blue supergiant star was observed in the Large Magellanic Cloud, producing the first supernova to be visible to the naked eye in 400 years, called SN 1987A, and the first ever detection of neutrinos from stellar collapse (Arnett, Bahcall, Kirshner, and Woosley, 1989; Murdin, 1989; Schramm, 1987; Trimble, 1988; Woosley and Phillips, 1988; and Mc Cray, 1993). The Japanese Kamiokande II and American Irvine-Michigan-Brookhaven (IMB) Čerenkov water detectors recorded the release of about 2×10^{53} ergs of neutrino energy that briefly provided a luminous output equal to that of all the stars in the visible Universe (Bionata et al., 1987; Hirata et al., 1987).

The neutrinos released during core collapse were detected about 3 hours before the visible explosion. They were detected in the form of electron antineutrinos, $\bar{\nu}_e$, absorbed by protons, p, in the reaction

$$\bar{\nu}_e + p \rightarrow n + e^+ \ , \tag{4.249}$$

where neutrons, n, are formed together with positrons, e^+, that then annihilate with electrons, e^-, to produce the Čerenkov radiation recorded by the light detectors. The two detectors contain several kilotons of water, corresponding to several times 10^{32} protons. (The cross section for the antineutrino-proton reaction is about two orders of magnitude larger than the cross section for the neutrino-electron scattering used to detect solar neutrinos.) Characteristics of the neutrino burst and visible explosion of SN 1987A are given in Table 4.10.

As discussed in the previous Sect. 4.3.8, these observations have led to an upper limit to the mass, m_ν, of the electron neutrino of $m_\nu \leq 16$ eV (Burrows, 1988). The extra time, Δt, that a finite mass neutrino requires to reach the Earth compared to a zero mass particle is (Bahcall, 1989)

$$\Delta t_i = 2.57(D/50\text{kpc})(10\text{MeV}/E_i)^2(m_{\nu e}/10\text{eV})^2\text{seconds} \ , \tag{4.250}$$

for a neutrino of energy E_i, mass $m_{\nu e}$, and distance D. Here the distance is given in units of kiloparsecs, or kpc, where $1\text{kpc} = 3.086 \times 10^{21}\text{cm} = 3,261.6$ light years.

4.4 Nucleosynthesis of the Elements

4.4.1 Stellar Nucleosynthesis and the Abundances of the Elements

One of the most intriguing astronomical inquiries has been the origin of the chemical elements of which the material world is composed. The relative abundances of the elements provide important clues for their origin. William D. Harkins, for example, noticed that elements of low atomic weight are more abundant than those of high atomic weight, suggesting that the heavy elements were synthesized from light ones. Harkins also noticed that elements with even atomic numbers are about 10 times more abundant than those of odd atomic number of similar value, indicating that the relative abundances of the elements depends on nuclear rather than chemical properties (Harkins, 1917; 1931).

The abundances of the elements in a star can be inferred from absorption lines in their radiation. Such spectral investigations showed that the Sun contains the same chemical elements as those found on Earth, with the same relative abundances (Payne, 1929), but that the lightest element, hydrogen, is by

Table 4.10. Supernova 1987A (SN 1987A) Neutrino Burst and Visible Explosion[1]

Neutrino Burst	
Kamiokande II Detector	
Event Start Time	$t_0 = 07^h\ 35^m\ 35^s$ UT ($\pm$ 1 min) on 23 Feb 1987
Number of Anti-Neutrinos Detected	$N = 11$
Energy Range of Anti-Neutrinos	$\Delta E = 7.5$ to 36 MeV
Burst Duration	$\Delta t = 12.5$ s
Irvine-Michigan-Brookhaven (IMB) Detector	
Event Start Time	$t_0 = 07^h\ 35^m\ 41.37^s$ UT ($\pm$ 10 ms) on 23 Feb 1987
Number of Anti-Neutrinos Detected	$N = 8$
Energy Range of Anti-Neutrinos	$\Delta E = 20$ to 40 MeV
Burst Duration	$\Delta t = 5.6$ s
Neutrinos and Core Collapse	
Total Anti-Neutrino Energy	$E \approx 3 \times 10^{52}$ erg
Total Neutrino Energy	$E \approx 2.5 \times 10^{53}$ erg
Total Neutrino Luminosity	$L \approx 10^{55}$ erg s^{-1}
Average Neutrino Temperature	$T = 4$ MeV $\approx 10^{10}$ K
Number of Neutrinos Produced	$N = 10^{58}$ neutrinos
Neutrino Flux at Earth	$F \approx 5 \times 10^{10}$ cm^{-2}
Inferred Neutrino Mass Limit	$m \leq 16$ eV
Core Collapse Time	$\Delta t = 0.01$ s $= 10$ ms
Collapsed Remnant Mass	$M = 1.4 M_\odot$, for solar mass $M_\odot$
Collapsed Remnant Radius	$R = 10$ km
Gravitational Potential Energy Released	$E = 2.5 \times 10^{53}$ erg
Visible Explosion*	
Optical Discovery	$V = 5.0$ mag on 24.122 Feb 1987 ($t \approx 19$ hours) (Ian Shelton)
Explosive Optical Brightening	$V = 6.36$ mag on 23.444 Feb 1987 ($t \approx 3$ hours) $M_{bol} = -12.8$ mag $L = 4.2 \times 10^{40}$ erg s$^{-1} \approx 10^7 L_\odot$, for solar luminosity $L_\odot$.
Secondary Visual Maximum	$V = 2.97$ mag on 20 May 1987 ($t \approx 90$ days) $M_{bol} = -16.2$ mag $L \approx 10^{42}$ erg s$^{-1} \approx 10^8 L_\odot$, for solar luminosity $L_\odot$

*Time t from neutrino burst at t_0 = 23.316 Feb 1987

[1] Adapted from Lang (1992).

far the most abundant element in the outer atmospheres of the Sun and other bright stars (Unsöld, 1928; Mc Crea, 1929; Russell, 1929) as well as in their interiors (Strömgren, 1932, 1933). Hydrogen is too light to be retained in the Earth's atmosphere, and the second-most abundant element in the Sun, helium, is so rare on the Earth that it was first discovered on the Sun. We now know that hydrogen, ^{1}H, and helium, ^{4}He, as well as trace amounts of the light elements deuterium, ^2D, helium-3, ^3He, and lithium-7, ^7Li, were formed in the immediate aftermath of the big-bang explosion that produced the expanding Universe. Such a big-bang nucleosynthesis of the elements is discussed in Volume II, together with its implications for the baryon density of the Universe.

By 1920 Arthur Stanley Eddington had pointed out that hydrogen might be transformed to helium, with the resultant mass difference used to power the stars (Eddington, 1920), but the details of the nuclear reactions were lacking. Even at the enormous interior stellar temperatures, the mean velocities of particles were far smaller than those needed to penetrate the Coulomb field of the nucleus. The development of the quantum mechanical theory for the probability of the penetration of a nucleus by another nucleus provided a resolution to this difficulty (Gamow, 1928; Gurney and Condon, 1929). Robert Atkinson and Fritz G. Houtermans applied and extended this tunneling theory to the nuclei within stars, providing the first attempt at stellar nuclear energy generation (Atkinson and Houtermans, 1929). The most effective nuclear interactions were those involving light nuclei with low charge, so only hydrogen-induced fusion of helium would be important, which is in accord with the great stellar abundance of hydrogen.

Atkinson (1931) then discussed the additional nucleosynthesis of heavier elements, and next showed that the most likely nuclear reaction within stars is the collision of two protons, ^1H, to form a deuteron, $^2H = {}^2D$, and a positron, e^+ (Atkinson, 1936). This nonresonant proton-proton reaction, given by $^1H + {}^1H \rightarrow {}^2D + e^+ + \nu_e$ where ν_e denotes an electron neutrino, provides the starting point for a subsequent chain of reactions leading to the synthesis of helium in stars. The quantitative aspects were developed by Hans Bethe and Carl von Weizsäcker, paving the way for our understanding of the synthesis of heavier elements inside stars (Weizsäcker, 1937, 1938; Bethe and Critchfield, 1938; Bethe, 1939).

Significant amounts of elements heavier than helium cannot be synthesized in the early stages of the big bang. By the time that the temperature of the expanding Universe had cooled enough for helium to form, the material was too rarefied to permit the formation of the next most abundant element, carbon or ^{12}C, by the triple collision of helium, ^4He. Instead, carbon had to be synthesized in the hot, dense cores of evolved, giant stars where the triple helium collisions are frequent enough (Öpik, 1951; Salpeter, 1952). It seemed natural then, to suppose that all the elements heavier than carbon have been formed over long time intervals during successive static burning stages in stars and that the exponential decline in abundance with increasing weight is explained by the rarity of stars that have evolved to later stages of life.

Hans E. Suess and Harold Clayton Urey then provided a detailed discussion of the element and isotopic abundances in the Sun and similar stars (Suess and Urey, 1956). This discussion, which called attention to the many fluctuations that appear in the general trend of an exponential decline of abundance with increasing atomic weight, served as the major stimulus for current ideas concerning the synthesis of elements in stars. Nowadays, the detailed solar system abundances of the elements are based primarily on primitive carbonaceous meteorites (Cameron, 1973, 1982; Anders and Grevesse, 1989; Lang, 1992; Grevesse and Noels, 1993).

In 1956 Margaret Burbidge, Geoffrey Burbidge, William Fowler and Fred Hoyle delineated the element-producing reactions in stars that account for the details of the abundance curve shown in Fig. 4.6 (Burbidge, Burbidge, Fowler and Hoyle, 1956, often called B^2FH after the first initials of the last names of the authors). At about the same time Alastair Cameron independently discussed most of the same topics and emphasized strongly that nuclear fuels might burn explosively during supernova explosions, accounting for some of the heaviest elements (Cameron, 1957).

The early history of ideas concerning element formation, in the big-bang and in stars, has been reviewed by Alpher and Herman (1950), Penzias (1979) and Fowler (1984). The basic scheme of stellar nucleosynthesis, in which heavy elements are built from light ones inside stars, has held up well since it was first proposed by B^2FH.

Thermonuclear astrophysics was reviewed by Clayton and Woosley (1974). Nucleosynthesis and stellar evolution has been discussed in a series of lectures by Woosley (1986). Advances in element burning processes in stars and explosive nucleosynthesis in supernova explosions have been reviewed by Trimble (1991) in the context of the origin and abundance of the chemical

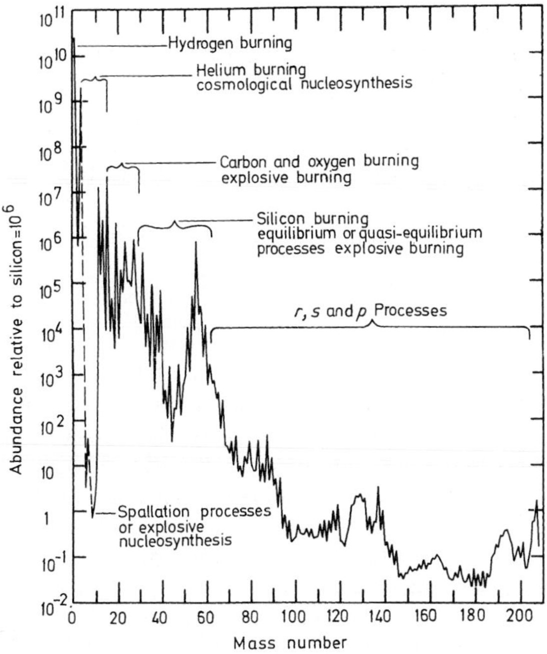

Fig. 4.6. Abundances of the nuclides plotted as a function of mass number. The figure also illustrates the different stellar nuclear processes which probably produce the characteristic features.

elements, first reviewed by her in Trimble (1975). Other related reviews are given by Arnett (1995) for explosive nucleosynthesis, Meyer (1994) for r-, s-, and p- processes in nucleosynthesis, Trimble (1982, 1983, 1988) for supernovae and their remnants, and Woosley and Weaver (1986), Bethe (1990) and Wheeler and Harkness (1990) for the physics of supernova explosions.

The major nuclear burning stages in stars are discussed in the monograph by Rolfs and Rodney (1988), and related to stellar evolution in the text by Kippenhahn and Weigert (1990). Nuclear energy generation in the solar interior is reviewed by Parker and Rolfs (1991).

4.4.2 Nucleosynthetic Processes in Ordinary Stars – Energy Generation Stages and Reaction Rates

An examination of the abundance of the elements leads to a description of a series of stellar nucleosynthetic processes which are nearly the same as those postulated by Burbidge, Burbidge, Fowler and Hoyle (1957). These fundamental processes are shown on the nuclide abundance curve in Fig. 4.6 and are as follows:

a) Hydrogen burning (conversion of hydrogen to helium) temperature $T > 10^7\,°K$ and $\approx 10^{10}$ years duration.
b) Helium burning (conversion of helium to carbon, oxygen, etc.) temperature $T \gtrsim 10^8\,°K$ and $\approx 10^7$ years duration.
c) Carbon burning, temperature $T \gtrsim 6 \times 10^8\,°K$, and oxygen burning, temperature $T \gtrsim 10^9\,°K$ (production of $16 \lesssim A \lesssim 28$) $\approx 10^5$ years duration, unless nucleosynthesis is explosive, then few seconds duration.
d) Silicon burning (production of $28 \lesssim A \lesssim 60$) temperature $T > 3$ or $4 \times 10^9\,°K$ and for the quasi-equilibrium and the e process ≈ 1 sec duration.
e) The s process (production of $A \geq 60$) temperature $T > 10^8\,°K$ and $10^3 - 10^7$ years duration.
f) The r process (production of $A \geq 60$) temperature $T > 10^{10}\,°K$ and 10–100 seconds duration (uncertain).
g) The p process (production of the low abundance proton-rich heavy nuclei) temperature $T > 2$ to $3 \times 10^9\,°K$ and 10–100 seconds duration.

The hydrogen and helium burning stages govern must of the evolution paths observed in the Hertzsprung-Russell diagram. The relation of thermonuclear reactions to stellar evolution is reviewed by Hayashi, Hoshi, and Sugimoto (1962) and Tayler (1966), and presented in the monograph by Kippenhahn and Weigert (1990). The detailed formulae for evaluating these reaction rates are given in Section 4.2, as well as in the books by Clayton (1965, 1983) and Rolfs and Rodney (1988). Reaction rate constants for many of the important thermonuclear reactions in stars are given by Fowler, Caughlan and Zimmerman (1967, 1975), Harris, Fowler, Caughlan and Zimmerman (1983), Caughlan, Fowler, Harris and Zimmerman (1985), and Caughlan and Fowler (1988). The reaction rates for hydrogen burning in the Sun are reviewed by Parker and Rolfs (1991). Nuclear partition functions were discussed by Fowler, Engelbrecht and Woosley (1978).

In what follows, reactions are specified for the nuclear processes suggested above. The component reactions for many of these processes will be specified together with the appropriate energy release and reaction rate constants. The reaction rate for the forward reaction can then be determined from the equation (cf. Eqs. (4.47), (4.53), (4.64), (4.65), and (4.67)).

$$\begin{aligned} N_A\langle 12\rangle &= C_1 T_9^{-2/3} \exp[-C_2 T_9^{-1/3} - (T_9/T_0)^2]\{1 + C_3 T_9^{1/3} + C_4 T_9^{2/3} \\ &\quad + C_5 T_9 + C_6 T_9^{4/3} + C_7 T_9^{5/3}\} + \sum C_8 T_9^{-3/2} \exp(-C_9/T_9) \\ &\quad + C_{10} \exp(-C_{11}/T_9) \quad \mathrm{cm^3\,sec^{-1}(g\text{-}mole)^{-1}} \ , \end{aligned} \tag{4.251}$$

and the reverse reaction rates from the equations

$$N_A\langle 34\rangle = C_{12}[\exp(-C_{13}/T_9)]N_A\langle 12\rangle \quad \mathrm{cm^3\,sec^{-1}(g\text{-}mole)^{-1}} \tag{4.252}$$

or

$$\lambda_\gamma(3) = C_{14} T_9^{3/2}[\exp(-C_{13}/T_9)]N_A\langle 12\rangle \quad \mathrm{sec^{-1}} \ . \tag{4.253}$$

Here Avogadro's number $N_A = 6.022136 \times 10^{23}\,(\text{g-mole})^{-1}$, $\langle 12\rangle$ is the reaction rate $\langle\sigma v\rangle$ in the forward direction, $T_9 = T/10^9$, where T is the temperature in °K, the constants $C_1 - C_{14}$ are given in the tables, $\langle 34\rangle$ is the reaction rate $\langle\sigma v\rangle$ in the reverse direction, and $\lambda_\gamma(3) = [\tau_\gamma(3)]^{-1}$ is the inverse of the lifetime for photodisintegration.

Hydrogen burning: the proton-proton chain. This process starts with the proton-proton reaction which has been discussed by Weizsäcker (1937), Bethe and Critchfield (1938), Bethe (1939), Salpeter (1952), Fowler (1959), Reeves (1963), and Bahcall and May (1969). Subsequent reactions were suggested by Schatzman (1951) and Fowler (1958); and the roles of various parts of the chain have been discussed by Parker, Bahcall, and Fowler (1964). The complete chain is

$$\begin{aligned}
&\mathrm{H^1 + H^1 \to D^2 + e^+ + \nu_e + 1.442\,MeV - 0.263\,MeV} \\
&\mathrm{D^2 + H^1 \to He^3 + \gamma + 5.493\,MeV} \\
&\left\{\begin{array}{ll} \mathrm{He^3 + He^3 \to He^4 + 2H^1 + 12.859\,MeV} & (86\%) \\ \text{or} \\ \mathrm{He^3 + He^4 \to Be^7 + \gamma + 1.586\,MeV} & (14\%) \end{array}\right. \\
&\left\{\begin{array}{ll} \mathrm{Be^7 + e^- \to Li^7 + \nu_e + 0.861\,MeV - 0.80\,MeV} & (14\%) \\ \mathrm{Li^7 + H^1 \to He^4 + He^4 + 17.347\,MeV} & \\ \text{or} \\ \mathrm{Be^7 + H^1 \to B^8 + \gamma + 0.135\,MeV} & (0.02\%) \end{array}\right. \\
&\mathrm{B^8 \to Be^8 + e^+ + \nu_e + 17.98\,MeV - 7.2\,MeV} \\
&\mathrm{Be^8 \to 2He^4 + 0.095\,MeV} \ .
\end{aligned} \tag{4.254}$$

Here the energy release in MeV includes the annihilation energy of the positron when it is a reaction product, and the negative energy release denotes the average neutrino energy loss. Hydrogen burning by the proton-proton chain is reviewed in detail in the monograph by Rolfs and Rodney (1988), while Parker and Rolfs

(1991) provide an overview of hydrogen burning in the Sun. The overall proton-proton chain is given by the reaction $4p \rightarrow {}^4He + 2e^+ + 2\nu_e + Q_{eff}$, where the proton $p = {}^1H$, the helium nucleus is ${}^4He$, the positron is e^+, the electron neutrino is ν_e, and Q_{eff} is the energy release. Most of the time (86%), the chain is completed by the ${}^3He + {}^3He$ reaction with $Q_{eff} = 26.20$ MeV (a total $Q = 26.731$ MeV and $q = 0.53$ MeV due to neutrinos). The less frequent terminations of ${}^3He + {}^7Be + e^- \rightarrow {}^7Li$ (14%) and ${}^3He + {}^7Be + {}^1H \rightarrow B^8$(0.02%) have $Q_{eff} = 25.66$ MeV and $Q_{eff} = 19.17$ MeV, respectively.

The proton-proton chain is driven by the nonresonant initiating proton-proton reaction $p + p \rightarrow {}^2H + e^+ + \nu_e$, where ${}^2H = {}^2D =$ the deuterium nucleus or deuteron. The low energy cross section factor (Eqs. 4.34, 4.46, 4.47) is $S_{pp}(0) = 3.89 \times 10^{-25}(1.0000 \pm 0.0011)$ MeV barns (Kamionkowski and Bahcall, 1994), and the $S'(0) = dS(0)/dE = 4.2 \times 10^{-24}$ barns. At the center of the Sun with $T_9 = 15$, the $\langle \sigma v \rangle_{pp} = 1.19 \times 10^{-43}$ cm^3 s^{-1} and the mean proton lifetime, τ_p, against combustion into deuterium inside the Sun is 10^{10} years (Rolfs and Rodney, 1988), demonstrating that nuclear reactions proceed very slowly inside stars. Cross section factors for the proton-proton chain are given in Table 4.11, while the reaction rate constants are given by Caughlan and Fowler (1988).

Table 4.11. Cross section factors $S(0)$ and $S'(0) = dS(0)/dE$ for the proton-proton chain[1]

Reaction	$S(0)$ MeV barns	$S'(0)$ barns
${}^1H(p, e^+ \nu){}^2D$	3.89×10^{-25}	4.2×10^{-24}
${}^2D(p, \gamma){}^3He$	0.25×10^{-6}	0.75×10^{-5}
${}^3He({}^3He, 2p){}^4He$	5.0 ± 0.3	-3.1
${}^3He(\alpha, \gamma){}^7Be$	$(0.553 \pm 0.017) \times 10^{-3}$	-3.1×10^{-4}
${}^7Be(p, \gamma){}^8B$	$(0.0243 \pm 0.0018) \times 10^{-3}$	-3×10^{-5}

[1] Adapted from Parker and Rolfs (1991) and Kamionkowski and Bahcall (1994).

Hydrogen burning: the C–N–O bi-cycle. For stars slightly more massive than the Sun, hydrogen will be fused into helium by the faster C–N cycle provided that carbon, nitrogen, or oxygen are present to act as a catalyst. These reactions are (Weizsäcker, 1938; Bethe, 1939)

$$\begin{aligned}
C^{12} + H^1 &\rightarrow N^{13} + \gamma + 1.944\ \text{MeV} \\
N^{13} &\rightarrow C^{13} + e^+ + \nu_e + 2.221\ \text{MeV} - 0.710\ \text{MeV} \\
C^{13} + H^1 &\rightarrow N^{14} + \gamma + 7.550\ \text{MeV} \\
N^{14} + H^1 &\rightarrow O^{15} + \gamma + 7.293\ \text{MeV} \\
O^{15} &\rightarrow N^{15} + e^+ + \nu_e + 2.761\ \text{MeV} - 1.000\ \text{MeV} \\
N^{15} + H^1 &\rightarrow C^{12} + He^4 + 4.965\ \text{MeV}.
\end{aligned} \tag{4.255}$$

Additional proton capture reactions, which may take place to form the complete C–N–O bi-cycle, are (Fowler, 1954; Salpeter, 1955; Marion and Fowler, 1957)

$$\begin{aligned} N^{15} + H^1 &\to O^{16} + \gamma + 12.126\,\text{MeV} \\ O^{16} + H^1 &\to F^{17} + \gamma + 0.601\,\text{MeV} \\ F^{17} &\to O^{17} + e^+ + \nu_e + 2.762\,\text{MeV} - 0.94\,\text{MeV} \\ O^{17} + H^1 &\to N^{14} + He^4 + 1.193\,\text{MeV} \end{aligned} \tag{4.256}$$

Here the energy release includes positron annihilation, and the negative energy release is the average neutrino energy loss. It is possible that the CNO cycle produces most of the N^{14} found in nature, and the details of nucleosynthesis by the CNO process are given by Caughlan and Fowler (1962) and Caughlan (1965). During supernovae explosions, a rapid CNO cycle might take place in which the (n, p) reactions replace the beta decays in the cycle.

Hydrogen burning by the CNO cycle is also reviewed in the text by Rolfs and Rodney (1988); the reaction rate constants are given by Caughlan and Fowler (1988).

Helium burning: the triple alpha and alpha capture processes. The reactions assigned to the triple alpha process are (Salpeter, 1952, 1953, 1957; Öpik, 1951, 1954; Hoyle, 1954; Fowler and Greenstein, 1956; Cook, Fowler, Lauritsen, and Lauritsen, 1957)

$$\begin{aligned} He^4 + He^4 &\to Be^8 - 0.0921\,\text{MeV} \\ Be^8 + He^4 &\to C^{12*} - 0.286\,\text{MeV} \\ C^{12*} &\to C^{12} + \gamma + 7.656\,\text{MeV} \ . \end{aligned} \tag{4.257}$$

The total reaction, $3He^4 \to C^{12} + \gamma + 7.274\,\text{MeV}$, results in the mean lifetime, $\tau_{3\alpha}$, for the destruction of He^4 by the 3α process given by (Fowler, Caughlan, and Zimmerman, 1967; Clayton, 1968; Barnes, 1971)

$$\begin{aligned} \lambda_{3\alpha} = \frac{1}{\tau_{3\alpha}} = \frac{3r_{3\alpha}}{\mathrm{N}_\alpha} &= 6.65 \times 10^{-10} (\rho X_\alpha)^2 T_9^{-3} \exp(-4.405/T_9) \\ &+ 1.66 \times 10^{-8} (\rho X_\alpha)^2 T_9^{-3} \exp(-27.443/T_9) \quad \text{sec}^{-1} \ , \end{aligned} \tag{4.258}$$

or for the forward triple alpha reaction:

$$\begin{aligned} N_\mathrm{A}\langle\sigma v\rangle = 2.05 \times 10^{-8} T_9^{-3} \exp(-4.405/T_9) \\ + 5.315 \times 10^{-7} T_9^{-3} \exp(-27.433/T_9) \quad \text{cm}^3\,\text{sec}^{-1}\,\text{g-mole}^{-1} \ , \end{aligned}$$

where the first term holds for $0.03 \leq T_9 \leq 8$ and the second term is added for $4 \leq T_9 \leq 8$. Here $r_{3\alpha}$ is the reaction rate, N_α is the density of α particles, X_α is their mass fraction, and ρ is the gas mass density. It is now believed that helium burning results in the production of approximately equal amounts of ^{12}C and ^{16}O in stars in the wide range of masses from 0.5 to 50 $M_\odot$.

Once C^{12} is formed, O^{16} will be the product of the α capture process

$$C^{12} + He^4 \to O^{16} + \gamma + 7.161\,\text{MeV}. \tag{4.259}$$

This is another special reaction with the rate (Fowler, Caughlan, and Zimmerman, 1975)

$$N_{\mathrm{A}}\langle\sigma v\rangle = 1.90\times10^{8}T_9^{-2}(1+0.046\,T_9^{-2/3})^{-2}\exp[-32.12\,T_9^{-1/3}-(T_9/3.270)^2] + 3.338\times10^{2}\exp(-26.316/T_9)\quad \mathrm{cm^3\,sec^{-1}\,g\text{-}mole^{-1}}\ . \tag{4.260}$$

The rate of the $^{12}\mathrm{C}(\alpha,\gamma)^{16}\mathrm{O}$ reaction is given by Caughlan and Fowler (1988); a value of 1.7 ± 0.5 times this rate is suggested by Weaver and Woosley (1993). Further α capture processes which then follow are

$$\begin{aligned}
&\mathrm{O}^{16} + \mathrm{He}^4 \rightarrow \mathrm{Ne}^{20} + \gamma + 4.730\ \mathrm{MeV}\\
&\mathrm{Ne}^{20} + \mathrm{He}^4 \rightarrow \mathrm{Mg}^{24} + \gamma + 9.317\ \mathrm{MeV}\\
&\mathrm{Mg}^{24} + \mathrm{He}^4 \rightarrow \mathrm{Si}^{28} + \gamma + 9.981\ \mathrm{MeV}\\
&\mathrm{Si}^{28} + \mathrm{He}^4 \rightarrow \mathrm{S}^{32} + \gamma + 6.948\ \mathrm{MeV}\\
&\mathrm{S}^{32} + \mathrm{He}^4 \rightarrow \mathrm{Ar}^{36} + \gamma + 6.645\ \mathrm{MeV}\ .
\end{aligned} \tag{4.261}$$

The reaction products of the C–N–O cycle might also produce neutrons by the α-capture reactions (Cameron, 1954)

$$\mathrm{C}^{13} + \mathrm{He}^4 \rightarrow \mathrm{O}^{16} + \mathrm{n} + 2.214\ \mathrm{MeV}$$

or (Cameron, 1960)

$$\begin{aligned}
&\mathrm{N}^{14} + \mathrm{He}^4 \rightarrow \mathrm{F}^{18} + \gamma + 4.416\ \mathrm{MeV}\\
&\mathrm{F}^{18} \rightarrow \mathrm{O}^{18} + \mathrm{e}^+ + \nu_{\mathrm{e}}\\
&\mathrm{O}^{18} + \mathrm{He}^4 \rightarrow \mathrm{Ne}^{22} + \gamma + 9.667\ \mathrm{MeV}\\
&\mathrm{O}^{18} + \mathrm{He}^4 \rightarrow \mathrm{n} + \mathrm{Ne}^{21} - 0.699\ \mathrm{MeV}\\
&\mathrm{Ne}^{22} + \mathrm{He}^4 \rightarrow \mathrm{n} + \mathrm{Mg}^{25} - 0.481\ \mathrm{MeV}\ .
\end{aligned} \tag{4.262}$$

Helium burning is reviewed in detail in the monograph by Rolfs and Rodney (1988), and placed within the context of stellar evolution of both low mass and high mass stars in the text by Kippenhahn and Weigert (1990). Reaction rate constants are given by Caughlan and Fowler (1988).

Carbon and oxygen burning. At the conditions of helium burning, the predominant nuclei are C^{12} and O^{16}. When temperatures greater than 8×10^8 °K are reached, carbon will begin to react with itself according to the reactions

$$\begin{aligned}
\mathrm{C}^{12} + \mathrm{C}^{12} &\rightarrow \mathrm{Mg}^{24} + \gamma + 13.930\ \mathrm{MeV}\\
&\rightarrow \mathrm{Na}^{23} + \mathrm{p} + 2.238\ \mathrm{MeV}\\
&\rightarrow \mathrm{Ne}^{20} + \mathrm{He}^4 + 4.616\ \mathrm{MeV}\\
&\rightarrow \mathrm{Mg}^{23} + \mathrm{n} - 2.605\ \mathrm{MeV}\\
&\rightarrow \mathrm{O}^{16} + 2\mathrm{He}^4 - 0.114\ \mathrm{MeV}\ .
\end{aligned} \tag{4.263}$$

Resonance energies and partial widths of the compound states formed by two carbon nuclei are discussed by Imanishi (1968, 1969).

At about 2×10^9 °K, oxygen will also react with itself according to the reactions

$$\begin{aligned} O^{16} + O^{16} &\rightarrow S^{32} + \gamma + 16.539 \text{ MeV} \\ &\rightarrow P^{31} + p + 7.676 \text{ MeV} \\ &\rightarrow S^{31} + n + 1.459 \text{ MeV} \\ &\rightarrow Si^{28} + He^4 + 9.593 \text{ MeV} \\ &\rightarrow Mg^{24} + 2He^4 - 0.393 \text{ MeV} . \end{aligned} \tag{4.264}$$

Patterson, Winkler, and Spinka (1968), Patterson, Winkler, and Zaidins (1969), and Spinka and Winkler (1972) have measured the cross sections for several of these reactions. They find that the large interaction radius of the heavy ions requires an extra term in the cross section, so that the average cross section $\langle\sigma\rangle$ is given by

$$\langle\sigma\rangle = \frac{\langle S\rangle}{E} \exp[-(E_G/E)^{1/2} - gE] , \tag{4.265}$$

where E is the kinetic energy in the center-of-mass system, the Gamow energy, E_G, is given by Eq. (4.33) and

$$g = 0.122(AR_f^3/Z_1Z_2)^{1/2}\text{MeV}^{-1} , \tag{4.266}$$

where A is the reduced mass of the two interacting particles in a.m.u., Z_1 and Z_2 are the particle charges in units of the proton charge, and the interaction radius, R_f, is given in Fermis. For the $C^{12} + C^{12}$ reaction $\langle S\rangle = 2.9 \times 10^{16}$ MeV-barns and $g = 0.46$ MeV^{-1}, whereas reaction $\langle S\rangle = 6 \times 10^{27}$ MeV-barns and $g = 0.84$ MeV^{-1} for the $O^{16} + O^{16}$ reactions above 7 MeV. These values result in a $C^{12} + C^{12}$ reaction rate of (Arnett and Truran, 1969)

$$\rho N_A\langle\sigma v\rangle = 3.24 \times 10^{26} T_9^{-2/3} \exp[-84.17\, T_9^{-1/3}(1 + 0.037\, T_9)^{1/3}] \quad \text{sec}^{-1} , \tag{4.267}$$

and an $O^{16} + O^{16}$ reaction rate of (Truran and Arnett, 1970)

$$\begin{aligned} N_A\langle\sigma v\rangle = 6.6 \times 10^{37} T_9^{-2/3} \exp[-135.79\, T_9^{-1/3} \\ (1 + 0.0530\, T_9)^{1/3}] \quad \text{cm}^3\,\text{sec}^{-1}(\text{g-mole})^{-1} . \end{aligned} \tag{4.268}$$

Hydrostatic oxygen burning is treated by Arnett (1972) and Woosley, Arnett and Clayton (1972), whereas explosive oxygen burning is considered by Woosley, Arnett, and Clayton (1973). Such explosive nucleosynthesis is discussed in the monograph by Rolfs and Rodney (1988), reviewed by Arnett (1995) and delineated by Woosley and Weaver (1995). Also see earlier work by Arnett (1974, 1978) and Cameron (1979).

The α particles, protons, and neutrons which are products of the carbon and oxygen burning reactions given in Eqs. (4.263) and (4.264) will interact with the other products of the burning to form many of the other nuclides with $16 \leq A \leq 28$.

It is now thought that most of the carbon, oxygen, and silicon burning, which accounts for the observed solar system abundances for $20 \leq A \leq 64$, occurs during fast explosions, and these explosive burning processes are discussed in Sect. 4.4.4.

Silicon burning. At the conclusion of carbon and oxygen burning, the most abundant nuclei will be S^{32} and Si^{28} with significant amounts of Mg^{24}. Because the binding energies for protons, neutrons and α particles in S^{32} are smaller than those in Si^{28}, the nuclide S^{32} will be the first to photodisintegrate according to the reactions

$$\begin{aligned} &S^{32} + \gamma \to P^{31} + p - 8.864 \text{ MeV} \\ &P^{31} + \gamma \to Si^{30} + p - 7.287 \text{ MeV} \\ &Si^{30} + \gamma \to Si^{29} + n \\ &Si^{29} + \gamma \to Si^{28} + n \ . \end{aligned} \tag{4.269}$$

The resulting reactions will leave little but Si^{28}. Silicon will then begin to photodisintegrate at temperatures greater than 3×10^9 °K according to the reactions

$$\begin{aligned} &Si^{28} + \gamma \to Al^{27} + p - 11.583 \text{ MeV} \\ &Si^{28} + \gamma \to Mg^{24} + \alpha - 9.981 \text{ MeV} \ . \end{aligned} \tag{4.270}$$

As the (γ, α) reaction has the lower threshold, it is the dominant reaction at low temperatures $T < 2 \times 10^9$ °K; whereas the (γ, p) reaction has the shorter lifetime at higher temperatures. Further photodisintegrations lead to the build-up of lighter elements according to the reactions

$$\begin{aligned} &Mg^{24} + \gamma \to Na^{23} + p - 11.694 \text{ MeV} \\ &Mg^{24} + \gamma \to Ne^{20} + \alpha - 9.317 \text{ MeV} \\ &Ne^{20} + \gamma \to O^{16} + \alpha - 4.730 \text{ MeV} \\ &O^{16} + \gamma \to C^{12} + \alpha - 7.161 \text{ MeV} \ . \end{aligned} \tag{4.271}$$

Resonance strengths for the various resonance energies of the $^{24}Mg(\alpha, \gamma)^{28}Si$ and $^{27}Al(p, \gamma)^{28}Si$ reactions are tabulated, respectively, by Lyons (1969) and Lyons, Toevs, and Sargood (1969). Analytic fits to experimental data for the $^{20}Ne(\alpha, \gamma)^{24}Mg$ and $^{23}Na(p, \gamma)^{24}Mg$ reactions may lead to the rates (Couch and Shane, 1971; Fowler, Caughlan, and Zimmerman, 1974).

For $Ne^{20} + \alpha \to Mg^{24} + \gamma$

$$\begin{aligned} N_A \langle \sigma v \rangle_{\alpha,\gamma} = &\frac{1.235 \times 10^{18}}{T_9^{2/3}} \exp\left[\frac{-46.765}{T_9^{1/3}} - \left(\frac{T_9}{0.849}\right)^2\right] \\ &+ 8.947 \times 10^3 \exp\left(\frac{-15.576}{T_9}\right) \quad \text{cm}^3\,\text{sec}^{-1}(\text{g-mole})^{-1} \ . \end{aligned} \tag{4.272}$$

For $Na^{23} + p \to Mg^{24} + \gamma$

$$\begin{aligned} N_A \langle \sigma v \rangle_{p,\gamma} = &\frac{2.263 \times 10^{13}}{T_9^{2/3}} \exp\left[\frac{-20.766}{T_9^{1/3}} - \left(\frac{T_9}{0.315}\right)^2\right] \\ &+ 4.938 \times 10^4 \exp\left(\frac{-3.564}{T_9}\right) \quad \text{cm}^3\,\text{sec}^{-1}(\text{g-mole})^{-1} \ . \end{aligned} \tag{4.273}$$

The corresponding photodisintegration rates for $1.0 \leq T_9 \leq 5.0$ are tabulated by Couch and Shane (1971). For the $^{20}Ne(\alpha, \gamma)^{24}Mg$ reaction, we have $C_{13} = 108.118$ and $C_{14} = 6.011 \times 10^{10}$, whereas for the $^{23}Na(p, \gamma)^{24}Mg$ reaction, we have $C_{13} = 135.708$ and $C_{14} = 7.489 \times 10^{10}$. These constants can be used in Eq. (4.253) to determine photodisintegration rates.

The abundances of some of the nuclei in the range $28 \lesssim A \lesssim 60$ are thought to be determined by equilibrium or quasi-equilibrium processes in which the importance of many individual reaction rates is diminished, and these processes are discussed in the next Sect. 4.4.3. Abundances of the elements with $28 \leq A \leq 60$ have been discussed by Hoyle (1946), Burbidge, Burbidge, Fowler, and Hoyle (1957), Fowler and Hoyle (1964), Clifford and Tayler (1965), Truran, Cameron, and Gilbert (1966), Bodansky, Clayton, and Fowler (1968), Truran (1968), and Michaud and Fowler (1971). Most nuclear species between ^{28}Si and ^{59}Co, except the neutron-rich species (^{36}S, ^{40}Ar, ^{43}Ca, ^{46}Ca, ^{48}Ca, ^{51}Ti, ^{54}Cr, and ^{58}Fe), are generated by a quasi-equilibrium process in which the only important thermonuclear reaction rates are thought to be those of the "bottle-neck" nuclei ^{44}Ca, ^{45}Sc, and ^{45}Ti. The reaction rates for these nuclei may be estimated by the methods given by Michaud and Fowler (1970). The abundances of the neutron-rich species could well be determined by the *s* or *r* processes discussed in the next section (Seeger, Fowler, and Clayton, 1965). They may, however, be formed in the explosive carbon burning stage (Howard, Arnett, Clayton, and Woosley, 1972).

The s, r and p processes. Because the binding energy per nucleon decreases with increasing A for nuclides beyond the iron peak ($A \gtrsim 60$), and because these elements have large Coulomb barriers, they are not likely to be formed by fusion or alpha and proton capture. It is thought that most of these heavier elements are formed by neutron capture reactions which start with the iron group nuclei (Cr, Mn, Fe, Co, and Ni). If the flux of neutrons is weak, most chains of neutron captures will include only a few captures before the beta decay of the product nucleus. Because the neutron capture lifetime is slower (*s*) than the beta decay lifetime, this type of neutron capture is called the *s* process. When there is a strong neutron flux, as it is believed to occur during a supernova explosion, the neutron-rich elements will be formed by the rapid (*r*) neutron capture process. In this case, the neutron capture lifetime is much less than that of beta decay, and the neutron-rich elements only beta decay to beta stable elements after the capture processes are over. The details of the *s* and *r* processes are discussed by Burbidge, Burbidge, Fowler, and Hoyle (1957), and the appropriate neutron capture rates and nuclide abundance equations are given in Sect. 4.4.5. The *r* and *s* processes are discussed in the monograph by Rolfs and Rodney (1988), and nucleosynthesis by the *r*, *s* and *p* processes is reviewed by Meyer (1994). Earlier papers on these processes include those by Butcher (1975, 1976), Schramm (1973) and Schramm and Tinsley (1974). Although the source of neutrons for the *s* process is uncertain, the most likely site of the *s* process synthesis is provided by the convective mixing of hydrogen and helium layers following a helium flash (Schwarzschild and Härm, 1962, 1967). In this case free neutrons are formed by the reactions (Sanders, 1967;

Cameron and Fowler, 1971)

$$C^{12}(p,\gamma)N^{13}(e^{+}\nu)C^{13}$$

and

$$C^{13}(\alpha,n)O^{16} \ . \qquad (4.274)$$

The site of r process synthesis is thought to be related to the formation of a neutron star, and, perhaps, the formation of a supernova explosion (Truran, Arnett, Tsuruta, and Cameron, 1968). The proton rich heavy elements are much less abundant than the elements thought to be produced by the r and s processes, and are thought to be formed by a rapid proton capture process. A possible occasion for this process is the passage of a supernova shock wave through the hydrogen outer layer of a pre-supernova star. Possible reactions for this process include the (p, γ), (p, n), (γ, n) and (n,2n) reactions and positron capture (Burbidge, Burbidge, Fowler, and Hoyle, 1957; Ito, 1961; Reeves and Stewart, 1965; Macklin, 1970). The p process in explosive nucleosynthesis has been discussed by Truran and Cameron (1972). Explosive r-process nucleosynthesis is considered by Schramm (1973).

4.4.3 Equilibrium Processes

Under conditions of thermodynamic equilibrium, the composition of matter in a star may be calculated without determining individual reaction rates, and only the binding energies and partition functions of the various nuclear species need to be specified. Although this condition greatly simplifies calculations, it is only possible if the matter is in equilibrium with the radiation, and if every nucleus is transformable into any other nucleus. Hoyle (1946) showed that matter is in equilibrium with radiation for temperatures $T \approx 10^9$ °K, and that all known nuclei may be transformed into any other nucleus by nuclear reactions for $T \gtrsim 2 \times 10^9$ °K. Early investigations by Tolman (1922), Urey and Bradley (1931), Pokrowski (1931), Sterne (1933), and Chandrasekhar and Henrich (1942) ruled out an equilibrium origin for most of the elements. Nevertheless, the abundance peak in the iron group, where the binding energy per nucleon is a maximum, can be accounted for by an equilibrium process (Hoyle, 1946; Burbidge, Burbidge, Fowler, and Hoyle, 1957). More detailed calculations of the equilibrium concentrations in this region ($28 \leq A \leq 60$) have been carried out by Fowler, and Hoyle (1964), Clifford and Tayler (1965), and Truran, Cameron, and Gilbert (1966) for strict equilibrium conditions; and by Bodansky, Clayton, and Fowler (1968), Truran (1968), Michaud and Fowler (1972) and Woosley, Arnett, and Clayton (1972), for quasi-equilibrium conditions.

Under conditions of statistical equilibrium, the number density, N_i, of particles of the ith kind is given by

$$N_i = \frac{1}{V}\sum_{r}^{n} \mu_i[\pm\mu_i + \exp(\varepsilon_{ir}/kT)]^{-1} \ , \qquad (4.275)$$

where V is the volume, μ_i is the chemical potential of the ith particle, the plus and minus signs refer to Fermi–Dirac and Bose–Einstein statistics, respectively, and the summation is over all energy states, ε_{ir}, which includes both internal energy levels and the kinetic energy. If an internal level has spin, J, then $2J+1$ states of the same energy must be included in the sum. When the nuclides are non-degenerate and non-relativistic, Maxwellian statistics can be employed to give

$$N_i = \frac{\mu_i}{V}\left[\sum(2J_r+1)\exp\left(-\frac{\varepsilon_r}{kT}\right)\right]\left[\frac{4\pi V}{h^3}\int_0^\infty p^2\exp\left(-\frac{p^2}{2M_ikT}\right)dp\right]$$

$$= \mu_i\omega_i\left(\frac{2\pi M_ikT}{h^2}\right)^{3/2}, \tag{4.276}$$

where p is the particle momentum, M_i is its mass, the partition function $\omega_i = \sum(2J_r+1)\exp(-\varepsilon_r/kT)$, and here ε_r refers to internal energy states only. For particles $p_i, p_j, \ldots$ which react according to

$$\alpha p_i + \beta p_j + \cdots \leftrightarrows \xi p_r + \eta p_s + \cdots \ , \tag{4.277}$$

the chemical potentials are related by the equation

$$\mu_i^\alpha \mu_j^\beta \cdots = \mu_r^\xi \mu_s^\eta \cdots \exp(-Q/kT), \tag{4.278}$$

where

$$Q = c^2[\alpha M_i + \beta M_j + \cdots - \xi M_r - \eta M_s - \cdots] \ . \tag{4.279}$$

Hoyle (1946) and Burbidge, Burbidge, Fowler, and Hoyle (1957) considered the condition of statistical equilibrium between the nuclei, (A, Z), and free protons, p, and neutrons, n. For a nucleus, there are Z protons and ($A - Z$) neutrons, and the statistical weight of both protons and neutrons is two. It then follows from Eqs. (4.276) to (4.279) that for equilibrium between nuclides, protons, and neutrons, the number density, $N(A, Z)$, of the nucleus, (A, Z), is given by

$$N(A,Z) = \omega(A,Z)\left(\frac{AM_\mu kT}{2\pi\hbar^2}\right)^{3/2}\left(\frac{2\pi\hbar^2}{M_\mu kT}\right)^{3A/2}\frac{N_{\mathrm{n}}^{(A-Z)}N_{\mathrm{p}}^Z}{2^A}\exp\left[\frac{Q(A,Z)}{kT}\right], \tag{4.280}$$

where the partition function, $\omega(A,Z)$, of the nucleus, (A,Z), is given by

$$\omega(A,Z) = \sum_r(2I_r+1)\exp\left(-\frac{E_r}{kT}\right), \tag{4.281}$$

where I_r and E_r are, respectively, the spin and energy of the rth excited level, the binding energy, $Q(A, Z)$, of the nucleus, (A, Z), is given by

$$Q(A,Z) = c^2[(A-Z)M_{\mathrm{N}} + ZM_{\mathrm{p}} - M(A,Z)] \ , \tag{4.282}$$

where $M_{\mathrm{N}}, M_{\mathrm{p}}$ and $M(A, Z)$ are, respectively, the masses of the free neutrons, free protons, and the nucleus, (A, Z), the factor

$$\left(\frac{2\pi\hbar^2}{M_\mu kT}\right)^{3/2} \approx 1.6827\times10^{-34}T_9^{-3/2} \quad \mathrm{cm}^{-3} \ , \tag{4.283}$$

the atomic mass unit is M_μ, and N_{n} and N_{p} denote, respectively, the number densities of free neutrons and protons. Conservation of mass requires that the mass density, ρ, be given by

$$\rho = N_p M_p + N_n M_N + \sum_{A,Z} N(A,Z) M(A,Z) \ . \tag{4.284}$$

Conservation of charge further implies that

$$N_p + \sum_{A,Z} Z N(A,Z) = N_{e^-} - N_{e^+} \ , \tag{4.285}$$

where N_{e^-} and N_{e^+} denote, respectively, the number densities of electrons and positrons. Clifford and Tayler (1965) use the additional parameter

$$R = \frac{N_P + \sum_{A,Z} Z N(A,Z)}{N_n + \sum_{A,Z} (A-Z) N(A,Z)} = \frac{\langle Z \rangle}{\langle N \rangle} \tag{4.286}$$

to denote the ratio of the total number of protons to the total number of neutrons. They use the above equations to calculate nuclidic abundances as a function of ρ, T, and R for conditions with and without beta decay. Tsuruta and Cameron (1965) consider nuclear statistical equilibrium at high densities and use T and the Fermi energy, E_F, as free parameters.

$$\begin{aligned} E_F &= (3\pi^2)^{2/3} \frac{\hbar^2}{2m_e} N_e^{2/3} \quad \text{(nonrelativistic)} \\ E_F &= (3\pi^2)^{1/3} \hbar c N_e^{1/3} \quad \text{(relativistic)} \ , \end{aligned} \tag{4.287}$$

where N_e is the electron density. When positive and negative beta decay and electron capture rates are considered together with the steady state condition that the rate of neutrino emission is equal to the antineutrino emission rate, a specification of E_F gives the values of N_n and N_p to be used in Eq. (4.280). The appropriate beta decay and electron capture rates were given in Sect. 4.3.

Truran, Cameron, and Gilbert (1966) noted that at temperatures greater than 3×10^9 °K the photodisintegration of silicon proceeds rapidly, releasing protons, neutrons and alpha particles. They calculated the synthesis of iron peak nuclei from ^{28}Si nuclei by solving thermonuclear reaction rates for all the neutron, proton and alpha particle reactions. Bodansky, Clayton, and Fowler (1968) then showed that the nuclei are in a quasi-equilibrium during silicon burning. It is a partial equilibrium in that ^{28}Si is not itself in equilibrium with the light particles, and the heavy nuclei are not in equilibrium among themselves; but nuclei heavier than ^{28}Si are in equilibrium with ^{28}Si under the exchange of photons, nucleons, and alpha particles. The light particles are ejected at a significant rate by the $^{28}\mathrm{Si}(\gamma,\alpha)^{24}\mathrm{Mg}$ and the $^{28}\mathrm{Si}(\gamma,\mathrm{p})(\gamma,\mathrm{p})(\gamma,\mathrm{n})(\gamma,\mathrm{n})^{24}\mathrm{Mg}$ reactions. The concentration of heavy nuclei builds up to such a value that the nuclei consume and liberate alpha particles at almost the same rate, and a quasi-equilibrium is established with the equilibrium reactions

$$\begin{aligned} &^{28}\mathrm{Si} + {^4}\mathrm{He} \leftrightarrows {^{32}}\mathrm{S} + \gamma \\ &^{32}\mathrm{S} + {^4}\mathrm{He} \leftrightarrows {^{36}}\mathrm{Ar} + \gamma \\ &\vdots \\ &^{52}\mathrm{Fe} + {^4}\mathrm{He} \leftrightarrows {^{56}}\mathrm{Ni} + \gamma \ . \end{aligned} \tag{4.288}$$

It follows from Eqs. (4.276) to (4.279) that the equilibria are characterized by Saha equations such as

$$\frac{N(^{32}\mathrm{S})}{N(^{28}\mathrm{Si})} = N_\alpha \frac{\omega(^{32}\mathrm{S})}{\omega(^{28}\mathrm{Si})} \left[\frac{A(^{32}\mathrm{S})}{A(^{28}\mathrm{Si})A(^{4}\mathrm{He})}\right]^{3/2} \left(\frac{2\pi\hbar^2}{M_\mu kT}\right)^{3/2} \exp\left[\frac{B_\alpha(^{32}\mathrm{S})}{kT}\right] \tag{4.289}$$

for

$$^{28}\mathrm{Si} + {}^{4}\mathrm{He} \leftrightarrows {}^{32}\mathrm{S} + \gamma \ .$$

Here N_α is the alpha particle number density, the partition functions, ω, are approximately unity for the alpha particle nuclei, the A's are atomic masses in units of the atomic mass unit, M_μ, and $B_\alpha(^{32}\mathrm{S})$ is the separation energy of an alpha particle from the ground state of $^{32}\mathrm{S}$. The equilibrium abundances, $N(A, Z)$, of other nuclei, (A, Z), depend on the number densities of $^{28}\mathrm{Si}$, as well as those of the alpha particles, neutrons, and protons. The abundance formula for this case is given in Sect. 4.4.5. Bodansky, Clayton, and Fowler (1968) have shown that quasi-equilibrium distributions account for the observed abundances of the $4n$ nuclei ($A = 4n$) for $A = 28$ through $A = 56$, as well as for the dominant nuclei in the lower part of the iron group ($A = 49$ through $A = 57$). Truran (1968) examined the effect of a variable initial composition on this silicon burning process, and Michaud and Fowler (1972) examined the dependence of the synthesized abundances on the initial composition before silicon burning, as well as the effect of adding products from different burning zones. The latter authors show that with an initial neutron enhancement of 4×10^{-3} the natural abundances for nuclei with $28 \lesssim A \lesssim 59$ may be accounted for by quasi-equilibrium burning. In this case, a quasi-equilibrium between elements with $24 \leq A \leq 44$ and a separate equilibrium for elements with $46 \leq A \leq 60$ is assumed, and detailed nuclear reactions are given for the "bottleneck" at $A = 45$. This quasi-equilibrium silicon burning process must have taken place in a short time, $t \lesssim 1$ sec, and at high temperatures, $T \gtrsim 4.5 \times 10^9$ °K, suggesting the explosive burning processes discussed in the next section.

4.4.4 Explosive Burning Processes

As explained by Burbidge, Burbidge, Fowler, and Hoyle (1957), the successive cycles of static nuclear burning and contraction which successfully account for much of stellar evolution, must end when the available nuclear fuel is exhausted. They showed that the unopposed action of gravity in a helium exhausted stellar core leads to violent instabilities and to rapid thermonuclear reactions in the stellar envelope. Schwarzschild and Härm (1962, 1967) showed that even for a hydrogen exhausted core the thermonuclear reactions runaway in a helium flash. Arnett (1968) then showed that when cooling by neutrino emission in a highly degenerate gas is considered, the $\mathrm{C}^{12} + \mathrm{C}^{12}$ reaction will ignite explosively at core densities of about $2 \times 10^9 \mathrm{g\ cm}^{-3}$. The stellar material is instantaneously heated and then expands adiabatically so that the density, ρ, and temperature, T, are related by

$$\rho(t) \propto [T(t)]^3 \ , \tag{4.290}$$

for a $\Gamma_3 = 4/3$ adiabat, and a time variable, t. The appropriate expansion time is the hydrodynamic time scale, τ_{HD}, given by

$$\tau_{\mathrm{HD}} = (24\pi G\rho)^{-1/2} \approx 446\rho^{-1/2} \quad \mathrm{sec} \ . \tag{4.291}$$

The initial temperature and density must be such that the mean lifetime, τ_{R}, for a nucleus undergoing an explosive reaction, R, must be close to τ_{HD}. For the interaction of nucleus 1 with a nucleus 2,

$$\tau_{\mathrm{R}} = \tau_2(1) = [N_2\langle\sigma v\rangle]^{-1} = \left[\rho N_{\mathrm{A}} \frac{X_2}{A_2} \langle\sigma v\rangle\right]^{-1} \ , \tag{4.292}$$

where the mass density is ρ, the X_2, A_2 and N_2 are, respectively, the mass fraction, mass number, and number density of nucleus 2, and the $N_{\mathrm{A}}\langle\sigma v\rangle$ were given in Sect. 4.4.2 for various reactions (cf. Eq. (4.251)). Arnett (1969) used a mean carbon nucleus lifetime, $\tau_{\mathrm{C}^{12}}$, given by

$$\log \tau_{\mathrm{C}^{12}} \approx 37.4\, T_9^{-1/3} - 25.0 - \log_{10}\rho \approx \log_{10}\tau_{\mathrm{HD}} \tag{4.293}$$

for carbon burning to determine the initial conditions of explosive carbon burning. Known reaction rates were then used with Eqs. (4.290) and (4.293) to calculate expected abundances using the numerical scheme given in Sect. 4.4.5. Abundance ratios which closely approximate those of the solar system were found for

$$^{20}\mathrm{Ne},\ ^{23}\mathrm{Na},\ ^{24}\mathrm{Mg},\ ^{25}\mathrm{Mg},\ ^{26}\mathrm{Mg},\ ^{27}\mathrm{Al},\ ^{29}\mathrm{Si},\ \text{and}\ ^{30}\mathrm{Si} \ , \tag{4.294}$$

when it was assumed that a previous epoch of helium burning produced equal amounts of C^{12} and O^{16}, and that

$$\begin{aligned} T_{\mathrm{p}} &= 2 \times 10^9\ {}^\circ\mathrm{K} \\ \rho_{\mathrm{p}} &= 1 \times 10^5\ \mathrm{g\ cm^{-3}} \end{aligned} \tag{4.295}$$

and

$$\eta = 0.002 \ .$$

Here T_{p} and ρ_{p} denote, respectively, the peak values of temperature and mass density in the shell under consideration, and the neutron excess, η, is given by

$$\eta = \frac{N_{\mathrm{n}} - N_{\mathrm{p}}}{N_{\mathrm{n}} + N_{\mathrm{p}}} \ , \tag{4.296}$$

where N_{n} and N_{p} denote, respectively, the number densities of free neutrons and protons. Hansen (1971) used a full energy generation equation to obtain similar results for explosive carbon burning. Arnett, Truran, and Woosley (1971) have integrated the composition changes in a ^{12}C-detonation to show that most of the solar system iron group abundances can be explained by explosive burning.

Truran and Arnett (1970) have considered explosive oxygen burning for which the mean lifetime, $\tau_{\mathrm{O}^{16}}$, of a ^{16}O nucleus is given by

$$\tau_{\mathrm{O}^{16}} = [X_{16}\rho N_{\mathrm{A}}\langle\sigma v\rangle/16]^{-1} \ , \tag{4.297}$$

where X_{16} is the fractional abundance by mass of ^{16}O,

$$N_{\rm A}\langle\sigma v\rangle \approx \exp(86.804 - \xi)T_9^{-2/3} \ , \tag{4.298}$$

where $T_9 = T/10^9$, and

$$\xi = 135.79(1 + 0.053\,T_9)^{1/3}T_9^{-1/3} \ . \tag{4.299}$$

By equating $\tau_{\rm O^{16}}$ to $\tau_{\rm HD}$ for initial conditions, assuming an adiabatic expansion, and calculating the appropriate relative abundances, they have shown that nuclei of intermediate mass ($28 \lesssim A \lesssim 42$) are created by explosive oxygen burning with abundance ratios close to those of the solar system. The appropriate conditions are

$$\begin{aligned} &3.5 \times 10^9 \lesssim T \lesssim 3.7 \times 10^9\ ^\circ\mathrm{K} \\ &10^5 \lesssim \rho \lesssim 10^6\ \mathrm{g\ cm^{-3}} \end{aligned} \tag{4.300}$$

and

$$\eta \approx 0.002 \ .$$

Similarly, Arnett (1969) has used the $\tau_{\rm Si^{28}}$ of Bodansky, Clayton, and Fowler (1968) to calculate explosive silicon burning, and to account for abundance ratios in the range $44 \le A \le 62$. Howard, Arnett, and Clayton (1971) have also considered explosive burning in helium zones and indicate that the nuclei ^{15}N,^{18}O,^{19}F, and ^{21}Ne are created primarily as ^{15}O,^{18}F,^{19}Ne and ^{21}Ne in the explosive nuclear processes in helium zones. The general success of explosive carbon, oxygen and silicon burning in accounting for solar system abundance ratios for $20 \le A \le 64$ has been summarized by Arnett and Clayton (1970).

Explosive nucleosynthesis has been incorporated in models of galactic evolution by Arnett (1971) and Truran and Cameron (1971). Woosley, Arnett, and Clayton (1973) give a thorough treatment of explosive oxygen and silicon burning together with a treatment of the equilibrium process at high and low freeze-out densities. A general review of explosive nucleosynthesis in stars is given by Arnett (1973). Explosive nucleosynthesis has been revisited in a review by Arnett (1995). The explosive hydrodynamics and nucleosynthesis of massive stars is discussed by Woosley and Weaver (1995) and explosive burning is included in the monograph by Rolfs and Rodney (1988).

4.4.5 Nuclide Abundance Equations

The equation governing the change in the number density, $N(A, Z)$, of the nucleus, (A, Z), is of the form

$$\frac{d}{dt}(N_i) = -\sum_j N_i N_j \langle\sigma v\rangle_{ij} + \sum_{kl} N_k N_l \langle\sigma v\rangle_{kl} \ , \tag{4.301}$$

where N_i is the number density of the ith species, $\langle\sigma v\rangle_{ij}$ is the product of cross section and the relative velocity for an interaction involving species i and j, the $N_m N_n$ is replaced by $N_n^2/2$ for identical particles, and the summations are over all reactions which either create or destroy the species, i. For numerical work it is convenient to deal with the parameter

$$Y_i = Y(A,Z) = \frac{N(A,Z)}{\rho N_{\mathrm{A}}} = \frac{N_i}{\rho N_{\mathrm{A}}} \quad , \tag{4.302}$$

where ρ is the mass density of the gas under consideration, and Avogadro's number, $N_{\mathrm{A}} = 6.022137 \times 10^{23}(\text{g-mole})^{-1}$. The differential equation linking all reactions which create or destroy the ith nucleus is then given by

$$\frac{d}{dt}(Y_i) = -\sum_j f_{ij} + \sum_{kl} f_{kl} \quad , \tag{4.303}$$

where the vector flow, f_{ij}, which contains nuclei i and j in the entrance channel, is given by

$$f_{ij} = \frac{N_i N_j \langle\sigma v\rangle_{ij}}{\rho N_{\mathrm{A}}} = Y_i Y_j \rho N_{\mathrm{A}} \langle\sigma v\rangle_{ij} \quad , \tag{4.304}$$

and the $N_{\mathrm{A}}\langle\sigma v\rangle_{ij}$ are given in the previous sections. When all of the different types of reactions are taken into account, Eq. (4.303) becomes

$$\begin{aligned}\frac{dY(A,Z)}{dt} = &-\big[\lambda_{\beta}^{-}(A,Z) + \lambda_{\beta+}(A,Z) + \lambda_K(A,Z) + \lambda_\alpha(A,Z)\\ &+ \lambda_\gamma(A,Z) + 2.48 \times 10^8 \sigma_T N_{\mathrm{n}} + \sum_j Y(A_j Z_j)\rho N_{\mathrm{A}}\langle\sigma v\rangle_j\big] Y(A,Z)\\ &+ \lambda_{\beta^-}(A,Z-1)Y(A,Z-1) + \lambda_{\beta^+}(A,Z+1)Y(A,Z+1)\\ &+ \lambda_K(A,Z+1)Y(A,Z+1) + \lambda_\alpha(A+4,Z+2)Y(A+4,Z+2)\\ &+ 2.48\times10^8\sigma_T N_{\mathrm{n}} Y(A-1,Z) + \sum_{ik} Y(A_i,Z_i)Y(A_k,Z_k)\rho N_{\mathrm{A}}\langle\sigma v\rangle_{ik} \quad ,\end{aligned} \tag{4.305}$$

where the symbol λ denotes the decay rate or the inverse mean lifetime, the subscripts β^-, β^+, K, α and γ denote, respectively, negative beta decay, positive beta decay, electron capture, alpha decay, and photodisintegration, σ_T is the cross section for neutron capture in cm^2, N_{n} is the number density of neutrons, the summation $\sum_j$ denotes all reactions between the nucleus (A,Z) and any other nucleus, the summation $\sum_{ik}$ denotes all reactions between two nuclei which have (A,Z) as a product, ρ is the gas mass density, and the reaction rate $N_{\mathrm{A}}\langle\sigma v\rangle$ is given by the superposition of all rates given in the previous sections. When a reaction occurs between n nuclei of the same type, the reaction rates given in Sect. 4.4.2 must be multiplied by $n!$ when used in Eq. (4.305). For example, the reaction rates for the $C^{12} + C^{12}$ and $O^{16} + O^{16}$ reactions would be multiplied by two, whereas the reaction rate for the triple α process would be multiplied by six.

The numerical solution to the complex set of nuclear reaction networks has been given by Arnett and Truran (1969). Assuming that the change in composition over some time interval $\Delta t = t^{n+1} - t^n$ is sufficiently small, the vector flows may be linearized to give

$$f_{ij}^{n+1} = Y_i^{n+1} Y_j^{n+1} [ij]^{n+1} \tag{4.306}$$

$$\approx (Y_i^n Y_j^n + \Delta_i Y_j^n + \Delta_j Y_i^n)[ij]^{n+1} \quad , \tag{4.307}$$

where the beginning and end of the time interval are denoted by superscripts n and $n+1$,

$$\Delta_i = Y_i^{n+1} - Y_i^n \ ,$$

and

$$[ij] = \rho N_{\mathrm{A}} \langle \sigma v \rangle_{ij} \ . \tag{4.308}$$

Similarly, the approximate expression for the vector flow which is symmetric in time is given by

$$f_{ij}^{n+1/2} \approx \left(Y_i^n Y_j^n + \frac{\Delta_i}{2} Y_j^n + \frac{\Delta_j}{2} Y_i^n \right) [ij]^{n+1/2} \ . \tag{4.309}$$

When the time derivative of Y_i is replaced by

$$\frac{d}{dt}(Y_i) = \frac{\Delta_i}{\Delta t} \ , \tag{4.310}$$

we have the coupled set of linear equations (Arnett and Truran, 1969)

$$Y_i^{n+1} a_{ii} + Y_j^{n+1} a_{ij} + Y_k^{n+1} a_{ik} + Y_l^{n+1} a_{il} = -Y_k^n a_{ik} - Y_i^n Y_j^n [ij] \ , \tag{4.311}$$

where

$$a_{ii} = \frac{1}{\Delta t} + Y_j^n [ij]$$

$$a_{ij} = Y_i^n [ij]$$

$$a_{ik} = -Y_l^n [kl]$$

and

$$a_{il} = -Y_k^n [kl] \ .$$

The solution of Eq. (4.311) by iteration is possible when $|a_{ii}| > |a_{ij}| + |a_{ik}| + |a_{il}|$.

In the buildup of nuclei by the s and the r processes, the important reactions are the (n, γ), (γ, n) and beta decay processes with respective inverse lifetimes of

$$\lambda_{\mathrm{n}} = \sigma_{\mathrm{n}} v_{\mathrm{n}} N_{\mathrm{n}} \approx 2.48 \times 10^8 \, \sigma_T N_{\mathrm{n}} \quad \mathrm{sec}^{-1} \ , \tag{4.312}$$

$$\lambda_\gamma = \sigma_\gamma c N_\gamma \ , \tag{4.313}$$

and

$$\lambda_\beta = \frac{f \ln 2}{ft} \approx 10^{-5.5} f \quad \mathrm{sec}^{-1} \ , \tag{4.314}$$

where σ_{n} and σ_γ are, respectively, the cross sections for the (n,γ) and (γ,n) reactions, v_{n} and N_{n} are the velocity and number density of the neutrons responsible for the (n,γ) process, σ_T is given in units of cm^2 and is assumed to be measured at 30 keV, N_γ is the density of photons, γ, and ft and f values are tabulated in the references given in Sect. (4.3.2). The general equation for the s process is

$$\begin{aligned} \frac{dN(A,Z)}{dt} &= \lambda_{\mathrm{n}}(A-1,Z)N(A-1,Z) - \lambda_{\mathrm{n}}(A,Z)N(A,Z) \\ &\quad + \lambda_\beta(A,Z-1)N(A,Z-1) - \lambda_\beta(A,Z)N(A,Z) \\ &\quad + \text{termination terms due to alpha decay at } A > 209 \ . \end{aligned} \tag{4.315}$$

Remembering that $\lambda_n \ll \lambda_\beta$, we have for the *s* process

$$\frac{dN(A,Z)}{dt} = -\lambda_n(A,Z)N(A,Z) + \lambda_n(A-1,Z)N(A-1,Z) \ . \tag{4.316}$$

The general equation for the *r* process is

$$\begin{aligned}\frac{dN(A,Z)}{dt} &= \lambda_n(A-1,Z)N(A-1,Z) - \lambda_n(A,Z)N(A,Z)\\ &+ \lambda_\beta(A,Z-1)N(A,Z-1) - \lambda_\beta(A,Z)N(A,Z)\\ &+ \lambda_\gamma(A+1,Z)N(A+1,Z) - \lambda_\gamma(A,Z)N(A,Z)\\ &+ \text{ termination terms due to fission for } A \geq 260 \ .\end{aligned} \tag{4.317}$$

Assuming that equilibrium is reached between the rapid (n,γ) and (γ,n) processes, and that $\lambda_n \gg \lambda_\beta$, we have for the *r* process

$$\frac{dN(A,Z)}{dt} = -\lambda_\beta(A,Z)N(A,Z) + \lambda_\beta(A,Z-1)N(A,Z-1) \ . \tag{4.318}$$

The general equation for the statistical balance of the *r* process is

$$\log\frac{N(A+1,Z)}{N(A,Z)} = \log N_n - 34.07 - \frac{3}{2}\log T_9 + \frac{5.04}{T_9}Q_n \ , \tag{4.319}$$

where the neutron binding energy, Q_n, is given by

$$Q_n(A,Z) = c^2[M_N + M(A,Z) - M(A+1,Z)] \ , \tag{4.320}$$

and is expressed in MeV, the $T_9 = T/10^9$, and N_n is the number density of neutrons.

At sufficiently high temperatures, $T \gtrsim 3 \times 10^9$ °K, the reactions are so profuse that nearly all nuclei, (A,Z), are converted into other nuclei, (A',Z'), even when Z and Z' are large. When the rates of all nuclear reactions (excepting beta decays) are exactly equal to the rates of the inverse reactions, the nuclear abundances may be determined from statistical considerations similar to those which led to the Saha ionization equation of Sect. 3.3.1.4. In the condition of statistical equilibrium between the nuclei and free protons and neutrons, detailed reaction rates become unnecessary and the number density, $N(A,Z)$, of the nuclide, (A,Z), is given by (Hoyle, 1946; Burbidge, Burbidge, Fowler, and Hoyle, 1957)

$$N(A,Z) = \omega(A,Z)\left(\frac{2\pi\hbar^2}{M_\mu kT}\right)^{3(A-1)/2} A^{3/2}\frac{N_p^Z N_n^{A-Z}}{2^A}\exp\left[\frac{E_B(A,Z)}{kT}\right] \ , \tag{4.321}$$

where the partition function or statistical weight factor, $\omega(A,Z)$, is given by

$$\omega(A,Z) = \sum_r (2I_r + 1)\exp\left(-\frac{E_r}{kT}\right) \ , \tag{4.322}$$

where E_r is the energy of the excited state measured above the ground level and I_r is the spin, M_μ is the atomic mass unit, the factor $[2\pi^2\hbar^2/M_\mu kT]^{3/2} = 1.6827 \times 10^{-34} T_9^{-3/2}\text{cm}^{-3}$, the N_p and N_n denote, respectively, the densities of free protons and neutrons, and the binding energy, $E_B(A,Z)$, of the ground level of the nucleus, (A, Z), is given by

$$E_{\mathrm{B}}(A,Z) = c^2[(A-Z)M_{\mathrm{N}} + ZM_{\mathrm{p}} - M(A,Z)] \ , \tag{4.323}$$

where $M_{\mathrm{N}}, M_{\mathrm{p}}$ and $M(A,Z)$ are the masses of the free neutron, free proton, and the nucleus (A,Z), respectively. Eq. (4.321) can be rewritten in the form

$$\log N(A,Z) = \log \omega(A,Z) + 33.77 + \frac{3}{2}\log(AT_9) + \frac{5.04}{T}E_{\mathrm{B}}(A,Z) + A\left(\log N_{\mathrm{n}} - 34.07 - \frac{3}{2}\log T_9\right) + Z\log\left(\frac{N_{\mathrm{p}}}{N_{\mathrm{n}}}\right) \ , \tag{4.324}$$

where $T_9 = T/10^9$. Provided that nuclear equilibrium is achieved faster than any of the relevant decay rates, the ratio

$$\frac{\langle Z\rangle}{\langle N\rangle} = \frac{\sum ZN(A,Z) + N_{\mathrm{p}}}{\sum(A-Z)N(A,Z) + N_{\mathrm{n}}} \tag{4.325}$$

must be preserved so that the equilibrium $N(A,Z)$ is determined by the density, ρ, temperature, T, and $\langle Z\rangle/\langle N\rangle$. The equilibrium process has been used to determine abundances of the iron group nuclei ($46 \le A \le 60$) by Fowler and Hoyle (1964) and Clifford and Tayler (1965).

Bodansky, Clayton, and Fowler (1968) have shown that if equilibrium is reached between ^{28}Si and the iron group, but not between ^{28}Si and the alpha particles, a quasi-equilibrium condition prevails in which nuclei heavier than ^{28}Si are in equilibrium with ^{28}Si under exchange of protons, neutrons and alpha particles. In this case, the equilibrium number density, $N(A,Z)$, relative to that of ^{28}Si is given by

$$N(A,Z) = C(A,Z)N(^{28}\mathrm{Si})N_{\alpha}^{\delta_{\alpha}}N_{\mathrm{p}}^{\delta_{\mathrm{p}}}N_{\mathrm{n}}^{\delta_{\mathrm{n}}} \ , \tag{4.326}$$

where

$$C(A,Z) = \frac{\omega(A,Z)}{\omega(^{28}\mathrm{Si})}2^{-(\delta_{\mathrm{p}}+\delta_{\mathrm{n}})}[U(A,Z)]^{3/2}\left(\frac{2\pi\hbar^2}{M_{\mu}kT}\right)^{3(\delta_{\alpha}+\delta_{\mathrm{p}}+\delta_{\mathrm{n}})/2} \times \exp\left[\frac{Q(A,Z)}{kT}\right],$$

and

$$U(A,Z) = \frac{A(A,Z)}{A(^{28}\mathrm{Si})}A(^4\mathrm{He})^{-\delta_{\alpha}}A_{\mathrm{p}}^{-\delta_{\mathrm{p}}}A_{\mathrm{n}}^{-\delta_{\mathrm{n}}} \ ,$$

where $A(A,Z)$ is the atomic mass of the nuclide, (A,Z), in a.m.u., A_i is the atomic weight of element, i,

$$Q(A,Z) = E_{\mathrm{B}}(A,Z) - E_{\mathrm{B}}(^{28}\mathrm{Si}) - \delta_{\alpha}E_{\mathrm{B}}(^4\mathrm{He}) \ , \tag{4.327}$$

where $E_{\mathrm{B}}(A,Z)$ is the binding energy of nuclide, (A,Z), given by Eqs. (4.10) or (4.19), and $\delta_{\alpha}, \delta_{\mathrm{p}}$ and δ_{n} specify, respectively, the number of free alpha-particles, protons, or neutrons in (A,Z) in excess of the number in ^{28}Si. If the largest alpha-particle nucleus contains N^1 neutrons and Z^1 protons, then

$$\delta_{\alpha} = \tfrac{1}{4}(N^1 + Z^1 - 28), \quad \delta_{\mathrm{p}} = Z - Z^1, \quad \text{and} \quad \delta_{\mathrm{n}} = N - N^1 \ . \tag{4.328}$$

The partition function, $\omega(A,Z)$, for the alpha-particles is unity, whereas it is two for the neutron and proton. Bodansky, Clayton, and Fowler (1968) have tabulated $\delta_{\alpha}, \delta_{\mathrm{p}}$ and $\delta_{\mathrm{n}}, Q(A,Z), \omega(A,Z)$ and $C(A,Z)$ for $24 \le A \le 62$ and for $T_9 = 3.8, 4.4$ and 5.0.

4.4.6 Formation of the Rare Light Elements – Spallation Reactions

The rare light nuclei, lithium, beryllium, and boron, are not the products of stellar nucleosynthesis, and are, in fact, destroyed in hot stellar interiors. This condition is reflected in the comparatively low abundances of these nuclei (see Fig. 4.6). The fact that these light elements do exist, however, means that they must arise in nuclear reactions of a non-thermonuclear character in relatively cool and moderately dense regions.

Lithium, beryllium and boron (Li, Be, B), sometimes collectively referred to as the *l*-elements (*l* for light), are thought to be produced through the shattering or fragmentation of larger, more abundant interstellar nuclei, such as carbon and oxygen, by high-energy cosmic ray particles. (The most abundant and energetic cosmic-ray particles are protons and α-particles or helium nuclei; cosmic rays are discussed in greater detail in Sect. 4.5.3.1). Such a spallation origin of the rare light elements was initially suggested by Bradt and Peters (1950), and applied to proton collisions on stellar surfaces by Fowler, Burbidge and Burbidge (1955). Then Reeves, Fowler and Hoyle (1970), Mitler (1970) and Meneguzzi, Audouze and Reeves (1971) showed that the stellar and solar system abundances of ^{6}Li, ^{7}Li, ^{9}Be, ^{10}B and ^{11}B, might be accounted for if they are produced by the spallation reactions of cosmic ray particles with interstellar carbon, ^{12}C, nitrogen, ^{14}N, and oxygen ^{16}O, over the lifetime of the Galaxy (also see Reeves, Audouze, Fowler and Schramm, 1973; Audouze and Tinsley, 1974; Reeves and Meyer, 1978; and Walker, Mathews and Viola, 1985). The relevant nuclear reactions, in which several particles are emitted, are called spallation reactions. Since they take place in the relatively cool, low density interstellar medium, the lithium, beryllium and boron products can survive after formation.

Reeves (1994) reviews the origin of the light elements, including their production by galactic cosmic-ray bombardment of the interstellar gas. The observed abundances of the lithium-6, beryllium-9 and boron-10 nuclei are explained by cosmic-ray spallation of interstellar carbon and oxygen (Reeves, 1994; Walker, Mathews and Viola, 1985). As discussed in Volume II, the lithium-7 nucleus owes its abundance to big-bang nucleosynthesis, as well as to subsequent cosmic-ray spallation reactions in interstellar space. Standard primordial big-bang nucleosynthesis is unable to produce significant yields of beryllium, Be, or boron, B, resulting in abundances relative to hydrogen of Be/H and B/H of $< 10^{-17}$ (Thomas, Schramm, Olive and Fields, 1993).

There was a problem with boron, since the current abundance ratio of ^{11}B/^{10}B is about twice the amount predicted by spallation theory, suggesting the need for another source of the boron-11 nucleus. This difficulty can be overcome if there is some extra source of low-energy cosmic rays, above the inverse power law spectrum used in the calculations (Walker, Mathews and Viola, 1985). The detection of an excess of gamma rays in the direction of the star-forming region in the Orion cloud (Bloemen et al., 1994) provides a natural explanation. These gamma rays are interpreted as arising from carbon and oxygen nuclei ejected from supernovae; these nuclei are excited when they collide with the surrounding gas, which is primarily molecular and atomic hydrogen, and emit gamma rays upon de-excitation. Such supernova-driven interactions offer a way to make lithium, beryllium and boron that is independent of the abundance of heavy elements in the surrounding medium (Cassé, Lehoucq and Vangioni-Flam (1995). When combined with the spallation effects of galactic cosmic rays, they can explain the observed solar system abundances of these light elements (see Ramaty, Kozlovsky and Lingenfelter (1998), and the references contained therein, for a review).

The present-day solar system and galactic abundances of lithium, Li, beryllium, Be, and boron, B, are (Anders and Grevesse, 1989; Reeves, 1994; Wilson and Rood, 1994):

$$\begin{aligned} \mathrm{Li/H} &= 2 \times 10^{-9} \\ \mathrm{Be/H} &= 1.3 \times 10^{-11} \\ \mathrm{B/H} &= 2 \times 10^{-10} \ . \end{aligned} \tag{4.329}$$

These ratios do not appear to have varied by more than a factor of two in the last five billion years or so.

The present abundance ratios of the light nuclei are (Reeves, 1994):

$$^{7}\mathrm{Li}/^{6}\mathrm{Li} = 12.5 \quad \text{(within five percent at solar birth)}$$

or

$$\begin{aligned} ^{6}\mathrm{Li/H} &= 2 \times 10^{-10} \\ ^{11}\mathrm{B}/^{10}\mathrm{B} &= 4.05 \ \text{(within one percent at solar birth)} \end{aligned} \tag{4.330}$$

or

$$^{10}\mathrm{B/H} = 4 \times 10^{-11} \ .$$

The spallation calculations depend on the fluxes and energy spectra of cosmic rays, which can be extrapolated from those observed in the solar system taking into account the effect of solar modulation (Ormes and Prothero, 1983; Webber et al., 1992; also see Sect. 4.5.3.1). Meteoritic data on the production rate of nuclei by spallation reactions shows that the average cosmic-ray flux entering the solar system has not varied significantly in the last few billion years (Lal and Peters, 1967; Forman and Schaeffer, 1979). A discussion of the relevant spallation cross sections and techniques for their measurement are reviewed by Austin (1981). The calculations also depend on the known interstellar target abundances of carbon and oxygen (Anders and Grevesse, 1989; Lang, 1992; Grevesse and Noels, 1993), which have remained constant for the last ten billion years or so (Spite, 1992).

Table 4.12. Production rate ratios of lithium, beryllium and boron by cosmic-ray spallation of interstellar carbon and oxygen[1]

$^{6}\mathrm{Li}/^{9}\mathrm{Be} = 5$	$^{7}\mathrm{Li}/^{9}\mathrm{Be} = 7$	$\mathrm{Li/Be} = 12$
$^{10}\mathrm{B}/^{9}\mathrm{Be} = 5$	$^{11}\mathrm{B}/^{9}\mathrm{Be} = 12$	$\mathrm{B/Be} = 17$
$^{7}\mathrm{Li}/^{6}\mathrm{Li} = 1.4$	$^{11}\mathrm{B}/^{10}\mathrm{B} = 2.5$	

[1] Adopted from Reeves (1994).

The production rates of lithium, beryllium and boron isotopes by cosmic-ray spallation reactions with abundant interstellar nuclei has been computed by Reeves and Meyer (1978) and Walker, Mathews and Viola (1985). These rates, which are given in Table 4.12, are uncertain by a factor of less than two.

As an example of the relevant spallation reactions, if carbon, ^{12}C, nuclei in the interstellar medium are bombarded by cosmic-ray protons, p:

$$\begin{aligned} p + {}^{12}C &\rightarrow {}^{11}B + 2p \quad (Q = -16.0\ \text{MeV}) \\ &\rightarrow {}^{10}B + 2p + n \quad (Q = -27.4\ \text{MeV}) \\ &\rightarrow {}^{10}B + {}^{3}He \quad (Q = -19.7\ \text{MeV}) \ , \end{aligned} \tag{4.331}$$

$$\begin{aligned} p + {}^{12}C &\rightarrow {}^{9}Be + 3p + n \quad (Q = -34.0\ \text{MeV}) \\ &\rightarrow {}^{9}Be + {}^{3}He + p \quad (Q = -26.3\ \text{MeV}) \ , \end{aligned} \tag{4.332}$$

$$\begin{aligned} p + {}^{12}C &\rightarrow {}^{7}Li + 4p + 2n \quad (Q = -52.9\ \text{MeV}) \\ &\rightarrow {}^{7}Li + {}^{4}He + 2p \quad (Q = -24.6\ \text{MeV}) \\ &\rightarrow {}^{6}Li + 4p + 3n \quad (Q = -60.2\ \text{MeV}) \\ &\rightarrow {}^{6}Li + {}^{4}He + 2p + n \quad (Q = -31.9\ \text{MeV}) \\ &\rightarrow {}^{6}Li + {}^{4}He + {}^{3}He \quad (Q = -24.2\ \text{MeV}) \ , \end{aligned} \tag{4.333}$$

where p denotes the proton, n is a neutron, ^{3}He is the helium-3 nucleus, ^{4}He is an α-particle, the nucleus of the helium atom, and the negative energy release Q denotes the energy threshold required for the reaction to proceed.

Although the abundances of elements formed by spallation reactions are dependent upon both reaction cross sections and the energy spectrum of the incident high energy particles, abundance ratios are relatively insensitive to the energy spectrum. For example, if only $p + {}^{16}O$ reactions are considered, measured cross sections give (Yiou, Seide, and Bernas, 1969)

$$\begin{aligned} \frac{^{7}Li}{^{6}Li} &= 0.98 \pm 0.13 \\ \frac{^{9}Be}{^{7}Be} &= 0.34 \pm 0.08 \\ \frac{^{10}Be}{^{9}Be} &= 0.30 \pm 0.08 \end{aligned} \tag{4.334}$$

and

$$\frac{^{11}B + {}^{11}C}{^{10}B + {}^{10}C} = 2.1 \pm 0.6 \ ,$$

for proton energies larger than 135 MeV. Furthermore, a proton flux capable of generating $Li/H = 10^{-9}$ will generate $D/H = 3 \times 10^{-7}$ and $^{3}He/^{4}He = 3 \times 10^{-6}$. Comparison with the observed abundances shows that spallation reactions might account for the ^{6}Li, Be and B, but cannot account for the higher solar system abundances of ^{7}Li and D. For detailed calculations, the formation rate of an isotope, L, from the spallation of the M elements by a flux of protons with energy spectrum, $\varphi(E)$, is given by

$$\frac{dN_L}{dt} = \sum_M N_M \int \sigma(M, L, E)\varphi(E)dE \tag{4.335}$$

where M is any of the M elements, C^{12}, N^{14}, O^{16}, and Ne^{20}, the number densities of the M and L elements are, respectively, N_M and N_L, the time variable is t, the proton energy is E, and the spallation reaction cross section is $\sigma(M, L, E)$. Observations of solar flares give a proton energy spectrum

$$\varphi(E) \propto E^{-\gamma} \ , \tag{4.336}$$

where γ ranges from 2 to 5, or more exactly,

$$\varphi(E) \propto \exp(-R/R_0) \ , \tag{4.337}$$

where the rigidity $R = pc/(Ze)$ is the ratio of momentum to charge, and R_0 is an experimentally determined parameter ranging from 40 to 400 MeV. If the proton rest mass energy is Mc^2, and its kinetic energy is E, then

$$R = \frac{A}{Ze}[2Mc^2E + E^2]^{1/2} \ , \tag{4.338}$$

where A is the atomic number and Ze is the charge of the nucleus.

For cosmic ray protons, Meneguzzi, Audouze, and Reeves (1971) adopt

$$\varphi(E) = 12.5(E_0 + E)^{-2.6} \quad \mathrm{cm^{-2}\,sec^{-1}\,GeV^{-1}}, \tag{4.339}$$

where $E_0 = 0.931$ GeV, and E is the proton kinetic energy in GeV.

Production rates involve the integration of the spallation cross sections over the incident spectrum, and these rates are given in Table 4.13 for the cosmic ray spectrum.

Detailed formulae for the energetics of cosmic ray spallation reactions are given by Meneguzzi, Audouze, and Reeves (1971), and Silberberg and Tsao (1973), whereas those for spallation reactions on stellar surfaces are given by

Table 4.13. Production rates of rare light elements by the spallation reactions of abundant nuclei and cosmic ray protons (H) or alpha particles (He)[1]

Nucleus	Production rate in $g^{-1}\,sec^{-1}$			
	C^{12}	N^{14}	O^{16}	Ne^{20}
Li^6 (H)	21.4×10^{-6}	9.9×10^{-6}	39.3×10^{-6}	1.26×10^{-6}
Li^6 (He)	5.39	1.59	9.53	1.23
Li^7 (H)	54.3	15.4	57.0	1.80
Li^7 (He)	8.63	2.43	14.54	1.87
Be^9 (H)	11.3	1.58	7.36	0.24
Be^9 (He)	2.39	0.96	5.75	0.74
Be^{10} (H)	47.6	2.90	43.8	1.40
Be^{10} (He)	2.70	1.14	6.85	0.88
Be^{11} (H)	170.7	23.9	85.0	2.70
Be^{11} (He)	3.65	1.13	6.78	0.87

[1] After Mitler (1970). All values are $\times 10^{-6}$ as indicated for the first member of each column.

Ryter, Reeves, Gradsztajn, and Audouze (1970). Approximate relations may be derived using the formula

$$\frac{N_L}{N_M} = \langle\varphi\rangle\langle\sigma\rangle T \ , \tag{4.340}$$

where $\langle\varphi\rangle$ and $\langle\sigma\rangle$ denote, respectively, average fluxes and cross sections, and T denotes the duration of the spallation reaction. For energies greater than 30 MeV, for example, the galactic cosmic ray fluxes are (Freier and Waddington, 1968; Comstock, Fan, and Simpson, 1969)

$$\begin{aligned} \varphi_{\mathrm{H}} &= 3.6\,\mathrm{cm}^{-2}\,\mathrm{sec}^{-1} \\ \varphi_{\mathrm{He}} &= 0.4\,\mathrm{cm}^{-2}\,\mathrm{sec}^{-1} \\ \varphi_{\mathrm{C,N,O}} &= 25 \times 10^{-3}\,\mathrm{cm}^{-2}\,\mathrm{sec}^{-1} \\ \varphi_{\mathrm{N}} &= 4 \times 10^{-3}\,\mathrm{cm}^{-2}\,\mathrm{sec}^{-1} \ , \end{aligned} \tag{4.341}$$

and the appropriate $T = 10^{10}$ years, the age of the Galaxy. Meneguzzi, Audouze, and Reeves (1971) have assumed a constant cosmic ray energy spectrum ($E^{-2.6}$) over 10^{10} years and obtain the abundance ratios

$$\begin{aligned} &\frac{^{6}\mathrm{Li}}{\mathrm{H}} = 8 \times 10^{-11} && \frac{^{10}\mathrm{B}}{\mathrm{H}} = 8.7 \times 10^{-11} \\ &\frac{^{7}\mathrm{Li}}{\mathrm{H}} = 1.2 \times 10^{-10} && \frac{^{11}\mathrm{B}}{\mathrm{H}} = 2 \times 10^{-10} \\ &\frac{^{7}\mathrm{Li}}{^{6}\mathrm{Li}} = 1.5 && \frac{^{11}\mathrm{B}}{^{10}\mathrm{B}} = 2.4 \\ &\frac{\mathrm{Li}}{\mathrm{H}} = 2 \times 10^{-10} && \frac{\mathrm{B}}{\mathrm{H}} = 3.0 \times 10^{-10} \\ &\frac{^{9}\mathrm{Be}}{\mathrm{H}} = 2 \times 10^{-11} && \frac{\mathrm{B}}{\mathrm{Be}} = 15 \\ &\frac{\mathrm{Li}}{\mathrm{Be}} = 10 && \frac{\mathrm{B}}{\mathrm{Li}} = 1.4 \ . \end{aligned} \tag{4.342}$$

The most important spallation reactions are those involving cosmic-ray protons, p, and interstellar carbon, C, and oxygen, O. Since the spallation cross sections do not depend strongly on energy, we can remove them from under the integral sign in Eq. (4.335) to obtain (Rolfs and Rodney, 1988)

$$\frac{dN_i}{dt} \approx \left(N_{\mathrm{C}}\sigma^i_{\mathrm{pC}} + N_{\mathrm{O}}\,\sigma^i_{\mathrm{pO}}\right)\Phi_{\mathrm{p}} \ , \tag{4.343}$$

where $\Phi_{\mathrm{p}} = 8.3\,\mathrm{cm}^2\,\mathrm{s}^{-1}$ is the total flux of cosmic-ray protons above the reaction threshold near 30 MeV, and σ^i denotes the spallation cross section, given in Table 4.14, for the production of element i by the reaction given in the subscript. Assuming a constant production rate between the time of formation of our Galaxy, t_{G}, and the formation of the Sun, t_{S}, the abundance ratios relative to hydrogen, with number density H, are:

$$\frac{N_i}{H} \approx \left[\left(\frac{N_{\mathrm{C}}}{H}\right)\sigma^i_{\mathrm{pC}} + \left(\frac{N_{\mathrm{O}}}{H}\right)\sigma^i_{\mathrm{pO}}\right]\Phi_{\mathrm{p}}(t_{\mathrm{G}} - t_{\mathrm{S}}) \ . \tag{4.344}$$

Using $t_{\mathrm{G}} - t_{\mathrm{S}} = 9 \times 10^9$ years with $N_{\mathrm{C}}/H = 0.048\%$ and $N_{\mathrm{O}}/H = 0.085\%$ (Austin, 1981), one obtains the spallation production adundances given in Table 4.14.

Table 4.14. Production of lithium, Li, beryllium, Be, and boron, B, by spallation reactions during the interaction of cosmic-ray protons, p, and interstellar carbon, C, and oxygen, O[1]

Nuclide, i	σ^i_{pC} (mb)	σ^i_{pO} (mb)	$(N_i/H)_{th}$ ($\times 10^{-12}$)	$(N_i/H)_{obs}$ ($\times 10^{-12}$)	$(N_i/H)_{th}/(N_i/H)_{obs}$
^{6}Li	14.8	13.9	45	70	0.64
^{7}Li	20.5	21.2	66	900	0.07
^{9}Be	6.2	4.4	16	14	1.14
^{10}B	22.7	12.7	51	30	1.70
^{11}B	57.0	26.5	118	120	0.98

[1] Adapted from Austin (1981) and Rolfs and Roney (1988). Here the spallation reaction cross sections are given in millibarns and the theoretical abundance ratios $(N/H)_{th}$ calculated from Eq. (4-344) are compared to the observed ones $(N/H)_{obs}$.

4.4.7 Rapid Thermonuclear Reactions in Supernovae Explosions

Burbidge, Burbidge, Fowler, and Hoyle (1957) argued that the gravitational collapse of a highly evolved star might lead to an imploding core whose heat might be sufficient to ignite the potentially explosive light nuclei such as C^{12} at temperatures of a few times 10^9 °K. The subsequent explosion of the stellar envelope was thought to coincide with nucleosynthesis by the equilibrium, r, and p processes. Hoyle and Fowler (1960) then suggested two origins for supernovae explosions. The Type I explosions were thought to originate from the ignition of degenerate material in the core of stars of intermediate mass. More massive stars with nondegenerate cores are the site of Type II explosions which result from the implosion-explosion process. Hayakawa, Hayashi, and Nishida (1960), Tsuji (1963), and Tsuda and Tsuji (1963) postulated that rapid thermonuclear reactions such as the rapid CNO process, p capture, n processes, and α capture could synthesize many of the elements in the range $20 \leq A \leq 60$ during supernovae explosions. Fowler and Hoyle (1964) presented a detailed analysis of nucleosynthesis during the Type II supernova process. Detailed hydrodynamic models of the implosion-explosion process were then given by Colgate and White (1966) and Arnett (1966, 1967).

The physics of supernova explosions has been reviewed more recently by Woosley and Weaver (1986), Bethe (1990) and Wheeler and Harkness (1990). Explosive nucleosynthesis was reviewed by Arnett (1973) and Arnett (1995). The nucleosynthesis of elements during the evolution and explosion of massive stars has been considered in detail by Woosley and Weaver (1995), and neutron capture processes in nucleosynthesis have been reviewed by Meyer (1994).

Before a supernova explosion, the mass density, ρ, and temperature, T, of a star are related by the equations of state which were discussed in detail in Chap. 3. For a completely degenerate relativistic gas we have a pressure, P, given by (Hoyle and Fowler, 1960)

$$P \approx 1.243 \times 10^{15} \left(\frac{\rho}{\mu_e}\right)^{4/3} \left[1 + \frac{1}{x^2}\left(\frac{2\pi^2 k^2 T^2}{m^2 c^4} - 1\right)\right] \quad \text{dynes cm}^2 \ , \qquad (4.345)$$

where

$$\rho = 9.74 \times 10^5 \, \mu_e x^3 \quad \text{g cm}^{-3} \; . \tag{4.345}$$

Here the constant $1.243 \times 10^{15} = (3\pi^2)^{1/3} c\hbar/(4M_\mu^{4/3})$, where M_μ is the atomic mass unit, the constant $9.74 \times 10^5 = M_\mu m^3 c^3/(3\pi^2\hbar^3)$, and $x = \hbar(3\pi^2 N_e)^{1/3}/(mc)$ where N_e is the electron density. The pressure of a degenerate gas is relatively insensitive to temperature, and the temperature release by expansion or neutrino processes is insufficient to prohibit a temperature rise to explosive values during gravitational contraction. For example, the temperature and density of a degenerate gas are related by (Hoyle and Fowler, 1960)

$$\left(\frac{T_9}{1.33}\right)^2 - 1 \approx \left(\frac{\rho}{9.74 \times 10^5 \mu_e}\right)^{2/3} \left[\left(\frac{M}{M_{cr}}\right)^{2/3} - 1\right] \; , \tag{4.346}$$

where $T_9 = T/10^9$, and the critical mass, M_{cr}, in units of the Sun's mass, $M_\odot$, is

$$M_{cr} = \frac{5.836}{\mu_e^2} M_\odot \; . \tag{4.347}$$

For a degenerate relativistic gas, $\rho > 7.3 \times 10^6 \mu_e$ g cm^{-3}, explosive temperatures for carbon burning, $T_9 \approx 1$, are realizable for M on the order of M_{cr}. Arnett (1969) has followed the evolution of a carbon-oxygen stellar core for stars of intermediate mass, $4M_\odot \leq M \leq 9M_\odot$, and shows that explosive ignition of carbon burning does indeed occur for degenerate core densities on the order of 10^9 g cm^{-3}. The explosion generates a strong shock wave, called a detonation wave, which progresses outward through the unburned stellar envelope momentarily increasing its temperature as well. Prior to ignition of this fuel, the pressure and specific volume before (P_1, V_1) and after (P_2, V_2) the shock front are related by the Rankine–Hugoniot relations (Sect. 3.5.9).

$$\frac{V_2}{V_1} = \left(\frac{2}{M_1^2} + \gamma - 1\right) / (\gamma + 1) \; , \tag{4.348}$$

$$\frac{P_2}{P_1} = (2\gamma M_1^2 - \gamma + 1)/(\gamma + 1) \; , \tag{4.349}$$

where

$$M_1^2 = \left(\frac{V_1}{S_1}\right)^2 = 1 + \left[1 + (1 + 2\beta)^{1/2}\right] / \beta \; ,$$

$$\beta = \gamma P_1 V_1 / \left[Q(\gamma^2 - 1)\right] \; ,$$

Q is the energy release per gram for the reaction, $M_1 = (V_1/S_1)$ is the ratio of the shock front speed to the velocity of sound before the front, and γ is the adiabatic index. Typical velocities are $V \approx 20{,}000$ km sec^{-1}, and the raise in temperature, T_2, can be calculated from P_2, V_2 and the equation of state. Explosive temperatures are found for T_2, and even higher temperatures are found to result from the ignition of stellar fuel after the passage of the detonation wave. These temperatures are sufficiently high to eventually set up complete nuclear statistical equilibrium following the passage of the wave. Initial nuclear abundances are then determined from the equilibrium equations (Sect. 4.4.5). During the subsequent expansion, the temperature and density are related by the adiabatic condition, and abundances depend only on the neutron-proton ratio and the rate of expansion. The same conditions also follow for the nondegenerate objects exploding from temperatures sufficiently high to initially establish nuclear equilibrium (cf. Wagoner, 1969).

The initial condition of a nondegenerate, nonrelativistic stellar core is, of course, different from the degenerate case, with a pressure, P, given by (Fowler and Hoyle, 1964)

$$P = \left[\frac{3R^4(1-\beta)}{a\mu^4\beta^4}\right]^{1/3} \rho^{4/3} , \tag{4.350}$$

and a mass density, ρ, given by

$$\rho = \frac{a\mu\beta}{3R(1-\beta)} T^3 . \tag{4.351}$$

Here $R = 8.3145 \times 10^{16}$ erg mole^{-1} $(10^9\ {}^\circ\mathrm{K})^{-1}$ is the gas constant, the radiation constant $a = 7.5659 \times 10^{21}$ erg cm$^{-3}(10^9\ {}^\circ\mathrm{K})^{-4}$, the mean molecular weight is μ, and β, the ratio of gas pressure to total pressure, satisfies the relation

$$1 - \beta = 0.0030\left(\frac{2M}{3M_\odot}\right)^2 \mu^4\beta^4 . \tag{4.352}$$

For a massive star, $M \approx 30M_\odot$, we have $\beta \approx 0.40$, and with $\mu \approx 2.1$ and $\beta \approx 0.4$ we have

$$\rho \approx 4.3 \times 10^4\ T_9^3 \quad \mathrm{g\ cm^{-3}} \tag{4.353}$$

and in general

$$\rho \approx 1.016 \times 10^7 \left(\frac{M_\odot}{M_\mathrm{c}}\right)^2 \left(\frac{T}{\mu\beta}\right)^3 \quad \mathrm{g\ cm^{-3}} \tag{4.354}$$

for a core of mass M_c. The implosion-explosion phenomenon which follows from a contracting core is described by Fowler and Hoyle (1964); whereas the details of the hydrodynamic processes following implosion are given by Colgate and White (1966), Arnett (1966, 1967, 1968), and Arnett and Cameron (1967). As pointed out by Fowler and Hoyle (1964), the raise in temperature following the outward moving shock wave is sufficient to cause explosive oxygen and hydrogen burning, and other rapid processes on the explosion time scale of ≈ 100 sec.

The explosive hydrodynamics and nucleosynthesis of isotopes lighter than mass number $A = 66$ during supernova explosions of massive stars has been discussed in detail by Woosley and Weaver (1995). Recent progress in our understanding of the formation of heavy nuclei, with mass numbers $A \geq 70$, by neutron capture processes is discussed by Meyer (1994). Predictions of nucleosynthesis yields (up to atomic number $Z = 36$) from massive stars have been compared and reviewed by Arnett (1995). An earlier review by Arnett (1973) provided an overview of many of the relevant pioneering papers.

4.5 High Energy Particles and High Energy Radiation

In this section we discuss the creation of energetic radiation, or photons, and high-energy charged particles, together with energy loss mechanisms for both the photons and particles. Energetic electromagnetic radiation is observed throughout the cosmos at X-ray and gamma ray wavelengths. This radiation is often expressed in terms of its photon energy in keV, MeV, or GeV, where $1\ \text{eV} = 1.6021773 \times 10^{-12}$ ergs, $1\ \text{keV} = 10^3$ eV, and $1\ \text{MeV} = 10^6$ eV, and $1\ \text{GeV} = 10^9$ eV. High energy electrons, accelerated to nearly the velocity of light, are inferred from their synchrotron radiation, and directly sampled as cosmic ray particles entering the terrestrial atmosphere. The primary cosmic rays mainly consist of protons, α-particles, or helium nuclei, and heavier nuclei. The energies of cosmic rays are usually expressed in units of MeV or GeV.

High-energy electrons can emit radiation as bremsstrahlung, synchrotron radiation and inverse Compton radiation. These processes were respectively discussed in Sects. 1.29, 1.25 and 1.38. They have been reviewed by Blumenthal and Gould (1970) and discussed in the monographs by Rybicki and Lightman (1979) and Longair (1992).

When high-energy, charged particles interact with matter they can lose energy by ionization losses and by bremsstrahlung. High-energy photons that interact with matter also lose energy by photoelectric absorption, Compton scattering and electron-positron pair production. All of these processes are discussed in the following parts of Sect. 4.5, as well in the book by Longair (1992). The energy loss of relativistic electrons and positrons traversing cosmic matter is discussed by Gould (1975).

Plasmas with thermal energy $kT_e \geq 1\text{MeV}$ will create pairs of electrons, e^-, and positrons, e^+, as well as bremsstrahlung photons, and a sufficiently luminous and compact astrophysical object will also create e^-, e^+ pairs by collisions of photons, γ, through the reaction $\gamma + \gamma \rightarrow e^- + e^+$. Such electron-positron pairs annihilate, producing radiation at the 0.511 MeV line that has been observed in the Sun (Murphy et al., 1985) and from the nucleus of our galaxy (Ramaty and Lingenfelter, 1983). Electron-positron pair creation and annihilation have been discussed by Begelman, Blandford and Rees (1984) in their review of extragalactic radio sources.

Photons with energies $E \geq 0.511$ MeV, the rest mass energy of the electron, are called gamma rays. They have been discussed in the monographs by Chupp (1976) and Ramana Murthy and Wolfendale (1993) and in the article by Gehrels and Paul (1998). Galactic gamma-ray sources are reviewed by Bignami and Hermensen (1983) and tabulated by Lang (1992). Gamma-ray line astronomy is reviewed by Ramaty and Lingenfelter (1979, 1983), Murphy and Ramaty (1984), and Ramaty and Murphy (1987), while very high energy gamma-ray astronomy in the range 10 GeV to 100 TeV, has been reviewed by Weekes (1988).

4.5.1 Creation of High Energy Particles and Energetic Radiation

4.5.1.1 Creation of Electron–Positron Pairs by Gamma Ray Absorption in the Presence of a Nucleus

A photon of energy, $h\nu$, greater than the threshold energy $2mc^2 = 1.022$ MeV, may form an electron-positron pair when passing through the Coulomb field of a nucleus of charge eZ. When all energies under consideration are large compared with mc^2, the cross section, $\sigma(E_0)$, for the pair creation of a positron and electron of respective energies, E_0 and E_1, is given by (Bethe and Heitler, 1934)

$$\sigma(E_0)dE_0 = 4\alpha Z^2 r_0^2 \frac{\left[E_0^2 + E_1^2 + \frac{2}{3}E_0E_1\right]}{(h\nu)^3}\left[\ln\left(\frac{2E_0E_1}{h\nu mc^2}\right) - \frac{1}{2}\right]dE_0$$

for relativistic energies, and no screening

$$\frac{2E_0E_1}{h\nu} \ll \frac{mc^2}{\alpha Z^{1/3}},$$

$$\sigma(E_0)dE_0 = \frac{4\alpha Z^2 r_0^2}{(h\nu)^3}\left[\left(E_0^2 + E_1^2 + \frac{2}{3}E_0E_1\right)\ln\left(\frac{183}{Z^{1/3}}\right) - \frac{E_1E_0}{9}\right]dE_0 \tag{4.355}$$

for relativistic energies and complete screening

$$\frac{2E_0E_1}{h\nu} \gg \frac{mc^2}{\alpha Z^{1/3}},$$

where the fine structure constant $\alpha \approx 1/137.036$, and the classical electron radius $r_0 = e^2/(mc^2) \approx 2.818 \times 10^{-13}$ cm. The total cross section, $\sigma(h\nu)$, for the creation of electron-positron pairs is obtained by integrating Eq. (4.355) from $E_0 = mc^2$ to $E_0 = h\nu - mc^2$.

$$\begin{aligned}\sigma(h\nu) &= 4\alpha Z^2 r_0^2\left[\frac{7}{9}\ln\left(\frac{2h\nu}{mc^2}\right) - \frac{109}{54}\right] \quad \text{for } \frac{2E_0E_1}{h\nu} \ll \frac{mc^2}{\alpha Z^{1/3}} \\ \sigma(h\nu) &= 4\alpha Z^2 r_0^2\left[\frac{7}{9}\ln\left(\frac{183}{Z^{1/3}}\right) - \frac{1}{54}\right] \quad \text{for } \frac{2E_0E_1}{h\nu} \gg \frac{mc^2}{\alpha Z^{1/3}},\end{aligned} \tag{4.356}$$

where $4\alpha r_0^2 \approx 2.318 \times 10^{-27}$ cm^{-2}. When electron velocities, v, are so small that $Ze^2/(h\nu) \gg 1$, a Coulomb correction factor, $C(Z) \approx (\alpha Z)^2$, must be subtracted from the terms in the square brackets in Eq. (4.356). The cross sections are also suppressed at very high energies in a very dense medium (Landau and Pomeranchuk, 1953; Migdal, 1956), and in a crystalline medium (Überall, 1956, 1957). For a completely ionized gas, Eq. (4.355) may be used.

Also see Chupp (1976), Ramana Murthy and Wolfendale (1993) and Longair (1992). Pair production in intense magnetic fields is discussed by Zaumen (1976).

4.5.1.2 Creation of Electron–Positron Pairs by Charged Particles

The electric field of a fast charged particle has an associated "virtual" photon which in turn may create an electron-positron pair. When a heavy charged particle of mass, M_0, charge eZ_0, and kinetic energy, E_0, collides with a heavy particle of mass, M, and charge, eZ, the total cross section, σ, for the creation of electron-positron pairs of any energy is given by (Heitler and Nordheim, 1934)

$$\sigma \approx \frac{(\alpha Z Z_0 r_0 mc^2)^2}{M_0 c^2 E_0} \left[\frac{Z M_0 - Z_0 M}{M} \right]^2 \quad \text{for } 2mc^2 < E_0 \ll M_0 c^2 \ , \tag{4.357}$$

where the fine structure constant $\alpha = 1/137.036$, the classical electron radius $r_0 = e^2/(mc^2) \approx 2.818 \times 10^{-13}$ cm, and mc^2 is the rest mass energy of the electron.

When the energy, E_0, of the incident particle is greater than $M_0 c^2$, and the other particle is at rest, the total cross section is given by (Bhabha, 1935)

$$\sigma \approx \frac{28}{27\pi} (\alpha Z Z_0 r_0)^2 \ln^3 \left(\frac{E_0}{M_0 c^2} \right) \quad \text{for } E_0 \gg M_0 c^2 \ . \tag{4.358}$$

If the particle at rest is an atom, and $E_0 \gg M_0 c^2 / (\alpha Z^{1/3})$ for complete screening, the total cross section is (Nishina, Tomonaga, and Kobayasi, 1935; Heitler, 1954)

$$\sigma = \frac{28}{27\pi} (\alpha Z Z_0 r_0)^2 \ln \left(\frac{1}{\alpha Z^{1/3}} \right) \left[3 \ln \left(\frac{E_0}{M_0 c^2} \right) \ln \left(\frac{E_0 Z^{1/3}}{191\, M_0 c^2} \right) + \ln^2 \left(\frac{191}{Z^{1/3}} \right) \right]$$

$$\text{for} \quad E_0 > M_0 c^2 / \alpha Z^{1/3} \ . \tag{4.359}$$

Eqs. (4.358) and (4.359) are thought to be accurate to within a factor of two. More complicated expressions accurate to twenty percent are given by Murota, Ueda, and Tanaka (1956) and Hayakawa (1969). When the incident particle is an electron, Eqs. (4.358) and (4.359) are valid with $Z_0 = 1$ and $M_0 = m$, the electron mass.

4.5.1.3 Creation of Electron–Positron Pairs by Two Photon Collision

An electron-positron pair may be produced in the collision of a photon of energy, E_1, with a photon of energy, E_2, provided that $E_1 E_2 > (mc^2)^2$, where $mc^2 \approx 0.511$ MeV is the rest mass energy of the electron. The pair creation cross section, $\sigma(E_1, E_2)$, is given by (Dirac, 1930; Heitler, 1954)

$$\sigma(E_1, E_2) = \frac{\pi r_0^2}{2} (1 - \beta^2) \left[2\beta(\beta^2 - 2) + (3 - \beta^4) \ln \left(\frac{1 + \beta}{1 - \beta} \right) \right] ,$$

where

$$\beta = \left[1 - \frac{(mc^2)^2}{E_1 E_2} \right]^{1/2} , \tag{4.360}$$

the velocity of the outgoing electron in the center-of-mass system is βc, and the classical electron radius $r_0 = e^2/(mc^2) \approx 2.818 \times 10^{-13}$ cm. Applications of Eq. (4.360) to gamma rays are given by Nikishov (1962), Goldreich and Morrison (1964), Gould and Schréder (1966), and Jelley (1966).

Gamma rays produced in a sufficiently luminous and compact astrophysical object will create electron-positron pairs. This process both depletes the escaping radiation at high energies and also changes the composition and properties of the radiating gas through the injection of new particles. Pair creation will become significant when the radiation energy exceeds 0.511 MeV and the compactness parameter, l, exceeds 10, where

$$l = \frac{L\sigma_T}{Rm_e c^3} = 2\pi\left(\frac{m_p}{m_e}\right)\left(\frac{L}{L_E}\right)\left(\frac{R_S}{R}\right) , \tag{4.361}$$

where L is the absolute luminosity, R is a characteristic source dimension, σ_T is the Thomson scattering cross section, m_p/m_e is the ratio of the proton to electron mass, $L_E = 4\pi GMcm_p/\sigma_T$ is the Eddington limit for a mass M, and $R_S = 2GM/c^2$ is the Schwarzschild radius. Pair production and Compton scattering in compact sources such as galactic nuclei are discussed by Lightman and Zdziarski (1987).

4.5.1.4 Creation of μ-Meson Pairs by Gamma Rays in the Presence of a Nucleus

A photon, γ, of energy, $h\nu$, greater than the threshold energy $2m_\mu c^2 \approx 211$ MeV may form a μ^- meson pair, μ^+ and μ^-, when passing through the Coulomb field of a nucleus. The cross section for pair creation will be given by Eq. (4.356) for the electron-positron pair creation, with m_e replaced by m_μ, and a slight modification due to the large momentum transfer to the nucleus during meson pair production. The cross section is reduced below that given by Eq. (4.356) by the ratio $(m/m_\mu)^2 \approx (1/207)^2$. Detailed calculations of meson pair production cross sections are given by Rawitscher (1956). Electron-positron pairs will be formed by the decay of the muon pairs, whose decay modes and lifetimes are given in Table 4.4.

4.5.1.5 Creation of Recoil (Knock-on) Electrons by Charged Particle Collision

The cross section, $\sigma(E_0, W_r)$, for the production of a recoil electron of kinetic energy, W_r, by the collision of a charged particle of total energy, E_0, with another charged particle of charge, eZ, is given by (Bhabha, 1936; Hayakawa, 1969)

$$\begin{aligned} \sigma(E_0, W_r)dW_r & \\ &= 2\pi Z^2 r_0^2 \frac{mc^2}{\beta_0^2}\frac{dW_r}{W_r^2} \times \begin{cases} \left(1 - \beta_0^2 \frac{W_r}{W_m}\right) & \text{for spin } 0 , \\ \left[1 - \beta_0^2 \frac{W_r}{W_m} + \frac{1}{2}\left(\frac{W_r}{E_0}\right)^2\right] & \text{for spin } \tfrac{1}{2} , \\ \left[\left(1 - \beta_0^2 \frac{W_r}{W_m}\right)\left(1 + \frac{1}{3}\frac{mW_r}{M^2}\right) \right. & \\ \left. + \frac{1}{3}\left(\frac{W_r}{E_0}\right)^2\left(1 + \frac{1}{2}\frac{mW_r}{M^2}\right)\right] & \text{for spin } 1 , \end{cases} \\ &\approx 2\pi r_0^2 \frac{mc^2 Z^2}{\beta_0^2 W_r^2} dW_r , \end{aligned} \tag{4.362}$$

where the spin is that of the incident particle, the classical electron radius $r_0 = e^2/(mc^2) \approx 2.82 \times 10^{-13}$ cm, the velocity of the incident particle is $c\beta_0$, the electron mass is m, the mass of the incident particle is M, and the maximum energy that can be transferred in a direct collision to a free electron, $W_{\rm m}$, is given by

$$W_{\rm m} = \frac{2mp^2}{m^2 + M^2 + 2mE_0} \approx p \quad \text{for } m \ll M \quad \text{and} \quad E_0 \gg \frac{M^2}{m}$$
$$\approx 2m\beta_0^2\gamma^2 = 2m\left(\frac{p}{M}\right)^2 \quad \text{for } m \ll M \quad \text{and} \quad p \ll \frac{M^2}{m} \ ,$$

where the momentum and total energy of the incident particle are, respectively p and E_0.

For the special case of relativistic protons of total energy, $E_{\rm p}$, and velocity, $c\beta_{\rm p}$, Eq. (4.361) becomes

$$\sigma(E_{\rm p}, W_{\rm r})dW_{\rm r} = 2\pi r_0^2 \frac{mc^2}{\beta_{\rm p}^2 W_{\rm r}^2}\left[1 - \frac{m^2c^2W_{\rm r}}{2mE_{\rm p}^2} + \frac{1}{2}\left(\frac{W_{\rm r}}{E_{\rm p}}\right)^2\right] dW_{\rm r} \ . \tag{4.363}$$

4.5.1.6 Creation of Photons by Electron–Positron Annihilation

A positron, e^+, may collide with an electron, e^-, to produce two gamma ray photons according to the reaction $e^+ + e^- \rightarrow \gamma + \gamma$. One photon will have a high energy and, if the electron is at rest, the other photon will have an energy on the order of $mc^2 = 0.511$ MeV. If the energy of the positron is given by γmc^2, where here γ is taken to denote an energy factor, the cross section, σ, for two photon annihilation with a free electron at rest is given by (Dirac, 1930)

$$\sigma = \frac{\pi r_0^2}{\gamma + 1}\left[\frac{\gamma^2 + 4\gamma + 1}{\gamma^2 - 1}\ln(\gamma + \sqrt{\gamma^2 - 1}) - \frac{\gamma + 3}{\sqrt{\gamma^2 - 1}}\right]$$
$$\approx \frac{\pi r_0^2}{\gamma}[\ln(2\gamma) - 1] \quad \text{for } \gamma \gg 1$$
$$\approx \frac{\pi r_0^2}{\beta} \quad \text{for } \beta \ll 1 \ , \tag{4.364}$$

where the classical electron radius $r_0 = e^2/(mc^2) \approx 2.818 \times 10^{-13}$ cm, and $\beta = v/c$, where v is the velocity of the positron.

The positron may also be annihilated by emitting only one photon when colliding with an electron which is bound to an atom. For the collision of a positron with an electron in the K shell of an atom, the one photon annihilation cross section is given by (Fermi and Uhlenbeck, 1933; Bethe and Wills, 1935)

$$\sigma = \frac{4\pi Z^5\alpha^4 r_0^2}{\beta\gamma(\gamma + 1)^2}\left[\gamma^2 + \frac{2\gamma}{3} + \frac{4}{3} - \frac{\gamma + 2}{\beta\gamma}\ln\left[(1 + \beta)\gamma\right]\right]$$
$$\approx \frac{4\pi Z^5\alpha^4 r_0^2}{\gamma} \quad \text{for } \gamma \gg 1$$
$$\approx \frac{4\pi}{3} Z^5\alpha^4 r_0^2 \beta \quad \text{for } \beta \ll 1 \ , \tag{4.365}$$

where the positron has an energy γmc^2 and velocity $v = \beta c$, the atom has charge, eZ, the fine structure constant $\alpha = e^2/(\hbar c) \approx 1/137.036$, and the classical electron radius, $r_0 = e^2/(mc^2) \approx 2.818 \times 10^{-13}$ cm.

Radiation of the 0.511 MeV line during electron-positron annihilation has been detected during solar flares and from a source in the vicinity of the center of the Galaxy (Ramaty and Lingenfelter, 1983).

4.5.1.7 Creation of π-Mesons, μ-Mesons, Positrons, Electrons, Photons, and Neutrinos by Nuclear Interaction

The important reactions for the production of π-mesons are

$$\begin{aligned} &\mathrm{p} + \mathrm{p} \rightarrow \mathrm{p} + \mathrm{p} + \mathrm{n}_1(\pi^+ + \pi^-) + \mathrm{n}_2\pi^0 \\ &\mathrm{p} + \mathrm{p} \rightarrow \mathrm{p} + \mathrm{n} + \pi^+ + \mathrm{n}_3(\pi^+ + \pi^-) + \mathrm{n}_4\pi^0 \\ &\mathrm{p} + \mathrm{p} \rightarrow \mathrm{n} + \mathrm{n} + 2\pi^+ + \mathrm{n}_5(\pi^+ + \pi^-) + \mathrm{n}_6\pi^0 \\ &\mathrm{p} + \mathrm{p} \rightarrow \mathrm{D} + \pi^+ + \mathrm{n}_7(\pi^+ + \pi^-) + \mathrm{n}_8\pi^0 \ , \end{aligned} \tag{4.366}$$

where p denotes a proton, n a neutron, D a deuteron, and $\mathrm{n}_1 - \mathrm{n}_8$ are positive integers. Once π-mesons are produced, μ-mesons, electrons, positrons, photons, and neutrinos may also be produced through the decay reactions

$$\begin{aligned} &\pi^{\pm} \rightarrow \mu^{\pm} + \nu_\mu/\bar{\nu}_\mu \\ &\pi^0 \rightarrow \gamma + \gamma \\ &\mu^{\pm} \rightarrow \mathrm{e}^{\pm} + \nu_\mathrm{e}/\bar{\nu}_\mathrm{e} + \bar{\nu}_\mu/\nu_\mu \ . \end{aligned} \tag{4.367}$$

In the proton energy region from 290 MeV to 1 GeV, only the reactions $\mathrm{p} + \mathrm{p} \rightarrow \mathrm{p} + \mathrm{n} + \pi^+$ or $\mathrm{p} + \mathrm{p} + \pi_0$ or $\mathrm{D} + \pi^+$ are important. Other pion reactions are given in Sect. 4.5.1.12 on solar flares.

The minimum kinetic energy, W_{min}, for a nuclear proton to produce a total of Y π-mesons is given by

$$W_{\mathrm{min}} = \frac{Y^2 m_\pi^2 c^4}{2 m_\mathrm{p} c^2} + 2Y m_\pi c^2 \approx Y(280 + 10Y)\,\mathrm{MeV} \ , \tag{4.368}$$

where $m_\pi c^2$ and $m_\mathrm{p} c^2$ are, respectively, the rest mass energies of the π meson and the proton.

Experimental measurements (Pollack and Fazio, 1963; Ramaty and Lingenfelter, 1966), indicate that the cross section, $\sigma(E_\mathrm{p})$, for the production of π-mesons by protons of energy E_p is

$$\sigma(E_\mathrm{p}) \approx 2.7 \times 10^{-26}\,\mathrm{cm}^2 \quad \text{for } E_\mathrm{p} > 2\,\mathrm{GeV} \ , \tag{4.369}$$

with multiplicity

$$\begin{aligned} m_\pm &= 2E_\mathrm{p}^{0.25} \\ m_0 &= 0.5 m_\pm \ . \end{aligned}$$

4.5.1.8 Emission of a High Energy Photon by the Inverse Compton Effect

When a free electron with kinetic energy $E = \gamma mc^2$ collides with a photon of energy, $h\nu$, the Compton scattering cross section, σ_c, is given by (Klein and Nishina, 1929)

$$\sigma_c = \frac{3}{4}\sigma_T\left\{\frac{1+q}{q^3}\left[\frac{2q(1+q)}{1+2q} - \ln(1+2q)\right] + \frac{1}{2q}\ln(1+2q) - \frac{1+3q}{(1+2q)^2}\right\}$$

$$\approx \sigma_T\left(1 - \frac{2\gamma h\nu}{mc^2}\right) \quad \text{for } \gamma h\nu \ll mc^2 \quad \text{or} \quad q \ll 1$$

$$\approx \frac{3}{8}\sigma_T\frac{mc^2}{\gamma h\nu}\left[\frac{1}{2} + \ln\left(\frac{2\gamma h\nu}{mc^2}\right)\right] \quad \text{for } \gamma h\nu \gg mc^2 \quad \text{or} \quad q \gg 1 \ , \tag{4.370}$$

where $q = \gamma h\nu/(mc^2)$, the classical electron radius $r_0 = e^2/(mc^2) \approx 2.818 \times 10^{-13}$ cm, and the Thomson (1903) scattering cross section $\sigma_T = 8\pi e^4/(3m^2c^4) \approx 6.65 \times 10^{-25}$ cm^2. Tables and graphs of the Compton cross section are given by Davisson and Evans (1952) and Nelms (1953).

The frequency, ν_0, of the emitted photon is given by

$$\nu_0 \approx \gamma^2\nu \quad \text{for } \gamma h\nu \ll mc^2$$

and

$$\nu_0 \approx \gamma mc^2/h \quad \text{for} \quad \gamma h\nu \gg mc^2 . \tag{4.371}$$

This process has been discussed in detail by Feenberg and Primakoff (1948), Donahue (1951), Felten and Morrison (1963, 1966) and Blumenthal and Gould (1970). When relativistic electrons produce synchrotron radiation and then undergo Compton scattering from it, the resulting Compton-synchrotron radiation may have photon energies nearly as high as that of the electrons, and a spectrum the same as that of the synchrotron radiation (Gould, 1965). This process is discussed in detail in Sect. 4.5.2.3.

4.5.1.9 High Energy Photon Emission by the Bremsstrahlung of a Relativistic Electron or Muon

In the collision of a relativistic electron of energy, E, with a nucleus of charge, eZ, the differential cross section, $\sigma_B(E, E_\gamma)\, dE_\gamma$, for the bremsstrahlung of a photon of energy, E_γ, is given by (Bethe and Heitler, 1934)

$$\sigma_B(E, E_\gamma)dE_\gamma \approx \frac{4\alpha r_0^2 Z^2}{E_\gamma}\ln\left(\frac{2E}{mc^2}\right)dE_\gamma \quad \text{for } mc^2 \ll E \ll \frac{mc^2}{\alpha Z^{1/3}} \tag{4.372}$$

for the relativistic case and no screening. Here the fine structure constant $\alpha = e^2/(\hbar c) \approx 1/137.036$, the classical electron radius $r_0 = e^2/(mc^2) \approx 2.818 \times 10^{-13}$ cm, and the rest mass energy of the electron is $mc^2 \approx 0.511$ MeV. In the extreme relativistic but complete screening case we have (Tsai, 1974)

$$\sigma_B(E, E_\gamma)dE_\gamma \approx \frac{M}{XE_\gamma}dE_\gamma \quad \text{for } E \gg \frac{mc^2}{\alpha Z^{1/3}} \ , \tag{4.373}$$

where M is the mass of the target nucleus in grams, and the radiation length, X, in gram cm^{-2} is given by

$$\frac{1}{X} = 4\alpha r_0^2 Z^2 N_A A^{-1}\ln(183Z^{-1/3}) \ , \tag{4.374}$$

where Avogadro's number $N_A \approx 6.022 \times 10^{23}$ (gram mole)$^{-1}$, and the mass number is A. Values of the radiation length in gm cm^{-2} are (Nishimura, 1967)

$$\begin{aligned} X_H &= 6.28, \quad X_{He} = 9.31, \quad X_{Li} = 83.3 \ , \\ X_C &= 43.3, \quad X_N = 38.6, \quad X_O = 34.6 \ , \\ X_{Al} &= 24.3, \quad X_{Si} = 22.2, \quad X_{Fe} = 13.9 \ , \end{aligned} \tag{4.375}$$

where the subscript denotes the element under consideration. The cross sections for electron-atom bremsstrahlung are given by Eqs. (4.372) and (4.373) if Z^2 is replaced by $Z(Z+1)$.

When accuracies greater than a few percent are required, the bremsstrahlung cross section may be calculated from the expressions (Wheeler and Lamb, 1939; Rossi, 1952; Tsai, 1974):

$$\sigma_B(E, E_\gamma) dE_\gamma = \frac{4\alpha r_0^2 Z^2}{E_\gamma} F(E, E_\gamma) dE_\gamma \ , \tag{4.376}$$

where the function $F(E, E_\gamma)$ is a slowly varying function of the relativistic electron energy, E, and the photon energy E_γ. Extreme values of $F(E, E_\gamma)$ are

$$F(E, E_\gamma) = \left\{1 + \left(1 - \frac{E_\gamma}{E}\right)^2 - \frac{2}{3}\left(1 - \frac{E_\gamma}{E}\right)\right\} \left\{\ln\left[\frac{2E}{mc^2} \frac{(E - E_\gamma)}{E_\gamma}\right] - \frac{1}{2}\right\} \tag{4.377}$$

for a bare nucleus ($\xi \gg 1$), and

$$F(E, E_\gamma) = \left[1 + \left(1 - \frac{E_\gamma}{E}\right)^2 - \frac{2}{3}\left(1 - \frac{E_\gamma}{E}\right)\right] \ln\left(\frac{183}{Z^{1/3}}\right) + \frac{1}{9}\left(1 - \frac{E_\gamma}{E}\right) \tag{4.378}$$

for complete screening ($\xi \approx 0$). Here the screening parameter, ξ, is given by

$$\xi = \frac{mc^2}{\alpha E} \frac{E_\gamma}{E - E_\gamma} Z^{-1/3} \ ,$$

where the fine structure constant $\alpha = e^2/(\hbar c) \approx 1/137.036$, and the rest mass energy of the electron is $mc^2 \approx 0.511$ MeV.

For muons of mass, m_μ, the bremsstrahlung cross section for collisions with a bare nucleus is given by

$$\begin{aligned} \sigma_B(E, E_\gamma) dE_\gamma &= \frac{4\alpha r_0^2 Z^2}{E_\gamma} \left(\frac{m}{m_\mu}\right)^2 \left\{1 + \left(1 - \frac{E_\gamma}{E}\right)^2 - \frac{2}{3}\left(1 - \frac{E_\gamma}{E}\right)\right\} \\ &\times \left\{\ln\left[\frac{2E}{m_\mu c^2} \frac{\hbar}{m_\mu R c} \left(\frac{E - E_\gamma}{E_\gamma}\right)\right] - \frac{1}{2}\right\} dE_\gamma \ , \end{aligned} \tag{4.379}$$

where m is the electron mass, and the nuclear radius $R \approx 1.2 \times 10^{-13} A^{1/3}$ cm, where A is the mass number.

4.5.1.10 Photon Emission by the Synchrotron Radiation of a Relativistic Electron (Magnetobremsstrahlung)

When a relativistic electron of energy $E = \gamma mc^2$ moves through a region with a uniform magnetic field of strength H, it will emit high energy photons by synchrotron radiation, which is discussed in detail in Sect. 1.25. Here we note that although the emitted photon spectrum is continuous, in most situations it is sufficient to assume that the emitted photon has a frequency

$$\nu = \frac{3\gamma^2}{4\pi} \frac{eH}{mc} \approx 4.6 \times 10^{-6} HE^2 \text{ Hz} \;, \tag{4.380}$$

where E is in electron volts and H is in Gauss.

4.5.1.11 Photon Emission from Nuclear Reactions

A nucleus (Z, A) can emit a photon by the following reactions

$$\begin{aligned} &(Z, A) + \mathrm{n} \rightarrow (Z, A+1) + \gamma \\ &(Z, A) + \mathrm{p} \rightarrow (Z+1, A) + \gamma \\ &(Z, A) + \alpha \rightarrow (Z+2, A+4) + \gamma \\ &(Z, A)^* \rightarrow (Z, A) + \gamma \;. \end{aligned} \tag{4.381}$$

where * denotes an excited state.

Gamma-ray line astronomy is reviewed by Ramaty and Lingenfelter (1979, 1983), and discussed within the solar context in the next Sect. 4.5.1.12 – also see Murphy and Ramaty (1984) and Ramaty and Murphy (1987).

4.5.1.12 Energetic Particles and Radiation from Solar Flares

Solar flares are sudden, brief and powerful outbursts, releasing an energy from 10^{28} to 10^{34} erg on time scales of a second to several tens of minutes. Such short-lived solar flares release their energy within active regions, the magnetized atmosphere in, around and above sunspots, and vary in intensity and frequency of occurrence with the 11-year solar activity cycle. Many of the physical processes involved in solar flares are similar to those that produce billion-ton magnetic bubbles, called Coronal Mass Ejections or CMEs (Gosling, 1993; Hudson, Haisch and Strong, 1995; Lang, 1995); these ejections also vary in step with the activity cycle.

Although first reported from observations of visible sunlight, by Carrington (1859) and Hodgson (1859), flares usually produce only minor perturbations in the combined colors, or white light, of the Sun. In contrast, the flaring radio and X-ray emission is frequently several thousands of times more energetic than the Sun's normal radiation at these wavelengths. This invisible radiation indicates the presence of energetic electrons, accelerated to very high speeds approaching that of light, and gas heated to temperatures of about 10 million degrees. The most intense, high-energy radiation of solar flares, X-rays and γ-rays, comes from magnetic loops in solar active regions.

An early overview of solar flares is given in the monograph by Svestka (1976). The physics of solar flares is reviewed in the book by Tandberg-Hanssen and Emslie (1988) and in the collection of articles edited by Culhane and Jordan (1991). Solar flares and Coronal Mass Ejections are reviewed by Kahler (1992). High-energy neutral radiation from the Sun was reviewed by Chupp (1984), and overviews of solar radio emission are given by Dulk (1985) and Mc Lean and Labrum (1985). The theory and diagnostics of energetic particles in solar flares is reviewed by Brown (1991) and evidence for high-energy flare particles is reviewed by Hudson and Ryan (1995). Flares on the Sun and other stars are reviewed by Haisch, Strong and Rodono (1991), and discussed in the collection of articles edited by Haisch and Rodono (1989).

Intense radio emission during solar flares was detected in the earliest days of radio astronomy (Southworth, 1945; Hey, 1946; Payne-Scott, Yabsley and Bolton, 1947; also see Hey, 1973, and Christiansen, 1984). The solar flare radiation at radio wavelengths is often called a radio burst to emphasize its brief, energetic and eruptive characteristics. Burst radiation at millimeter and microwave wavelengths demonstrates the existence of nonthermal electrons accelerated to nearly the velocity of light. Such a powerful acceleration cannot occur by thermal processes. The nonthermal electrons produce synchrotron or gyrosynchrotron radiation at microwave wavelengths when spiraling about the intense magnetic fields of coronal loops. These magnetic loops reside within active regions and are anchored in underlying sunspots. Radio bursts at the longer meter wavelengths provide signatures of energetic particles travelling out through, or down into, the lower solar corona. Radio observations of solar flares have been reviewed by Wild, Smerd and Weiss (1962), Kundu (1965), Kundu (1982), Kundu and Vlahos (1982), Dulk (1985), Kundu and Lang (1985) and Mc Lean and Labrum (1985). Radio evidence for nonthermal particle acceleration on other stars of late spectral type is reviewed by Lang (1994).

Because solar flares have very high temperatures of up to 10 million degrees, the bulk of their radiation is emitted at X-ray and ultraviolet wavelengths that must be observed from above the Earth's absorbing atmosphere. Balloon-borne observations provided the first evidence for energetic X-ray radiation from solar flares (Peterson and Winkler, 1959). X-rays with relatively long wavelengths and low energies of about 10 keV are called soft X-rays to distinguish them from the more energetic, about 100 keV, hard X-rays. Satellites that have investigated the X-ray radiation of solar flares include the Skylab mission, which imaged solar flares at ultraviolet and X-ray wavelengths in 1973–74 (Eddy, 1979), the Solar Maximum Mission, or SMM, launched by NASA in 1980, and the Japanese Hinotori (fire-bird) spacecraft, launched in 1981. Flare observations from SMM and Hinotori have been respectively reviewed by Sakurai (1991) and Strong (1991). The SMM instruments included spectrometers at soft and hard X-ray wavelengths, designed to image the bremsstrahlung of energetic flares, as well as temperature-sensitive and density-sensitive spectral lines. Hinotori had the temporal and spatial resolution required to position and resolve the flaring hard X-ray emission. New data on the X-ray and gamma-ray radiation produced by solar flares has been provided by the Yohkoh spacecraft at both soft X-ray and hard X-ray wavelengths and by the Compton Gamma-Ray Observatory, both launched in 1991 (see Hudson and Ryan, 1995, for a review). An overview of the first five years of Yohkoh observations of the Sun is also given by Shimizu (1996).

The radio and X-ray radiation of solar flares can be interpreted in terms of a canonical model involving the release of stored magnetic energy in coronal loops, or arcades of loops, that link photospheric magnetism (sunspots) of opposite magnetic polarity. A general description of this model is given by Lang (1995), which is illustrated in Fig. 4.7. The energy appears initially as electrons and protons that are impulsively accelerated near the apex of the coronal loops on time scales of seconds or less. Nonthermal, high-speed relativistic electrons, that can reach an energy of 1 MeV, emit intense microwave radiation via gyrosynchrotron or synchrotron radiation near the loop tops during the impulsive phase of the solar flare. Energy is transported downward by the electrons and protons which follow the coronal loop's arching magnetism into the lower, denser reaches of the solar chromosphere. As the accelerated, nonthermal electrons stream from the corona toward and through the chromosphere, they produce nonthermal hard X-ray bremsstrahlung via interactions with the ambient protons. Interactions of flare-accelerated protons and alpha particles with the ambient heavier nuclei produce gamma-ray lines and energetic neutrons. The chromospheric plasma is heated very

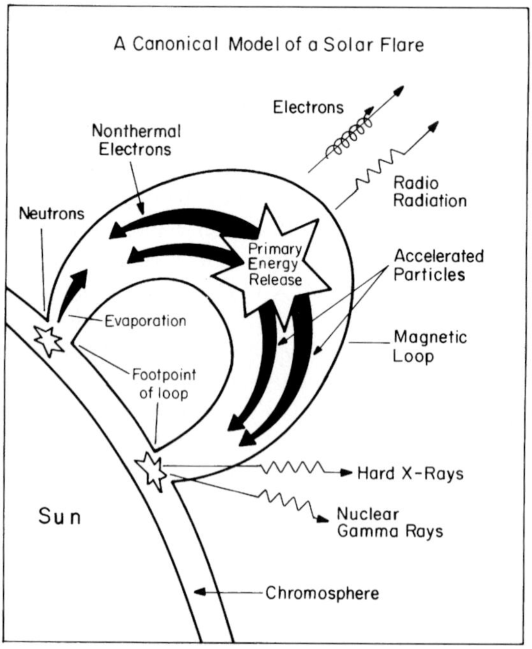

Fig. 4.7. In this canonical model, the impulsive flare begins with a primary energy release at the top of a coronal loop, where electrons and protons are accelerated to speeds approaching the velocity of light. The high-energy electrons emit microwave and radio radiation by the gyrosynchrotron and synchrotron process near the loop top, and they produce hard X-ray nonthermal bremsstrahlung near the loop footpoints in the chromosphere. Beams of protons and alpha particles stream downward along the curved magnetic channel, producing gamma-ray spectral lines and energetic neutrons when they enter the dense, lower atmosphere. This is followed by the rise of heated material, called chromospheric evaporation, into the loop, accompanied by a gradual increase in soft X-ray radiation. (Adapted from Lang, 1995.)

rapidly (in seconds) by the accelerated particles that slam into it, and is driven upward and "evaporates" along the guiding magnetic field by the large pressure gradients. Relatively long-lived (tens of minutes) soft X-ray radiation is then emitted by thermal bremsstrahlung as the flaring loop is filled with the rising evaporated material.

A comprehensive discussion of electron and proton beams in solar flares is given by Brown, Karlicky, Mac Kinnon and Van Den Oord (1990). Impulsive solar X-ray bursts from electrons beamed toward the photosphere are discussed by Langer and Petrosian (1977). Impulsive hard X-ray emission comes largely from the chromospheric footpoints of coronal magnetic loops (Hoyng et al., 1981; Duijveman et al., 1982). The microwave (8,000 to 17,000 MHz) and hard X-ray (10 to 100 keV) bursts have a similar time profile, suggesting a common origin from nonthermal electrons. Solar hard X-ray bursts are reviewed by Dennis (1985, 1988) and Tanaka (1987). The hard X-ray footpoints appear nearly simultaneous in time (Hudson et al., 1994; Sakao, 1994; Masuda, 1994), implying excitation in the chromosphere by energetic electron beams in coronal loops that connect the footpoints. Remarkably bright hard X-ray sources have also been observed in the corona above solar flares, with time variations that match those of the impulsive footpoint sources (Masuda et al., 1994; Takakura et al., 1993).

Nuclear gamma-ray lines, emitted at energies between 0.5 and 7 MeV during solar flares, have been observed from deuterium formation, electron–positron annihilation, and excited carbon, nitrogen, oxygen and heavier nuclei (Chupp et al., 1973; Chupp, 1984; Murphy et al., 1991; see Table 4.15). The narrow deexcitation lines (≤ 100 keV width) result from the interaction of ambient nuclei with flare-accelerated protons and alpha particles having

Table 4.15. Some important gamma-ray lines from solar flares[1]

Element	Energy (MeV)	Element	Energy (MeV)
Delayed Lines		*Proton Excitation*	
$e^+ + e^-$	0.511*	^{14}N	5.105*
(pair annihilation)			2.313
^{2}H = ^{2}D	2.223*	^{20}Ne	1.634*
(neutron capture)			2.613
			3.34
Spallation			
^{12}C	4.438*	^{24}Mg	1.369*
			2.754
^{16}O	6.129*	^{28}Si	1.779*
	6.917*		6.878
	7.117*		
	2.741	^{56}Fe	0.847*
Alpha Excitation			1.238*
^{7}Be	0.431*		1.811
^{7}Li	0.478*		

[1] The most prominent lines are marked with an asterisk * and have been detected in the flare of 27 April 1981 (Murphy et al., 1985); also see Longair (1992).

energies between 1 and 100 MeV per nucleon (Lingenfelter and Ramaty, 1967, 1973; Lingenfelter, 1969; Ramaty and Lingenfelter, 1979, 1983; Ramaty and Murphy, 1987).

Chromospheric evaporation during solar flares is discussed by Antonucci et al. (1982) and by Mariska et al. (1993), while the subsequent thermal soft X-ray emission is described by Pallavicini et al. (1977) and Doschek et al. (1993). The slow, smooth rise of the soft X-ray radiation resembles the integral of the rapid, impulsive time profile of the microwave and hard X-ray radiation, a relationship known as the Neupert effect (Neupert, 1968; Dennis and Zarro, 1993). This implies that much of the thermal decay phase of the flare is an atmospheric response to energetic particles or to the process that accelerates the particles.

The theory of thermal bremsstrahlung, which accounts for the soft X-ray radiation, was discussed in detail in Sect. 1.30. The photon energy, $h\nu$, is given by

$$h\nu \approx kT \approx 8.6 \times 10^{-8} T \text{ keV} \quad , \tag{4.382}$$

where ν is the frequency of the radiation and T is the temperature of the plasma in °K. The free–free flux, η_{FF}, per unit volume of the solar corona, at the Earth's distance, is given by (cf. Eq. (1.219))

$$\eta_{\mathrm{FF}} = 7.15 \times 10^{-50} N_e \sum_z N_z Z^2 \left[\exp\left(-\frac{143.89}{\lambda T} \right) \right] \times \frac{\langle g(Z, T, c/\lambda) \rangle}{T^{1/2} \lambda^2} \, d\lambda \quad \text{erg cm}^{-2} \text{ sec}^{-1} \text{ Å}^{-1} \quad . \tag{4.383}$$

Here N_e is the free electron density, N_z is the number density of ions of charge Z, the photon wavelength $\lambda = c/(2\pi\nu)$ is assumed to be given in Å (1 Å $= 10^{-8}$ cm), the electron temperature, T, is in units of 10^6 °K, and the $\langle g(Z, T, c/\lambda) \rangle \approx 1$ is the temperature averaged free–free Gaunt factor. In different units, the free-free X-ray flux at the Earth is given by

$$F_{\mathrm{FF}}(h\nu) = \frac{4.78 \times 10^{-39}}{T^{1/2}} \frac{\exp[-h\nu/(kT)]}{h\nu} \left[\int_v N_e N_i \, dv \right] \times \langle g(T, \nu) \rangle \quad \text{photons cm}^{-2} \text{ sec}^{-1} \text{ keV}^{-1} \quad , \tag{4.384}$$

where $h\nu$ is the photon energy, and the emission measure, $\int N_e N_i dv$ is the volume integral of the product of the free electron and ion densities. The free-free Gaunt factor, as well as the free–bound radiation formula, is given in Sect. 1.31 and 1.30, and numerical evaluations of the thermal emissivity are given in the references listed in these sections. For wavelengths in the range 1 Å to 30 Å and at temperatures in the range 0.8×10^6 to 10^8 °K, the free–free and free–bound emissivity have been evaluated by Culhane (1969).

X-ray spectroscopy of high-temperature solar flare plasmas using temperature-sensitive and density-sensitive line ratios indicate electron temperatures of $T_e = (2 \text{ to } 20) \times 10^6$ °K and electron densities of $N_e = 10^{11}$ to 10^{12} cm^{-3} (Phillips, 1991).

Detailed formulae for the microwave gyrosynchrotron radiation from solar flares have been provided by Takakura (1967), Ramaty (1969) and Ramaty and Petrosian (1972), reviewed by Dulk (1985), and given in Sect. 1.26. The radiation flux for an electron of energy E spiraling about a magnetic field of strength H, is a maximum at about twice the gyrofrequency, ν_H, given by:

$$\nu_H = \frac{eH}{2\pi mc\gamma}\sin\psi \approx 2.8 \times 10^6 H \ \sin\ \psi \ \text{Hz} \ , \tag{4.385}$$

where $\gamma = E/mc^2 = [1-(v/c)^2]^{-1/2} \approx 1$ for an electron of velocity, v, and kinetic energy, $E \approx mc^2$, and ψ is the pitch angle between the direction of the magnetic field and the direction of the electron motion. To a first approximation, the total gyrosynchrotron emission, P, from one electron is given by

$$P = \frac{8\pi^2}{3}\frac{e^2}{c}\nu_H^2\langle\sin^2\psi\rangle[\gamma^2 + 2\gamma] \tag{4.386}$$

for an electron with kinetic energy $E = \gamma mc^2$. The change in electron kinetic energy due to both gyrosynchrotron radiation and collisions with thermal electrons is given by (Takakura, 1967)

$$\frac{d\gamma}{dt} = \frac{1}{mc^2}\frac{dE}{dt} = -3.8\times 10^{-9}H^2\langle\sin\psi\rangle\left[\frac{\gamma^2}{2}+\gamma\right] + 1.5\times 10^{-16}N_e\gamma^{-3/2} \ , \tag{4.387}$$

where H is the magnetic field intensity in Gauss and N_e is the electron number density in cm^{-3}.

The hard X-ray radiation from solar flares is attributed to the nonthermal bremsstrahlung of suprathermal electron beams interacting with ambient protons. In thin-target bremsstrahlung, the nonthermal electrons radiate in the corona and lose only a fraction of their energy while radiating. For the thick-target bremsstrahlung, favored by the canonical flare model, the electrons radiate where they come to rest and thermalize, losing all their suprathermal energy there. The two models for nonthermal bremsstrahlung have been specified by Brown (1971) and Hudson (1972). Here we will adopt the discussion given by Tandberg-Hanssen and Emslie (1988).

The hydrodynamics of beam-heated solar flare atmospheres is provided by Brown and Emslie (1989), and electron and proton beams in solar flares are discussed by Brown, Karlicky, Mac Kinnon and Van Den Oord (1990). A comparison of the millimeter-wave and gamma-ray emissions from at least one flare has been interpreted in terms of the thick-target bremsstrahlung produced by nonthermal relativistic electrons (Ramaty et al., 1994).

Let $I(\varepsilon)$ denote the hard X-ray bremsstahlung flux at energy ε observed at the Earth, in units of photons $cm^{-2}\ s^{-1}\ keV^{-1}$, resulting from the injection of a beam of suprathermal energetic electrons with a differential energy spectrum $F(E_0)$, in units of electrons $cm^{-2}\ s^{-1}\ keV^{-1}$, over a flare area S. For the thin-target situation, no significant modification of the injected spectrum occurs, and:

$$I_{\text{thin}}(\varepsilon) = \frac{S\Delta N}{4\pi R^2}\int_{\varepsilon}^{\infty} F(E_0)\sigma_B(\varepsilon, E_0)dE_0 \ , \tag{4.388}$$

where $R = 1$ AU $\approx 1.496 \times 10^{13}$ cm is the mean distance between the Earth and the Sun, $\Delta N = \int_{\mathrm{source}} n_p(s)ds$ is the column density of the source observed, and $n_p(s)$ is the ambient proton density as a function of the distance, s, along the injected electron's path. The nonrelativistic Bethe–Heitler expression for the bremsstrahlung cross section is used:

$$\sigma(\varepsilon, E) = \frac{8\alpha}{3} r_0^2 \frac{m_e c^2}{\varepsilon E} \ln\left[\frac{1 + (1 - \varepsilon/E)^{1/2}}{1 - (1 - \varepsilon/E)^{1/2}}\right] ,$$

$$\sigma(\varepsilon, E) = \frac{7.9 \times 10^{-25}}{\varepsilon E} \ln\left[\frac{1 + (1 - \varepsilon/E)^{1/2}}{1 - (1 - \varepsilon/E)^{1/2}}\right] \mathrm{cm}^2\ \mathrm{keV}^{-1} ,$$

where ε and E are in keV in the numerical approximation, the classical electron radius is $r_0 = e^2/(m_e c^2) \approx 2.82 \times 10^{-13}$ cm and the fine structure constant is $\alpha = 2\pi e^2/(hc) \approx 7.3 \times 10^{-3}$. The cross section is differential in photon energy ε, but includes all possible directions of the outgoing electron and all possible directions and polarizations of the outgoing photon. The bremsstrahlung cross sections have been tabulated by Koch and Motz (1959), and become highly anisotropic at high values for the ratio of the photon energy ε to electron energy E (Gluckstern and Hull, 1953; Tseng and Pratt, 1973).

For a thick-target situation, the electrons thermalize in the bremsstrahlung source, and a target-averaged flux spectrum $F(E)$ should be used in place of the injected spectrum $F(E_0)$, both in units of electrons cm^{-2} s^{-1} keV^{-1}. In this case:

$$I_{\mathrm{thick}}(\varepsilon) = \frac{S}{4\pi R^2} \frac{1}{C} \int_{E_0=\varepsilon}^{\infty} F(E_0) \int_{\varepsilon}^{E_0} E \sigma_{\mathrm{B}}(\varepsilon, E) dE\, dE_0 , \tag{4.389}$$

where $C = 2\pi e^4 \ln \Lambda$ and typical values of the Coulomb logarithm in the solar situation are $\ln \Lambda = 20$ to 30. An effective column density ΔN_{eff} for the thick target, which permits comparison to the thin-target expression, is:

$$\Delta N_{\mathrm{eff}} = \frac{1}{C \sigma_{\mathrm{B}}(\varepsilon, E)} \int_{\varepsilon}^{E_0} E \sigma_{\mathrm{B}}(\varepsilon, E) dE .$$

This effective column density varies roughly with E_0^2 and corresponds approximately to the column density required to stop an electron of injected energy E_0. The principal contributions to the emission of photons with energy ε are from electrons with energy $E \geq \varepsilon$, particularly when $F(E_0)$ is a rapidly decreasing function of E_0. The thin-target approximation will be a good one when the actual column density $\Delta N \ll \Delta N_{\mathrm{eff}}(\varepsilon)$; whereas the thick-target expression is appropriate when $\Delta N \gg \Delta N_{\mathrm{eff}}(\varepsilon)$.

Eqs. (4.388) and (4.389) are integral equations, relating the injected spectrum $F(E_0)$, often called the source function, to the hard X-ray spectrum, $I(\varepsilon)$, termed the data function. Brown (1971) showed how these equations can be reduced to Abel's integral equation, permitting analytic inversion to yield the source function in terms of the properties of $I(\varepsilon)$.

If we assume a source function of the form

$$F(E_0) = AE_0^{-\delta} , \tag{4.390}$$

we obtain (Tandberg-Hanssen and Emslie, 1988):

$$I_{\text{thin}}(\varepsilon) = \frac{S\Delta NA}{4\pi R^2} \kappa_{\text{BH}} \overline{Z^2} \frac{B(\delta, 1/2)}{\delta} \varepsilon^{-(\delta+1)} , \tag{4.391}$$

where $\kappa_{\text{BH}} = (8\alpha/3)r_0^2 m_e c^2 = 7.9 \times 10^{-25}$ cm^2 keV is the constant in the Bethe-Heitler cross section, an abundance weighted value of $\overline{Z^2} = 1.4$ is appropriate for solar conditions, and $B(a, b)$ can be evaluated using the identity

$$B(a, b) \equiv \frac{\Gamma(a)\Gamma(b)}{\Gamma(a+b)} ,$$

where $\Gamma(a)$ is the gamma (factorial) function

$$\Gamma(a) = \int_0^\infty e^{-x} x^{a-1} \, dx ,$$

with the properties $\Gamma(a) = (a-1)\Gamma(a-1)$ and $\Gamma(1) = 1$. Eq. (4.391) shows that the thin-target hard X-ray spectrum is itself a power law with

$$I_{\text{thin}}(\varepsilon) = a\varepsilon^{-\gamma}, \tag{4.392}$$

with $\gamma = \delta + 1$, and

$$a = \frac{S\Delta NA}{4\pi R^2} \kappa_{\text{BH}} \overline{Z^2} \frac{B(\gamma - 1, 1/2)}{(\gamma - 1)} .$$

For the thick-target case we obtain (Tandberg-Hanssen and Emslie, 1988):

$$I_{\text{thick}}(\varepsilon) = \frac{SA}{4\pi R^2 C} \kappa_{\text{BH}} \overline{Z^2} \frac{B(\delta - 2, 1/2)}{(\delta - 1)(\delta - 2)} \varepsilon^{1-\delta} . \tag{4.393}$$

This is also a power law $I(\varepsilon) = a\varepsilon^{-\gamma}$ with the exponent $\gamma = \delta - 1$, and

$$a = \frac{SA\kappa_{\text{BH}}\overline{Z^2}}{4\pi R^2 C} \frac{B(\gamma - 1, 1/2)}{\gamma(\gamma - 1)} .$$

For the constant a in both the thin-target and thick-target case, an abundance weighted average of $\overline{Z^2} = 1.4$ can be used in the solar situation.

The thick-target hard X-ray yield is that of a thin target with an effective injected flux spectrum

$$F_{\text{eff}}(E_0) = \frac{A}{(\delta - 1)} \left[\frac{E_0^2}{C\Delta N} \right] E_0^{-\delta},$$

which is two powers of E_0 harder (flatter) than the injected spectrum $F(E_0) = AE_0^{-\delta}$.

Eqs. (4.392) and (4.393) can be used to infer properties of the injected electron flux, the parameters A and δ of Eq. (4.390), from the observed hard X-ray parameters a and γ, once a thin-target or thick-target model has been chosen. Related quantities include the total electron flux, F_1, above some reference energy E_1:

$$F_1 = \int_{E_1}^{\infty} AE_0^{-\delta}\, dE_0 = \frac{A}{(\delta-1)} E_1^{1-\delta} \ ,$$

and the energy flux, $\mathscr{F}_1$, above the same reference energy: (4.394)

$$\mathscr{F}_1 = \int_{E_1}^{\infty} AE_0^{-\delta} E_0\, dE_0 = \frac{A}{(\delta-2)} E_1^{2-\delta} \ .$$

The high energy processes responsible for gamma-ray emission from solar flares have been discussed by Ramaty and Lingenfelter (1983) and Murphy et al. (1987), and reviewed by Chupp (1984), Murphy and Ramaty (1984) and Ramaty and Murphy (1987). As illustrated in Table 4.15, interactions of flare accelerated particles with the ambient solar atmosphere produce gamma-ray lines in the energy range 0.5 to 10 MeV.

Flare-accelerated protons interact with abundant elements in the solar atmosphere, exciting nuclei that emit gamma-ray lines on de-excitation (Lingenfelter and Ramaty, 1967; Chupp, 1971, 1974; Ramaty and Lingenfelter, 1979, 1983). The inelastic scattering process of the nucleus, N, with a proton, p, can be written:

$$\mathrm{N} + \mathrm{p} \rightarrow \mathrm{N}^* + \mathrm{p}' \qquad (4.395)$$

where the unprimed and primed sides respectively denote the incident and scattered proton, p, and the excited nucleus N^* reverts to its former state emitting a gamma-ray line, γ, by:

$$\mathrm{N}^* \rightarrow \mathrm{N} + \gamma \ .$$

Elements and energies of prominent gamma-ray lines resulting from proton excitation of abundant elements are given in Table 4.15.

Excited nuclei can also be generated by the fusion of flare-associated particles with atmospheric nuclei. Examples include beryllium, $^7\mathrm{Be}^*$, and lithium, $^7\mathrm{Li}^*$, produced by flaring alpha particles, α, and ambient helium, $^4\mathrm{He}$ (Kozlovsky and Ramaty, 1977; Share and Murphy, 1997).

$$^4\mathrm{He} + \alpha \rightarrow {}^7\mathrm{Be}^* + \mathrm{n}$$

and

$$^4\mathrm{He} + \alpha \rightarrow {}^7\mathrm{Li}^* + \mathrm{p} \qquad (4.396)$$

with the excited nuclei reverting to their former unexcited state accompanied by the emission of a gamma ray line, γ, of energy, $h\nu$, by:

$$^7\mathrm{Be}^* \rightarrow {}^7\mathrm{Be} + \gamma \quad (h\nu = 0.431\ \mathrm{MeV})$$

and

$$^7\mathrm{Li}^* \rightarrow {}^7\mathrm{Li} + \gamma \quad (h\nu = 0.478\ \mathrm{MeV}) \ .$$

The collision of a flare-associated proton, p, or alpha-particle, α, with a heavy ambient nucleus may cause the nucleus to break up into lighter fragments that are left in excited states that subsequently deexcite to emit gamma-ray lines.

Important examples are the production of excited carbon, $^{12}C^*$, and oxygen, $^{16}O^*$, by flaring protons, p, that break up oxygen, ^{16}O, or neon, ^{20}Ne, nuclei by the reactions:

$$^{16}O + p \rightarrow ^{12}C^* + \alpha$$

and

$$^{20}Ne + p \rightarrow ^{16}O^* + \alpha \ , \tag{4.397}$$

with deexcitation:

$$^{12}C^* \rightarrow ^{12}C + \gamma \quad (h\nu = 4.438\,\text{MeV})$$

and

$$^{16}O^* \rightarrow ^{16}O + \gamma \quad (h\nu = 6.129\,\text{MeV}) \ .$$

Additional gamma-ray lines that are observed during solar flares (also see Table 4.15) include those associated with electron–positron annihilation at 0.511 MeV

$$e^- + e^+ \rightarrow \gamma + \gamma \quad (h\nu = 0.511 \text{ MeV}) \ , \tag{4.398}$$

and the feature at 2.223 MeV associated with neutron, n, capture by hydrogen nuclei, ^1H, to produce excited deuterium $^2H^*$, that relaxes with emission of gamma rays, γ.

$$^1H + n \rightarrow {}^2H^* \tag{4.399}$$

with

$$^2H^* \rightarrow {}^2H + \gamma \quad (h\nu = 2.223\,\text{MeV}) \ .$$

The neutrons are released in spallation induced reactions by particles accelerated in the flare. The line emission is usually delayed from the time of flare acceleration in order to allow the neutrons to thermalize before they are captured.

High-energy neutrons are also produced when flare-accelerated protons and alpha-particles interact with ambient hydrogen, ^1H, and helium, ^4He (see Table 4.16). Neutrons with energies of 20 to 100 MeV have been observed directly in space during solar flares (Chupp et al., 1982, 1987; Chupp, 1990; Ryan et al., 1994); they are in turn produced by flare protons with energies of up to 1 GeV (Ramaty and Mandzhavidze, 1994). Neutrons with very high

Table 4.16. Neutron producing reactions for the abundant nuclei

Reaction	Threshold energy (MeV/nucleon)
H^1 (p, nπ^+) H^1	292.3
He^4 (p, pn) He^3	25.9
He^4 (p, 2pn) H^2	32.8
He^4 (p, 2p2n) H^1	35.6
C^{12} (p, n...)	19.8
N^{14} (p, n ...)	6.3
O^{16} (p, pn...)	16.5
Ne^{20} (p, pn...)	17.7

energy of about 1 GeV have been detected by ground-based neutron monitors following exceptionally intense solar flares (Debrunner et al., 1983, 1990).

For detailed calculations, the proton energy spectrum is taken to be

$$N(P) = P_0^{-1} \exp\left(-\frac{P}{P_0}\right) , \tag{4.400}$$

where for a nucleus of mass number, A, and charge, Ze, the rigidity, $P = A[2M_p c^2 E_p + E_p^2]^{1/2}/(Ze)$ for protons of energy, E_p, and rest mass, M_p. The rigidity is the ratio of momentum to charge, P_0 is characteristic rigidity, $P_0 \approx 80$ to 200 MeV, and $N(P)$ may be normalized at the Sun to one particle of rigidity greater than zero. The most important neutron producing reactions are given in Table 4.16. The total neutron production cross sections, σ_n, are (Lingenfelter and Ramaty, 1967) $\sigma_n \approx 20$, 200, and 1000 mb, respectively, for the pp, pα, and pCNONe reactions when the proton energy, E_p, is greater than 1 BeV. For reference, one millibarn = 1 mb = 10^{-27} cm^2. The differential energy spectrum, $f(E_n, E_p)$, for incident protons of energy, E_p, and secondary neutrons of energy, E_n, is given by

$$f(E_n, E_p) = \frac{1}{E_p(1-K)} \quad \text{for } 0 < E_n < E_p(1-K) \text{ and the (p, p) reactions}$$

$$f(E_n, E_p) = \frac{1}{E_p - Q} \quad \text{for } 0 < E_n < E_p - Q \text{ and the (p, }\alpha\text{) reactions}$$

$$f(E_n, E_p) = E_n \theta^{-2} \exp(-E_n/\theta) \text{ for the pCNONe reactions} . \tag{4.401}$$

Here $K \approx 0.25$ is the inelasticity defined as the fraction of the incident proton energy going into total pion energy, Q is the kinetic energy converted into rest mass in the (p, α) reaction, and $\theta \approx 1.5$ MeV is the temperature of the excited nucleus. The neutrons so produced interact with hydrogen according to the reaction $n + p \rightarrow D + \gamma$ to produce the 2.23 MeV gamma ray line. The relative probability of neutron capture by other nuclei in the solar atmosphere is less than 10^{-3} of that of hydrogen.

The principal pion producing reactions are

$$\begin{aligned}
p + p \rightarrow & D + \pi^+ \\
& p + p + a(\pi^+ + \pi^-) + b\pi^0 \\
& p + n + \pi^+ + a(\pi^+ + \pi^-) + b\pi^0 \\
& 2n + 2\pi^+ + a(\pi^+ + \pi^-) + b\pi^0 \\
p + He^4 \rightarrow & p + He^4 + a(\pi^+ + \pi^-) + b\pi^0 \\
& p + He^3 + n + a(\pi^+ + \pi^-) + b\pi^0 \\
& 2p + H^2 + n + a(\pi^+ + \pi^-) + b\pi^0 \\
& 4p + n + \pi^- + a(\pi^+ + \pi^-) + b\pi^0 \\
& 3p + 2n + a(\pi^+ + \pi^-) + b\pi^0 \\
& 2p + 3n + \pi^+ + a(\pi^+ + \pi^-) + b\pi^0 \\
& p + 4n + 2\pi^+ + a(\pi^+ + \pi^-) + b\pi^0
\end{aligned} \tag{4.402}$$

The measured cross sections for these reactions have been summarized by Lingenfelter and Ramaty (1967), and approximate values for the production of π pions are $\sigma_{\pi^+} \approx 20$ and 100 mb for the pp and pα reactions, and $\sigma_{\pi^0} \approx 9E_\mathrm{p}$ and $90E_\mathrm{p}$ mb for a proton energy, E_p, in GeV and greater than one GeV. The differential energy spectrum, $f(E_\pi, E_\mathrm{p})$, for incident protons of energy, E_p, and secondary pions of energy, E_π, is given by

$$f(E_\pi, E_\mathrm{p}) = \delta(E_\pi - \langle E_\pi \rangle), \tag{4.403}$$

where δ is the Dirac delta function and

$$\langle E_\pi \rangle = 0.985 E_\mathrm{p}^{3/4} \text{ in MeV for the (pp) reaction } ,$$

and

$$\langle E_\pi \rangle = 0.754 E_\mathrm{p}^{3/4} \text{ in MeV for the (p}\alpha\text{) reaction } .$$

Lingenfelter and Ramaty (1967) assumed that protons are accelerated in an isotropic manner and interact with the solar atmosphere isotropically. They calculated the total number of secondary components, $Q(E_\mathrm{s})$, of energy, E_s, produced during the acceleration of protons, and also when the protons are released in the solar material and stop. The acceleration and slowing down phases may be taken, respectively, to be equivalent to the thin and thick target models. Assuming that the spectral energy density, $\langle N(E) \rangle$, of accelerated particles is constant over the acceleration time, t_1, the total number of secondaries produced during the acceleration phase is (Lingenfelter and Ramaty, 1967; Chupp, 1971)

$$Q(E_\mathrm{s}) = N t_1 \int_0^\infty \sigma(E) f(E_\mathrm{s}, E) V(E) \langle N(E) \rangle dE \quad \mathrm{MeV}^{-1} , \tag{4.404}$$

where $Q(E_\mathrm{s})$ is in particles MeV^{-1} when the number density of target nuclei, N, is in cm^{-3}, the acceleration time, t_1, is in seconds, the cross section, $\sigma(E)$, at energy, E, for secondary production is in cm^2, $f(E_\mathrm{s}, E)$ is the normalized differential energy spectrum, $V(E)\langle N(E) \rangle$ is the flux of accelerated nuclei in cm^{-2} sec^{-1} MeV^{-1}, and the proton kinetic energy and the proton velocity are, respectively, E and $V(E)$. The $f(E_\mathrm{s}, E)$ is the probability that a neutron or gamma ray which results from interaction with a proton of energy, E, will have an energy, dE_s, around E_s. The proton energy spectrum, $\langle N(E) \rangle$, is given by Eq. (4.400) and is normalized so that it gives the total number of accelerated particles at energy, E.

In the slowing down phase, the total number of particles per MeV is given by

$$Q(E_\mathrm{s}) = \frac{N}{\rho} \int_0^\infty \sigma(E) f(E_\mathrm{s}, E) \frac{dR(E)}{dE} \, dE \int_E^\infty N(E_1, t_1) \times \exp\left(- \frac{[R(E_1) - R(E)]}{L} \right) dE_1 \, \mathrm{MeV}^{-1} , \tag{4.405}$$

where N/ρ is the number of target atoms per gram of solar material, ρ is the mass density in g cm^{-3} seen in slowing down, $dR(E)/dE$ is the slope of the stopping range, $R(E)$, versus energy, E, curve, E_1 is the energy the proton must have at t_1 in order to slow down to an energy, E, at time, t, and L is the mean attenuation length of protons over energy, E_1, to energy, E. Both $R(E)$ and L are in units of g cm^{-2}.

If during acceleration one half of all the neutrons produced escape the Sun, and the neutron production rate is an impulse function in time, the time dependent neutron flux at the Earth, $\varphi(t)$, is given by

$$\varphi(t) = \frac{Q(E_n)}{2\pi R^2}\frac{dE_n}{dt} P_e(E_n) P_s(E_n) \quad \text{cm}^{-2}\ \text{sec}^{-1}\ , \tag{4.406}$$

where $Q(E_n)$ is given by Eqs. (4.404) or (4.405), R is the Sun-Earth distance $\approx 1\ A.\ U.$, dE_n/dt is the rate of change of neutron energy at 1 $A.\ U.$ for a delta function production at the Sun, $P_e(E_n) = 0.5$, the neutron survival probability at 1 $A.\ U.$ is $P_s(E_n)$ and $Q(E_n)$ is evaluated at $E_n(t) = mc^2\{[1-(R/ct)^2]^{-1/2} - 1\}$ where $R = vt$ and v is the velocity of the neutron.

Gamma-ray flares exhibit evidence for ions up to 100 MeV per nucleon, and some appear capable of accelerating protons up to 1 or 10 GeV. Ions that escape from the impulsive solar flare and are observed directly in interplanetary space have energies up to 100 MeV per nucleon (Reames et al., 1992; Mazur et al., 1992). Shock waves responsible for accelerating interplanetary MeV protons may be related to coronal mass ejections rather than solar flares (Kahler, 1992). Both the electrons and ions in solar flares can reach relativistic velocities promptly (in a few seconds) and nearly simultaneously (Forrest and Chupp, 1983; Forrest et al., 1986; Chupp et al., 1987).

A number of mechanisms that might account for particle acceleration during solar flares have been reviewed by Miller et al. (1997) and by De Jager (1986). Particle acceleration on the Sun and in other cosmic objects is also discussed in the collection of articles edited by Trottet and Pick (1987), Zank and Gaiser (1992) and Chupp and Benz (1994). Flare acceleration processes include stochastic acceleration by electromagnetic and electrostatic waves, drift and diffusive shock acceleration and acceleration by large-scale, quasi-static electric fields. Both shock and high-speed plasma jets can be produced during magnetic reconnection; pinch sheets and magnetic reconnection have been reviewed by Syrovatskii (1981). One global reconnection scenario occurs at the top of a coronal loop or arcade of loops (Carmichael, 1964; Sturrock, 1968; Kopp and Pneuman, 1976; Cargill and Priest, 1983; Forbes et al., 1989).

Observations at radio, soft X-ray and ultraviolet wavelengths, respectively using the Very Large Array and the Yohkoh and SOHO spacecraft, indicate that solar flares may be due to forced magnetic reconnection in the corona (Kundu and Lang, 1985; Tsuneta et al., 1992; Masuda et al., 1994; Tsuneta, 1995, 1996; Manoharan et al., 1996; Shimizu, 1996; and Innes et al., 1997).

4.5.2 Energy Loss Mechanisms for High Energy Particles and High Energy Radiation

4.5.2.1 Charged Particle Energy Loss by Ionization

When a charged particle passes through matter, it loses energy by exciting and ionizing atoms. The loss in the energy, E, of a particle of charge, eZ, is given by the Bethe–Bloch formula for ionization loss of protons and nuclei in passing through matter (Bethe, 1930, 1932; Bloch, 1933).

$$-\frac{dE}{dx} = \frac{2\pi Z^2 e^4 N}{mv^2}\left[\ln\left(\frac{2mv^2\gamma^2 W_{\rm m}}{I^2}\right) - 2\beta^2 + f\right]$$

$$\approx 2.54 \times 10^{-19} Z^2 N \sqrt{2/(\gamma - 1)}[\ln(\gamma - 1) + 11.8] \quad \text{eV cm}^{-1}$$

for atomic hydrogen and $\gamma \ll 1$

$$\approx 2.54 \times 10^{-19} Z^2 N \left[3 \ln \gamma + \ln\left(\frac{M}{m}\right) + 19.5\right] \quad \text{eV cm}^{-1}$$

for atomic hydrogen and $\gamma \gg M/m$, (4.407)

where dx is an element of unit length in the direction of particle motion, m is the electron mass, M is the mass of the incident particle whose velocity is v, the

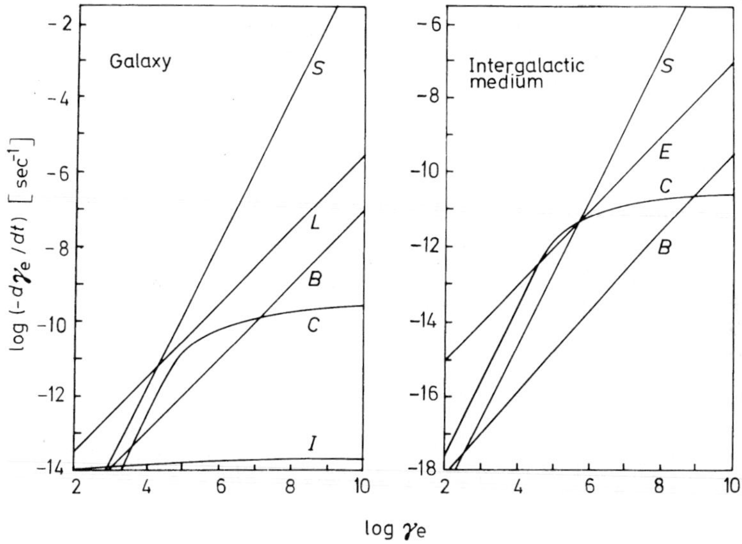

Fig. 4.8. Electron energy loss rates in the Galaxy and the intergalactic medium by synchrotron emission (S), cosmic expansion (E), leakage out of the halo (L), bremsstrahlung emission (B), Compton scattering (C), and ionization (I), (after Burbidge, 1966, by permission of Academic Press, Inc.). Here $\gamma = E/mc^2$ where E is the total energy of the electron. All energy loss rates are given in Sect. 4.5 except for the leakage loss rate $d\gamma/dt = \gamma lc/R^2$, and the expansion loss rate $d\gamma/dt = \gamma/H_0$. Here the radius of the galactic halo is $R \approx 5 \times 10^{22}$ cm, the mean free path for Brownian motion of interstellar gas clouds is $l \approx 1$ kiloparsec $= 3 \times 10^{21}$ cm, and the Hubble constant gives an age of $H_0^{-1} \approx 10^{17}$ sec

$\beta = v/c$, the $\gamma = (1-\beta^2)^{-1/2} = E/(Mc^2)$, the number density of electrons in the material is N, the ionization potential is I, and the maximum energy transfer, W_m, is given after Eq. (4.362). The density effect term, f, was first suggested by Fermi (1939, 1940), and is tabulated together with other constants in Eq. (4.407) by Sternheimer (1956) and Hayakawa (1969). Ionization potentials were given in Table 3.2.

For a completely ionized gas and an ultrarelativistic incident particle, we have (Ginzburg, 1969)

$$f = \ln(1-\beta^2) + \ln\left(\frac{I^2}{\hbar^2\omega_p^2}\right) + 1 \ ,$$

where the plasma frequency, ω_p, is given by

$$\omega_p = (4\pi e^2 N_e/m)^{1/2} \ ,$$

and N_e is the free electron density. For the case of an ionized gas and an ultrarelativistic incident particle, Eq. (4.407) becomes

$$-\frac{dE}{dx} \approx 2.54\times10^{-19} Z^2 N_e \left[\ln\left(\frac{W_m}{mc^2}\right) - \ln N_e + 74.1\right] \quad \text{eV cm}^{-1} \ , \qquad (4.408)$$

where $W_m = E$ if $\gamma \gg (M/m)$ and $W_m = 2E^2/(mc^2)$ for $1 \ll \gamma \ll (M/m)$.

For the case of electrons, the ionization loss is (Longair, 1992)

$$-\frac{dE}{dx} = \frac{2\pi e^4 NZ}{m_e v^2}\left\{\left[\ln\left(\frac{\gamma^2 m_e v^2 T_{max}}{2I^2}\right)\right] - \left(\frac{2}{\gamma} - \frac{1}{\gamma^2}\right)\ln 2 + \frac{1}{\gamma^2} + \frac{1}{8}\left(1-\frac{1}{\gamma}\right)^2\right\} \ , \qquad (4.409)$$

where N, Z and I denote the number density, atomic number, and ionization potential of the atoms in the material through which the electron passes, v is the velocity of the electron, and the maximum kinetic energy of a particle with mass M and momentum $M\beta\gamma$ is given by

$$T_{max} = \frac{2m_e c^2\beta^2\gamma^2}{1 + 2\gamma m_e/M + (m_e/M)^2} = \frac{2\gamma^2 M^2 m_e v^2}{m_e^2 + M^2 + 2\gamma m_e M} \ , \qquad (4.410)$$

where $\beta = v/c$ and the incident particle energy $E = \gamma Mc^2$. In the case of electron-electron collisions, $M = m_e$ and T_{max} has the value:

$$T_{max} = \frac{\gamma^2 m_e v^2}{1+\gamma} \ .$$

For the special case of an electron incident upon neutral atoms,

$$-\frac{dE}{dx} = \frac{2\pi e^4 N}{mv^2}\left[\ln\left(\frac{mv^2\gamma^2 W_m}{I^2}\right) + \frac{9}{8} - \beta^2 + f\right]$$
$$\approx 2.54\times10^{-19} N[3\ln\gamma + 20.2] \quad \text{eV cm}^{-1} \text{ for hydrogen and } \gamma \ll 1. \qquad (4.411)$$

When ultrarelativistic electrons are incident upon a fully ionized gas,

$$-\frac{dE}{dx} \approx 2.54 \times 10^{-19} N_e[\ln \gamma - \ln N_e + 73.4] \quad \text{eV cm}^{-1} \ . \tag{4.412}$$

Electron energy losses by ionization and other processes are illustrated in Fig. 4.8 for the Galaxy and the intergalactic medium.

Ionization losses for non-relativistic and relativistic particles are reviewed by Longair (1992). A practical form of the Bethe–Bloch equation for the ionization loss of protons and nuclei is expressed in terms of the total mass per unit cross-section traversed by the particle. Using $\xi = \rho x$ for a mass density, ρ, the mean rate of energy loss, $-dE/d\xi$, called the stopping power of the material, is given by (Review of Particle Properties, Phys. Rev. **D50**, 1173, 1994):

$$-\frac{dE}{d\xi} = Kz^2 \frac{Z}{A} \frac{1}{\beta^2} \left[\frac{1}{2} \ln \left(\frac{2m_e c^2 \beta^2 \gamma^2 T_{\max}}{I^2} \right) - \beta^2 - \frac{\delta}{2} \right] ,$$

or

$$-\frac{dE}{d\xi} \approx z^2 \frac{NZ}{\rho} f(v) = z^2 \frac{Z}{m} f(v) \ , \tag{4.413}$$

for protons and nuclei incident on material of nuclear mass, m. Here the constant $K = 4\pi N_A r_e^2 m_e c^2 = 0.307\,075$ MeVg^{-1}cm^2 for an atomic mass A of 1 g mol^{-1}, the quantity ze is the charge of the incident particle, Z and A are the atomic number and mass of the material, the $\beta = v/c$ for an incident particle of velocity v, the velocity of light is c, the incident particle energy is $E = \gamma Mc^2$ for a particle mass M, the N is the number density of atoms, the ionization potential of the relevant atom is I, the electron density is $N_e = ZN_m$, the $f(v)$ is a function of the incident particle velocity v, the maximum kinetic energy, $T_{\max}$, is given above in Eq. (4.410), the δ is the density effect correction to the ionization energy loss, and at very high energies

$$\frac{\delta}{2} = \left[\ln \left(\frac{h\nu_p}{I} \right) \right] + \ln(\beta\gamma) - \frac{1}{2} ,$$

for a plasma energy of $h\nu_p$ and plasma frequency ν_p. The energy loss rate $-(dE/d\xi)/Z^2$ has been provided by Enge (1966) and Hillas (1972).

4.5.2.2 Electron Energy Loss by Bremsstrahlung

When a relativistic electron of energy, E, collides with nuclei of charge, eZ, it emits photons of energy, E_γ, and loses energy according to the equation (Bethe and Heitler, 1934)

$$-\frac{dE}{dx} = N \int_0^{E-mc^2} E_\gamma \sigma_B(E, E_\gamma) dE_\gamma \ , \tag{4.414}$$

when N is the number density of the nuclei, and the differential cross section $\sigma_B(E, E_\gamma)$ is given by Eqs. (4.372), (4.373), or (4.376). Using Eqs. (4.372) and (4.373) in Eq. (4.414), we obtain

$$-\frac{dE}{dx} \approx 4N\alpha r_0^2 E Z^2 \ln\left(\frac{2E}{mc^2}\right) \quad \text{for } mc^2 \ll E \ll \frac{mc^2}{\alpha Z^{1/3}} ,$$
$$-\frac{dE}{dx} \approx \frac{MNE}{X} \quad \text{for } E \gg \frac{mc^2}{\alpha Z^{1/3}} . \tag{4.415}$$

Here N is the number density of target nuclei, the fine structure constant $\alpha = e^2/(\hbar c) \approx 1/137.036$, the classical electron radius $r_0 = e^2/(mc^2) \approx 2.818 \times 10^{-13}$ cm, the rest mass energy of the electron is $mc^2 \approx 0.511$ MeV, the mass of the target nucleus is M, and the radiation length X is given by Eqs. (4.374) or (4.375). For electron-atom bremsstrahlung, the energy losses are given by Eqs. (4.414) and (4.415) with Z^2 replaced by $Z(Z+\xi)$ where (Wheeler and Lamb, 1939; Tsai, 1974):

$$\xi = \frac{\ln(183Z^{-1/3})}{\ln(1440Z^{-2/3})} . \tag{4.416}$$

For the special case of hydrogen atoms, $Z = 1, M = 1.67 \times 10^{-24}$ grams, $X = 62.8$ gram cm^{-2}, and

$$-\frac{dE}{dx} \approx 2.6 \times 10^{-26} NE \text{ eV cm}^{-1} \quad \text{for } E \gg \frac{mc^2}{\alpha} \tag{4.417}$$
$$\approx 2.3 \times 10^{-27} NE \ln\left(\frac{2E}{mc^2}\right) \text{ eV cm}^{-1} \quad \text{for } mc^2 \ll E \ll \frac{mc^2}{\alpha Z^{1/3}} .$$

When the gas is completely ionized, the no screening effect equations are appropriate. Bremsstrahlung, synchrotron radiation, and Compton scattering of high energy electrons traversing a dilute gas are reviewed by Blumenthal and Gould (1970).

4.5.2.3 Electron Energy Loss by Compton Scattering – The Inverse Compton Effect

When a relativistic electron of energy, $E = \gamma mc^2$, collides with photons of energy, $h\nu$, it produces a photon of energy, E_γ, and loses energy according to the equation

$$-\frac{dE}{dx} = \int \sigma_c(E_\gamma, h\nu) N(h\nu) E_\gamma \, dE_\gamma \, d(h\nu) , \tag{4.418}$$

where the Klien–Nishina (1929) scattering cross section, $\sigma_c(E_\gamma, h\nu)$, is given in Eqs. (4.370), and $N(h\nu)$ is the total number density of incident photons of energy, $h\nu$. From Eqs. (4.370) and (4.371) we obtain

$$E_\gamma \approx \gamma^2 h\nu \quad \text{and} \quad \sigma_c \approx \sigma_T \quad \text{for } \gamma h\nu \ll mc^2$$

and

$$E_\gamma \approx \gamma^2 mc^2 \quad \text{and} \quad \sigma_c \approx \frac{3}{8}\sigma_T \left(\frac{mc^2}{\gamma h\nu}\right)\left[\ln\left(\frac{2\gamma h\nu}{mc^2}\right) + \frac{1}{2}\right] \quad \text{for } \gamma h\nu \gg mc^2 , \tag{4.419}$$

where the Thomson scattering cross section, $\sigma_T = 8\pi e^4/(3m^2c^4) \approx 6.65 \times 10^{-25}$cm^2. Substituting Eqs. (4.419) into Eq. (4.418) we obtain

$$-\frac{dE}{dx} \approx \sigma_T U \gamma^2 \quad \text{for} \quad \gamma h\nu \ll mc^2$$

and

$$-\frac{dE}{dx} \approx \frac{3}{8}\sigma_T U \left(\frac{mc^2}{h\nu}\right)^2 \ln\left(\frac{2\gamma h\nu}{mc^2}\right) \quad \text{for } \gamma h\nu \gg mc^2 \ , \tag{4.420}$$

where the photon energy density, U, is given by

$$U = \int h\nu N(h\nu)\, d(h\nu) \approx h\nu N(h\nu)$$

and the photon energy is $h\nu$. Various contributions to U have been discussed by Feenberg and Primakoff (1948), Donahue (1951), and Felten and Morrison (1963, 1966) for the Galaxy and for intergalactic objects. The energy density of electromagnetic radiation in various spectral regions is illustrated in Fig. 4.9. Bremsstrahlung, synchrotron radiation, and Compton scattering of high energy electrons traversing a dilute gas are reviewed by Blumenthal and Gould (1970).

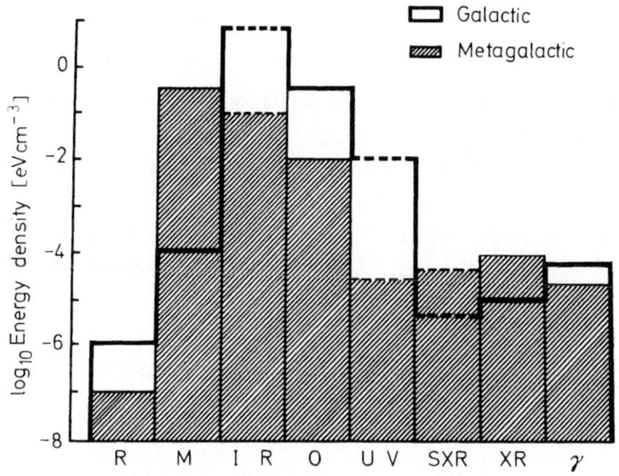

Fig. 4.9. Estimates of the flux of electromagnetic radiation in different spectral regions, plotted in histogram form and in energy density units of eV cm^{-3} = 1.6×10^{-12} erg cm^{-3}. The spectral regions considered are the radio ($R \approx 10^8 - 10^{10}$ Hz), microwave ($M \approx 10^{10} - 10^{12}$ Hz), infrared ($I - R \approx 10^{12} - 10^{14}$ Hz), optical ($0 \approx 10^{13} - 10^{15}$ Hz), ultraviolet ($UV \approx 10^{15} - 10^{17}$ Hz), soft X-ray ($SXR \approx 10^{17}$ Hz), X-ray ($XR \approx 10^{17} - 10^{19}$ Hz) and γ-ray ($\gamma \approx 10^{18} - 10^{20}$ Hz) frequencies. Fluxes are shown in the thatched regions for the galactic contribution in the solar neighbourhood, and under the heavy line for an average point in intergalactic space. This histogram compares, in effect, the surface brightness of the Milky Way (after subtracting the surface brightness of the galactic poles) with that at the galactic poles. Estimates for those spectral regions where considerable uncertainty still exists are shown as dashed lines. Only in the microwave region, where the 2.7 °K black body photon radiation contributes ≈ 0.3 eV cm^{-3}, and in the X-ray region, where there is $\approx 10^{-4}$ eV cm^{-3} in hard photons, does the metagalactic contribution greatly exceed the galactic contribution (after Silk, 1970, by permission of D. Reidel Publishing Co.)

4.5.2.4 Electron Energy Loss by Synchrotron (Magneto-Bremsstrahlung) Radiation

When a relativistic electron of energy $E = \gamma mc^2$ moves at a velocity v in a magnetic field of strength H, it loses energy according to the equation (Schott, 1912) (cf. Eq. (1.155)).

$$-\frac{dE}{dx} = \frac{1}{c}\frac{dE}{dt} = \frac{2e^4}{3m^2c^4}\gamma^2 H^2 \approx 6.6. \times 10^{-14} E^2 H^2 \quad \text{erg cm}^{-1} , \tag{4.421}$$

where $\beta = v/c$, E is in ergs, and H is in Gauss. For conversion, 1 eV = 1.6021×10^{-12} erg.

4.5.2.5 Photon Energy Loss by the Photoelectric Effect, Compton Scattering, and Pair Formation

A beam of photons of original intensity, I_0, will have an intensity, I, given by

$$I = I_0 \exp[-\tau] \tag{4.422}$$

after passing through a given distance, L, of a medium. The optical depth, τ, is given by

$$\tau = \int_0^L N\sigma \, dL \approx N\sigma L , \tag{4.423}$$

where N is the number density of scattering or absorbing particles and σ is the cross section for the scattering or absorbing process.

When the quantum energy, $h\nu$, of a proton is large compared with the ionization energy, I, of the K shell electrons of an atom, K shell electrons will be ejected when the photon and atom collide. The cross section, σ_K, for this photoelectric effect, in the nonrelativistic case, $h\nu \ll mc^2$, and the ejection of two K shell electrons, is given by (Stobbe, 1930; Heitler, 1954)

$$\begin{aligned}\sigma_K &= \frac{128\pi\sigma_T}{\alpha^3 Z^2}\left(\frac{I}{h\nu}\right)^4 \frac{\exp(-4\eta\cot^{-1}\eta)}{[1-\exp(-2\pi\eta)]} \quad \text{for } \nu \gtrsim \nu_K \text{ and } h\nu \ll mc^2 \\ &= \frac{64\sigma_T}{\alpha^3 Z^2}\left(\frac{I}{h\nu}\right)^{7/2} = 4\sqrt{2}\sigma_T \alpha^4 Z^5 \left(\frac{mc^2}{h\nu}\right)^{7/2} \quad \text{for } \nu \gg \nu_K \text{ and } h\nu \ll mc^2 ,\end{aligned} \tag{4.424}$$

where the Thomson scattering cross section, σ_T, is given by (Thomson, 1903)

$$\sigma_T = \frac{8\pi}{3}\left(\frac{e^2}{mc^2}\right)^2 = 6.6524616(18) \times 10^{-25} \text{ cm}^2 ,$$

the constant, η, is given by

$$\eta = \left(\frac{I}{h\nu - I}\right)^{1/2} = \frac{Ze^2}{\hbar v} ,$$

where v is the velocity of the ejected electron, Z is the atomic number of the atom, and the ionization energy, I, is related to the frequency, ν_K, of the absorption edge and the frequency, ν, by

$$I = h\nu_K = h\nu - mv^2/2 = mc^2\alpha^2 Z^2/2 \ . \tag{4.425}$$

Cross sections for other shells may be evaluated by changing Z and I in Eq. (4.424). Ionization potentials are given in Table 3.2 of Sect. 3.3. The total cross section, σ, due to the photoelectric effect of all shells is $\sigma \approx 5\sigma_K/4$ for frequencies $\nu > \nu_K$, the frequency of the K shell absorption edge. Empirical values of photoelectric cross sections are given by Davisson and Evans (1952). More detailed cross sections with appropriate Gaunt factors are given by Karzas and Latter (1961). The combination of the dependence $\nu^{-7/2}$ on the frequency, ν, and the Z^5 dependence on the atomic number, Z, means that cosmically rare, heavy elements can make important contributions to the photoelectric absorption cross section at hard ultraviolet and X-ray energies.

For the special case of the hydrogen-like atom, the cross section, σ_n, for photoionization of the nth energy level is given by (Kramers, 1923) (cf. Eq. (1.230))

$$\sigma_n = \frac{32\pi^2 e^6 R_\infty Z^4}{3^{3/2} h^3 \nu^3 n^5} \approx 2.8 \times 10^{29} \frac{Z^4}{\nu^3 n^5} \ \mathrm{cm}^2 \ , \tag{4.426}$$

where the frequency, ν, is larger than $2\pi^2 me^4 Z^2/(h^2 n^2) \approx 3.3 \times 10^{15} Z^2 n^{-2}$ Hz, and the Rydberg constant for infinite mass, $R_\infty = 2\pi^2 me^4/(ch^3) \approx 1.097 \times 10^5 \ \mathrm{cm}^{-1}$.

In the relativistic case, the photoelectric cross section, σ_K, for the two K shell electrons is given by (Sauter, 1931; Heitler, 1954)

$$\begin{aligned}\sigma_K &= \frac{3\sigma_T Z^5 \alpha^4}{2}\left(\frac{mc^2}{h\nu}\right)^5 (\gamma^2 - 1)^{3/2} \\ &\quad \times \left\{\frac{4}{3} + \frac{\gamma(\gamma-2)}{\gamma+1}\left[1 - \frac{1}{2\gamma\sqrt{\gamma^2-1}} \ln\left(\frac{\gamma + \sqrt{\gamma^2-1}}{\gamma - \sqrt{\gamma^2-1}}\right)\right]\right\} \\ &\approx \frac{3\sigma_T Z^5 \alpha^4}{2}\left(\frac{mc^2}{h\nu}\right) \quad \text{for } h\nu \gg mc^2 \ ,\end{aligned} \tag{4.427}$$

where

$$\gamma = \left[1 - \left(\frac{v}{c}\right)^2\right]^{-1/2} = \frac{h\nu + mc^2}{mc^2} \ ,$$

the velocity of the ejected electron is v, the fine structure constant $\alpha = e^2/(\hbar c) \approx 1/(137.036)$, the rest mass energy of the electron is $mc^2 \approx 0.511$ MeV, the atomic number of the atom is Z, and $h\nu$ is the quantum energy of the photon of frequency, ν.

The total Compton scattering cross section, σ_c, for photon scattering from a free electron of energy, $E = \gamma mc^2$, is given by Eq. (4.370) with $q = \gamma h\nu/(mc^2)$. The asymptotic forms of this equation are (Klein and Nishina, 1929)

$$\begin{aligned}\sigma_c &\approx \sigma_T \quad \text{for } \gamma h\nu \ll mc^2 \\ \sigma_c &\approx \frac{3}{8}\sigma_T \frac{mc^2}{\gamma h\nu}\left[\frac{1}{2} + \ln\left(\frac{2\gamma h\nu}{mc^2}\right)\right] \quad \text{for } \gamma h\nu \gg mc^2 \ ,\end{aligned} \tag{4.428}$$

where a photon of frequency, ν, has quantum energy, $h\nu = 4.106 \times 10^{-21} \nu$ MeV, the rest mass energy of the electron is $mc^2 \approx 0.511$MeV, and the Thomson (1906) scattering cross section $\sigma_T = 8\pi e^4/(3m^2c^4) \approx 6.65 \times 10^{-25}\,\mathrm{cm}^2$. The photon is, of course, not completely absorbed but scattered with the new frequency, ν_0, given by (Compton, 1923)

$$\begin{aligned} \nu_0 &= \nu\left[1 + \left(\frac{h\nu}{mc^2}\right)(1 - \cos\varphi)\right]^{-1} \quad \text{for } \gamma \ll 1 \text{ and } \gamma h\nu \ll mc^2 \\ &= 4\gamma^2\nu/3 \quad \text{for } \gamma \gg 1 \text{ and } \gamma h\nu \ll mc^2 \ , \\ &= \gamma mc^2/h \quad \text{for } \gamma \gg 1 \text{ and } \gamma h\nu \gg mc^2 \ , \end{aligned} \tag{4.429}$$

where ν is the frequency of the incident photon and φ is its deflection angle.

A photon may be absorbed in creating electron-positron pairs when it collides with another photon or when it passes through the Coulomb field of a nucleus. The total cross sections for these two processes are given by Eqs. (4.360) and (4.356). Discussions of various photon absorption processes in astrophysical objects have been made by Feenberg and Primakoff (1948), Donahue (1951), Felten and Morrison (1963, 1966), and Goldreich and Morrison (1964). Absorption of X-ray photons in the energy range 0.1 to 8 keV is dominated by photoionization and optical depths as large as unity may arise in the interstellar medium. Calculations of photoelectric absorption of X-rays in the interstellar medium are given by Brown and Gould (1970). For X-ray energies above 8 keV, Compton scattering predominates with a maximum cross section of $5 \times 10^{-25}\,\mathrm{cm}^2$ at 100 keV, but optical depths for galactic and metagalactic space are less than unity. Pair production interactions with starlight photons give sufficient optical depth to absorb gamma ray photons in the energy range 100 GeV to 5000 GeV (Nikishov, 1962), whereas all metagalactic gamma rays with energies larger than 10^5 GeV will be absorbed by pair production interaction with the universal 3 °K black body photon gas (Gould and Schréder, 1966; Jelley, 1966).

In the extreme relativistic limit with velocities close to that of light, c, the cross section for electron-positron annihilation is:

$$\sigma = \frac{\pi r_e^2}{\gamma}[\ln 2\gamma - 1] \ ,$$

and for thermal electrons and positrons it becomes:

$$\sigma \approx \frac{\pi r_e^2}{(v/c)} \ .$$

4.5.3 The Origin of High Energy Particles

4.5.3.1 Energy Spectrum of Cosmic Ray Electrons, Protons, and Positrons

Cosmic rays are energetic charged particles that rain down on the Earth's atmosphere, coming from all directions in outer space and from beyond the solar system. They were discovered by Victor Hess, an ardent amateur balloonist, who recorded an unexpected increase in ionizing radiation at high altitudes (Hess, 1911, 1912). Further balloon flights by Werner Kolhörster showed that the ionization continued to increase with height to values as large as fifty times that at sea level (Kolhörster, 1913). High-altitude measurements by Robert Millikan confirmed that the radiation did come from beyond the terrestrial atmosphere, and he incidentally gave the radiation its present name cosmic rays (Millikan, 1926; Millikan and Cameron, 1926).

Cosmic rays were believed to be high-energy radiation, in the form of gamma-ray photons, until the world-wide measurements of the systematic increase of cosmic ray intensities at increasing latitude, with a greater latitude increase at high altitudes than at sea level, showed that cosmic rays are charged particles deflected by the Earth's magnetic field toward the magnetic poles (Clay, 1927; Compton, 1933). Geiger counters were also used in coincidence to discriminate charged particles from photons, showing that cosmic rays near the ground consist of high-speed charged particles (Bothe and Kolhörster, 1929).

The Wilson cloud chamber was next used to detect primary cosmic ray particles (Skobeltzn, 1929) and air showers of secondary particles produced when the primaries interact with the Earth's atmosphere (Blackett and Occhialini, 1933). Such cloud chamber observations led to the discovery of several fundamental particles, including the positron (Anderson, 1932, 1933) and a meson with a mass intermediate between that of the electron and proton (Street and Stevenson, 1937; Anderson and Neddermeyer, 1936, 1937; Neddermeyer and Anderson, 1937). A second meson was discovered from high-altitude observations of cosmic rays using photographic emulsions (Lattes, Occhialini and Powell, 1947).

These early investigations of cosmic rays led to several Nobel Prizes in Physics – Charles T. R. Wilson in 1927 for his cloud chamber, Victor F. Hess in 1936 for the discovery of cosmic radiation, Carl D. Anderson in 1936 for the discovery of the positron, Patrick M. S. Blackett in 1948 for his development of the Wilson cloud chamber and discoveries made therewith, and Cecil F. Powell in 1950 for the discovery regarding mesons.

Subsequent measurements with instruments carried by high-altitude balloons established that cosmic rays consist of protons, alpha-particles (helium nuclei) and heavier nuclei (Schein, Jesse and Wollan, 1941; Bradt and Peters, 1948; Freier et al., 1948). Electrons were not discovered as a primary component of cosmic rays until 1961 (Earl, 1961; Meyer and Vogt, 1961), although their presence had been inferred earlier from the synchrotron radio emissions of our Galaxy (Kiepenheuer, 1950; Ginzburg, 1956). In the same year, cosmic rays with a very high energy of $E \approx 10^{19}$ eV were inferred from the detection of a huge (1.8 km diameter) air shower (Linsley, Scarsi and Rossi, 1961).

Reviews of cosmic rays are given by Rossi (1948) and Simpson (1983). Details on the history can be found in the books by Rossi (1952), Hayakawa (1969), Hillas (1972) and Friedlander (1989), and in the collection of early papers edited by Rosen (1969). The book by Pais (1986) places cosmic ray discoveries in the context of the fundamental particles of matter. Books that emphasize the physics of cosmic rays are those by Berenzinskii et al. (1990), Gaisser (1990) and Longair (1992).

The high-energy cosmic ray particles have a smooth, continuous energy spectrum, with fewer particles at higher energies, from about 1 MeV $= 10^6$ eV to more than 10^{11} GeV $= 10^{14}$ MeV $= 10^{20}$ eV. (High-energy physicists use the electron volt, of eV, unit of energy where 1 eV $= 1.6021773 \times 10^{-12}$ erg.) Protons, nuclei and electrons in the energy range 10^9 eV to 10^{12} eV account for most of the cosmic ray energy flux, and the local cosmic ray energy density, ρ_{E}, is $\rho_{\mathrm{E}} \approx 1$ eV cm^{-3}. In the peak energy range at about 10^9 eV, cosmic rays consist of approximately 87% hydrogen (protons), 12% helium nuclei (alpha particles) and 1% heavier nuclei. At a given energy, the cosmic ray electrons are of much lower flux then the cosmic ray protons. Data from meteorites indicate that the flux of cosmic rays entering the solar system has been nearly constant over the past 4 billion years (Forman and Schaeffer, 1979).

When solar activity is at the peak of its 11-year cycle, the flux of low-energy cosmic rays detected at the Earth is least, and vice versa. This anticorrelation between the cosmic ray flux at the top of the Earth's atmosphere and the Sun's magnetic activity is often called the Forbush effect after its discoverer (Forbush, 1950, 1962, 1966); enhanced magnetism in the solar wind at the peak of the solar activity cycle deflects cosmic rays from their Earth-bound paths (Meyer, Parker and Simpson, 1956). Energetic protons are also emitted from the Sun during solar flares and coronal mass ejections (see Sect. 4.5.1.12). They were discovered and described in the late 1950s (Freier, Ney and Winckler, 1959). Such particles have been called solar cosmic rays, a somewhat confusing nomenclature.

When observing high energy particles, the differential energy flux, $J(E)$, is often measured. If $dJ(E)$ denotes the number of particles in the energy range, dE, which pass a unit area per unit time per unit solid angle, then

$$J(E) = \frac{dJ(E)}{dE} \quad \text{particles (cm}^2 \text{ sec ster MeV or GeV)}^{-1} \ . \tag{4.430}$$

Detectors are often sensitive to particles whose energies lie above a certain threshold energy, and therefore the integral energy spectrum

$$J(>E) = \int_E^\infty J(E)dE \quad \text{particles (cm}^2 \text{ sec ster)}^{-1} \ , \tag{4.431}$$

is also often measured. The differential energy spectrum of the cosmic ray electrons (first observed by Earl, 1961) is shown in Fig. 4.10). This spectrum is well represented by the following approximations:

$$\begin{aligned} J(E) &= 1.32 \times 10^{-2} E^{-1.75} \, (\text{cm}^2 \text{ sec ster MeV})^{-1} \text{ for } 3 \text{ MeV} \le E \le 20 \text{ MeV} \\ J(E) &= 1.00 \times 10^{-5} \, (\text{cm}^2 \text{ sec ster MeV})^{-1} \text{ for } 30 \text{ MeV} \le E \le 300 \text{ MeV} \\ J(E) &= 3.5 \times 10^{-3} E^{-1.6} \, (\text{cm}^2 \text{ sec ster GeV})^{-1} \text{ for } 0.5 \text{ GeV} \le E \le 3 \text{ GeV} \\ J(E) &= 1.16 \times 10^{-2} E^{-2.6} \, (\text{cm}^2 \text{ sec ster GeV})^{-1} \text{ for } 3 \text{ GeV} \le E \le 300 \text{ GeV} \ . \end{aligned} \tag{4.432}$$

The spectrum of cosmic ray electrons with energies between 6 and 100 GeV is given by Meegan and Earl (1975). Their energy spectra has also been

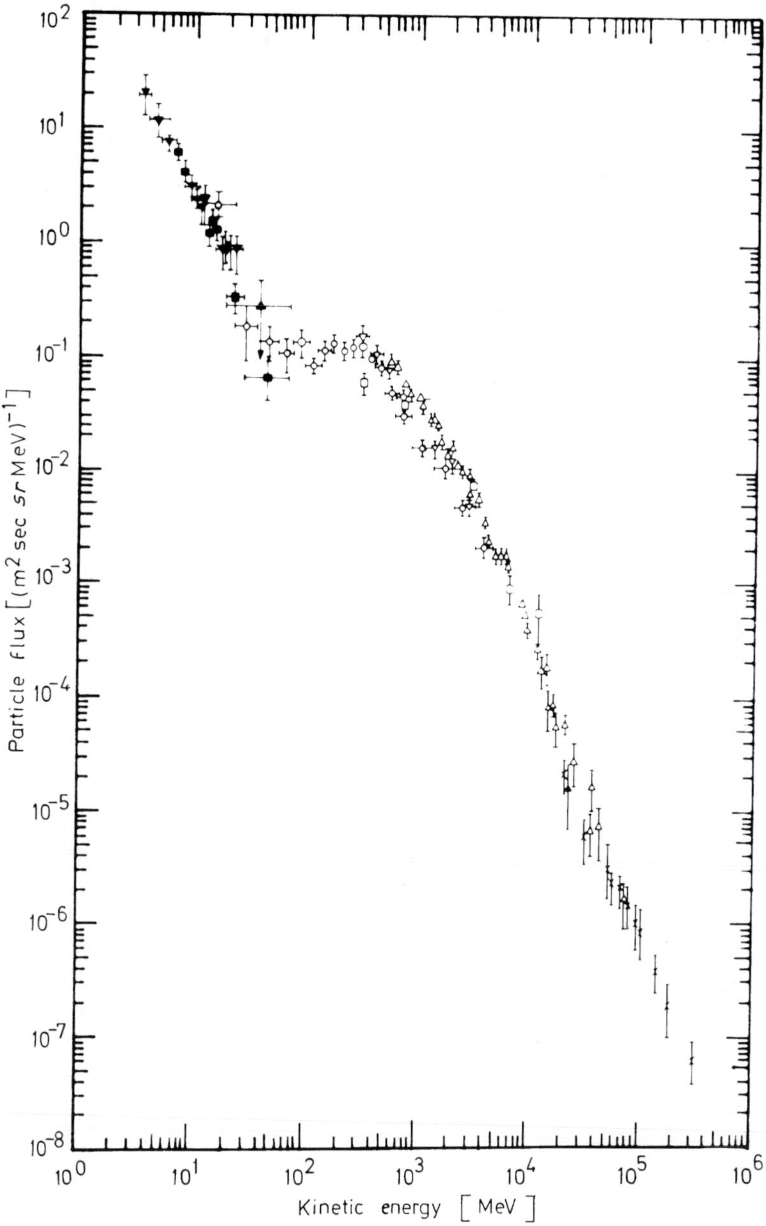

Fig. 4.10. Energy spectrum of the primary cosmic ray electrons (after Fanselow et al., 1969, by permission of the American Astronomical Society and the University of Chicago Press). Each symbol denotes a different set of observations, and these are referenced by Fanselow et al.

summarized by Webber (1983), obtaining a power-law differential energy spectrum proportional to $E^{-3.3}$ for energies $E \geq 10$ GeV, which is somewhat steeper than that obtained for cosmic ray protons and nuclei. At high energies the spectrum has probably been significantly influenced by the effects of synchrotron radiation losses (Longair, 1992). The observed electron energy spectrum is strongly influenced by the solar wind at energies $E \leq 1$ GeV.

The primary proton spectrum is parallel to the high energy ($3 \leq E \leq 300$ GeV) electron spectrum with a displacement to the right by a factor of about twenty in energy. The differential energy spectrum of protons and nuclei (first observed by Freier et al., 1948 and Bradt and Peters, 1948) is given by (Ryan, Ormes, and Balasubrahmanyan, 1972)

$$J(E) = 2.0 \pm 0.2E^{-2.75\pm0.03} \quad \text{protons (cm}^2\text{ sec ster GeV)}^{-1}$$
$$J(E) = 8.6 \pm 1.4 \times 10^{-2}E^{-2.77\pm0.05} \quad \text{He (cm}^2\text{ sec ster GeV/nucleon)}^{-1}$$
$$\text{for } 10\,\text{GeV} \leq E \leq 10^3\,\text{GeV} \qquad (4.433)$$

for protons and helium (He) nuclei. Here the energy, E, is in GeV. The integral energy spectrum is given by (Hayakawa, 1969)

$$J(>E) = 1.6 \times 10^{-5}\left(\frac{E}{10^3}\right)^{-1.6} \quad (\text{cm}^2\text{ sec ster})^{-1}$$
$$\text{for } 10\text{ GeV} \leq E \leq 10^4\text{ GeV}$$
$$J(>E) = 2.0 \times 10^{-14}\left(\frac{E}{10^8}\right)^{-2.2} \quad (\text{cm}^2\text{ sec ster})^{-1} \qquad (4.434)$$
$$\text{for } 10^6\text{ GeV} \leq E \leq 10^9\text{ GeV}$$

where E is in GeV. The spectrum and composition of the primary cosmic radiation and the heavy cosmic ray nuclei are reviewed by Webber (1967) and Shapiro and Silberberg (1970), respectively.

The differential energy spectra of cosmic ray protons, helium nuclei, carbon nuclei and iron nuclei are provided in the review given by Simpson (1983), and reproduced by Longair (1992). They can be represented by a power law distribution in energy E, proportional to E^{-x} between 1 and 10^3 GeV, where E is the kinetic energy per nucleon and the exponent $x = 2.5$ to 2.7. At higher energies it is only possible to measure the total energy of the particles. At lower energies, below about 1 GeV per nucleon, the observed flux of these cosmic rays is also modulated by the 11-year solar activity cycle (Webber and Lezniak, 1974), with a greater cosmic ray intensity reaching the Earth at a minimum in the solar activity cycle when the modulation by the solar wind is less.

Cosmic rays with energies above 10^{14} eV $= 10^5$ GeV are too rare to be detected with instruments lofted above the atmosphere, and are instead inferred from extensive air showers they produce when interacting with the atmosphere (Watson, 1991; Cronin, 1997). The spectrum of energy, E, above $E = 10^{14}$ eV is a power law, falling off as $E^{-2.7}$ up to an energy of $E = 5 \times 10^{15}$ eV and then steepening to a power law $E^{-3.07}$. The slight kink or knee in the spectrum coincides with the limit in energy predicted for supernova shock acceleration in our Galaxy. The differential energy spectrum of the cosmic-ray primary energy, inferred from extensive air showers produced by the primary particles, above 10^{17} eV is (Bird et al., 1994):

$$J(E) = 10^{-33.55}\left(\frac{E}{10^{18}\,\mathrm{eV}}\right)^{-3.07\pm0.01} \mathrm{cm}^{-2}\,\mathrm{s}^{-1}\,\mathrm{sr}^{-1}\,\mathrm{eV}^{-1}$$

$$\text{for } 10^{17}\,\mathrm{eV} \leq E \leq 10^{20}\,\mathrm{eV}\ . \qquad (4.435)$$

The measured cosmic ray flux at 10^{19} eV is 2 to 3.4×10^{-37} cm^{-2} s^{-1} sr^{-1} eV^{-1}.

The origin of ultra-high-energy cosmic rays is reviewed by Hillas (1984). Those of energy $E \approx 10^{19}$ eV will interact strongly with the 2.7 °K microwave background (Greisen, 1966; Zatsepin and Kuzmin, 1966). However, observations of cosmic rays with energies beyond 10^{20} eV suggest the absence of a 2.7 °K cutoff in the energy spectrum, or a possible new component at very high energies (Takeda et al. 1998). Cosmic rays with energies in excess of 10^{20} eV are exceedingly rare, with only a few being detected (Bird et al., 1994; Bird, 1995; Hayashida et al., 1994). The possible origin of these highest energy cosmic rays is discussed by Elbert and Sommers (1995) and Waxman (1995).

In the low energy range (3 to 20 MeV), the cosmic ray electrons lose most of their energy by ionization of interstellar hydrogen, and are thought to originate from the knock-on collisions of higher energy protons (Abraham, Brunstein, and Cline, 1966; Simnett and McDonald, 1969). Observations of the electron-positron positive fraction, R, indicate that higher energy electrons are directly accelerated in sources and do not originate as secondaries of the protons. The positive fraction, R, is given by

$$R = \frac{J^{+}(E)}{J^{+}(E) + J^{-}(E)}\ , \qquad (4.436)$$

where the superscripts + and – denote, respectively, positrons and electrons. Observed values of R are given in Table 4.17. Measurements of the positron-electron ratio in the primary cosmic rays for energies of 5 to 50 GeV are given by Buffington, Orth and Smoot (1974). The high energy electrons must be produced within our Galaxy because otherwise the X-ray background would

Table 4.17. Fraction of positrons in the primary cosmic ray electron component[1]

Energy interval (BeV)	$R = \frac{J^{+}(E)}{J^{+}(E)+J^{-}(E)}$
0.053–0.088	0.31 ± 0.19
0.088–0.173	0.45 ± 0.08
0.173–0.44	0.29 ± 0.09
0.44–0.86	0.10 ± 0.07
0.86–1.70	0.083 ± 0.024
1.70–4.2	0.046 ± 0.018
4.2–8.4	0.013 ± 0.05
8.4–14.3	0.15 ± 0.18

[1]After Fanselow et al. (1969).

be much larger than that observed (Felten and Morrison, 1966). The question of the origin of most of the high energy particles reduces, then, to a suitable mechanism for accelerating particles within our Galaxy.

4.5.3.2 Acceleration Mechanisms for High Energy Particles

For the development of theories of the origin and acceleration of cosmic rays there are the collection of original papers by Rosen (1969), the reviews by Morrison (1961) and Ginzburg and Ptuskin (1976), and the monographs by Ginzburg and Syrovatskii (1964) and Hayakawa (1969).

As an example, Baade and Zwicky (1934) demonstrated that the enormous energy released during the supernova explosions of massive stars can account for the observed flux of cosmic rays, even if they occur only once a millennium in our Galaxy. The idea that cosmic rays are energized by supernovae was developed by Ginzburg (1953, 1956) and Ginzburg and Syrovatskii (1964), showing that they can account for the observed cosmic ray energy density of $\rho_{\mathrm{E}} \approx 1\mathrm{eV\ cm}^{-3}$ spread uniformly throughout the Galaxy (also see Gaisser, 1990). This idea received additional support from the observation of radio emission from supernova remnants, attributed to the synchrotron radiation of electrons accelerated to speeds approaching that of light (Shklovskii, 1953). Recent arguments for the origin and acceleration of cosmic rays in supernova explosions have been given by Ellison, Drury and Meyer (1997), Meyer, Drury and Ellison (1997), Ramaty, Kozlovsky and Lingenfelter (1998) and Tanimori et al. (1998). As discussed later in this section, strong shock waves generated by supernovae provide a plausible mechanism for accelerating cosmic rays for energies, E, of $E \leq 10^{15}$ eV. Shock waves have also been discussed in Section 3.5.9; the physics of supernova explosions has been reviewed by Woosley and Weaver (1986), Bethe (1990) and Wheeler and Harkness (1990); supernovae and supernova remnants are reviewed by Trimble (1982, 1983), Reynolds (1988), and Weiler and Sramek (1988); detailed physical information about the radio and X-ray radiation from supernova remnants is given by Lang (1992); and SN 1987A, the first supernova visible to the naked eye in 400 years, is reviewed by Trimble (1988), Woosley and Phillips (1988), Arnett, Bahcall, Kirshner and Woosley (1989), Murdin (1989), and Mc Cray (1993) with physical data summarized by Lang (1992) and in Table 4.10.

The acceleration of cosmic rays has been reviewed by Drury (1983) and Blandford and Eichler (1987), and the origin of ultra-high-energy cosmic rays has been discussed by Hillas (1984). The origin and acceleration of cosmic rays are also described in the monographs by Berenzinskii et al. (1990), Gaisser (1990) and Longair (1994).

The acceleration mechanism for high-energy cosmic ray particles must account for their power-law differential spectrum which varies roughly with energy, E, as $E^{-2.7}$. There are two well-known processes that can produce a power-law spectrum of particle energies, the classical Fermi acceleration mechanism, introduced by Fermi (1949, 1954) and acceleration by strong shocks, described in several independent papers in the late 1970s by Axford, Leer and Skadron (1977), Krymskii (1977), Bell (1978), and Blandford and

Ostriker (1978, 1980). Shock wave acceleration of cosmic rays has also been described in the books by Gaisser (1990) and Longair (1994), and extensively reviewed by Axford (1981), Drury (1983), and Blandford and Eichler (1987). A third acceleration mechanism involves pulsars with their strong magnetic fields and rapid rotation (also see Volume II). Cosmic ray particles might be accelerated by the low frequency waves of the rotating magnetic neutron star (Gunn and Ostriker, 1969; Ostriker and Gunn, 1969), with the particles originating from the surface of the neutron star (Goldreich and Julian, 1969).

Fermi imagined that cosmic rays are scattered by interstellar clouds containing turbulent magnetic fields and moving at velocity, V. For charged particles with velocities, v, approaching that of light, c, the average energy gain, ΔE, per collision is:

$$\left\langle \frac{\Delta E}{E} \right\rangle = \frac{8}{3}\left(\frac{V}{c}\right)^2 . \tag{4.437}$$

The number of particles, $N(E)dE$, with an energy between E and $E + dE$ is given by:

$$N(E)dE \propto E^{-x}\, dE \tag{4.438}$$

where the spectral index, x, is given by:

$$x = \left[1 + \frac{1}{\alpha \tau_{esc}}\right] , \tag{4.439}$$

when the particle remains in the acceleration region for an average time τ_{esc}, and the typical rate of energy increase, dE/dt, is

$$\frac{dE}{dt} = \frac{4}{3}\left(\frac{V^2}{cL}\right)E = \alpha E ,$$

for a mean free path between clouds of L, and an average time between collisions with clouds of $2L/c$. Early variants of the Fermi acceleration mechanism employing turbulence, hydrodynamic waves, and plasma waves are discussed by Morrison, Olbert and Rossi (1954), Parker (1955, 1958), Davis (1956), Burbidge (1957), Wentzel (1963, 1964) and Tsytovich (1964, 1966). A full treatment taking into account the stochastic nature of the acceleration process and the spreading of the energy spectrum by scattering is described in the review by Blandford and Eichler (1987). They obtain a spectral index of:

$$x = \frac{3}{2}\left(1 + \frac{4cL}{3\langle V\rangle^2 \tau_{esc}}\right) - \frac{1}{2} \tag{4.440}$$

for an isotropic distribution of scatterers with mean square velocity $\langle V\rangle^2$ and a mean free path between clouds of L.

A more efficient acceleration mechanism is provided by strong shock waves. Due to the turbulent magnetic fields on either side of the shock front, a cosmic ray particle bounces back and forth across it. The average fractional energy gain in crossing a shock wave moving at a non-relativistic velocity, V, is:

$$\left\langle \frac{\Delta E}{E} \right\rangle = \frac{4}{3}\frac{V}{c} \ . \tag{4.441}$$

Since this process varies as the first order in V/c, it has been called the first order Fermi acceleration process, in contrast with the second order, $(V/c)^2$, Fermi acceleration mechanism originally proposed by Fermi. The strong shock acceleration mechanism also leads to a differential energy spectrum with (Bell, 1978; Longair, 1994)

$$N(E)dE \propto E^{-x}\, dE$$

where the spectral index, x, is

$$x = 1 - \ln P / \ln \beta = 2.0 \ , \tag{4.442}$$

where $P = 1 + 4V/(3c)$ is the probability that the particle remains in the shock accelerating region after one collision and $\beta = 1 + \Delta E/E = 1 + 4V/(3c)$. A spectral index of $x = 2.0$ is somewhat flatter than the typical spectra of cosmic rays or extragalactic radio sources.

The steepening of the energy spectra, beyond that attributed to the standard strong shock model, is described in Longair (1994), with references that include Baring and Kirk (1991), Begelman and Kirk (1990), Biermann and Strittmatter (1987), Bregman (1985), Heavens and Meisenheimer (1987), and Kirk and Schneider (1987).

It appears that acceleration by strong shocks, generated during supernova explosions in our Galaxy, might account for most cosmic rays. The maximum energy, E_{max}, per particle that can be achieved by this mechanism is limited by the eventual disintegration of the expanding supernova remnant. Lagange and Cesarsky (1983) showed that most of the acceleration occurs by the time the supernova has swept up its own mass, limiting the energy to

$$E_{\mathrm{max}} \approx 10^{15} \text{ eV} \quad \text{for supernova acceleration.} \tag{4.443}$$

At this energy, the gyroradii of the cosmic-ray particles are of the same size as the supernova remnant. This threshold energy also corresponds to the knee in the cosmic-ray energy spectrum where the spectrum steepens in power law dependence from $E^{-2.7}$ to $E^{-3.0}$.

Cosmic rays have been observed with energies up to 10^{20} eV, and those with energies below about 10^{19} eV are thought to originate in our Galaxy. Between 10^{15} and 10^{19} eV the origin is uncertain, but one possibility is pulsars. Due to the conservation of magnetic flux, HR^2, the magnetic fields are amplified to $H_{\mathrm{N}} \approx 10^{12}$ Gauss when a normal star of radius $R = 10^{11}$ cm collapses to a neutron star of radius $R_{\mathrm{N}} = 10^6$ cm. The pulsars are also spinning rapidly, with periods of one minute or less, and the low frequency waves generated by the rotating magnetic dipole could accelerate cosmic ray particles (Gunn and Ostriker, 1969; Ostriker and Gunn, 1969).

This mechanism makes use of the fact that, in sufficiently strong wave fields, a charged particle is accelerated in the propagation direction to nearly the velocity of light in a small fraction of a wavelength, and thereafter rides the wave at essentially constant phase, slowly taking energy from the wave. The maximum energy, $E_{\max}$, available for an electron is given by

$$E_{\max} \approx mc^2 \left[2\omega_B r_0 c^{-1} \ln \frac{r_c}{r_0} \right]^{2/3} , \tag{4.444}$$

where r_c is the radius at which the wave energy is exhausted, r_0 is the inner wave radius, $r_0 \approx c/\omega$, and ω_B is the cyclotron frequency, $eB/(mc)$, in the magnetic field at maximum amplitude at r_0. For example, if $B = 10^6$ Gauss at $r_0 = 10^8$ cm, then $E_{\max} = 10^{13}$ eV if $r_c = 10^{12}$ cm. For a particle of atomic mass, A, and atomic number, Z,

$$E_{\max} \approx 3.6 \times 10^{13} A \left(\frac{Z}{A}\right)^{2/3} \left(\frac{\omega}{10^2}\right)^{4/3} \left(\frac{B}{10^{12}}\right)^{2/3} \left(\frac{R}{10^6}\right)^2 \text{ eV} , \tag{4.445}$$

where R is the radius of the star. Goldreich and Julian (1969) first showed that cosmic ray particles could initially come from the surface of a neutron star, and that substantial acceleration might take place on the surface as well. They showed that the potential difference, V, between the pole and equator of the neutron star is given by (cf. Eq. (1.98))

$$V \approx \frac{\omega R^2 B}{c} \approx 10^{16} \omega \left(\frac{B}{10^{12}}\right) \left(\frac{R}{10^6}\right)^2 \text{ volts} , \tag{4.446}$$

an equation first applied to a rotating star by Alfvén (1938). Thus a potential difference as large as 10^{14} volts is possible. This acceleration is more than sufficient to allow the particles to escape the surface gravity of the star and to flow out into space along the magnetic field lines, along which the particle is accelerated as well.

The rate of injection of particles, r, is given by

$$r = \frac{B\omega^2 R^3}{Zec} \approx 7 \times 10^{32} \left(\frac{\omega}{10^2}\right)^2 \left(\frac{B}{10^{12}}\right) \left(\frac{R}{10^6}\right)^3 Z^{-1} \text{ particles sec}^{-1} . \tag{4.447}$$

It follows that the number of particles, $N(E)dE$, with an energy between E and $E + dE$, is given by

$$N(E)dE = r \left[\left(\frac{dE}{d\omega}\right) \left(\frac{d\omega}{dt}\right) \right]^{-1} dE \propto E^{-2.5} \, dE , \tag{4.448}$$

a spectrum remarkably close to the observed spectral index of -2.6 for the high energy cosmic rays. The number density of cosmic rays depends upon the birthrate of pulsars, and therefore the birthrate of supernovae which is thought to be about one every thirty years. Within the uncertainty of this number, the observed number of high energy cosmic ray particles may be accounted for.

The origin and acceleration of the highest energy cosmic rays has been reviewed by Hillas (1984). For any gradual mode of acceleration, where the particle traverses the site many times while gaining energy, the size, R, of the accelerating region must be greater than the gyroradius of the particle in the magnetic field. This results in a limit on the product of the size, R, and the magnetic field strength, H, given by:

$$R(\text{kiloparsec}) \times H(10^{-6}\text{Gauss}) \geq \frac{2E(10^{18}\text{eV})}{Z\beta} , \tag{4.449}$$

where the units of R, H, and the cosmic-ray particle energy, E, are given in parenthesis, the particle charge is Ze, and the factor $\beta \leq 1$ is the fraction of the maximum energy that can be obtained and is related to the acceleration mechanism. Only highly condensed objects with very intense magnetic fields or enormously extended objects with weaker magnetism have a product $R \times H$ sufficient to produce cosmic rays of the highest energies. Pulsars remain a possibility, as do extragalactic objects such as active galactic nuclei (Protheroe and Szabo, 1992).

4.5.3.3 The Origin of High Energy Photons

Following the prediction that cosmic X-rays and γ-rays might be observed (Morrison, 1958), several groups have observed a diffuse, isotropic, extragalactic X and γ ray background together with discrete, galactic X-ray sources. The differential energy spectrum of the diffuse background, shown in Fig. 4.11, is given by (Friedman, 1973)

$$\begin{aligned}
J(E) &\propto E^{-1} \text{ to } E^{-2}\,(\text{cm}^2 \text{ sec ster keV})^{-1} \text{ for } 100 \text{ eV} \leq E \leq 1 \text{ keV} \\
J(E) &\propto E^{-0.4}\,(\text{cm}^2 \text{ sec ster keV})^{-1} \text{ for } 1 \text{ keV} \leq E \leq 10 \text{ keV} \\
J(E) &\propto E^{-0.75}\,(\text{cm}^2 \text{ sec ster keV})^{-1} \text{ for } 20 \text{ keV} \leq E \leq 40 \text{ keV} \\
J(E) &\propto E^{-1.3}\,(\text{cm}^2 \text{ sec ster keV})^{-1} \text{ for } 40 \text{ keV} \leq E \leq 1 \text{ MeV} \\
J(E) &\propto E^{-2}\,(\text{cm}^2 \text{ sec ster keV})^{-1} \text{ for } 1 \text{ MeV} \leq E \leq 10 \text{ MeV} ,
\end{aligned} \tag{4.450}$$

or in integral form (Silk, 1973)

$$\begin{aligned}
dJ(E) &= 90 \quad \text{keV}(\text{cm}^2 \text{ sec ster keV})^{-1} \quad \text{for } E = 0.25 \text{ keV} \\
dJ(E) &= 12E^{-0.5} \quad \text{keV}(\text{cm}^2 \text{ sec ster keV})^{-1} \quad \text{for } 1 \text{ keV} \leq E \leq 20 \text{ keV} \\
dJ(E) &= 40E^{-1.2} \quad \text{keV}(\text{cm}^2 \text{ sec ster keV})^{-1} \quad \text{for } 20 \text{ keV} \leq E \leq 1000 \text{ keV} .
\end{aligned}$$

The diffuse X-rays between 10 and 100 keV are isotropic to better than five percent over angular sizes of ten degrees (Schwartz, 1970), and are probably extragalactic. The soft X-ray flux below 1 keV rises in intensity towards the galactic pole, which may be due to absorption effects.

The most intense X-ray sources include X-ray binaries within our Galaxy and active galactic nuclei outside it; these objects are reviewed in Volume II. Normal galaxies and clusters of galaxies also emit X-ray radiation (Forman and Jones, 1982; Fabbiano, 1989, see Volume II). There is no definite explanation for the origin of the diffuse X-ray background (Mc Cammon and Sanders, 1990; Fabian and Barcons, 1992), although active galactic nuclei and quasars probably dominate the X-ray radiation near 1 keV and may produce

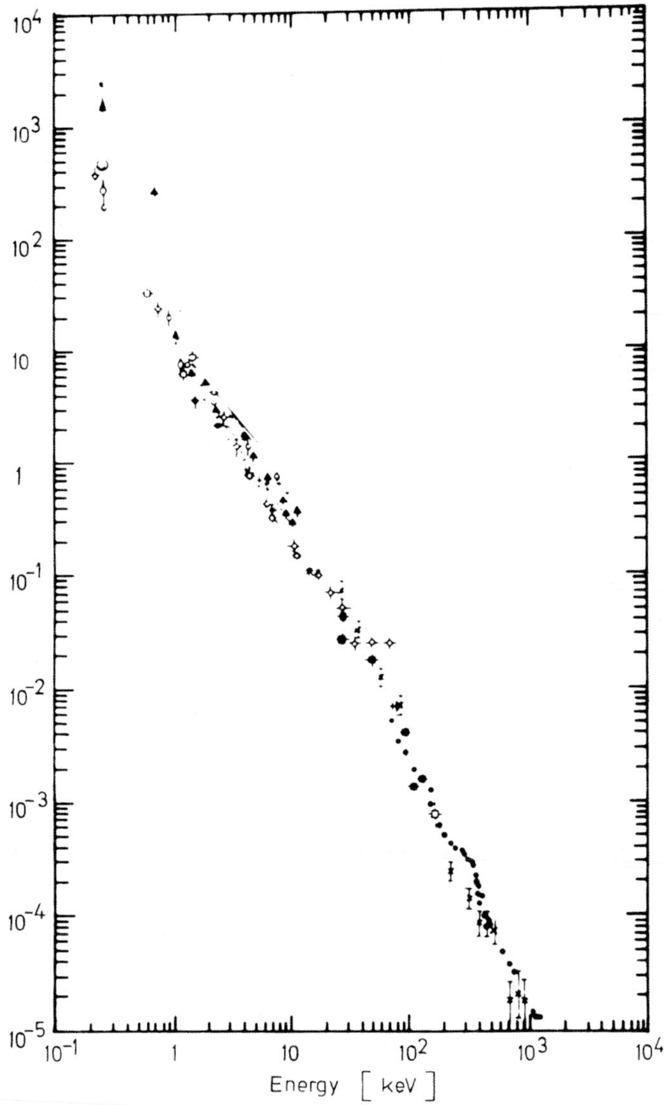

Fig. 4.11. The differential energy spectrum of the diffuse X- and γ-ray background from 0.27 keV to 1000 keV (after Oda, 1970, by permission of the International Astronomical Union). The data in the energy range 1–100 keV are represented by a power law spectrum $E^{-1.7 \pm 0.2}$, where E is the energy. Each symbol denotes a different set of observations, and these are referenced by Oda

the entire X-ray background from 2 to 100 keV (Zdziarski, Zycki and Krolik, 1993).

The differential spectrum for gamma rays has the form (Fichtel et al., 1975):

$$\frac{dJ}{dE} = (2.7 \pm 0.5) \times 10^{-7} \left(\frac{E}{100}\right)^{-2.4\pm0.2} \gamma - \text{rays cm}^{-2}\ \text{sr}^{-1}\ \text{s}^{-1}\ \text{MeV}^{-1}$$

$$\text{for } 35\ \text{MeV} \leq E \leq 170\ \text{MeV} \ , \qquad (4.451)$$

where E is in MeV.

Gamma rays in the high energy range around 100 MeV are produced during the interaction of cosmic ray electrons and protons with the highly-structured interstellar medium. This diffuse band of gamma rays, concentrated in the plane of our Galaxy, has been reviewed by Bloemen (1989) and modeled by Bertsch et al. (1993) in terms of cosmic rays, matter and photon interactions. Gamma-ray astronomy is reviewed in the monograph by Ramana Murthy and Wolfendale (1993); very high energy gamma-ray astronomy in the range 10 GeV to 100 GeV was reviewed by Weekes (1988).

Gamma ray bursts are amongst the most intriguing discoveries in the high energy part of the electromagnetic spectrum. They were accidentally discovered at energies of $E \approx 1$ MeV with the Vela satellites, intended for monitoring nuclear explosions (Klebesadel, Strong and Olson, 1973; Cline, Desai, Klebesadel and Strong, 1973; and Wheaton et al., 1973).

Reviews of gamma ray bursts are provided by Higdon and Lingenfelter (1990), Fishman and Meegan (1995) and Piran (1997). They are intense gamma-ray transients with fluxes at energies $E > 30$ keV as large as 10^{-2} erg cm^{-2} s^{-1}, orders of magnitude greater than that of the most intense steady-state X-ray or gamma-ray sources. The observed durations of gamma-ray bursts range from several microseconds to several hundred seconds, and their rise times can be as short as 10^{-4} seconds. The observed photon energy generally peaks in the several hundred keV range. The spectrum is power law,

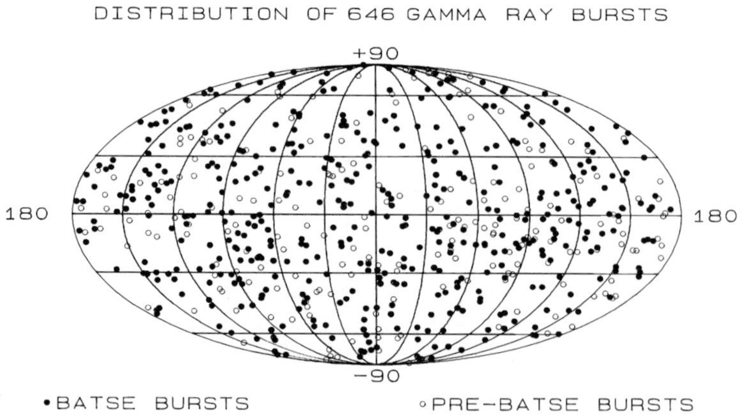

Fig. 4.12. The isotropic distribution of cosmic gamma-ray bursts, plotted in galactic coordinates, from the BATSE instrument aboard the Compton Gamma Ray Observatory (solid circles, from Fishman and Meegan, 1995) and from pre-Batse experiments (open circles) mainly from the Konus experiment (Mazets et al., 1981) and the interplanetary network (Atteia et al., 1987). The plane of our Galaxy extends horizontally across the center of this diagram and the galactic poles are at the central top and bottom. (Courtesy of Kevin Hurley.)

varying with energy, E, as $E^{-\alpha}$ where the spectral index $\alpha \approx 2$. The observed fluence is $F \approx 10^{-7}$ erg cm^{-2}.

Gamma ray bursts are distributed isotropically across the sky (Meegan et al., 1992; Fishman and Meegan 1995, see Fig. 4.12). The dimmest bursts are longer by a factor of about 2.3 compared to bright ones (Nemiroff et al., 1994; Norris et al., 1995).

Gamma ray bursts involve the release of about 10^{50} erg of radiation into a small volume with a radius of roughly 10^8 cm, generating photons with an energy of about 1 MeV. They may originate in a fireball of dense radiation and electron-positron pairs (Goodman, 1986; Paczynski, 1986, 1990; Shemi and Piran, 1990; Rees and Mészáros, 1992; Mészáros and Rees, 1997; Piran, 1997). The violent fireball bursts are followed by faint afterglows in progressively longer, less energetic wavelengths: X-rays, visible light, and radio waves. Studies of these afterglows, lasting for weeks or months after the gamma-ray burst, suggest that the gamma-ray bursts emerge from gigantic, distant energetic fireballs that are at cosmological distances, billions of light years away (Costa et al., 1997; Frail et al., 1997; Groot et al., 1997; Kulkarni et al., 1998; Metzger et al., 1997; Sari, Piran and Narayan, 1998; Van Paradijs et al., 1997; Vietri, 1997; Waxman, 1997). A summary of the arguments in favor of a cosmological origin for the gamma ray bursts was given by Paczynski (1995); an overview of arguments for a galatic origin was provided by Lamb (1995). Investigations of the radio afterglow of one gamma ray burst implied that this fireball was about a tenth of a light year across, moving outward at close to the speed of light (Frail et al., 1997).

Merging neutron stars may provide the cataclysm needed to create such a brief, powerful fireball with its gamma ray bursts (Eichler et al., 1989; Narayan, Piran and Shemi, 1991; Piran, 1995). However, one recent gamma ray burst was so powerful that even this mechanism might not be able to supply enough energy. When combined with the optically-determined redshift of $z = 3.42$, the observed gamma ray flux implies an energy of 3×10^{53} erg in gamma rays alone if the emission is isotropic (Kulkarni et al., 1998).

Soft gamma ray repeaters are a different class of high-energy transient sources, that originate within our Galaxy. Unlike gamma ray bursts, these objects repeat their outburts. They have been associated with very strongly magnetized neutron stars, dubbed magnetars, with magnetic fields of strength $H = 10^{14}$ to 10^{15} Gauss. Theoretical speculations about the existence of magnetars (Usov, 1984; Duncan and Thomson, 1992), were confirmed when the soft gamma ray repeaters were identified with young supernova remnants (Kulkarni and Frail, 1993; Vasisht et al., 1994), and with intense, periodic X-ray flashes (Kouveliotou et al., 1993, 1998).

References

Aannestad, P. A. (1973): Molecule formation: I. In normal H I clouds; II. In interstellar shock waves. Ap. J. Suppl. No. 217, **25**, 205.

Aannestad, P. A. (1975): Absorptive properties of silicate core-mantle grains. Ap. J. **200**, 30.

Aannestad, P. A., Purcell, E. M. (1973): Interstellar grains. Ann. Rev. Astron. Astrophys. **11**, 309.

Aarons, J., Whitney, H. E., Allen, R. S. (1971): Global morphology of ionospheric scintillations. Proc. I.E.E.E. **59** (2),159.

Abazov, A. I., et al. (1991): Search for neutrinos from the Sun using gallium. Phys. Rev. Lett. **67**, 3332.

Abney, W. DE W. (1877): Effect of a star's rotation on its spectrum. M.N.R.A.S. **37**, 278.

Abraham, P. B., Brunstein, K. A., Cline, T. L. (1966): Production of low-energy cosmic-ray electrons. Phys. Rev. **150**, 1088.

Abt, H. A., Biggs, E. S. (1972): Bibliography of stellar radial velocities. New York: Latham Press and Kitt Peak Nat. Obs.

Abt, H. A., Hunter, J. H. (1962): Stellar rotation in galactic clusters. Ap. J. **136**, 381.

Abt, H. A., et al. (1969): An atlas of low dispersion grating stellar spectra. Kitt Peak Nat. Obs., Yerkes Obs.

Acker, A., et al. (1990): The Strassbourg-ESO catalogue of galactic planetary nebulae.

Acquista, C., Anderson, J. L. (1974): Radiative transfer of partially polarized light. Ap. J. **191**, 567.

Adams, F. C., et al. (1997): Constraints on the intergalactic transport of cosmic rays. Ap. J. **491**, 6.

Adams, J. B., Ruderman, M. A., Woo, C. H. (1963): Neutrino pair emission by a stellar plasma. Phys. Rev. **129**, 1383.

Adams, W. S. (1914): An A-type star of very low luminosity. P.A.S.P. **26**, 198. Reproduced in: A source book in astronomy and astrophysics 1900–1975 (eds. K. R. Lang and O. Gingerich). Cambridge, Mass.: Harvard University Press 1979.

Adams, W. S. (1915): The spectrum of the companion of Sirius. P.A.S.P. **27**, 136. Reproduced in: A source book in astronomy and astrophysics 1900–1975 (eds. K. R. Lang and O. Gingerich). Cambridge, Mass.: Harvard University Press 1979.

Adams, W. S. (1925): The relativity displacement of the spectral lines in the companion of Sirius. Proc. Nat. Acad. Sci. (Wash.) **11**, 382.

Adams, W. S., Dunham, T. (1932): Absorption bands in the infra-red spectrum of Venus. P.A.S.P. **44**, 243.

Aggarwal, H. R., Oberbeck, V. R. (1974): Roche limit of a solid body. Ap. J. **191**, 577.

Airy, G. B. (1835): On the diffraction of an object-glass with circular aperture. Trans. Camb. Phil. Soc. **5**, 283.

Aizenman, M. L., Cox, J. P. (1974): Pulsational stability of stars in thermal imbalance. IV. Direct solution of differential equation. Ap. J. **194**, 663.

Aizenman, M. L., Cox, J. P. (1975): Pulsational stability of stars in thermal imbalance. VI. Physical mechanisms and extension to nonradial oscillations. Ap. J. **195**, 175.

Aizenman, M. L., Hansen, C. J., Ross, R. R. (1975): Pulsation properties of upper-main-sequence stars. Ap. J. **201**, 387.

Ajzenberg-Selove, F. (1960): Nuclear spectroscopy. New York: Academic Press.

Akasofu, S.-I. (1982): Interaction between a magnetized plasma flow and a strongly magnetized celestial body with an ionized atmosphere: Energetics of the magnetosphere. Ann. Rev. Astron. Astrophys. **20**, 117.

Alazraki, G., Couturier, P. (1971): Solar wind acceleration caused by the gradient of Alfvén wave pressure. Astron. Astrophys **13**, 380.

Alexander, D. R. (1975): Low-temperature Rosseland opacity tables. Ap. J. Supp. **29,** 363.

Alfvén, H. (1938): On the sidereal time variation of the cosmic radiation. Phys. Rev. **54**, 97.

Alfvén, H. (1942): On the existence of electromagnetic-hydrodynamic waves. Arkiv. f. Mat., Astron., Physik. **29**, 1.

Alfvén, H. (1942): The existence of electromagnetic-hydrodynamic waves. Nature **150**, 405.

Alfvén, H. (1947): Granulation, magneto-hydrodynamic waves and the heating of the solar corona. M.N.R.A.S. **107**, 211.

Alfvén, H.: Cosmical electrodynamics. Oxford: Oxford University Press 1950.

Alfvén, H., Falthammer, C. G.: Cosmical electrodynamics. Oxford: Oxford at the Clarendon Press 1963.

Alfvén, H., Herlofson, N. (1950): Cosmic radiation and radio stars. Phys. Rev. **78**, 616. Reproduced in: Source book in astronomy and astrophysics 1900-1975 (eds. K. R. Lang and O. Gingerich). Cambridge, Mass.: Harvard University Press 1979.

Ali, A. W., Griem, H. R. (1965): Theory of resonance broadening of spectral lines by atom-atom impacts. Phys. Rev. **140 A**, 1044 **144**, 366 (1966).

Allen, B. J., Gibbons, J. H., Macklin, R. L.: Nucelo-synthesis and neutron-capture cross sections. In: Advances in nuclear physics, vol 4 (ed. M. BARANGER and E. VOGT). New York: Plenum Press 1971.

Allen, C. W. (1947): Interpretation of electron densities from corona brightness. M.N.R.A.S. **107**, 426.

Allen, C. W.: Astrophysical quantities. University of London, London: Athlone Press 1963.

Allen, M., Robinson, G. W. (1975): Formation of molecules on small interstellar grains. Ap. J. **195**, 81.

Allen, M., Robinson, G. W. (1976): Molecular hydrogen in interstellar dark clouds. Ap. J. **207**, 745.

Allen, M., Robinson, G. W. (1977): Molecular hydrogen in interstellar dark clouds: erratum. Ap. J. **214**, 955.

Allen, R. J. (1969): Intergalactic hydrogen along the path to Virgo A. Astron. Astrophys. **3**, 382.

Aller, L. H.: Gaseous nebulae. London: Chapman and Hall 1956.

Aller, L. H.: The abundance of the elements. New York: Interscience Publ. 1961.

Aller, L. H.: Atoms, stars and nebulae – Third edition. New York: Cambridge University Press, 1991.

Aller, L. H., Bowen, I. S., Wilson, O. C. (1963): The spectrum of NGC 7027. Ap. J. **138**, 1013.

Aller, L. H., Czyzak, S. S.: The chemical composition of planetary nebulae. In: Planetary nebulae (ed. D. E. Osterbrock, C. R. O'Dell). I.A.U. Symp. No. 34. Dordrecht, Holland: D. R. Reidel. Berlin-Heidelberg-New York: Springer 1968.

Aller, L. H., Faulker, D. J.: Spectrophotometry of 14 southern planetary nebulae. In: The Galaxy and the Magellanic clouds – I.A.U. Symp. No. 20 Aust. Acad. Sci., Canberra, Australia. 1964.

Aller, L. H., Greenstein, J. L. (1960): The abundances of the elements in G-type subdwarfs. Ap. J. Supp. No. 46, **5**, 139.

Aller, L. H., Liller, W.: Planetary nebulae. In: Nebulae and interstellar matter – Stars and stellar systems VII (ed. B. M. Middlehurst, L. H. Aller). Chicago, Ill.: Univ. of Chicago Press 1960.

Alpar, M. A., Chau, H. F., Cheng, K. S., Pines, D. (1993): Postglitch relaxation of Vela pulsar after its first eight large glitches: A reevaluation with the vortex creep model. Ap. J. **409**, 345.

Alpher, R. A., Bethe, H. A., Gamow, G. (1948): The origin of chemical elements. Phys. Rev. **73**, 803.

Alpher, R. A., Herman, R. C. (1950): Theory of the origin and relative abundance distribution of the elements. Rev. Mod. Phys. **22**, 153.

Altenhoff, W., Mezger, P. G., Strassl, H., Wendker, H., Westerhout, G. (1960): Radio astronomical measurements at 2.7 KMHz.Veroff Sternwarte, Bonn **59**, 48.

Ambartsumian, V. A. (1955): Stellar systems of positive total energy. Observatory **75**, 72.

Ampere, A. M. (1827): Sur l'action mutuelle d'un aiment et d'un conduct voltaique. Ann. Chim. Phys. **37**.

Anders, E., Grevesse, N. (1989): Abundances of the elements – Meteoritic and solar. Geochimica et Cosmochimica **53**, 297. Abundances reproduced in: Lang, K. R., Astrophysical data: planets and stars. New York: Springer-Verlag 1992.

Anderson, C. D. (1932): Energies of cosmic-ray particles. Phys. Rev. **41**, 405.

Anderson, C. D. (1932): The apparent existence of easily deflectable positives. Science **76**, 238.

Anderson, C. D. (1933): The positive electron. Phys. Rev. **43**, 491.

Anderson, C. D., Neddermeyer, S. H. (1936): Cloud chamber observations of cosmic rays at 4300 meters elevation and near sea-level. Phys. Rev. **50**, 263.

Anderson, C. D., Neddermeyer, S. H. (1937): Cosmic ray particles of intermediate mass. Phys. Rev. **54**, 88.

Anderson, J. L., Spiegel, E. A. (1975): Radiative transfer through a flowing refractive medium. Ap. J. **202**, 454.

Anderson, L. S., Athay, R. G. (1989): Model solar chromosphere with prescribed heating. Ap. J. **346**, 1010.

Anderson, P. W. (1949): Pressure broadening in the microwave and infra-red regions. Phys. Rev. **76**, 647.

Anderson, P. W., Itoh, N. (1975): Pulsar glitches and restlessness as a hard superfluidity phenomenon. Nature **256**, 25.

Anderson, W. Von (1929): Gewöhnliche Materie und strahlende Energie als verschiedene "Phasen" eines und desselben Grundstoffes (Common matter and radiated energy as different "phases" of the same chemical element). Z. Physik **54**, 433.

Ando, H., Osaki, Y. (1975): Nonadiabatic nonradial oscillations – an application to the five-minute oscillation of the Sun. Publ. Astron. Soc. Japan **27**, 581.

Andrews, M. H., Hjellming, R. M. (1969): Intensities of radio recombination lines (II). Astrophys. Lett. **4**, 159.

Angel, J. R. P. (1978): Magnetic white dwarfs. Ann. Rev. Astron. Astrophys. **16**, 487.

Angel, J. R. P., Borra, E. F., Landstreet, J. D. (1981): The magnetic fields of white dwarf stars Ap. J. Supp. **45**, 457.

Angel, J. R. P., Landstreet, J. D. (1971): Detection of circular polarization in a second white dwarf. Ap. J. (Letters) **164**, L15.

Angel, J. R. P., Landstreet, J. D. (1972): Discovery of circular polarization in the red degenerate star G 99-47. Ap. J. (Letters) **178**, L21.

Antonucci, E., et al. (1982): Impulsive phase of flares in soft X-ray emission. Solar Phys. **78**, 107.

Apollonio, M., et al. (1998): Initial results from the CHOOZ long baseline reactor neutrino oscillation experiment Phys. Lett. **B420**, 397.

Appleton, E. V. (1932): Wireless studies of the ionosphere. J. Inst. Elec. Engrs. (London) **71**, 642.

Arnett, W. D. (1966): Gravitational collapse and weak interactions. Can. J. Phys. **44**, 2553.

Arnett, W. D. (1967): Mass dependence in gravitational collapse of stellar cores. Can. J. Phys. **45**, 1621.

Arnett, W. D. (1968): On supernova hydrodynamics. Ap. J. **153**, 341.

Arnett, W. D. (1969): A possible model of supernovae: Detonation of C^{12}. Astrophys. and Space Sci. **5**, 180.

Arnett, W. D. (1969): Explosive nucleosynthesis in stars. Ap. J. **137**, 1369.

Arnett, W. D. (1971): Galactic evolution and nucleosynthesis. Ap. J. **166**, 153.

Arnett, W. D. (1972): Hydrostatic oxygen burning in stars: I. Oxygen stars. Ap. J. **173**, 393.
Arnett, W. D. (1973): Explosive nucleosynthesis in stars. Ann. Rev. Astron. Astrophys. **11**, 73.
Arnett, W. D. (1974): Advanced evolution of massive stars. V. Neon burning. Ap. J. **193**, 169.
Arnett, W. D. (1974): Advanced evolution of massive stars. VI. Oxygen burning. Ap. J. **194**, 373.
Arnett, W. D. (1977): Advanced evolution of massive stars. VII. Silicon burning. Ap. J. Suppl. **35**, 145.
Arnett, W. D. (1977): Neutrino trapping during gravitational collapse of stars. Ap. J. **218**, 815.
Arnett, W. D. (1978): On the bulk yields of nucleosynthesis from massive stars. Ap. J. **219**, 1008.
Arnett, W. D. (1995): Explosive nucleosynthesis revisited: Yields. Ann. Rev. Astron. Astrophys. **33**, 115.
Arnett, W. D., Cameron, A. G. W. (1967): Supernova hydrodynamics and nucleosynthesis. Can. J. Phys. **45**, 2953.
Arnett, W. D., Clayton, D. D. (1970): Explosive nucleosynthesis in stars. Nature **227**, 780.
Arnett, W. D., et al. (1989): Supernova 1987A. Ann. Rev. Astron. Astrophys. **27**, 629.
Arnett, W. D., Truran, J. W. (1969): Carbon-burning nucleosynthesis at constant temperature. Ap. J. **157**, 339.
Arnett, W. D., Truran, J. W., Woosley, S. E. (1971): Nucleosynthesis in supernova models II. The ^{12}C detonation model. Ap. J. **165**, 87.
Arnould, M. (1972): Influence of the excited states of target nuclei in the vincinity of the iron peak on stellar reaction rates. Astron. Astrophys. **19**, 92.
Aston, F. W. (1927): A new mass-spectrograph and the whole number rule. Proc. Roy. Soc. London **A115**, 487.
Aström, E. (1950): On waves in an ionized gas. Arkiv. Physik **2**, 443.
Athay, R. G., White, O. R. (1978): Chromospheric and coronal heating by sound waves. Ap. J. **226**, 1135.
Athay, R. G., White, O. R. (1979): Chromospheric oscillations observed with OSO 8 IV. Power and phase spectra for CIV. Ap. J. **229**, 1147.
Atkinson, R. d'E. (1931): Atomic synthesis and stellar energy I, II. Ap. J. **73**, 250, 308. Reproduced in: A source book in astronomy and astrophysics 1900–1975 (eds. K. R. Lang and O. Gingerich). Cambridge, Mass.: Harvard University Press 1979.
Atkinson, R. d'E. (1936): Atomic synthesis and stellar energy III. Ap. J. **84**, 73.
Atkinson, R. d'E., Houtermans, F. G. (1929): Zur Frage der Aufbaumöglichkeit der Elemente in Sternen. Zeits. f. Phys. **54**, 656.
Atteia, J.-L., et al. (1987): A second catalog of gamma ray bursts 1978–1980. Localizations from the interplanetary network. Ap. J. Supp. **64**, 305.
Audouze, J., Epherre, M., Reeves, H.: Survey of experimental cross sections for proton-induced spallation reactions in He^4, C^{12}, N^{14}, and O^{16}. In: High energy nuclear reactions in astrophysics (ed. B. Shen). New York: W. A. Benjamin 1967.
Audouze, J., Lequeux, J., Reeves, H. (1973): On the cosmic boron abundance. Astron. and Astrophys. **28**, 85.
Audouze, J., Tinsley, B. M. (1974): Galactic evolution and the formation of the light elements. Ap. J. **192**, 487.
Audouze, J., Tinsley, B. M. (1976): Chemical evolution of galaxies. Ann. Rev. Astron. Astrophys. **14**, 43.
Aumann, H. H. (1985): IRAS observations of matter around nearby stars. P.A.S.P. **97**, 885.
Aumann, H. H., et al. (1984): Discovery of a shell around Alpha Lyrae. Ap. J. (Letters) **278**, L23.
Austin, S. M. (1981): The creation of light elements – cosmic rays and cosmology. Prog. Part. Nucl. Phys. **7**, 1.

Avogadro, A. (1811): Essay on a manner of determining the relative masses of the elementary molecules of bodies and the proportions in which they enter into these compounds. J. de. Physique **73**, 58. Engl. trans. in: Foundations of the molecular theory. Alemic Club. Repr. No. 4. Edinburgh: Bishop 1950.

Axford, W. I. (1961): Ionization fronts in interstellar gas: The structure of ionization fronts. Phil. Trans. Roy. Soc. London **A253**, 301.

Axford, W. I. (1964): The initial development of H II regions. Ap. J. **139**, 761.

Axford, W. I. (1981): Acceleration of cosmic rays by shock waves. In: Proc. 17th International Cosmic Ray Conference (Paris) **12**, 155.

Axford, W. I., Leer, E., Skadron, G. (1977): In: Proc. 15th International Cosmic Ray Conference **11**, 132.

Baade, W., Zwicky, F. (1934): On super-novae. Proc. Nat. Acad. Sci. (Wash.) **20**, 254. Reproduced in: A source book in astronomy and astrophysics 1900-1975 (eds. K. R. Lang and O. Gingerich). Cambridge, Mass.: Harvard University Press 1979.

Baade, W., Zwicky, F. (1934): Cosmic rays from super-novae. Proc. Nat. Acad. Sci. (Wash.) **20**, 259. Reprod. in: Selected papers on cosmic ray origin theories (ed. S. Rosen). New York: Dover 1969.

Baade, W., Zwicky, F. (1934): Remarks on super-novae and cosmic rays. Phys. Rev. **46**, 76. Reproduced in: Selected papers on cosmic ray origin theories (ed. S. Rosen). New York: Dover 1969.

Babcock, H. W. (1947): Zeeman effect in stellar spectra. Ap. J. **105**, 105.

Babcock, H. W. (1960): The 34-kilogauss magnetic field of HD 215441. Ap. J. **132,** 521.

Babcock, H. W., Babcock, H. D. (1955): The Sun's magnetic field, 1952–1954. Ap. J. **121**, 349.

Bachiller, R. (1996): Bipolar molecular outflows from young stars and protostars. Ann. Rev. Astron. Astrophys. **34**, 111.

Back, E. Von, Goudsmit, S. (1928): Kernmoment und Zeeman-Effekt von Wismut (Nuclear moment and Zeeman effect in bismuth). Z. Physik **47**, 174.

Bahcall, J. N. (1962): Beta decay in stellar interiors. Phys. Rev. **126**, 1143.

Bahcall, J. N. (1962): The exclusion principle and photobeta reactions in nucleogenesis. Ap. J. **136**, 445.

Bahcall, J. N. (1964): Electron capture in stellar interiors. Ap. J. **139**, 318.

Bahcall, J. N. (1964): Solar neutrino cross sections and nuclear beta decay. Phys. Rev. **135**, B 137.

Bahcall, J. N. (1964): Neutrino opacity: I. Neutrino-lepton scattering. Phys. Rev. **136**, B1164.

Bahcall, J. N. (1964): Solar neutrinos I. Theoretical. Phys. Rev. Letters **12**, 300. Reproduced in: A source book in astronomy and astrophysics 1900–1975 (eds. K. R. Lang and O. Gingerich). Cambridge, Mass.: Harvard University Press 1979.

Bahcall, J. N. (1965): Observational neutrino astronomy. Science **147**, 115.

Bahcall, J. N. (1968): A systematic method for identifying absorption lines as applied to PKS 0237 – 23. Ap. J. **153**, 679.

Bahcall, J. N. (1978): Solar neutrino experiments. Rev. Mod. Phys. **50**, 881.

Bahcall, J. N. (1978): Masses of neutron stars and black holes in X-ray binaries. Ann. Rev. Astron. Astrophys. **16**, 241.

Bahcall, J. N. (1984): K giants and the total amount of matter near the sun. Ap. J. **287**, 926.

Bahcall, J. N.: Neutrino astrophysics. New York: Cambridge University Press, 1989.

Bahcall, J. N. (1991): Shapes of solar-neutrino spectra: Unconventional tests of the standard electroweak model. Phys. Rev. **D44**, 1644.

Bahcall, J. N. (1994): Two solar neutrino problems. Phys. Lett. **B338**, 276.

Bahcall, J. N. (1996): Solar neutrinos: Where we are, where we are going. Ap. J. **467**, 475.

Bahcall, J. N.: Solar neutrinos: Solved and unsolved problems. In: Unsolved problems in astrophysics (eds. J. N. Bahcall, J. P. Ostriker). Princeton, N. J.: Princeton University Press 1997.

Bahcall, J. N., Basu, S., Kumar, P. (1997): Localized helioseismic constraints on solar structure. Ap. J. (Letters) **485**, L91.

Bahcall, J. N., Bethe, H. A. (1990): A solution of the solar neutrino problem. Phys. Rev. Letters **65**, 2233.
Bahcall, J. N., et al. (1997): Are standard solar models reliable? Phys. Rev. Letters **78**, 171.
Bahcall, J. N., Frautschi, S. C. (1964): Neutrino opacity: II. Neutrino-nucleon interactions. Phys. Rev. **136**, B1547.
Bahcall, J. N., Frautschi, S. C. (1969): Lepton non-conservation and solar neutrinos. Phys. Lett **29B**, 623.
Bahcall, J. N., Krastev, P. I. (1996): How well do we (and will we) know solar neutrino fluxes and oscillation parameters? Phys. Rev. **D53**, 4211.
Bahcall, J. N., Krastev, P. I. (1997): Does the Sun appear brighter at night in neutrinos? Phys. Rev. C. Nuclear Physics **56**, No. 5., 2839.
Bahcall, J. N., Lande, K., Lanou, R. E., Jr., Learned, J. G., Hamish Robertson, R. G., Wolfenstein, L. (1995): Progress and prospects in neutrino astrophysics. Nature **375**, 29.
Bahcall, J. N., May, R. M. (1969): The rate of the proton-proton reaction and some related reactions. Ap. J. **155**, 501.
Bahcall, J. N., Ostriker, J. P. (eds.): Unsolved problems in astrophysics. Princeton, New Jersey: Princeton University Press 1997.
Bahcall, J. N., Pinsonneault, M. H. (1992): Helium diffusion in the Sun. Ap. J. (Letters) **395**, L119.
Bahcall, J. N., Pinsonneault, M. H. (1992): Standard solar models, with and without helium diffusion, and the solar neutrino problem. Rev. Mod. Phys. **64**, 885.
Bahcall, J. N., Salpeter, E. E. (1965): On the interaction of radiation from distant sources with the intervening medium. Ap. J. **142**, 1677.
Bahcall, J. N., Sears, R. L. (1972): Solar neutrinos. Ann. Rev. Astron. Astrophys. **10**, 25.
Bahcall, J. N., Tremaine, S. (1981): Methods for determining the masses of spherical systems I. Test particles around a point mass. Ap. J. **244**, 805.
Bahcall, J. N., Ulmer, A. (1996): Temperature dependence of solar neutrino fluxes. Phys., Rev. **D53**, 4202.
Bahcall, J. N., Ulrich, R. K. (1971): Solar neutrinos: III. Composition and magnetic field effects and related inferences. Ap. J. **170**, 593.
Bahcall, J. N., Ulrich, R. K. (1988): Solar models, neutrino experiments, and helioseismology. Rev. Mod. Phys. **60**, 297.
Bahcall, J. N., Wolf, R. A. (1965): Neutron stars. Phys. Rev. Lett. **14**, 343.
Bahcall, J. N., Wolf, R. A. (1965): Neutron stars: I. Properties at absolute zero temperature. Phys. Rev. **140**, B1445.
Bahcall, J. N., Wolf, R. A. (1965): Neutron stars: II. Neutrino cooling and observability. Phys. Rev. **140**, B1452.
Bahcall, J. N., Wolf, R. A. (1968): Fine-structure transitions. Ap. J. **152**, 701.
Bahcall, N. A. (1979): The X-ray luminosity function of clusters of galaxies: predictions from a thermal bremsstrahlung model. Ap. J. (Letters) **232**, L83.
Bahcall, N. A., Fowler, W. A. (1969): The effect of excited nuclear states on stellar reaction rates. Ap. J. **157**, 645.
Bahcall, N. A., Fowler, W. A. (1970): Nuclear partition functions for stellar reaction rates. Ap. J. **161**, 119.
Bailes, M., Johnston, S.: Recent pulsar discoveries. In: The review of radio science 1990–1992 (ed. W. Ross Stone). Oxford: Oxford University Press 1993.
Bailey, S. I. (1899): The periods of the variable stars in the cluster Messier 5. Ap. J. **10**, 255.
Bailey, V. A. (1948): Plane waves in an ionized gas with static electric and magnetic fields present. Austr. J. Sci. Res. **A1**, 351.
Baker, J. G., Menzel, D. H. (1938): Physical processes in gaseous nebulae III. The Balmer decrement. Ap. J. **88**, 52. Reprod. in: Selected papers on physical processes in ionized plasmas (ed. D. H. Menzel). New York: Dover 1962.
Baker, N., Kippenhahn, R. (1962): The pulsation of models of δ Cephei stars. Z. Ap. **54**, 114.

Baker, N., Kippenhahn, R. (1965): The pulsations of models of Delta Cephei stars II. Ap. J. **142**, 868.
Balbus, S. A., Hawley, J. F. (1991): A powerful local shear instability in weakly magnetized disks. I. Linear analysis. Ap. J. **376**, 214.
Baliunas, S. L. (1985): Stellar activity cycles. Ann. Rev. Astron. Astrophys. **23**, 379.
Balmer, J. J. (1885): Notiz über die Spectrallinien des Wasserstoffs (A note on the spectral lines of hydrogen). Ann. Phys. Chem. **25**, 80.
Baranger, M. (1958): Problem of overlapping lines in the theory of pressure broadening. Phys. Rev. **111**, 494.
Baranger, M. (1958): General impact theory of pressure broadening. Phys. Rev. **112**, 855.
Baranger, M.: Spectral line broadening in plasmas. In: Atomic and molecular processes (ed. D. R. Bates). New York: Academic Press 1962.
Barbosa, D. D. (1982): A note on Compton scattering. Ap. J. **254**, 301.
Barcons, X., Fabian, A. C. (eds.): The X-ray background. New York: Cambridge University Press 1992.
Bardeen, J. M., Cooper, L. N., Schrieffer, J. R. (1957): Theory of superconductivity. Phys. Rev. **108**, 1175.
Bardeen, J. M., et al. (1977): A new criterion for secular instability of rapidly rotating stars. Ap. J. (Letters) **217**, L49.
Bardeen, J. M., Friedman, J. L., Schutz, B. F., Sorkin, R. (1977): A new criterion for secular instability of rotating stars. Ap. J. (Letters) **217**, L49.
Bardin, R. K., Barnes, C. A., Fowler, W. A., Seeger, P. A. (1962): ft value of O^{14} and the universality of the Fermi interaction. Phys. Rev. **127**, 583.
Baring, M. G., Kirk, J. G. (1991): The modification of relativistic shock fronts by accelerated particles. Astron. Astrophys. **241**, 329.
Barkat, Z. (1975): Neutrino processes in stellar interiors. Ann. Rev. Astron. Astrophys. **13**, 45.
Barkat, Z., Buchler, J. R., Ingber, L. (1972): Equation of state of neutron-star matter at subnuclear densities. Ap. J. **176**, 723.
Barlow, M. J., Silk, J. (1976): H_2 recombination on interstellar grains. Ap. J. **207**, 131.
Barnard, E. E. (1919): On the dark markings of the sky with a catalogue of 182 such objects. Ap. J. **49**, 1.
Barnard, E. E.: A photographic atlas of regions of the milky way (ed. E. B. Frost and M. R. Calvert). Washington, D.C.: Carnegie Institution 1927.
Barnes, A. (1968): Collisionless heating of the solar-wind plasma: I. Theory of the heating of collisionless plasma by hydromagnetic waves. Ap. J. **154**, 751.
Barnes, A. (1969): Collisionless heating of the solar-wind plasma: II. Application of the theory of plasma heating by hydromagnetic waves. Ap. J. **155**, 311.
Barnes, C. A.: Nucleosynthesis by charged-particle reactions. In: Advances in nuclear physics, vol 4 (ed. M. Baranger, E. Vogt). New York: Plenum Press 1971.
Batchelor, G. K., The theory of homogeneous turbulence. Cambridge: Cambridge University Press 1967.
Bateman, G.: MHD instabilities. Cambridge, Mass.: MIT Press 1978, 1980.
Bates, D. R. (1951): Rate of formation of molecules by radiative association. M.N.R.A.S. **111**, 303.
Bates, D. R., Dalgarno, A.: Electronic recombination. In: Atomic and molecular processes (ed. D.R. Bates). New York: Academic Press 1962.
Bates, D. R., Damgaard, A. (1949): The calculation of the absolute strengths of spectral lines. Phil. Trans. Roy. Soc. London **A242**, 101.
Bates, D. R., Spitzer, L. (1951): The density of molecules in interstellar space. Ap. J. **113**, 441.
Baumbach, S. Von (1937): Strahlung, Ergiebigkeit und Elektronendichte der Sonnenkorona (Radiation, abundance, and electron density of the solar corona). Astron. Nach. **263**, 120.
Bavassano, B., Smith, E. J. (1986): Radial variation of interplanetary Alfvénic fluctuations: Pioneer 10 and 11 observations between 1 and 5 AU. J. Geophys. Res. **91**, 1706.
Baym, G., Bethe, H. A., Pethick, C. J. (1971): Neutron star matter. Nuclear Phys. **A175**, 225.

Baym, G., Pethick, C. (1975): Neutron stars. Ann. Rev. Nucl. Sci. **25**, 27.

Baym, G., Pethick, C. (1979): Physics of neutron stars. Ann. Rev. Astron. Ap. **17**, 415.

Baym, G., Pethick, C., Pines, D., Ruderman, M. (1969): Spin up in neutron stars: The future of the Vela pulsar. Nature **224**, 872.

Baym, G., Pethick, C., Sutherland, P. (1971): The ground state of matter at high densities: Equation of state and stellar models. Ap. J. **170**, 299.

Baym, G., Pines, D. (1971): Neutron starquakes and pulsar speedup. Ann. Phys. **66**, 816.

Beaudet, G., Petrosian, V., Salpeter, E. E. (1967): Energy losses due to neutrino processes. Ap. J. **150**, 979.

Beaudet, G., Salpeter, E. E., Silvestro, M. L. (1972): Rates for URCA neutrino processes. Ap. J. **174**, 79.

Beaujardiere, O. De La (1966): L'èvolution du spectre de rayonnement synchrotron. Ann. Astrophys. **29**, 345.

Beck, R., et al. (1996): Galactic magnetism: recent developments and perspectives. Ann. Rev. Astron. Astrophys. **34**, 155.

Beck, R., Gräve, R. (eds.): Interstellar magnetic fields. Observation and theory. New York: Springer-Verlag 1987.

Beckwith, S. V. W., Sargent, A. I. (1996): Circumstellar disks and the search for neighbouring planetary systems. Nature **383**, 139.

Becquerel, H. (1896): Sur les radiations émises par phosphorescence (On the radiation emitted by phosphorescence). Compt. Rend. **122**, 420.

Begelman, M. C., Blandford, R. D., Rees, M. J. (1984): Theory of extragalactic radio sources. Rev. Mod. Phys. **56**, 255.

Begelman, M. C., Kirk, J. G. (1990): Shock-drift particle acceleration in superluminal shocks: A model for hot spots in extragalactic radio sources. Ap. J. **353**, 66.

Begelman, M. C., Rees, M.: Gravity's fatal attraction – black holes in the universe. New York: W. H. Freeman 1996.

Beichman, C. A. (1987): The IRAS view of the Galaxy and the solar system. Ann. Rev. Astron. Astrophys. **25**, 521.

Bekefi, G.: Radiation processes in plasmas. New York: John Wiley and Sons 1966.

Belcher, J. W., Davis, L. Jr. (1971): Large amplitude Alfvén waves in the interplanetary medium. J. Geophys. Res. **76**, 3534.

Belcher, J. W., Davis, L. Jr., Smith, E. J. (1969): Large-amplitude waves in the interplanetary medium: Mariner 5. J. Geophys. Res. **74**, 2302.

Bell, A. R. (1978): The acceleration of cosmic rays in shock fronts – I., II. M.N.R.A.S. **182**, 147, 443.

Bélopolsky, A. (1897): Researches on the spectrum of the variable star η Aquilae. Ap. J. **6**, 393.

Benjamini, R., Londrillo, P., Setti, G. (1967): The cosmic black-body radiation and the inverse Compton effect in the radio galaxies: The X-ray background. Nuovo Cimento **B52,** 495.

Bennett, W. H. (1934): Magnetically self-focusing streams. Phys. Rev. **45**, 890.

Benz, A.O.: Plasma astrophysics: Kinetic processes in solar and stellar coronae. Boston: Kluwer Academic 1993.

Benz, A. O., Güdel, M. (1994): The soft X-ray/microwave ratio of solar and stellar flares and coronae. Astron. Astrophys. **285**, 621.

Berezinskii, V. S., Bulanov, S. V., Dogiel, V. A., Ginzburg, V. L., Ptuskin, V.S.: Astrophysics of cosmic rays. New York: North Holland 1990.

Berge, G. L., Greisen, E. W. (1969): High-resolution interferometry of Venus at 12-cm wavelength. Ap. J. **156**, 1125.

Bergeron, J. (1969): Étude des possibilités d'existence d'un plasma intergalactique dense. Astron. Astrophys. **3**, 42.

Berman, L. (1936): The effect of reddening on the Balmer decrement in planetary nebulae. M.N.R.A.S. **96**, 890.

Bernas, R., et al. (1967): On the nucleosynthesis of lithium, beryllium, and boron, Ann. Phys. (N.Y.) **44**, 426.
Bernoulli, D.: Hydrodynamia Argentorati (Hydrodynamics) 1738.
Bertin, G., Lin, C.-C.: Spiral structure in galaxies: A density wave theory. Cambridge, Mass.: Massachusetts Institute of Technology Press 1996.
Bertin, G., Radicati, L. A. (1976): The bifurcation from the Maclaurin to the Jacobi sequence as a second-order phase transition. Ap. J. **206**, 815.
Bertout, C. (1989): T Tauri stars: Wild as dust. Ann. Rev. Astron. Astrophys. **27**, 351.
Bertsch, D. L., et al. (1993): Diffuse gamma-ray emission in the galactic plane from cosmic-ray, matter, and photon interactions. Ap. J. **416**, 587.
Bethe, H. A. (1930): The influence of screening on the creation and stopping of electrons. Proc. Camb. Phil. Soc. **30**, 524.
Bethe, H. A. (1930): Zur Theorie des Durchgangs schneller Korpuskularstrahlen durch Materie (On the theory of the penetration of matter by fast (nuclear) particle beams). Ann. Phys. **5**, 325.
Bethe, H. A. (1932): Bremsformel für Elektronen relativistischer Geschwindigkeit (A braking formula for relativistic electrons). Z. Phys. **76**, 293.
Bethe, H. A.: Quantenmechanik der Ein- und Zwei-Elektronen-Probleme (Quantum mechanics of one- and two-electron problems). In: Handbuch der Physik (ed. H. Geiger, K. Scheel), Vol. XXIV, p. 273. Berlin: Springer 1933.
Bethe, H. A. (1937): Nuclear physics B. Nuclear dynamics, theoretical. Rev. Mod. Phys. **9**, 69.
Bethe, H. A. (1939): Energy production in stars. Phys. Rev. **55**, 434. Reproduced in: A source book in astronomy and astrophysics 1900–1975 (eds. K. R. Lang and O. Gingerich). Cambridge, Mass.: Harvard University Press 1979.
Bethe, H. A. (1986): Possible explanation of the solar-neutrino puzzle. Phys. Rev. Letters **56**, 1305.
Bethe, H. A. (1989): Solar-neutrino experiments. Phys. Rev. Letters **63**, 837.
Bethe, H. A. (1990): Supernova mechanisms. Rev. Mod. Phys. **62**, 801.
Bethe, H. A., Critchfield, C. L. (1938): The formation of deutrons by proton combination. Phys. Rev. **54**, 248.
Bethe, H. A., Heitler, W. (1934): On the stopping of fast particles and on the creation of positive electrons. Proc. Roy. Soc. London **A146**, 83.
Bethe, H. A., Johnson, M. B. (1974): Dense baryon matter calculations with realistic potentials. Nucl. Phys. **A230**, 1.
Bethe, H. A., Salpeter, E. E.: Quantum mechanics of one and two electron atoms. Berlin-Göttingen Heidelberg: Springer 1957.
Bethe, H. A., Wills, H. H. (1935): On the annihilation radiation of positrons. Proc. Roy. Soc. London **A150**, 129.
Bhabha, H. J. (1935): On the calculation of pair creation by fast charged particles and the effect of screening. Proc. Camb. Phil. Soc. **31**, 394.
Bhabha, H. J. (1935): The creation of electron pairs by fast charged particles. Proc. Roy. Soc. London **A152**, 559.
Bhabha, H. J. (1936): The scattering of positrons by electrons with exchange on Dirac's theory of the positron. Proc. Roy. Soc. London **A154**, 195.
Biermann, L. (1946): Zur Deutung der chromosphärischen Turbulenz und des Exzesses der UV-Strahlung der Sonne (An explanation of chromospheric turbulence and the UV excess of solar radiation). Naturwiss. **33**, 118.
Biermann, L. (1947): Über die Ursache der chromosphärischen Turbulenz und des UV-Exzesses der Sonnenstrahlung (About the cause of the chromospheric turbulence and the UV excess of the solar radiation). Z. Ap. **25**, 161.
Biermann, L. (1951): Kometenschweife und solare Korpuskular-Strahlung (The tails of comets and the solar corpuscular radiation). Z. Ap. **29**, 274.

Biermann, L. (1957): Solar corpuscular radiation and the interplanetary gas. Observatory **77**, 109. Reproduced in: Source book in astronomy and astrophysics 1900–1975 (eds. K. R. Lang and O. Gingerich). Cambridge, Mass.: Harvard University Press 1979.

Biermann, P. L., Strittmatter, P. A. (1987): Synchrotron emission from shock waves in active galactic nuclei. Ap. J. **322**, 643.

Biganami, G. F., Hermsen, W. (1983): Galactic gamma-ray sources. Ann. Rev. Astron. Astrophys. **21**, 67.

Bilenky, S. M., Petcov, S. T. (1987): Massive neutrinos and neutrino oscillations. Rev. Mod. Phys. **59**, 671.

Bilenky, S. M., Pontecorvo, B. M. (1978): Lepton mixing and neutrino oscillations. Sov. Phys. Usp. **20**, 776.

Binney, J. (1982): Dynamics of elliptical galaxies and other spheroidal components. Ann. Rev. Astron. Astrophys. **20**, 399.

Binney, J., Tremaine, S.: Galactic dynamics. Princeton, New Jersey: Princeton University Press 1987.

Bionta, R. M., et al. (1987): Observation of a neutrino burst in coincidence with supernova 1987A in the Large Magellanic Cloud. Phys. Rev. Letters **58**, 1494.

Biot, J. B., Savart, F. (1820): Sur le magnétisme de la pile de Volta. Ann. Chimie **15**, 222.

Bird, D. J., et al. (1994): The cosmic-ray energy spectrum observed by the fly's eye. Ap. J. **424**, 491.

Bird, D. J., et al. (1995): Detection of a cosmic ray with measured energy well beyond the expected spectral cutoff due to cosmic microwave radiation. Ap. J. **441**, 144.

Birkeland, K.: The Norwegian aurora polar expedition 1902–1903: I. On the cause of magnetic storms and the origin of terrestrial magnetism. Christiana: H. Aschehong 1908.

Blaauw, A. (1964): The O associations in the solar neighborhood. Ann. Rev. Astron. Ap. **2**, 213.

Black, D. C., Bodenheimer, P. (1976): Evolution of rotating interstellar clouds. II. The collapse of proto-stars of 1, 2, and 5 M_O. Ap. J. **206**, 138.

Black, D. C., Matthews, M. S. (eds.): Protostars and planets II. Tucson, Arizona: University of Arizona Press 1985.

Black, J. H.: Diffuse interstellar clouds. In: Spectroscopy of astrophysical plasmas (eds. A. Dalgarno and D. Layzer). New York: Cambridge University Press 1987.

Black, J. H., Dalgarno, A. (1973): The cosmic abundance of deuterium. Ap. J. (Letters) **184**, L101.

Black, J. H., Dalgarno, A. (1976): Interstellar H_2 : the population of excited rotational states and the infrared response to ultraviolet radiation. Ap. J. **203**, 132.

Black, J. H., Dalgarno, A. (1977): Models of interstellar clouds I. The zeta ophiuchi cloud. Ap. J. Supp. **34**, 405.

Black, J. H., Dalgarno, A., Oppenheimer, M. (1975): The formation of CH^+ in interstellar clouds. Ap. J. **199**, 633.

Blackett, P. M. S., Occchialini, G. P. S. (1933): Some photographs of the tracks of penetrating radiation. Poc. Roy. Soc. (London) **A139**, 699.

Blades, J. C., Turnshek, D. A., Norman, C. A. (eds.): QSO absorption lines: Probing the universe. New York: Cambridge University Press 1988.

Blake, J. B., Schramm, D. N. (1975): A consideration of the neutron capture time scale in the s-process. Ap. J. **197**, 615.

Blake, J. B., Schramm, D. N. (1976): A possible alternative to the r-process. Ap. J. **209**, 846.

Blamont, J. E., Roddier, F. (1961): Precise observation of the profile of the Fraunhofer strontium resonance line. Evidence for the gravitational redshift of the Sun. Phys. Rev. Lett. **7**, 437.

Blandford, R. D., Eichler, D. (1987): Particle acceleration at astrophysical shocks: A theory of cosmic ray origin. Physics Reports **154**, 1.

Blandford, R. D., Königl, A. (1979): Relativistic jets as compact radio sources. Ap. J. **232**, 34.

Blandford, R. D., Ostriker, J. P. (1978): Particle acceleration by astrophyscial shocks. Ap. J. (Letters). **221**, L29.

Blandford, R. D., Ostriker, J. P. (1980): Supernova shock acceleration of cosmic rays in the Galaxy. Ap. J. **237**, 793.

Blatt, J. M., Weisskopf, V. F.: Theoretical nuclear physics. New York: John Wiley and Sons 1962.

Blin-Stoyle, R. J., Freeman, J. M. (1970): Coupling constants and electromagnetic radiative corrections in beta-decay and the mass of the intermediate vector boson. Nucl. Phys. **A150**, 369.

Bloch, F. Von (1933): Bremsvermögen von Atomen mit mehreren Elektronen (Braking capabilities of multi-electron atoms). Z. Phys. **81**, 363.

Bloemen, H. (1989): Diffuse galactic gamma-ray emission. Ann. Rev. Astron. Astrophys. **27**, 469.

Bloemen, H., et al. (1994): COMPTEL observations of the Orion complex: evidence for cosmic-ray induced gamma-ray lines. Astron. Astrophys. **281**, L5.

Bludman, S. A., Van Riper, K. A. (1977): Equation of state of an ideal Fermi gas. Ap. J. **212**, 859.

Blumenthal, G. R. (1974): The Poynting-Robertson effect and Eddington limit for electrons scattering with hard photons. Ap. J. **188**, 121.

Blumenthal, G. R., Gould, R. J. (1970): Bremsstrahlung, synchrotron radiation, and Compton scattering of high-energy electrons traversing dilute gases: Rev. Mod. Phys. **42** (2), 237,.

Bobrovnikoff, N. T. (1942): Physical theory of comets in the light of spectroscopic data. Rev. Mod. Phys **14**, 164.

Bodansky, D., Clayton, D. D., Fowler, W. A. (1968): Nuclear quasi-equilibrium during silicon burning. Ap. J. Suppl. No. 148, **16**, 299.

Bodansky, D., Clayton, D. D., Fowler, W. A. (1968): Nucleosynthesis during silicon burning. Phys. Rev. Lett. **20**, 161.

Bodenheimer, P. (1965): Studies in stellar evolution: II. Lithium depletion during the pre-main sequence contraction. Ap. J. **142**, 451.

Bodenheimer, P. (1966): Depletion of deuterium and beryllium during pre-main sequence evolution. Ap. J. **144**, 103.

Bodenheimer, P., Sweigart, A. (1968): Dynamic collapse of the isothermal sphere. Ap. J. **152**, 515.

Bogdan, T. J., Braun, D. C.: Active region seismology. In: Helioseismology – Proc. Fourth Soho Workshop. ESA SP-376 (1995) – June.

Bogolyubov, N. N.: Problems of a dynamical theory in statistical physics. State Tech. Press, Moscow 1946.

Böhm, D., Aller, L. H. (1947): The electron velocity distribution, in gaseous nebulae and stellar envelopes. Ap. J. **105**, 131. Reprod. In: Selected papers on physical processes in ionized plasmas (ed. D. H. Menzel). New York: Dover 1962.

Bohm, D., Gross, E. P. (1949): Theory of plasma oscillations: A. Origin of medium-like behavior; B. Excitation and damping of oscillations. Phys. Rev. **75**, 1851, 1854.

Böhm, K. H.: Basic theory of line formation. In: Stars and stellar systems (ed. J. Greenstein) vol. 6. Chicago, Ill.: Univ. of Chicago Press 1960.

Böhm, K. H.: The atmospheres and spectra of central stars. In: Planetary nebulae-I.A.U. Symp. No. 34. Dordrecht, Holland: D. Reidel. Berlin-Heidelberg-New York: Springer 1968.

Böhm, K.-H., et al. (1977): Some properties of very low temperature, pure helium surface layers of degenerate dwarfs. Ap. J. **217**, 521.

Böhm-Vitense, E. (1953): Die Wasserstoffkonvektionszone der Sonne (The hydrogen convection zone of the Sun). Z. Ap. **32**, 135.

Böhm-Vitense, E. (1958): Über die Wasserstoffkonvektionszone in Sternen verschiedener Effektivtemperaturen und Leuchtkräfte (About the hydrogen convection zone in stars of various effective temperatures and luminosities). Z. Ap. **46**, 108.

Böhm-Vitense, E. (1981): The effective temperature scale. Ann. Rev. Astron. Astrophys. **19**, 295.

Böhm-Vitense, E.: Introduction to stellar astrophysics. Volume 1. Basic stellar observations and data. Volume 2. Stellar Atmospheres. Volume 3. Stellar Structure and Evolution. New York: Cambridge University Press, 1989, 1989, 1992.

Bohr, N. (1913): On the constitution of atoms and molecules. Phil. Mag. **26**, 1, 10, 476, 857.

Bohr, N. (1915): On the decrease of velocity of swiftly moving electrified particles in passing through matter. Phil. Mag. **30**, 581.

Bolton, J. G., Wild, J. P. (1957): On the possibility of measuring interstellar magnetic fields by 21-cm Zeeman splitting. Ap. J. **125**, 296.

Boltzmann, L. (1868): Studien über das Gleichgewicht der lebendigen Kraft zwischen bewegten materiellen Punkten (Studies of the equilibrium and the life force between material points). Wien. Ber. **58**, 517.

Boltzmann, L. (1872): Weitere Studien über das Wärmegleichgewicht unter Gasmolekülen (Further studies on the thermal equilibrium of gas molecules). Sitz. Acad. Wiss. **66**, 275. Engl. trans. Brush (1966).

Boltzmann, L. (1877): Über die Beziehung eines allgemeinen mechanischen Satzes zum zweiten Hauptsatz der Wärmetheorie (On the relation of a thermal mechanical theorem to the second law of thermodynamics). Sitz. Acad. Wiss. **75**, 67. Engl. trans. Brush (1966).

Boltzmann, L. (1884): Über eine von Hrn. Bartoli entdeckte Beziehung der Wärmestrahlung zum zweiten Hauptsatze (On a relation between thermal radiation and the second law of thermodynamics discovered by Bartoli). Ann. Phys. **22**, 31.

Boltzmann, L.: Vorlesungen über Gastheorie (Lectures in gas theory). 1896. Eng. trans. by Brush. Berkeley: Univ. of Calif. Press 1967.

Bond, G. P. (1863): Light of the moon and of Jupiter. Mem. Amer. Ac. **8**, 221.

Bond, J. W., Watson, K. M., Welch, J. A.: Atomic theory of gas dynamics. Reading, Mass.: Addison-Wesley 1965.

Bondi, H. (1952): On spherically symmetrical accretion. M.N.R.A.S. **112**, 195.

Bondi, H., Hoyle, F. (1944): On the mechanism of accretion by stars. M.N.R.A.S. **104**, 273.

Bonner, W. B. (1956): Boyle's law and gravitational instability. M.N.R.A.S. **116**, 351.

Booker, H. G. (1956): A theory of scattering by nonisotropic irregularities with application to radar reflections from the aurora. J. Atmos. Terr. Phys. **8**, 204.

Booker, H. G., Gordon, W. E. (1950): A theory of radio scattering in the troposphere. Proc. I. R. E. **38**, 401.

Booker, H. G., Ratcliffe, J. A., Shinn, D. H. (1950): Diffraction from an irregular screen with applications to ionospheric problems. Phil. Trans. Roy. Soc. London **A242**, 75.

Borderies, N., Goldreich, P., Tremaine, S. (1982): Sharp edges of planetary rings. Nature **299**, 209.

Born, M., Green, H. S.: A general kinetic theory of liquids. Cambridge: Cambridge University Press 1949.

Born, M., Oppenheimer, R. (1927): Zur Quantentheorie der Molekeln (On the quantum theory of molecules). Ann. Phys. **84**, 457.

Born, M., Wolf, E.: Principles of optics. New York: Pergamon Press 1964, 1970.

Borra, E. F., Landstreet, J. D., Mestel, L. (1982): Magnetic stars. Ann. Rev. Astron. Astrophys. **20**, 191.

Bose, S. N. (1924): Planck's Gesetz und Lichtquantenhypothese (Planck's law and the light quantum hypothesis). Z. Physik **26**, 178.

Boss, A.: Looking for Earths: The race to find new solar systems. New York: John Wiley and Sons, 1998.

Bothe, W., Kolhörster, W. (1929): Das Wesen der Höhenstrahlung . Zeits. f. Phys. **56**, 751. English translation in: Cosmic Rays (A. M. Hillas). Oxford: Pergamon Press 1972.

Boussinesq, J.: Théorie analytique de la chaleur (Analytic theory of heat). **2**, 172. Paris: Gauthier-Villars 1903.

Bowen, I. S. (1928): The origin of the nebular lines and the structure of the planetary nebulae. Ap. J. **67**, 1. Reproduced in: Source book in astronomy and astrophysics 1900–1975 (eds. K. R. Lang and O. Gingerich). Cambridge, Mass.: Harvard University Press 1979.

Bowen, I. S. (1935): The spectrum and composition of the gaseous nebulae. Ap. J. **81**, 1.

Bowen, I. S. (1960): Wave lengths of forbidden nebular lines II. Ap. J. **132**, 1.

Bowers, R. L., Deeming, T.: Astrophysics I, Stars. Boston: Jones and Bartlett Publishers 1984.

Bowers, R. L., Deeming, T.: Astrophysics II. Interstellar matter and galaxies. Boston: Jones and Bartlett 1984.

Bowers, R. L., et al. (1975): A realistic lower bound for the maximum mass of neutron stars. Ap. J. **196**, 639.

Bowyer, S. (1991): The cosmic far ultraviolet background. Ann. Rev. Astron. Astrophys. **29**, 59.

Boyarchuk, A. A., Gershberg, R. E., Godovnikov, N. V. (1968): Formulae, graphs, and tables for a quantitative analysis of the hydrogen radiation of emission objects. Bull. Crimean. Ap. Obs. **38**, 208.

Boyarchuk, A. A., Gershberg, R. E., Godovnikov, N. V., Pronik, V. I. (1969): Formulae and graphs for a quantitative analysis of the forbidden line radiation of emission objects. Bull. Crimean Ap. Obs. **39**, 147.

Boyarchuk, A. A., Kopyloc, I. M. (1964): A general catalogue of rotational velocities of 2558 stars. Bull. Crimean, Ap. Obs. **31**, 44.

Boyd, T. J. M., Sanderson, J. J.: Plasma dynamics. New York: Barnes and Noble 1969.

Boyle, R. (1660): New experiments physico-mechanical, touching the spring of air and its effects, made for the most part in a new pneumatical engine. Oxford, 1660 – reprod. Brush.

Boyle, R. (1662): A defense of the doctrine touching the spring and weight of the air,....Oxford 1662. Repr. in Boyle's Works (ed. T. Birch). London 1772.

Bracewell, R. N., Preston, G. W. (1956): Radio reflection and refraction phenomena in the high solar corona. Ap. J. **123**, 14.

Brackett, F. S. (1922): Visible and infra-red radiation of hydrogen. Ap. J. **56**, 154.

Bradt, H. L., Peters, B. (1948): Investigation of the primary cosmic radiation with nuclear photographic emulsion. Phys. Rev. **74**, 1828.

Bradt, H. L., Peters, B. (1950): The heavy nuclei of the primary cosmic radiation. Phys. Rev. **77**, 54.

Bradt, H. V. D., Mc Clintock, J. E. (1983): The optical counterparts of compact galactic X-ray sources. Ann. Rev. Astron. Astrophys. **21**, 13.

Brecher, K., Chanmugam, G. (1978): Why do collapsed stars rotate so slowly? Ap. J. **221**, 969.

Brecher, K., Morrison, P. (1969): Leakage electrons from normal galaxies: The diffuse cosmic X-ray source. Phys. Rev. Lett. **23**, 802.

Breger, M., Bregman, J. N. (1975): Period-luminosity-color relations and pulsation modes of pulsating variable stars. Ap. J. **200,** 343.

Bregman, J. N. (1985): Diffusive shock acceleration and quasar photospheres. Ap. J. **288**, 32.

Brewster, D. (1815): On the laws which regulate the polarization of light from transparent bodies. Phil. Trans. **15**, 125.

Briet, G., Teller, E. (1940): Metastability of hydrogen and helium levels. Ap. J. **91**, 215.

Briet, G., Wigner, E. (1936): Capture of slow neutrons. Phys. Rev. **49**, 519.

Briggs, B. H., Phillips, G. J., Shinn, D. H. (1950): The analysis of observations on spaced receivers of the fading of radio signals. Proc. Roy. Soc. London **B63**, 106.

Brilloin, L. (1926): Les principes de la nouvelle mécanique ondulatoire (Principles of the undulatory mechanics). J. Phys. et le Radium **7**, 321. Remarques sur la mecanique ondulatoire (Notes on undulatory mechanics). J. Phys. Radium **7**, 353.

Brocklehurst, M. (1970): Level populations of hydrogen in gaseous nebulae, M.N.R.A.S. **148**, 417.

Brocklehurst, M., Leeman, S. (1971): The pressure broadening of radio recombination lines. Astrophys. Lett. **9**, 35.

Brocklehurst, M., Seaton, M. J. (1971): The profiles of radio recombination lines. Astrophys. Lett. **9**, 139.

Brocklehurst, M., Seaton, M. J. (1972): On the interpretation of radio recombination line observations. M.N.R.A.S. **157**, 179.

Brown, J. C. (1971): The deduction of energy spectra of non-thermal electrons in flares from the observed dynamic spectra of hard X-ray bursts. Solar Phys. **18**, 489.

Brown, J. C. (1972): The decay characteristics of models of solar hard X-ray bursts. Solar Phys. **25**, 158.

Brown, J. C. (1972): The directivity and polarization of thick target X-ray bremsstrahlung from solar flares. Solar Phys. **26**, 441.

Brown, J. C. (1973): Thick target X-ray bremsstrahlung from partially ionized targets in solar flares. Solar Phys. **28**, 151.

Brown, J. C. (1991): Energetic particles in solar flares: theory and diagnostics. Phil. Trans. Roy. Soc. (London) **A336**, 413.

Brown, J. C., Emslie, A. G. (1989): Self-similar Lagrangian hydrodynamics of beam-heated solar flare atmospheres. Ap. J. **339**, 1123.

Brown, J. C., Karlicky, M., Mac Kinnon, A. L., Van Den Oord, G. H. J. (1990): Beam heating in solar flares: electrons or protons? Ap. J. Supp. **73**, 343.

Brown, J. C., Smith, D. F. (1980): Solar flares. Rep. Prog. Phys. **43**, 125.

Brown, R. L. (1971): On the photoionization of hydrogen and helium. Ap. J. **164**, 387.

Brown, R. L.: Radio observations of H II regions. In: Spectroscopy of astrophysical plasmas (eds. A. Dalgarno and D. Layzer). New York: Cambridge University Press 1987.

Brown, R. L., Gomez-Gonzalez, J. (1975): The transfer of radio recombination line radiation through a cold gas. I. Hydrogen and helium lines in compact H II regions. Ap. J. **200**, 598.

Brown, R. L., Gould, R. J. (1970): Interstellar absorption of cosmic X-rays. Phys. Rev. D 1, **1**, 2252.

Brown, R. L., Liszt, H. S. (1984): Sagittarius A and its environment. Ann. Rev. Astron. Astrophys. **22**, 223.

Brown, R. L., Mathews, W. G. (1970): Theoretical continuous spectra of gaseous nebulae. Ap. J. **160**, 939.

Brown, T. M., et al. (1989): Inferring the Sun's internal angular velocity from observed p-mode frequency splittings. Ap. J. **343**, 526.

Brown, T. M., Morrow, C. A. (1987): Depth and latitude dependence of solar rotation. Ap. J. (Letters) **314**, L21.

Browning, P. K., Priest, E. R. (1986): Heating of coronal arcades by magnetic tearing turbulence, using the Taylor-Heyvaerts hypothesis. Astron. Astrophys. **159**, 129.

Browning, P. K., Sakurai, T., Priest, E. R. (1986): Coronal heating in closely-packed flux tubes: a Taylor-Heyvaerts relaxation theory. Astron. Astrophys. **158**, 217.

Brueckner, G. E., Bartoe, J.-D. F. (1983): Observations of high-energy jets in the corona above the quiet sun, the heating of the corona and the acceleration of the solar wind. Ap. J. **272**, 329.

Bruenn, S. W. (1972): The effect of URCA shells on the density of carbon ignition in degenerate stellar cores. Ap. J. **177**, 459.

Brush, S. G.: Kinetic theory, vol. I. New York: Pergamon Press 1965.

Brush, S. G.: Kinetic theory, vol. II. New York: Pergamon Press 1966.

Brussard, P. J., Hulst, H. C. Van De (1962): Approximation formulas for non-relativistic bremsstrahlung and average Gaunt factors for a Maxwellian electron gas. Rev. Mod. Phys. **34** (3), 507.

Bryan, G. H. (1889): The waves on a rotating liquid spheroid of finite ellipticity. Phil Trans. Roy. Soc. (London) **A180**, 187.

Buchler, J. R. (1978): On the vibrational stability of stars in thermal imbalance. Ap. J. **220**, 629.

Buchler, J. R., Barkat, Z. (1971): Properties of low density neutron star matter. Phys. Rev. Lett. **27**, 48.

Buchler, J. R., Coon, S. A. (1977): The interacting neutron gas at high density and temperature. Ap. J. **212**, 807.

Buchler, J. R., Ingber, L. (1971): Properties of the neutron gas and applications to neutron stars. Nucl. Phys. **A170**, 1.

Buffington, A., Orth, C. D., Smoot, G. F. (1974): Measurement of the positron-electron ratio in the primary cosmic rays from 5 to 50 GeV. Phys. Rev. Lett. **33**, 34.

Buneman, O. (1958): Instability, turbulence, and conductivity in current-carrying plasma. Phys. Rev. Lett. **1**, 8.

Buneman, O. (1959): Dissipation of currents in ionized media. Phys. Rev. **115**, 503.

Buneman, O. (1962): Scattering of radiation by the fluctuations in a nonequilibrium plasma. J. Geophys. Res. **67** (5), 2050.

Burbidge, E. M., Burbidge, G. R. (1972): Optical observations of southern radio sources. Ap. J. **172**, 37.

Burbidge, E. M., et al. (1957): Synthesis of the elements in stars. Rev. Mod. Phys. **29**, 547. Reproduced in: Source book in astronomy and astrophysics 1900–1975 (eds. K. R. Lang and O. Gingerich). Cambridge, Mass.: Harvard University Press 1979.

Burbidge, E. M., Strittmatter, P. A. (1972): Redshifts of twenty radio galaxies. Ap. J. (Letters) **172**, L37.

Burbidge, G. R. (1957): Acceleration of cosmic-ray particles among extragalactic nebulae. Phys. Rev. **107**, 269. Reprod. in: Selected papers on cosmic ray origin theories (ed. S. Rosen). New York: Dover 1969.

Burbidge, G. R. (1959): Estimates of the total energy in particles and magnetic field in the non-thermal radio sources. Ap.J. **129**, 849.

Burbidge, G. R.: X-ray and γ-ray sources. In: High energy astrophysics – Int. school of physics Enrico Fermi course 35 (ed. L. Gratton). New York: Academic Press 1966.

Burbidge, G. R. (1970): Nuclei of galaxies. Ann. Rev. Astron. Astrophys. **8**, 369.

Burbidge, G. R., Burbidge, E. M.: Quasi-stellar objects. San Francisco, Calif.: W. H. Freeman 1967.

Burbidge, G. R., Burbidge, E. M. (1969): Red-shifts of quasi-stellar objects and related extragalactic systems. Nature **222**, 735.

Burbidge, G. R., Crowne, A. H. (1979): An optical catalogue of radio galaxies. Ap. J. Supp. **40**, 583.

Burbidge, G. R., Hoyle, F. (1964): On cosmic rays as an extra-galactic phenomenon. Proc. Phys. Soc. **84**, 141.

Burbidge, G. R., Jones, T. W., O'Dell, S. L. (1974): Physics of compact nonthermal sources. III. energetic considerations. Ap. J. **193**, 43.

Burger, H. C. Von, Dorgelo, H. B. (1924): Beziehung zwischen inneren Quantenzahlen und Intensitäten von Mehrfachlinien (Relations between inner quantum numbers and intensities of multiple lines). Z. Physik **23**, 258.

Burgess, A. (1958): The hydrogen recombination spectrum. M.N.R.A.S. **118**, 477.

Burgess, A. (1964): Dielectronic recombination and the temperature of the solar corona. Ap. J. **139**, 776.

Burgess, A. (1965): A general formula for the estimation of dielectronic recombination coefficients in low density plasmas. Ap. J. **141**, 1588.

Burgess, A., Seaton, M. J. (1964): The ionization equilibrium for iron in the solar corona. M.N.R.A.S. **127**, 355.

Burke, B. F., Franklin, K. L. (1955): Observations of a variable radio source associated with the planet jupiter. J. Geophys. Res. **60**, 213. Reproduced in: A source book in astronomy and astrophysics 1900–1975 (eds. K. R. Lang and O. Gingerich). Cambridge, Mass.: Harvard University Press 1979.

Burrows, A. (1988): A statistical derivation of an upper limit to the electron neutrino mass from the SN 1987A neutrino data. Ap. J. (Letters) **328**, L51.

Burrows, A., Lattimer, J. M. (1986): The birth of neutron stars. Ap. J. **307**, 178.

Burrows, A., Liebert, J. (1993): The science of brown dwarfs. Rev. Mod. Phys. **65**, 301.

Burrows, C. J., et al. (1996): Hubble Space Telescope observations of the disk and jet of HH 30. Ap. J. **473**, 437.

Burton, W. B.: The structure of our Galaxy derived from observations of neutral hydrogen. In: Galactic and extragalactic radio astronomy, second edition (eds. G. L. Verschuur and K. I. Kellermann). New York: Springer-Verlag 1988.

Burton, W. B.: Distribution and observational properties of the ISM. In: The galactic interstellar medium. Saas-Fee advanced course 21 (eds. D. Pfenniger and P. Bartholdi). New York: Springer Verlag 1992.

Butcher, H. R. (1975): Studies of heavy-element synthesis in the galaxy. II. A survey of e-, r-, and s-process abundances. Ap. J. **199**, 710.

Butcher, H. R. (1976): On s-process abundance evolution in the galactic disk. Ap. J. **210**, 489.

Callaway, J. (1985): Scattering of electrons by atomic hydrogen at intermediate energies: Elastic scattering and $n = 2$ excitation from 12 to 54 eV. Phys. Rev. **A32**, 775.

Callaway, J., Unnikrishnan, K., Oza, D. H. (1987): Optical-potential study of electron-hydrogen scattering at intermediate energies. Phys. Rev. **A36**, 2576.

Cameron, A. G. W. (1954): Origin of anomalous abundances of the elements in giant stars. Phys. Rev. **93**, 932.

Cameron, A. G. W. (1955): Origin of anomalous abundances of the elements in giant stars. Ap. J. **121**, 144.

Cameron, A. G. W. (1957): Nuclear reactions in stars and nucleogenesis. P.A.S.P. **69**, 201.

Cameron, A. G. W.: Stellar evolution, nuclear astrophysics and nucleogenesis. Atomic Energy Chalk River Report No. CRL-454. Ontario, Canada: Atomic Energy of Canada Chalk River Project 1957.

Cameron, A. G. W. (1959): Photobeta reactions in stellar interiors. Ap. J. **130**, 452.

Cameron, A. G. W. (1959): Pycnonuclear reactions and nova explosions. Ap. J. **130**, 916.

Cameron, A. G. W. (1960): New neutron sources of possible astrophysical interest. Astron. J. **65**, 485.

Cameron, A. G. W. (1973): Abundances of the elements in the solar system. Space Sci. Rev. **15**, 121.

Cameron, A. G. W. (1979): The neutron-rich silicon-burning and equilibrium processes of nucleosynthesis. Ap. J. (Letters) **230**, L53.

Cameron, A. G. W.: Element and nuclidic abundances in the solar system. In: Essays in nuclear astrophysics (eds. C. A. Barnes, D. D. Clayton and D. N. Schramm). New York: Cambridge University Press 1982.

Cameron, A. G. W., Fowler, W. A. (1971): Lithium and the s-process in red giant stars. Ap. J. **164**, 111.

Cameron, R. C.: The magnetic and related stars. Baltimore, Maryland: Mono Book 1967.

Canal, R., Isern, J., Labay, J. (1990): The origin of neutron stars in binary systems. Ann. Rev. Astron. Astrophys. **28**, 183.

Canfield, R. C. (1974): A simplified method for calculation of radiative energy loss due to spectral lines. Ap. J. **194**, 483.

Canuto, V. (1975): Equation of state ultrahigh densities. Ann. Rev. Astron. Ap. **13**, 335.

Canuto, V., Chiu, H. Y. (1970): Nonrelativistic electron bremsstrahlung in a strongly magnetized plasma. Phys. Rev. **A2**, 518.

Canuto, V., Chiu, H. Y. (1971): Intense magnetic fields in astrophysics. Space Sci. Rev. **12**, 3.

Canuto, V., Chiu, H. Y., Chou, C. K. (1970): Neutrino bremsstrahlung in an intense magnetic field. Phys. Rev. **D2**, 281.

Canuto, V., Chiu, H. Y., Fassio-Canuto, L. (1969): Electron bremsstrahlung in intense magnetic fields. Phys. Rev. **185**, 1607.

Canuto, V., Chiuderi, C., Chou, C. K. (1970): Plasmon neutrino emission in a strong magnetic field: I. Transverse plasmons; II. Longitudinal plasmons. Astrophys. and Space Sci. **7**, 407, **9**, 453.

Canuto, V., Chou, C. K. (1971): Neutrino luminosity by the ordinary URCA process in an intense magnetic field. Astrophys. and Space Sci. **10**, 246.

Canuto, V., Datta, B., Kalman, G. (1978): Superdense neutron matter. Ap. J. **221**, 274.

Canuto, V., et al. (1970): Neutrino bremsstrahlung in a strong magnetic field. Phys. Rev. **D2**, 281.

Canuto, V., Lodenquai, J., Ruderman M. (1971): Thomson scattering in a strong magnetic field. Phys. Rev. **D3,** 2303.

Capriotti, E. R. (1964): The hydrogen radiation spectrum in gaseous nebulae. Ap. J. **139**, 225.

Capriotti, E. R., Daub, C. T. (1960): Hβ and [0 III] fluxes from planetary nebulae. Ap. J. **132**, 677.

Carathéodory, C. (1909): Untersuchungen über die Grundlagen der Thermodynamik (Investigation of the foundations of thermodynamics). Math. Ann. **67**, 355.

Cargill, P. J., Priest, E. R. (1983): The heating of post-flare loops. Ap. J. **266**, 383.

Carmichael, H.: A process for flares. In: AAS-NASA symposium on the physics of solar flares (ed. W. N. Hess). Washington, D. C.: National Aeronautics and Space Administration Special Publication NASA SP-50 1964.

Carnot, Par S.: Reflexions sur la puissance motrice du feu et sur les machines (Reflections on the motivating force of fire and the machines). Bachelier, Paris 1824. Eng. trans. in: Reflections on the motive power of fire,...(ed. E. Mendoza). New York: Dover 1960.

Carrington, R. C. (1860): Description of a singular appearance seen in the sun on september 1, 1859. M.N.R.A.S. **20**, 13. Reproduced in: Early solar physics (ed. A. J. Meadows). Oxford: Pergamon Press 1970.

Carroll, T. D., Salpeter, E. E. (1966): On the abundance of interstellar OH. Ap. J. **143**, 609.

Carruthers, G. R. (1970): Rocket observation of interstellar molecular hydrogen. Ap. J. (Letters) **161**, L81.

Carson, T. R. (1976): Stellar opacity. Ann. Rev. Astron. Ap. **14**, 95.

Cartan, H. (1924): Sur la stabilite ordinare des ellipsoides de Jacobi (On the ordinary stability of Jacobian ellipsoids). Proc. Int. Math. Cong. Toronto **2**, 2 (cf. Lyttleton 1953).

Cassé, M., Lehoucq, R., Vangioni-Flam, E. (1995): Production and evolution of light elements in active star-forming regions. Nature **373**, 318.

Cassinelli, J. P. (1979): Stellar winds. Ann. Rev. Astron. Ap. **17**, 275.

Catalán, M. A. (1922): Series and other regularities in the spectrum of manganese. Phil. Trans. Roy. Soc. London **A223**, 127.

Caughlan, G. R. (1965): Approach to equilibrium in the CNO bi-cycle. Ap. J. **141**, 688.

Caughlan, G. R., et al. (1985): Tables of thermonuclear reactions rates for low-mass nuclei ($1 \leq Z \leq 14$). Atomic Data and Nuclear Data Tables. **32**, 197.

Caughlan, G. R., Fowler, W. A. (1962): The mean lifetimes of carbon, nitrogen, and oxygen nuclei in the CNO bi-cycle. Ap. J. **136**, 453.

Caughlan, G. R., Fowler, W. A. (1988): Thermonuclear reaction rates V. Atomic Data and Nuclear Data Tables **40**, 283.

Čerenkov, P. A. (1937): Visible radiation produced by electrons moving in a medium with velocities exceeding that of light. Phys. Rev. **52**, 378. (Cf. J. V. Jelley: Čerenkov Radiation and its Applications. New York: Pergamon Press 1958.)

Cesarsky, C. J. (1980): Cosmic-ray confinement in the Galaxy. Ann. Rev. Astron. Astrophys. **18**, 289.

Cesarsky, D. A., Moffet, A. T., Pasachoff, J. M. (1973): 327 MHz observations of the galactic center: Possible detection of a deuterium absorption line. Ap. J. (Letters) **180**, L1.
Chadwick, J. (1932): The existence of a neutron. Proc. Roy. Soc. London **A136**, 692.
Chakrabarti, S. K.: Theory of transonic astrophysical flows. London: World Scientific 1990.
Chamberlain, O., Segre, E., Wiegand, C., Ypsilantis, T. (1955): Observation of antiprotons. Phys. Rev. **100**, 947.
Chanan, G. A., Middleditch, J., Nelson, J. E. (1976): The geometry of the eclipse of a pointlike star by a Roche-lobefilling companion. Ap. J. **208**, 512.
Chandrasekhar, S. (1931): The maximum mass of ideal white dwarfs. Ap. J. **74**, 81.
Chandrasekhar, S. (1934): Stellar configurations with degenerate cores. Observatory **57**, 373.
Chandrasekhar, S. (1935): The highly collapsed configurations of a stellar mass (second paper). M.N.R.A.S. **95**, 207.
Chandrasekhar, S. (1935): Stellar configurations with degenerate cores. M.N.R.A.S. **95**, 226.
Chandrasekhar, S.: An introduction to the study of stellar structure. Chicago, Ill.: University of Chicago Press 1939, and New York: Dover.
Chandrasekhar, S. (1943): Dynamical friction: I. General considerations: The coefficient of dynamical friction. Ap. J. **97**, 255.
Chandrasekhar, S. (1943): Stochastic problems in physics and astronomy. Rev. Mod. Phys. **15**, 1. Reprod. in: Selected papers on noise and stochastic processes (ed. N. Wax). New York: Dover 1954.
Chandrasekhar, S. (1949): On Heisenberg's elementary theory of turbulence. Proc. Roy. Soc. (London) **A200**, 20.
Chandrasekhar, S. (1949): Turbulence – A physical theory of astrophysical interest. Ap. J. **110**, 329.
Chandrasekhar, S. (1951): The fluctuations in density in isotropic turbulence. Proc. Roy. Soc. (London) **A210**, 18.
Chandrasekhar, S. (1951): The gravitational instability of an infinite homogeneous turbulent medium. Proc. Roy. Soc. (London) **A210**, 26.
Chandrasekhar, S. (1952): A statistical basis for the theory of stellar scintillation. M.N.R.A.S. **112**, 475.
Chandrasekhar, S. (1952): On the inhibition of convection by a magnetic field. Phil. Mag. **43**, 501.
Chandrasekhar, S. (1953): The instability of a layer of fluid heated from below and subject to coriolis force. Proc. Roy. Soc. (London) **A217,** 306.
Chandrasekhar, S. (1953): The virial theorem in hydromagnetics. M.N.R.A.S. **113**, 667.
Chandrasekhar, S. (1954): The gravitational instability of an infinite homogeneous medium when coriolis force is acting and when a magnetic field is present. Ap. J. **119**, 7.
Chandrasekhar, S. (1955): Hydrodynamic turbulance: II. An elementary theory. Proc. Roy. Soc. (London) **A233**, 330.
Chandrasekhar, S. (1955): The character of the equilibrium of an incompressible heavy viscous fluid of variable density. Proc. Camb. Phil. Soc. **51**, 162.
Chandrasekhar, S. (1959): The oscillations of a viscous liquid globe. Proc. London Math. Soc. **9**, 141.
Chandrasekhar, S.: Radiative transfer. New York: Dover 1960.
Chandrasekhar, S.: Hydrodynamics and hydromagnetic stability. Oxford: Oxford at the Clarendon Press 1961.
Chandrasekhar, S. (1964): A general variational principle governing the radial and non-radial oscillations of gaseous masses. Ap. J. **139**, 664.
Chandrasekhar, S. (1964): Dynamical instability of gaseous masses approaching the Schwarzschild limit in general relativity. Phys. Rev. Letters **12**, 114.
Chandrasekhar, S. (1964): The dynamical instability of gaseous masses approaching the Schwarzschild limit in general relativity. Ap. J. **140**, 417.
Chandrasekhar, S.: Ellipsoidal figures of equilibrium. New Haven, Conn.: Yale University Press 1969.

Chandrasekhar, S. (1974): On a criterion for the onset of dynamical instability by a nonaxisymmetric mode of oscillation along a sequence of differentially rotating configurations. Ap. J. **187**, 169.

Chandrasekhar, S. (1984): On stars, their evolution and their stability. Rev. Mod. Phys. **56**, 137.

Chandrasekhar, S., Breen, F. H. (1946): On the continuous absorption coefficient of the negative hydrogen ion. III. Ap. J. **104**, 430.

Chandrasekhar, S., Elbert, D. D. (1955): The instability of a layer of fluid heated from below and subject to coriolis force II. Proc. Roy. Soc. (London) **A231**, 198.

Chandrasekhar, S., Fermi, E. (1953): Problems of gravitational stability in the presence of a magnetic field. Ap. J. **118**, 116.

Chandrasekhar, S., Henrich, L. R. (1942): An attempt to interpret the relative abundances of the elements and their isotopes. Ap. J. **95**, 288.

Chandrasekhar, S., Kaufman, A. N., Watson, K. M. (1958): The stability of the pinch. Proc. Roy. Soc. (London) **A245**, 435.

Chandrasekhar, S., Lebovitz, N. R. (1962): On the oscillations and the stability of rotating gaseous masses. Ap. J. **135**, 248.

Chandrasekhar, S., Lebovitz, N. R. (1962): On the oscillations and the stability of rotating gaseous masses: II. The homogeneous compressible model. Ap. J. **136**, 1069.

Chandrasekhar, S., Lebovitz, N. R. (1963): Non-radial oscillations and the convective instability of gaseous masses. Ap. J. **138**, 185.

Chandrasekhar, S., Lebovitz, N. R. (1964): Non-radial oscillations of gaseous masses. Ap. J. **140**, 1517.

Chandrasekhar, S., Limber, D. N. (1954): On the pulsation of a star in which there is a prevalent magnetic field. Ap. J. **119**, 10.

Chang, M. (1977): Mean lives of some astrophysically important excited levels in carbon, nitrogen, oxygen. Ap. J. **211**, 300.

Chanmugam, G. (1992): Magnetic fields of degenerate stars. Ann. Rev. Astron. Astrophys. **30**, 143.

Chao, N. C., Clark, J. W., Yang, C. H. (1972): Proton superfluidity in neutron star matter. Nucl. Phys. **179**, 320.

Chapman, S. (1916): On the law of distribution of molecular velocities and the theory of viscosity and thermal conductivity in a non-uniform simple monatomic gas. Phil. Trans. Roy. Soc. London **A216,** 279.

Chapman, S. (1917): On the kinetic theory of a gas: Part. II. A composite monatomic gas, diffusion, viscosity, and thermal conduction. Phil. Trans. Roy. Soc. London **A217**, 115.

Chapman, S. (1918): The energy of magnetic storms. M.N.R.A.S. **79**, 70.

Chapman, S. (1919): An outline of a theory of magnetic storms. Proc. Roy. Soc. London **A95**, 61.

Chapman, S. (1957): Noise in the solar corona and the terrestrial ionosphere. Smithsonian Contr. Astrophys. **2**, 1.

Chapman, S. (1959): Interplanetary space and the earth's outermost atmosphere. Proc. Roy. Soc. London **A253**, 462.

Chapman, S., Cowling, T. G.: The mathematical theory of non-uniform gases. Cambridge: Cambridge University Press 1953.

Chashei, V. I., Shishov, V. I. (1983): Turbulence in the solar atmosphere and the formation of the solar wind. Sov. Phys. Dokl. **28**, 702.

Chashei, I. V., Shishov, V. I. (1986): Origin of low-frequency Alfvén waves in the solar wind. Sov. Astron. **30**, 322.

Chashei, I. V., Shishov, V. I. (1987): Influence of the outflow of material on the thermal regime of the corona. Sov. Astron. **31**, 61.

Chen, K., Ruderman, M. (1993): Origin and radio pulse properties of millisecond pulsars. Ap. J. **408**, 179.

Cheung, A. C., Rank, D. M., Townes, C. H., Thornton, D. D., Welch, W. J. (1968): Detection of NH_3 molecules in the interstellar medium by their microwave emission. Phys. Rev. Letters **21**, 1701.

Chevalier, R. A. (1983): Blast waves with cosmic-ray pressure. Ap. J. **272**, 765.

Chiosi, C., Bertelli, G. Bressan, A. (1992): New developments in understanding the HR diagram. Ann. Rev. Astron. Astrophys. **30**, 235.

Chiosi, C., Maeder, A. (1986): The evolution of massive stars with mass loss. Ann. Rev. Astron. Astrophys. **24**, 329.

Chiu, H. Y. (1961): Neutrino emission processes, stellar evolution, and supernovae, part I, part II. Ann. Phys. (N.Y.) **15**, 1, **16**, 321.

Chiu, H. Y. (1961): Annihilation process of neutrino production in stars. Phys. Rev. **123**, 1040.

Chiu, H. Y.: Stellar physics. Waltham Mass.: Blaisdell 1968.

Chiu, H. Y., Canuto, V., Fassio-Canuto, L. (1969): Nature of radio and optical emissions from pulsars. Nature **221**, 529.

Chiu, H. Y., Fassio-Canuto, L. (1969): Quantized synchrotron radiation in intense magnetic fields. Phys. Rev. **185**, 1614.

Chiu, H. Y., Morrison, P. (1960): Neutrino emission from black-body radiation at high stellar temperatures. Phys. Rev. Lett. **5**, 573.

Chiu, H. Y., Salpeter, E. E. (1964): Surface X-ray emission from neutron stars. Phy. Rev. Lett. **12**, 413.

Chiu, H. Y., Stabler, R. C. (1961): Emission of photoneutrinos and pair annihilation neutrinos from stars. Phys. Rev. **122**, 1317.

Christensen, C. J., et al. (1967): The half-life of the free neutron. Phys. Lett. **26B**, 11.

Christensen-Dalsgaard, J., et al. (1985): Speed of sound in the solar interior. Nature **315**, 378.

Christensen-Dalsgaard, J., Gough, D. O., Thompson, M. J. (1991): The depth of the solar convection zone. Ap. J. **378**, 413.

Christensen-Dalsgaard, J., Gough, D., Toomre, J. (1985): Seismology of the Sun. Science **229**, 923.

Christiansen, W. N.: The first decade of solar radio astronomy in Australia. In: The early years of radio astronomy (ed. W. T. Sullivan, III). New York: Cambridge University Press 1984.

Christy, R. F. (1966): A study of pulsation in RR Lyrae models. Ap. J. **144**, 108.

Christy, R. F. (1966): Pulsation theory. Ann. Rev. Astron. Astrophys. **4**, 353.

Chupp, E. L. (1971): Gamma ray and neutron emissions from the Sun. Space Sci. Rev. **12**, 486.

Chupp, E. L.: Gamma-ray astronomy: Dordrecht, Holland: D. Reidel 1976.

Chupp, E. L. (1984): High energy neutral radiations from the Sun. Ann. Rev. Astron. Astrophys. **22**, 359.

Chupp, E. L. (1990): Emission characteristics of three intense solar flares observed in cycle 21. Ap. J. Supp. **73**, 213.

Chupp, E. L., Benz, A. O. (eds) (1994): Particle acceleration phenomena in astrophysical plasmas. Proceedings of the International Astronomical Union (IAU) colloquium 142. Ap. J. Supp. **90**, 511–983

Chupp, E. L., et al. (1982): A direct observation of solar neutrons following the 0118 UT flare on 1980 June 21. Ap. J. (Letters) **263**, L95.

Chupp, E. L., et al. (1987): Solar neutron emissivity during the large flare on 1982 June 3. Ap. J. **318**, 913.

Chupp, E. L., Forrest, D. J., Higbie, P. R., Suri, A. N., Tsai, C., Dunphy, P. P. (1973): Solar gamma ray lines observed during the solar activity of August 2 to August 11, 1972. Nature (GB) **241**, 335.

Churchwell, E., Mezger, P. G. (1970): On the determination of helium abundance from radio recombination lines. Astrophys. Lett. **5**, 227.

Clark, B. G. (1965): An interferometer investigation of the 21-centimeter hydrogen-line absorption. Ap. J. **142**, 1398.

Clark, B. G., Kuz'min, A. D. (1965): The measurement of the polarization and brightness distribution of Venus at 10.6-cm wavelength. Ap. J. **142**, 23.

Clark, J. W., Heintzmann, H., Hillebrandt, W., Grewing, M. (1971): Nuclear forces, compressibility of neutron matter, and the maximum mass of neutron stars. Astrophys. Lett. **10**, 21.

Clarke, R. W., et al. (1969): Long baseline interferometer observations at 408 and 448 MHz II – The interpretation of the observations. M.N.R.A.S. **146**, 381.

Clausius, R. Von (1850): Über die bewegende Kraft der Wärme und die Gesetze, die sich daraus für die Wärmelehre selbst ableiten lassen (On the moving force of heat and the laws of thermodynamics that can be deduced from it). Ann. Phys. **79**, 368. Engl. trans. in Phil. Mag. **2**, 1 (1851) and in: Reflections on the motive power of fire,...(ed. E. Mendoza). New York: Dover 1960.

Clausius, R. Von (1857): Über die Art der Bewegung, welche wir Wärme nennen (The nature of the motion which we call heat). Ann. Phys. **100**, 353. Eng. trans. in Phil. Mag. **14**, 108 (1857) and Brush (1965).

Clausius, R. Von (1858): Über die mittlere Länge der Wege, welche bei Molekularbewegung gasförmigen Körper von den einzelnen Molekülen zurückgelegt werden, nebst einigen anderen Bemerkungen über die mechanischen Wärmetheorie (On the mean length of the paths described by the separate molecules of gaseous bodies). Ann. Phys. **105,** 239. Eng. trans. in Phil. Mag. **17**, 81 (1859) and Brush (1965).

Clausius, R. Von (1865): Über verschiedene für die Anwendung bequeme Formen der Hauptgleichungen der mechanischen Wärmetheorie (On different, convenient to use, forms of the main equations of mechanical heat theory). Ann. Phys. u. Chem. **125**, 353.

Clausius, R. Von (1870): Über einen auf die Wärme anwendbaren mechanischen Satz (On a mechanical theorem applicable to heat). Sitz. Nidd. Ges. 114. Engl. trans. in Phil. Mag. **40**, 122 (1870) and Brush (1965).

Claverie, A., et al. (1979): Solar structure from global studies of the five-minute oscillation. Nature **282**, 591.

Clay, J. (1927): Penetrating radiation. Proc. Acad. Sci. Amsterdam **30**, 1115.

Clayton, D. D. (1964): Cosmoradiogenic chronologies of nucleosynthesis. Ap. J. **139**, 637.

Clayton, D. D. (1969): Istopic composition of cosmic importance. Nature **224**, 56.

Clayton, D. D. (1971): New prospect for gamma-ray line astronomy. Nature **234**, 291.

Clayton, D. D.: Principles of stellar evolution and nucleosynthesis. New York: McGraw-Hill 1968. Chicago: University of Chicago Press, 1983.

Clayton, D. D., Colgate, S. A., Fishman, G. J. (1969): Gamma-ray lines from young supernova remnants. Ap. J. **155**, 75.

Clayton, D. D., Craddock, W. L. (1965): Radioactivity in supernova remnants. Ap. J. **142**, 189.

Clayton, D. D., Silk, J. (1969): Measuring the rate of nucleosynthesis with a gamma-ray detector. Ap. J. (Letters) **158**, L43.

Clayton, D. D., Woosley, S. E. (1974): Thermonuclear astrophysics. Rev. Mod. Phys. **46**, 755.

Clifford, F. E., Tayler, R. J. (1965): The equilibrium distribution of nuclides in matter at high temperatures. Mem. R.A.S. **69**, 21.

Cline, T. L., et al. (1973): Energy spectra of cosmic gamma-ray bursts. Ap. J. (Letters) **185**, L1.

Cocke, W. J. (1975): On the production of power-law spectra and the evolution of cosmic synchrotron sources: a model for the Crab Nebula. Ap. J. **202**, 773.

Cogan, B. C. (1977): The pulsation periods of stars with convection zones. Ap. J. **211**, 890.

Cohen, M. H. (1961): Radiation in a plasma. I. Čerenkov effect. Phys. Rev. **123** (3), 711.

Cohen, M. H., et al. (1967): Interplanetary scintillations, II. Observations. Ap. J. **147**, 449.

Cohen, M. H., Shaffer, D. B. (1971): Positions of radio sources from long-baseline interferometry. Astron. J. **76**, 91.

Coles, W. A., Rickett, B. J., Rumsey, V. H.: In Solar wind three (ed. C. T. Russell) Los Angeles: University of California 1974.

Colgate, S. A., Johnson, M. H. (1960): Hydrodynamic origin of cosmic rays. Phys. Rev. Lett. **5**, 235.

Colgate, S. A., White, R. H. (1966): The hydrodynamic behavior of supernovae explosions. Ap. J. **143**, 626.

Combes, F. (1991): Distribution of CO in the milky way. Ann. Rev. Astron. Astrophys. **29**, 195.

Compton, A. H. (1923): A quantum theory of the scattering of X-rays by light elements. Phys. Rev. **21**, 207, 483.

Compton, A. H. (1923): The spectrum of scattered X-rays. Phys. Rev. **22** (5), 409.

Compton, A. H. (1933): A geographic study of cosmic rays. Phys. Rev. **43**, 387.

Comstock, G. M., Fan, C. Y., Simpson, J. A. (1969): Energy spectra and abundances of the cosmic-ray nuclei helium to iron from the OGOI satellite experiment. Ap. J. **155**, 609.

Condon, E. U. (1926): A theory of intensity distribution in band systems. Phys. Rev. **28**, 1182.

Condon, E. U. (1928): Nuclear motions associated with electron transitions in diatomic molecules Phys. Rev. **32**, 858.

Condon, E. U., Shortley, G. H.: The theory of atomic spectra. Cambridge, England: Cambridge University Press 1963.

Condon, J. J., Backer, D. C. (1975): Interstellar scintillation of extragalactic radio sources. Ap. J. **197**, 31.

Cook, A. H. (1978): Physics of celestial masers. Q.J.R.A.S. **19**, 255.

Cook, C. W., et al. (1957): B^{12}, C^{12}, and the red giants. Phys. Rev. **107**, 508.

Cooper, J. (1966): Plasma spectroscopy. Rpt. Prog. Phys. **29**, 2.

Cord, M. S., Lojko, M. S., Petersen, J. D.: Microwave spectral tables – Spectral lines listing. Nat. Bur. Stands. (Wash.) Mon. 79, 1968.

Cordes, J. M., et al. (1991): The galactic distribution of free electrons. Nature **354**, 121.

Corliss, C. H., Bozman, W. R.: Experimental transition probabilities for spectral lines of seventy elements. Nat. Bur. Stands. (Wash.) Mon. 53, 1962.

Corliss, C. H., Warner, B. (1964): Absolute oscillator strengths for Fe I. Ap. J. Suppl. **8**, 395.

Costa, E., et al. (1997): Discovery of an X-ray afterglow associated with the γ-ray burst of 28 February 1997. Nature **387**, 783.

Couch, R. G., Arnett, W. D. (1974): On the thermal properties of the convective URCA process. Ap. J. **194**, 537.

Couch, R. G., Arnett, W. D. (1975): Carbon ignition and burning in degenerate stellar cores. Ap. J. **196**, 791.

Couch, R. G., Loumos, G. L. (1974): The URCA process in dense stellar interiors. Ap. J. **194**, 385.

Couch, R. G., Schmiedekamp, A. B., Arnett, W. D. (1974): s-Process nucleosynthesis in massive stars: core helium burning. Ap. J. **190**, 95.

Couch, R. G., Shane, K. C. (1971): The Photodisintegration rate of ^{24}Mg. Ap. J. **169**, 413.

Coulomb, C. A. (1785): Sur l'électricité et le magnétisme. Mem. l'Acad

Cowie, L. L., Mc Kee, C. F., Ostriker, J. P. (1981): Supernova remnant evolution in an inhomogeneous medium. I. Numerical models. Ap. J. **247**, 908.

Cowie, L. L., Songaila, A. (1986): High-resolution optical and ultraviolet absorption-line studies of interstellar gas. Ann. Rev. Astron. Astrophys. **24**, 499.

Cowley, C. R., Cowley, A. P. (1964): A new solar curve of growth. Ap. J. **140**, 713.

Cowling, T. G. (1941): The non-radial oscillations of polytropic stars. M.N.R.A.S. **101**, 367.

Cowling, T. G. (1945): On the Sun's general magnetic field. M.N.R.A.S. **105**, 166.

Cowling, T. G. (1945): The electrical conductivity of an ionized gas in a magnetic field, with applications to the solar atmosphere and the ionosphere. Proc. Roy. Soc. London **A183**, 453.

Cowling, T. G. (1946): The growth and decay of the sunspot magnetic field. M.N.R.A.S. **106**, 218.

Cowling, T. G.: Magnetohydrodynamics. First edition. New York: Interscience 1957. Second edition. Bristol, England: A. Hilger 1976.

Cowling, T. G. (1981): The present status of dynamo theory. Ann. Rev. Astron. Astrophys. **19**, 115.

Cowling, T. G., Newing, R. A. (1949): The oscillations of a rotating star. Ap. J. **109**, 149.

Cowsik, R., Kobetich, E. J. (1972): Comment on inverse Compton models for the isotropic X-ray background and possible thermal emission from a hot intergalactic gas. Ap. J. **177**, 585.

Cox, A. N., Livingston, W. C., Matthews, M. S. (eds.): Solar interior and atmosphere. Tucson: University of Arizona Press 1991.

Cox, A. N., Stewart, J. N. (1970): Rosseland opacity tables for population II compositions. Ap. J. Suppl. No. 174, **19**, 201.

Cox, A. N., Stewart, J. N., Eiler. D. D. (1965): Effects of bound-bound absorption on stellar opacities. Ap. J. Suppl. **11**, 1.

Cox, D. P., Mathews, W. G. (1969): Effects of self-absorption and internal dust on hydrogen-line intensities in gaseous nebulae. Ap. J. **155**, 859.

Cox, D. P., Reynolds, R. J. (1987): The local interstellar medium. Ann. Rev. Astron. Astrophys. **25**, 303.

Cox, J. P. (1963): On second helium ionization as a cause of pulsational instability in stars. Ap. J. **138**, 487.

Cox, J. P. (1974): Effects of thermal imbalance on the pulsational stability of stars undergoing thermal runaways. Ap. J. (Letters) **192**, L85.

Cox, J. P. (1974): Pulsating stars. Rep. Prog. Phys. **37**, 563.

Cox, J. P. (1976): Nonradial oscillations of stars: theories and observations. Ann. Rev. Astron. Ap. **14**, 247.

Cox, J. P.: Theory of stellar pulsation. Princeton, New Jersey: Princeton University Press 1980.

Cox, J. P., Davey, W. R., Aizenman, M. L. (1974): Pulsational stability of stars in thermal imbalance. III. Analysis in terms of absolute variations. Ap. J. **191**, 439.

Cox, J. P., Giuli, R. T.: Principles of stellar structure. Volume 2. Applications to stars. New York: Gordon and Breach 1968.

Cox, J. P., Whitney, C. (1958): Stellar pulsation: IV. A semitheoretical period-luminosity relation for classical Cepheids. Ap. J. **127**, 561.

Crandall, W. E., Millburn, G. P., Pyle, R. W., Birnbaum, W. (1956): C^{12} (x, xn) C^{11} and A^{27} (x, x2pn) Na^{24} cross sections at high energies. Phys. Rev. **101**, 329.

Cronin, J. W.: The highest energy cosmic rays. In: Unsolved problems in astrophysics (eds. J. N. Bahcall and J. P. Ostricker). Princeton, New Jersey: Princeton University Press 1997.

Cropper, M. (1990): The polars. Space Sci. Rev. **54**, 195.

Croswell, K.: Planet Quest: The epic discovery of alien solar systems. New York: The Free Press 1997.

Crutcher, R. M., et al. (1975): OH Zeeman observations of interstellar dust clouds. Ap. J. **198**, 91.

Culhane, J. L. (1969): Thermal continuum radiation from coronal plasmas at soft X-ray wavelengths. M.N.R.A.S. **144**, 375.

Culhane, J. L., Jordan, C. (eds.) (1991): The physics of solar flares. Phil. Trans. Roy. Soc. (London) **A336**, 321–495. Reprinted London: The Royal Society 1991.

Curie, I., Joliot, F. (1932): Émission de protons de grande vitesse par les substances hydrogénées sous l'influence des rayons γ trés pénétrants (Emission of very fast protons by hydrogen substances under the influence of very penetrating γ-rays). Comp. Rend. **194**, 273.

Curtis, H. D. (1918): Descriptions of 762 nebulae and clusters, The planetary nebulae. Lick Obs. Bull. **13**, 11, 57.

Czyzak, S. J., Krueber, T. K., Martins, P. De A. P., Saraph, H. E., Seaton, M. J., Shemming, J.: Collision strengths for excitation of forbidden lines. In: Planetary nebulae (ed. D. E.

Osterbrock, C. R. O'Dell). I.A.U. Symp. No. 34. Dordrecht, Holland: D. Reidel. Berlin-Heidelberg-New York: Springer 1968.
D'Antonia, F., Mazzitelli, I. (1990): Cooling of white dwarfs. Ann. Rev. Astron. Astrophys. **28**, 139.
Dalgarno, A., Layzer, D.: Spectroscopy of astrophysical plasmas. New York: Cambridge University Press 1987.
Dalgarno, A., Mc Cray, R. A. (1973): The formation of molecules from negative ions. Ap. J. **181**, 85.
Danby, G., et al. (1962): Observation of high-energy neutrino reactions and the existence of two kinds of neutrinos. Phys. Rev. Lett. **9**, 36.
Danby, J. M. A., Bray, T. A. (1967): Density of interstellar matter near a star. Astron. J. **72**, 219.
Danby, J. M. A., Camm, G. L. (1957): Statistical dynamics and accretion. M.N.R.A.S. **117**, 50.
Däppen, W.: Solar models and oscillation theory. In: Helioseismology – Proc. Fourth Soho Workshop. ESA SP-376 (1995) – June.
Darwin, Sir C. (1949): Source of the cosmic rays. Nature **164**, 1112. Reprod. in: Selected papers on cosmic ray origin theories (ed. S. ROSEN). New York: Dover 1966.
Davey, W. R. (1974): Pulsational stability of stars in thermal imbalance. V. Eigensolutions for quasi-adiabatic oscillations. Ap. J. **194**, 687.
Davey, W. R., Cox, J. P. (1974): Pulsational stability of stars in thermal imbalance. II. An energy approach. Ap. J. **189**, 113.
Davidson, K., Fesen, R. A. (1985): Recent developments concerning the crab nebula. Ann. Rev. Astron. Astrophys. **23**, 119.
Davidson, K., Netzer, H. (1979): The emission lines of quasars and similar objects. Rev. Mod. Phys. **51**, 715.
Davidson, K., Ostriker, J. P. (1973): Neutron star accretion in a stellar wind: Model for a pulsed X-ray source. Ap. J. **179**, 585.
Davies, P. C. W., Seaton, M. J. (1969): Radiation damping in the optical continuum. J. Phys. B. (Atom. Molec. Phys.) **2**, 757.
Davis, L. (1956): Modified Fermi mechanism for the acceleration of cosmic rays. Phys. Rev. **101**, 351.
Davis, L., Greenstein, J. L. (1951): The polarization of starlight by aligned dust grains. Ap. J. **114**, 506.
Davis, R., Harmer, D. S., Hoffman, K. C. (1968): Search for neutrinos from the Sun. Phys. Rev. Lett. **20**, 1205. Reproduced in: Source book in astronomy and astrophysics 1900–1975 (eds. K. R. Lang and O. Gingerich). Cambridge, Mass.: Harvard University Press 1979.
Davisson, C. M., Evans, R. D. (1952): Gamma-ray absorption coefficients. Rev. Mod. Phys. **24**, 79.
Davy, Sir H. (1821): Further researches on the magnetic phenomena produced by electricity. Phil. Trans. **111**, 425.
De Broglie, L. (1923): Waves and quanta. Nature **112**, 540.
De Broglie, L. (1925): Recherches sur la théorie des quanta. Ann. Physique **3**, 22.
De Jaeger, C.: Structure and dynamics of the solar atmosphere. In: Handbuch der Physik, vol. LII: Astrophysics IV: The solar system (ed. S. Flügge). Berlin-Heidelberg-New York: Springer 1959.
De Jager, C. (1986): Solar flares and particle acceleration. Space Science Reviews **44**, 43.
De Jager, C., Kundu, M. R. (1963): A note on bursts of radio emission and high energy (>20 keV) X-rays from solar flares. Space Res. **3**, 836.
De Jong, T. (1972): The density of H_2 molecules in dark interstellar clouds. Astron. Astrophys. **20**, 263.
De Vaucouleurs, G., De Vaucouleurs, A.: Reference catalogue of bright galaxies. Austin, Texas: Univ. of Texas Press 1964.
De Veny, J. B., Osborn, W. H., Janes, K. (1971): A catalogue of quasars. P.A.S.P. **83**, 611.

De Witt, H. E., Graboske, H. C., Cooper, M. S. (1973): Screening factors for nuclear reactions: I. General theory. Ap. J. **181**, 439.

De Young, D. S. (1976): Extended extragalactic radio sources. Ann. Rev. Astron. Ap. **14**, 447.

Debrunner, H., Flückiger, E., Chupp, E. L., Forrest, D. J. (1983): The solar cosmic ray neutron event on 1983 June 3. Proc. 18th Int. Cosmic Rays Conf. **4**, 75.

Debrunner, H., Flückiger, E. O., Lockwood, J. A. (1990): Signature of the solar cosmic-ray event on 1982 June 3. Ap. J. Supp. **73**, 259.

Debye, P. Von (1909): Der Lichtdruck auf Kugeln von beliebigem Material (The light pressure on spheres of arbitrary material). Ann. Phys. **30**, 57.

Debye, P. Von, Hückel, E. (1923): Zur Theorie der Elektrolyte: I. Gefrierpunktserniedrigung und verwandte Erscheinungen; II. Das Grenzgesetz für die elektrische Leitfähigkeit (On the theory of electrolytes: I. Lowering of the freezing point and related phenomena; II. The limiting laws for the electrical conductivity). Phys. Z. **24**, 185.

Dennis, B. R. (1985): Solar hard X-ray bursts. Solar Phys. **100**, 465.

Dennis, B. R. (1988): Solar flare hard X-ray observations. Solar Phys. **118**, 49.

Dennis, B. R., Zarro, D. M. (1993): The Neupert effect: what can it tell us about the impulsive and gradual phases of solar flares? Solar Phys. **146**, 177.

Dennison, D. M., Uhlenbeck, G. E. (1932): The two-minima problem and the ammonia molecule Phys. Rev. **41**, 313.

Dennison, P. A., Hewish, A. (1967): The solar wind outside the plane of the ecliptic. Nature **213**, 343.

Denskat, K. U., Neubauer, F. M., Schwenn, R.: Properties of "Alfvénic" fluctuations near the sun: Helios-1 and Helios-2. In: Solar wind four (ed. H. Rosenbauer). Katlenburg-Lindau: Max Planck Institut für Aeronomie 1981, Garching: Max Planck Institut für Extraterrestrishe Physik 1981.

Dent, W. A., Aller, H. D., Olsen, E. T. (1974): The evolution of the radio spectrum of Cassiopeia A. Ap. J. (Letters) **188**, L11.

Dere, K. P., et al. (1997): Chianti an atomic database for emission lines. Astron. Astrophys. Supp. **125**, 149.

Despain, K. H. (1977): Convective neutron and s-process element production in deeply mixed envelopes. Ap. J. **212**, 774.

Detweiler, S. L., Lindblom, L. (1977): On the evolution of the homogeneous ellipsoidal figures. Ap. J. **213**, 193.

Deubner, F.-L. (1975): Observations of low wavenumber nonradial eigenmodes of the Sun. Astron. Astrophys. **44**, 371.

Deubner, F.-L., Gough, D. (1984): Helioseismology: Oscillations as a diagnostic of the solar interior. Ann. Rev. Astron. Astrophys. **22**, 593.

Dickel, J. R., Degioanni, J. J., Goodman, G. C. (1970): The microwave spectrum of Jupiter. Radio Science **5** (2), 517.

Dickey, J. M., Lockman, F. J. (1990): H I in the Galaxy. Ann. Rev. Astron. Astrophys. **28**, 215.

Dickman, R. L., Snell, R. L., Young, J. S. (eds.): Molecular clouds in the milky way and external galaxies. Lecture notes in physics 315. New York: Springer-Verlag 1988.

Didelon, P. (1983): Catalog of magnetic field measurements. Astron. Astrophys. Supp. **53**, 119.

Dirac, P. A. M. (1925): The effect of Compton scattering by free electrons in a stellar atmosphere. M.N.R.A.S. **85**, 825.

Dirac, P. A. M. (1926): The elimination of the nodes in quantum mechanics, Sec. 10. The relative intensities of the lines of a multiplet. Proc. Roy. Soc. London **A111**, 302.

Dirac, P. A. M. (1926): On the theory of quantum mechanics. Proc. Roy. Soc. London **A112**, 661.

Dirac, P. A. M. (1930): On the annihilation of electrons and protons. Proc. Camb. Phil. Soc. **26**, 361.

Dirac, P. A. M. (1931): Quantised singularities in the electromagnetic field. Proc. Roy. Soc. (London) **A133**, 60.

Dismuke, N., et al. (1952): Tables for the analysis of beta spectra. Nat. Bur. Stands. (Wash) App. Math. Ser. **13**.

Ditchburn, R. W., Öpik, U.: Photoionization processes. In: Atomic and molecular processes (ed. D.R. Bates). New York: Academic Press 1962.

Dombrovski, V. A. (1954): On the nature of the radiation from the Crab nebula. Dokl. Acad. Nauk. S.S.S.R. **94**, 1021.

Donahue, T. M. (1951): The significance of the absence of primary electrons for theories of the origin of the cosmic radiation. Phys. Rev. **84** (5), 972.

Doppler, C. (1842): Über das. farbige Licht d. Doppelsterne usw. (On the colored light of double stars, etc.). Abhandlungen der. königlichen. Böhmischen Gesellschaff. d. Wissenschaften. **2**, 467.

Doschek, G. A., et al. (1993): The 1992 January 5 flare at 13.3 UT: Observations from Yohkoh. Ap. J. **416**, 845.

Doschek, G. A., Mariska, J. T., Sakao, T. (1996): Soft X-ray flare dynamics. Ap. J. **459**, 823.

Dougherty, J. P., Farley, D. T. (1960): A theory of incoherent scattering of radio waves by a plasma. Proc. Roy. Soc. London **A259**, 79.

Draine, B. T. (1980): Interstellar shock waves with magnetic precursors. Ap. J. **241**, 1021.

Draine, B. T., Salpeter, E. E. (1979): Destruction mechanisms for interstellar dust. Ap. J. **231**, 438.

Dravskikh, Z. V., Dravskikh, A. F. (1964): An attempt of observation of an excited hydrogen radio line. Astron. Tsirk **282**, 2.

Dreicer, H. (1960): Electron and ion runaway in a fully ionized gas II. Phys. Rev. **117**, 329.

Drury, L. O'C (1983): An introduction to the theory of diffusive shock acceleration of energetic particles in tenuous plasmas. Rep. Prog. Phys. **46**, 973.

Duijveman, A., Hoyng, P., Machado, M. E. (1982): X-ray imaging of three flares during the impulsive phase. Solar Phys. **81**, 137.

Dulk, G. A. (1985): Radio emission from the sun and other stars. Ann. Rev. Astron. Astrophys. **23**, 169.

Dulk, G. A., Marsh, K. A. (1982): Simplified expressions for the gyrosynchrotron radiation from mildly relativistic, nonthermal and thermal electrons. Ap. J. **259**, 350.

Duncan, R. C., Thompson, C. (1992): Formation of very strongly magnetized neutron stars: Implications for gamma-ray bursts. Ap. J. (Letters) **392**, L9.

Dungey, J. W. (1953): Conditions for the occurrence of electrical discharges in astrophysical systems. Phil. Mag. **44**, 725.

Dunham, T. (1933): The spectra of venus, mars, jupiter and saturn under high dispersion. P.A.S.P. **45**, 202. Reproduced in: A source book in astronomy and astrophysics 1900–1975 (eds. K. R. Lang and O. Gingerich). Cambridge, Mass.: Harvard University Press 1979.

Dunham, T. (1937): Interstellar neutral potassium and neutral calcium. P.A.S.P. **49**, 26.

Dunn, R. B., Evans, J. W., Jefferies, J. T., Orrall, F. Q., White, O. R., Zirker, J. B. (1968): The chromospheric spectrum at the 1962 eclipse. Ap. J. Suppl. No. 139, **15**, 275.

Dupree, A. K. (1969): Radiofrequency recombination lines from heavy elements: Carbon. Ap. J. **158**, 491.

Dupree, A. K. (1986): Mass loss from cool stars. Ann. Rev. Astron. Astrophys. **24**, 377.

Durisen, R. H. (1973): Viscous effects in rapidly rotating stars with application to white-dwarf models. II. Numerical results. Ap. J. **183**, 215.

Durisen, R. H. (1975): Upper mass limits for stable rotating white dwarfs. Ap. J. **199**, 179.

Durisen, R. H., Imamura, J. N. (1981): Improved secular stability limits for differentially rotating polytropes and degenerate dwarfs. Ap. J. **243**, 612.

Duval, P., Karp, A. H. (1978): The combined effects of expansion and rotation on spectral lines shapes. Ap. J. **222**, 220.

Duvall, T. L. Jr. (1982): A dispersion law for solar oscillations. Nature **300**, 242.

Duvall, T. L., Jr., D'Silva, S., Jefferies, S. M., Harvey, J. W., Schou, J. (1996): Downflows under sunspots detected by helioseismic tomography. Nature **379**, 235.
Duvall, T. L., Jr., Dziembowski, W. A., Goode, P. R., Gough, D. O., Harvey, J. W., Leibacher, J. W. (1984): Internal rotation of the Sun. Nature **310**, 22.
Duvall, T. L. Jr., et al. (1993): Time-distance helioseismology. Nature **362**, 430.
Duvall, T. L., Jr., et al. (1997): Time-distance helioseismology with the MDI instrument: Initial results. Solar Phys. **170**, 63.
Duvall, T. L. Jr., Harvey, J. W. (1983): Observations of solar oscillations of low and intermediate degree. Nature **301**, 24.
Duvall, T. L. Jr., Harvey, J. W. (1984): Rotational frequency splitting of solar oscillations. Nature **310**, 19.
Duvall. T. L. Jr., Harvey, J. W., Pomerantz, M. A. (1986): Latitude and depth variation of solar rotation. Nature **321**, 500.
Dwek, E., Arendt, R. G. (1992): Dust-gas interactions and the infrared emission from hot astrophysical plasmas. Ann. Rev. Astron. Astrophys. **30**, 11.
Dyson, J. E., Williams, D. A.: Physics of the interstellar medium. New York: John Wiley and Sons 1980.
Dziembowski, W., Goode, P. R. (1985): Frequency splitting in Ap stars. Ap. J. (Letters) **296**, L27.
Earl, J. A. (1961): Cloud chamber observations of primary cosmic ray electrons. Phys. Rev. Letters **6**, 125.
Easson, I., Pethick, C. J. (1979): Magnetohydrodynamics of neutron star interiors. Ap. J. **227**, 995.
Eberhardt, P., et al. (1970): Trapped solar wind gases, Kr^{81}/Kr exposure ages, and K/Ar ages in Apollo 11 lunar material. Science **167**, 558.
Ebert, R. (1957): Zur instabilität kugelsymmetrischer Gasverteilungen. Zeits. f. Astrophysik **42**, 263.
Ecker, G., Müller, K. G. (1958): Plasmapolarisation und Trägerwechselwirkung (Plasma polarization and carrier interaction). Z. Physik **153**, 317.
Eddington, A. S. (1916): The kinetic energy of a star cluster. M.N.R.A.S. **76**, 525. Reproduced in: Source book in astronomy and astrophysics 1900–1975 (eds. K. R. Lang and O. Gingerich). Cambridge, Mass.: Harvard University Press 1979.
Eddington, A. S. (1917): On the radiative equilibrium of the stars M.N.R.A.S. **77**, 16. Reproduced in: Source book in astronomy and astrophysics 1900–1975 (eds. K. R. Lang and O. Gingerich). Cambridge, Mass.: Harvard University Press 1979.
Eddington, A. S. (1918): On the pulsations of a gaseous star and the problem of Cepheid variables. M.N.R.A.S. **79**, 2, 177. Reproduced in: A source book in astronomy and astrophysics 1900–1975 (eds. K. R. Lang and O. Gingerich). Cambridge, Mass.: Harvard University Press 1979.
Eddington, A. S.: Report on the relativity theory of gravitation. London: Physical Society of London 1918.
Eddington, A. S. (1919): The total eclipse of 1919 May 29 and the influence of gravitation on light. Observatory **42**, 119.
Eddington, A. S. (1919): The pulsations of a gaseous star and the problem of the Cepheid variables. M.N.R.A.S. **79**, 177. Reproduced in: Source book in astronomy and astrophysics 1900–1975 (eds. K. R. Lang and O. Gingerich). Cambridge, Mass.: Harvard University Press 1979.
Eddington, A. S.: Space, time and gravitation. Cambridge: Cambridge University Press 1920.
Eddington, A. S. (1920): The internal constitution of the stars. Nature **106**, 14. Observatory **43**, 341. Reproduced in: A source book in astronomy and astrophysics 1900–1975 (eds. K. R. Lang and O. Gingerich). Cambridge, Mass.: Harvard University Press 1979.
Eddington, A. S.: The mathematical theory of relativity. Cambridge, England: Cambridge Univ. Press 1923.

Eddington, A. S. (1924): A comparison of Whitehead's and Einstein's formulae. Nature **113**, 192.

Eddington, A. S. (1924): On the relation between the masses and luminosities of the stars. M.N.R.A.S. **84**, 308. Reproduced in: A source book in astronomy and astrophysics 1900–1975 (eds. K. R. Lang and O. Gingerich). Cambridge, Mass.: Harvard University Press 1979.

Eddington, A. S. (1926): Diffuse matter in interstellar space. Proc. Roy. Soc. (London) **A111**, 424.

Eddington, A. S.: The internal constitution of the stars. Cambridge, England: Cambridge University Press 1926. Republished New York: Dover Publ. 1959.

Eddington, A. S. (1937): Interstellar matter. Observatory **60,** 99.

Eddington, A. S. (1941): On the cause of Cepheid pulsation. M.N.R.A.S. **101**, 182.

Eddington, A. S. (1942): Conditions in the hydrogen convection zone. M.N.R.A.S. **102**, 154.

Eddy, J. A.: A new sun, the solar results from Skylab. Washington, D. C.: National Aeronautics and Space Administration NASA SP-402 1979.

Edlén, B. (1941): An attempt to identify the emission lines in the spectrum of the solar corona. Arkiv. mat. astron. fysik **28B**, 1. Reproduced in: A source book in astronomy and astrophysics 1900–1975 (eds. K. R. Lang and O. Gingerich). Cambridge, Mass.: Harvard University Press 1979.

Edlén, B. (1942): Die Deutung der Emissionslinien im Spektrum der Sonnenkorona (The interpretation of the emission line spectrum of the solar corona). Z. Ap. **22**, 30.

Edlen, B. (1953): The dispersion of standard air. J. Opt. Soc. Amer. **43**, 339.

Eggen, O. J. (1951): Photoelectric studies: V. Magnitudes and colors of classical Cepheid variable stars. Ap. J. **113**, 367.

Eggen, O. J., Greenstein, J. L. (1965): Spectra, colors, luminosities, and motions of the white dwarfs. Ap. J. **141**, 83.

Eggen, O. J., Greenstein, J. L. (1965): Observations of proper-motion stars II. Ap. J. **142**, 925.

Eggen, O. J., Greenstein, J. L. (1967): Observations of proper-motion stars III. Ap. J. **150**, 927.

Eggleton, P. P. (1983): Approximations to the radii of Roche lobes. Ap. J. **268**, 368.

Eichler, D., et al. (1989): Nucleosynthesis, neutrino bursts and γ-rays from coalescing neutron stars. Nature **340**, 126.

Eidman, V. Ya. (1958): Investigation of the radiation of an electron moving in a magnetoactive medium. Sov. Phys. J.E.T.P. **7**, 91. (Corrections in Sov. Phys. J.E.T.P. **9**, 947 (1959)).

Einstein, A. (1905): Über einen die Erzeugung und Verwandlung des Lichtes betreffenden heuristischen Gesichtspunkt (On a heuristic point of view concerning the generation and transformation of light). Ann. Phys. **17**, 132.

Einstein, A. (1905): Ist die Trägheit eines Körpers von seinem Energieinhalt abhängig? (Does the inertia of a body depend on its energy?). Ann Phys. **18**, 639. Eng. trans. in: Source book in astronomy and astrophysics 1900–1975 (eds. K. R. Lang and O. Gingerich). Cambridge, Mass.: Harvard University Press 1979.

Einstein, A. (1906): Das Prinzip von der Erhaltung der Schwerpunktsbewegung und die Trägheit der Energie (Conservation of the motion of the mass center). Ann. Phys. **20**, 627. Eng. trans. in: Source book in astronomy and astrophysics 1900–1975 (eds. K. R. Lang and O. Gingerich). Cambridge, Mass.: Harvard University Press 1979.

Einstein, A. (1907): Über die vom Relativitätsprinzip geforderte Trägheit der Energie (Variation of intertia with energy on the principle of relativity). Ann. Phys. **23**, 371.

Einstein, A. (1909): Zum gegenwärtigen Stand des Strahlungsproblems (Additional new opinions on radiation problems). Phys. Z. **10**, 185.

Einstein, A. (1911): Über den Einfluß der Schwerkraft auf die Ausbreitung des Lichtes (On the influence of gravity on the propagation of light). Ann. Phys. **35**, 898. Trans. in: The principle of relativity. H. A. Lorentz, A. Einstein, H. Minkowski, H. Weyl. New York: Dover 1952.

Einstein, A. (1916): Die Grundlage der allgemeinen Relativitätstheorie (The foundations of the general theory of relativity). Ann. Physik **49**, 769. Trans. in: The principle of relativity. H. A. Lorentz, A. Einstein, H. Minkowski, H. Weyl. New York: Dover 1952.

Einstein, A. (1917): Zur Quantentheorie der Strahlung (On the quantum theory of radiation). Phys. Z. **18**, 121.

Einstein, A. (1924): Quantentheorie des einatomigen idealen Gases (The quantum theory of the monatomic perfect gas). Preuss. Acad. Wiss. Berl. Berlin Sitz. **22**, 261; **1**, 3, 18 (1925).

Ekers, J. A. (1969): The Parkes catalogue of radio sources – declination zone +20° to −90°. Austr. J. Phys. Suppl. No. 7.

Elander, N., Smith, W. H. (1973): Predissociation in the $C^2\Sigma^+$ state of CH and its astrophysical importance. Ap. J. **184**, 663.

Elbert, J. W., Sommers, P. (1995): In search of a source for the 320 EeV fly's eye cosmic ray. Ap. J. **441**, 151.

Elder, F. R., Langmuir, R. V., Pollock, H. C. (1948): Radiation from electrons accelerated in a synchrotron. Phys. Rev. **74**, 52.

Elitzur, M. (1992): Astronomical masers. Ann. Rev. Astron. Astrophys. **30**, 75.

Ellison, D. C., Drury, L. O'C., Meyer, J.-P. (1997): Galactic cosmic rays from supernova remnants II. Shock acceleration of gas and dust. Ap. J. **487**, 197.

Elmegreen, B. G.: Molecular clouds and star formation: An overview. In: Protostars and planets II (eds. D. C. Black and M. S. Matthews). Tucson, Arizona: University of Arizona Press 1985.

Elmegreen, B. G.: Large scale dynamics of the interstellar medium. In: The galactic interstellar medium (eds. W. B. Burton, B. G. Elmegreen and R. Genzel). New York: Springer-Verlag 1992.

Elmegreen, B. G., Lada, C. J. (1977): Sequential formation of subgroups in OB associations. Ap. J. **214**, 725.

Elsasser, W. M. (1933): Sur le principe de Pauli dans les noyaux (On Pauli's principle within the nucleus). J. Phys. Radium **4**, 549.

Elsner, R. F., Lamb, F. K. (1977): Accretion by magnetic neutron stars. I. Magnetospheric structure and stability. Ap. J. **215**, 897.

Elsworth, Y. P., et al. (1990): Variation of low-order acoustic solar oscillations over the solar cycle. Nature **345**, 322.

Elsworth, Y., et al. (1995): Slow rotation of the Sun's interior. Nature **376**, 669.

Elwert, G. (1954): Die weiche Röntgenstrahlung der ungestörten Sonnenkorona (The soft X-ray radiation of the undisturbed solar corona). Z. Naturforschung, **9**, 637.

Emden, R.: Gaskugeln (Gas spheres). Leipzig: Teuber 1907.

Endal, A. S. (1975): Carbon-burning nucleosynthesis with convection. Ap. J. **195**, 187.

Enge, H. A.: Introduction to nuclear physics. London: Addison-Wesley 1966.

Enskog, D.: Kinetische Theorie der Vorgänge in Mässing verdünnten Gasen (Kinetic theory of processes in massive, dilute gases). Uppsala: Almqvist and Wiksell 1917.

Epstein, I. (1950): Pulsation properties of giant-star models. Ap. J. **112**, 6.

Epstein, P. S. (1916): Zur Theorie des Starkeffektes (The theory of the Stark effect). Ann. Physik **50**, 489.

Epstein, P. S. (1926): The Stark effect from the point of view of Schroedinger's quantum theory. Phys. Rev. **28**, 695.

Epstein, R. I., Arnett, W. D., Schramm, D. N. (1976): Synthesis of the light elements in supernovae. Ap. J. Suppl. **31**, 111.

Epstein, R. I., Feldman, P. A. (1967): Synchrotron radiation from electrons in helical orbits. Ap. J. (Letters) **150**, L109.

Erber, T. (1966): High-energy electromagnetic conversion processes in intense magnetic fields. Rev. Mod. Phys. **38**, 626.

Erickson, W. C. (1964): The radio-wave scattering properties of the solar corona. Ap. J. **139**, 1290.

Erman, P. (1977): Experimental oscillator strengths of molecular ions. Ap. J. (Letters) **213**, L89.

Euler, L.: Principes generaux du movement des fluides (General principles of the movement of fluids). Hist. de l'acad. de Berlin 1755.

Evans, A.: The dusty universe. New York: Ellis Horwood 1993.

Evans, J. V. (1969): Radar studies of planetary surfaces. Ann. Rev. Astron. Astrophys. **7**, 201.

Evans, N. J.: Star formation: An overview. In: Protostars and planets II (eds. D. C. Black and M. S. Matthews). Tucson, Arizona: University of Arizona Press 1985.

Ewart, G. M., Guyer, R. A., Greenstein, G, (1975): Electrical conductivity and magnetic field decay in neutron stars. Ap. J. **202**, 238.

Ewen, H. I., Purcell, E. M. (1951): Radiation from galactic hydrogen at 1,420 MHz. Nature **168**, 356. Reproduced in: Source book in astronomy and astrophysics 1900–1975 (eds. K. R. Lang and O. Gingerich). Cambridge, Mass.: Harvard University Press 1979.

Fabbiano, G. (1989): X rays from normal galaxies. Ann. Rev. Astron. Astrophys. **27**, 87.

Fabian, A. C., Barcons, X. (1992): The origin of the X-ray background. Ann. Rev. Astron. Astrophys. **30**, 429.

Fabricius, D. (1594): Vierteljahrsschrift (Fourth year book). **4**, 290.

Fanselow, J. L., et al. (1969): Charge composition and energy spectrum of primary cosmic-ray electrons. Ap. J. **158**, 771.

Faraday, M. (1843): On static electrical inductive action. Phil. Mag. **22**, 200.

Faraday, M.: Experimental researches in electricity. London: R. Taylor 1844. Repr. New York: Dover 1952.

Farley, D. T. (1966): A theory of incoherent scattering of radio waves by plasma: 4. The effect of unequal ion and electron temperatures. J. Geophys. Res. **17**, 4091.

Fassio-Canuto, L. (1969): Neutron beta decay in a strong magenetic field. Phys. Rev. **187**, 2141.

Feenberg, E. (1932): The scattering of slow electrons by neutral atoms. Phys. Rev. **40**, 40.

Feenberg, E., Primakoff, H. (1948): Interaction of cosmic-ray primaries with sunlight and starlight. Phys. Rev. **73** (5), 449.

Feenberg, E., Trigg, G. (1950): The interpretation of compartive half-lives in the Fermi theory of beta-decay. Rev. Mod. Phys. **22**, 399.

Feingold, A. M. (1951): Table of ft values in beta-decay. Rev. Mod. Phys. **23**, 10.

Fejer, J. A. (1953): The diffraction of waves in passing through an irregular refracting medium. Proc. Roy. Soc. London **A220**, 455.

Fejer, J. A. (1960): Radio-wave scattering by an ionized gas in thermal equilibrium. J. Geophys. Res. **65**, 2635.

Fejer, J. A. (1960): Scattering of radio waves by an ionized gas in thermal equilibrium. Can. J. Phys. **38**, 1114.

Feldman, W. C., Asbridge, J. R., Bame, S. J., Gosling, J. T.: Plasma and magnetic fields in the Sun. In: The solar output and its variations (ed. O. R. White). Boulder, Colorado: Colorado Associated University Press 1977.

Felten, J. E., Morrison, P. (1963): Recoil photons from scattering of starlight by relativistic electrons. Phys. Rev. Lett. **10** (10), 453.

Felten, J. E., Morrison, P. (1966): Omnidirectional inverse Compton and synchrotron radiation from cosmic distributions of fast electrons and thermal photons. Ap. J. **146**, 686.

Fermat, P. (1891): Oeuvres de Fermat. Paris **2**, 354.

Fermi, E. (1926): Zur Quantelung des idealen einatomigen Gases (On quantisation of the ideal monatomic gas). Z. Physik **36**, 902.

Fermi, E. (1928): Eine statistische Methode zur Bestimmung einiger Eigenschaften des Atoms und ihre Anwendung auf die Theorie des periodischen Systems der Elemente (A statistical method for determining the eigenstate of atoms and its application to the theory of the periodical system of the elements). Z. Physik **48**, 73.

Fermi, E. (1930): Über die magnetischen Momente der Atomkerne (On the magnetic moments of atomic nuclei). Z. Physik **60**, 320.

Fermi, E. (1934): Versuch einer Theorie der β-strahlen I. (An attempt at the theory of β-rays). Z. Physik **88**, 161. English translation in Am. J. Phys. **36**, 1150 (1968).
Fermi, E. (1939): The absorption of mesotrons in air and in condensed materials. Phys. Rev. **56**, 1242.
Fermi, E. (1940): The ionization loss of energy in gases and in condensed materials. Phys. Rev. **57**, 485.
Fermi, E. (1949): On the origin of cosmic radiation. Phys. Rev. **75**, 1169. Reprod. in: Selected papers on cosmic ray origin theories (ed. S. Rosen). New York: Dover 1969.
Fermi, E. (1954): Galactic magnetic fields and the origin of cosmic radiation. Ap. J. **119**, 1. Reprod. in: Selected papers on cosmic ray origin theories (ed. S. Rosen). New York: Dover 1969. Reproduced in: Source book in astronomy and astrophysics 1900–1975 (eds. K. R. Lang and O. Gingerich). Cambridge, Mass.: Harvard University Press 1979.
Fermi, E., Uhlenbeck, G. E. (1933): On the recombination of electrons and positrons. Phys. Rev. **44**, 510.
Festa, G. G., Ruderman, M. A. (1969): Neutrino-pair bremsstrahlung from a degenerate electron gas. Phys. Rev. **180**, 1227.
Feynman, R. P., Gell-Mann, M. (1958): Theory of the Fermi interaction. Phys. Rev. **109**, 193.
Fichtel, C. E., et al. (1975): High-energy gamma-ray results from the second small astronomy satellite. Ap. J. **198**, 163.
Field, G. B. (1958): Excitation of the hydrogen 21-cm line. Proc. I.R.E. **46**, 240.
Field, G. B. (1959): The spin temperature of intergalactic neutral hydrogen. Ap. J. **129**, 536.
Field, G. B. (1965): Thermal instability. Ap. J. **142**, 531.
Field, G. B. (1974): Interstellar abundances: gas and dust. Ap. J. **187**, 453.
Field, G. B., Goldsmith, D. W., Habing, H. J. (1969): Cosmic-ray heating in the interstellar gas. Ap. J. (Letters) **155**, L149.
Field, G. B., Henry, R. C. (1964): Free-free emission by intergalactic hydrogen. Ap. J. **140**, 1002.
Field, G. B., Somerville, W. B., Dressler, K. (1966): Hydrogen molecules in astronomy. Ann. Rev. Astron. Astrophys. **4**, 207.
Field, G. B., Steigman, G. (1971): Charge transfer and ionization equilibrium in the interstellar medium. Ap. J. **166**, 59.
Finn, G. D., Mugglestone, D. (1965): Tables of the line broadening function H (a,v). M.N.R.A.S. **129**, 221.
Finn, L. S. (1994): Observational constraints on the neutron star mass distribution. Phys. Rev. Lett. **73**, 1878.
Finzi, A. (1965): Cooling of a neutron star by the URCA process. Phys. Rev. **137**, B 472.
Finzi, A. (1965): Vibrational energy of neutron stars and the exponential light curves of type I supernovae. Phys. Rev. Lett. **15**, 599.
Finzi, A., Wolf, R. A. (1967): Type I supernovae. Ap. J. **150**, 115.
Finzi, A., Wolf, R. A. (1968): Hot, vibrating neutron stars. Ap. J. **153**, 835.
Fishbone, L. G. (1975): The relativistic Roche problem. II. Stability theory. Ap. J. **195**, 499.
Fisher, A. J., et al. (1976): The isotopic composition of cosmic rays with $5 \leq Z \leq 26$. Ap. J. **205**, 938.
Fishman, G. J., Clayton, D. D. (1972): Nuclear gamma rays from ^{7}Li in the galactic cosmic radiation. Ap. J. **178**, 337.
Fishman, G. J., Meegan, C. A. (1995): Gamma-ray bursts. Ann. Rev. Astron. Astrophys. **33**, 415.
Fleischer, R. L., Hart, H. R. Jr., Renshaw, A. (1974): Composition of heavy cosmic rays from 25 to 180 MeV per atomic mass unit. Ap. J. **193**, 575.
Flowers, E., Ruderman, M. A. (1977): Evolution of pulsar magnetic fields. Ap. J. **215**, 302.

Flowers, E., Ruderman, M., Sutherland, P. (1976): Neutrino pair emission from finite-temperature neutron superfluid and the cooling of young neutron stars. Ap. J. **205**, 541,.

Fokker, A. D. (1914): Die mittlere Energie rotierender elektrischer Dipole im Strahlungsfeld (Mean energy of a rotating electric dipole in a radiation field). Ann. Physik **43**, 810.

Foley, H. M. (1946): The pressure broadening of spectral lines. Phys. Rev. **69**, 616.

Fomalont, E. B., Moffet, A. T. (1971): Positions of 352 small diameter sources. Astron. J. **76**, 5.

Fontaine, G., et al. (1974): The effects of differences in composition, equation of state, and mixing length upon the structure of white-dwarf convection zones. Ap. J. **193**, 205.

Fontenla, J. M., Avrett, E. H., Loeser, R. (1991): Energy balance in the solar transition region. II. Effects of pressure and energy input on hydrostatic models. Ap. J. **377**, 712.

Fontenla, J. M., Avrett, E. H., Loeser, R. (1993): Energy balance in the solar transition region. III. Helium emission in hydrostatic, constant-abundance models with diffusion. Ap. J. **406**, 319.

Forbes, T. G. (1986): Fast-shock formation in line-tied magnetic reconnection models of solar flares. Ap. J. **305**, 553.

Forbes, T. G., Malherbe, J.-M., Priest, E. R. (1989): The formation of flare loops by magnetic reconnection and chromospheric ablation. Solar Phys. **120**, 285.

Forbes, T. G., Priest, E. R. (1987): A comparison of analytical and numerical models for steadily driven reconnection. Rev. Geophys. **25**, 1583.

Forbush, S. E. (1950): Cosmic-ray intensity variations during two solar cycles. J. Geophys. Res. **63**, 651.

Forbush, S. E.: Solar cycle variation of cosmic ray intensity, cosmic ray activity and geomagnetic activity 1937–1961. Semaine d'Etude sur le Probleme du Rayonnement Cosmique dans l'Espace Interplanetaire. Rome: Pontificial Academy 1962.

Forbush, S. E.: Time-variations of cosmic rays. In: Handbuch der Physik, vol. XLIX/1: Geophysics III. Berlin-Heidelberg-New York: Springer 1966.

Ford, A. L., Docken, K. K., Dalgarno, A. (1975): The photoionization and dissociative photoionization of H_2, HD, and D_2. Ap. J. **195**, 819.

Ford, A. L., Docken, K. K., Dalgarno, A. (1975): Cross sections for photoionization of vibrationally excited molecular hydrogen. Ap. J. **200**, 788.

Ford, W. K., Rubin, V. C., Roberts, M. S. (1971): A comparison of 21-cm radial velocities and optical radial velocities of galaxies. Astron. J. **76**, 22.

Forman, M. A., Schaeffer, O. A. (1979): Cosmic ray intensity over long time scales. Rev. Geophys. Space Phys. **17**, 552.

Forman, W., Jones, C. (1982): X-ray-imaging observations of clusters of galaxies. Ann. Rev. Astron. Astrophys. **20**, 547.

Forman, W., Jones, C. A., Liller, W. (1972): Optical studies of UHURU sources: III. Optical variations of the X-ray eclipsing system HZ Herculis. Ap. J. (Letters) **177**, L103.

Forrest, D. J., Chupp, E. L. (1983): Simultaneous acceleration of electrons and ions in solar flares. Nature **305**, 291.

Forrest, D. J., et al. (1986): Very energetic gamma rays from the 3 June 1982 solar flare. Adv. Space Res. **6**, No. 6, 115.

Fossat, E., Grec, G., Pomerantz, M. A. (1981): Solar pulsations observed from the geographic South Pole - initial results. Solar Phys. **74**, 59.

Foster, P. N., Chevalier, R. A. (1993): Gravitational collapse of an isothermal sphere. Ap. J. **416**, 303.

Fowler, R. H. (1926): On dense matter. M.N.R.A.S. **87**, 114. Reproduced in: Source book in astronomy and astrophysics 1900–1975 (eds. K. R. Lang and O. Gingerich). Cambridge, Mass.: Harvard University Press 1979.

Fowler, W. A. (1954): Experimental and theoretical results on nuclear reactions in stars. Mem. Soc. Roy. Sci. Liege **14**, 88.

Fowler, W. A. (1958): Completion of the proton-proton reaction chain and the possibility of energetic neutrino emission by hot stars. Ap. J. **127**, 551.
Fowler, W. A. (1959): Experimental and theoretical results on nuclear reactions in stars II. Mem. Soc. Roy. Sci. Liege Ser 5, **3**, 207.
Fowler, W. A. (1972): What cooks with solar neutrinos. Nature **238**, 24.
Fowler, W. A. (1984): Experimental and theoretical astrophysics: the quest for the origin of the elements. Rev. Mod. Phys. **56**, 149.
Fowler, W. A., Burbidge, G. R., Burbidge, E. M. (1955): Nuclear reactions and element synthesis in the surfaces of stars. Ap. J. Suppl. No. 17, **2**, 167.
Fowler, W. A., Caughlan, G. R., Zimmerman, B. A. (1967): Thermonuclear reaction rates. Ann. Rev. Astron. Astrophys. **5**, 525.
Fowler, W. A., Caughlan, G. R., Zimmerman, B. A. (1975): Thermonuclear reaction rates, II. Ann. Rev. Astron. Astrophys. **13**, 69.
Fowler, W. A., Engelbrecht, C. A., Woosley, S. E. (1978): Nuclear partition functions. Ap. J. **226**, 984.
Fowler, W. A., Greenstein, J. L. (1956): Element building reactions in stars. Proc. Nat. Acad. Sci. (Wash.) **42**, 173.
Fowler, W. A., Hoyle, F. (1964): Neutrino processes and pair formation in massive stars and supernovae. Ap. J. Suppl. **9**, 201.
Fowler, W. A., Reeves, H., Silk, J. (1970): Spallation limits on interstellar fluxes of low-energy cosmic rays and nuclear gamma rays. Ap. J. **162**, 49.
Frail, D. A., et al. (1997): The radio afterglow from the γ-ray burst of 8 May 1997. Nature **389**, 261.
Franck, J. (1925): Elementary processes of photochemical reactions. Trans. Far. Soc. **21**, 536.
Frank, J., King, A., Raine, D.: Accretion power in astrophysics. Second edition. New York: Cambridge Univeristy Press 1992.
Fraunhofer, J. (1817): Auszug davon in. Gilb. Ann. Physik **56**, 264.
Frautschi, S., et al. (1971): Ultra-dense matter. Comm. Ap. and Space Phys. **3**, 121.
Frazier, E. N. (1968): A spatio-temporal analysis of velocity fields in the solar photosphere. Zeits. f. Ap. **68**, 345.
Freeman, J. M., Clark, G. J., Ryder, J. S., Burcham, W. E., Sguier, G. T. A., Draper, J. E.: Present values of the weak interaction coupling constants. U.K. Atom. Energy Res. Group, 1972.
Freeman, J., et al. (1977): The local interstellar helium density. Ap. J. **215**, L 83.
Freier, P. S., Ney, E. P., Winckler, J. R. (1959): Balloon observations of solar cosmic rays on March 26, 1958. J. Geophys. Res. **64**, 685.
Freier, P. S., Waddington, C. J. (1968): Singly and doubly charged particles in the primary cosmic radiation. J. Geophys. Res. **73**, 4261.
Freier, P., et al. (1948): Evidence for heavy nuclei in the primary cosmic radiation. Phys. Rev. **74**, 213.
Freier, P., et al. (1948): The heavy component of primary cosmic rays. Phys. Rev. **74**, 1818.
Fresnel, A. J. (1822): Explication de la réfraction dans le systeme des ondes. Mem. de l' Acad. **11**, 893.
Friedlander, M. W.: Cosmic rays. Cambridge, Mass.: Harvard University Press 1989.
Friedman, B., Pandharipande, V. R. (1981): Hot and cold, nuclear and neutron matter. Nucl. Phys. **A361**, 502.
Friedman, H. (1973): Cosmic X-ray sources: A progress report. Science **181**, 395.
Friman, B. L., Maxwell, O. V. (1979): Neutrino emissivities of neutron stars. Ap. J. **232**, 541.
Fristrom, C. C. (1987): The equation of state for degenerate dwarf stars. Fund. Cosmic Phys. **12**, 119.
Fröberg, C. E. (1955): Numerical treatment of Coulomb wave functions. Rev. Mod. Phys. **27**, 399.
Fuchs, V., Cairns, R. A., Lashmore-Davies, C. N. (1986): Velocity space structure of runaway electrons. Phys. Fluids **29**, 2931.

Fukuda, Y., et al. (1998): Evidence for oscillation of atmospheric neutrinos. Phys. Rev. **81**, 1562.

Fuller, G. M., Fowler, W. A., Newman, M. J. (1980): Stellar weak-interaction rates for sd-shell nuclei. I. Nuclear matrix element systematics with application to ^{26}Al and selected nuclei of importance to the supernova problem. Ap. J. Supp. **42**, 447.

Fuller, G. M., Fowler, W. A., Newman, M. J. (1982): Stellar weak interaction rates for intermediate-mass nuclei. II. A = 21 to A = 60. Ap. J. **252**, 715.

Fuller, G. M., Fowler, W. A., Newman, M. J. (1982): Stellar weak interaction rates for intermediate mass nuclei. III. Rate tables for the free nucleons and nuclei with A = 21 to A = 60. Ap. J. Supp. **48**, 279.

Furth, H. P., Killeen, J., Rosenbluth, M. N. (1963): Finite resistivity instabilities of a sheet pinch. Phys. Fluids **6**, 459.

Gaisser, T. K.: Cosmic rays and particle physics. New York: Cambridge University Press 1990.

Gallagher, J. S., Starrfield, S. (1978): Theory and observations of classical novae. Ann. Rev. Astron. Ap. **16**, 171.

Galli, D., Shu, F. H. (1993): Collapse of magnetized molecular cloud cores. I. Semianalytical solution. Ap. J. **417**, 220.

Gamow, G. (1928): Zur Quantentheorie des Atomkernes (On the quantum theory of the atomic nucleus). Z. Physik **51**, 204.

Gamow, G. (1946): Expanding universe and the origin of the elements. Phys. Rev. **70**, 572.

Gamow, G., Schoenberg, M. (1941): Neutrino theory of stellar collapse. Phys. Rev. **59**, 539. Reproduced in: Source book in astronomy and astrophysics 1900–1975 (eds. K. R. Lang and O. Gingerich). Cambridge, Mass.: Harvard University Press 1979.

Gandelman, G. M., Pinaev, V. A. (1960): Emission of neutrino pairs by electrons and the role played by it in stars. Sov. Phys. JETP **10**, 764.

Garcia-Munoz, M., Mason, G. M., Simpson, J. A. (1975): The isotopic composition of galactic cosmic-ray lithium, beryllium, and boron. Ap. J. (Letters) **201**, L145.

Garcia-Munoz, M., Mason, G. M., Simpson, J. A. (1977): The age of the galactic cosmic rays derived from the abundance of ^{10}Be. Ap. J. **217**, 859.

Gardner, E., Lattes, C. M. G. (1948): Production of mesons by the 184 inch Berkeley cyclotron. Science **107**, 270.

Garstang, R. H.: Forbidden transitions. In: Atomic and molecular processes (ed. D. R. Bates). New York: Academic Press 1962.

Garstang, R. H.: Transition probabilities for forbidden lines. In: Planetary nebulae (ed. D. E. Osterbrock, C. R. O'Dell). I.A.U. Symp. No. 34. Dordrecht, Holland: D. Reidel. Berlin-Heidelberg-New York: Springer 1968.

Garvey, G. T., et al. (1969): Set of nuclear-mass relations and a resultant mass table. Rev. Mod. Phys. **41**, S 1.

Gaunt, J. A. (1930): Continuous absorption. Phil. Trans. Roy. Soc. London **A229**, 163.

Gautschy, A., Saio, H. (1995): Stellar pulsations across the HR diagram. Part 1. Ann. Rev. Astron. Astrophys. **33**, 75.

Gautschy, A., Saio, H. (1996): Stellar pulsations across the HR diagram: Part 2. Ann. Rev. Astron. Astrophys. **34**, 551.

Gay-Lussac, M. (1809): Memoir on the combination of gaseous substances with each other. Mem. de la société d' arcueil **2**, 207.Eng. trans. in: Foundations of the molecular theory. Alemic Club Repr. No. 4. Edinburgh: Bishop 1950.

Geballe, T. R., Oka, T. (1996): Detection of H_3^+ in interstellar space. Nature **384**, 334.

Gehrels, N., Paul, J. (1998): The new gamma-ray astronomy. Physics Today **51**, 26 – February.

Gelb, J. M., Kwong, W., Rosen, S. P. (1992): Implications of new GALLEX results for the Mikheyev-Smirnov-Wolfenstein solution of the solar neutrino problem. Phys. Rev. Letters **69**, 1864.

Genzel, R.: Physics and chemistry of molecular clouds. In: The galactic interstellar medium. Saas-Fee advanced course 21 (eds. D. Pfenniger and P. Bartholdi). New York: Springer-Verlag 1992.
Genzel, R., Stutzki, J. (1989): The Orion molecular cloud and star-forming region. Ann. Rev. Astron. Astrophys. **27**, 41.
Genzel, R., Townes, C. H. (1987): Physical conditions, dynamics, and mass distribution in the center of the Galaxy. Ann. Rev. Astron. Astrophys. **25**, 377.
Gerwin, R. A. (1968): Stability of the interface between two fluids in relative motion. Rev. Mod. Phys. **40**, 652.
Ghosh, P., Lamb, F. K. (1979): Accretion by rotating magnetic neutron stars. II. Radial and vertical structure of the transition zone in disk accretion. Ap. J. **232**, 259.
Ghosh, P., Lamb, F. K. (1979): Accretion by rotating magnetic neutron stars. III. Accretion torques and period changes in pulsating X-ray sources. Ap. J. **234**, 296.
Ghosh, P., Lamb, F. K., Pethick, C. J. (1977): Accretion by rotating magnetic neutron stars. I. Flow of matter inside the magnetosphere and its implications for spin-up and spin-down of the star. Ap. J. **217**, 578.
Giacconi, R., et al. (1972): The Uhuru catalog of X-ray sources. Ap. J. **178**, 281.
Gibbs, J. W.: Elementary principles in statistical mechanics. New York 1902.
Giclas, H. L. Burnham, R., Thomas, N. G.: A list of white dwarf suspects I, II, III. Lowell Obs. Bull. **125, 141, 153** (1965, 1967, 1970)
Gilbert, A., Cameron, A. G. W. (1965): A composite nuclear-level density formula with shell corrections. Can. J. Phys. **43**, 1446.
Giles, P. M., et al. (1997): A subsurface flow of material from the Sun's equator to its poles. Nature **390**, 52.
Gilman, R. C. (1974): Free-free and free-bound emission in low-surface-gravity stars. Ap. J. **188**, 87.
Gingerich, O., De Jager, C. (1968): The Bilderberg model of the photosphere and low chromosphere. Solar Phys. **3**, 5.
Gingerich, O., et al. (1971): The Harvard-Smithsonian reference atmosphere. Solar Phys. **18**, 347.
Ginzburg, V. L. (1946): On solar radiation in the radio spectrum. C. R. Acad Sci. USSR **52**, 487.
Ginzburg, V. L. (1953): The origin of cosmic rays and radio astronomy. Usp. Fiz. Nauk. **51**, 343.
Ginzburg, V. L. (1953): The origin of cosmic rays. Dokl. Akad. Nauk SSSR **92**, 1133.
Ginzburg, V. L. (1956): The nature of cosmic radio emission and the origin of cosmic rays. Nuovo Cimento Supplement **3**, 38. Reproduced in: A source book in astronomy and astrophysics 1900–1975 (eds. K. R. Lang and O. Gingerich). Cambridge, Mass.: Harvard University Press 1979.
Ginzburg, V. L.: Propagation of electromagnetic waves in a plasma. New York: Gordon and Breach 1961 .
Ginzburg, V. L. (1969): Superfluidity and superconductivity in astrophysics. Comments Astrophys. and Space Phys. **1**, 81.
Ginzburg, V. L.: Elementary processes for cosmic ray astrophysics. New York: Gordon and Breach 1969.
Ginzburg, V. L., Ptuskin, V. S. (1976): On the origin of cosmic rays: Some problems in high-energy astrophysics. Rev. Mod. Phys. **48**, 161.
Ginzburg, V. L., Syrovatskii, S. I.: The origin of cosmic rays. New York: Macmillan 1964.
Ginzburg, V. L., Syrovatskii, S. I. (1965): Cosmic magnetobremsstrahlung (Synchrotron radiation). Ann. Rev. Astron. Astrophys. **3**, 297.
Ginzburg, V. L., Zhelezniakov, V. V. (1958): On the possible mechanisms of sporadic solar radio emission (Radiation in an isotropic plasma). Sov. Astron. **2**, 653.
Ginzburg, V. L., Zhelezniakov, V. V.: On the mechanisms of sporadic solar radio emission. In: Paris symposium on radio astronomy (ed. R. N. Bracewell). Stanford, Calif.: Stanford University Press 1959.

Gioumousis, G., Stevenson, D. P. (1958): Reactions of gaseous molecule ions with gaseous molecules V. Theory. J. Chem. Phys. **29**, 294.

Giovanelli, R. G. (1949): A note on heat transfer in the upper chromosphere and corona. M.N.R.A.S. **109**, 372.

Giovanelli, R., Haynes, M. P.: Extragalactic neutral hydrogen. In: Galactic and extragalactic radio astronomy, second edition (eds. G. L. Verschuur and K. I. Kellermann). New York: Springer-Verlag 1988.

Glasser, M. L. (1975): Ground state of electron matter in high magnetic fields. Ap. J. **199**, 206.

Glennon, B. M., Wiese, W. L.: Bibliography on atomic transition probabilities. Nat. Bur. Stands. (Wash.). Mon. 50, 1962.

Gliese, W.: Catalogue of nearby stars. Veroff-Astron. Rechen-Institut Heidelberg: Braun, Karlsruhe 1969.

Gluckstern, R. L., Hull, M. H., Jr. (1953): Polarization dependence of the integrated bremsstrahlung cross section. Phys. Rev. **90**, 1030.

Gold, T.: The origin of cosmic radio noise. Proc. Conf. Dynamics of Ionized Media. University College, London: Dept. of Physics 1951. Reproduced in: A source book in astronomy and astrophysics 1900–1975 (eds. K. R. Lang and O. Gingerich). Cambridge, Mass.: Harvard University Press 1979.

Gold, T.: Magnetic energy shedding in the solar atmosphere. In: AAS-NASA Symposium on the Physics of Solar Flares NASA SP-50 (ed. W. N. Hess). Washington, DC: National Aeronautics and Space Administration 1964, pp. 389–395.

Gold, T. (1968): Rotating neutron stars as the origin of the pulsating radio sources. Nature **218**, 731. Reproduced in: Source book in astronomy and astrophysics 1900–1975 (eds. K. R. Lang and O. Gingerich). Cambridge, Mass.: Harvard University Press 1979.

Gold, T., Hoyle, F. (1960): On the origin of solar flares. M.N.R.A.S. **120**, 89.

Goldberg, H. S., Scadron, M. D.: Physics of stellar evolution and cosmology. New York: Gordon and Breach, 1981.

Goldberg, L. (1935): Relative multiplet strengths in LS coupling. Ap. J. **82**, 1.

Goldberg, L. (1966): Stimulated emission of radio-frequency lines of hydrogen. Ap. J. **144**, 1225.

Goldberg, L., Dupree, A. K. (1967): Population of atomic levels by dielectronic recombination. Nature **215**, 41.

Goldberg, L., Müller, E. A., Aller, L. H. (1960): The abundances of the elements in the solar atmosphere. Ap. J. Suppl. No. 45, **5**, 1.

Goldreich, P., Julian, W. H. (1969): Pulsar electrodynamics. Ap. J. **157**, 869.

Goldreich, P., Julian, W. H. (1970): Stellar winds. Ap. J. **160**, 971.

Goldreich, P., Kumar, P. (1990): Wave generation by turbulent convection. Ap. J. **363**, 694.

Goldreich, P., Morrison, P. (1964): On the absorption of gamma rays in intergalactic space. Soviet Phys. JETP **18**, 239.

Goldreich, P., Tremaine, S. (1979): Towards a theory for the uranian rings. Nature **277**, 97.

Goldreich, P., Tremaine, S. (1982): The dynamics of planetary rings. Ann. Rev. Astron. Astrophys. **20**, 249.

Goldschmidt, V. M. (1937): Geochemische Verteilungsgesetze der Elemente: IX. Die Mengenverhältnisse der Elemente und der Atom-Arten (Geochemical distribution laws of the elements: IX. The abundances of elements and of types of atoms). Skrift. Norske Videnskaps-Akad. Oslo I. Mat. Nat. No. 4.

Goldsmith, D.: Worlds unnumbered: The search for extrasolar planets. Sausalito, California: University Science Books, 1997.

Goldsmith, P. F.: Molecular clouds – an overview. In: Interstellar processes (eds. D. J. Hollenbach and H. A. Thronson, Jr.). Boston: D. Reidel 1987.

Goldsmith, P. F., Langer, W. D. (1978): Molecular cooling and thermal balance of dense interstellar clouds. Ap. J. **222**, 881.

Goldwire, H. C. (1968): Oscillator strengths for electric dipole transitions of hydrogren. Ap. J. Suppl. No. 152, **17**, 445.

Goode, P. R., et al. (1991): What we know about the Sun's internal rotation from oscillations. Ap. J. **367**, 649.

Goodman, F. O. (1978): Formation of hydrogen molecules on interstellar grain surfaces. Ap. J. **226**, 87.

Goodman, J. (1986): Are gamma-ray bursts optically thick? Ap. J. (Letters) **308**, L47.

Goodman, J. A., et al. (1979): Composition of primary cosmic rays above 10^{13} eV from the study of time distributions of energetic Hadrons near air-shower cores. Phys. Rev. Lett. **42**, 854.

Goodricke, J. (1783): A series of observations on, and a discovery of, the period of the variation of the light of the bright star in the head of Medufa, called Algol. Phil. Trans. **73**, 474.

Gordon, M. A.: H II regions and radio recombination lines. In: Galactic and extragalactic radio astronomy, second edition (eds. G. L. Verschuur and K. I. Kellermann). New York: Springer-Verlag 1988.

Gordon, M. A., Burton, W. B. (1976): Carbon monoxide in the galaxy. I. The radial distribution of CO, H_2, and nucleons. Ap. J. **208**, 346.

Gosling, J. T. (1993): The solar flare myth. J. Geophys. Res. **98**, 18937.

Goss, W. M., Field, G. B. (1968): Collisional excitation of low-energy permitted transitions by charged particles. Ap. J. **151**, 177.

Gough, D. O. (1976): Random remarks on solar hydrodynamics. In: The energy balance and hydrodynamics of the solar chromosphere and corona. Proc. I.A.U. Colloq. No.36. (eds. R.-M. Bonnet and Ph. Delache) Paris: G. De Bussac, Claremont-Ferrand.

Gough, D. O. (1995): The future of helioseismology. In: Helioseismology – Proc. Fourth Soho Workshop. ESA SP-376 – June.

Gough, D.O., et al. (1996): The seismic structure of the Sun. Science **272**, 1296.

Gough, D. O., Leibacher, J. W., Scherrer, P. H., Toomre, J. (1996): Perspectives in helioseismology. Science **272**, 1281.

Gough, D. O., Thompson, M. J. (1990): The effect of rotation and a buried magnetic field on stellar oscillations. M.N.R.A.S. **242**, 25.

Gough, D. O., Toomre, J. (1991): Seismic observations of the solar interior. Ann. Rev. Astron. Astrophys. **29**, 627.

Gould, R. J. (1965): High-energy photons from the Compton-synchrotron process in the Crab nebula. Phys. Rev. Lett. **15**, 577.

Gould, R. J. (1975): Energy loss of relativistic electrons and positrons traversing cosmic matter. Ap. J. **196**, 689.

Gould, R. J. (1978): Radiative recombination of complex ions. Ap. J. **219**, 250.

Gould, R. J., Gold, T., Salpeter, E. E. (1963): The interstellar abundance of the hydrogen molecule II. Galactic abundance and distribution. Ap. J. **138**, 408.

Gould, R. J., Salpeter, E. E. (1963): The interstellar abundance of the hydrogen molecule I. Basic processes. Ap. J. **138**, 393.

Gould, R. J., Schréder, G. (1966): Opacity of the universe to high-energy photons. Phys. Rev. Lett. **16** (6), 252.

Graboske, H. C., De Witt, H. E., Grossman, A. S., Cooper, M. S. (1973): Screening factors for nuclear reactions: II. Intermediate screening and astrophysical applications. Ap. J. **181**, 457.

Gradsztajn, E.: Production of Li, Be, and B isotopes in C, N, and O. In: High energy nuclear reactions in astrophysics (ed. B. Shen). New York: W.A. Benjamin 1967.

Gray, D. F.: The observation and analysis of stellar photospheres. Second edition. New York: Cambridge University Press 1992.

Grec, G., Fossat, E., Pomerantz, M. A. (1980): Solar oscillations – full disk observations from the geographic South Pole. Nature **288**, 541.

Grec, G., Fossat, E., Pomerantz, M. A. (1983): Full-disk observations of solar oscillations from the geographic South Pole – latest results. Solar Phys. **82**, 55.
Green, A. E. S. (1954): Coulomb radius constant from nuclear masses. Phys. Rev. **95**, 1006.
Green, H. S.: The molecular theory of fluids. New York: Interscience 1952.
Green, L. C., Rush, P. P., Chandler, C. D. (1957): Oscillator strengths and matrix elements for the electric dipole moment of hydrogen. Ap. J. Suppl. No. 26, **3**, 37.
Green, P. E.: Radar measurements of target scattering properties. In: Radar astronomy (ed. J. V. Evans, T. Hagfors). New York: McGraw-Hill 1968.
Green, S. (1975): Rotational excitation of molecular ions in interstellar clouds. Ap. J. **201**, 366.
Green, S., Thaddeus, P. (1976): Rotational excitation of CO by collisions with He, H, and H_2 under conditions in interstellar clouds. Ap. J. **205**, 766.
Greenberg, J. M. (1963): Interstellar grains. Ann. Rev. Astron. Astrophys. **1**, 267.
Greenberg, J. M., Van De Hulst, H. C.: Interstellar dust and related topics, IAU symposium no. 52. Boston: D. Reidel 1973.
Greenstein, G. (1976): Superfluidity in neutron stars. II. After a period jump. Ap. J. **208**, 836.
Greenstein, J. L. (1951): A search for He^3 in the Sun. Ap. J. **113**, 531.
Greenstein, J. L. (1969): The Lowell suspect white dwarfs. Ap. J. **158,** 281.
Greenstein, J. L. (1970): Some new white dwarfs with peculiar spectra VI. Ap. J. (Letters) **162**, L55.
Greenstein, J. L., et al. (1977): The rotation and gravitational redshift of white dwarfs. Ap. J. **212**, 186.
Greenstein, J. L., Oke, J. B., Shipman, H. L. (1971): Effective temperature, radius, and gravitational redshift of Sirius B. Ap. J. **169**, 563.
Greenstein, J. L., Richardson, R. S. (1951): Lithium and the internal circulation of the Sun. Ap. J. **113**, 536.
Greenstein, J. L., Trimble, V. L. (1967): The Einstein redshift in white dwarfs. Ap. J **149**, 283.
Greenstein, J. L., Trimble, V. L. (1972): The gravitational redshift of 40 Eridani B. Ap. J. (Letters) **175**, L1.
Greisen, K. (1966): End to the cosmic-ray spectrum? Phys. Rev. Lett. **16**, 748.
Grevesse, N. (1968): Solar abundances of lithium, beryllium, and boron. Solar Phys. **5**, 159.
Grevesse, N., Noels, A.: Cosmic abundances of the elements. In: Origin and Evolution of the Elements (eds. N. Prantzos, E. Vangioni-Flam, and M. Cassé). New York: Cambridge University Press 1993.
Gribov, V., Pontecorvo, B. (1969): Neutrino astronomy and lepton charge. Phys. Lett. **28B**, 493.
Griem, H. R. (1960): Stark broadening of higher hydrogen and hydrogen-like lines by electrons and ions. Ap. J. **132**, 883.
Griem, H. R. (1962): Wing formulae for Stark-broadened hydrogenic lines. Ap. J. **136**, 422.
Griem, H. R.: Plasma spectroscopy. New York: McGraw-Hill 1964.
Griem, H. R. (1967): Corrections to the asymptotic Holtsmark formula for hydrogen lines broadened by electrons and ions in a plasma. Ap. J. **147**, 1092.
Griem, H. R. (1967): Stark broadening by electron and ion impacts of $n\alpha$ hydrogen lines of large principal quantum number. Ap. J. **148**, 547.
Griem, H. R., Kolb, A. C., Shen, K. Y. (1959): Stark broadening of hydrogen lines in a plasma. Phys. Rev. **116**, 4.
Groot, P. J., et al. (1998): The rapid decay of the optical emission from GRB 980326 and its possible implications. Ap. J. (Letters) **502**, L123.
Grotrian, W.: Gesetzmäßigkeiten in den Serienspektren (Regularities in the series spectra). In: Handbuch der Astrophysik. Berlin: Springer 1930.
Grotrian, W. (1939): On the question of the significance of the lines in the spectrum of the solar corona. Naturwiss. **27**, 214. English translation in: A source book in astronomy

and astrophysics 1900–1975 (eds. K. R. Lang and O. Gingerich). Cambridge, Mass.: Harvard University Press 1979.
Güdel, M., et al. (1993): A tight correlation between radio and X-ray luminosities of M dwarfs. Ap. J. **415**, 236.
Güdel, M., Schmitt, J. H. M. M., Benz, A. O. (1994): Discovery of microwave emission from four nearby solar-type G stars. Science **265**, 933.
Guenther, D. B., Demarque, P. (1997): Seismic tests of the Sun's interior structure, composition, and age, and implications for solar neutrinos. Ap. J. **484**, 937.
Gundermann, E. J.: Observations of the interstellar hydroxyl radical. Ph.D. thesis, Harvard University, 1965.
Gunn, J. E., Ostriker, J. P. (1969): Acceleration of high-energy cosmic rays by pulsars. Phys. Rev. Letters **22**, 728.
Gunn, J. E., Peterson, B. E. (1965): On the density of neutral hydrogen in intergalactic space. Ap. J. **142**, 1633.
Gurney, R. W., Condon, E. U. (1928): Wave mechanics and radioactive disintegration. Nature **122**, 493.
Gurney, R. W., Condon, E. U. (1929): Quantum mechanics and radioactive disintegration. Phys. Rev. **33**, 127.
Habing, H. J. (1968): The interstellar radiation density between 912 Å and 2400 Å. B. A. N. **19**, 421.
Hack, M.: The evolution of close binary systems. In: Star evolution – Proc. Int. Sch. Phys. Enrico Fermi – Course 38 (ed. L. Gratton). New York: Academic Press 1963.
Hagan, L. (1977): Bibliography on atomic energy levels and spectra, NBS SP 363, Wash..
Hagen, F. A., Fisher, A. J., Ormes, J. F. (1977): ^{10}Be abundance and the age of cosmic rays: A balloon measurement. Ap. J. **212**, 262.
Hagfors, T. (1961): Some properties of radio waves reflected from the moon and their relation to the lunar surface. J. Geophys. Res. **66** (3), 777.
Hagfors, T. (1961): Density fluctuations in a plasma in a magnetic field, with applications to the ionosphere. J. Geophys. Res. **66** (6), 1699.
Hagfors, T. (1964): Backscattering from an undulating surface with applications to radar returns from the moon. J. Geophys. Res. **69** (18), 3779.
Hagfors, T. (1970): Remote probing of the moon by infrared and microwave emissions and by radar. Radio Science **5**, 189.
Hainebach, K. L., et al. (1974): On the e-process: its components and their neutron excesses. Ap. J. **193**, 157.
Hainebach, K. L., Norman, E. B., Schramm, D. N. (1976): Consistency of cosmic-ray composition, acceleration mechanism, and supernova models. Ap. J. **203**, 245.
Hainebach, K. L., Schramm, D. N., Blake, J. B. (1976): Cosmic-ray spallative origin of the rare odd-odd nuclei, consistent with light-element production. Ap. J. **205**, 920.
Haisch, B., Rodono, M.: Solar and stellar flares. Solar Physics **121**, Nos. 1–2 (1989). Reprinted Boston: Kluwer Academic 1989.
Haisch, B., Strong, K. T., Rodono, M. (1991): Flares on the sun and other stars. Ann. Rev. Astron. Astrophys. **29**, 275.
Hale, G. E. (1908): On the probable existence of a magnetic field in sun-spots. Ap. J. **28**, 315. Reproduced in: A source book in astronomy and astrophysics 1900–1975 (eds. K. R. Lang and O. Gingerich). Cambridge, Mass.: Harvard University Press 1979.
Hale, G. E., Ellerman, F., Nicholson, S. B., Joy, A. H. (1919): The magnetic polarity of sun-spots. Ap. J. **49**, 153.
Hall, J. S. (1949): Observations of the polarized light from stars. Science **109**, 166. Reproduced in: A source book in astronomy and astrophysics 1900–1975 (eds. K. R. Lang and O. Gingerich). Cambridge, Mass.: Harvard University Press 1979.
Halpern, P.: The quest for alien planets: Exploring worlds outside the solar system. New York: Plenum Trade, 1997.

Hamada, T., Salpeter, E. E. (1961): Models for zero-temperature stars. Ap. J. **134**, 683.
Hansen, C. J. (1968): Some weak interaction processes in highly evolved stars. Astrophys. and Space Sci. **1**, 499.
Hansen, C. J. (1971): Explosive carbon-burning nucleosynthesis. Astrophys. and Space Sci. **14**, 389.
Harding, D., Guyer, R. A., Greenstein, G. (1978): Superfluidity in neutron stars. III. Relaxation processes between the superfluid and the crust. Ap. J. **222**, 991.
Harkins, W. D. (1917): The evolution of the elements and the stability of complex atoms – A new periodic system which shows a relation between the abundance of the elements and the structure of the nuclei of atoms. J. Am. Chem. Soc. **39**, 856.
Harkins, W. D. (1931): The periodic system of atomic nuclei and the principle of regularity and continuity of series. Phys. Rev. **38**, 1270.
Harris, D. E., Romanishin, W. (1974): Inverse Compton radiation and the magnetic field in clusters of galaxies. Ap. J. **188**, 209.
Harris, D. E., Zeissig, G. A., Lovelace, R. V. (1970): The minimum observable diameter of radio sources. Astron. Astrophys. **8**, 98.
Harris, D. L. (1948): On the line-absorption coeffcient due to Doppler effect and damping. Ap. J. **108**, 112.
Harris, M. J., et al. (1983): Thermonuclear reactions rates, III. Ann. Rev. Astron. Astrophys. **21**, 165.
Harrison, B. K., et al..: Gravitation theory and gravitational collapse. Chicago, Ill.: University Chicago Press 1964.
Harrison, E. R., Noonan, T. W. (1979): Interpretation of extragalactic redshifts. Ap. J. **232**, 18.
Harrison, R. A. (1991): Coronal mass ejection. Phil. Trans. Roy. Soc. (London) **A336**, 401.
Hartle, J. B. (1978): Bounds on the mass and moment of inertia of non-rotating neutron stars. Physics Reports **46**, 201.
Hartle, J. B., Munn, M. W. (1975): Slowly rotating relativistic stars. V. Static stability analysis of $n = 3/2$ polytropes. Ap. J. **198**, 467.
Hartle, J. B., Sabbadine, A. G. (1977): The equation of state and bounds on the mass of nonrotating neutron stars. Ap. J. **213**, 831.
Hartmann, J. F. (1904): Investigations of the spectrum and orbit of delta orionis. Ap. J. **19**, 268. Reproduced in: A source book in astronomy and astrophysics 1900–1975 (eds. K. R. Lang and O. Gingerich). Cambridge, Mass.: Harvard University Press 1979.
Hartree, D. R. (1928): The wave mechanics of an atom with a non-Coulomb central field. Part. 1. Theory and methods. Proc. Camb. Phil. Soc. **24**, 89.
Hartree, D. R. (1931): The propagation of electromagnetic waves in a refracting medium in a magnetic field. Proc. Camb. Phil. Soc. **27**, 143.
Harvey, J. W. (1995): Helioseismology. Phys. Today **48**, 32 – October.
Harvey, J. W.: Helioseismology: The state of the art. In: Helioseismology – Proc. Fourth Soho Workshop. ESA SP-376 (1995) – June.
Harvey, J. W., et al. (1996): The global oscillation network group (GONG) project. Science **272**, 1284.
Harwit, M.: Astrophysical concepts. Second Edition. New York: Springer-Verlag, 1988.
Hasegawa, A. (1971): Plasma instabilities in the magnetosphere. Rev. of Geophys. and Space Sci. **9**, 703.
Haubold, H. J., John, R. W. (1978): On the evaluation of an integral connected with the thermonuclear reaction rate in closed form. Astron. Nach. **299**, 225.
Haubold, H. J., John, R. W. (1979): On resonant thermonuclear reaction rate integrals - closed-form evaluation and approximation considerations. Astron. Nach. **300**, 63.
Haubold, H. J., John, R. W.: A new approach to the analytic evaluation of thermonuclear reaction rates. In: Fundamental problems in the theory of stellar evolution (eds. D. Sugimoto, D. Q. Lamb and D. N. Schramm). Boston: D. Reidel 1981.
Haubold, H. J., Mathai, A. M. (1984): On the nuclear energy generation rate in a simple analytic stellar model. Ann. Phys. **41**, 372.

Haubold, H. J., Mathai, A. M. (1984): On nuclear reaction rate theory. Ann. Phys. **41**, 380.
Haubold, H. J., Mathai, A. M. (1986): Analytic solution to the problem of nuclear energy generation rate in a simple stellar model. Astron. Nach. **307**, 9.
Haubold, H. J., Mathai, A. M. (1986): The resonant thermonuclear reaction rate. J. Math. Phys. **27**, 2203.
Haxton, W. C. (1995): The solar neutrino problem. Ann. Rev. Astron. Astrophys. **33**, 459.
Hayakawa, S.: Cosmic ray physics. New York: Wiley Interscience 1969.
Hayakawa, S., Hayashi, C., Nishida, M. (1960): Rapid thermonuclear reactions in supernova explosions. Suppl. Prog. Theor. Phys. (Japan) **16**, 169.
Hayashi, C. (1961): Stellar evolution in early phases of gravitational contraction. Pub. Aston. Soc. Japan **13**, 450. Reproduced in: A source book in astronomy and astrophysics 1900–1975 (eds. K. R. Lang and O. Gingerich). Cambridge, Mass.: Harvard University Press 1979.
Hayashi, C. (1966): Evolution of protostars. Ann. Rev. Astron. Ap. **4**, 171.
Hayashi, C., Hoshi, R., Sugimoto, D. (1962): Evolution of stars. Prog. Theor. Phys. (Japan) Suppl. **22**, 1.
Hayashida, N., et al. (1994): Observation of a very energetic cosmic ray well beyond the predicted 2.7 K cutoff in the primary energy spectrum. Phys. Rev. Lett. **73**, 3491.
Heavens, A. F., Meisenheimer, K. (1987): Particle acceleration in extragalactic sources: the role of synchrotron losses in determining the spectrum. M.N.R.A.S. **225**, 335.
Heaviside, O. (1889): On the electromagnetic effects due to the motion of electrification through a dielectric. Phil. Mag. **27**, 324.
Heaviside, 0. (1902): The waste of energy from a moving electron. Nature **67**, 6.
Heaviside, 0. (1904): The radiation from an electron moving in an elliptic, or any other orbit. Nature **69**, 342.
Hebb, M. H., Menzel, D. H. (1940): Physical processes in gaseous nebulae X. collisional excitation of nebulium. Ap. J. **92**, 408. Reprod. in: Selected papers on physical processes in ionized plasmas (ed. D. H. Menzel). New York: Dover 1962.
Heeger, K. M., Robertson, R. G. H. (1996): Probability of a solution to the solar neutrino problem with the minimal standard model. Phys. Rev. Letters **77**, 3720.
Hegyi, D. J. (1977): The upper mass limit for neutron stars including differential rotation. Ap. J. **217**, 244.
Heiles, C. E., Drake, F. D. (1963): The polarization and intensity of thermal radiation from a planetary surface. Icarus **2**, 291.
Heiles, C. (1974): An almost complete survey of 21 centimeter line radiation for $|b| = 10°$ II. Data on machine readable tape. Astron. and Astrophys, Suppl. **14**, 557.
Heiles, C., Habing, H. J. (1974): An almost complete survey of 21 centimeter line radiation for $|b| \geq 10°$ I. Atlas of contour maps. Astron. and Astrophys. Suppl. **14**, 1.
Heisenberg, W. (1932): Über den Bau der Atomkerne I (On the structure of atomic nuclei I). Z. Physik **77**, 1.
Heisenberg, W. (1948): Zur statischen Theorie der Turbulenz (The statistical theory of turbulence). Z. Phys. **124**, 628.
Heisenberg, W. (1949): On the theory of statistical and isotropic turbulence. Proc. Roy. Soc. **195**, 402.
Heitler, W.: The quantum theory of radiation. Oxford: Oxford University Press 1954.
Heitler, W., Nordheim, L. (1934): Sur la production des paires par des chocs de particules lourdes (On the production of pairs by the collision of heavy particles). J. Phys. Radium **5**, 449.
Held, E. F. M. Van Der (1931): Intensity and natural widths of spectral lines. Z. Phys. **70**, 508.
Helfand, D. J., Tademaru, E. (1977): Pulsar velocity observations: correlations, interpretations, and discussion. Ap. J. **216**, 842.
Helfand, D. J., Taylor, J. H., Manchester, R. N. (1977): Pulsar proper motions. Ap. J. **213**, L 1.

Helmholtz, H. Von: Über die Erhaltung der Kraft (The conservation of force). Berlin: G. Riemer 1847. Engl. trans. in Brush, 1965.
Helmholtz, H. Von (1854): Popular lectures.
Helmholtz, H. Von (1868): Über diskontinuierliche Flüssigkeitsbewegungen (On discontinuities in moving fluids). Wiess Abhandlungen 146, J. A. Barth, 1882. Engl. trans. in Phil. Mag. **36**, 337.
Hempelmann, A., et al. (1995): Coronal X-ray emission and rotation of cool main-sequence stars. Astron. Astrophys. **294**, 515.
Hénon, M. (1982): Vlasov equation? Astron. Astrophys. **114**, 211.
Henry, J. P. et al. (1979): Detection of X-ray emission from distant clusters of galaxies. Ap. J. (Letters) **234**, L15.
Henry, R. C. (1991): Ultraviolet background radiation. Ann. Rev. Astron. Astrophys. **29**, 89.
Henry, R. C., Fritz, G., Meekins, J. F., Friedman, H., Byram, E. T. (1968): Possible detection of a dense intergalactic plasma. Ap. J. (Letters) **153**, L11.
Henry, R. J. W. (1970): Photoionization cross-sections for atoms and ions of carbon, nitrogen, oxygen, and neon. Ap. J. **161**, 1153.
Henyey, L. G. (1940): The Doppler effect in resonance lines. Proc. Nat. Acad. Sci. **26**, 50.
Henyey, L., Vardya, M. S. Bodenheimer, P. (1965): Studies in stellar evolution: III. The calculation of model envelopes. Ap. J. **142**, 841.
Herbst, E.: Gas phase chemical processes in molecular clouds. In: Interstellar processes (ed. D. J. Hollenbach and H. A. Thronson, Jr.). Boston: D. Reidel 1987.
Herbst, E., Klemperer, W. (1973): The formation and depletion of molecules in dense interstellar clouds. Ap. J. **185**, 505.
Herbst, E., Leung, C. M. (1986): Synthesis of complex molecules in dense interstellar clouds via gas-phase chemistry: Model update and sensitivity analysis. M.N.R.A.S. **222**, 689.
Herlofson, N. (1950): Magneto-hydrodynamic waves in a compressible fluid conductor. Nature **165**, 1020.
Hernanz, M., et al.: The age of the Galaxy obtained from white dwarfs. In: Astrophysical ages and dating methods (eds. E. Vangioni-Flam, M. Cassé, J. Audouze and J. Tran Thanh Van). Paris: Editions Frontieres 1990.
Hershberg, R. E., Pronik, V. I. (1959): The theory of the Strömgren zone. Astron. Zh. **36**, 902.
Hertz, H. (1889): Die Kräfte elektrischer Schwingungen, behandelt nach der Maxwell'schen Theorie (The force of electrical oscillations treated with the Maxwell theory). Ann. Phys. **36**, 1.
Hertz, H. (1889): Über die Beziehungen zwischen Licht und Elektrizität (On the relations between light and electricity). Gesammelte Werke **1**, 340.
Hertzsprung, E. (1905): Zur Strahlung der Sterne (Giants and dwarfs). Z. Wiss. Photog. 3. Eng. trans. in: Source book in astronomy (ed. H. Shapley). Cambridge, Mass.: Harvard University Press 1960.
Hertzsprung, E. (1926): On the relation between period and form of the light curve of variable stars of the δ Cephei type. B.A.N. **3**, 115.
Herzberg, G.: Molecular spectra and molecular structure I. Spectra of diatomic molecules. New York: Van Nostrand 1950.
Herzberg, G.: Atomic spectra and atomic structure. New York: Dover 1964.
Hess, V. F. (1911): Über die Absorption der γ-Strahlen in der Atmosphäre (On the absorption of γ-rays in the atmosphere). Phys. Z. **12**, 998.
Hess, V. F. (1912): Über Beobachtungen der durchdringenden Strahlung bei sieben Freiballonfahrten (On observations of the penetrating radiation during seven balloon flights). Phys. Zeits. **13**, 1084. English translation in: A source book in astronomy and astrophysics 1900–1975 (eds. K. R. Lang and O. Gingerich). Cambridge, Mass.: Harvard University Press 1979.
Hewish, A. (1951): The diffraction of radio waves in passing through a phase-changing ionosphere. Proc. Roy. Soc. London **A209**, 81.

Hewish, A. (1975): Pulsars and high density physics. Rev. Mod. Phys. **47**, 567.
Hewish, A. (1975): Pulsars and high density physics. Science **188**, 1079.
Hewish, A., Bell, S. J., Pilkington, J. D. H., Scott, P. F., Collins, R. A. (1968): Observation of a rapidly pulsating radio source. Nature **217**, 709. Reproduced in: Source book in astronomy and astrophysics 1900–1975 (eds. K. R. Lang and O. Gingerich). Cambridge, Mass.: Harvard University Press 1979.
Hewitt, A., Burbidge, A. (1980): A revised optical catalogue of quasi-stellar objects. Ap. J. Supp. **43**, 57; **46**, 113 (1981).
Hey, J. S. (1945): Solar radiations in the 2–6 metre radio wavelength band. Nature **157**, 47.
Hey, J. S.: The evolution of radio astronomy. New York: Neale Watson Academic Pub. 1973.
Heyvaerts, J., Priest, E. R. (1984): Coronal heating by reconnection in DC current systems, a theory based on Taylor's hypothesis. Astron. Astrophys. **137**, 63.
Hide, R. (1956): The character of the equilibrium of a heavy, viscous, incompressible fluid of variable density: I. General theory; II. Two special cases. Quart. J. Math. Appl. Math. **9**, 22, 35.
Hietler, W.: The quantum theory of radiation. London: Oxford at the Clarendon Press 1960.
Higdon, J. C. (1979): Distribution of cosmic rays and magnetic field in the Galaxy as deduced from synchrotron radio and gamma-ray observations. Ap. J. **232**, 113.
Higdon, J. C., Lingenfelter, R. E. (1990): Gamma-ray bursts. Ann. Rev. Astron. Astrophys. **28**, 401.
Higgs, L. A.: A catalogue of radio observations of planetary nebulae and related optical data. Nat. Res. Council (Canada), NRC 12129, Astrophys. Branch, Ottawa, 1971.
Higgs, L. A. (1971): A survey of microwave radiation from planetary nebulae. M.N.R.A.S. **153**, 315.
Hikasa, K., et al. (1992): Review of particle properties. Phys. Rev. **D45**, 1.
Hill, F.: Local probes of the solar interior. In: Helioseismolgy: Proc. Fourth SOHO Workshop. ESA SP-376 (1995) - June.
Hill, J. K., Hollenbach, D. J. (1976): H_2 molecules and the intercloud medium. Ap. J. **209**, 445.
Hill, J. K., Hollenbach, D. J. (1978): Effects of expanding compact H II regions upon molecular clouds: molecular dissocation waves, shock waves, and carbon ionization. Ap. J. **225**, 390.
Hillas, A. M.: Cosmic rays. Oxford: Pergamon Press 1972.
Hillas, A. M. (1984): The origin of ultra-high-energy cosmic rays. Ann. Rev. Astron. Astrophys. **22**, 425.
Hills, J. G. (1978): The rate of formation of white dwarfs in stellar systems. Ap. J. **219,** 550.
Hills, J. G. (1978): An upper limit to the rate of formation of neutron stars in the galaxy. Ap. J. **221**, 973.
Hiltner, W. A. (1949): Polarization of light from distant stars by the interstellar medium. Science **109**, 165. Reproduced in: A source book in astronomy and astrophysics 1900-1975 (eds. K. R. Lang and O. Gingerich). Cambridge, Mass.: Harvard University Press 1979.
Hinata, S. (1979): The role of turbulent heating in the solar atmosphere. Ap. J. **232**, 915.
Hinata, S. (1980): Electrostatic ion-cyclotron heating of solar atmosphere. Ap. J. **235**, 258.
Hindman, J. V., Kerr, F. J., Mc Gee, R. X. (1963): A low resolution hydrogen line survey of the Magellanic system. Austr. J. Phys. **16**, 570.
Hines, C. O. (1960): Internal atmospheric gravity waves at ionospheric heights. Can. J. Phys. **38**, 1441.
Hinteregger, H. E. (1965): Absolute intensity measurements in the extreme ultraviolet spectrum of solar radiation. Space Sci. Rev. **4**, 461.
Hirata, K. S., et al. (1987): Observation of a neutrino burst from the Supernova SN1987A. Phys. Rev. Letters **58**, 1490.

Hirata, K. S., et al. (1989): Observation of boron-8 solar neutrinos in the Kamiokande-II detector. Phys. Rev. Letters **63**, 16.

Hirata, K. S., et al. (1990): Constraints on neutrino-oscillation parameters from the Kamiokande-II solar-neutrino data. Phys. Rev. Letters **65**, 1301.

Hirayama, T. (1974): Theoretical model of flares and prominences. I. Evaporating flare model. Solar Phys. **34**, 323.

Hirshfeld, A., Sinnott, R. W.: Sky catalogue 2000.0. Volume 2. Double stars, variable stars and nonstellar objects. New York: Cambridge University Press 1985.

Hjellming, R. M. (1966): Physical processes in H II regions. Ap. J. **143**, 420.

Hjellming, R. M.: The effects of star formation and evolution on the evaluation of H II regions, and theoretical determinations of temperatures in H II regions. In: Interstellar ionized hydrogen (ed. Y. Terzian). New York: W.A. Benjamin 1968.

Hjellming, R. M., Andrews, M. H., Sejnowksi, T. J. (1969): Intensities of radio recombination lines. Astrophys. Lett. **3**, 111.

Hjellming, R. M., Churchwell, E. (1969): An analysis of radio recombination lines emitted by the Orion nebula. Astrophys. Lett. **4**, 165.

Hjellming, R. M., Gordon, C. P., Gordon, K. J. (1969): Properties of interstellar clouds and the intercloud medium. Astron. Astrophys. **2**, 202.

Hjellming, R. M., Gordon, M. A. (1971): Radio recombination lines and non-LTE theory: A reanalysis. Ap. J. **164**, 47.

Hjerting, F. (1938): Tables facilitating the calculation of line absorption coefficients. Ap. J. **88**, 508.

Ho, P. T. P., Townes, C. H. (1983): Interstellar ammonia. Ann. Rev. Astron. Astrophys. **21**, 239.

Hodgson, R. (1860): On a curious appearance seen in the sun. M.N.R.A.S. **20**, 15. Reproduced in: Early solar physics (ed. A. J. Meadows). Oxford: Pergamon Press 1970.

Hoffmeister, C., Richter, G., Wenzel, W.: Variable stars. New York: Springer-Verlag 1985.

Hogan, C., Layzer, D. (1977): Origin of the X-ray background. Ap. J. **212**, 360.

Hollenbach, D. J., Mc Kee, C. F. (1979): Molecule formation and infrared emission in fast interstellar shocks I. Physical processes. Ap. J. Supp. **41**, 555.

Hollenbach, D. J., Salpeter, E. E. (1970): Surface adsorption of light gas atoms. J. Chem. Phys. **53**, 79.

Hollenbach, D. J. Salpeter, E. E. (1971): Surface recombination of hydrogen molecules. Ap. J. **163**, 155.

Hollenbach, D. J., Thronson, H. A. Jr.: Interstellar processes. Boston: D. Reidel 1987.

Hollenbach, D. J., Werner, M. W., Salpeter, E. E. (1971): Molecular hydrogen in HI regions. Ap. J. **163**, 165.

Hollweg, J. V. (1972): Alfvénic motions in the solar atmosphere. Ap. J. **177**, 255.

Hollweg, J. V. (1978): Alfvén waves in the solar atmosphere. Solar Phys. **56**, 305.

Holstein, T. (1950): Pressure broadening of spectral lines. Phys. Rev. **79**, 744.

Holt, S. S., Mc Cray, R. (1982): Spectra of cosmic X-ray sources. Ann. Rev. Astron. Astrophys. **20**, 323.

Holtsmark, J. (1919): Über die Verbreiterung von Spektrallinien (On the broadening of spectral lines). Ann. Physik **58**, 577.

Holtsmark, J. (1925): Über die Absorption in Na-Dampf (On the absorption in sodium vapor). Z. Physik **34**, 722.

Holweger, H., Müller, E. A. (1974): The photospheric barium spectrum: Solar abundance and collision broadening of Ba II lines by hydrogen. Solar Physics **39**, 19.

Holzer, T. E. (1989): Interaction between the solar wind and the inerstellar medium. Ann. Rev. Astron. Astrophys. **27**, 199.

Hönl, H. Von (1925): Die Intensitäten der Zeeman-Komponenten (The intensity of the Zeeman components). Z. Physik **31**, 340.

Hooper, C. F. (1968): Low-frequency component electric microfield distributions in plasmas. Phys. Rev. **165**, 215.

House, L. L., Stemitz, R. (1975): The non LTE transport equation for polarized radiation in the presence of magnetic fields. Ap. J. **195**, 235.
Howard, W. M., Arnett, W. D., Clayton, D. D. (1971): Explosive nucleosynthesis in helium zones. Ap. J. **165**, 495.
Howard, W. M., et al. (1972): Nucleosynthesis of rare nuclei from seed nuclei in explosive carbon burning. Ap. J. **175**, 201.
Hoyle, F. (1946): The synthesis of elements from hydrogen. M.N.R.A.S. **106**, 343.
Hoyle, F. (1954): On nuclear reactions occurring in very hot stars: I. The synthesis of elements from carbon to nickel. Ap. J. Suppl. **1**, 121.
Hoyle, F., Burbidge, G. R. Sargent, W. L. W. (1966): On the nature of the quasi-stellar sources. Nature **209**, 751.
Hoyle, F., Clayton, D. D. (1974): Nucleosynthesis in white-dwarf atmospheres. Ap. J. **191**, 705.
Hoyle, F., Fowler, W. A. (1960): Nucleosynthesis in supernovae. Ap. J. **132**, 565.
Hoyle, F., Lyttleton, R. A. (1939): The evolution of the stars. Proc. Camb. Phil. Soc. **35**, 592.
Hoyle, F., Tayler, R. J. (1964): The mystery of the cosmic helium abundance. Nature **203**, 1108.
Hoyng, P., et al. (1981): Origin and location of the hard X-ray emission in a two-ribbon flare. Ap. J. (Letters) **246**, L155.
Huang, S. S., Struve, O. (1955): Study of Doppler velocities in stellar atmospheres. The spectrum of Alpha Cygni, Ap. J. **121**, 84.
Huang, S. S., Struve, O.: Stellar rotation and atmospheric turbulence. In: Stars and stellar systems, vol. 6 (ed. J. Greenstein). Chicago, Ill.: Univ. of Chicago Press 1960.
Hubble, E. (1922): A general study of diffuse galactic nebulae. Ap. J. **56**, 162.
Hubble, E. (1922): The source of luminosity in galactic nebulae. Ap. J. **56**, 400.
Huchtmeier, W. K., Richter, O.-G.: A general catalog of H I observations of galaxies. The reference catalog. New York: Springer-Verlag 1989.
Hudson, H. S. (1972): Thick-target processes and white-light flares. Solar Phys. **24**, 414.
Hudson, H. S., et al. (1994): Impulsive behavior in solar soft X-radiation. Ap. J. (Letters) **422**, L25.
Hudson, H. S., Haisch, B, Strong, K. T. (1995): Comment on "the solar flare myth" by J. T. Gosling. J. Geophys. Res. **100**, 3473.
Hudson, H. S., Ryan, J. (1995): High-energy particles in solar flares. Ann. Rev. Astron. Astrophys. **33**, 239.
Hughes, M. P., Thompson, A. R., Colvin, R. S. (1971): An absorption line study of the galactic neutral hydrogen at 21 centimeters wavelength. Ap. J. Suppl. No. 200, **23**, 323.
Hugoniot, Par H. (1887): Sur la propagation du mouvement dans les corps et specialement dans les gaz parfarts (On the propagation of the movement of bodies, and especially of the perfect gas). J. de l'Ecole Polytechnique **57**, 1, **59**, 1 (1889).
Hull, M. H., Briet, G.: Coulomb wave functions. In: Handbuch der Physik, vol. XLI: Nuclear reactions; II: Theory (ed. S. Flügge), p. 408. Berlin-Göttingen-Heidelberg: Springer 1959.
Hulst, H. C. Van De (1945): Radiogloven mit hat wereldrium: I. Ontvangst der radiogloven; II. Herkomst der radiogloven (Radio waves from space: I. Reception of radiowaves; II. Origin of radiowaves). Ned. Tijdsch. Natuurk **11**, 201. Engl. trans. in: Source book in astronomy and astrophysics 1900-1975 (eds. K. R. Lang and O. Gingerich). Cambridge, Mass: Harvard University Press 1979).
Hulst, H. C. Van De (1949): On the attenuation of plane waves by obstacles of arbitrary size and form. Physica **15**, 740.
Hulst, H. C. Van De (1949): The solid particles in interstellar space.Rech. Astr. Obs. Utrecht XI, part 2. Reproduced in: Source book in astronomy and astrophysics 1900-1975 (eds. K. R. Lang and O. Gingerich). Cambridge, Mass.: Harvard University Press 1979.
Hulst, H. C. Van De: Light scattering by small particles. New York: Wiley 1957.
Humason, M. L., Mayall, N. U., Sandage, A. R. (1956): Redshifts and magnitudes of extragalactic nebulae. Astron. J. **61**, 97.

Hummer, D. G. (1988): A fast and accurate method for evaluating the nonrelativistic free-free Gaunt factor for hydrogenic ions. Ap. J. **327**, 477.
Hummer, D. G., Seaton, M. J. (1963): The ionization structure of planetary nebulae: I. Pure hydrogen nebulae. M.N.R.A.S. **125**, 437.
Hummer, D. G., Storey, P. J. (1987): Recombination-line intensities for hydrogenic ions - I. Case B calculations for H I and He II. M.N.R.A.S. **224**, 801.
Humphreys, C. J. (1953): The sixth series in the spectrum of atomic hydrogen. J. Res. Nat. Bur. Stands., **50**, 1.
Hund, F. Von (1927): Zur Deutung der Molekelspektren II (On the interpretation of molecular spectra II). Z. Physik **42**, 93.
Hunger, K. (1956): The theory of curves of growth. Z. Ap. **39**, 38.
Hunter, C. (1977): On secular stability, secular instability, and points of bifurcation of rotating gaseous masses. Ap. J. **213**, 497.
Hunter, C. (1977): The collapse of unstable isothermal spheres. Ap. J. **218**, 834.
Hunter, D. A., Watson. W. D. (1978): The translational and rotational energy of hydrogen molecules after recombination on interstellar grains. Ap. J. **226**, 477.
Huntress, W. T., Jr., Mitchell, G. F. (1979): The synthesis of complex molecules in interstellar clouds. Ap. J. **231**, 456.
Hutchins, J. B. (1976): The thermal effects of H_2 molecules in rotating and collapsing spheroidal gas clouds. Ap. J. **205**, 103.
Iben, I. Jr. (1969): The Cl^{37} solar neutrino experiment and the solar helium abundance. Ann. Phys. (N.Y.) **54**, 164.
Iben, I. Jr. (1975): Thermal pulses; p-capture, α-capture, s-process nucleosynthesis; and convective mixing in a star of intermediate mass. Ap. J. **196**, 525.
Iben, I. Jr. (1975): Neon-22 as a neutron source, light elements as modulators, and s-process nucleosynthes in a thermally pulsating star. Ap. J. **196**, 549.
Iben, I. Jr. (1978): URCA neutrino-loss rates under conditions found in the carbon-oxygen cores of intermediate-mass stars. Ap. J. **219**, 213.
Iben, I, Jr. (1978): Thermal oscillations during carbon burning in an electron-degenerate stellar core. Ap. J. **226**, 996.
Iben, I. Jr., Renzini, A. (1983): Asymptotic giant branch evolution and beyond. Ann. Rev. Astron. Astrophys. **21**, 271.
Iben, I. Jr., Renzini, A. (1984): Single star evolution I. Massive stars and early evolution of low and intermediate mass stars. Phys. Rpt. **105**, 329.
Iben, I. Jr., Rood, R. T. (1970): Metal-poor stars I. evolution from the main sequence to the giant branch. Ap. J. **159**, 605.
Iben, I. Jr., Rood, R. T. (1970): Metal-poor stars III. on the evolution of horizontal-branch stars. Ap. J. **161**, 587.
Iben, I. Jr., Truran, J. W. (1978): On the surface composition of thermally pulsing stars of high luminosity and on the contribution of such stars to the element enrichment of the interstellar medium. Ap. J. **220**, 980.
Iben, I. Jr., Tuggle, R. S. (1972): Comments on a PLC relationship for Cepheids and on the comparison between pulsation and evolution masses for Cepheids. Ap. J. **178**, 441.
Ichimaru, S. (1982): Strongly coupled plasmas: High-density classical plasmas and degenerate electron liquids. Rev. Mod. Phys. **54**, 1017.
Ichimaru, S., Utsumi, K. (1984): Screening potential and enhancement of thermonuclear reaction rate due to relativistic degenerate electrons in dense multi-ionic plasmas. Ap. J. **278**, 382.
Imanishi, B. (1968): Resonance energies and partial widths of quasimolecular states formed by the two carbon nuclei. Phys. Lett. **27B**, 267.
Imanishi, B. (1969): Quasimolecular states in the interaction between ^{12}C and ^{12}C nuclei. Nuclear Phys. **A125**, 33.
Inglis, D. R. (1981): Dynamo theory of the earth's varying magnetic field. Rev. Mod. Phys. **53**, 481.

Inglis, D. R., Teller, E. (1939): Ionic depression of series limits in one-electron spectra. Ap. J. **90**, 439.
Inman, C. L., Ruderman, M. A. (1964): Plasma neutrino emission from a hot, dense electron gas. Ap. J. **140**, 1025.
Innes, D. E., et al. (1997): Bi-directional plasma jets produced by magnetic reconnection on the sun. Nature **386**, 811.
Irvine, W. M., et al.: The chemical state of dense interstellar clouds: An overview. In: Protostars and planets II (eds. D. C. Black and M. S. Matthews). Tucson, Arizona: The University of Arizona Press 1985.
Isenberg, P. A. (1977): Adiabatic self-similar blast waves, their radial instabilities, and their application to supernova remnants. Ap. J. **217**, 597.
Ito, K. (1961): Stellar synthesis of the proton-rich heavy elements. Prog. Theor. Phys. (Japan) **26**, 990.
Jackson, J. D. (1960): Longitudinal plasma oscillation. J. Nucl. Energy C, **1**, 171.
Jackson, J. D.: Classical electrodynamics. New York: Wiley 1962.
Jackson, J. D.: Classical electrodynamics, second edition. New York: John Wiley and Sons 1975.
Jacobi, C. G. J. (1834): Über die Figur des Gleichgewichts (On the figure of objects of the same specific gravity). Ann. Phys. u. Chem. **33**, 229.
Jacobs, V. L., et al. (1977): The influence of autoionization accompanied by excitation on dielectronic recombination and ionization equilibrium. Ap. J. **211**, 605.
Jacobs, V. L., et al. (1979): Dielectric recombination rates, ionization equilibrium, and radiative energy-loss rates for neon, magnesium, and sulfur ions in low-density plasmas. Ap. J. **230**, 627.
Jansky, K. G. (1935): A note on the source of interstellar interference. Proc. Inst. Radio Eng. **23**, 1158. Reproduced in: A source book in astronomy and astrophysics 1900–1975 (eds. K. R. Lang and O. Gingerich). Cambridge, Mass.: Harvard University Press 1979.
Jaschek, C., Jaschek, M.: The behavior of chemical elements in stars. New York: Cambridge University Press 1995.
Jauch, J. M., Rohrlich, F.: The theory of protons and electrons. Reading, Mass.: Addison-Wesley 1955.
Jeans, Sir J. H. (1902): On the stability of a spherical nebula. Phil. Trans. Roy. Soc. London **199**, 1. Reproduced in: Source book in astronomy and astrophysics 1900-1975 (eds. K. R. Lang and O. Gingerich). Cambridge, Mass.: Harvard University Press 1979.
Jeans, Sir J. H. (1905): On the partition of energy between matter and aether. Phil. Mag. **10**, 91.
Jeans, Sir J. H. (1909): Temperature-radiation and the partition of energy in continuous media. Phil. Mag. **17**, 229.
Jeans, Sir J. H. (1917): The motion of tidally-distorted masses, with special reference to theories of cosmogony. Mem. R.A.S. London **62**, 1.
Jeans, Sir J. H.: Problems of cosmogony and stellar dynamics. Cambridge, England: Cambridge University Press 1919.
Jeans, Sir J. H.: Astronomy and cosmogony. Cambridge, England: Cambridge University Press 1929.
Jefferts, K. B., Penzias, A. A., Wilson, R. W. (1973): Deuterium in the Orion nebula. Ap. J. (Letters) **179**, L57.
Jeffreys, H. (1926): The stability of a layer of fluid heated below. Phil. Mag. **2**, 833.
Jeffreys, H. (1930): The instability of a compressible fluid heated below. Proc. Çamb. Phil. Soc. **26**, 170.
Jeffries, J. T.: Spectral line formation. Waltham, Mass.: Blaisdell 1968.
Jelley, J. V. (1966): High-energy γ-ray absorption in space by a 3.5 °K microwave field. Phys. Rev. Lett. **16**, 479.
Jenkins, F.A., Segre, É. (1939): The quadratic Zeeman effect. Phys. Rev. **55**, 52.

Johns, O., Reeves, H. (1975): The r-process production ratios of long-lived radionuclides. Ap. J. **202**, 214.

Jokipii, J. R. (1973): Turbulence and scintillations in the interplanetary plasma. Ann. Rev. Astron. Astrophys. **11**, 1.

Jokipii, J. R. (1976): Consequences of a lifetime greater than 10^7 years for galactic cosmic rays. Ap. J. **208**, 900.

Jokipii, J. R., Hollweg, J. V. (1970): Interplanetary scintillations and the structure of solar-wind fluctuations. Ap. J. **160**, 745.

Jokipii, J. R., Kóta, J. (1989): The polar heliospheric magnetic field. Geophysical Research Letters **16**, 1.

Jones, W., O'Dell, S. L. (1977): Transfer of polarized radiation in self-absorbed synchrotron sources. Ap. J. **214**, 522.

Jones, W., O'Dell, S. L. (1977): Transfer of polarized radiation in self-absorbed synchrotron sources. II. Treatment of inhomogeneous media and calculation of emergent polarization. Ap. J. **215**, 236.

Jones, T. W., O'Dell, S. L., Stein, W. A. (1974): Physics of compact nonthermal sources. I. Theory of radiation processes. Ap. J. **188**, 353.

Jones, T. W., O'Dell, S. L., Stein, W. A. (1974): Physics of compact nonthermal sources. II. Determination of physical parameters. Ap. J. **192**, 261.

Joss, P. C., Rappaport, S. A. (1984): Neutron stars in interacting binary systems. Ann. Rev. Astron. Astrophys. **22**, 537.

Joule, J. P. (1847): On matter, living force, and heat., lecture repr. in Joule's scientific papers and Brush (1965).

Joy, A. H. (1939): Rotation effects, interstellar absorption, and certain dynamical constants of the Galaxy determined from Cepheid variables. Ap. J. **89**, 356. Reproduced in: A source book in astronomy and astrophysics 1900–1975 (eds. K. R. Lang and O. Gingerich). Cambridge, Mass.: Harvard University Press 1979.

Julienne, P. S., Krauss, M.: Molecule formation by inverse predissociation. In: Molecules in the galactic environment (ed. M. A. Gordon, L. E. Snyder). New York: Wiley 1973.

Julienne, P. S., Krauss, M., Donn, B. (1971): Formation of OH through inverse predissociation. Ap. J. **170**, 65.

Jura, M. (1974): Formation and destruction rates of interstellar H_2. Ap. J. **191**, 375.

Kafatos, M. C., Harrington, R. S., Maran, S. P.: Astrophysics of brown dwarfs. New York: Cambridge University Press 1986.

Kahler, S. W. (1992): Solar flares and coronal mass ejections. Ann. Rev. Astron. Astrophys. **30**, 113.

Kahn, F. D. (1961): Sound waves trapped in the solar atmosphere. Ap. J. **134**, 343.

Kaler, J. B. (1970): Chemical composition and the parameters of planetary nebulae. Ap. J. **160**, 887.

Kaler, J. B. (1978): The [0 III] lines as a quantitative indicator of nebular central-star temperature. Ap. J. **220**, 887.

Kaler, J. B. (1978): Galactic abundances of neon, argon, and chlorine derived from planetary nebulae. Ap. J. **225**, 527.

Kaler, J. B. (1985): Planetary nebulae and their central stars. Ann. Rev. Astron. Astrophys. **23**, 89.

Kaler, J. B. (1986): Electron temperatures in planetary nebulae. Ap. J. **308**, 322.

Kaler, J. B. (1986): C^{+2} electron temperatures in planetary nebulae. Ap. J. **308**, 337.

Källén, G. (1967): Radioactive corrections to β-decay and nucleon form factors. Nucl. Phys. **B1**, 225.

Kallman, T., Mc Cray, R. (1980): Efficiency of the Bowen fluorescence mechanism in static nebulae. Ap. J. **242**, 615.

Kalogera, V., Baym, G. (1996): The maximum mass of a neutron star. Ap. J. (Letters) **470**, L61.

Kamionkowski, M., Bahcall, J. N. (1994): The rate of the proton-proton reaction. Ap. J. **420**, 884.

Kane, S. R., Anderson, K. A. (1976): Characteristics of cosmic X-ray bursts observed with the OGO-5 satellite. Ap. J. **210**, 875.

Kaplan, S. A. (1954): A system of spectral equations of magneto-gas-dynamic isotropic turbulence. Dokl. Acad. Nauk. SSSR **94**, 33.

Kaplan, S. A., Pikelner, S. B.: The interstellar medium. Cambridge, Mass.: Harvard University Press 1970.

Kapteyn, J. C. (1905): Star streaming. Rep. British Assoc. Adv. Sci. 237. Reproduced in: A source book in astronomy and astrophysics 1900-1975 (eds. K. R. Lang and O. Gingerich). Cambridge, Mass.: Harvard University Press 1979.

Kardashev, N. S. (1959): On the possibility of detection of allowed lines of atomic hydrogen in the radio frequency spectrum. Sov. Astron. **3**, 813.

Kardashev, N. S. (1962): Nonstationariness of spectra of young sources of nonthermal radio emission. Sov. Astron. **6**, 317.

Kármán, T. (1937): On the statistical theory of turbulence. Proc. Nat. Acad. Sci. **23**, 98.

Kármán, T. , Howarth, L. (1938): On the statistical theory of isotropic turbulence. Proc. Roy. Soc. London **A164**, 192.

Karp, A. H. (1975): Hydrodynamic models of a cepheid atmosphere. I. Deep envelope calculations. Ap. J. **199**, 448.

Karp, A. H. (1975): Hydrodynamic models of a cepheid atmosphere. II. Continuous spectrum. Ap. J. **200**, 354.

Karp, A. H. (1975): Hydrodynamic models of a cepheid atmosphere. III. Line spectrum and radius determinations. Ap. J. **201**, 641.

Karp, A. H., et al. (1977): The opacity of expanding media: the effect of spectral lines. Ap. J. **214**, 161.

Karzas, W. J., Latter, R. (1961): Electron radiative transitions in a Coulomb field. Ap. J. Suppl. **6**, 167.

Katz, J. I. (1976): The origin of X-ray sources in clusters of galaxies. Ap. J. **207**, 25.

Katz, J., Horwitz, G., Klapisch, M. (1975): Thermodynamic stability of relativistic stellar clusters. Ap. J. **199**, 307.

Kaufman, F. (1969): Elementary gas processes. Ann. Rev. Phys. Chem. **20**, 46.

Kazanas, D., Schramm, D. N. (1977): Neutrino damping of nonradial pulsations in gravitational collapse. Ap. J. **214**, 819.

Kellermann, K. I. (1966): The radio source 1934-63. Austr. J. Phys. **19**, 195.

Kellermann, K. I. (1970): Thermal radio emission from the major planets. Radio Science **5** (2), 487.

Kellermann, K. I., Owen, F. N.: Radio galaxies and quasars. In: Galactic and extragalactic radio astronomy, second edition (eds. G. L. Verschuur and K. I. Kellermann). New York: Springer-Verlag 1988.

Kellerman, K. I., Pauliny-Toth, I. I. K. (1969): The spectra of opaque radio sources. Ap. J. (Letters) **155**, L71.

Kellerman, K. I., Pauliny-Toth, I. I. K. (1981): Compact radio sources. Ann. Rev. Astron. Astrophys. **19**, 373.

Kellermann, K. I., Pauliny-Toth, I. I. K., Williams, P. J. S. (1969): The spectra of radio sources in the revised 3 C catalogue, Ap. J. **157**, 1.

Kelvin, Lord (W. Thomson) (1835): Transient electric currents. Proc. Glasgow Phil. Soc. **3**, 281.

Kelvin, Lord (W. Thomson) (1848): On an absolute thermometric scale founded on Carnot's theory of the motive power of heat. Proc. Camb. Phil. Soc. **1**, 66.

Kelvin, Lord (W. Thomson) (1850): A mathematical theory of magnetism. Phil. Mag. **37**, 241.

Kelvin, Lord (W. Thomson) (1862): Physical considerations regarding the possible age of the Sun's heat. Phil. Mag. **23**, 158. Brit. Assoc. Rpt. 27 (1861), Math. and Phys. Papers **5**, 141 (1861).

Kelvin, Lord (W. Thomson) (1862): On the convective equilibrium of temperature in the atmosphere. Math. and Phys. Papers **3**, 255.
Kelvin, Lord (W. Thomson) (1863): Sur le refroidissement seculaire du soleil. De la température actuelle due soleil. De l'origine et de la somme totale de la chaleur solaire (On the age of the Sun's heat, the actual temperature of the Sun, and the origin of the sum total of the Sun's heat). Les Mondes, **3**, 473.
Kelvin, Lord (W. Thomson) (1871): Hydrokinetic solutions and observations, on the motion of free solids through a liquid. Phil. Mag. **42**, 362.
Kemp, J. C. (1970): Circular polarization of thermal radiation in a magnetic field. Ap. J. **162**, 169.
Kemp, J. C., et al. (1970): Discovery of circularly polarized light from a white dwarf star. Ap. J. (Letters) **161**, L77. Reproduced in: A source book in astronomy and astrophysics 1900–1975 (eds. K. R. Lang and O. Gingerich). Cambridge, Mass.: Harvard University Press 1979.
Kemp, J. C., Swedlund, J. B., Wolstencroft, R. D. (1971): Confirmation of the magnetic white dwarf G 195-19. Ap. J. (Letters) **164**, L17.
Kepple, P. C. (1972): Improved Stark-profile calculations for the He II lines at 256, 304, 1085, 1216, 1640, 3203 and 4686 Å. Phys. Rev. **A6**, 1.
Kepple, P. C., Griem, H. R. (1968): Improved Stark profile calculations for the hydrogen lines Hα, Hβ, Hγ and Hδ. Phys. Rev. **A173**, 317.
Kerr, F. J.: Radio line emission and absorption by the interstellar gas. In: Stars and stellar systems, vol. 7 (ed. G. P. Kuiper), Chicago, Ill.: Univ. of Chicago Press 1968.
Khandelwal, G. S., Khan, F., Wilson, J. W. (1989): An asymptotic expression for the dipole oscillator strength for transitions of the He sequence. Ap. J. **336,** 504.
Khintchine, A. (1934): Korrelationstheorie der stationären stochastischen Prozesse (Correlation theory of a stationary stochastic process). Math. Annalen **109**, 604.
Kiepenheuer, K. O. (1950): Cosmic rays as the source of general galactic radio emission. Phys. Rev. **79**, 738. Reproduced in: A source book in astronomy and astrophysics 1900–1975 (eds. K. R. Lang and O. Gingerich). Cambridge, Mass.: Harvard University Press 1979.
Kinahan, B. F., Härm, R. (1975): Chemical composition and the Hertzsprung gap. Ap. J. **200**, 330.
King, D., et al. (1975): Applications of linear pulsation theory to the cepheid mass problem and the double-mode cepheids. Ap. J. **195**, 467.
Kinman, T. D. (1956): An attempt to detect deuterium in the solar atmosphere. M.N.R.A.S. **116**, 77.
Kippenhahn, R., Weigert, A.: Stellar structure and evolution. New York: Springer-Verlag 1990.
Kirchhoff, H. (1860): On the simultaneous emission and absorption of rays of the same definite refrangibility. Phil. Mag. **19**, 193.
Kirchhoff, H. (1860): On a new proposition in the theory of heat. Phil. Mag. **21**, 240. See also: Gesammelte Abhandlungen. Leipzig: J. A. Barth 1882.
Kirchhoff, G., Bunsen, R. (1861): Chemical analysis by spectrum-observations. Phil. Mag. **22**, 329, 498.
Kirk, J. G., Schneider, P. (1987): Particle acceleration at shocks: A monte carlo method. Ap. J. **322**, 256.
Kirkwood, J. G. (1947): The statistical mechanical theory of transport processes: II. Transport in gases. J. Chem. Phys. **15**, 72, 155.
Kirshner, R. P., Kwan, J. (1975): The envelopes of Type II supernovae. Ap. J. **197**, 415.
Kislinger, M. B., Morley, P. D. (1978): Asymptotic freedom and dense stellar matter. II. The equation of state for neutron stars. Ap. J. **219**, 1017.
Kitchen, C. R.: Stars, nebulae and the interstellar medium. Boston: Adam Hilger, 1987.
Klebesadel, R. W., Strong, I. B., Olson, R. A. (1973): Observations of gamma-ray bursts of cosmic origin. Ap.J. (Letters) **182**, L85.

Klein, O., Nishina, Y. (1929): Über die Streuung von Stahlung durch freie Elektronen nach der neuen relativistischen Quantendynamik von Dirac (On the scattering of radiation by free electrons according to the new relativistic quantum dynamics by Dirac). Z. Physik **52**, 853.

Klemperer, W. (1971): Interstellar molecule formation, radiative association and exchange reactions. In: Highlights of astronomy, vol 2 (ed. C. De Jager). Dordrecht, Holland: D. Reidel.

Knapp, H. F. P., Van Den Meijdenberg, C. J. N., Beenakker, J. J. M., Van De Hulst, H. C. (1966): Formation of molecular hydrogen in interstellar space. B.A.N. **18**, 256.

Koch, H. W., Motz, J. W. (1959): Bremsstrahlung cross-section formulas and related data. Rev. Mod. Phys. **31**, 920.

Kodama, T. (1971): β-stability line and liquid-drop mass formulas. Prog. Theor. Phys. (Japan) **45**, 1112.

Koehler, J. A. (1966): Intergalactic atomic neutral hydrogen II. Ap. J. **146**, 504.

Koester, D. (1987): Gravitational redshift in white dwarfs. Mem. Societa Italiana, Nuovo Serie **58**, 45.

Kohyama, Y., Itoh, N., Obama, A., Mutoh, H. (1993): Neutrino energy loss in stellar interiors. V. Recombination neutrino process. Ap. J. **415**, 267.

Kolb, A. C., Griem, H. (1958): Theory of line broadening in multiplet spectra. Phys. Rev. **111**, 514.

Kolhörster, W. (1913): Messungen der durchdringenden Strahlung im Freiballon in größeren Höhen (Measurements of the penetrating radiation from a balloon at greater altitudes). Phys. Z. **14**, 1153.

Kolmogoroff, A. N. (1941): The local structure of turbulence in incompressible viscous fluids for very large Reynolds numbers. Compt. Rend. Acad. Sci. (SSSR) **30**, 301.

Kolmogoroff, A. N. (1941): Dissipation of energy in the locally isotropic turbulence. Compt. Rend. Acad. Sci. (SSSR) **32**, 16.

Konopinski, E. J. (1959): The experimental clarification of the laws of β-radioactivity. Ann. Rev. Nucl. Sci. **9**, 99.

Konopinski, E. J.: The theory of beta radioactivity. Oxford: Oxford at the Clarendon Press 1966.

Kopp, R. A., Pneuman, G. W. (1976): Magnetic reconnection in the corona and the loop prominence phenomenon. Solar Phys. **50**, 85.

Kosloff, R., Kafri, A., Levine, R. D. (1977): Rotational excitation of interstellar OH molecules. Ap. J. **215**, 497.

Kosovichev, A. G. (1996): Tomographic imaging of the Sun's interior. Ap. J. (Letters) **461**, L55.

Kosovichev, A. G. (1996): Helioseismic constraints on the gradient of angular velocity at the base of the solar convection zone. Ap. J. (Letters) **469**, L61.

Kosovichev, A. G., et al. (1997): Structure and rotation of the solar interior: Initial results from the MDI medium-l program. Solar Phys. **170**, 43.

Kothari, D. S. (1938): The theory of pressure–ionization and its applications. Proc. Roy. Soc. London **A165**, 486.

Kouveliotou, C., et al. (1993): Recurrent burst activity from the soft γ-ray repeater SGR 1900+14. Nature **362**, 728.

Kouveliotou, C., et al. (1998): An X-ray pulsar with a superstrong magnetic field in the soft γ-ray repeater SGR 1806-20. Nature **393**, 235.

Kozlovsky, B, Ramaty, R. (1977): Narrow lines from alpha–alpha reactions. Astrophysical Letters **19**, 19.

Kraft, R. P. (1965): Studies in stellar rotation I. Comparison of rotational velocities in the Hyades and Coma clusters. Ap. J. **142**, 681.

Kraft, R. P.: Stellar rotation. In: Spectroscopic astrophysics – An assessment of the contributions of Otto Struve. Berkeley: University of California Press 1970.

Kragh, H.: Dirac - a scientific biography. New York: Cambridge University Press 1990.

Kramers, H. A. (1923): On the theory of X-ray absorption and of the continuous X-ray spectrum. Phil. Mag. **46**, 836.

Kramers, H. A. (1924): The law of dispersion and Bohr's theory of spectra. Nature **113**, 783.

Kramers, H. A. (1926): Wellenmechanik und halbzahlige Quantisierung (Wave mechanics and semi-numerical quantization). Z. Phys. **39**, 828.

Kramers, H. A., Ter Haar, D. (1946): Condensation in interstellar space. B.A.N. **10**, 137.

Krieger, A. S., Timothy, A. F., Roelof, E. C. (1973): A coronal hole and its identification as the source of a high velocity solar wind stream. Solar Phys. **29**, 505.

Kristian, J.: Optical identification of 3CR sources. In: Galaxies and the universe (eds. A. Sandage, M. Sandage and J. Kristian). Chicago: University of Chicago Press 1975.

Kronberg, P. P. (1994): Extragalactic magnetic fields. Rep. Prog. Phys. **57**, 325.

Kronig, R. De L. (1925): Über die Intensität der Mehrfachlinien und ihrer Zeeman-Komponenten (On the intensity of multiple lines and their Zeeman components). Z. Physik **31**, 885.

Kronig, R. De L., Kramers, H. A. (1928): Zur Theorie der Absorption und Dispersion in den Röntgenspektren (Theory of absorption and dispersion in X-ray spectra). Z. Physik **48**, 174.

Krotscheck, E. (1972): Superfluidity in neutron matter. Z. Phys. **251**, 135.

Kruskal, M., Schwarzschild, M. (1954): Some instabilities of a completely ionized plasma. Proc. Roy. Soc. London **A223**, 348.

Kruskal, M., Schwarzschild, M., Härm, R. (1977): An instability due to the local mixing-length approximation. Ap. J. **214**, 498.

Kruskal, M., Tuck, J. L. (1958): The instability of a pinched fluid with a longitudinal magnetic field. Proc. Roy. Soc. London A**245**, 222.

Krymskii, G. F. (1977): A regular mechanism for the acceleration of charged particles on the front of a shock wave. Sov. Phys. Dokl. **22**, 327 (1978). Dokl. Akad. Nauk SSSR **234**, 1306.

Kuchowicz, B. (1963): Neutrino gas statistics. Bull. Acad. Pol. Sci. Serv. Sci. Mat. Astr. et Phys. **11**, 317.

Kuchowicz, B. (1963): Neutrinos in superdense matter: I. A tentative statistical approach. Inst. Nucl. Res. Warsaw Rpt. 384.

Kuchowicz, B.: Nuclear astrophysics: A bibliographical survey. New York: Gordon and Breach 1967.

Kuhn, W. : Über die Gesamtstärke der von einem Zustande ausgehenden Absorptionslinien (On the total intensity of the absorption lines originating at one level). Z. Physik **33**, 408.

Kuiper, T. B. H., Rodriguez Kuiper, E. N., Zuckerman, B. (1978): Spectral line shapes in spherically symmetric radially moving clouds. Ap. J. **219**, 129.

Kukarin, B. V. Parenago, P. P.: General catalogue of variable stars. 3rd ed. Moscow, 1969.

Kulkarni, S. R., et al. (1998): Identification of a host galaxy at redshift $z = 3.42$ for the γ-ray burst of 14 december 1997. Nature **393**, 35.

Kulkarni, S. R., Frail, D. A. (1993): Identification of a supernova remnant coincident with the soft γ-ray repeater SGR 1806-20. Nature **365**, 33.

Kulkarni, S. R., Heiles, C.: Neutral hydrogen and the diffuse interstellar medium. In: Galactic and extragalactic radio astronomy, second edition (eds. G. L. Verschuur and K. I. Kellermann). New York: Springer-Verlag 1988.

Kulsrud, R. M.: Plasma instabilities. In: Plasma astrophysics – Proc. Int. Sch. Phys. Enrico Fermi - Course 39 (ed. P. A. Sturrock) New York: Academic Press 1967.

Kulsrud, R. M., Ostriker, J. P., Gunn, J. E. (1972): Acceleration of cosmic rays in supernova remnants. Phys. Rev. Lett. **28**, 636.

Kundu, M. R.: Solar radio astronomy. New York: Interscience 1965.

Kundu, M. R. (1982): Advances in solar radio astronomy. Rep. Prog. Phys. **45**, 1435.

Kundu, M. R., Lang, K. R. (1985): The sun and nearby stars: Microwave observations at high resolution. Science **228**, 9.

Kundu, M. R., Vlahos, L. (1982): Solar microwave bursts – a review. Space Sci. Rev. **32**, 405.

Kuo, T. K., Pantaleone, J. (1989): Neutrino oscillations in matter. Rev. Mod. Phys. **61**, 937.

Kuperus, M., Ionson, J. A., Spicer, D. S. (1981): On the theory of coronal heating mechanisms. Ann. Rev. Astron. Astrophys. **19**, 7.

Kurtz, D. W. (1990): Rapidly oscillating Ap stars. Ann. Rev. Astron. Astrophys. **28**, 607.

Kurucz, R. L. (1979): Model atmospheres for G, F, A, B, and O stars. Ap. J. Supp. **40**, 1.

Kutsenko, A. B., Stepanov, K. N. (1960): Instability of plasma with anisotropic distributions of ion and electron velocities. Sov. Phys. Jetp **11**, 1323.

Laan, H. Van der (1963): Radio galaxies: I. The interpretation of radio source data. M.N.R.A.S. **126**, 519.

Lada, C. J. (1985): Cold outflows, energetic winds and enigmatic jets around young stellar objects. Ann. Rev. Astron. Astrophys. **23**, 267.

Ladenburg, R. Von (1921): Die Quantentheoretische Deutung der Zahl der Dispersionselektronen (The quantum theory interpretation of the number of scattering electrons). Z. Physik **4**, 451.

Lagage, P. O., Cesarsky, C. J. (1983): Cosmic-ray shock acceleration in the presence of self-excited waves. Astron. Astrophys. **118**, 223.

Lagage, P. O., Cesarsky, C. J. (1983): The maximum energy of cosmic rays accelerated by supernova shocks. Astron. Astrophys. **125**, 249.

Laing, R. A., et al. (1978): Investigations of the optical fields of 3CR radio sources to faint limiting magnitudes - II. M.N.R.A.S. **183**, 547.

Lal, D., Peters, B.: Cosmic ray produced radioactivity on the earth. In: Handbuch der Physik 46/2, Cosmic Rays II (ed. K. Sitte). Berlin: Springer-Verlag 1967.

Lamb, D. Q. (1995): The distance scale to gamma-ray bursts. P.A.S.P. **107**, 1152.

Lamb, D. Q., et al. (1978): Hot dense matter and stellar collapse. Phys. Rev. Lett. **41**, 1623.

Lamb, D. Q., Lamb, F. K. (1978): Nuclear burning in accreting neutron stars and X-ray bursts. Ap. J. **220**, 291.

Lamb, F. K., et al. (1977): A model for bursting X-ray sources: time-dependent accretion by magnetic neutron stars and degenerate dwarfs. Ap. J. **217**, 197.

Lamb, F. K., Pethick, C. J., Pines, D. (1973): A model for compact X-ray sources: Accretion by rotating magnetic stars. Ap. J. **184**, 271.

Lamb, H.: Hydrodynamics. Cambridge, England: Cambridge University Press 1879, 1895, 1906, 1916, 1924, 1932. New York: Dover Pub. 1932, 1945.

Lamb, H. (1881): On the oscillations of a viscous spheroid. Proc. London Math. Soc. **13**, 51.

Lamb, H. (1909): On the theory of waves propagated vertically in the atmosphere. Proc. London Math. Soc. **7**, 122.

Lamb, S. A., et al. (1977): Neutron-capture nucleosynthesis in the helium-burning cores of massive stars. Ap. J. **217**, 213.

Lamb, S. A., Iben, I. Jr., Howard, W. M. (1976): On the evolution of massive stars through the core carbon-burning phase. Ap. J. **207**, 209.

Lambert, D. L., Warner, B. (1968): The abundances of the elements in the solar photosphere II., III. M.N.R.A.S. **138**, 181, 213.

Lambert, J. H.: Photometria, sive de mensura et gradibus luminis, colorum et umbrae. Augsburg 1760.

Landau, L. D. (1932): On the theory of stars. Phys. Z. Sowjetunion **1**, 285. Reproduced in: Source book in astronomy and astrophysics 1900–1975 (eds. K. R. Lang and O. Gingerich). Cambridge, Mass.: Harvard University Press 1979.

Landau, L. D. (1946): On the vibrations of the electronic plasma. J. Phys. (U.S.S.R,) **10**, 25.

Landau, L. D., Lifshitz, E.M.: Fluid mechanics. New York: Pergamon Press 1959.

Landau, L. D., Lifshitz, E. M.: The classical theory of fields. Reading, Mass.: Addison-Wesley 1962.

Landau, L. D., Lifshitz, E. M.: Statistical physics. Reading, Mass.: Addison-Wesley 1969.

Landau, L. D., Lifshitz, E. M.: The classical theory of fields. New York: Pergamon Press 1971.

Landau, L.D., Pomeranchuk, I. (1953): Limits of applicability of the theory of bremsstrahlung electron and pair production for high energies. Dokl. Akad. Nauk. USSR **92**, 535.

Landau, L. D., Rumer, G. (1938): The cascade theory of electronic showers. Proc. Roy Soc. (London) **A116**, 213.
Landé, A. (1920): Über den anomalen Zeeman-Effekt (On the anomalous Zeeman effect). Z. Physik **5**, 231.
Landé, A. (1923): Termstruktur und Zeeman-Effekt der Multipletts (Term structure and Zeeman effect in multiplets) Z. Physik **19**, 112.
Landini, M., Monsignori Fossi, B. C. (1972): Ionization balance for ions of Na, Al, P, Cl, A, K, Ca, Cr, Mn, Fe and Ni. Astron. Astrophys. Supp. **7**, 291.
Landolt, A. U. (1968): A new short-period blue variable. Ap. J. **153**, 151.
Landstreet, J. D. (1967): Synchrotron radiation of neutrinos and its astrophysical significance. Phys. Rev. **153**, 1372.
Landstreet, J. D. (1980): The measurement of magnetic fields in stars. Astron. J. **85**, 611.
Landstreet, J. D., Angel, J. R. P. (1971): Discovery of circular polarization in the white dwarf G 99-37. Ap. J. (Letters) **165**, L67.
Lane, J. H. (1870): On the theoretical temperature of the Sun, under the hypothesis of gaseous mass maintaining its volume by its internal heat and depending on the laws of gases as known to terrestrial experiment. Amer. J. Sci **50**, 57.
Lang, K. R. (1971): Interstellar scintillation of pulsar radiation. Ap. J. **164**, 249.
Lang, K. R.: Astrophysical data – planets and stars. New York: Springer-Verlag, 1992.
Lang, K. R. (1994): Radio evidence for nonthermal particle acceleration on stars of late spectral type. Ap. J. **90**, 753.
Lang, K. R.: Sun, earth and sky. New York: Springer-Verlag 1995.
Lang, K. R.: Radio evidence for nonthermal magnetic activity on main-sequence stars of late spectral type. In: Magnetodynamic phenomena in the solar atmosphere. Prototypes of stellar magnetic activity (eds. Y. Uchida, T. Kosugi and H. S. Hudson). Boston: Kluwer Academic Pub. 1996).
Lang, K. R. (1996): Unsolved mysteries of the Sun, Parts 1 and 2, Sky and Telescope **92**, 38-August, **92**, 24-September.
Lang, K. R. (1997): SOHO reveals the secrets of the Sun. Scientific American **276**, 32-March.
Lang, K. R., Gingerich, O. (eds.): A source book in astronomy and astrophysics, 1900–1975. Cambridge, Mass.: Harvard University Press, 1979.
Lang, K. R., Rickett, B. J. (1970): Size and motion of the interstellar scintillation pattern from observations of CP 1133. Nature **225**, 528.
Langanke, K.: Nuclear reaction rates from laboratory experiments and in the stellar plasma. In: Nuclei in the cosmos (ed. H. Oberhummer). New York: Springer-Verlag 1991.
Langer, S. H., Petrosian, V. (1977): Impulsive solar X-ray bursts. III. Polarization, directivity, and spectrum of the reflected and total bremsstrahlung radiation from a beam of electrons directed toward the photosphere. Ap. J. **215**, 666.
Langer, W. D. (1978): The stability of interstellar clouds containing magnetic fields. Ap. J. **225**, 95.
Langer, W. D. (1978): The formation of molecules in interstellar clouds from singly and multiply ionized atoms. Ap. J. **225**, 860.
Laplace, P. S. Marquis DE(1816): Sur la vitesse du son dans l'air et dan l'eau (On the velocity of sound in the air and the water). Ann. Chem. Phys. **3**, 238.
Laporte, O. (1924): Die Struktur des Eisenspektrums (The structure of the iron spectrum). Z. Physik **23**, 135.
Larmor, Sir J. (1897): On the theory of the magnetic influence on spectra; and on the radiation from moving ions. Phil. Mag. **44**, 503.
Larson, R. B. (1969): Numerical calculations of the dynamics of a collapsing proto-star. M.N.R.A.S. **145**, 271.
Larson, R. B. (1973): Processes in collapsing interstellar clouds. Ann. Rev. Astron. Ap. **11**, 219.
Lasker, B. M. (1966): An investigation of the dynamics of old H II regions. Ap. J. **143**, 700.

Lasker, B. M. (1966): Ionization fronts for H II regions with magnetic fields. Ap. J. **146**, 471.

Lasker, B. M. (1967): The energization of the interstellar medium by ionization limited H II regions. Ap. J. **149**, 23.

Lasker, B. M., Hesser, J. E. (1969): High frequency stellar oscillations: II. G44-32 A new short period blue variable star. Ap. J. (Letters) **158**, L171.

Lasker, B. M., Hesser, J. E. (1971): High frequency stellar oscillations: VI. R 548 A periodically variable white dwarf. Ap. J. (Letters) **163**, L89.

Lattes, C. M. G., et al. (1947): Processes involving charged mesons. Nature **159**, 694.

Lattes, C. M. G., Occhialini, G. P. S., Powell, C. F. (1947): Observations on the tracks of slow mesons in photographic emulsions. Nature **160**, 453.

Lattimer, J. M., et al. (1991): Direct URCA process in neutron stars. Phys. Rev. Letters **66**, 2701.

Le Roux, E. (1961): Étude théorique de rayonnement synchrotron des radiosources. Ann. Astrophys. **24**, 71.

Lean, J. (1997): The sun's variable radiation and its relevance for earth. Ann. Rev. Astron. Astrophys. **35**, 33.

Leavitt, H. S. (1912): Periods of 25 variable stars in the small Magellanic cloud. Harvard Circular No. 173. Reproduced in: Source book in astronomy (ed. H. Shapley). Cambridge, Mass.: Harvard University Press 1960.

Lebovitz, N. R., Russell, G. W. (1972): The pulsations of polytropic masses in rapid, uniform rotation. Ap. J. **171**, 103.

Lederer, C. M., Hollander, J. M., Perlman, J.: Table of istopes. New York: Wiley 1967.

Ledoux, P. (1945): On the radial pulsation of gaseous stars. Ap. J. **102**, 143.

Ledoux, P.: Stellar stability and stellar evolution. In: Star evolution – Proc. Int. Sch. Phys. "Enrico Fermi" Course 28. New York: Academic Press 1963.

Ledoux, P.: Stellar stability. In: Stellar structure – Stars and stellar systems VIII (ed. L. H. Aller and D. B. Mc Laughlin). Chicago, Ill.: University of Chicago Press 1965).

Ledoux, P., Perkeris, C. L. (1941): Radial pulsations of stars. Ap. J. **94**, 124.

Ledoux, P. Renson, P. (1966): Magnetic stars, Ann. Rev. Astron. Astrophys. **4**, 326.

Ledoux, P., Walraven, T.: Variable stars. In: Handbuch der Physik, vol. LI: Astrophysics II – Stellar structure (ed. S. Flügge). Berlin-Heidelberg-New York: Springer 1958.

Lee, L. C., Jokipii, J. R. (1975): Strong scintillations in astrophysics. I. the Markov approximation, its validity and application to angular broadening. Ap. J. **196**, 695.

Lee, L. C., Jokipii, J. R. (1975): Strong scintillations in astrophysics. II. A theory of temporal broadening of pulses. Ap. J. **201**, 532.

Lee, L. C., Jokipii, J. R. (1975): Strong scintillations in astrophysics. III. The fluctuations in intensity. Ap. J. **202**, 439.

Lee, L. C., Jokipii, J. R. (1976): The iregularity spectrum in interstellar space. Ap. J. **206**, 735.

Legg, M. P. C., Westfold, K. C. (1968): Elliptic polarization of synchrotron radiation. Ap. J. **154**, 499.

Leibacher, J., Stein, R. F. (1971): A new description of the solar five-minute oscillation. Ap. Lett. **7**, 191.

Leighton, R. B. (1961): in Aerodynamic phenomena in stellar atmospheres. Proc. Fourth Symp. Cosm. Gas Dyn. Supp. Nuovo Cimento **22**, 321.

Leighton, R. B.: Principles of modern physics. New York: McGraw-Hill 1959.

Leighton, R. B., Noyes, R. W., Simon, G. W. (1962): Velocity fields in the solar atmosphere: I. Preliminary report. Ap. J. **135**, 474.

Lemaitre, G. E., Vallarta, M. S. (1933): On Compton's latitude effect of cosmic radiation. Phys. Rev. **43**, 87.

Lèna, P.: Observational astrophysics. New York: Springer-Verlag, 1988.

Leung, C. M., Liszt, H.S. (1976): Radiation transport and non-LTE analysis of interstellar molecular lines. I. Carbon monoxide. Ap. J. **208**, 732.

Leung, Y. C., Wang, C. G. (1971): Properties of hadron matter: II. Dense baryon matter and neutron stars. Ap. J. **170**, 499.

Leventhal, M., Maccallum, C. J., Stang. P. D. (1978): Detection of 511 keV positron annihilation radiation from the galactic center direction. Ap. J. (Letters) **225**, L11.

Levine, R. H. (1974): A new theory of coronal heating. Ap. J. **190**, 457.

Levy, E. H.: The solar cycle and dynamo theory. In: The solar cycle – ASP conference series vol. 27 (ed. K. L. Harvey). San Farancisco: Astronomical Society of the Pacific 1992.

Lewin, J. H. G., Van Den Heuvel, E. P. J. (eds): Accretion-driven stellar X-ray sources. New York: Cambridge University Press, 1983.

Lewis, M., Margenau, H. (1958): Statistical broadening of spectral lines emitted by ions in a plasma. Phys. Rev. **109**, 842.

Libbrecht, K. G. (1989): Solar p-mode frequency splittings. Ap. J. **336**, 1092.

Libbrecht, K. G., Woodard, M. F. (1990): Solar-cycle effects on solar oscillation frequencies. Nature **345**, 779.

Libbrecht, K. G., Woodard, M. F. (1991): Advances in helioseismology. Science **253**, 152.

Libbrecht, K. G., Woodard, M. F., Kaufman, J. M. (1990): Frequencies of solar oscillations. Ap. J. Supp. **74**, 1129.

Lichtenstadt, I., et al. (1978): Effects of neutrino degeneracy and of downscatter on neutrino radiation from dense stellar cores. Ap. J. **226**, 222.

Liebert, J. (1980): White dwarf stars. Ann. Rev. Astron. Astrophys. **18**, 363.

Liemohn, H. B. (1965): Radiation from electrons in a magnetoplasma. J. Res. Nat. Bur. Stands. USNC-URSI Radio Science **69D**, 741.

Liénard, A. M. (1898): Théorie de Lorentz, Théorie de Larmor et celle de Lorentz. Éclairage électr. **14**, 16.

Lieu, R., Axford, W. I. (1993): Synchrotron radiation: An inverse Compton effect. Ap. J. **416**, 700.

Lighthill, M. J. (1952): On sound generated aerodynamically: I. General theory. Proc. Roy. Soc. London **A211**, 564.

Lighthill, M. J. (1954): On sound generated aerodynamically: II. Turbulence as a source of sound. Proc. Roy. Soc. London **A222**, 1.

Lightman, A. P., Press, W. H., Odenwald, S. F. (1978): Present and past death rates for globular clusters. Ap. J. **219**, 629.

Lightman, A. P., Shapiro, S. L. (1975): Spectrum and polarization of X-rays from accretion disks around black holes. Ap. J. (Letters) **198**, L73.

Lightman, A. P., Shapiro, S. L. (1978): The dynamical evolution of globular clusters. Rev. Mod. Phys. **50**, 437.

Lightman, A. P., Zdziarski, A. A. (1987): Pair production and Compton scattering in compact sources and comparison to observations of active galactic nuclei. Ap. J. **319**, 643.

Liller, W. (1955): The photoelectric photometry of planetary nebulae. Ap. J. **122**, 240.

Liller, W., Aller, L. H. (1955): Photoelectric spectrophotometry of planetary nebulae. Ap. J. **120**, 48.

Lilley, A. E., Palmer, P. (1968): Tables of radio-frequency recombination lines. Ap. J. Suppl. No. 144, **16**, 143.

Lin, D. N. C., Papaloizou, J. C. B. (1996): Theory of accretion disks II. Application to observed systems. Ann. Rev. Astron. Astrophys. **34**, 703.

Lin, R. P., Hudson, H. S. (1971): 10-100 keV electron acceleration and emission from solar flares. Solar Phys. **17**, 412.

Lindblad, B. (1935): A condensation theory for meteoric matter and its cosmological significance. Nature **135**, 133.

Lindblom, L., Detweiler, S. L. (1977): On the secular instabilities of the Maclaurin spheroids. Ap. J. **211**, 565.

Lindblom, L., Detweiler, S. (1979): The role of neutrino dissipation in gravitational collapse. Ap. J. **232**, L 101.

Lindholm, E. (1941): Zur Theorie der Verbreiterung von Spektrallinien (On the theory of broadening of spectral lines). Arkiv. Mat. Astron. Physik **28** B, No. 3.

Lingenfelter, R. E. (1969): Solar flare optical, neutron, and gamma-ray emission. Solar Phys. **8**, 341.

Lingenfelter, R. E., Ramaty, R.: High energy nuclear reactions in solar flares. In: High energy nuclear reactions in astrophysics (ed. B. Shen). New York: W.A. Benjamin 1967.

Linsley, J., Scarsi, L., Rossi, B. (1961): Extremely energetic cosmic ray event. Phys. Rev. Letters **6**, 485.

Liszt, H. S., Leung, C. M. (1977): Radiation transport and non-LTE analysis of interstellar molecular lines. II. Carbon monosulfide. Ap. J. **218**, 396.

Litvak, M. M. (1969): Infrared pumping of interstellar OH. Ap. J. **156**, 471.

Lockman, F. J., Brown, R. L. (1976): On the derivation of nebular electron temperatures from radio recombination line observations. Ap. J. **207**, 436.

Longair, M. S.: Theoretical concepts in physics. New York: Cambridge University Press, 1984.

Longair, M. S.: High energy astrophysics, volume 1, particles, photons and their detection, second edition. New York: Cambridge University Press 1992.

Longair, M. S.: High energy astrophysics, volume 2, stars, the Galaxy and the interstellar medium, second edition. New York: Cambridge University Press 1994.

Longair, M. S., Gunn, J. E. (1975): An investigation of the optical fields of 35 3CR radio sources to faint limiting optical magnitudes. M.N.R.A.S. **170**, 121.

Lorentz., H. A. (1880): Über die Beziehung zwischen der Fortpflanzungsgeschwindigkeit de Lichtes und der Körperdichte (Concerning the relation between the velocity of propagation of light and the density and composition of media). Ann. Phys. **9**, 641.

Lorentz, H. A. (1892): La théorie électromagnétique de Maxwell et son application aux corps mouvants. Archives Neerl, **25**, 363.

Lorentz, H. A. (1898): Influence du champ magnétique sur l'emission lumineuse (On the influence of magnetic forces on light emission). Rev. d'electricité **14**, 435 and also Ann. Physik **63**, 278 (1897).

Lorentz, H. A. (1904): Electromagnetic phenomena in a system moving with any velocity smaller than that of light. Proc. Am. Acad. Sci. **6**, 809. Reprod. in: The principle of relativity (H.A. Lorentz, A. Einstein, H. Minkowski, H. Weyl). New York: Dover 1952.

Lorentz, H. A. (1906): The absorption and emission lines of gaseous bodies. Proc. Am. Acad. Sci. **8**, 591.

Lorentz, H. A.: The theory of electrons. 1909. Republ. New York: Dover 1952.

Lorenz, C. P., Ravenhall, D. G., Pethick, C. J. (1993): Neutron star crusts. Phys. Rev. Lett. **70**, 379.

Lorenz, L. (1881): Über die Refractionskonstante (On the constant of refraction). Wiedem. Ann. **11**, 70.

Loschmidt, J. (1865): Zur Größe der Luftmoleküle (The size of atmospheric molecules). Wien. Ber. **52**, 395.

Lovas, F. J. (1986): Recommended rest frequencies for observed interstellar molecular microwave transitions – 1985 revision. J. Phys. Chem. Ref. Data **15**, 251.

Lovas, F. J. (1992): Nist recommended rest frequencies for observered interstellar molecular microwave transitions – 1991 revision. J. Phys. Chem. Ref. Data **21**, 181. Also on the World Wide Web see http://physics.nist.gov/ under Physical Reference Data.

Lovas, F. J., Snyder, L. E., Johnson, D. R. (1979): Recommended rest frequencies for observed interstellar molecular transitions. Ap. J. Supp. **41**, 451.

Love, S. G., Brownlee, D. E. (1993): A direct measurement of the terrestrial mass accretion rate of cosmic dust. Science **262**, 550.

Lovelace, R. V. E., et al. (1971): Analysis of observations of interplanetary scintillations. Ap. J. **159**, 1047.

Lumley, J., Panofsky, H.: The structure of atmospheric turbulence. New York: Interscience Publ. 1964.

Luo, D., Pradhan, A. K., Shull, J. M. (1988): Oscillator strengths for Si II. Ap. J. **335**, 498.

Lüst, R.: The solar wind. In: Interstellar gas dynamics (ed. H. J. Habing). Dordrecht, Holland: D. Reidel 1970.

Lutz, B. L. (1972): Radiative association and formation of diatomic molecules. Ann. N.Y. Acad. Sci. **194**, 29.

Lyman, T. (1906): The spectrum of hydrogen in the region of extremely short wave-length. Ap. J. **23**, 181.

Lynden-Bell, D., Ostriker, J. P. (1967): On the stability of differentially rotating bodies. M.N.R.A.S. **136**, 293.

Lynds, B. T. (1974): An atlas of dust and H II regions in galaxies. Ap. J. Suppl. **28**, 391.

Lynds, B. T., Wickramasinghe, N. C. (1968): Interstellar dust. Ann. Rev. Astron. Astrophys. **6**, 215.

Lyne, A., Graham-Smith, F.: Pulsar Astronomy. New York: Cambridge University Press 1990.

Lyons, P. B. (1969): Total yield measurements in $^{24}Mg(\alpha, \gamma)^{28}Si$. Nucl. Phys. **A130**, 25.

Lyons, P. B., Toevs, J. W., Sargood, J. W. (1969): Total yield measurements in $^{27}Al(p,\gamma)^{28}Si$. Phys. Rev. **C2**, 22041.

Lyttleton, R. A.: Theory of rotating fluid masses. Cambridge, England: Cambridge University Press 1953.

Lyttleton, R. A.: The stability of rotating liquid masses. Cambridge, England: Cambridge University Press 1955.

Machado, M. E. (1991): Flare transport by conduction and radiation. Phil. Trans. R. Soc. Lond. **A336**, 425.

Macklin, R. L. (1970): Were the lightest stable isotopes produced by photodissociation? Ap. J. **162**, 353.

Maclaurin, C.: A treatise on fluxions. (1742. Cf.: History of the mathematical theories of attraction, and the figure of the Earth by I. Todhunter. Macmillan 1873. Repr. New York: Dover 1962.)

Maddalena, R. J. et al. (1986): The system of molecular clouds in Orion and Monoceros. Ap. J. **303**, 375.

Mahaffy, J. H., Hansen, C. J. (1975): Carbon detonations in rapidly rotating stellar cores. Ap. J. **201**, 695.

Major, J .K., Biedenharn, L. C. (1954): Sargent diagram and comparative half-lives for electron capture transitions. Rev. Mod. Phys. **26**, 321.

Malik, G. P., Malik, U., Varma, V. S. (1991): Recalculation of radial matrix integrals for the electric dipole transitions in hydrogen. Ap. J. **371**, 418.

Malone, R. C., Johnson, M. B., Bethe, H. A. (1975): Neutron star models with realistic high-density equations of state. Ap. J. **199**, 741.

Maltby, P., et al. (1986): A new sunspot umbral model and its variation with the solar cycle. Ap. J. **306**, 284.

Manchester, R. N. (1974): Structure of the local galactic magnetic field. Ap. J. **188**, 637.

Manoharan, P. K., Van Driel-Gesztelyi, L., Pick, M., Démoulin, P. (1996): Evidence for large-scale solar magnetic reconnection from radio and X-ray measurements. Ap. J. (Letters) **468**, L73.

Marcus, P. S., Press, W. H., Teukolsky, S. A. (1977): Stablest shapes for an axisymmetric body of gravitating, incompressible fluid. Ap. J. **214**, 584.

Margenau, H. (1939): Van der Waals forces. Rev. Mod. Phys. **11**, 1.

Margenau, H. (1946): Conduction and dispersion of ionized gases at high frequencies. Phys. Rev. **69**, 508.

Margenau, H., Lewis, M. (1959): Structure of spectral lines from plasmas. Rev. Mod. Phys. **31**, 579.

Marion, J. B., Fowler, W. A. (1957): Nuclear reactions with the neon isotopes in stars. Ap. J. **125**, 221.

Mariska, J. T., Doschek, G. A., Bentley, R. D. (1993): Flare plasma dynamics observed with the Yohkoh Bragg crystal spectrometer I. Properties of the Ca XIX resonance line. Ap. J. **419**, 418.
Marscher, A. P. (1983): Accurate formula for the self-Compton X-ray flux density from a uniform spherical, compact radio source. Ap. J. **264**, 296.
Marshak, R. E., Bethe, H. A. (1947): On the two-meson hypothesis. Phys. Rev. **72**, 506.
Martin, P. G. (1975): On the Kramers-Kronig relations for interstellar polarization. Ap. J. **202**, 389.
Martin, P. G.: Cosmic dust. Oxford: Clarendon Press 1978.
Martin, P. G., Campbell, B. (1976): Circular polarization observations of the interstellar magnetic field. Ap. J. **208**, 727.
Martin, W. C.: Sources of atomic spectroscopic data for astrophysics. In: Atomic and molecular data for space astronomy: Needs and availability. Lecture Notes in Physics 407 (eds. C. P. Smith and W. L. Wiese). New York: Springer-Verlag 1992.
Martyn, D. F. (1946): Temperature radiation from the quiet sun in the radio spectrum. Nature **158**, 632.
Martyn, D. F. (1948): Solar radiation in the radio spectrum. Proc. Roy. Soc. (London) **A193**, 44.
Mason, H. E. (1991): Atomic physics calculations relevant to solar flare spectra. Phil Trans. R. Soc. Lond. **A336**, 471.
Masuda, S.: Hard X-ray sources and the primary energy release site in solar flares. Ph. D. thesis, Univ. of Tokyo, Tokyo, Japan, 1994.
Masuda, S., et al. (1994): A loop-top hard X-ray source in a compact solar flare as evidence for magnetic reconnection. Nature **371**, 495.
Mathews, W. G. (1965): The time evolution of an H II region. Ap. J. **142**, 1120.
Mathews, W. G., O'dell, C. R. (1969): Evolution of diffuse nebulae. Ann. Rev. Astron. Astrophys. **7**, 67.
Mathewson, D. S., Ford, V. L. (1970): Polarization observations of 1800 stars. Mem. R.A.S. **74**, 139.
Mathis, J. S. (1985): Ionization correction factors for low-excitation gaseous nebulae. Ap. J. **291**, 247.
Mathis, J. S., Rumpl, W., Nordsieck, K. H. (1977): The size distribution of interstellar grains. Ap. J. **217**, 425.
Matinyan, S. G., Tsilosani, N. N. (1962): Transformation of photons into neutrino pairs and its significance in stars. Sov. Phys. Jetp **14**, 1195.
Maxwell, J. C. (1860): Illustrations of the dynamical theory of gases: Part I. On the motions and collisions of perfectly elastic spheres; Part II. On the process of diffusion of two or more kinds of moving particles among one another. Phil. Mag. **19**, 19, **20**, 21, reprod. Brush (1965).
Maxwell, J. C. (1861): On physical lines of force: Part. I. The theory of molecular vortices applied to magnetic phenomena; Part II. The theory of molecular vortices applied to electric currents. Phil. Mag. **21**, 161, 281.
Maxwell, J. C.: (1866): Viscosity or internal friction of air and other gases. Phil. Trans. Roy. Soc. London **156**, 249.
Maxwell, J. C.: A treatise on electricity and magnetism. 1873. Republ. New York: Dover 1954.
Maxwell, J. C. (1899): The theory of anomalous dispersion. Phil. Mag. **48**, 151.
Mayall, N. U., De Vaucouleurs, A. (1962): Redshifts of 92 galaxies. Astron. J. **67**, 363.
Mayer, C. H.: Thermal radio emission of the planets and moon. In: Surface and interiors of planets and satellites (ed. A. Dollfuss). New York: Academic Press 1970.
Mayer, C. H., Mc Cullough, T. P. (1971): Microwave radiation of Uranus and Neptune. Icarus **14**, 187.
Mayer, M. G. (1931): Über Elementarakte mit zwei Quantensprüngen (On elementary processes with two quantum emissions). Ann. Phys. **9**, 273.
Mayer, M. G. (1948): On closed shells in nuclei. Phys. Rev. **74**, 235.

Mayer, R. (1842): The forces of inorganic nature. Ann. Chem. and Pharm. **42**, 233. Engl. trans. in Phil. Mag. **24**, 371 (1862) reprod. Brush (1965).
Mayle, R., Wilson, J. R., Schramm, D. N. (1987): Neutrinos from gravitational collapse. Ap. J. **318**, 288.
Maza, J., Van Den Bergh, S. (1976): Statistics of extragalactic supernovae. Ap. J. **204**, 519.
Mazets, E., et al. (1981): Catalog of cosmic gamma-ray bursts from the KONUS experiment. Parts 1 and 2, 3, 4. Astrophys. Space Sci. **80**, 3, 85, 119.
Mazur, J. E., Mason, G. M., Klecker, B., Mc Guire, R. E. (1992): The energy spectra of solar flare hydrogen, helium, oxygen, and iron: Evidence for stochastic acceleration. Ap. J. **401**, 398.
Mazurek, T. J., Lattimer, J. M., Brown, G. E. (1979): Nuclear forces, partition functions, and dissociation in stellar collapse. Ap. J. **229**, 713.
Mc Cammon, D., Bunner, A. N., Coleman, P. L., Kraushaar, W. L. (1971): A search for absorption of the soft X-ray diffuse flux by the small Magellanic cloud. Ap. J. (Letters) **168**, L33.
Mc Cammon, D., Sanders, W. T. (1990): The soft X-ray background and its origins. Ann. Rev. Astron. Astrophys. **28**, 657.
Mc Cook, G. P., Sion, E. M. (1987): A catalog of spectroscopically identified white dwarfs. Ap. J. Supp. **65**, 603.
Mc Cray, R. (1969): Synchrotron radiation losses in self-absorbed radio sources. Ap. J. **156**, 329.
Mc Cray, R. (1993): Supernova 1987A revisited. Ann. Rev. Astron. Astrophys. **31**, 175.
Mc Cray, R., Snow, T. P. Jr. (1979): The violent interstellar medium. Ann. Rev. Astron. Ap. **17**, 213.
Mc Crea, W. H. (1929): The hydrogen chromosphere. M.N.R.A.S. **89**, 483.
Mc Crea, W. H. (1957): The formation of population I stars: Part I. Gravitational contraction. M.N.R.A.S. **117**, 562.
Mc Dougall, J., Stoner, E. C. (1938): The computation of Fermi-Dirac functions. Phil. Trans. Roy. Soc. London **237**, 67.
Mc Kee, C. F.: Astrophysical shocks in diffuse gas. In: Spectroscopy of astrophysical plasmas (eds. A. Dalgarno and D. Layzer). New York: Cambridge University Press 1987.
Mc Kee, C. F., Cowie, L. L. (1977): The evaporation of spherical clouds in a hot gas. II. Effects of radiation. Ap. J. **215**, 213.
Mc Kee, C. F., Hollenbach, D. J. (1980): Interstellar shock waves. Ann. Rev. Astron. Astrophys. **18**, 219.
Mc Kee, C. F., Ostriker, J. P. (1977): A theory of the interstellar medium: three components regulated by supernova explosions in a inhomogeneous substrate. Ap. J. **218**, 148.
Mc Lean, D. J., Labrum, N. R. (eds.): Solar radiophysics. Cambridge, England: Cambridge University Press 1985.
Mc Nalley, D.: Interstellar molecules. In: Advances in astronomy and astrophysics, vol. 6 (ed. Z. Kopal). New York: Academic Press 1968.
Meegan, C. A., Earl, J. A. (1975): The spectrum of cosmic electrons with energies between 6 and 100 GeV. Ap. J. **197**, 219.
Meegan, C. A., et al. (1992): Spatial distribution of γ-ray bursts observed by BATSE. Nature **355**, 143.
Meggers, W. F., Corliss, C. H., Scribner, B. F.: Tables of spectral line intensities. Nat. Bur. Stands. Mon. 32 (Wash.), 1961.
Meier, D. L., et al. (1976): Magnetohydrodynamic phenomena in collapsing stellar cores. Ap. J. **204**, 869.
Melrose, D. B. (1971): On the degree of circular polarization of synchrotron radiation. Astrophys.Space Sci. **12**, 172.
Melrose, D. B. (1991): Collective plasma radiation processes. Ann. Rev. Astron. Astrophys. **29**, 31.

Mendoza, C.: Recent advances in atomic calculations and experiments of interest in the study of planetary nebulae. In: Planetary nebulae. Proc. IAU Symp. No. 103 (ed. D. R. Flower). Boston: D. Reidel 1983).

Mendoza, C., Zeippen, C. J. (1982): Transition probabilities for forbidden lines in the $3p^2$ configuration-II. M.N.R.A.S. **199**, 1025.

Meneguzzi, M., Audouze, J., Reeves, H. (1971): The production of the elements Li, Be, B by galactic cosmic rays in space and its relation with stellar observations. Astron. Astrophys. **15**, 337.

Menzel, D. H. (1926): The planetary nebulae. P.A.S.P. **38**, 295. Reproduced in: A source book in astronomy and astrophysics 1900–1975 (eds. K. R. Lang and O. Gingerich). Cambridge, Mass.: Harvard University Press 1979.

Menzel, D. H. (1936): The theoretical interpretation of equivalent breadths of absorption lines. Ap. J. **84**, 462.

Menzel, D. H. (1947): Extended sum rules for transition arrays. Ap. J. **105**, 126.

Menzel, D. H. (1969): Oscillator strengths, f, for high-level transitions in hydrogen. Ap. J. Suppl. No. 161, **18**, 221.

Menzel, D. H. (ed.): Selected papers on physical processes in ionized plasmas. New York: Dover Pub. 1962.

Menzel, D. H., Goldberg, L. (1936): Multiplet strengths for transitions involving equivalent electrons. Ap. J. **84**, 1.

Menzel, D. H., Pekeris, C. L. (1935): Absorption coefficients and hydrogen line intensities. M.N.R.A.S. **96**, 77. Reproduced in: Selected papers on physical processes in ionized plasmas (ed. D. H. Menzel). New York: Dover 1962.

Mercier, R. P. (1962): Diffraction by a screen causing large random phase fluctuations. Proc. Camb. Phil. Soc. **58**, 382.

Mercier, R. P. (1964): The radio-frequency emission coefficient of a hot plasma. Proc. Phys. Soc. **83**, 819.

Merrill, P. W.: Lines of chemical elements in astronomical spectra. Washington, D.C.: Carnegie Inst. of Wash. Publ. 610, 1958.

Merzbacher, E.: Quantum mechanics. New York: John Wiley and Sons 1961.

Mestel, L.: The theory of white dwarfs. In: Stellar structure – Stars and stellar systems VIII (ed. L. H. Aller and D. Mc Laughlin). Chicago, Ill.: University of Chicago Press 1965.

Mestel, L. (1968): Magnetic braking by a stellar wind - I. M.N.R.A.S. **138**, 359.

Mestel, L.: Magnetic fields. In: Protostars and planets II (eds. D. C. Black and M. S. Matthews). Tucson, Arizona: University of Arizona Press 1985.

Meszarós, P.: High energy from magnetized neutron stars. Chicago, Illinois: University of Chicago Press 1992.

Mészáros, P., Rees, M. J. (1997): Optical and long-wavelength afterglow from gamma-ray bursts. Ap. J. **476**, 232.

Metzger, M. R., et al. (1997): Spectral constraints on the redshift of the optical counterpart of the γ-ray burst of 8 May 1997. Nature **387**, 878.

Mewaldt, R. A., Stone, E. C., Vogt, R. E. (1976): The isotopic composition of hydrogen and helium in low-energy cosmic rays. Ap. J. **206**, 616.

Meyer, B. S. (1994): The r-, s-, and p-processes in nucleosynthesis. Ann. Rev. Astron. Astrophys. **32**, 153.

Meyer, J.-P., Drury, L. O'C, Ellison, D. C. (1997): Galactic cosmic rays from supernova remnants I. A cosmic-ray composition controlled by volatility and mass-to-charge ratio. Ap. J. **487**, 182.

Meyer, P., Parker, E. N., Simpson, J. A. (1956): Solar cosmic rays of February, 1956 and their propagation through interplanetary space. Phys. Rev. **104**, 768.

Meyer, P., Vogt, R. (1961): Electrons in the primary cosmic radiation. Phys. Rev. Letters **6**, 193.

Mezger, P. G., Henderson, A. P. (1967): Galactic H II regions I. Observations of their continuum radiation at the frequency 5 GHz. Ap. J. **147**, 471.

Michaud, G., Fowler, W. A. (1970): Thermonuclear reaction rates at high temperatures. Phys. Rev. **C2**, 22041.

Michaud, G., Fowler, W. A. (1972): Nucleosynthesis in silicon burning. Ap. J. **173**, 157.

Michel, F. C. (1972): Accretion of matter by condensed objects. Astrophys. Space Sci. **15**, 153.

Michel, F. C. (1977): Accretion magnetospheres: General solutions. Ap. J. **213**, 836.

Michel, F. C. (1982): Theory of pulsar magnetospheres. Rev. Mod. Phys. **54**, 1.

Michel, F. C.: Theory of neutron star magnetospheres. Chicago, Illinois: University of Chicago Press 1991.

Mie, G. Von (1908): Beiträge zur Optik trüber Medien, speziell Kolloidaler Metallösungen (Contributions to the optics of opaque media, especially colloide metal solutions). Ann. Physik **25**, 377.

Migdal, A. B. (1956): Bremsstrahlung and pair production in condensed media at high energies. Phys. Rev. **103**, 1811.

Mihalas, D.: Stellar atmospheres. San Francisco, Calif.: W. H. Freeman 1970.

Mikheyev, S. P., Smirnov, A. Y. (1987): Neutrino oscillations in a variable-density medium and neutrino bursts due to the gravitational collapse of stars. Sov. Phys. JETP **64**, 4.

Mikheyev, S. P., Smirnov, A. Y. (1986): Resonance enhancement of oscillations in matter and solar neutrino spectroscopy. Sov. J. Nucl. Phys. **42**, 913.

Mikheyev, S. P., Smirnov, A. Y. (1986): Resonant amplification of neutrino oscillations in matter and solar-neutrino spectroscopy. Nuovo Cimento **9C**, 17.

Miley, G. K. (1971): The radio structure of quasars – a statistical investigation. M.N.R.A.S. **152**, 477.

Miley, G. (1980): The structure of extended extragalactic radio sources. Ann. Rev. Astron. Astrophys. **18**, 165.

Miller, J. A., et al. (1997): Critical issues for understanding particle acceleration in impulsive solar flares. J. Geophys,. Res. **102**, 14,631.

Miller, J. A., Guessoum, N., Ramaty, R. (1990): Stochastic Fermi acceleration in solar flares. Ap. J. **361**, 701.

Millikan, R. A. (1926): High frequency rays of cosmic origin. Proc. Nat. Acad. Sci. **12**, 48.

Millikan, R. A., Cameron, G. H. (1926): High frequency rays of cosmic origin III. Measurements in snow-fed lakes at high altitudes. Phys. Rev. **28**, 851.

Milne, D. K. (1970): Nonthermal galactic radio sources. Austr. J. Phys. **23**, 425.

Milne, E. A. (1921): Radiative equilibrium in the outer layers of a star: The temperature distribution and the law of darkening. M.N.R.A.S. **81**, 361.

Milne, E. A. (1930): Thermodynamics of the stars. Handbuch der Astrophysik **3**, 80. Reproduced in: Selected papers on the transfer of radiation (ed. D. H. Menzel). New York: Dover 1966.

Minkowski, R. (1942): The crab nebula. Ap. J. **96**, 199. Reproduced in: A source book in astronomy and astrophysics 1900–1975 (eds. K. R. Lang and O. Gingerich). Cambridge, Mass.: Harvard University Press 1979.

Misner, C. W., Thorne, K. S., Wheeler, J. A.: Gravitation. San Francisco: W. H. Freeman 1973.

Mitchell, G. F., Ginsburg, J. L., Kuntz, P. J. (1977): The formation of molecules in diffuse interstellar clouds. Ap. J. **212**, 71.

Mitler, H. E. (1970): The solar light element abundances and primeval helium. Smithsonian Astrophys. Obs. Rpt. 323.

Mitler, H. E. (1970): Cosmic-ray production of deuterium, He^3, lithium, beryllium, and boron in the Galaxy. Smithsonian Ap. Obs. Spec. Rpt. 330.

Moffet, A.: Strong nonthermal radio emission from galaxies. In: Galaxies and the universe – Stars and stellar systems, vol. IX (ed. A. R. Sandage). Chicago, Ill.: Univ. of Chicago Press 1974.

Moiseev, S. S., Sagdeev, R. Z. (1963): On the Bohm diffusion coefficient. Sov. Phys. JETP **17**, 515.

Möller, C. (1937): On the capture of orbital electrons by nuclei. Phys. Rev. **51**, 84.

Monaghan, J. J. (1992): Smoothed particle hydrodyanamics. Ann. Rev. Astron. Astrophys. **30**, 543.

Montmerle, T. (1977): On the possible existence of cosmological cosmic rays. I. The framework for light-element and gamma-ray production. Ap. J. **216**, 177.

Montmerle, T. (1977): On the possible existence of cosmological cosmic rays. II. The observational constraints set by the γ-ray background spectrum and the lithium and deuterium abundances. Ap. J. **216**, 620.

Montmerle, T. (1977): Light-element production by cosmological cosmic rays. Ap. J. **217**, 878.

Montmerle, T. (1977): On the possible existence of cosmological cosmic rays. III. Nuclear γ-ray production. Ap. J. **218**, 263.

Moore, C. E.: An ultraviolet multiplet table. Nat. Bur. Stands. (Wash.) 1950–69.

Moore, C. E.: Atomic energy levels – Vol. I, II, III. Nat. Bur. Stands (Wash.) 1949, 1952, 1959.

Moore, C. E. (1970): Ionization potentials and ionization limits derived from the analysis of optical spectra. Nat. Bur. Stands. (Wash.) rpt. NSRDS-NBS 34.

Moore, C. E., Merrill, P. W.: Partial Grotrian diagrams of astrophysical interest. Nat. Bur. Stands. (Wash.) NSRDS-NBS. **23**, 1968.

Moore, C. E., Minnaert, M. G. J., Houtgast, J.: The solar spectrum 2935 Å to 8770 Å. Nat. Bur. Stands. (Wash.) Mon. **61**, 1966.

Moran, J. M. et al. (1968): OH interferometry; long baseline OH interferometry. Aston. J. **73**, S27; S108.

Morgan, W. W., Keenan, P. C., Kellman, E.: An atlas of stellar spectra. Chicago, Ill.: University of Chicago Press 1943.

Morris, M., Rickard, L. J. (1982): Molecular clouds in galaxies. Ann. Rev. Astron. Astrophys. **20**, 517.

Morrison, D. (1970): Thermophysics of the planet Mercury. Space Sci. Rev. **11**, 271.

Morrison, P. (1958): On gamma-ray astronomy. Nuovo Cimento **7**, 858.

Morrison, P.: The origin of cosmic rays. In: Handbuch der Physic 46/1. Berlin: Springer-Verlag 1961.

Morrison, P., Olbert, S., Rossi, B. (1954): The origin of cosmic rays. Phys. Rev. **94**, 440.

Morse, P. M. (1929): Diatomic molecules according to the wave mechanics: II. Vibrational levels. Phys. Rev. **34**, 57.

Mosengeil, K. (1907): Theorie der stationären Strahlung in einem gleichförmig bewegten Hohlraum (Theory of stationary radiation in a uniformly moving cavity). Ann. Phys. **22**, 867.

Mott, N. F., Massey, H. S. W.: The theory of atomic collisions. Oxford: Oxford at the Clarendon Press 1965.

Mott-Smith, H. M. (1951): The solution of the Boltzmann equation for a shock wave. Phys. Rev. **82**, 885.

Mottemann, J.: Thesis, Dept of Astron., U.C.L.A. 1972.

Mouschovias, T. Ch. (1974): Static equilibria of the interstellar gas in the presence of magnetic and gravitational fields. Ap. J. **192**, 37.

Mouschovias, T. Ch. (1976): Nonhomologous contraction and equilibria of self-gravitating, magnetic interstellar clouds embedded in an intercloud medium: star formation. I. formulation of the problem and method of solution. Ap. J. **206**, 753.

Mouschovias, T. Ch. (1976): Nonhomologous contraction and equilibria of self-gravitating, magnetic interstellar clouds embedded in an intercloud medium: star formation. II. Results. Ap. J. **207**, 141.

Mouschovias, T. Ch. (1979): Magnetic braking of self-gravitating, oblate interstellar clouds. Ap. J. **228**, 159.

Mouschovias, T. Ch., Spitzer, L. Jr. (1976): Note on the collapse of magnetic interstellar clouds. Ap. J. **210**, 326.

Mozer, B., Baranger, M. (1960): Electric field distributions in an ionized gas. Phys. Rev. **118**, 626.

Mufson, S. L. (1974): The structure and stability of shock waves in a multiphase interstellar medium. Ap. J. **193**, 561.

Muhleman, D. O. (1969): Microwave opacity of the Venus atmosphere. Astron. J. **74**, 57.

Muller, C. A., Oort, J. H. (1951): The interstellar hydrogen line at 1,420 MHz and an estimate of galactic rotation. Nature **168**, 356. Reproduced in: A source book in astronomy and astrophysics 1900-1975 (eds. K. R. Lang and O. Gingerich). Cambridge, Mass.: Harvard University Press 1979.

Murdin, P. (1989): The LMC supernova in order of magnitude. Q.J.R.A.S. **30**, 419.

Murota, T., Ueda, A. (1956): On the foundation and the applicability of Williams-Weizsäcker method. Prog. Theor. Phys. **16**, 497.

Murota, T., Ueda, A., Tanaka, H. (1956): The creation of an electron pair by a fast charged particle. Prog. Theor. Phys. **16**, 482.

Murphy, R. J., Dermer, C. D., Ramaty, R. (1987): High-energy processes in solar flares. Ap. J. Supp. **63**, 721.

Murphy, R. J., et al. (1985): Solar flare gamma-ray line spectroscopy. In: Proc.19th International Cosmic Ray Conference (La Jolla) **4**, 253.

Murphy, R. J., et al. (1991): Solar abundances from gamma-ray spectroscopy: Comparisons with energetic particle, photospheric, and coronal abundances. Ap. J. **371**, 793.

Murphy, R. J., Ramaty, R. (1984): Solar flare neutrons and gamma rays. Adv. Space Res. **4**, No. 7, 127.

Myers, P. C. (1978): A compilation of interstellar gas properties. Ap. J. **225**, 380.

Myers, W. D. (1970): Droplet model nuclear density distributions and single-particle potential wells. Nucl. Phys. **A145**, 387.

Myers, W. D., Swiatecki, W. J. (1966): Nuclear masses and deformations. Nucl. Phys. **81**, 1.

Nakagawa, M., Kohyama, Y., Itoh, N. (1987): Relativistic free-free Gaunt factor of the dense high-temperature stellar plasma. Ap. J. Supp. **63**, 661.

Narayan, R. (1987): The birthrate and initial spin period of single radio pulsars. Ap. J. **319**, 162.

Narayan, R., Piran, T., Shemi, A. (1991): Neutron star and black hole binaries in the Galaxy. Ap. J. (Letters) **379**, L17.

Navier, C. L. M. H. (1822): Mem. de l'acad. Sci. **6**.

Neddermeyer, S. H., Anderson, C. D. (1937): Note on the nature of cosmic-ray particles. Phys. Rev. **51**, 884.

Nelms, A. T. (1953): Graphs of the Compton energy-angle relationship and the Klein-Nishina formula from 10 keV to 500 MeV. Cir. Nat. Bur. Stands. No. 542.

Nelson, R. D., Lide, D. R., Maryott, A. A.: Selected values of electric dipole moments for molecules in the gas phase, NSRDS-NBS 10. Nat. Bur. Stands. (Wash.), 1967.

Nemiroff, R. J., et al. (1994): Gamma-ray bursts are time-asymmetric. Ap. J. **423**, 432.

Nernst, W. (1906): Über die Beziehungen zwischen Wärmeentwicklung und maximaler Arbeit bei kondensierten Systemen (About the heat and maximum work of a condensed system). Sitz. Berl. **1**, 933.

Nernst, W.: The new heat theorem – Its foundations in theory and experiment. 1926. Repr. New York: Dover 1969.

Nesme-Ribes, E., Baliunas, S. L., Sokoloff, D. (1996): Sunspot cycles – on other stars – are helping astronomers study the Sun's variations and the ways they might affect the earth. Scientific American **275**, 47 – August.

Ness, N. F., Scearce, C. S., Seek, J. B. (1964): Initial results of the Imp 1 magnetic field experiment. J. Geophys. Res. **69**, 3531.

Neupert, W. M. (1968): Comparison of solar X-ray line emission with microwave emission during solar flares. Ap. J. (Letters) **153**, L59.

Newkirk, G. (1967): Structure of the solar corona. Ann. Rev. Astron. Astrophys. **5**, 213.

Newman, J. J., Fowler, W. A. (1976): Maximum rate for the proton-proton reaction compatible with conventional solar models. Phy. Rev. Lett. **36**, 895.

Newman, M. J. (1978): S-process studies: the exact solution. Ap. J. **219**, 676.

Newton, H. W., Nunn, M. L. (1951): The Sun's rotation derived from sunspots 1934–1944 and additional results M.N.R.A.S. **111**, 413.

Nicholls, R. W. (1977): Transition probability data for molecules of astrophysical interest. Ann. Rev. Astron. Ap. **15**, 197.

Nikishov, A. I. (1962): Absorption of high-energy photons in the universe. Soviet Phys. J.E.T.P. **14**, 393.

Nishimura, J.: Theory of cascade showers. In: Handbuch der Physik, vol. XLVI/2, p.1. Berlin-Heidelberg-New York: Springer 1967.

Nishina, Y., Takeuchi, M., Ichimiya, T. (1937): On the nature of cosmic ray particles. Phys. Rev. **52**, 1198.

Nishina, Y., Tomonaga, S., Kobayasi, M. (1935): On the creation of positive and negative electrons by heavy charged particles. Sci. Pap. Inst. Phys. Chem. Res. Japan **27**, 137.

Noerdlinger, P. D. (1978): Positrons in compact radio sources. Phys. Rev. Lett. **41**, 135.

Nolte, J. T., et al. (1976): Coronal holes as sources of solar wind. Solar Phys. **46**, 303.

Norman, E. B., Schramm, D. N. (1979): On the conditions required for the r-process. Ap. J. **228**, 881.

Norris, J. P., et al. (1995): Duration distributions of bright and dim BATSE gamma-ray bursts. Ap. J. **439**, 542.

Novaco, J. C., Brown, L. W. (1978): Nonthermal galactic emission below 10 megahertz. Ap. J. **221**, 114.

Novikov, I. D., Syunyaev, R. A. (1967): An explanation of the anolmalous helium abundance in the star 3 Cen A. Sov. Astron. A. J. **11**, 2, 252.

Noyes, R. W., Hall, D. N. B. (1972): Thermal oscillations in the high solar photosphere. Ap. J. (Letters) **176**, L89.

Nussbaumer, H., Storey, P. J. (1983): Dielectronic recombination at low temperatures. Astron. Astrophys. **126**, 75.

O'Connell, R. F. (1975): Internal magnetic fields of pulsars, white dwarfs, and other stars. Ap. J. **195**, 751.

O'Dell, C. R. (1962): A distance scale for planetary nebulae based on emission line fluxes. Ap. J. **135**, 371.

O'Dell, C. R. (1963): Photoelectric spectrophotometry of planetary nebulae. Ap. J. **138**, 1018.

O'Dell, S. L., Sartori, L. (1970): Limitation on synchrotron models with small pitch angles. Ap. J. (Letters) **161**, L63.

Oda, M.: Observational results on diffuse cosmic X-rays. In: Non-solar X-and gamma ray astronomy – I.A.U. Symp. No. 37 (ed. L. Gratton). Dordrecht, Holland: D. Reidel 1970.

Odegard, N. (1985): Determination of nebular density and temperature from radio recombination lines. Ap. J. Supp. **57**, 571.

Ögelman, H. B., Maran, S. P. (1976): The origin of OB associations and extended regions of high-energy activity in the Galaxy through supernova cascade processes. Ap. J. **209**, 124.

Ohm, G. S. (1826): Versuch einer Theorie der durch galvanische Kräfte hervorgebrachten elektroskopischen Erscheinungen (An attempt to a theory of the galvanic forces generated through electroscopic phenomena). Ann. Phys. **6**, 459; **7**, 45.

Olson, B. I. (1975): On the ratio of total-to-selective absorption. P.A.S.P. **87**, 349.

Onsager, L. (1945): The distribution of energy in turbulence. Phys. Rev. **68**, 286.

Oort, J. H. (1946): Some phenomena connected with interstellar matter. M.N.R.A.S. **106**, 159.

Oort, J. H. (1954): Outline of a theory of the origin and acceleration of interstellar clouds and O associations. B.A.N. **12**, 177.

Oort, J. H., Kerr, F. J., Westerhout, G. (1958): The galactic system as a spiral nebula. M.N.R.A.S. **118**, 379. Reproduced in: A source book in astronomy and astrophysics 1900–1975 (eds. K. R. Lang and O. Gingerich). Cambridge, Mass.: Harvard University Press 1979.

Oort, J. H., Van De Hulst, H. C. (1946): Gas and smoke in interstellar space. B.A.N. **10**, 187.

Oort, J. H., Walraven, T. (1956): Polarization and composition of the Crab nebula. B.A.N. **12**, 285.

Öpik, E. (1922): An estimate of the distance of the Andromeda nebula. Ap. J. **55**, 406.

Öpik, E. J. (1951): Stellar models with variable composition II. Sequences of models with energy generation proportional to the fifteenth power of temperature. Proc. Roy. Irish Academy **54**, No. 4, 7 and Contributions from the Armagh Obs. **1**, No. 3, 1 (1951).

Öpik, E. J. (1953): Stellar associations and supernovae. Irish Astron. J. **2**, 219.

Öpik, E. J. (1954): The chemical composition of white dwarfs. Mem. Soc. Roy. Sci. Liege **14**, 131.

Oppenheimer, J. R., Serber, R. (1937): Note on the nature of cosmic-ray particles. Phys. Rev. **51**, 1113.

Oppenheimer, J. R., Snyder, H. (1939): On continued gravitational contraction. Phys. Rev. **56**, 455.

Oppenheimer, J. R., Volkoff, G. M. (1939): On massive neutron cores. Phys. Rev. **55**, 374. Reproduced in: A source book in astronomy and astrophysics 1900–1975 (eds. K. R. Lang and O. Gingerich). Cambridge, Mass.: Harvard University Press 1979.

Oppenheimer, M., Dalgarno, A. (1975): The formation of carbon monoxide and the thermal balance in interstellar clouds. Ap. J. **200**, 419.

Orear, J., Salpeter, E. E. (1973): Black holes and pulsars in the introductory physics course. Am. J. Phys. **41**, 1131.

Ormes, J., Freier, P. (1978): On the propagation of cosmic rays in the Galaxy. Ap. J. **222**, 471.

Ormes, J. F., Protheroe, R. J. (1983): Implications of HEAO 3 data for the acceleration and propagation of galactic cosmic rays. Ap. J. **272**, 756.

Ornstein, L. S., Burger, H. C. (1924): Strahlungsgesetz und Intensität von Mehrfachlinien (The radiation laws and intensities of multiple lines). Z. Physik **24**, 41.

Ornstein, L. S., Burger, H. C. (1924): Nachschrift zu der Arbeit Intensität der Komponenten im Zeeman Effekt (Postscript to the work on intensities of the components in the Zeeman effect). Z. Physik **29**, 241.

Osaki, J. (1974): An excitation mechanism for pulsation in beta cephei stars. Ap. J. **189**, 469.

Oster, L. (1960): Effects of collisions on the cyclotron radiation from relativistic particles. Phys. Rev. **119**, 1444.

Oster, L. (1961): Emission, absorption, and conductivity of a fully ionized gas at radio frequencies. Rev. Mod. Phys. **33** (4), 525.

Oster, L. (1970): The free-free emission and absorption coefficients in the radio frequency range at very low temperatures. Astron. Astrophys. **9**, 318.

Osterbrock, D. E. (1961): The heating of the solar chromosphere, plages and corona by magnetohydrodynamic waves. Ap. J. **134**, 347.

Osterbrock, D. E. (1963): Expected ultraviolet emission spectrum of a gaseous nebulae. Planet Space Sci. **11**, 621.

Osterbrock, D. E. (1965): Temperature in H II regions and planetary nebulae. Ap. J. **142**, 1423.

Osterbrock, D. E.: Astrophysics of gaseous nebulae. San Franciso: W. H. Freeman 1974.

Osterbrock, D. E.: Astrophysics of galactic nebulae and active galactic nuclei. Mill Valley, California: University Science Books, 1989.

Osterbrock, D. E., Mathews, W. G. (1986): Emission-line regions of active galaxies and qsos. Ann. Rev. Astron. Astrophys. **24**, 171.

Ostriker, J. P., Bodenheimer, P. (1968): Rapidly rotating stars. II. Massive white dwarfs. Ap. J. **151**, 1089.

Ostriker, J. P., Bodenheimer, P. (1973): On the oscillations and stability of rapidly rotating stellar models. III. Zero-viscosity polytropic sequences. Ap. J. **180**, 171. Erratum. Ap. J. **182**, 1037 (1973).

Ostriker, J. P., Bodenheimer, P., Lynden-Bell, D. (1966): Equilibrium models of differentially rotating zero-temperature stars. Phys. Rev. Lett. **17**, 816.

Ostriker, J. P., et al. (1976): A new luminosity limit for spherical accretion onto compact X-ray sources. Ap. J. (Letters) **208**, L61.

Ostriker, J. P., Gunn, J. E. (1969): On the nature of pulsars: I. Theory. Ap. J. **157**, 1395.

Ostriker, J. P., Hesser, J. E. (1968): Ultrashort-period stellar oscillations: II. The period and light curve of HZ 29. Ap. J. (Letters) **153**, L151.

Ostriker, J. P., Mc Kee, C. F. (1988): Astrophysical blastwaves. Rev. Mod. Phys. **60**, 1.

Ostriker, J. P., Tassoul, J. L. (1969): On the oscillations and stability of rotating stellar models. II. Rapidly rotating white dwarfs. Ap. J. **155**, 987.

Owen, F. N., Helfand, D. J., Spangler, S. R. (1981): The correlation of X-ray emission with strong millimeter activity in extragalactic sources. Ap. J. (Letters) **250**, L55.

Owens, A. J., Jokipii, J. R. (1977): Cosmic rays in a dynamical halo. I. Age and matter traversal distributions and anisotropy for nuclei. Ap. J. **215**, 677.

Owens, A. J., Jokipii, J. R. (1977): Cosmic rays in a dynamical halo. II. Electrons. Ap. J. **215**, 685.

Ozernoy, L. M., Chibisov, G. V. (1971): Galactic parameters as a consequence of cosmological turbulence. Astrophys. Lett. **7**, 201.

Pacholczyk, A. G.: Radio astrophysics. San Francisco: W. H. Freeman 1979.

Pacholczyk, A. G., Stodolkiewicz, J. S. (1960): On the gravitational instability of some magnetohydrodynamical systems of astrophysical interest. Acta. Astron. (Polska Akad. Nauk) **10**, 1.

Pacini, F. (1967): Energy emission from a neutron star. Nature **216**, 567. Reproduced in: Source book in astronomy and astrophysics 1900–1975 (eds. K. R. Lang and O. Gingerich). Cambridge, Mass.: Harvard University Press 1979.

Paczynski, B. (1971): Evolutionary processes in close binary systems. Ann. Rev. Astron. Astrophys. **9**, 183.

Paczynski, B. (1972): Carbon ignition in degenerate stellar cores. Astrophys. Lett. **11**, 53.

Paczynski, B. (1974): Helium-shell flashes in population I stars. Ap. J. **192**, 483.

Paczynski, B. (1977): Helium shell flashes. Ap. J. **214**, 812.

Paczynski, B. (1977): A model of accretion disks in close binaries. Ap. J. **216**, 822.

Paczynski, B. (1986): Gamma-ray bursters at cosmological distances. Ap. J. (Letters) **308**, L43.

Paczynski, B. (1990): Super-Eddington winds from neutron stars. Ap. J. **363**, 218.

Paczynski, B. (1995): How far away are gamma-ray bursters? P.A.S.P. **107**, 1167.

Paczynski, B., Zytkow, A. N. (1978): Hydrogen shell flashes in a white dwarf with mass accretion. Ap. J. **222**, 604.

Pagel, B. E. J. (1965): Revised abundance analysis of the halo red-giant HD 122563. Roy. Obs. Bul..No. 104.

Pais, A.: Inward bound: Of matter and forces in the physical world. New York: Oxford Univesity Press 1986.

Pallavicini, R. (1991): The role of magnetic loops in solar flares. Phil. Trans. R. Soc. Lond. **A336**, 389.

Pallavicini, R., et al. (1981): Relations among stellar X-ray emission observed from Einstein, stellar rotation and bolometric luminosity. Ap. J. **248**, 279.

Pallavicini, R., Serio, S., Vaiana, G. S. (1977): A survey of soft X-ray limb flare images.: The relation between their structure in the corona and other physical parameters. Ap. J. **216**, 108.

Pallé, P. L., Régulo, C., Roca-Cortés T. (1989): Solar cycle induced variations of the low *l* solar acoustic spectrum. Astron. Astrophys. **224**, 253.

Palmer, P., et al. (1967): Detection of a new microwave spectral line. Nature **215**, 40.
Panagia, N., Felli, M. (1975): The spectrum of the free–free radiation from extended envelopes. Astron. Astrophys. **39**, 1.
Pandharipande, V. R. (1971): Hyperonic matter. Nucl. Phys. **A178**, 123.
Pandharipande, V. R., Pines, D., Smith, R. A. (1976): Neutron star structure: Theory, observation and speculation. Ap. J. **208**, 550.
Pandharipande, V. R., Smith, R. A. (1975): A model neutron solid with π^o condensate. Nucl. Phys. **A237**, 507.
Pandharipande, V. R., Smith, R. A. (1975): Nuclear matter calculations with mean scalar fields. Phys. Lett. **B59**, 15.
Pardo, R. C., Couch, R. G., Arnett, W. D. (1974): A study of nucleosynthesis during explosive carbon burning. Ap. J. **191**, 711.
Parke, S. J. (1986): Nonadiabatic level crossing in resonant neutrino oscillations. Phys. Rev. Letters **57**, 1275.
Parker, E. N. (1952): Gravitational instability of a turbulent medium. Nature **170**, 1030.
Parker, E. N. (1955): Dynamics of the interplanetary gas and magnetic fields. Ap. J. **125**, 668.
Parker, E. N. (1955): Hydromagnetic waves and the acceleration of cosmic rays. Phys. Rev. **99**, 241. Reprod. in: Selected papers on cosmic ray origin theories (ed. S. Rosen). New York: Dover 1969.
Parker, E. N. (1958): Dynamics of the interplanetary gas and magnetic fields. Ap. J. **128**, 664.
Parker, E. N. (1958): Origin and dynamics of cosmic rays. Phys. Rev. **109**, 1328. Reprod. in: Selected papers on cosmic ray origin theories (ed. S. Rosen). New York: Dover 1969.
Parker, E. N. (1958): Cosmic-ray modulation by solar wind. Phys. Rev. **110**, 1445.
Parker, E. N. (1959): Extension of the solar corona into interplanetary space. J. Geophys. Res. **64**, 2675.
Parker, E. N. (1960): The hydrodynamic theory of solar corpuscular radiation and stellar winds. Ap. J. **132**, 821.
Parker, E. N. (1963): The solar-flare phenomenon and the theory of reconnection and annihilation of magnetic fields. Ap. J. Supp. **77**, no. 8, 177.
Parker, E. N.: Interplanetary dynamical processes. New York: John Wiley and Sons 1963.
Parker, E. N. (1965): Dynamical theory of the solar wind. Space Sci. Rev. **4**, 666.
Parker, E. N. (1974): Hydraulic concentration of magnetic fields in the solar photosphere. I. Turbulent pumping. Ap. J. **189**, 563.
Parker, E. N. (1974): Hydraulic concentration of magnetic fields in the solar photosphere. II. Bernoulli effect. Ap. J. **190**, 429.
Parker, E. N. (1988): Nanoflares and the solar X-ray corona. Ap. J. **330**, 474.
Parker, P. D. Mac D., Rolfs, C. E.: Nuclear energy generation in the solar interior. In: Solar interior and atmosphere (eds. A. N. Cox, W. C. Livingston and M. S. Matthews). Tucson, Arizona: University of Arizona Press 1991.
Parker, P. D., Bahcall, J. N., Fowler, W. A. (1964): Termination of the proton-proton chain in stellar interiors. Ap. J. **139**, 602.
Paschen, F. (1908): Zur Kenntnis ultraroter Linienspektra I (On the knowledge of infrared line spectra I). Ann. Physik **27**, 537.
Paschen, F., Back, E. (1912): Normale und anomale Zeeman-Effekte (Normal and anomalous Zeeman effect). Ann. Physik **39**, 929, Ann. Physik **40**, 960 (1913).
Patterson, J. R., Winkler, H., Spinka, H. M. (1968): Experimental investigation of the stellar nuclear reaction $^{16}O + ^{16}O$ at low energies. Bull. Am. Phys. Soc. **13**, 1495.
Patterson, J. R., Winkler, H., Zaidins, C. S. (1969): Experimental investigation of the stellar nuclear reaction $^{12}C + ^{12}C$ at low energies. Ap. J. **157**, 367.
Paul, J., Cassé, J., Cerarsky, C. J. (1976): Distribution of gas, magnetic fields, and cosmic rays in the Galaxy, Ap. J. **207**, 62.

Pauli, W. (1925): Über den Zusammenhang des Abschlusses der Elektronengruppen im Atom mit der Komplexstruktur der Spektren (On the connection of the termination of electron groups in atoms with the complex structure of spectra). Z. Physik **31**, 765.

Pauli, W. (1927): Über Gasentartung und Paramagnetismus (Gas degeneration and paramagnetism). Z. Physik **41**, 81.

Pauli, W. (1927): Zur Quantenmechanik des magnetischen Elektrons (On the quantum mechanics of the magnetic electron). Z. Physik **43**, 601.

Pauli, W.: Les théories quantities du magnétisive l'électron magnetique (The theory of magnetic quantities: The magnetic electron). Rpt. Septiene Couseil. Phys. Solvay, Bruxelles, 1930.

Pauli, W.: Remarks at the Seventh Solvay Conference, October 1933. Reproduced in the original French in: Collected scientific papers of Wolfgang Pauli, Vol. 2 (eds. R. Kronig and V. F. Weisskopf). New York: Wiley Interscience 1964.

Pawsey, J. L. (1946): Observation of million degree radiation from the sun at a wavelength of 1.5 metres. Nature **158**, 633.

Payne, C. H.: The relative abundances of the elements.In: Stellar atmospheres (C. H. Payne). Cambridge, Mass.: Harvard University Press 1925. Reproduced in: A source book in astronomy and astrophysics 1900–1975 (eds. K. R. Lang and O. Gingerich). Cambridge, Mass.: Harvard University Press 1979.

Payne-Scott, R., Yabsley, D. E., Bolton, J. G. (1947): Relative times of arrival of bursts of solar noise on different radio frequencies. Nature **160**, 256.

Peach, G. (1972): The broadening of radio recombination lines by electron collisions. Astrophys. Lett. **10**, 129.

Pedlar, A., Davies, R. D. (1971): Stark broadening in high quantum number recombination lines of hydrogen. Nature-Phys. Sci. **231**, 49.

Peebles, P. J. E.: Physical cosmology. Princeton, N. J.: Princeton University Press 1971.

Peebles, P. J. E., Wilkinson, D. T. (1968): Comment on the anisotropy of the primeval fireball. Phys. Rev. **174**, 2168.

Pengelly, R. M. (1964): Recombination spectra I. Calculations for hydrogenic ions in the limit of low densities. M.N.R.A.S. **127**, 145.

Penston, M. V. (1969): Dynamics of self-gravitating gaseous spheres III. Analytical results in the free-fall of isothermal cases. M.N.R.A.S. **144**, 425.

Penzias, A. A. (1979): The origin of the elements. Rev. Mod. Phys. **51**, 425.

Penzias, A. A., Wilson, R. W. (1969): Intergalactic H I emission at 21 centimeters. Ap. J. **156**, 799.

Perek, L. (1971): Photometry of southern planetary nebulae. Czech. Astr. Bull. No.22, 103.

Perek, L., Kohoutek, L.: Catalogue of planetary nebulae. Prague: Czech. Acad. Sci. 1967.

Perl, M. L., et al. (1975): Evidence for anomalous lepton production in e^+ - e^- annihilation. Phys. Rev. Lett. **35**, 1489.

Perl, M. L., et al. (1976): Properties of anomalous $e\mu$ events produced in e^+ e^- annihilation. Physics Letters **63B**, 466.

Perrin, J.-M. (1994): Corrections to an expression for the oscillator strength of hydrogen. Astron. Astrophys. **283**, 1025.

Peterson, B. A., Bolton, J. G. (1972): Redshifts of southern radio sources. Ap.J. (Letters) **173**, L19.

Peterson, F. W., King, C. (1975): A model for simultaneous synchrotron and inverse Compton fluxes. Ap. J. **195**, 753.

Peterson, L. E., Winckler, J. R. (1959): Gamma ray burst from a solar flare. J. Geophys. Res. **64**, 697.

Peterson, V. L., Bahcall, J. N. (1963): Exclusion principle inhibition of beta decay in stellar interiors. Ap. J. **138**, 437.

Pethick, C. J. (1992): Cooling of neutron stars. Rev. Mod. Phys. **64**, 1133.

Petit, M.: Variable stars. New York: John Wiley and Sons 1987.

Petrosian, V., Beaudet, G., Salpeter, E. E. (1967): Photoneutrino energy loss rates. Phys. Rev. **154**, 1445.

Petschek, H. E.: Annihilation of magnetic fields. Proc. AAS-NASA conference on physics of solar flares (ed. W. N. Hess). NASA SP-50, Wash. D.C. 1964.

Pfund, A. H. (1924): The emission of nitrogen and hydrogen in the infrared. J. Opt. Soc. Am. **9**, 193.

Phillips, J. L., et al. (1995): Ulysses solar wind plasma observations at high southerly latitudes. Science **268**, 1030.

Phillips, J. L., et al. (1995): Ulysses solar wind plasma observations from pole to pole. Geophys. Res. Lett. **22**, No. 23, 3301.

Phillips, K. J. H. (1991): Spectroscopy of high-temperature solar flare plasmas. Phil. Trans. Roy. Soc. (London) **A336**, 461.

Pickering, E. C. (1896): Stars having peculiar spectra, new variable stars in Crux and Cygnus. Ap.J. **4**, 369.

Pickering, E. C. (1897): The spectrum of ξ Puppis. Ap.J. **5**, 92.

Piddington, J. H. (1956): Solar atmospheric heating by hydromagnetic waves. M.N.R.A.S. **116**, 314.

Pierce, J. R. (1948): Possible fluctuations in electron streams due to ions. J. Appl. Phys. **19**, 231.

Pikelner, S. (1968): Ionization and heating of the interstellar gas by subcosmic rays, and the formation of clouds. Sov. Astron. **11**, 737.

Pinaev, V. S. (1964): Some neutrino pair production processes in stars. Sov. Phys. JETP **18**, 377.

Pines, D.: Inside neutron stars. In: Proc. 12th int. conf. on low temperature physics. Academic Press Japan 1970.

Pines, D., Bohm, D. (1952): A collective description of electron interactions: II. Collective vs. individual particle aspects of the interactions. Phys. Rev. **85**, 338.

Piran, T. (1995): Binary neutron stars. Scientific American **272**, 52 – May.

Piran, T.: Toward understanding gamma-ray bursts. In: Unsolved problems in astrophysics (eds. J. N. Bahcall and J. P. Ostriker). Princeton, New Jersey: Princeton University Press 1997.

Planck, M. (1901): Über das Gesetz der Energieverteilung im Normalspectrum (On the law of energy distribution in a normal spectrum). Ann. Physik **4**, 553.

Planck, M. (1910): Zur Theorie der Wärmestrahlung (On the theory of thermal radiation). Ann. Physik **31**, 758.

Planck, M.: The theory of heat radiation. 1913. Reprod. New York: Dover 1959.

Plummer, H. C. (1913): Note on the orbit of ζ Geminorum. M.N.R.A.S. **73**, 661.

Plummer, H. C. (1914): Note on the velocity of light and Doppler's principle. M.N.R.A.S. **74**, 660.

Pneuman, G. W., Kopp, R. A. (1970): Coronal streamers III. Energy transport in streamer and interstreamer regions. Solar Phys. **13**, 176.

Pneuman, G. W., Kopp, R. A. (1971): Gas-magnetic field interactions in the solar corona. Solar Phys. **18**, 258.

Poincaré, H.: Lecons sur les hypothèses cosmogoniques (Lessons on cosmological hypothesis). Paris: Librarie Scientifique, A. Hermann 1811.

Poincaré, H. (1855): Lecons l'équilibre d'une masse fluide animeé d'un mouvement de rotation (Lesson on the equilibrium of rotating fluid masses). Acta. Math. **7**, 259.

Poincaré, H. (1905): Électricité - sur la dynamique de l'électron. Comptes Rendus Acad. Sci. Paris **140**, 1504.

Poisson, S. D. (1813): Remarques sur une equation qui se présente das la theorie des attractions des spheroides (Remarks on an equation which presents itself in the theory of spheroidal attractions) Bull. de la Soc. Philomatique **3**, 388.

Pokrowski, G. I. (1931): Versuch der Anwendung einiger thermodynamischer Gesetzmä*ß*igkeiten zur Beschreibung von Erscheinungen in Atomkernen (Thermodynamical principles of nuclear phenomena). Phys. Z. **32**, 374.

Polanyi, J. C.: Chemical processes. In: Atomic and molecular processes (ed. D. R. Bates). New York: Academic Press 1962.

Pollack, J. B., Fazio, G. G. (1963): Production of π mesons and gamma radiation in the Galaxy by cosmic rays. Phys. Rev. **131**, 2684.

Pollack, J. B., Morrison, D. (1970): Venus: Determination of atmospheric parameters from the microwave spectrum. Icarus **12**, 376.

Pontecorvo, B. (1958): Mesonium and antimesonium. Sov. Phys. JETP **6**, 429.

Pontecorvo, B. (1958): Inverse beta processes and nonconservation of lepton charge. Sov. Phys. JETP **7**, 172.

Pontecorvo, B. (1959): Electron and muon neutrinos. Sov. Phys. JETP **37**, 1236 (1960). J. Exp. Theor. Phys. USSR **37**, 1751.

Pontecorvo, B. (1959): The universal Fermi interaction and astrophysics. Sov. Phys. JETP **9**, 1148.

Pontecorvo, B. (1968): Neutrino experiments and the problem of conservation of leptonic charge. Sov. Phys. JETP **26**, 984.

Popper, D. M. (1954): Red shift in the spectrum of 40 Eridani B. Ap. J. **120**, 316.

Power, H. (1663): Experimental philosophy in three books containing new experiments, microscopical, mercurial, magnetical. London (1663), cf. I. B. Cohen. Newton, Hooke, and Boyle's law. Nature **204**, 618 (1964).

Poynting, J. H. (1884): On the transfer of energy in the electromagnetic field. Phil. Trans. **175**, 343.

Prandtl, L. (1905): Verhandlungen des Dulten Internationalen Mathematiker-Kongresses (Transactions of the international mathematical congress). 484, see also The physics of solids and fluids by P. P. Ewald and L. Prandtl. London: Blakie 1930.

Prandtl, L. (1925): Bericht über Untersuchungen zur ausgebildeten Turbulenz. Z. Angew. Math. Mech. **5**, 136.

Prandtl, L.: Essentials of fluid dynamics. London: Blakie 1952.

Prendergast, K. H., Burbidge, G. R. (1968): On the nature of some galactic X-ray sources. Ap. J. (Letters) **151**, L83.

Press, W. H., Wiita, P. J., Smarr, L. L. (1975): Mechanism for inducing synchronous rotation and small eccentricity in close binary systems. Ap. J. **202**, L135.

Preston, G. W. (1970): The quadratic Zeeman effect and large magnetic fields in white dwarfs. Ap. J. (Letters) **160**, L143.

Preston, M. A.: Physics of the nucleus. Reading, Mass.: Addison-Wesley 1965.

Preston, T. (1898): Radiation phenomena in the magnetic field. Phil. Mag. **45**, 325.

Priest, E. R., Solar flare magnetohydrodynamics. New York: Gordon and Breach 1981.

Priest, E. R.: Solar magnetohydrodynamics. Dordrecht, Holland: D. Reidel 1982.

Priest, E. R. (1991): The magnetohydrodynamics of energy release in solar flares. Phil. Trans. Roy. Soc. (London) **A336**, 363.

Pringle, J. E. (1981): Accretion discs in astrophysics. Ann. Rev. Astron. Astrophys. **19**, 137.

Pringle, J. E., Rees, M. J. (1972): Accretion disc models for compact X-ray sources. Astron. and Astrophys **21**, 1.

Protheroe, R. J., Szabo, A. P.: High energy cosmic rays from active galactic nuclei. Phys. Rev. Lett. **69**, 2885 (1992).

Proudman, J. (1916): On the motion of solids in a liquid possessing vorticity. Proc. Roy. Soc. London **A92**, 408.

Proudman, J. (1952): The generation of noise by isotropic turbulence. Proc. Roy. Soc. London **A214**, 119.

Puget, J. L., Léger, A. (1989): A new component of the interstellar matter: Small grains and large aromatic molecules. Ann. Rev. Astron. Astrophys. **27**, 161.

Purcell, E. M. (1952): The lifetime of the $2^2S_{1/2}$ state of hydrogen in an ionized atmosphere. Ap. J. **116**, 457.

Rabi, I. I. Von (1928): Das freie Elektron im homogenen Magnetfeld nach der Diracschen Theorie (The free electron in a homogeneous magnetic field according to the Dirac theory). Z Physik **49**, 507.

Rabin, D. M., et al.: The solar activity cycle. In: Solar interior and atmosphere (eds. A.N. Cox, W. C. Livingston, M. S. Matthews. Tucson: University of Arizona Press 1991.

Radhakrishnan, V., Brooks, J. W., Goss, W. M., Murray, J. D., Schwarz, U. J. (1972): The Parkes survey of 21-centimeter absorption in discrete-source spectra. Ap. J. Suppl. No. 203, **24**, 1.

Radin, J. (1970): Cross section for $C^{12}(\alpha, \alpha n)C^{11}$ at 920 MeV. Phys. Rev. **C 2**, 793.

Ramana Murthy, P. V., Wolfendale, A. W.: Gamma-ray astronomy, Second edition. New York: Cambridge University Press 1993.

Ramaty, R. (1969): Gyrosynchrotron emission and absorption in a magnetoactive plasma. Ap. J. **158**, 753.

Ramaty, R., Kozlovsky, B., Lingenfelter, R. E. (1979): Nuclear gamma rays from energetic particle interactions. Ap. J. Supp. **40**, 487.

Ramaty, R., Kozlovsky, B., Lingenfelter, R. (1998): Cosmic rays, nuclear gamma rays and the origin of the light elements. Physics Today **51**, 30 – April.

Ramaty, R., Lingenfelter, R. E. (1966): Galactic cosmic-ray electrons. J. Geophys. Res. **71**, 3687.

Ramaty, R., Lingenfelter, R. E. (1979): γ-ray line astronomy. Nature **278**, 127.

Ramaty, R., Lingenfelter, R. E. (1983): Gamma-ray line astronomy. Space Science Reviews **36**, 305.

Ramaty, R., Mandzhavidze, N.: Theoretical models for high-energy solar flare emissions. In: High energy solar phenomena - a new era of spacecraft measurements (eds. J. M. Ryan and W. T. Vestrand). New York: American Institute of Physics 1994.

Ramaty, R., Mandzhavidze, N., Kozlovsky, B., Skibo, J. G. (1993): Acceleration in solar flares: interacting particles versus interplanetary particles. Adv. Space Res. **13**, No. 9, 275.

Ramaty, R., Murphy, R. J. (1987): Nuclear processes and accelerated particles in solar flares. Space Science Reviews **45**, 213.

Ramaty, R., Petrosian, V. (1972): Free-free absorption of gyrosynchrotron radiation in solar microwave bursts. Ap. J. **178**, 241.

Ramaty, R., Schwartz, R. A., Enome, S., Nakajima, H. (1994): Gamma-ray and millimeter-wave emissions from the 1991 june x-class flares. Ap. J. **436**, 941.

Ramaty, R., Stecker, F. W., Misra, D. (1970): Low-energy cosmic ray positrons and 0.51-MeV gamma rays from the Galaxy. J. Geophys. Res. **75**, 1141.

Rana, N. C. (1991): Chemical evolution of the Galaxy. Ann. Rev. Astron. Astrophys. **29**, 129.

Rank, D. M, Townes, C. H., Welch, W. J. (1971): Interstellar molecules and dense clouds. Science **174**, 1083.

Rankine, W. J. M. (1870): On the thermodynamic theory of waves of finite longitudinal disturbance. Phil. Trans. Roy. Soc. London **160**, 277.

Rapp, D., Francis, W. E. (1962): Charge exchange between gaseous ions and atoms. J. Chem. Phys. **37**, 2631.

Rastall, P. (1977): The maximum mass of a neutron star. Ap. J. **213**, 234.

Ratcliffe, J. A. (1956): Some aspects of diffraction theory and their applications to the ionosphere. Rep. Prog. Phys. **19**, 188.

Ratcliffe, J. A.: The magneto-ionic theory and its applications to the ionosphere. Cambridge, England: Cambridge University Press 1959.

Rau, A. R. P., Spruch, L. (1976): Energy levels of hydrogen in magnetic fields of arbitrary strength. Ap. J. **207**, 671.

Ravenhall, D. G., Bennett, C. D., Pethick, C. J. (1972): Nuclear surface energy and neutron star matter. Phys. Rev. Lett. **78**, 978.

Rawitscher, G. H. (1956): Effect of the finite size of the nucleus on μ-pair production by gamma rays. Phys. Rev. **101**, 423.
Rayleigh, Lord (1871): On the light from the sky, its polarization and colour. Phil. Mag. **41**, 107, 274.
Rayleigh, Lord (1879): Investigations in optics, with special reference to the spectroscope. Phil. Mag. **8**, 403.
Rayleigh, Lord (1883): Investigation of the character of the equilibrium of an incompressible heavy fluid of variable density. Proc. London Math. Soc. **14**, 170.
Rayleigh, Lord (1889): On the limit to interference when light is radiated from moving molecules. Phil. Mag. **27**, 298.
Rayleigh, Lord (1889): On the visibility of faint interference bands. Phil. Mag. **27**, 484.
Rayleigh, Lord (1899): On the transmission of light through an atmosphere containing small particles in suspension, and on the origin of the blue of the sky. Phil. Mag. **47**, 375.
Rayleigh, Lord (1900): Remarks upon the law of complete radiation. Phil. Mag. **49**, 539.
Rayleigh, Lord (1905): The dynamical theory of gases and of radiation. Nature **72**, 54.
Rayleigh, Lord (1915): On the widening of spectrum lines. Phil. Mag. **29**, 274.
Rayleigh, Lord (1916): On convective currents in a horizontal layer of fluid when the higher temperature is on the under side. Phil. Mag. **32**, 529.
Raymond, J. C. (1978): On dielectric recombination and resonances in excitation cross sections. Ap. J. **222**, 1114.
Raymond, J. C. (1979): Shock waves in the interstellar medium. Ap. J. Suppl. **39**, 1.
Razin, V. A. (1960): The spectrum of nonthermal cosmic radio emission. Radiophysica **3**, 584, 921.
Reames, D. V., Richardson, I. G., Wenzel, K.-P. (1992): Energy spectra of ions from impulsive solar flares. Ap. J. **387**, 715.
Reber, G. (1944): Cosmic static. Ap. J. **100**, 279. Reproduced in: A source book in astronomy and astrophysics (eds. K. R. Lang and O. Gingerich). Cambridge, Mass.: Harvard University Press 1979.
Rees, M. J. (1966): Appearance of relativistically expanding radio sources. Nature **211**, 468.
Rees, M. J. (1976): Opacity-limited hierarchical fragmentation and the masses of protostars. M.N.R.A.S. **176**, 483.
Rees, M. J. (1984): Black hole models for active galactic nuclei. Ann. Rev. Astron. Astrophys. **22**, 471.
Rees, M. J., Mészáros, P. (1992): Relativistic fireballs: energy conversion and time-scales. M.N.R.A.S. **258**, 41P.
Reeves, H.: Stellar energy sources. In: Stellar structure - Stars and stellar systems VIII (ed. L. H. Aller and D. B. Mc Laughlin). Chicago, Ill.: University of Chicago Press 1963.
Reeves, H.: Nuclear reactions in stellar surfaces and their relations with stellar evolution. New York: Gordon and Breach 1971.
Reeves, H. (1994): On the origin of the light elements ($Z < 6$). Rev. Mod. Phys. **66**, 193.
Reeves, H., Audouze, J., Fowler, W. A., Schramm, D. N. (1973): On the origin of the light elements. Ap. J. **179**, 909.
Reeves, H., Fowler, W. A., Hoyle, F. (1970): Galactic cosmic ray origin of Li, Be and B in stars. Nature **226**, 727.
Reeves, H., Johns, O. (1976): The long-lived radioisotopes as monitors of stellar, galactic, and cosmological phenomena. Ap. J. **206**, 958.
Reeves, H., Meyer, J.-P. (1978): Cosmic-ray nucleosynthesis and the infall rate of extragalactic matter in the solar neighborhood. Ap. J. **226**, 613.
Reeves, H., Stewart, P. (1965): Positron-capture processes are a possible source of the p elements. Ap. J. **141**, 1432.
Reid, M. J., Moran, J. M. (1981): Masers. Ann. Rev. Astron. Astrophys. **19**, 231.
Reifenstein, E. C., et al. (1970): A survey of H 109 α recombination line emission in galactic H II regions of the northern sky. Astron. Astrophys. **4**, 357.

Reilman, R. F., Manson, S. T. (1979): Photoabsorption cross sections for positive atomic ions with $Z \leq 30$. Ap. J. Supp. **40**, 815.

Reines, F., Cowan, C. L., Jr. (1953): Detection of the free neutrino. Phys. Rev. **92**, 830.

Reiss, A. G., Press, W. H., Kirshner, R. P. (1996): A precise distance indicator: Type Ia supernova multicolor light-curve shapes. Ap. J. **473**, 88.

Reynolds, O. (1876): On the force caused by the communication of heat between a surface and a gas and on a new photometer. Phil. Mag. **23**, 1.

Reynolds, O. (1883): An experimental investigation of the circumstances which determine whether the motion of water shall be direct or sinuous, and of the law of resistance in parallel channels. Phil. Trans. Roy. Soc. London **174**, 935.

Reynolds, S. P.: Supernova remnants. In: Galactic and extragalactic radio astronomy, second edition (eds. G. L. Verschuur and K. I. Kellermann). New York: Springer-Verlag 1988.

Rhoades, C. E. Jr., Ruffini, R. (1974): Maximum mass of a neutron star. Phys. Rev. Lett. **32**, 324.

Rhodes, E. J. Jr., Ulrich, R. K., Simon, G. W. (1977): Observations of nonradial p-mode oscillations on the Sun. Ap. J. **218**, 901.

Richardson, R. S., Schwarzschild, M. (1950): On the turbulent velocities of solar granules. Ap. J. **111**, 351.

Rickett, B. J. (1970): Interstellar scintillation and pulsar intensity variations. M.N.R.A.S. **150**, 67.

Rickett, B. J. (1977): Interstellar scattering and scintillation of radio waves. Ann. Rev. Astron. Astrophys. **15**, 479.

Rickett, B. J. (1990): Radio propagation through the turbulent interstellar plasma. Ann. Rev. Astron. Astrophys. **28**, 561.

Rickett, B. J., Coles, W. A. (1991): Evolution of the solar wind structure over a solar cycle: Interplanetary scintillation velocity measurements compared with coronal observations. J. Geophys. Res. **96**, 1717.

Rickett, B. J., Lang, K. R. (1973): Two-station observations of the interstellar scintillation from pulsars. Ap. J. **185**, 945.

Riddle, R. K. (1970): First catalogue of trigonometric parallaxes of faint stars. Publ. U.S. Naval Obs. **120**, part 3.

Riley, J. M., Longair, M. S., Gunn, J. E. (1980): Investigations of the optical fields of 3CR radio sources to faint limiting magnitudes - III. M.N.R.A.S. **192**, 233.

Ritter, A. Von (1880): Untersuchungen über die Höhe der Atmosphäre und die Konstitution gasförmiger Welkörper (Investigations on the height of the atmosphere and the constitution of gaseous celestial bodies). Ann. Phys. u. Chem **8**, 157; **13**, 360 (1881).

Ritus, V. I. (1962): Photoproduction of neutrinos on electrons and neutrino radiation from stars. Sov. Phys. JETP **14**, 915.

Ritz, W. (1908): On a new law of series spectra. Ap. J. **28**, 237.

Roberts, D. A. (1989): Interplanetary observational constraints on Alfvén wave acceleration of the solar wind. J. Geophys. Res. **94**, 6899.

Roberts, D. A., Klein, L. W., Goldstein, M. L., Mathaeus, W. H. (1987): The nature and evolution of magnetohydrodynamic fluctuations in the solar wind: Voyager observations. J. Geophys. Res. **92**, 11021.

Roberts, M. S. (1968): Neutral hydrogen observations of the binary galaxy system NGC 4631/4656. Ap. J. **151**, 171.

Roche, M. (1847): Mémoire sur la figure d'une masse fluide (Soumise à l'attraction d'un point eloigne), Memoir on the figure of a fluid mass (subject to the attraction of a distant point). Acad. des Sci. de Montpellier **1**, 243, 333.

Rogers, A. E. E., Barrett, A. H. (1968): Excitation temperature of the 18 cm line of OH in HI regions. Ap. J. **151**, 163.

Rogerson, J. (1969): On the abundance of iron in the solar photosphere. Ap. J. **158**, 797.

Rohrlich, F. (1959): Sum rules for multiplet strengths. Ap. J. **124**, 449.

Rolfs, C. E., Rodney, W. S.: Cauldrons in the cosmos – nuclear astrophysics. Chicago: University of Chicago Press, 1988.
Röntgen, W. C. (1896): On a new kind of rays. Nature **103**, 274.
Rosen, J., Rosen, N. (1975): The maximum mass of a cold neutron star. Ap. J. **202**, 782.
Rosen, S. (ed.): Selected papers on cosmic rays origin theories . New York: Dover 1969.
Rosenberg, L. (1963): Electromagnetic interactions of neutrinos. Phys. Rev. **129**, 2786.
Rosenbluth, M., Mac Donald, W. M., Judd, D. L. (1957): Fokker-Planck equation for an inverse square force. Phys. Rev. **107**, 1.
Rosenfeld, A. H., et al. (1968): Data on particles and resonant states. Rev. Mod. Phys. **40, 77**.
Rosner, R., et al. (1978): Heating of coronal plasma by anomalous current dissipation. Ap. J. **222**, 317.
Rosner, R., (1985): On stellar X-ray emission. Ann. Rev. Astron. Astrophys. **23**, 413.
Rosner, R., Tucker, W. H., Vaiana, G. S. (1978): Dynamics of the quiescent solar corona. Ap. J. **220**, 643.
Rosseland, S. (1924): Note on the absorption of radiation within a star, M.N.R.A.S. **84**, 525. Reprod. in: Selected papers on the transfer of radiation (ed. D. H. Menzel). New York: Dover 1966.
Rosseland, S.: The pulsation theory of variable stars. Oxford: Clarendon Press 1949. New York: Dover 1964.
Rosseland, S. (1929): Viscosity in the stars. M.N.R.A.S. **89**, 49.
Rossi, B. (1939): The disintegration of mesotrons. Rev. Mod. Phys. **11**, 296.
Rossi, B. (1948): Interpretation of cosmic-ray phenomena. Rev. Mod. Phys. **20**, 537.
Rossi, B.: High energy particles. New York: Prentice Hall 1952.
Roughton, N. A., et al. (1976): Thermonuclear reaction rates derived from thick target yields. Ap. J. **205**, 302.
Routly, P. M. (1972): Second catalogue of trigonometric parallaxes of faint stars. Publ. U.S. Naval Obs. **20**, part 6.
Rowan-Robinson, M. (1979): Clouds of dust and molecules in the galaxy. Ap. J. **234**, 111
Rowan-Robinson, M.: The cosmological distance ladder. New York: W. H. Freeman 1985.
Rowan-Robinson, M. (1988): The extragalactic distance scale. Space Sci. Rev. **48**, 1.
Russell, H. N. (1925): The intensitites of lines in multiplets. Nature **115**, 835.
Roxburgh, I. W. (1965): On models of nonspherical stars: II. Rotating white dwarfs. Z. Ap. **62**, 134.
Rubin, R. H. (1968): The structure and properties of H II regions. Ap. J. **153**, 761.
Rubin, R. H. (1985): Models of H II regions: Heavy element opacity, variation of temperature. Ap. J. Supp. **57**, 349.
Ruderman, M. (1969): Superdense matter in stars. J. Phys. Suppl. 11, **30**, 152.
Ruderman, M. (1976): Crust-breaking by neutron superfluids and the vela pulsar glitches. Ap. J. **203**, 213.
Ruderman, M. (1991): Neutron star crustal plate tectonics. II. Evolution of radio pulsar magnetic fields. Ap. J. **382**, 576.
Ruderman, M. (1991): Neutron star crustal plate tectonics. III. Cracking, glitches, and gamma-ray bursts. Ap. J. **382**, 587.
Ruderman, M.: In and around neutron stars. In: Unsolved problems in astrophysics (eds. J. N. Bahcall and J. P. Ostriker). Princeton, New Jersey: Princeton University Press 1997.
Ruderman, M. A., Sutherland, P. G. (1974): Rotating superfluid in neutron stars. Ap. J. **190**, 137.
Russell, H. N. (1916): On the albedo of the planets and their satellites. Ap. J. **43**, 173.
Russell, H. N. (1921): Comments on the excitation of planetary nebulae. Observatory **44**, 72.
Russell, H. N. (1929): On the composition of the sun's atmosphere. Ap. J. **70**, 11. Reproduced in: A source book in astronomy and astrophysics 1900-1975 (eds. K. R. Lang and O. Gingerich). Cambridge, Mass.: Harvard University Press 1979.
Russell, H. N.. (1934): Molecules in the sun and stars. Ap. J. **69**, 317.
Russell, H. N. (1936): Tables for intensities of lines in multiplets. Ap. J. **83**, 129.

Russell, H. N., Saunders, F. A. (1925): New regularities in the spectra of the alkaline earths. Ap. J. **61**, 38.

Rutherford, E. (1899): Uranium radiation and the electrical conduction produced by it. Phil. Mag. **47**, 109.

Rutherford, E. (1911): The scattering of α and β particles by matter and the structure of the atom. Phil. Mag. **21**, 669.

Rutherford, E. (1914): The structure of the atom. Phil. Mag. **27**, 488.

Rutherford, E., Chadwick, J. (1921): The disintegration of elements by α-particles. Nature **107**, 41.

Rutherford, E., Soddy, F. (1902): The cause and nature of radioactivity, part I, part II. Phil. Mag. **4**, 370, 569.

Rutherford, E., Soddy, F. (1903): The radioactivity of uranium. A comparative study of the radioactivity of radium and thorium. Condensation of the radioactive emanations. The products of radioactive change and their specific material nature. Phil. Mag. **5**, 441, 445, 561, 576.

Ryan, J., et al.: Neutron and gamma-ray measurements of the solar flare of 1991 June 9. In: High-Energy Solar Phenomena – A New Era of Spacecraft Measurements. AIP Conference Proceedings 294 (eds. J. M. Ryan and W. T. Vestrand). New York: American Institute of Physics 1994.

Ryan, M. J., Ormes, J. F., Balasubrahmanyam, V. K. (1972): Cosmic-ray proton and helium spectra above 50 GeV. Phys. Rev. Lett. **28**, 985.

Rybicki, G. B., Lightman, A. P.: Radiative processes in astrophysics. New York: Wiley-Interscience 1979.

Rydberg, J. R. (1890): On the structure of the line-spectra of the chemical elements. Phil. Mag. **29**, 331.

Rydberg, J. R. (1896): The new elements of Cleveite gas. Ap. J. 4, 91.

Rydgren, A. E., Cohen, M.: Young stellar objects and their circumstellar dust: An overview. In: Protostars and planets II (eds. D. C. Black and M. S. Matthews). Tucson, Arizona: University of Arizona Press 1985.

Ryter, C., et al. (1970): The energetics of L nuclei formation in stellar atmospheres and its relevance to X-ray astronomy. Astron. Astrophys. **8**, 389.

Sackmann, I.-J. (1977): What quenches the helium shell flashes? Ap. J. **212**, 159.

Sackur, O. (1911): Die Anwendung der Kinetischen Theorie der Gase auf chemische Probleme (The application of the kinetic theory of gases to chemical problems). Ann. Phys. **36**, 958.

Sackur, O. (1913): Die universelle Bedeutung des sog. elementaren Wirkungsquantums (Universal significance of the elementary working-quantum). Ann. Phys. **40**, 67.

Saenz, R. A. (1977): Maximum mass of neutron stars: Dependence on the assumptions. Ap. J. **212**, 816.

Saenz, R. A., Shapiro, S. L. (1979): Gravitational and neutrino radiation from stellar core collapse: Improved ellipsoidal model calculations. Ap. J. **229**, 1107.

Sagdeev, R. Z. (1979): The 1976 Oppenheimer lectures: critical problems in plasma astrophysics. I. Turbulence and nonlinear waves. Rev. Mod. Phys. **51**, 1.

Saha, M. N. (1920): Ionization in the solar chromosphere. Phil. Mag. **40**, 479. Reproduced in: Source book in astronomy and astrophysics 1900-1975 (eds. K. R. Lang and O. Gingerich). Cambridge, Mass.: Harvard University Press 1979.

Saha, M. N. (1921): On the physical theory of stellar spectra. Proc. Roy. Soc. London **A99**, 135.

Saika, D. J., Salter, C. J. (1988): Polarization properties of extragalactic radio sources. Ann. Rev. Astron. Astrophys. **26**, 93.

Sakao, T.: Characteristics of solar flare hard X-ray sources as revealed with the hard X-ray telescope aboard the Yohkoh satellite. Ph. D. Thesis, Univ. of Tokyo, Tokyo, Japan, 1994.

Sakata, S., Inoue, T. (1946): On the correlations between mesons and Yukawa particles. Prog. Theor. Phys. **1**, 143.

Sakurai, T. (1991): Observations from the Hinotori mission. Phil. Trans. Roy. Soc. (London) **A336**, 339.

Salem, M., Brocklehurst, M. (1979): A table of departure coefficients from thermodynamic equilibrium (b_n factors) for hydrogenic ions. Ap. J. Suppl. **39**, 633.

Salpeter, E. (1952): Nuclear reactions in stars without hydrogen. Ap. J. **115**, 336. Reproduced in: Souce book in astronomy and astrophysics 1900–1975 (eds. K. R. Lang and O. Gingergich). Cambridge, Mass.: Harvard University Press 1979.

Salpeter, E. E. (1952): Nuclear reactions in the stars: I. Proton-proton chain. Phys. Rev. **88**, 547.

Salpeter, E. E. (1953): Energy production in stars. Ann. Rev. Nucl. Sci. **2**, 41.

Salpeter, E. E. (1954): Electron screening and thermonuclear reactions. Austr. J. Phys. **7**, 373.

Salpeter, E. E. (1955): Nuclear reactions in stars: II. Protons on light nuclei. Phys. Rev. **97**, 1237.

Salpeter, E. E. (1957): Nuclear reactions in stars. Buildup from helium. Phys. Rev. **107**, 516.

Salpeter, E. E. (1960): Electron density fluctuations in a plasma. Phys. Rev. **120**, 1528.

Salpeter, E. E. (1961): Energy and pressure of a zero temperature plasma. Ap. J. **134**, 669.

Salpeter, E. E. (1964): Accretion of interstellar matter by massive objects. Ap. J. **140**, 796.

Salpeter, E. E. (1967): Interplanetary scintillations I. Theory. Ap. J. **147**, 433.

Salpeter, E. E. (1974): Nucleation and growth of dust grains. Ap. J. **193**, 579.

Salpeter, E. E. (1974): Formation and flow of dust grains in cool stellar atmospheres. Ap. J. **193**, 585.

Salpeter, E. E. (1974): Dying stars and reborn dust. Rev. Mod. Phys. **46**, 433.

Salpeter, E. E. (1976): Planetary nebulae, supernova remnants, and the interstellar medium. Ap. J. **206**, 673.

Salpeter, E. E. (1977): Formation and destruction of dust grains. Ann. Rev. Astron. Astrophys. **15**, 267.

Salpeter, E. E., Van Horn, H. M. (1969): Nuclear reaction rates at high densities. Ap. J. **155**, 183.

Salter, C. J., Brown, R. L.: Galactic nonthermal continuum emission. In: Galactic and extragalactic radio astronomy, second edition (eds. G. L. Verschuur and K. I. Kellermann). New York: Springer-Verlag 1988.

Sampson, D. H. (1974): Electron-impact excitation cross-sections for complex ions. I. Theory for ions with one and two valence electrons. Ap. J. Suppl. **28**, 309.

Sampson, R. A. (1894): On the rotation and mechanical state of the Sun. Mem. R.A.S. **51**, 123.

Sandage, A. (1975): On the ratio of extinction to reddening for interstellar matter using galaxies. I. A limit on the neutral extinction from photometry of the 3C 129 group. P.A.S.P. **87**, 853.

Sanders, R. H. (1967): S-process nucleosynthesis in thermal relaxation cycles. Ap. J. **150**, 971.

Sargent, A. I., Beckwith, S. (1987): Kinematics of the circumstellar gas of HL Tauri and R. Monocerotis. Ap. J. **323**, 294.

Sargent, W. L. W. (1964): The atmospheres of the magnetic and metallic-line stars. Ann. Rev. Astron. Astrophys. **2**, 297.

Sargent, W. L. W. (1973): Redshifts for 51 galaxies identified with radio sources in the 4C catalog. Ap. J. (Letters) **182**, L13.

Sargent, W. L. W., Jugaku, J. (1961): The existence of He^3 in 3 Centauri A. Ap. J. **134**, 777.

Sari, R., Piran, T., Narayan, R. (1998): Spectra and light curves of gamma-ray burst afterglows. Ap. J. (Letters) **497**, L17.

Saslaw, W. C.: Gravitational physics of stellar and galactic systems. Cambridge, England: Cambridge University Press 1985.

Sastri, V. K., Stothers, R. (1974): Influence of opacity on the pulsational stability of massive stars with uniform chemical composition. II. Modified Kramers opacity. Ap. J. **193**, 677.

Saunders, R. L., et al. (1992): Magellan at venus. J. Geophys. Res. **97**, 13,603, 15,921.
Sauter, F. (1931): Über den atomaren Photoeffekt bei groβer Härte der anregenden Strahlung (Atomic photoelectric effect excited by very hard rays). Ann. Phys. **9**, 217.
Sauter, F. (1933): Zur unrelativistischen Theorie des kontinuierlichen Röntgenspektrums (On the non-relativistic theory of the continuous X-ray spectrum). Ann. Physik **18**, 486.
Savage, B. D., et al. (1977): A survey of interstellar molecular hydrogen, I. Ap. J. **216**, 291.
Savage, B. D., Mathis, J. S. (1979): Observed properties of interstellar dust. Ann. Rev. Astron. Astrophys. **17**, 73.
Sawyer R. F., Soni, A. (1977): Neutrino transport in pion-condensed neutron stars. Ap. J. **216**, 73.
Sawyer, R. F., Soni, A. (1979): Transport of neutrinos in hot neutron-star matter. Ap. J. **230**, 859.
Schalen, C. (1975): On the value of $R = A_V / E_{B-V}$. Astron. Astrophys. **42**, 251.
Scharleman, E. T. (1978): The fate of matter and angular momentum in disk accretion onto a magnetized neutron star. Ap. J. **219**, 617.
Schatzman, E. (1948): Les reactions thermonucléaires aux grandes densités, gaz dégénérés et non dégénérés (Thermonuclear reactions at large densities, degenerate and nondegenerate gases). J. Phys. Radium **9**, 46.
Schatzman, E. (1949): The heating of the solar corona and chromosphere. Ann. Astrophys. **12**, 203.
Schatzman, E. (1951): L'isotope ^{3}He das les étoiles. Application a la théorie des novae et des naines blanches (The helium three isotope in stars. Application of the theory of novae and of white dwarfs). Compt. Rend. **232**, 1740.
Schatzman, E.: White dwarfs. Amsterdam: North Holland 1958.
Scheffler, H., Elsässer, H.: Physics of the galaxy and interstellar matter. New York: Springer-Verlag, 1987.
Schein, M., Jesse, W. P., Wollan, E. O. (1941): The nature of the primary cosmic radiation and the origin of the mesotron. Phys. Rev. **59**, 615.
Scheuer, P. A. G. (1960): The absorption coefficient of a plasma at radio frequencies. M.N.R.A.S. **120**, 231.
Scheuer, P. A. G. (1965): A sensitive test for the presence of atomic hydrogen in intergalactic space. Nature **207**, 963.
Scheuer, P. A. G. (1968): Synchrotron radiation formulae. Ap. J. (Letters) **151**, L139.
Schild, R. E. (1977): Interstellar reddening law. Astron. J. **82**, 337.
Schlesinger, B. M. (1977):The hydrogen profile, previous mixing, and loops in the H-R diagram during core helium burning. Ap. J. **212**, 507.
Schlesinger, F. (1909): The algol-variable δ Libre. Pub. Allegheny Obs. **1**, 125.
Schmidt, G. D.: Magnetic fields in white dwarfs. In: White dwarfs, proceeding of IAU colloquium no. 114 (ed. G. Wegner). New York: Springer-Verlag 1988.
Schneider, S., Elmegreen, B. G. (1979): A catalog of dark globular filaments. Ap. J. Suppl. **41**, 87.
Schott, G. A.: Electromagnetic radiation and the mechanical reactions arising from it. Cambridge, England: Cambridge University Press 1912.
Schramm, D. N. (1973): Explosive r-process nucleosynthesis. Ap. J. **185**, 293.
Schramm, D. N. (1987): Neutrinos from Supernova 1987A. Comm. Nucl. Part. Phys. **17**, 239.
Schramm, D. N., Arnett, W. D. (1975): The weak interaction and gravitational collapse. Ap. J. **198**, 629.
Schramm, D. N., Tinsley, B. M. (1974): On the origin and evolution of s-process elements. Ap. J. **193**, 151.
Schreier, E., et al. (1972): Evidence for the binary nature of Centaurus X-3 from UHURU X-ray observations. Ap. J. (Letters) **172**, L79.
Schrödinger, E. (1925, 1926): Quantisierung als Eigenwertproblem (Quantisation as an eigenvalue problem). Ann. Physik **79**, 361, **80**, 437, **81**, 109.

Schuster, A. (1905): Radiation through a foggy atmosphere. Ap. J. **21**, 1 Reprod. in: Selected papers on the transfer of radiation (ed. D. H. Menzel). New York: Dover 1966.
Schwartz, D. A. (1970): The isotropy of the diffuse cosmic X-rays determined by OSO-III. Ap. J. **162**, 439.
Schwartz, M. (1960): Feasibility of using high-energy neutrinos to study the weak interactions. Phys. Rev. Lett. **4**, 306.
Schwartz, R. A., Stein. R. F. (1975): Waves in the solar atmosphere. IV. Magneto-gravity and acoustic-gravity modes. Ap. J. **200**, 499.
Schwarzschild, K. Von (1906): Über das Gleichgewicht der Sonnenatmosphäre (On the equilibrium of the solar atmosphere.) Nach. Ges. Gott. **195**, 41. Engl. trans. in: Selected papers on the transfer of radiation (ed. D. H. Menzel). New York: Dover 1966.
Schwarzschild, K. Von (1914): Über Diffusion und Absorption in der Sonnenatmosphäre (On scattering and absorption in the solar atmosphere). Sitz. Akad. Wiss. 1183, Engl. trans. in: Selected papers on the transfer of radiation (ed. D. H. Menzel). New York: Dover 1966.
Schwarzschild, K. Von (1916): Zur Quantenhypothese (On the quantum hypothesis). Sitz. Akad. Wiss. 548.
Schwarzschild, M. (1941): Overtone pulsations for the standard model. Ap. J. **94**, 245.
Schwarzschild, M. (1948): On noise arising from the solar granulation. Ap. J. **107**, 1.
Schwarzschild, M.: Structure and evolution of stars. Princeton, New Jersey: Princeton University Press 1958.
Schwarzschild, M., Härm, R. (1962): Red giants of population II. Ap. J. **136**, 158.
Schwarzschild, M., Härm, R. (1967): Hydrogen mixing by helium-shell flashes. Ap. J. **150**, 961.
Schwinger, J. (1949): On the classical radiation of accelerated electrons. Phys. Rev. **75**, 1912.
Scoville, N. Z., Solomon, P. M. (1974): Radiative transfer, excitation, and cooling of molecular emission lines (CO and CS). Ap. J. (Letters) **187**, L67.
Scoville, N. Z., Solomon, P. M. (1975): Molecular clouds in the galaxy. Ap. J. (Letters) **199**, L105.
Seagrave, J. D. (1952): Radiative capture of protons by C^{13}. Phys. Rev. **85**, 197.
Seares, F. H. (1913): The displacement-curve of the sun's general magnetic field. Ap. J. **38**, 99.
Sears, R. L., Brownlee, R. R.: Stellar evolution and age determinations. In: Stellar structure - stars and stellar systems VIII (ed. L. H. Aller and D. B. Mc Laughlin). Chicago, Ill.: Univ. of Chicago Press 1965.
Seaton, M. J. (1951): The chemical composition of the interstellar gas. M.N.R.A.S. **111**, 368.
Seaton, M. J. (1954): Electron temperatures and electron densities in planetary nebulae. M.N.R.A.S. **114**, 154.
Seaton, M. J. (1955): The kinetic temperature of the interstellar gas in regions of neutral hydrogen. Ann. Astrophys. **18**, 188.
Seaton, M. J. (1958): Thermal inelastic collision processes. Rev. Mod. Phys. **30**, 979.
Seaton, M. J. (1960): Planetary nebulae. Rep. Prog. Phys. **23**, 314.
Seaton, M. J.: The theory of excitation and ionization by electron impact. In: Atomic and molecular processes (ed. D. R. Bates). New York: Academic Press 1962.
Seaton, M. J. (1964): Recombination spectra. M.N.R.A.S. **127**, 177.
Seaton, M. J.: Atomic collision processes in gaseous nebulae. In: Advances in atomic and molecular physics, volume 4 (eds. D. R. Bates and I. Estermann). New York: Academic Press 1968.
Seaton, M. J.: Electron impact excitation of positive ions. In: Advances in atomic and molecular physics, volume 11 (eds. D. R. Bates and B. Benderson). New York: Academic Press 1975.

Sedov, L. I.: Similarity and dimensional methods in mechanics. New York: Academic Press 1959 and Course in continuum mechanics, vol. 1 to 4. Groningen: Wolter-Noordhoff 1971.

Seeger, P. A., Fowler, W. A., Clayton, D. D. (1965): Nucleosynthesis of heavy elements by neutron capture. Ap. J. Suppl. No. 9, **11**, 121.

Seeger, P. A., Schramm, D. N. (1970): R-process production ratios of chronologic importance. Ap. J. (Letters) **160**, L157.

Sejnowski, T. J., Hjellming, R. M. (1969): The general solution of the b_n problem for gaseous nebulae. Ap. J. **156**, 915.

Sekido, Y., Elliot, H. (eds.): Early history of cosmic ray studies. Dordrecht: D. Reidel 1985.

Sellwood, J. A. (1987): The art of N-body building. Ann. Rev. Astron. Astrophys. **25**, 151.

Service, A. T. (1977): Concise approximation formulae for the Lane-Emden functions. Ap. J. **211**, 908.

Shajn, G., Struve, O. (1929): On the rotation of the stars. M.N.R.A.S. **89**, 222. Reproduced in: Source book in astronomy and astrophysics 1900–1975 (eds. K. R. Lang and O. Gingerich). Cambridge, Mass.: Harvard University Press 1979.

Shakura, N. I., Sunyaev, R. A. (1973): Black holes in binary systems. Observational appearance. Astron. Astrophys. **24**, 337.

Shapiro, I. I. (1967): Theory of the radar determination of planetary rotations. Astron. J. **72**, 1309.

Shapiro, M. M., Silberberg, R. (1970): Heavy cosmic ray nuclei. Ann. Rev. Nucl. Sci. **20**, 323.

Shapiro, P. R., Field, G. B. (1976): Consequences of a new hot component of the interstellar medium. Ap. J. **205**, 762.

Shapiro, S. L., Teukolsky, S. A. (1976): On the maximum gravitational redshift of white dwarfs. Ap. J. **203**, 697.

Shapiro, S. L., Teukolsky, S. A.: Black holes, white dwarfs, and neutron stars. New York: John Wiley and Sons 1983.

Shapley, H. (1914): On the nature and cause of Cepheid variation. Ap. J. **40**, 448.

Shapley, H. (ed.): Source book in astronomy 1900-1950. Cambridge, Mass.: Harvard University Press 1960.

Shapley, H., Nicholson, S. B. (1919): On the spectral lines of a pulsating star. Comm. Nat. Acad. Sci. Mt. Wilson Obs. **2**, 65.

Share, G. H., Murphy, R. J. (1997): Intensity and directionality of flare-accelerated α-particles at the sun. Ap. J. **485**, 409.

Shaver, P. A. (1980): Accurate electron temperatures from radio recombination lines. Astron. Astrophys. **91**, 279.

Shaver, P. A., et al. (1983): The galactic abundance gradient. M.N.R.A.S. **204**, 53.

Shaw, P. B., Clayton, D. D., Michel, F. C. (1965): Photon-induced beta decay in stellar interiors. Phys. Rev. **140**, B 1433.

Sheeley, N. R. Jr., Wang, Y.-M., Nash, A. G. (1992): A new determination of the solar rotation rate. Ap. J. **401**, 378.

Shemi, A., Piran, T. (1990): The appearance of cosmic fireballs. Ap. J. (Letters) **365**, L55.

Shields, G. A., Oke, J. B., Sargent, W. L. W. (1972): The optical spectrum of the Seyfert galaxy 3C 120. Ap. J. **176**, 75.

Shimizu, T. (ed.): Yohkoh views the sun. The first five years. Tokyo: Institute of Space and Astronautical Science, National Astronomical Observatory, Yohkoh Group 1996.

Shipman, H. L. (1972): Masses and radii of white dwarfs. Ap. J. **177**, 723.

Shipman, H. L. (1977): Masses, radii, and model atmospheres for cool white-dwarf stars. Ap. J. **213**, 138.

Shipman, H. L. (1979): Masses and radii of white-dwarf stars. III. Results for 110 hydrogen-rich and 28 helium-rich stars. Ap. J. **228**, 240.

Shipman, H. L., Mehan, R. G. (1976): The unimportance of pressure shifts in the measurement of gravitational redshifts in white dwarfs. Ap. J. **209**, 205.

Shipman, H. L., Provencal, J. L., Høg, E., Thejll, P. (1997): The mass and radius of 40 Eridani B from Hipparcos: An accurate test of stellar interior theory. Ap. J. (Letters) **488**, L43.

Shklovskii, I. S. (1953): On the nature of the luminescence of the Crab nebula. Dokl. Akad. Nauk. S.S.S.R. **90**, 983. Eng. trans. in: Source book in astronomy and astrophysics 1900–1975 (eds. K. R. Lang and O. Gingerich). Cambridge, Mass: Harvard University Press 1979.

Shklovskii, I. S. (1953): On the origin of cosmic rays. Dokl. Akad. Nauk (SSSR) **91**, 475.

Shklovskii, I. S. (1953): The possibility of observing monochromatic radio emissions from interstellar molecules. Dokl. Akad. Nauk SSSR **92**, No. 1, 25.

Shklovskii, I. S. (1960): Secular variations in the flux and intensity of radio emission from discrete sources. Sov. Astron. **6**, 317.

Shklovskii, I. S. (1964): Physical conditions in the gaseous envelope of 3C 173. Astron. Zh. **41**, 408. Sov. Astron. **8**, 638 (1965).

Shklovskii, I. S.: Supernovae. New York: Wiley Interscience 1968.

Shortley, G. H. (1935): Line strengths in intermediate coupling. Phys. Rev. **47**, 295, 419.

Shortley, G. H. (1940): The computation of quadrupole and magnetic-dipole transition probabilities. Phys. Rev. **57**, 225.

Shu, F. H. (1977): Self-similar collapse of isothermal spheres and star formation. Ap. J. **214**, 488.

Shu, F. H.: The physics of astrophysics, volume 1, radiation. Mill Valley, California: University Science Books 1991.

Shu, F. H.: The physics of astrophysics, volume 2, gas dynamics. Mill Valley, California: University Science Books 1992.

Shu, F. H., Adams, F. C., Lizano, S. (1987): Star formation in molecular clouds: Observation and theory. Ann. Rev. Astron. Astrophys. **25**, 23.

Shu, F. H., et al. (1972): Galactic shocks in an interstellar medium with two stable phases. Ap. J. **173**, 557.

Shukla, P. G., Paul, J. (1976): Gamma-ray production by the inverse Compton process in interstellar space. Ap. J. **208**, 893.

Shull, J. M. (1977): Grain disruption in interstellar hydromagnetic shocks. Ap. J. **215**, 805.

Shull, J. M., Beckwith, S. (1982): Interstellar molecular hydrogen. Ann. Rev. Astron. Astrophys. **20**, 163.

Shull, J. M., Van Steenberg, M. (1982): The ionization equilibrium of astrophysically abundant elements. Ap. J. Supp. **48**, 95.

Siemans, P. J., Pandharipande, V. R. (1971): Neutron matter computations in Brueckner and variational theory. Nucl. Phys. **A173**, 561.

Silberberg, R., Tsao, C. H. (1973): Partial cross-sections in high energy nuclear reactions and astrophysical applications: I. Targets with $Z \leq 28$. Ap. J. Suppl. No. 220, **25**, 315.

Silberberg, R., Tsao, C. H., Letaw, J. R. (1985): Improved cross section calculations for astrophysical applications. Ap. J. Supp. **58**, 873.

Silk, J. (1970): Diffuse cosmic X and gamma radiation: The isotropic component. Space Sci. Rev. **11**, 671

Silk, J. (1973): Diffuse X and gamma radiation. Ann. Rev. Astron. Astrophys. **11**, 269.

Silk, J. (1975): Hydromagnetic waves and shock waves as an interstellar heat source. Ap. J. (Letters) **198**, L77.

Silk, J. (1977): On the fragmentation of cosmic gas clouds. II. Opacity limited star formation. Ap. J. **214**, 152.

Silk, J. (1977): On the fragmentation of cosmic gas clouds. III. The initial stellar mass function. Ap. J. **214**, 918.

Silk, J., Arons, J. (1975): On the nature of the globular cluster X-ray sources. Ap. J. **200**, L131 .

Simnett, G. H., Mc Donald, F. B. (1969): Observations of cosmic-ray electrons between 2.7 and 21.5 MeV. Ap. J. **157**, 1435.

Simon, A. (1955): Diffusion of like particles across a magnetic field. Phys. Rev. **100**, 1557.

Simon, M. (1969): Asymptotic form for synchrotron spectra below Razin cutoff. Ap. J. **156**, 341.

Simpson, J. A. (1983): Elemental and isotopic composition of the galactic cosmic rays. Ann. Rev. Nucl. Part. Sci. **33**, 323.

Simpson, J. A., et al. (1995): Cosmic ray and solar particle investigations over the south polar regions of the sun. *Science* **268**, 1019.

Simpson, J. P., Rubin, R. H., Erickson, E. F., Haas, M. R. (1986): The ionization structure of the Orion nebula: Infrared line observations and models. Ap. J. **311**, 895.

Singh, M., Chaturvedi, J. P. (1987): One hundred and fifty-three diatomic molecules, molecular ions, and radicals of astrophysical interest. Astrophys. Space Sci. **135**, 1.

Sion, E. M., Acierno, M. J., Tomczyk, S. (1979): Hydrogen shell flashes in massive accreting white dwarfs. Ap. J. **230**, 832.

Sitenko, A. G., Stepanov, K. N. (1957): On the oscillations of an electron plasma in a magnetic field. Sov. Phys. J.E.T.P. **4**, 512.

Skobeltzyn, D. (1929): Über eine Neue Art sehr Schneller β-Strahlen. Zeits. f. Phys. **54**, 686.

Slettebak, A. (1949): On the axial rotation of the brighter O and B stars. Ap. J. **110**, 498.

Slettebak, A. (1954): The spectra and rotational velocities of the bright stars of Draper types B8-A2. Ap. J. **119**, 146.

Slettebak, A. (1955): The spectra and rotational velocities of the bright stars of Draper types A3-G0. Ap. J. **121**, 653.

Slettebak, A., ed.: Proceedings I.A.U. colloquium on stellar rotation. New York: Gordon and Breach 1970.

Slettebak, A., Howard, R. F. (1955): Axial rotation in the brighter stars of Draper types B2-B5. Ap. J. **121**, 102.

Slipher, V. M. (1912): On the spectrum of the nebula in the Pleiades. Lowell Obs. Bull. No. **55**, 26.

Slipher, V. M. (1916): On the spectrum of the nebula about Rho Ophiuchi. Lowell Obs. Bull. No. **75**, 155.

Slipher, V. M. (1918): The spectrum of NGC 7023. P.A.S.P. **30**, 63.

Slipher, V. M. (1919): On the spectra of the Orion nebulosities. P.A.S.P. **31**, 212.

Slish, V. I. (1963): Angular size of radio stars. Nature **199**, 682.

Smerd, S. F., Westfold, K. C. (1949): The characteristics of radio-frequency radiation in an ionized gas, with applications to the transfer of radiation in the solar atmosphere. Phil. Mag. **40**, 831.

Smerd, S. F., Wild, J. P., Sheridan, K. V. (1962): On the relative position and origin of harmonics in the spectra of solar radio bursts of spectral types II and III. Austr. J. Phys. **15**, 180.

Smith, A. M., Stecher, T. P. (1971): Carbon monoxide in the interstellar spectrum of Zeta Ophiuchi. Ap. J. (Letters) **164**, L43.

Smith, B. A., Terrile, R. J. (1984): A circumstellar disk around Beta Pictoris. Science **226**, 1421.

Smith, D., Adams, N. G. (1978): Molecular synthesis in interstellar clouds: Radiative association reactions of CH_3^+ ions. Ap. J. (Letters) **220**, L87.

Smith, D., Adams, N. G. (1984): Dissociative recombination coefficients for H_3^+, HCO^+, N_2H^+, and CH_5^+ at low temperature: Interstellar implications. Ap. J. (Letters) **284**, L13.

Smith, E. J., et al. (1995): Ulysses observations of Alfvén waves in the southern and northern solar hemispheres. Geophysical Res. Lett. **22**, 3381.

Smith, H. E., Spinrad, H., Smith, E. O. (1976): The revised 3C catalogue of radio sources: A review of optical identifications and spectroscopy. P.A.S.P. **88**, 621.

Smith, W. H., Liszt, H. S., Lutz, B. L. (1973): A reevaluation of the diatomic processes leading to CH and CH^+ formation in the interstellar medium. Ap. J. **183**, 69.

Smoot, G. F., Buffington, A., Orth, C. D. (1975): Search for cosmic-ray antimatter. Phys. Rev. Lett. **35**, 258.

Snell, W.: 1621, unpublished.

Snodgrass, H. B. (1983): Magnetic rotation of the solar photosphere. Ap. J. **270**, 288.

Snyder, L. E., et al. (1969): Microwave detection of interstellar formaldehyde. Phys. Rev. Letters **22**, 679.

Snyder, L. T., in: M.T.P. International review of science, sec. 1,3, ch. 6, 1973.

Sofue, Y., Fujimoto, M., Wielebinski, R. (1986): Global structure of magnetic fields in spiral galaxies. Ann. Rev. Astron. Astrophys. **24**, 459.

Sokolsky, P.: Introduction to ultrahigh energy cosmic ray physics. New York: Addison Wesley 1988.

Solomon, P., Klemperer, W. (1972): The formation of diatomic molecules in interstellar clouds. Ap. J. **178**, 389.

Sommerfeld, A. (1916): Zur Quantentheorie der Spektrallinien (On the quantum theory of spectral lines). Ann. Phys. **50**, 385, **51**, 1.

Sommerfeld, A. (1931): Über die Beugung und Bremsung der Elektronen (On the deflection and deceleration of electrons). Ann. Phys. **11**, 257.

Sommerfeld, A.: Thermodynamics and statistical mechanics. New York: Academic Press 1964.

Southworth, G. C. (1944): Microwave radiation from the sun. J. Franklin Inst. **239**, 285.

Spiegel, E. A.: The gas dynamics of accretion. In: Interstellar gasdynamics (ed. H. J. Habing). Dordrecht, Holland: D. Reidel 1970.

Spiegel, E. A. (1971): Convection in stars: Part. I. Basic Boussinesq convection. Ann. Rev. Astron. Astrophys. **9**, 223.

Spinka, H., Winkler, H. (1972): Experimental investigation of the $^{16}O + {}^{16}O$ total reaction cross section at astrophysical energies. Ap. J. **174**, 455.

Spite, M.: Trends of element abundances in the stars of our Galaxy. In: The Stellar Populations of Galaxies. Proc. IAU Symposium No. 149 (ed. B. Barbuy and A. Renzini). Boston: Kluwer Academic 1992.

Spitzer, L. (1940): The stability of isolated clusters. M.N.R.A.S. **100**, 396.

Spitzer, L. (1940): Impact broadening of spectral lines. Phys. Rev. **58**, 348.

Spitzer, L. (1948): The temperature of interstellar matter I. Ap. J. **107**, 6.

Spitzer, L. (1949): The temperature of interstellar matter II. Ap. J. **109**, 337.

Spitzer, L. (1976): Hydrogen molecules in space. Q.J.R.A.S. **17**, 97.

Spitzer, L.: Physics of fully ionized gases. New York: Wiley 1962.

Spitzer, L.: Physical processes in the interstellar medium. New York: John Wiley 1978.

Spitzer, L.: Dynamical evolution of globular clusters. Princeton: Princeton University Press 1987.

Spitzer, L. (1990): Theories of the hot interstellar gas. Ann. Rev. Astron. Astrophys. **28**, 71.

Spitzer, L., Drake, J. F., Jenkins, E. B., Morton, D. C., Rogerson, J. B., York, D. G. (1973): Spectrophotometric results from the Copernicus satellite IV molecular hydrogen in interstellar space. Ap. J. (Letters) **181**, L116.

Spitzer, L., Greenstein, J. L. (1951): Continuous emission from planetary nebulae. Ap. J. **114**, 407.

Spitzer, L., Härm, R. (1953): Transport phenomena in a completely ionized gas. Phys. Rev. **89**, 977.

Spitzer, L., Härm, R. (1958): Evaporation of stars from isolated clusters. Ap. J. **127**, 544.

Spitzer, L., Hart, M. H. (1971): Random gravitational encounters and the evolution of spherical systems I. Method. Ap. J. **164**, 399.

Spitzer, L., Savedoff, M. P. (1950): The temperature of interstellar matter III. Ap. J. **111**, 593. Reproduced in: A source book in astronomy and astrophysics 1900–1975 (eds. K. R. Lang and O. Gingerich). Cambridge, Mass.: Harvard University Press 1979.

Spitzer, L., Scott, E. H. (1969): Heating of H I regions by energetic particles: II. Interaction between secondaries and thermal electrons. Ap. J. **158**, 161.

Spitzer, L., Shull, J. M. (1975): Random gravitational encounters and the evolution of spherical systems. VI. Plummer's model. Ap. J. **200**, 339.
Spitzer, L., Shull, J. M. (1975): Random gravitational encounters and the evolution of spherical systems. VII. Systems with several mass groups. Ap. J. **201**, 773.
Spitzer, L., Tukey, J. W. (1951): A theory of interstellar polarization. Ap. J. **114**, 187.
Spitzer, L., Zweibel, E. G. (1974): On the theory of H_2 rotational excitation. Ap. J. (Letters) **191**, L127.
Spruch, L. (1991): Pedagogic notes on Thomas-Fermi theory (and on some improvements): atoms, stars, and the stability of bulk matter. Rev. Mod. Phys. **63**, 151.
Stark, J. Von (1913): Beobachtungen über den Effekt des elektrischen Feldes auf Spektrallinien (Observations of the effect of the electric field on spectral lines). Sitz. Akad. Wiss. **40**, 932.
Starrfield, S., Truran, J. W., Sparks, W. M. (1975): Novae, supernovae, and neutron sources. Ap. J. (Letters) **198**, L113.
Stauffer, J. R. (1987): Dynamical mass determinations for white dwarf components of HZ 9 and Case 1. Astron. J. **94**, 996.
Stebbins, J., Huffer, C. M., Whitford, A. E. (1940): The colors of 1332 B stars. Ap. J. **91**, 20.
Stecher, P., Williams, D. A. (1968): Interstellar molecule formation. Ap. J. **146**, 88.
Stefan, A. J. (1879): Beziehung zwischen Wärmestrahlung und Temperatur (Relation between thermal radiation and temperature). Wien. Ber. **79**, 397.
Stein, R. F. (1968): Waves in the solar atmosphere: I. The acoustic energy flux. Ap. J. **154**, 297.
Stein, R. F., Schwartz, R. A. (1972): Waves in the solar atmosphere: II. Large amplitude acoustic pulse propagation. Ap. J. **177**, 807.
Stein, W. A., Soifer, B. T. (1983): Dust in galaxies. Ann. Rev. Astron. Astrophys. **21**, 177.
Stenflo, J. O. (ed.): Solar photosphere: Structure, convection and magnetic fields. Dordrecht: Kluwer 1990.
Stern, D. P., Ness, N. F. (1982): Planetary magnetospheres. Ann. Rev. Astron. Astrophys. **20**, 139.
Sterne, T. E. (1933): The equilibrium theory of the abundance of the elements: A statistical investigation of assemblies in equilibrium in which transmutations occur. M.N.R.A.S. **93**, 736.
Sternheimer, R. M. (1956): Density effect for the ionization loss in various materials. Phys. Rev. **103**, 511.
Stevenson, D. J., Salpeter, E. E. (1977): The phase diagram and transport properties for hydrogen-helium fluid planets. Ap. J. Suppl. **35**, 221.
Stevenson, D. J., Salpeter, E. E. (1977): The dynamics and helium distribution in hydrogen-helium fluid planets. Ap. J. Suppl. **35**, 239.
Stibbs, D. W. N. (1950): A study of the spectrum and magnetic variable star HD 125248. M.N.R.A.S. **110**, 395.
Stief, L. J.: Photochemistry of interstellar molecules. In: Molecules in the galactic environment (ed. M. A. Gordon, L. E. Snyder). New York: Wiley 1973.
Stief, L. J., et al. (1972): Photochemistry and lifetimes of interstellar molecules. Ap. J. **171**, 21.
Stix, T. H.: The theory of plasma waves. New York: McGraw-Hill 1962.
Stobbe, M. Von (1930): Zur Quantenmechanik photoelektrischer Prozesse (On the quantum mechanics of photoelectric processes). Ann. Phys. **7**, 661.
Stokes, G. G. (1845): On the theories of the internal friction of fluids in motions, and of the equilibrium and motion of elastic solids. Trans. Camb. Phil. Soc. **8**, 287.
Stokes, G. G. (1852): On the composition and resolution of streams of polarized light from different sources. Trans. Camb. Phil. Soc. **9**, 399.
Stoner, E. C. (1930): The equilibrium of dense stars. Phil. Mag. **9**, 944.
Stothers, R. (1974): Influence of rotation on the maximum mass of pulsationally stable stars. Ap. J. **192**, 145.

Street, J. C., Stevenson, E. C. (1937): New evidence for the existence of a particle of mass intermediate between the proton and electron. Phys. Rev. **52**, 1003.

Striganov, A. R., Sventitskii, N. S.: Tables of spectral lines of neutral and ionized atoms. New York: IFI/Plenum 1968.

Strömgren, B. (1932): The opacity of stellar matter and the hydrogen content of the stars. Zeits. f. Ap. **4**, 118.

Strömgren, B. (1933): On the interpretation of the Hertzsprung-Russell diagram. Zeits. f. Ap. **7**, 222.

Strömgren, B. (1939): The physical state of interstellar hydrogen. Ap. J. **89**, 526. Reproduced in: Source book in astronomy and astrophysics 1900–1975 (eds. K. R. Lang and O. Gingerich). Cambridge, Mass.: Harvard University Press 1979.

Strong, K. T. (1991): Observations from the Solar Maximum Mission. Phil. Trans. Roy. Soc. (London) **A336**, 327.

Struve, O. (1929): The Stark effect in stellar spectra. Ap. J. **69**, 173.

Struve, O. (1930): On the axial rotation of stars. Ap. J. **72**, 1.

Struve, O., Elvey, C. T. (1934): The intensities of stellar absorption lines. Ap. J. **79**, 409.

Sturrock, P. A. (1958): Kinematics of growing waves. Phys. Rev. **112**, 1488.

Sturrock, P. A. (1966): Model of the high-energy phase of solar flares. Nature **211**, 695.

Sturrock, P. A.: A model for solar flares. In: Structure and development of solar active regions (ed. K. O. Kiepenheuer). Norwell, Mass.: D. Reidel 1968.

Sturrock, P. A.: Plasma physics: An introduction to the theory of astrophysical, geophysical and laboratory plasmas. New York: Cambridge University Press 1994.

Suess, H. E., Urey, H. C. (1956): Abundances of the elements. Rev. Mod. Phys. **28**, 53.

Surmelian, G. L., O'Connell, R. F. (1974): Energy spectrum of hydrogen-like atoms in a strong magnetic field. Ap. J. **190**, 741.

Surmelian, G. L., O'Cconnell, R. F. (1974): Quadratic Zeeman Effect in the hydrogen Blamer lines from a strong magnetic field. Ap. J. **193**, 705

Svestka, Z.: Solar Flares. Norwell, Mass.: Kluwer 1976.

Sweet, P. A.: The neutral point theory of solar flares. In: Proc. I.A.U. Symp. on electromagnetic phenomenon in cosmical physics (ed. B. Lehnert). Cambridge, England: Cambridge Univ. Press 1958.

Sweet, P. A. (1969): Mechanisms of solar flares. Ann. Rev. Astron. Astrophys. **7**, 149.

Sweigart, A. V. (1997): Effects of helium mixing on the evolution of globular cluster stars. Ap. J. (Letters) **474**, L23.

Swings, P. (1942): Considerations regarding cometary and interstellar molecules. Ap. J., **95**, 270.

Swings, P. (1948): Le spectre de la comete d'Encke, 1947. Annales d'Astrophysique **11**, 124.

Swings, P., Rosenfeld, L. (1937): Considerations regarding interstellar molecules. Ap. J. **86**, 483.

Syrovatskii, S. I. (1981): Pinch sheets and reconnection in astrophysics. Ann. Rev. Astron. Astrophys. **19**, 163.

Taam, R. E., Faulkner, J. (1975): Ultrashort-period binaries. III. The accretion of hydrogen-rich matter onto a white dwarf of one solar mass. Ap. J. **198**, 435.

Takakura, T. (1960): Synchrotron radiation from intermediate energy electrons and solar radio outbursts at microwave frequencies. Publ. Astron. Soc. Japan **12**, 325, 352.

Takakura, T. (1966): Implications of solar radio bursts for the study of the solar corona. Space Sci. Rev. **5**, 80.

Takakura, T. (1967): Theory of solar bursts. Solar Phys. **1**, 304.

Takakura, T. (1969): Interpretation of time characteristics of solar X-ray bursts referring to associated microwave bursts. Solar Phys. **6**, 133.

Takakura, T, et al. (1993): Time variation of the hard X-ray image during the early phase of solar impulsive bursts. Pub. Astron. Soc. Japan **45**, 737.

Takakura, T., Kai, K. (1966): Energy distribution of electrons producing microwave impulsive bursts and X-ray bursts from the Sun. Publ. Astron. Soc. Japan **18**, 57.

Takeda, M., et al. (1998): Extension of the cosmic-ray energy spectrum beyond the predicted Greisen-Zatasepin-Kuz'min cutoff. Phys. Rev. Lett. **81**, 1163.

Tanaka, K. (1987): Impact of X-ray observations from the Hinotori satellite on solar flares. Pub. Astron. Soc. Japan **39**, 1.

Tananbaum, H., Gursky, H., Kellogg, E. M., Levinson, R., Schreier, E., Giacconi, R. (1972): Discovery of a periodic pulsating binary X-ray source from UHURU. Ap. J. (Letters) **174**, L143.

Tandberg-Hanssen, E., Emslie, A. G.: The physics of solar flares. New York: Cambridge University Press 1988.

Tanimori, T., et al. (1998): Discovery of TeV gamma rays from SN 1006: Further evidence for the supernova remnant origin of cosmic rays., Ap. J. (Letters) **497**, L25.

Tarlé, G., Swordy, S. P. (1998): Cosmic antimatter. Scientific American **278**, 36 – April.

Tarter, C. B. (1969): Ultraviolet and forbidden-line intensities. Ap. J. Suppl. No. 154, **18**, 1.

Tassoul, J. L. (1968): Adiabatic pulsations and convective instability of gaseous masses III. M.N.R.A.S. **138**, 123.

Tassoul, J. L.: Theory of rotating stars. Princeton, New Jersey: Princeton University Press 1978.

Tassoul, J. L., Ostriker, J. P. (1968): On the oscillations and stability of rotating stellar models. I. Mathematical techniques. Ap. J. **154**, 613.

Tassoul, M., Tassoul, J. L. (1967): Adiabatic pulsations and convective instability of gaseous masses I, II. Ap. J. **150**, 213, 1031.

Tatarski, V. I.: Effects of the turbulent atmosphere on wave propagation. Washington, DC: Nat. Sci. Foundation 1971.

Tayler, R. J. (1957): Hydrodynamic instabilities of an ideally conducting fluid. Proc. Phys. Soc. London **B70**, 31.

Tayler, R. J. (1966): The origin of the elements. Prog. Theor. Phys. **29**, 490.

Taylor, B. N., Parker, W. H., Langenberg, D. N. (1969): Determination of e/h, using macroscopic quantum phase coherence in superconductors: Implications for quantum electrodynamics and the fundamental physical constants. Rev. Mod. Phys. **41**, 375.

Taylor, G. (1950): The formation of a blast wave by a very intense explosion II. The atomic explosion of 1945. Proc. Roy. Soc. (London) **A201**, 175.

Taylor, G. I. (1921): Experiments with rotating fluids. Proc. Roy. Soc. London **A100**, 114.

Taylor, G. I. (1935): Statistical theory of turbulence. Proc. Roy. Soc. London **A151**, 421.

Taylor, G. I. (1938): The spectrum of turbulence. Proc. Roy. Soc. London **A164**, 1476.

Taylor, G. I. (1950): The instability of liquid surfaces when accelerated in a direction perpendicular to their planes I. Proc. Roy. Soc. London **A201**, 192

Taylor, J. H., Stinebring, D. R. (1986): Recent progress in the understanding of pulsars. Ann. Rev. Astron. Astrophys. **24**, 285.

Tenorio-Tagle, G., Bodenheimer, P. (1988): Large-scale expanding superstructures in galaxies. Ann. Rev. Astron. Astrophys. **26**, 145.

Terzian, Y.: Observations of planetary nebulae at radio wavelengths. In: Planetary nebulae (ed. D. E. Osterbrock, C. R. O'dell), I.A.U. Symp. No. 34. Dordrecht, Holland: D. Reidel. Berlin-Heidelberg-New York: Springer 1968.

Terzian, Y., Parrish, A. (1970): Observations of the Orion nebula at low radio frequencies. Astrophys. Lett. **5**, 261.

Tetrode, H. Von (1912): Die chemische Konstante der Gase und das elementare Wirkungsquantum (Chemical gas constants and the elementary action quantum). Ann. Phys. **38**, 434.

Thirring, W.: A course in mathematical physics 4. Quantum mechanics of large systems. New York: Springer-Verlag 1983.

Thomas, D., Schramm, D. N., Olive, K. A., Fields, B. D. (1993): Primordial nucleosynthesis and the abundances of beryllium and boron. Ap. J. **406**, 569.

Thomas, L. H. (1926): The motion of the spinning electron. Nature **117**, 514.

Thomas, L. H. (1927): The calculation of atomic fields. Proc. Camb. Phil. Soc. **23**, 542.

Thomas, L. H. (1927): The kinematics of an electron with an axis. Phil. Mag. **1**, 1.

Thomas, W. (1925): Über die Zahl der Dispersionselektronen die einem stationären Zustande zugeordnet sind (On the number of scattering electrons which are associated with a stationary state). Naturwiss **13**, 627.

Thompson, M. J. (1990): A new inversion of solar rotational splitting data. Solar Phys. **125**, 1.

Thompson, M. J., et al. (1996): Differential rotation and dynamics of the solar interior. Science **272**, 1300.

Thompson, W. B. (1951): Thermal convection in a magnetic field. Phil. Mag. **42**, 1417.

Thomson, J. J. (1897): Conductivity of a gas through which cathode rays are passing. Phil. Mag. **44**, 298.

Thomson, J. J.: Conduction of electricity through gases. Cambridge, England: Cambridge University Press 1903. Republ. New York: Dover 1969.

Thorne, K. S. (1977): The relativistic equations of stellar structure and evolution. Ap. J. **212**, 825.

Thorne, K. S., Zytkow, A. N. (1977): Stars with degenerate neutron cores. I. Structure of equilibrium models. Ap. J. **212**, 832.

Thorsett, S. E., Arzoumanian, Z., Mc Kinnon, M. M., Taylor, J. H. (1993): The masses of two binary neutron star systems. Ap. J. (Letters) **405**, L29.

Tidman, D. A. (1958): Structure of a shock wave in fully ionized hydrogen. Phys. Rev. **111**, 1439.

Tidman, D. A., Krall, N. A.: Shock waves in collisionless plasmas. New York: John Wiley and Sons 1971.

Timmes, F. X., Woosley, S. E., Weaver, T. A. (1996): The neutron star and black hole initial mass function. Ap. J. **457**, 834.

Tiomno, J., Wheeler, J. A. (1949): Charge-exchange reaction of the μ-meson with the nucleus. Rev. Mod. Phys. **21**, 153.

Todhunter, I.: A history of the mathematical theories of attraction and the figure of the earth. New York: Dover 1962.

Toevs, J. W., et al. (1971): Stellar rates for the $^{28}Si(\alpha,\gamma)^{32}S$ and $^{16}O(\alpha,\gamma)^{20}Ne$ reactions. Ap. J. **169**, 421.

Tolman, R. C. (1922): Thermodynamic treatment of the possible formation of helium from hydrogen. J. Am. Chem. Soc. **44**, 1902.

Tomczyk, S., Schou, J., Thompson, M. J. (1995): Measurement of the rotation rate in the deep solar interior. Ap. J. (Letters) **448**, L57.

Tomisaka, K., Ikeuchi, S., Nakamura, T. (1988): The equilibria and evolutions of magnetized, rotating, isothermal clouds . I. Basic equations and numerical methods. Ap. J. **326**, 208.

Tonks, L., Langmuir, I. (1929): Oscillations in ionized gases. Phys. Rev. **33**, 195.

Totsuka, Y. (1991): Recent results on solar neutrinos from Kamiokande. Nucl. Phys. **B19**, 69.

Townely, R. (1662): cf. I. B. Cohen. Newton, Hooke, and Boyle's law. Nature **204**, 618 (1964).

Townes, C. H.: Microwave and radiofrequency resonance lines of interest to radio astronomy. In: I.A.U. Symp. No. 4 (ed H. C. Van De Hulst). Cambridge, England: Cambridge University Press 1957.

Townes, C. H. (1977): Interstellar molecules. Observatory **97**, 52.

Townes, C. H., Schawlow, A. L.: Microwave spectroscopy. New York: McGraw-Hill 1955.

Tremaine, S.: The centers of elliptical galaxies. In: Unsolved problems in astrophysics (eds. J. N. Bahcall and J. P. Ostriker). Princeton, New Jersey: Princeton University Press 1997.

Trimble, V. (1975): The origin and abundances of the chemical elements. Rev. Mod. Phys. **47**, 877.

Trimble, V. (1982): Supernovae. Part I: the events. Rev. Mod. Phys. **54**, 1183.

Trimble, V. (1983): Supernovae. Part II: the aftermath. Rev. Mod. Phys. **55**, 511.

Trimble, V. (1988): 1987A - the greatest supernova since Kepler. Rev. Mod. Phys. **60**, 859.
Trimble, V. (1991): The origin and abundances of the chemical elements revisited. Astron. Astrophys. Rev. **3**, 1.
Trimble, V. L., Greenstein, J. L. (1972): The Einstein redshift in white dwarfs III. Ap. J. **177**, 441.
Trimble, V. L., Thorne, K. S. (1969): Spectroscopic binaries and collapsed stars. Ap. J. **156**, 1013.
Troitskii, V. S. (1970): On the possibility of determining the nature of the surface material of Mars from its radio emission. Radio Science **5** (2), 481.
Troland, T. H., Heiles, C. (1977): The Zeeman effect in radio frequency recombination lines. Ap. J. **214**, 703.
Trottet, G., Pick, M.(eds.) (1987): Particle acceleration and trapping in solar flares. Solar Physics **111**, No. 1. Boston: Reidel, 1987.
Trubnikov, B. A. (1958): Plasma radiation in a magnetic field. Sov. Phys. "Doklady" **3**, 136.
Trubnikov, B. A. (1965): Particle interactions in a fully ionized plasma. Rev. of Plasma Phys. **1**, 105.
Trumpler, R. J. (1930): Absorption of light in the galactic system. P.A.S.P. **42**, 214.
Trumpler, R. J. (1930): Preliminary results on the distances, dimensions, and space distribution of open star clusters. Lick Obs. Bull. **14**, no. 420, 154. Reproduced in: A source book in astronomy and astrophysics 1900–1975 (eds. K. R. Lang and O. Gingerich). Cambridge, Mass.: Harvard University Press 1979.
Trumpler, R. J., Weaver, H. F.: Statistical astronomy. Berkeley, Calif.: University of Calif. Press 1953.
Truran, J. W. (1968): The influence of a variable initial composition on stellar silicon burning. Astrophys. and Space Sci. **2**, 384.
Truran, J. W. (1972): Charged particle thermonuclear reactions in nucleosynthesis. Astrophys. and Space Sci. **18**, 306.
Truran, J. W., Arnett, W. D. (1970): Nucleosynthesis in explosive oxygen burning. Ap. J. **160**, 181.
Truran, J. W., Cameron, A. G. W. (1971): Evolutionary models of nucleosynthesis in the Galaxy. Astrophys. and Space Sci. **14**, 179.
Truran, J. W., Cameron, A. G. W. (1972): The p process in explosive nucleosynthesis. Ap. J. **171**, 89.
Truran, J. W., Cameron, A. G. W., Gilbert, A. (1966): The approach to nuclear statistical equilibrium. Can. J. Phys. **44**, 576.
Truran, J. W., Cameron, A. G. W., Hilf, E. (1970): In: Proc. inst. conf. prop. nucl far from regular betastability Geneva report CERN 70-30.
Truran, J. W., et al. (1968): Rapid neutron capture in supernova explosions. Astrophys. and Space Sci. **1**, 129.
Tsai, Y-S. (1974): Pair production and bremsstrahlung of charged leptons. Rev. Mod. Phys. **46**, 815.
Tseng, H. K., Pratt, R. H. (1973): Polarization correlations in atomic-field bremsstrahlung. Phys. Rev. **A7**, 1502.
Tsuda, H., Tsuji, H. (1963): Synthesis of Fe-group elements by the rapid nuclear process. Prog. Theor. Phys. **30**, 34.
Tsuji, H. (1963): Synthesis of 4 N and their neighboring nuclei by the rapid nuclear process. Prog. Theor. Phys. **29**, 699.
Tsuji, T. (1986): Molecules in stars. Ann. Rev. Astron. Astrophys. **24**, 89.
Tsuneta, S. (1995): Particle acceleration and magnetic reconnection in solar flares. Pub. Astron. Soc. Japan **47**, 691.
Tsuneta, S. (1996): Structure and dynamics of magnetic reconnection in a solar flare. Ap. J. **456**, 840.
Tsuneta, S., et al. (1992): Observation of a solar flare at the limb with the Yohkoh soft X-ray telescope. Pub. Astron. Soc. Japan **44**, L63.

Tsuneta, S., et al. (1992): Global restructuring of the coronal magnetic fields observed with the Yohkoh soft X-ray telescope. Pub. Astron. Soc. Japan **44**, L211.

Tsuruta, S., Cameron, A. G. W. (1965): Composition of matter in nuclear statistical equilibrium at high densities. Can. J. Phys. **43**, 2056.

Tsuruta, S., Cameron, A. G. W. (1966): Some effects of nuclear forces on neutron-star models. Can. J. Phys. **44**, 1895.

Tsuruta, S., Cameron, A. G. W. (1970): URCA shells in dense stellar interiors. Astrophys. and Space Sci. **7**, 374.

Tsurutani, B. T., et al. (1994): The relationship between interplanetary discontinuities and Alfvén waves: Ulysses observations. Geophys. Res. Lett. **21**, No. 21, 2267.

Tsytovich, V. N. (1951): The problem of radiation by fast electrons in a magnetic field in the presence of a medium Vestn. Mosk. Univ. **11**, 27.

Tsytovich, V. N. (1964): Acceleration by radiation and the generation of fast particles under cosmic conditions. Sov. Astron. A. J. **7**, 471. Reprod. in: Selected papers on cosmic ray origin theories (ed. S. Rosen). New York: Dover 1969.

Tsytovich, V. N. (1966): Statistical acceleration of particles in a turbulent pasma. Usp. Fiz. Nauk. **89**, 89.

Tu, C.-Y., Marsch, E. (1995): Mhd structures, waves and turbulence in the solar wind: Observations and theories. Space. Sci. Rev. **73**, 1.

Tuchman, Y., Sack, N., Barkat, Z. (1978): Mass loss from dynamically unstable stellar envelopes. Ap. J. **219**, 183.

Tucker, W. H. (1973): Heating of solar active regions by magnetic energy dissipation : The steady-state case. Ap. J. **186**, 285.

Tully, R. B.: Nearby galaxies catalog. New York: Cambridge University Press 1980.

Turner, B. E. (1970): Fifty new OH sources associated with H II regions. Astrophys. Lett. **6**, 99.

Turner, B. E.: Molecules as probes of the interstellar medium and of star formation. In: Galactic and extragalactic radio astronomy, second edition (eds. G. L. Verschuur and K. I. Kellermann). New York: Springer-Verlag 1988.

Turner, B. E., Ziurys, L. M.: Interstellar molecules and astrochemistry. In: Galactic and extragalactic radio astronomy, second edition (eds. G. L. Verschuur and K. I. Kellermann). New York: Springer-Verlag 1988.

Turner. J., Kirby-Docken, K., Dalgarno, A. (1977): The quadrupole vibration-rotation transition probabilities of molecular hydrogen. Ap. J. Suppl. **35**, 281.

Twiss, R. Q. (1954): On the nature of discrete radio sources. Phil. Mag. **45**, 249.

Überall, H. (1956): High-energy interference effect of bremsstrahlung and pair production in crystals. Phys. Rev. **103**, 1055.

Überall, H. (1957): Polarization of bremsstrahlung from monocrystalline targets. Phys. Rev. **107**, 223.

Uhlenbeck, G. E., Goudsmit, S. (1925): Zuschriften und vorläufige Mitteilungen (Replacement of the hypothesis of nonmechanical constraint by a requirement referring to the inner properties of each individual electron). Naturwiss. **13**, 953.

Uhlenbeck, G. E., Goudsmit, S. (1926): Spinning electrons and the structure of spectra. Nature **117**, 264.

Ulmschneider, P., Priest, E. R., Rosner, R. (eds.): Mechanisms of chromospheric and coronal heating. New York: Springer-Verlag 1991.

Ulrich, M.-H., Maraschi, L., Urry, C. M. (1997): Variability of active galactic nuclei. Ann. Rev. Astron. Astrophys. **35**, 445.

Ulrich, R. K. (1970): The five-minute oscillations on the solar surface. Ap. J. **162**, 993.

Ulrich, R. K. (1976): A nonlocal mixing-length theory of convection for use in numerical calculations. Ap. J. **207**, 564.

Ulrich, R. K., Burger, H. L. (1976): The accreting component of mass-exchange binaries. Ap. J. **206**, 509.

Underhill, A. B., Waddell, J. H.: Stark broadening functions for the hydrogen lines. Nat. Bur. Stands. (Wash.), Circ. 603, 1959.
Ungerechts, H., Thaddeus, P. (1987): A CO survey of the dark nebulae in Perseus, Taurus and Auriga. Ap. J. Supp. **63**, 645.
Uno, W., et al.: Nonradial oscillations of stars. Tokyo: University of Tokyo Press 1989.
Unsöld, A. (1928): Über die Struktur der Fraunhoferschen Linien und die quantitative Spektralanalyse der Sonnenatmosphäre. Zeits. f. Phys. **46**, 765.
Unsöld, A. (1969): Stellar abundances and the origin of the elements. Science **163**, 1015.
Unsold, A., Baschek, B.: The new cosmos. Fourth Edition. New York: Springer-Verlag, 1991.
Urey, H. C., Bradley, C. A. (1931): On the relative abundances of the isotopes. Phys. Rev. **38**, 718.
Usov, V. V. (1984): Generation of γ-bursts by old magnetic neutron stars. Astrophys. Space Sci. **107**, 191.
Vallée, J. P. (1997): Observations of the magnetic fields inside and outside the milky way, starting with globules ($\sim$1 parsec), filaments, clouds, superbubbles, spiral arms, galaxies, superculusters, and ending with the cosmological universe's background surface (at $\sim$8 Teraparsecs). Fund. Cosmic Phys. **19**, 1.
Vallé, J. P. (1998): Observations of the magnetic fields inside and outside the solar system: from meteorites ($\sim$10 attoparsecs), asteroids, planets, stars, pulsars, masers, to protostellar cloudlets ($<$1 parsec). Fund. Cosmic Phys. **19**, 319.
Van De Hulst, H. C. (1945): Radio waves in space: Origin of radiowaves. Ned. tijds. natuur. **11**, 210. English translation in: A source book in astronomy and astrophysics 1900–1975 (eds. K. R. Lang and O. Gingerich). Cambridge, Mass.: Harvard University Press 1979.
Van De Hulst, H. C. (1949): The solid particles of interstellar space. Recherches astron. obs. utrecht **11**, pt. 2, 1. Reproduced in: A source book in astronomy and astrophysics 1900–1975 (eds. K. R. Lang and O. Gingerich). Cambridge, Mass.: Harvard University Press 1979.
Van De Hulst, H. C.: Interstellar polarization and magneto-hydrodynamic waves. In: Problems of cosmical aerodynamics Dayton, Ohio: Central Air Documents Office 1949.
Van De Hulst, H. C. (1950): The amount of polarization by interstellar grains. Ap. J. **112**, 1.
Van Den Berg, D. A., Stetson, P. B., Bolte, M. (1996): The age of the galactic globular cluster system. Ann. Rev. Astron. Astrophys. **34**, 461.
Van Den Bergh, S., Mc Clure, R. D., Evans, R. (1987): The supernova rate in Shapley-Ames galaxies. Ap. J. **323**, 44.
Van Der Klis, M. (1989): Quasi-periodic oscillations and noise in low-mass X-ray binaries. Ann. Rev. Astron. Astrophys. **27**, 517.
Van Dishoeck, E. F., Black, J. H. (1986): Comprehensive models of diffuse interstellar clouds: Physical conditions and molecular abundances. Ap. J. Suppl. **62**, 109.
Van Hieu, N., Shabalin, E. P. (1963): Role of the $\gamma + \gamma \rightarrow \gamma + \nu + \nu$ process in neutrino emission by stars. Sov. Phys. JETP **17**, 681.
Van Horn, H. M. (1991): Dense astrophysical plasmas. Science **252**, 384.
Van Horn, H. M., Salpeter, E. E. (1967): WKB approximation in three dimensions. Phys. Rev. **157**, 751.
Van Paradijs, J., et al. (1997): Transient optical emission from the error box of the γ-ray burst of 28 February 1997. Nature **386**, 686.
Van Riper, K. A. (1978): The hydrodynamics of stellar collapse. Ap. J. **221**, 304.
Van Riper, K. A. (1979): General relativistic hydrodynamics and the adiabatic collapse of stellar cores. Ap. J. **232**, 558.
Van Vleck, J. H. (1929): On σ-type doubling and electron spin in the spectra of diatomic molecules. Phys. Rev. **33**, 467.
Van Vleck, J. H.: Theory of electric and magnetic susceptibilities. London: Oxford University Press, 1932.

Van Vleck. J. H., Huber, D. L. (1977): Absorption, emission and linebreadths: a semihistorical perspective. Rev. Mod. Phys. **49**, 939.

Vandenberg, D. A., Stetson, P. B., Bolte, M. (1996): The age of the galactic globular cluster system. Ann. Rev. Astron. Astrophys. **34**, 461.

Vandervoort, P. O. (1963): The formation of H II regions. Ap. J. **137**, 381 **138**, 426.

Vandervoort, P. O. (1963): The stability of ionization fronts and the evolution of H II regions. Ap. J. **138**, 599.

Vandervoort, P. O. (1975): New applications of the equations of stellar hydrodynamics. Ap. J. **195**, 333.

Varsavsky, M. (1961): Some atomic parameters of ultraviolet lines. Ap. J. Suppl. **6**, 75.

Vasisht, G., Kulkarni, S. R., Frail, D. A., Greiner, J. (1994): Supernova remnant candidates for the soft γ-ray repeater 1900+14. Ap. J. (Letters) **431**, L35.

Vasyliunas, V. M. (1975): Theoretical models of magnetic field line merging. Rev. Geophys. **13**, 303.

Vauclair, S., Reeves, H. (1972): Spallation processes in stellar surfaces: Anomalous helium ratios. Astron. Astrophys. **18**, 215.

Vernazza, J. E., Avrett, E. H., Loeser, R. (1981): Structure of the solar chromosphere III. Models of the EUV brightness components of the quiet sun. Ap. J. Supp. **45**, 635.

Verschuur, G. L. (1968): Positive determination of an interstellar magnetic field by measurement of the Zeeman splitting of the 21-cm hydrogen line. Phys. Rev. Lett. **21**, 775.

Verschuur, G. L. (1971): Recent measurements of the Zeeman effect at 21-centimeter wavelength. Ap. J. **165**, 651.

Verschuur, G. L. (1979): Observations of the galactic magnetic field. Fund. Cosmic Phys. **5**, 113.

Verschuur, G. L., Kellermann, K. I.: Galactic and extragalactic radio astronomy. Second Edition. New York: Springer-Verlag, 1990.

Vidal, C. R., Cooper, J., Smith, E. W. (1973): Hydrogen Stark-broadening tables. Ap. J. Suppl. No. 214, **25**, 37.

Vietri, M. (1997): The afterglow of gamma-ray bursts: the cases of GRB 970228 and GRB 970508. Ap. J. (Letters) **488**, L105.

Vila, S. C., Sion, E. M. (1976): The pulsational properties of high-luminosity degenerate stars with hydrogen burning near the surface. Ap. J. **207**, 820.

Villard, M. P. (1900): Sur le rayonnement du radium (On the radiation of radium). Compt. Rend. **130**, 1178.

Vitello, P., Pacini, F. (1977): The evolution of expanding nonthermal sources. I. Nonrelativistic expansion. Ap. J. **215**, 452.

Vitello, P., Pacini, F. (1978): The evolution of expanding nonthermal sources. II. Relativistic expansions. Ap. J. **220**, 756.

Vladimirskii, V. V. (1948): Influence of the terrestrial magnetic field on large Auger showers. Zh. Exp. Teor. Fiz. **18**, 392.

Vlasov, A. A. (1938): The oscillation properties of an electron gas. Zhur. Eksp. Theor. Fiz. **8**, 291.

Vlasov, A. A. (1945): On the kinetic theory of an assembly of particles with collective interaction. J. Phys. (U.S.S.R.) **9**, 25.

Vogel, H. C. (1889): Spectrographische Beobachtungen an Algol (Spectroscopy of Algol). Astron. Nach. **123**, 289.

Voigt, W. (1913): Über die Intensitätsverteilung innerhalb einer Spektrallinie (Distribution of intensity within a spectrum line. Phys. Zeits. **14**, 377.

Von Hoerner, S., Saslaw, W. C. (1976): The evolution of massive collapsing gas clouds. Ap. J. **206**, 917.

Wadehra, J. M. (1978): Transition probabilities and some expectation values for the hydrogen atom in intense magnetic fields. Ap. J. **226**, 372.

Wagoner, R. V. (1969): Synthesis of the elements within objects exploding from very high temperatures. Ap. J. Suppl. No. 162, **18**, 247.
Wagoner, R. V., Fowler, W. A., Hoyle, F. (1967): On the synthesis of elements at very high temperatures. Ap. J. **148**, 3.
Walecka, J. D. (1974): A theory of highly condensed matter. Ann. Phys. **83**, 491.
Walker, M. F. (1956): Studies of extremely young clusters I: NGC 2264. Ap. J. Supp. **2**, 365. Reproduced in: A source book in astronomy and astrophysics 1900–1975 (eds. K. R. Lang and O. Gingerich). Cambridge, Mass.: Harvard University Press 1979.
Walker, T. P., Mathews, G. J., Viola, V. E. (1985): Astrophysical production rates for Li, Be, and B isotopes from energetic ^{1}H and ^{4}He reactions with HeCNO nuclei. Ap. J. **299**, 745.
Wallerstein, G., Conti, P. S. (1969): Lithium and beryllium in stars. Ann. Rev. Astr. Astrophys. **7**, 99.
Wampler, E. J., et al. (1973): Redshift of OQ 172. Nature **243**, 336.
Wang, C. G., Rose, W. K., Schlenker, S. L. (1970): Models for neutron-core stars based on realistic nuclear-matter calculations. Ap. J. (Letters) **160**, L17.
Wapstra, A. H., Audi, G. (1985): The 1983 atomic mass evaluation I. Atomic mass table. Nuclear Physis **A432**, 1.
Wapstra, A. H., Audi, G. (1985): The 1983 atomic mass evaluation II. Nuclear-reaction and separation energies. Nuclear Physics **A432**, 55.
Wapstra, A. H., Gove, N. B. (1971): The 1971 atomic mass evaluation. Nuclear Data Tables **9**, 265.
Warner, B. (1972): Observations of rapid blue variables X. G61-29. M.N.R.A.S. **159**, 315.
Warner, B., Nather, R. E. (1972): Observations of rapid blue variables II – HL Tau 76. M.N.R.A.S. **156**, 1.
Watson, A. A. (1991): The highest energy cosmic rays. Nucl. Phys. B (Proc. Suppl.) **22B**, 116.
Watson, W. D., Salpeter, E. E. (1972): Molecule formation on interstellar grains. Ap. J. **174**, 321.
Watson, W. D., Salpeter, E. E. (1972): On the abundances of interstellar molecules. Ap. J. **175**, 659.
Watson, W. D. (1974): Ion-molecule reactions, molecule formation, and hydrogen-isotope exchange in dense interstellar clouds. Ap. J. **188**, 35.
Watson, W. D. (1976): Interstellar molecule reactions. Rev. Mod. Phys. **49**, 513.
Watson, W. D. (1978): Gas phase reactions in astrophysics. Ann. Rev. Astron. Ap. **16**, 585.
Waxman, E. (1995): Cosmological origin for cosmic rays above 10^{19} eV. Ap. J. (Letters) **452**, L1.
Waxman, E. (1997): Gamma-ray-burst afterglow: supporting the cosmological fireball model, constraining parameters, and making predictions. Ap. J. (Letters) **485**, L5.
Waxman, E. (1997): γ-ray burst afterglow: confirming the cosmological fireball model. Ap. J. (Letters) **489**, L33.
Weaver, H. F., Williams, D. R. W. (1973): The Berkeley low latitude survey of neutral hydrogen part I. Profiles. Astron. Astrophys. Suppl. **8**, 1.
Weaver, H., Williams, D. R. W., Dieter, N. H., Lum, W. T. (1965): Observations of a strong unidentified microwave line and of emission from the OH molecule. Nature **208**, 29.
Weaver, T. A., Woosley, S. E. (1980): Evolution and explosion of massive stars. Ann. N. Y. Acad Sci. **336**, 335.
Weaver, T. A., Woosley, S. E. (1993): Nucleosynthesis in massive stars and the $^{12}C(\alpha, \gamma)^{16}O$ reaction rate. Phys. Reports **227**, 65.
Webber, W. R.: The spectrum and charge composition of the primary cosmic radiation. In: Handbuch der Physik, vol. XLVI/2: Cosmic Rays II, p. 181 (ed. K. Sitte). Berlin-Heidelberg-New York: Springer 1967.
Webber, W. R.: Cosmic ray electrons and positrons – A review of current measurements and some implications. In: Composition and Origin of Cosmic Rays (ed. M. M. Shapiro) Boston: D. Reidel 1983.

Webber, W. R., et al. (1992): Studies of the low-energy galactic cosmic-ray composition near 28 au at sunspot minimum: the primary-to-primary ratios. Ap. J. (Letters) **392**, L91.

Webber, W. R., Kish, J. C., Schrier, D. A. (1990): Individual charge changing fragmentation cross sections of relativistic nuclei in hydrogen, helium, and carbon targets. Phys. Rev. **C41**, 533.

Webber, W. R., Kish, J. C., Schrier, D. A. (1990): Individual isotopic fragmentation cross sections of relativistic nuclei in hydrogen, helium, and carbon targets. Phys. Rev. **C41**, 547.

Webber, W. R., Kish, J. C., Schrier, D. A. (1990): Total charge and mass changing cross sections of relativistic nuclei in hydrogen, helium, and carbon targets. Phys. Rev. **C41**, 520.

Webber, W. R., Kish, J. C., Schrier, D. A. (1990): Formula for calculating partial cross sections for nuclear reactions of nuclei with E $\geq$ 200 MeV/nucleon in hydrogen targets. Phys. Rev. **C41**, 566.

Webber, W. R., Lezniak, J. A. (1974): The comparative spectra of cosmic-ray protons and helium nuclei. Astrophys. Space Sci. **30**, 361.

Weber, E. J., Davis, L. Jr. (1967): The angular momentum of the solar wind. Ap. J. **148**, 217.

Weber, S. V. (1976): Oscillation and collapse of interstellar clouds. Ap. J. **208**, 113.

Weekes, T. C. (1988): Very high energy gamma-ray astronomy. Physics Reports **160**, 1.

Wefel, J. P.: The composition of the cosmic rays: An update. In: Cosmic Rays, Supernovae and the Interstellar Medium (eds. M. M. Shapiro, R. Silberberg and J. P. Wefel). Boston: Kluwer Academic 1991.

Wegner, G. (1980): A new gravitational redshift for the white dwarf σ^2 Eri B. Astron. J. **95**, 1255.

Weiler, K. W., Sramek, R. A. (1988): Supernovae and supernova remnants. Ann. Rev. Astron. Astrophys. **26**, 295.

Weinberg, S.: Gravitation and cosmology: principles and application of the general theory of relativity. New York: Wiley 1972.

Weinreb, S. (1962): A new upper limit to the galactic deuterium-to-hydrogen ratio. Nature **195**, 367.

Weinreb, S., et al. (1963): Radio observations of OH in the interstellar medium. Nature **200**, 829. Reproduced in: A source book in astronomy and astrophysics 1900–1975 (eds. K. R. Lang and O. Gingerich). Cambridge, Mass.: Harvard University Press 1979.

Weinreb, S., et al. (1965): Observations of polarized OH emission. Nature **208**, 440.

Weisheit, J. C. (1974): X-ray ionization cross-sections and ionization equilibrium equations modified by Auger transitions. Ap. J. **190**, 735.

Weisskopf, V. (1933): Die Breite der Spektrallinien in Gasen (The width of spectral lines in gases). Phys. Zeits. **34**, 1.

Weisskopf, V. (1933): The intensity and structure of spectral lines. Observatory **56**, 291.

Weisskopf, V. , Wigner, E. (1930): Berechnung der natürlichen Linienbreite auf Grund der Diracschen Lichttheorie (Calculation of the natural line width on the basis of Dirac's theory of light). Z. Phys. **63**, 54.

Weizsäcker, C. F. (1935): Zur Theorie der Kernmassen (On the theory of nuclear masses). Z. Physik **96**, 431.

Weizsäcker, C. F. (1937): Über Elementumwandlungen im Innern der Sterne I (On transformation of the elements in stellar interiors I). Phys. Z. **38**, 176.

Weizsäcker, C. F. (1938): Über Elementumwandlungen im Innern der Sterne II (On transformation of the elements in stellar interiors II). Phys. Z. **39**, 633. Eng. trans. in: A source book in astronomy and astrophysics 1900–1975 (eds. K. R. Lang and O. Gingerich). Cambridge, Mass.: Harvard University Press 1979.

Weizsäcker, C. F. (1951): The evolution of galaxies and stars. Ap. J. **114**, 165.

Wentzel, D. G. (1963): Fermi acceleration of charged particles. Ap. J. **137**, 135.

Wentzel, D. G. (1964): Motion across magnetic discontinuities and Fermi acceleration of charged particles. Ap. J. **140**, 1013.

Wentzel, D. G. (1967): An upper limit on the abundance of H_2 formed by chemical exchange Ap. J. **150**, 453.
Wentzel, D. G. (1974): Coronal heating by Alfvén waves. Solar Phys. **39**, 129.
Wentzel, D. G. (1976): Coronal heating by Alfvén waves, II. Solar Phys. **50**, 343.
Wentzel, G. (1926): Eine Verallgemeinerung der Quantenbedingungen für die Zwecke der Wellenmechanik (An overall description of the quantum requirements for the use of the wave mechanics). Z. Physik **38**, 518.
Werner, M. W. (1970): Ionization equlibrium of carbon in interstellar clouds. Astrophys. Lett. **6**, 81.
Westerhout, G.: Maryland-Greenbank 21-cm line survey. College Park, Maryland: Univ. of Maryland 1969.
Westfold, K. C. (1959): The polarization of synchrotron radiation. Ap. J. **130**, 241.
Weymann, R. (1966): The energy spectrum of radiation in the expanding universe. Ap. J. **145**, 560.
Weymann, R. (1967): Possible thermal histories of intergalactic gas. Ap. J. **147**, 887.
Weymann, R. J., Carswell, R. F., Smith, M.G. (1981): Absorption lines in the spectra of quasistellar objects. Ann. Rev. Astron. Astrophys. **19**, 41.
Weymann, R., Sears, R. L. (1965): The depth of the convective envelope on the lower main sequence and the depletion of lithium. Ap. J. **142**, 174.
Whang, Y. C., Liu, C. K., Chang, C. C. (1966): A viscous model of the solar wind. Ap. J. **145**, 255.
Wheaton, W. A., et al. (1973): The direction and spectral variability of a cosmic gamma-ray burst. Ap. J. (Letters) **185**, L57.
Wheeler, J. A. (1947): Mechanism of capture of slow mesons. Phys. Rev. **71**, 320.
Wheeler, J. A., Lamb, W. E., Jr. (1939): Influence of atomic electrons on radiation and pair production. Phys. Rev. **55**, 858.
Wheeler, J. C. (1974): Type I supernovae. Ap. J. **187**, 337.
Wheeler, J. C. (1977): X-ray bursts from magnetized accretion disks. Ap. J. **214**, 560.
Wheeler, J. C. (1978): Type I supernovae, R Coronae Borealis stars, and the Crab Nebula. Ap. J. **225**, 212.
Wheeler, J. C., Harkness, R. P. (1990): Type I supernovae. Rep. Prog. Phys. **53**, 1467.
Wheeler, J. C., Sneden, C., Truran, J. W. Jr. (1989): Abundance ratios as a function of metallicity. Ann. Rev. Astron. Astrophys. **27**, 279.
White, H. E., Eliason, A. Y. (1933): Relative intensity tables for spectrum lines. Phys. Rev. **44**, 753.
White, R. B.: Resistive instabilites and field line reconnection. In: Handbook of plasma physics. (eds. M. N. Rosenbluth and R. Z. Sagdeev). Vol. 1, 2. Basic Plasma Physics (eds. A. A. Galeev and R. N. Sudan). New York: Elsevier Science Publishers 1983, 1984.
Whitford, A. E. (1948): An extension of the interstellar absorption curve. Ap. J. **107**, 102.
Whitford, A. E. (1958): The law of interstellar reddening. Astron. J. **63**, 201.
Whittaker, W. A. (1963): Heating of the solar corona by gravity waves. Ap. J. **137**, 914.
Wiechert, J. E. (1901): Elektrodynam. Elementargesetze (Fundamental electrodynamic laws). Arch. Néerl **5**, 1.
Wien, W. (1893): Eine neue Beziehung der Strahlung schwarzer Körper zum zweiten Hauptsatz der Wärmetheorie (One new relation between the radiation of blackbodies and the second law of thermodynamics). Sitz. Acad. Wiss. Berlin **1**, 55.
Wien, W. (1894): On the division of energy in the emission-spectrum of a black body. Phil. Mag. **43**, 214.
Wiener, N. (1930): Generalized harmonic analysis. Acta. Math. Stockholm **55**, 117.
Wiese, W. L., Smith, M. W., Glennon, B. M.: Atomic transition probabilities, vol. 1. Hydrogen through helium. Nat. Bur. Stands. (Wash.), NSRDS-NBS4 1966.
Wigner, E., Seitz, F. (1934): On the constitution of metallic sodium II. Phys. Rev. **46**, 509.
Wiita, P. J., Press, W. H. (1976): Mass-angular-momentum regimes for certain instabilities of a compact, rotating stellar core. Ap. J. **208**, 525.

Wild, J. P. (1952): Radio-frequency line spectrum of atomic hydrogen and its applications in astronomy. Ap. J. **115**, 206.
Wild, J. P., Hill, E. R. (1971): Approximation of the general formulae for gyro and synchrotron radiation in a vacuum and isotropic plasma. Austr. J. Phys. **24**, 43.
Wild, J. P., Smerd, S. F., Weiss, A. A. (1963): Solar bursts. Ann. Rev. Astron. Astrophys. **1**, 291.
Wildhack, W. A. (1940): The proton-deuteron transformation as a source of energy in dense stars. Phys. Rev. **57**, 81.
Wilkinson, P. G. (1963): Diatomic molecules of astrophysical interest: Ionization potentials and dissociation energies. Ap. J. **138**, 778.
Williams, P. J. S. (1963): Absorption in radio sources of high brightness temperature. Nature **200**, 56.
Wilson, J. R. (1974): Coherent neutrino scattering and stellar collapse. Phys. Rev. Lett. **32**, 849.
Wilson, O. C. (1950): A survey of internal motions in the planetary nebulae. Ap. J. **111**, 279.
Wilson, O. C. (1958): Kinematic structures of gaseous envelopes - internal kinematics of the planetary nebulae. Rev. Mod. Phys. **30**, 1025.
Wilson, P. R., Burtonclay, D., Li, Y. (1996): Calculations of the solar internal angular velocity for 1986-1990. Ap. J. **457**, 440.
Wilson, R. W., et al. (1973): Interstellar deuterium: The hyperfine structure of DCN. Ap. J. (Letters) **179**, L107.
Wilson, R. W., Jefferts, K. B., Penzias, A. A. (1970): Carbon monoxide in the Orion nebula. Ap. J. (Letters) **161**, L43.
Wilson, T. L., et al. (1970): A survey of H 109 α recombination line emission in galactic H II regions of the southern sky. Astron. Astrophy. **6**, 364.
Wilson, T. L., Rood, R. T. (1994): Abundances in the interstellar medium. Ann. Rev. Astron. Astrophys. **32**, 191.
Winget, D. E., et al. (1987): An independent method for determining the age of the universe. Ap. J. (Letters) **315**, L77.
Withbroe, G. L., Noyes, R. W. (1977): Mass and energy flow in the solar chromosphere and corona. Ann. Rev. Astron. Astrophys. **15**, 363.
Wolf, M. (1923): On the dark nebula NGC 6960. Astron. Nach. **219**, 109. English translation in: A source book in astronomy and astrophysics 1900–1975 (eds. K. R. Lang and O. Gingerich). Cambridge, Mass.: Harvard University Press 1979.
Wolf, R. A. (1965): Rates of nuclear reactions in solid-like stars. Phys. Rev. **137**, B 1634.
Wolfendale, A. W.: Dust, gas and cosmic rays in the interstellar medium. In: Cosmic rays, supernovae and the interstellar medium (eds. M. M. Shapiro, R. Silberberg and J. P. Wefel). Boston: Kluwer Academic 1991.
Wolfenstein, L. (1978): Neutrino oscillations in matter. Phys. Rev. **D17**, 2369.
Wolfenstein, L. (1979): Neutrino oscillations and stellar collapse. Phys. Rev. **D20**, 2634.
Wolff, C. L. (1972): Free oscillations of the Sun and their possible stimulation by solar flares. Ap. J. **176**, 833.
Wolff, C. L. (1972): The five-minute oscillations as nonradial pulsations of the entire Sun. Ap. J. (Letters) **177**, L87.
Wolfson, R. (1977): Axisymmetric accretion near compact objects. Ap. J. **213**, 200.
Wolfson, R. (1977): Energy considerations in axisymmetric accretion. Ap. J. **213**, 208.
Woo, R.: Spacecraft radio scintillation and solar system exploration. In: Wave propagation in random media (scintillation) (eds. V. I. Tatarskii, A. Ishimaru and V. U. Zavorotny). Bellingham, Washington: SPIE 1993.
Woo, R.: Coronal structures observed by radio propagation measurements. In: Solar wind eight: AIP conference proceedings 382 (eds. D. Winterhalter, J. T. Gosling, S. R. Habbal, W.S. Kurth and N. Neugebauer) Woodbury, New York: American Institute of Physics 1996.
Wood, M. A. (1992): Constraints on the age and evoution of the Galaxy from the white dwarf luminosity function. Ap. J. **386**, 539.

Wood, P. R., Cahn, J. H. (1977): Mira variables, mass loss, and the fate of red giant stars. Ap. J. **211**, 499.
Woodard, M. O., Hudson, H. (1983): Solar oscillations observed in the total irradiance. Solar Phys. **82**, 67.
Woodward, P. R. (1976): Shock-driven implosion of interstellar gas clouds and star formation. Ap. J. **207**, 484.
Woodward, P. R. (1978): Theoretical models of star formation. Ann. Rev. Astron. Astrophys. **16**, 555.
Woolley, R. V. D. R. (1947): The solar corona. Suppl. Austr. J. Sci. **10**, 2.
Woosley, S. E.: Nucleosynthesis and stellar evolution. In: Nucleosynthesis and Chemical Evolution (ed. B. Hauck, A. Maeder and G. Meynet). Geneva Observatory: Swiss Society of Astrophysics and Astronomy 1986.
Woosley, S. E., Arnett, W. D., Clayton, D. D. (1972): Hydrostatic oxygen burning in stars: II. Oxygen burning at balanced power. Ap. J. **175**, 731.
Woosley, S. E., Arnett, W. D., Clayton, D. D. (1973): The explosive burning of oxygen and silicon. Ap. J. Supp. No. 231, **26**, 231.
Woosley, S. E., Phillips, M. M. (1988): Supernova 1987A! Science **240**, 750.
Woosley, S. E., Weaver, T. A. (1986): The physics of supernova explosions. Ann. Rev. Astron. Astrophys. **24**, 205.
Woosley, S. E., Weaver, T. A. (1995): The evolution and explosion of massive stars. II. Explosive hydrodynamics and nucleosynthesis. Ap. J. Supp. **101**, 181.
Woosley, S. E., Wilson, J. R., Mayle, R. (1986): Gravitational collapse and the cosmic antineutrino background. Ap. J. **302**, 19.
Wright, W. H. (1918): The wave lengths of the nebular lines and general observations of the spectra of the gaseous nebulae. Pub. Lick Obs. **13**, 193.
Wrubel, M. H. (1950): Exact curves of growth for the formation of absorption lines according to the Milne-Eddington model: II. Center of the disk. Ap. J. **111**, 157.
Wrubel, M. H. (1954): Exact curves of growth: III. The Schuster-Schwarzschild model. Ap. J. **119**, 51.
Wu, T. Y.: Quantum mechanics. New York: World Scientific 1986.
Wyndham, J. D. (1966): Optical identification of radio sources in the 3C revised catalogue. Ap. J. **144**, 459.
Yang, C. H., Clark, J. W. (1971): Superfluid condensation energy of neutron matter. Nucl. Phys. **A174**, 49.
Yiou, F., Seide, C., Bernas, R. (1969): Formation cross sections of lithium, beryllium, and boron isotopes produced by spallation of oxygen by high-energy protons. J. Geophys. Res. **74**, 2447.
Yokoi, K., Takahashi, K., Arnould, M. (1983): The ^{187}Re - ^{187}Os chronology and chemical evolution of the Galaxy. Astron. Astrophys. **117**, 65.
York, D. G., Rogerson, J. B. (1976): The abundance of deuterium relative to hydrogen in interstellar space. Ap. J. **203**, 378.
Yorke, H. W. (1986): The dynamical evolution of H II regions – recent theoretical developments. Ann. Rev. Astron. Astrophys. **24**, 49.
Young, J. S., Scoville, N. Z. (1991): Molecular gas in galaxies. Ann. Rev. Astron. Astrophys. **29**, 581.
Yukawa, H. (1937): On a possible interpretation of the penetrating component of the cosmic ray. Proc. Phys. Math. Soc. Japan **19**, 712.
Yukawa, H. (1938): On the interaction of elementary particles IV. Proc. Phys. Math. Soc. Japan **20**, 720.
Yvon, J.: La theorie des fluides et l'equation d'état: Actualites scientifiques et industrielles (The theory of fluids and the equation of state: Scientific and industrial actualities). Paris: Hermann and Cie 1935.
Zaidi, M. H. (1965): Emission of neutrino-pairs from stellar plasma. Nuovo Cimento **40 A**, 502.

Zank, G. P., Gaisser, T. K. (eds.): Particle acceleration in cosmic plasmas. New York: American Institute of Physics 1992.
Zanstra, H. (1927): An application of the quantum theory to the luminosity of diffuse nebulae. Ap. J. **65**, 50. Reproduced in: A source book in astronomy and astrophysics 1900–1975 (eds. K. R. Lang and O. Gingerich). Cambridge, Mass.: Harvard University Press 1979.
Zatsepin, G. T., Kuzmin, V. A. (1966): Upper limit of the spectrum of cosmic rays. Sov. Phys. JETP **4**, 78. ZhETF Pisma **4**, No. 3, 114.
Zaumen, W. T. (1976): Pair production in intense magnetic fields. Ap. J. **210**, 776.
Zdziarski, A. A., Zycki, P. T., Krolik, J. H. (1993): Active galactic nuclei make the cosmic X-ray background. Ap. J. (Letters) **414**, L81.
Zeeman, P. (1896): On the influence of magnetism on the nature of the light emitted by a substance. Phil. Mag. **43**, 226.
Zeeman, P. (1897): Doublets and triplets in the spectrum produced by external magnetic forces. I., II. Phil. Mag. **44**, 55, 255.
Zeippen, C. J. (1982): Transition probabilities for forbidden lines in the $2p^3$ configuration. M.N.R.A.S. **198**, 111.
Zeippen, C. J. (1987): Improved radiative transition probabilities for O II forbidden lines. Astron. Astrophys. **173**, 410.
Zeldovich, Y. B. (1958): Nuclear reactions in super-dense cold hydrogen. Sov. Phys. JETP **6**, 760.
Zeldovich, Y. B., Novikov, I. D.: Relativistic astrophysics, vol. 1: Stars and relativity. Chicago, Ill.: University of Chicago Press 1971.
Zeldovich, Y. B., Raizer, Y. P.: Physics of shock waves and high temperature hydrodynamic phenomena, Vols. 1, 2. New York: Academic Press 1966.
Zensus, J. A. (1997): Parsec-scale jets in extragalactic radio sources. Ann. Rev. Astron. Astrophys. **35**, 607.
Zhang, S. N., Cul, W., Chen, W. (1997): Black hole spin in X-ray binaries: Observational consequences. Ap. J. (Letters) **482**, L155.
Zheleznyakov, V. V.: Radio emission of the sun and planets. New York: Pergamon Press 1970.
Zheleznyakov, V. V., Zaitsev, V. V. (1970): Contribution to the theory of type III radio bursts I. Sov. Astron. A. J. **14**, 47.
Zhevakin, S. A. (1953): Theory of Cepheids. Astron. J. Sov. Union **30**, 161.
Zhevakin, S. A. (1963): Pulsation theory of variable stars. Ann. Rev. Astron. Astrophys. **1**, 367.
Zlotnik, E. Y. (1968): The theory of the slowly changing component of solar radio emission. I., II. Sov. Astron. A J **12**, No. 2, 245, No. 3, 464.
Zobel, W., et al. (1967): ONRL 4183.
Zuckerman, B., Ball, J. A. (1974): On microwave recombination lines from H I regions. Ap. J. 190, 35.
Zuckerman, B., Evans, N. J. (1974): Models of massive molecular clouds. Ap. J. **192**, L 149.
Zuckerman, B., Palmer, P. (1974): Radio radiation from interstellar molecules. Ann. Rev. Astron. Astrophys. **12**, 279.
Zwaan, C. (1987): Elements and patterns in the solar magnetic field. Ann. Rev. Astron. Astrophys. **25**, 83.

Author Index

M

Subject Index

C

D

E

G

I

M

O

Q

R

T

X

Y

Z

Springer and the environment

At Springer we firmly believe that an international science publisher has a special obligation to the environment, and our corporate policies consistently reflect this conviction.
We also expect our business partners – paper mills, printers, packaging manufacturers, etc. – to commit themselves to using materials and production processes that do not harm the environment. The paper in this book is made from low- or no-chlorine pulp and is acid free, in conformance with international standards for paper permanency.

Springer

Printing: Saladruck, Berlin
Binding: H. Stürtz AG, Würzburg